GEOMORPHOLOGY

The Harker Glacier at the head of Moraine Fjord, sub-Antarctic island of South Georgia. The glacier is nourished by snow avalanching from the high peaks that form the spine of the island. High debris turnover gives rise to the deposition of numerous inset lateral moraines that give the fjord its appropriate name. (Photograph © D.J.A. Evans. Reproduced with permission.)

Geomorphology: Critical Concepts in Geography

General Editor David J.A. Evans

Volume IV

Glacial Geomorphology

Edited by David J.A. Evans

Routledge
Taylor & Francis Group

LONDON AND NEW YORK

First published 2004
by Routledge
2 Park Square, Milton Park, Abingdon, OXON, OX14 4RN

Simultaneously published in the USA and Canada
by Routledge

711 Third Avenue, New York, NY 10017

Transferred to Digital Printing 2006

Routledge is an imprint of the Taylor & Francis Group

British Library Cataloguing in Publication Data
A catalogue record for this book is available from the British Library

Library of Congress Cataloging in Publication Data
A catalog record for this book has been requested

ISBN 10: 0-415-27608-X (Set)
ISBN 10: 0-415-27612-8 (Volume IV)
ISBN 13: 978-0-415-27608-5 (Set)
ISBN 13: 978-0-415-27612-2 (Volume IV)

Publisher's Note
References within each chapter are as they appear in the original complete work

CONTENTS

CONTENTS

CONTENTS

ACKNOWLEDGEMENTS

The publishers would like to thank the following for permission to reprint their material:

The University of Chicago Press for permission to reprint J.H. Bretz, 'The channeled scablands of the Columbia Plateau', *Journal of Geology* 31(8) (1923) 617–49. © 1923, The University of Chicago Press.

Blackwell Publishing for permission to reprint C.M. Mannerfelt, 'Marginal drainage channels as indicators of the gradients of Quaternary ice caps', *Geografiska Annaler* 31 (1949): 194–9.

The University of Chicago Press for permission to reprint W.R.B. Battle and W.V. Lewis, 'Temperature observations in bergschrunds and their relationship to cirque erosion', *Journal of Geology* 59 (1951): 537–45. © 1951, The University of Chicago Press.

American Journal of Science for permission to reprint J.W. Glen, J.J. Donner and R.G. West, 'On the mechanism by which stones in till become oriented', *American Journal of Science* 255 (1957): 194–205.

The University of Chicago Press for permission to reprint C.P. Gravenor and W.O. Kupsch, 'Ice-disintegration features in western Canada', *Journal of Geology* 67 (1959): 48–64. © 1959, The University of Chicago Press.

Blackwell Publishing for permission to reprint G. Östrem, 'Ice melting under a thin layer of moraine, and the existence of ice cores in moraine ridges', *Geografiska Annaler* 41 (1959): 228–30.

J. Weertman, 'Mechanism for the formation of inner moraines found near the edge of cold ice caps and ice sheets', *Journal of Glaciology* 3 (1961): 965–78. Reprinted from the *Journal of Glaciology* with permission of the International Glaciology Society.

I.J. Smalley and D.J. Unwin, 'The formation and shape of drumlins and their distribution and orientation in drumlin fields', *Journal of Glaciology* 7(51) (1968): 377–90. Reprinted from the *Journal of Glaciology* with permission of the International Glaciology Society.

Blackwell Publishing for permission to reprint R.J. Price, 'Moraines, sandar, kames and eskers near Breidamerkurjökull, Iceland', *Transactions of the Institute of British Geographers* 46 (1969): 17–43.

Aleksis Dreimanis for permission to reprint A. Dreimanis and U.J. Vagners, 'Bimodal distribution of rock and mineral fragments in basal tills', in R.P. Goldthwait (ed.) *Till: A Symposium*, Columbus: Ohio State University Press, 1971, pp. 237–50.

McGill University for permission to reprint M. Kälin, *The Active Push Moraine of the Thompson Glacier, Axel Heiberg Island, Canadian Arctic Archipelago*, Axel Heiberg Island Research Reports, Glaciology, no. 4, Montreal: McGill University, 1971, 68pp.

Geological Society of London for permission to reprint G.S. Boulton, 'Modern Arctic glaciers as depositional models for former ice sheets', *Quarterly Journal of the Geological Society of London* 128 (1972): 361–88. © Copyright Geological Society of London.

Blackwell Publishing for permission to reprint D.E. Sugden, 'Landscapes of glacial erosion in Greenland and their relationship to ice, topographic and bedrock conditions', *Institute of British Geographers, Special Publication* 7 (1974): 177–95.

SEPM (Society for Sedimentary Geology) for permission to reprint I. Banerjee and B.C. McDonald, 'Nature of esker sedimentation', in A.V. Jopling and B.C. McDonald, (eds) *Glaciofluvial and Glaciolacustrine Sedimentation*, Tulsa, Okla.: Society of Economic Paleontologists and Mineralogists, Special Publication 23, 1975, pp. 132–54.

SEPM (Society for Sedimentary Geology) for permission to reprint M. Church and R. Gilbert, 'Proglacial fluvial and lacustrine environments', in A.V. Jopling and B.C. McDonald (eds) *Glaciofluvial and Glaciolacustrine Sedimentation*, Tulsa, Okla.: Society of Economic Paleontologists and Mineralogists, Special Publication 23, 1975, pp. 22–100.

N. Eyles and R.J. Rogerson, 'A framework of the investigation of medial moraine formation: Austerdalsbreen, Norway, and Berendon Glacier, British Columbia, Canada', *Journal of Glaciology* 20(82) (1978): 99–113. Reprinted from the *Journal of Glaciology* with permission of the International Glaciology Society.

B. Hallet, 'A theoretical model of glacial abrasion', *Journal of Glaciology* 23(89) (1979): 39–50. Reprinted from the *Journal of Glaciology* with permission of the International Glaciology Society.

R.D. Powell, 'A model for sedimentation by tidewater glaciers', *Annals of Glaciology* 2 (1981): 129–34. Reprinted from the *Annals of Glaciology* with permission of the International Glaciology Society.

The American Geophysical Union for permission to reprint G.S. Boulton and R.C.A. Hindmarsh, 'Sediment deformation beneath glaciers: rheology and geological consequences', *Journal of Geophysical Research* 92(B9) (1987): 9059–82. Copyright by the American Geophysical Union.

The American Geophysical Union for permission to reprint R.B. Alley, D.D. Blankenship, C.R. Bentley and S.T. Rooney, 'Till beneath ice stream B: 3. Till deformation: evidence and implications', *Journal of Geophysical Research* 92(B9) (1987): 8921–9. Copyright by the American Geophysical Union.

The Canadian Association of Geographers for permission to reprint A.S. Dyke and T.F. Morris, 'Drumlin fields, dispersal trains and ice streams in Arctic Canada', *Canadian Geographer* 32(1): (1988): 86–90.

M. Sharp, J. Campbell Gemmell and J.-L. Tison, 'Structure and stability of the former subglacial drainage system of the Glacier de Tsanfleuron, Switzerland', *Earth Surface Processes and Landforms* 14 (1989): 119–34. Reproduced by permission of John Wiley & Sons Limited.

Reprinted from M.A. Paul and N. Eyles, 'Constraints on the preservation of diamict facies (melt-out tills) at the margins of stagnant glaciers', *Quaternary Science Reviews* 9(1) (1990): 51–69, with permission from Elsevier.

N.R. Iverson, 'Potential effects of subglacial water-pressure fluctuations on quarrying', *Journal of Glaciology* 37(125) (1991): 27–36. Reprinted from the *Journal of Glaciology* with permission of the International Glaciology Society.

Reprinted from S.M. Tulaczyk, R.P. Scherer and C.D. Clark, 'A ploughing model for the origin of weak tills beneath ice streams: a qualitative treatment', *Quaternary International* 86 (2001): 59–70, with permission from Elsevier.

Disclaimer

The publishers have made every effort to contact authors/copyright holders of works reprinted in *Geomorphology: Critical Concepts in Geography*. This has not been possible in every case, however, and we would welcome correspondence from those individuals/companies who we have been unable to trace.

Chronological Table of reprinted articles and chapters

Date	Author	Article/chapter	References	Chap.
1923	J.H. Bretz	The channeled scablands of the Columbia Plateau	*Journal of Geology* 31(8): 617–49	1
1949	C.M. Mannerfelt	Marginal drainage channels as indicators of the gradients of Quaternary ice caps	*Geografiska Annaler* 31: 194–9	2
1951	W.R.B. Battle and W.V. Lewis	Temperature observations in bergschrunds and their relationship to cirque erosion	*Journal of Geology* 59: 537–45	3
1957	J.W. Glen, J.J. Donner and R.G. West	On the mechanism by which stones in till become oriented	*American Journal of Science* 255: 194–205	4
1959	C.P. Gravenor and W.O. Kupsch	Ice-disintegration features in western Canada	*Journal of Geology* 67: 48–64	5
1959	G. Östrem	Ice melting under a thin layer of moraine, and the existence of ice cores in moraine ridges	*Geografiska Annaler* 41: 228–30	6
1961	J. Weertman	Mechanism for the formation of inner moraines found near the edge of cold ice caps and ice sheets	*Journal of Glaciology* 3: 965–78	7
1968	I.J. Smalley and D.J. Unwin	The formation and shape of drumlins and their distribution and orientation in drumlin fields	*Journal of Glaciology* 7(51): 377–90	8
1969	R.J. Price	Moraines, sandar, kames and eskers near Breidamerkurjökull, Iceland	*Transactions of the Institute of British Geographers* 46: 17–43	9
1971	A. Dreimanis and U.J. Vagners	Bimodal distribution of rock and mineral fragments in basal tills	R.P. Goldthwait (ed.) *Till: A Symposium*, Columbus: Ohio State University Press, pp. 237–50	10
1971	M. Kälin	The active push moraine of the Thompson Glacier: Axel Heiberg Island, Canadian Arctic Archipelago	Axel Heiberg Island Research Reports, Glaciology, No. 4, Montreal: McGill University, 68 pp.	11
1972	G. Stewart Boulton	Modern Arctic glaciers as depositional models for former ice sheets	*Quarterly Journal of the Geological Society of London* 128: 361–88	12
1974	D.E. Sugden	Landscapes of glacial erosion in Greenland and their relationship to ice, topographic and bedrock conditions	*Institute of British Geographers, Special Publication* 7: 177–95	13

Chronological Table continued

GENERAL EDITOR'S PREFACE

All areas of research benefit from periodic exercises in reflection – a time for assessing where we are and how we have arrived, who set the ball rolling and whether it still has momentum, and even whether or not we are re-inventing our own metaphorical wheels. Such reflection is especially valuable in these times of accelerating productivity and expanding volumes of literary output, because no geomorphologist could ever hope to keep abreast of every advance in all aspects of the discipline. For the academic this has just as much to do with shrinking library budgets as it has to do with the erosion of reading and processing time. Additionally, when caught up in the frenzy of pushing back research frontiers we can lose sight of the origins of the critical concepts in our science. We all use our relatively small set of core references when compiling papers for submission to journals but these necessarily mostly originate in the more modern literature and often focus on the novel application of a critical concept rather than its initial airing. In other words, we could all name a 'top ten' in our area of expertise but would they be specifically examples of excellent practice rather than benchmarks or breakthroughs? From beginning to end I have found the exercise of selection very much like repeatedly choosing the best ever England football team – 'So we've got Gordon Banks in goal . . .'

The seven volumes in this series contain reproductions of papers that the individual editors regarded as the initiators of critical concepts. It can, and undoubtedly will, be argued by the readership that someone somewhere thought of concept 'x' before author 'y' in Volume 'z'. We have hopefully covered these eventualities in our individual assessments by attempting to integrate the chosen paper with what came before and what has developed since its publication. Furthermore, we have all deliberated long and hard over our selections in the knowledge that our own volume must be objective in its reflection of the development of a subject area. Nonetheless, despite wide consultation with each other and with the wider geomorphic community, the compilations are ultimately personal and therefore reflect the sources of our inspiration. We hope that they coincide with those of other geomorphologists. As few libraries now house collections comprehensive enough to cover both the subject and age range represented in these seven volumes, we hope that they become a valuable reference source for both students and practitioners.

Whilst compiling this series I have been most ably assisted by Rebekah Taylor at Routledge, who possesses admirable levels of patience and the power of gentle persuasion necessary to get academics to deliver book manuscripts in times when their employers would frankly prefer to see them doing other things. I am glad that we persevered, for the exercise was informative and rewarding. Thanks also to Les Hill in Geography and Geomatics at the University of Glasgow for his continued excellent support in the reproduction of photographs that in many cases at the outset did not leave much margin for improvement! Finally, I would also like to thank each of the editors for their knowledgeable insight and hard labour in compiling the critical concept papers in geomorphology.

David J.A. Evans
University of Glasgow

INTRODUCTION

David J.A. Evans

Science typically progresses in a series of major breakthroughs, which then formulate new paradigms. Glacial geomorphology has advanced more gradually (see Evans in press) but is not immune to the odd outrageous hypothesis and paradigm shift (Davis 1926; Kuhn 1962; Shaw 1983a, 1996, 2002; Boulton 1986;). Because glacier beds are extremely difficult and dangerous to access, an actualistic approach to assessing process-form relationships often proves impossible. We invariably have to revert to what Shaw (1996) calls 'quasi-actualistic' studies on landforms emerging from receding glacier snouts or the erection of 'fantasy' theories (guesses, Gilbert 1896) which are then tested by inductive reasoning. This is a reasonably sound approach providing theories/hypotheses are falsifiable/testable and do not contradict the fundamental principles or laws of physics (determinism). It is because of the uncertainty surrounding exactly what takes place beneath glaciers that many of the critical concepts in glacial geomorphology evolve slowly, after much painstaking hypothesis formulation and testing. Several researchers often simultaneously move forward our understanding of glacial environments, communicating their ideas and findings after following different experimental procedures and often exploiting techniques refined in other disciplines. It is therefore difficult to identify exactly which paper represents *the* initial airing of a critical concept. Consequently, many of the papers selected for this volume do not communicate a completely original idea or explanation but instead constitute a major step forward based upon refinements of previous work. Advances in glacial geomorphology are often a direct consequence of developments in glaciology. As this volume is devoted to geomorphology the papers have been selected based upon their contributions to processes and forms rather than glacier physics. However, the impacts of glaciological research upon the themes reviewed below are recognized where pertinent.

An outrageous geological hypothesis

J.H. Bretz (1923) The channeled scablands of the Columbia Plateau. *Journal of Geology* 31(8): 617–49

The channeled scablands of Washington State are inextricably linked with the contrasting ostracisim and ultimate vindication of J. Harlen Bretz (Gould 1980) and have been proposed as one of the geological wonders of the world by Clarke (1986). In the 1920s Bretz presented his revolutionary ideas on the formation of the spectacular erosional forms of the region by a cataclysmic flood, released from an ice-dammed lake (Bretz 1923a, 1923b, 1925, 1927, 1928a, 1928b, 1928c, 1932). As Baker *et al.* (1987) explain in their review of Bretz's work, the geomorphological and geological community were not ready for such a radical overthrowing of accepted knowledge on fluvial landscape development, even though Davis (1926) had thrown down the challenge to geologists to do 'violence' to their accepted principles. This was despite the fact that the source of the 'Spokane flood' quickly became apparent in the shape of Glacial Lake Missoula (Pardee 1910, 1942). Vindication for Bretz's glacial flood and its immense power came slowly but he continued to produce seminal papers on the topic with increasing support from an initially wary discipline (Bretz *et al.* 1956; Bretz 1959, 1969). Although they remain the most evocative of the glacier flood landscapes on the planet they are far from a curiosity and similarly awesome erosional signatures of glacial lake drainage are now recognized routinely in deglaciated terrain (e.g. Malde 1968; Kehew and Lord 1986; Baker *et al.* 1993; Rudoy and Baker 1993).

The advent of the Scandinavian school of glacial landform analysis

C.M. Mannerfelt (1949) Marginal drainage channels as indicators of the gradients of Quaternary ice caps. *Geografiska Annaler* 31: 194–9

In the three decades that followed the Second World War, Scandinavian geomorphologists made considerable contributions to the understanding of glaciated landscapes. Particularly significant was the work of Carl Mannerfelt, who applied observations from contemporary glacierized landscapes to the interpretation of deglacial landforms in Scandinavia (Mannerfelt 1945, 1949, 1960). This work led to a paradigm shift in glacial geomorphology, especially in the British Isles where the influence of Percy Kendall's mapping of former ice-dammed lakes was driven by interpretations of meltwater channels as spillways (Kendall 1902; Dwerryhouse 1902; Harmer 1907; Jowett 1914; Charlesworth 1929). In contrast to these interpretations, Mannerfelt introduced the concept of meltwater channel incision in subglacial, marginal and proglacial settings. This was the catalyst for considerable advances in the study of glacifluvial features, with further seminal papers emerging from the Scandinavian work (e.g. Hoppe 1950; Gjessing 1960). Additionally the Scandinavian approach was then applied in Britain by Hollingworth (1952), Sissons (1958a, 1958b, 1958c, 1960a, 1960b, 1961a, 1961b, 1961c, 1963), Embleton (1961, 1964a, 1964b), Derbyshire (1962), Bowen and Gregory (1965) and Gregory (1965). In his 1949 paper, Mannerfelt identified the fact that meltwater drains along certain parts of receding glacier margins, thereby cutting lateral meltwater channels, and then plunges down under the glacier to produce features like engorged eskers. His concept that the abandoned, low gradient meltwater channels could be used to reconstruct the margins of receding glaciers is now applied routinely in polar regions, where cold-based ice margins have left remarkably intricate series of lateral meltwater channels (e.g. Schytt 1956; Maag 1969; Dyke 1993).

Early assessments of glacial erosion

W.R.B. Battle and W.V. Lewis (1951) Temperature observations in bergschrunds and their relationship to cirque erosion. *Journal of Geology* 59: 537–45

Any study that succeeds in collecting data from the beds of glaciers is destined for citation in all future literature on glacier process-form relationships, and the benchmark work of W.V. Lewis and W.R.B. Battle on glacial erosion and cirque development is a classic example. The impact of the contributions of Lewis and Battle are reviewed in Embleton and King (1975) in their chapter on cirques. A publication entitled *Norwegian Cirque Glaciers*, edited by Lewis (1960) and in which Battle presented bergschrund temperatures from a variety of settings, is widely regarded as a benchmark set of papers. However, it was the Battle and Lewis (1951) paper that set early standards for cirque temperature measurements that few researchers have since attempted to emulate (e.g. Gardner 1987; Mair and Kuhn 1994). Following the example of Johnson (1904), Battle descended into the bergschrunds of cirque glaciers in Norway, Greenland, Switzerland and Baffin Island and became the first researcher to collect data pertinent to possible freeze-thaw mechanisms at cirque backwalls. This work illustrated clearly that freeze-thaw was particularly ineffective in the mechanical breakdown of cirque backwalls at the base of bergschrunds, overturning the contemporary wisdom amongst glacial geomorphologists that frost shattering was extremely efficient at such locations. Battle met an untimely death while crossing a glacier on Baffin Island (Lewis 1954), leaving his colleagues to publish his results from the area at a later date (Thompson and Bonnlander 1956).

INTRODUCTION

Understanding till fabric

J.W. Glen, J.J. Donner and R.G. West (1957) On the mechanism by which stones in till become oriented. *American Journal of Science* 255: 194–205

The apparent alignment of clasts in tills appears to have been initially reported by Hind (1859; see Elson 1966) and later described by Miller (1884). It was not until the mid-twentieth century, however, that quantification techniques were applied to clast alignment in tills. The benchmark paper on this subject was that by Holmes (1941) who demonstrated that the majority of clast A-axes were orientated parallel to the former ice flow direction and possessed low dip angles. Additionally, a secondary population of clasts was aligned transverse to former ice flow (see also Richter 1932, 1933, 1936). Holmes proposed that the ice flow parallel clast A-axes were sliding whereas the transverse clasts were rolling prior to deposition. He also assessed the influence of clast characteristics (e.g. form, roundness, size, axial length) on their alignment by glacier ice and recognized that successive till layers could record shifts in glacial flow direction. Numerous till fabric studies followed on from Holmes (1941) but the most significant contribution was that of Glen *et al.* (1957). They provided explanation of the processes involved in the alignment of a clast in till. Developing the theory of Holmes (1941) they suggested that elongate clasts placed at random in a flowing liquid should rapidly align themselves so that their A-axes are parallel to the flow. The clasts then progressively become re-orientated so that their A-axes are transverse to the flow, the position at which the minimum energy is required for the flowing medium to move over the clasts. Glen *et al.* (1957) became the most influential paper on till sedimentology, although one aspect of clast attitudes in tills, the dip angles of the A-axes, was not considered in the study. The dip angles were incorporated, however, in another influential paper produced at the same time by Harrison (1957). Suggesting that the strong alignments of clasts and their up-glacier dips were due to inheritance from englacial structures (melt-out), Harrison linked clast alignment to shear planes in the ice but also acknowledged that tills and their fabrics could be remobilized by post-depositional mass flows. The concepts that were communicated by Glen *et al.* (1957) became the catalyst for a rapid expansion in till fabric analysis and its employment in a variety of benchmark papers on glacial geomorphology (e.g. Wright 1957, 1962; Hoppe 1959; Andrews 1963a, 1963b, 1965; McClintock and Dreimanis 1964; Andrews and Shimizu 1966; Andrews and Smithson 1966). Macro-fabric analysis remains a central and often controversial technique in the analysis of till genesis (e.g. Benn 1994a, 1995; Hicock *et al.* 1996; Evans *et al.* 1998; Bennett *et al.* 1999; Benn and Ringrose 2001).

4

Controlled moraines and other enigmatic forms

C.P. Gravenor and W.O. Kupsch (1959) Ice-disintegration features in western Canada. *Journal of Geology* 67: 48–64

In the two decades following the Second World War, glacial geomorphology enjoyed a phase of expansion characterized by the mapping and classification of landforms, enabled by the increasing availability of aerial photography. Several influential papers on deglacial landforms appeared as a result of this expansion. In Scandinavia, for example, Gunnar Hoppe produced papers on subaquatically formed 'annual moraines' (De Geer moraines; Hoppe 1948, 1957), submarginal crevasse-squeeze ridges (Hoppe 1959), subglacial hummocky moraine (Veiki moraines; Hoppe 1952, 1957, 1959) and "sediment plateaux' of subglacial cavity fill origin (Hoppe 1963). Similar features were the subject of close scrutiny also in North America (e.g. Lawrence and Elson 1953; Leighton 1959; Mackay 1960; Andrews 1963a, 1963b, 1963c; Andrews and Smithson 1966), where hummocky moraine was regarded as supraglacial ablation terrain (Gravenor 1955; Hartshorn 1958; Gravenor and Kupsch 1959; Kirby 1961; Winters 1961; but see Stalker 1960). Terms like 'prairie mounds', 'moraine plateaux' and 'crevasse fills' entered the North American glacial literature and became synonimous with ice stagnation. As analyses of the internal sediments and structures of landforms since the 1960s have gradually complemented assessments of their morphology, numerous refinements of the genetic classifications of earlier studies have been made. However, the papers cited above contributed to a major growth spurt in glacial geomorphology in the mid-twentieth century (see Evans in press). It is therefore difficult to identify one paper as the benchmark work. Probably the most lasting impact was made by Gravenor and Kupsch (1959) who introduced the concept of 'controlled' and 'uncontrolled' moraine deposition and brought the spectacular glacial landforms of the prairies to the attention of geomorphologists everywhere.

Moraines with ice cores

G. Östrem (1959) Ice melting under a thin layer of moraine, and the existence of ice cores in moraine ridges. *Geografiska Annaler* 41: 228–30

Research on modern glaciers and glacial processes in the 1950s highlighted the role of debris transport at glacier margins in the construction of supraglacial moraine assemblages (see the following section). Observations clearly demonstrated that the differential ablation imparted by the juxtaposition of debris-rich and debris-poor ice led to the development of supraglacial moraine whose morphology was controlled by debris concentrations in the glacier (e.g. Goldthwait 1951, 1960, 1961). Moreover, the advanced stages of deglaciation in polar regions were seen to be characterized by the wide-spread occurrence of moraines with substantial ice cores. Such ice-cored moraines are now widely recognized in both neoglacial (e.g. Miller 1973; Dyke, Andrews and Miller 1982) and Late Wisconsinan/Weichselian (e.g. St Onge and McMartin 1995; Dyke and Savelle 2000) glacial terrains at higher latitudes. The contemporary environmental conditions at these latitudes are not conducive to complete melt-out of buried glacier ice, implying that much of it could survive until it is re-incorporated by re-advancing glaciers (cf. Astakhov and Isayeva 1988; French and Harry 1990; Sugden *et al.* 1995; Evans 1989). The protection of glacier ice from melting is achieved by the accumulation of a supraglacial debris cover that exceeds 1 cm in thickness. This was demonstrated by the work of Gunnar Östrem based upon observations and measurements on ice-cored moraines in Norway, north Sweden and Baffin Island. Östrem (1959, 1962, 1963, 1964) provided details on the effect of an accumulating debris cover on ice ablation rates and identified contemporary ice-cored moraines from aerial photographs. This work was later verified by the observations of Loomis (1970) and Nakawa and Young (1981, 1982).

Geomorphological implications of thermal regimes

J. Weertman (1961) Mechanism for the formation of inner moraines found near the edge of cold ice caps and ice sheets. *Journal of Glaciology* 3: 965–78

A lasting influence of the work undertaken by Karl Gripp on the University of Hamburg expeditions to Spitsbergen in the 1920s related to process-form relationships in end moraine construction (Gripp and Todtmann 1925; Gripp 1938). In his benchmark paper, entitled 'Endmoränen', Gripp (1938) classified moraines according to process, recognizing that they were either 'depositional' or 'thrust' in origin. Not only did this recognize the importance of glacitectonics (see pp. 10–11) but also identified 'inner marginal' moraines on the glacier surface. Gripp interpreted these as the products of vertical debris transport between the boundary of active and stagnant ice, thereby identifying supraglacial moraines similar to those explained as the products of shearing by Chamberlin (1895) and Salisbury (1896).

Debris transport and supraglacial deposition were later elucidated by the seminal work of Sharp (1948, 1949, 1951) and R.P. Goldthwait (1951, 1960, 1961, 1971). Goldthwait's work in particular sparked further assessments of the vertical transfer of englacial debris by shearing to produce 'Thule-Baffin' or shear moraines on glacier surfaces (Bishop 1957; Swinzow 1962; Souchez 1967). Weertman (1961) questioned the englacial shearing process that was invoked by previous workers to explain the transport of debris to the glacier surface. His alternative mechanism followed on from an earlier benchmark contribution to the field of glaciology, the formation of regelation ice on the down ice side of bed perturbations (Weertman 1957; *Weertman regelation*). He proposed that meltwater and debris froze on to the glacier bed at the junction of warm and cold-based ice in the marginal zones of glaciers. This had implications far beyond the process-form relationships of Thule-Baffin moraines. Known as *net adfreezing*, it is valid today as a major process of basal ice formation and explains the concentration of debris in basal ice at the margins of glaciers (see Hubbard and Sharp 1989; Benn and Evans 1998, sections 5.4 and 11.4.2). Although support for Weertman's hypothesis was forthcoming (e.g. Hooke 1973a, 1973b; Hooke and Hudleston 1978), later work demonstrated that the transport of debris along englacial thrust faults was possible (e.g. Wakahama and Tusima 1981). Moreover, the thrusting mode of controlled moraine formation is preferred in some modern work (e.g. Hambrey *et al.* 1997; Bennett *et al.* 1998). More fundamentally Weertman (1961) laid the foundations for the development of glacier thermal regimes and their influence on debris entrainment, transport and deposition. This concept was later elaborated in a seminal paper by Boulton (1972b) in which zones of melting and freezing were linked to glacial processes and forms. The most recent development of the concept of bulk freezing on of meltwater and sediment is that of supercooling associated with subglacial overdeepenings (see Alley *et al.* 1998, 1999; Lawson *et al.* 1998; Evenson *et al.* 1999).

Sediment dilatancy and its subglacial implications

I.J. Smalley and D.J. Unwin (1968) The formation and shape of drumlins and their distribution and orientation in drumlin fields. *Journal of Glaciology* 7(51): 377–90

The concept of glacier flow aided by subglacially deforming materials is a relatively new concept (see pp. 17–20) but it had been acknowledged in earlier research on glacial landforms, particularly in the study of drumlins by Smalley and Unwin (1968; see also Smalley 1966). They applied existing knowledge on the dilatancy of granular materials (e.g. Reynolds 1885; Mead 1925; Andrade and Fox 1949), focusing on a variety of criteria, formation, shape distribution and orientation, previously highlighted as significant in understanding the 'drumlin problem'. It is in the drumlin literature that this paper receives its largest number of citations, often being regarded as the benchmark paper in a new era of sedimentological, rather than purely morphological (e.g. Chorley 1959), investigation of subglacial landforms (see Czechowna 1953; Gravenor 1953; Menzies 1979, 1984 for reviews). However, the implications for understanding sediment behaviour under stress are far reaching, and most research reports on subglacial till formation now routinely refer to the dilatancy of the materials. This is almost entirely due to the impact of the Smalley and Unwin (1968) paper, which placed dilatancy firmly as a central and critical issue in the understanding of the sediment/glacier interface. Later benchmark papers on the rheology of subglacial materials used the principle of dilatancy to classify and explain till formation (e.g. the 'A' and 'B' horizons of Boulton *et al.* 1974; Boulton and Hindmarsh 1987; Alley 1991; Benn 1995; Benn and Evans 1996; see pp. 17–20).

Dilatancy is a property that can be observed in any granular mass. In simple terms, when a granular mass is put under stress it will initially expand and individual grains will climb over one another (pervasive deformation). If the strain rate drops below that required to maintain dilatancy then the sediment will collapse into a denser, stable state. The sediment will not deform again until stresses rise and the strain rate reaches its critical threshold. In Smalley and Unwin's model, the flowing till moving into the area from up glacier will flow around the stable till already deposited at the glacier bed, streamlining it into a drumlin form. They related these critical stress levels to the distribution of drumlins by suggesting that stess levels beneath ice sheets are high enough to maintain sediment dilatancy except near the ice sheet margin. Here the reduction in stress levels allows the production of drumlins in a narrow belt that is located just up ice from marginal conditions of low stress and no drumlin formation. As further research on the nature and distribution of drumlins continued, it became clear that the dilatancy theory as applied by Smalley and Unwin (1968) could not explain various aspects of drumlin distribution and content. For example, their theory assumed that drumlins comprised only till but later observations of stratified cores became more widespread. However, Smalley and Unwin did acknowledge that pre-existing sediments could act as rigid cores, a concept developed more fully by Boulton (1987). Moreover, their appreciation that dilatant/deforming conditions are localized and vary spatially and temporally are concepts central to modern reconstructions of glacier beds (e.g. Piotrowski and Tulaczyk 1999). More recently the role of subglacial water, specifically in changing porewater pressures, has been cited as a possible driving force behind deformation and changing dilatant states in subglacial tills (e.g. Boulton and Jones 1979; Alley 1989; Murray and Dowdeswell 1992; Benn and Evans 1996).

Glacial landsystems at modern ice margins

R.J. Price (1969) Moraines, sandar, kames and eskers near Breidamerkurjökull, Iceland. *Transactions of the Institute of British Geographers* 46: 17–43

Early glacial geomorphological literature is replete with intensive investigations into individual landform types rather than whole assemblages. This type of approach often restricted the interpretations of landform genesis, because it failed to acknowledge the fact that landforms occupied a position on a spatial and temporal continuum and that some features, for example eskers and subglacial meltwater channels (or tunnel valleys) and moraines and flutings, were inextricably linked. Additional problems arose wherever genetic models were constructed in ignorance of modern processes. Breidamerkurjökull, a southern outlet of the Vatnajökull ice cap in Iceland, has featured significantly in some major advances in glacial research. The wealth of glacial geomorphological information in the foreland of the glacier became evident to an international audience following an intensive mapping programme undertaken by the University of Glasgow in the mid-1960s (see Evans and Twigg 2002 for review). By applying modern survey and mapping techniques to the recently uncovered glaciated terrains in Alaska and Iceland, R.J. Price and his colleagues provided the first comprehensive assessments of landform evolution associated with temperate glacier snouts (see also Price 1965, 1966; Petrie and Price 1966; Welch and Howarth 1968). At Breidamerkurjökull, where changes in proglacial geomorphology had been mapped since 1904, this approach facilitated accurate reconstructions of short-term landscape change that could be reconciled with observed processes and glacier dynamics. Price (1969) is a benchmark contribution to understanding the process-form relationships of actively receding temperate glacier margins (see also Price 1970). Of particular importance was the documented impact of melting buried glacier ice on the evolution of glacifluvial landforms, especially the well-developed esker networks that had clearly accumulated in englacial tunnels.

Terminal grade

A. Dreimanis and U.J. Vagners (1971) Bimodal distribution of rock
and mineral fragments in basal tills. In R.P. Goldthwait (ed.) *Till:
A Symposium*. Columbus: Ohio State University Press, pp. 237–50

In the high stress environment of the glacier bed rock fragments of all sizes are constantly brought into contact with one another, leading to substantial modification. These inter-particle stresses lead to fracture and abrasion, providing clear indications of former subglacial transport in the form of individual clasts (particle morphology; see Benn and Evans 1998, section 5.7.3 for a review). The production of stoss-and-lee shapes, facetting, rounding and striated surfaces on clasts due to glacial processes was noted by many early geologists (e.g. Giekie 1894), but it was the seminal work of Holmes (1960) that provided the first quantification of such features. Clearly, clasts underwent fracture and abrasion in the subglacial traction zone and debris pathways through glaciers were later quantified using patterns of change in clast form (e.g. Boulton 1978; Matthews and Petch 1982; Benn and Ballantyne 1994; Evans 1999). In addition to these wear patterns on larger clasts, changes in the matrix of subglacial tills due to fracture and abrasion have been identified using the grain size distributions of tills and experimentally crushed rock samples.

The progressive size reduction in subglacial rock fragments during transport was demonstrated by Dreimanis and Vagners (1971, 1972). They employed changes in till granulometry with distance from known bedrock sources to show that not only did grain size decrease with distance down glacier but also that there was a lower size limit beyond which no further fracturing or particle breakdown occurred. This they referred to as 'terminal grade', equivalent to what engineers call the 'limit of grindability'. The terminal grade was defined as the particle size at which the available energy is insufficient to cause further fracture. Because rock mineralogy varies considerably, so too does their terminal grade. Rocks composed of softer minerals will possess smaller terminal grades. Dreimanis and Vagners (1971) also reported grain size distributions for subglacial tills that differed significantly from those found in mechanically weathered or aqueously transported sediments and which matched those produced by experimental crushing (Gaudin 1926). The distributions were predominantly bi-modal to poly-modal, developing higher peaks in the finer grain sizes with distance down glacier (see also Beaumont 1971). This is thought to reflect gradual comminution of the sediment during its subglacial transport, the coarse peak representing the optimum rock fragment size and the fine peak or peaks representing the optimum mineral grain size(s) or terminal grade(s). The bi-modal and poly-modal grain size distributions were later explained as the products of crushing and abrasion, the two processes being responsible for coarse and fine particles respectively (Boulton 1978; Haldorsen 1981, 1983). The use of particle size distributions in assessing glacial processes and till genesis have been further developed by Hooke and Iverson (1995) and Iverson *et al.* (1996), who have applied the work of Sammis *et al.* (1987) on fault gouge, and by Hubbard *et al.* (1996). This recent research proposes that the large amount of fine-grained material in tills is a product of the slippage and abrasion between particles in addition to intergranular fracture, thereby highlighting the fact that tills will deform or shear internally. Words of caution need to be stressed here with respect to using grain size distributions as indicators of till maturity. Specifically, tills may include rafts of non-glacial sediment with their own grain size signature, and the grain size distribution of any till may not represent the subglacial wear signature of a mature till.

Proglacial tectonic processes and forms

Kälin M. (1971) *The Active Push Moraine of the Thompson Glacier, Axel Heiberg Island, Canadian Arctic Archipelago.* Axel Heiberg Island Research Reports, Glaciology no. 4, Montreal: McGill University, 68pp.

The recognition of disturbance of proglacial materials by advancing glacier snouts is as old as the glacial theory but the spectacular folding and faulting that is displayed in ancient and modern glacitectonic landforms lay largely unutilized by glacial researchers until recently. The wide variety of glacitectonic forms that exists in glaciated terrains, as displayed by Aber *et al.* (1989), provided a clear justification for their separate classification rather than merely variants of push moraine (see Benn and Evans 1998, section 11.3.1). A number of early researchers identified glacially deformed materials in a variety of settings (see Aber *et al.* 1989 and Benn and Evans 1998 for review and references), but the work of George Slater (1926, 1927a, 1927b, 1927c, 1927d, 1927e, 1931) and Karl Gripp (Gripp and Todtmann 1925; Gripp 1929, 1938) are the most widely cited. This led to some intensive work on ancient glacitectonic forms in Europe and North America and the realization that materials had been folded, thrust and stacked in a similar fashion to fold mountain construction (e.g. Gry 1940; Kozarski 1959; Mackay 1959; Kupsch 1962). With the exception of Gripp's work on Svalbard, very few observations on modern glacitectonic features were available at the time of Kälin's (1971) report on the moraine at the snout of the Thompson Glacier on Axel Heiberg Island in the Canadian arctic. To this day, Kälin's work is available only in its original relatively obscure format, a research report of McGill University, despite its regular citations in the glacial literature. It is remarkable in that it represents the first intensive survey and monitoring of an actively forming composite ridge; few similar attempts having been made since (e.g. Croot 1988a). The mechanics of moraine construction are elucidated mathematically by Kälin and related to stresses transmitted into the glacier foreland sediments by the advancing glacier. This concept of proglacial tectonic disturbance was later refined by the research of Rotnicki (1976), Aber (1982) and van der Wateren (1985) and led to later recognition of thrust moraines or composite ridges at contemporary glacier snouts elsewhere (e.g. Klassen 1982; Croot 1988b; Evans and England 1991).

Till genesis

G.S. Boulton (1972) Modern Arctic glaciers as depositional models for former ice sheets. *Quarterly Journal of the Geological Society of London* 128: 361–88

Interpretations of glacial sediments have little credibility unless they are reconciled with modern-day process observations. As glacier beds are notoriously difficult to access, many of our modern process observations are made at glacier margins where subglacial sediments are being freshly uncovered and the impacts of debris release mechanisms on sediment genesis can be monitored. In a series of papers reporting observations on Svalbard glacier margins, Geoffrey Boulton (1967, 1968, 1970a, 1970b, 1971, 1972a) provided models of glacial sedimentation that elucidated and refined our knowledge of glacigenic process-form relationships, clarified our nomenclature for till sedimentology, and offered explanations of sediment-landform associations in ancient glaciated terrain. It is difficult to identify which of Boulton's early papers had the most impact on glacial geomorphology. This includes his later seminal offerings on subglacial processes (e.g. Boulton 1972b, 1974, 1975, 1976, 1978, 1979, 1982; Boulton *et al.* 1974). However, Boulton (1972a) is significant as a synthesis of process-form observations and their application to reconstructing glacial depositional environments. Moreover, the till classification scheme that was refined in this paper to include 'melt-out till' has served as the accepted nomenclature, with the addition of deformed sediments, until the modern era.

Central to the model of glacial transport and deposition proposed in Boulton (1972a) was the fact that, in addition to lodgement, considerable volumes of debris were transferred vertically through the glacier due to basal freeze on and the folding and thrusting of debris-rich basal ice. Sediment was then released either by basal melting or supraglacial flowage once the glacier began receding. Melt-out tills (Boulton 1970b), passively released from the debris-rich ice, and flow tills (Hartshorn 1958), released and remobilized on the changing morphology of the glacier surface, were both interbedded with stratified sediments deposited by meltwater streams. The most significant message that emerged from Boulton's paper was that glaciers often deposit complex depositional sequences comprising tills separated by stratified sediments during one glacier advance. Such sequences had repeatedly been interpreted in ancient glaciated terrain as the products of several glaciations and therefore Boulton's findings led to the complete re-assessment of glacial stratigraphies, one of the most widely cited being that at Glanllynnau on the North Wales coast (Boulton 1977; Harris *et al.* 1990). The depositional model proposed by Boulton (1972a) complemented other benchmark publications such as those by Nobles and Weertman (1971) and Mickelson (1971) that addressed the role of subglacial topography and heat fluxes in the release of subglacial debris. Although Boulton's melt-out till was the subject of some critical research output in later years, for example Shaw (1982), Haldorsen and Shaw (1982) and Lawson (1979a, 1979b), some considerable doubt was later cast over its preservation potential in the geological record (Paul and Eyles 1990; see pp. 22–23).

The selectivity of glacial erosion

D.E. Sugden (1974) Landscapes of glacial erosion in Greenland and their relationship to ice, topographic and bedrock conditions. *Institute of British Geographers, Special Publication 7*: 177–95

In a series of papers on the regional impact of glacial erosion, D.E. Sugden identified that glacier ice was discharged through large ice streams within ice sheets, and that the contrast in erosion between those areas affected by ice streams and those that were not was remarkable (Sugden 1968, 1974). Based upon modern and ancient examples, Sugden went on to suggest that spatial variations in the amount of erosion were dictated by the thermal regime of the ice mass (e.g. Sugden 1976, 1977, 1978). This complemented the findings of Clayton (1965) on the Finger Lakes of New York and Gjessing (1966) in the Norwegian fjords. The contrast in erosional patterns over short distances in glaciated terrains had previously been noted and linked to possible nunataks by researchers like Linton (1949, 1950, 1952, 1963) and Sissons (1964). This fuelled ongoing debates about the occurrence and extent of full glacial refugia and nunataks during previous glaciations. Specifically, observations of deeply weathered blockfields and well developed tors on mountain summits at the continental margins of North America and Fennoscandinavia were interpreted as the remnants of Tertiary surfaces untouched by glacier ice during the Quaternary (e.g. Fernald 1925; Dahl 1946, 1955; Love and Love 1963; Ives 1966, 1978). This nunatak hypothesis has received renewed impetus recently in northwest Europe based upon the lowermost extent of well-developed periglacial features on mountain summits (e.g. Nesje *et al.* 1987, 1988; Nesje and Dahl 1990; Ballantyne 1997; Ballantyne *et al.* 1997, 1998; Ballantyne and McCarroll 1995).

In a model of glacial landscape development that encompassed the combined roles of ice dynamics and topography, Sugden (1974) proposed that upland surfaces with their preserved Tertiary features were juxtaposed with deeply eroded glacial troughs because they were covered by protective cold-based ice during full glacial conditions. Trough location is a function of positive feedbacks between subglacial topography, ice velocity basal temperature and erosion rates. The subglacial topography was initiated by preglacial fluvial and/or tectonic processes. Basal melting is accelerated beneath the thicker ice that accumulates in valley systems, thereby enhancing sliding rates and erosion. In contrast, the upland surfaces are covered by thin ice, which is more likely to be cold-based and non-erosive. The result over long periods of glaciation is a landscape of selective linear erosion. The selectivity of glacial erosion is now fully acknowledged in reconstructions of long-term landscape evolution in glaciated terrains. For example, the Palaeic surface on the interfjord plateaux in Norway contains Tertiary weathering residues that have survived numerous glaciations due to the concentration of glacial erosion in overdeepened preglacial valley systems (Nesje and Whillans 1994). Moreover, glacial landforms appear to have survived subsequent glaciations without being eroded due to the fact that they were located under cold-based ice (e.g. Kleman 1994; Kleman and Bergstrom 1994; see also pp. 8–9, vol. VII).

Esker genesis

I. Banerjee and B.C. McDonald (1975) Nature of esker sedimentation. In A.V. Jopling and B.C. McDonald (eds) *Glaciofluvial and Glaciolacustrine Sedimentation*. Tulsa, Okla.: SEPM Special Publication 23, pp. 132–54

The glacial geomorphological literature contains expansive descriptions of eskers and esker genesis (see Benn and Evans 1998, section 11.2.9) but a small number of papers serve as benchmarks of our improved understanding of esker formation and its implications for the operation of subglacial drainage networks. Of seminal status in this regard is the work of Shreve (1972), who provided the hydraulic theory to explain drainage pathways and esker patterns (see also Shreve 1985a, 1985b). Important contributions on esker genesis have come also from the interpretation of esker sedimentology and its association with esker morphology (e.g. Shaw 1972; Rust and Romanelli 1975; Saunderson 1975; Ringrose 1982; Hebrand and Amark 1989; Gorrell and Shaw 1991; Brennand 1994; Syverson *et al.* 1994). Esker classification schemes (e.g. Warren and Ashley 1994) now routinely incorporate information about the internal structure and general morphology, but the first attempt to undertake this type of assessment of esker process/form relationships was that of Bannerjee and McDonald (1975). Building upon the earlier theoretical and sedimentological work of Rothlisberger (1972), Shreve (1972) and Shaw (1972), the authors set out to provide models of esker sedimentation as dictated by the nature of the stream conduit and the site of deposition. From this they identified three models of esker sedimentation, including open channel, tunnel and deltaic systems. The latter was based upon one of the first exhaustive explanations of esker beads (see also Rust and Romanelli 1975).

Glacio-aqueous deposition

M. Church and R. Gilbert (1975) Proglacial fluvial and lacustrine environments. In A.V. Jopling and B.C. McDonald (eds) *Glaciofluvial and Glaciolacustrine Sedimentation*. Tulsa, Okla.: SEPM Special Publication 23, pp. 22–100

It is impossible in the space available, and beyond the scope of this volume, to highlight all the papers that would qualify as critical concept benchmarks in the understanding of fluvial and lacustrine depositional environments. Even the literature pertaining to glaciofluvial and glacilacustrine processes, forms and sediments contains numerous significant markers that could be identified for inclusion here (see Jopling 1975; Smith 1985; Smith and Ashley 1985; Ashley 1995; Benn and Evans 1998, Chapters 3, 8 and 10, for reviews).

It is therefore the widely cited overview of Church and Gilbert (1975) that is nominated, largely because of its lasting value as a synthesis of the principles of physical hydrology and their controls over sediment transport, deposition and classification in proglacial environments. Church (1972) had already delivered a seminal work in the shape of his intensive study of Baffin Island sandar and Gilbert (1971, 1972) had published some of the earliest reports on glacial lake sedimentation based on studies in the Canadian Rockies (see also Gustavson 1975). Their work therefore was at the vanguard of ground-breaking research efforts on proglacial sedimentation along with that of investigators such as Hjulstrom (1952, 1955), Krigstrom (1962), Fahnestock (1963), Smith (1970, 1974), McDonald and Banerjee (1971) and Bluck (1974). The thorough integration of hydrological parameters, field monitoring, sedimentology and morphology in the Church and Gilbert (1975) paper provided the foundations of concerted efforts to understand proglacial fluvial and lacustrine systems in the last quarter of the twentieth century. Moreover, it continues to be cited as a literary and directional benchmark in glacial research.

Medial moraine genesis and classification

N. Eyles and R.J. Rogerson (1978) A framework for the investigation of medial moraine formation: Austerdalsbreen, Norway, and Berendon Glacier, British Columbia, Canada. *Journal of Glaciology* 20(82): 99–113

Because they are striking features on the surfaces of most modern glaciers, medial moraines have been the subject of intensive research investigations since the classic work of Salisbury (1894). Even before Salisbury's work, a number of researchers had speculated on the origins of these supraglacial debris accumulations. As more examples were described and analysed it became clear that medial moraines were not solely the products of the joining of lateral moraines at valley glacier confluences. In the modern era of glacial research one classification scheme is regarded as the framework for medial moraine investigation and analysis, that of Eyles and Rogerson (1978a). Based upon systematic assessments of medial moraines on glaciers in Norway and Canada (Eyles and Rogerson 1978a, 1978b; Boulton and Eyles 1979) and incorporating Boulton's (1978) model of debris transport pathways in glacier systems, the Eyles and Rogerson classification recognized a wide variety of medial moraine origins. First, they identified *ablation-dominant moraines*, where englacial debris septa melt-out at the glacier surface, of which there are three sub-classifications. Specifically, these are the *below firn-line type* (AD1), the *above firn-line type* (AD2) and the *sub-glacial rock knob type* (AD3). Second, the *ice stream interaction* (ISI) type moraines are produced at the confluence of two valley glaciers and constitute the amalgamation of two lateral moraines. Third, the *avalanche type* (AT) moraines accumulate on the glacier surface below the locations of large rockslope failures where they become transported and attenuated by glacier flow. Acknowledgement of the general applicability and utility of this classification scheme is clearly illustrated by its repeated employment in terms of reference in medial moraine studies (e.g. Anderson 2000).

The abrasion process modelled

B. Hallet (1979) A theoretical model of glacial abrasion. *Journal of Glaciology* 23(89): 39–50

Striae, produced by the scoring of bedrock by debris carried in the base of a moving glacier, were some of the first features ever acknowledged as evidence of former glaciation (Agassiz 1838). Glacial geomorphologists have since routinely employed these abrasion damage tracks to map and reconstruct glacier flow histories (e.g. Viellette 1986; Sharp *et al.* 1989; Viellette *et al.* 1999), and evaluate the origins of glacially streamlined rock beds (e.g. Gjessing 1965; Boulton 1974; Goldthwait 1979; Rea *et al.* 2000). Glacial textbooks generally agree that six factors control the efficiency of the abrasion process: relative hardness of rock surface and impacting clast; force pressing the clast against the bed; velocity of clast relative to bed; concentration of debris in ice at the abrading surface; removal of wear products; and availability of basal debris. The first and most seminal attempts to quantify these factors and model the abrasion process were by Geoffrey Boulton (1974, 1975, 1979), based partly upon subglacial experiments at Breidamerkurjökull, Iceland, and Bernard Hallet (1979, 1981), based upon theoretical evaluations of abrasion products. Boulton's approach involved the application of the Coulomb equation and assumed that the friction between clasts in the ice and the rigid bed is proportional to the normal pressure pressing the surfaces together. So basal friction was thought to increase with ice thickness and to be inversely proportional to basal water pressure. It has hence become known as the *Coulomb friction model* (Schweizer and Iken 1992). Hallet (1979) proposed that ice will deform completely around subglacial clasts and that contact forces will be independent of ice thickness. This made the Coulomb friction model inapplicable to the abrasion problem because it relates to friction between rigid bodies. Alternatively, Hallet reasoned that the contact force between a clast and the bed is the sum of the bouyant weight of the particle and the drag force resulting from ice flow towards the bed. This became known as the *Hallet friction model*, and was later verified by laboratory experiment by Iverson (1990). A modified form of the Hallet friction model was later proposed by Schweizer and Iken (1992) in their sandpaper friction model, designed to account for particularly debris-rich basal ice.

A model for glacimarine sedimentation

R.D. Powell (1981) A model for sedimentation by tidewater glaciers. *Annals of Glaciology* 2: 129–134

Stratigraphic assessments of glacimarine sediments date back to the early work of Armstrong and Brown (1954), and the first systematic overview of glacimarine sedimentation was published by Carey and Ahmad (1961). This was followed by numerous descriptions of glacimarine sediment in a variety of ancient settings (see Andrew and Matsch 1983; Gravenor *et al.* 1984; Powell 1984; Eyles *et al.* 1985, for reviews and references). However, it was not until the work of Drewry and Cooper (1981), Nelson (1981) and Powell (1981) that models of glacimarine sedimentation accounted for spatial and temporal variability in processes and sediments associated with marine terminating glaciers. Powell's model (1981) was particularly significant in that it assimilated a growing body of knowledge on proximal glacimarine environments (e.g. Anderson *et al.* 1980; Elverhoi *et al.* 1980; Orheim and Elverhoi 1981; Domack 1982) and identified facies associations related to glacier snout morphology and dynamics, ice-proximity to site of deposition and water depth. Typical lateral variability in facies associations in stratigraphic sequences were then presented for use in interpretations of ancient glacimarine deposits, particularly in fjord settings. Powell (1984) later expanded his models of glacimarine sedimentation to account for the wide range of environmental settings and glacier types. A large literature on glacimarine processes and sediments has accumulated since the late 1970s and Powell's (1981) model provided the impetus for assessments of ice-marginal depo-centres in a wide variety of settings and linking spatial and temporal variability in sedimentary sequences due to the combined effects of glaciological and oceanographic parameters (e.g. Powell 1984, 1990; Bednarski 1988; Boulton 1990; Lønne 2001).

The subglacial paradigm shift

G.S. Boulton and R.C.A. Hindmarsh (1987) Sediment deformation beneath glaciers: rheology and geological consequences. *Journal of Geophysical Research* 92(B9): 9059–82

The most widely cited research output on subglacial processes is that of Geoffrey Boulton and his co-workers on the subglacial observations beneath Breidamerkurjökull, south Vatnajökull, Iceland. In a daring experiment that has never been duplicated, Boulton and his team excavated tunnels into the glacier margin and accessed the till at the glacier bed. They inserted segmented rods into the till which, when re-excavated several days later, had been displaced down glacier. This clearly demonstrated that the till was deforming subglacially via displacement in a two-tiered till. The upper and lower tiers were defined as the A and B horizons respectively, the A horizon being a porous, low density material and the B horizon being a dense, fissile layer. Moreover, the convex-upward displacement curve recognized in the A horizon was interpreted as the product of ductile flow, responsible for 80–95 per cent of the forward movement of the glacier. Deformation also occurred as brittle failure along discrete shear planes in the B horizon. Porewater pressures in the A horizon were generally high but fluctuated on a daily basis, driving changes in strain rates in the till. Although the concept of a subglacial deforming bed, based upon the two-tiered till sequence at Breidamerkurjökull, was proposed by Boulton et al. (1974), Boulton (1979) and Boulton and Jones (1979), it is the paper by Boulton and Hindmarsh (1987) that has had the most significant impact on the glaciological community. Initial impressions by Boulton and Jones (1979) that the Breidamerkurjökull till had a perfect plastic, Coulomb rheology were revised by Boulton and Hindmarsh (1987), who considered that the patterns of strain were more typical of a non-linearly viscous or Bingham material. Boulton and Hindmarsh (1987) also provided some thought-provoking theoretical concepts on the development of subglacial drainage channels or tunnel valleys in deforming sediment. Specifically, in locations where subglacial meltwater cannot be accommodated by Darcian flow through the deforming bed, piping failure initiates subglacial drainage conduits. Deforming sediment then creeps into the conduit where it is flushed out by meltwater and so the conduit cuts downwards into the bed, enlarging into a tunnel valley. This provided a theoretical explanation of tunnel valleys in areas of extensive deformation tills, although debate continues on subglacial drainage systems over soft substrates (e.g. Alley 1992; Walder and Fowler 1994).

Since the experiments at Breidamerkurjökull, evidence for subglacial deformation has been cited in numerous studies based upon field and laboratory experiments and interpretations of till sedimentology (see Phillips et al. in prep. for review). A lively debate on the rheology of tills has ensued, driven by the need to know more about the spatial and temporal behaviour of subglacial materials and their role in glacier dynamics. Hence, the Boulton and Hindmarsh (1987) paper marked the beginning of a new paradigm in glacial research, during which many benchmark research projects have contributed valuable knowledge on glacier-bed interactions (see the following section). The data collected at Breidamerkurjökull remain central to continuing debates on the subject of subglacial deformation, particularly because later experiments discovered different rheological properties in subglacial materials. Specifically, the position of deformation in subglacial sediments appeared to vary. Although the vertical deformation profile at Breidamerkurjökull was interpreted as typical of a non-linearly viscous material, later research documented failure at deeper levels in tills (e.g. Truffer et al. 2000; Tulaczyk et al. 2000a, 2000b) and supported a Coulomb-plastic rheology (e.g. Kamb 1991;

Iverson *et al.* 1998; Iverson and Iverson 2001). Most significantly, Iverson and Iverson (2001) reported a convex-upward displacement curve, thereby implying that such curves do not necessarily indicate a viscous rheology. Interestingly, more recent studies at Breidamerkurjökull have convinced Boulton *et al.* (2001) that variable responses by a till may be driven by temporal variations in water pressure/effective pressure. Consequently vertical variations in the locus of plastic failure can result in cumulative, distributed net strain in response to localized failure events.

Deformation confirmation

R.B. Alley, D.D. Blankenship, C.R. Bentley and S.T. Rooney (1987) Till beneath ice stream B: 3. Till deformation: evidence and implications. *Journal of Geophysical Research* 92(B9): 8921–9

Considerable efforts have been made in recent years to understand the dynamics of the Antarctic ice streams. Crucial to this understanding was the recovery of data on the nature of the ice-bed interface using high resolution seismic surveys beneath Ice Stream B. A series of influential papers were produced as a result of this research programme (see Alley *et al.* 1986, 1987a, 1987b and 1987c; Alley 1989; Blankenship *et al.* 1986, 1987; Rooney *et al.* 1987). The seismic data indicated that unfrozen till lay beneath the ice stream, an interpretation later confirmed by Engelhardt *et al.* (1990) when they recovered unfrozen till from the base of a borehole. It was proposed that most of the ice stream velocity is due to bed deformation, which takes place throughout the 6 m thick till layer. Implications for landform genesis were also provided by Alley *et al.* (1987a). They reported fluted bedrock, interpreted as the erosional products of the deforming till layer, and vast 'till deltas' at the grounding line of the ice stream. The latter are related to the partial decoupling of till and glacier sole at the outer margin of the ice stream where its surface slope is reduced and basal shear stresses fall.

The Ice Stream B research project proved to be a catalyst for further direct observations on ice stream dynamics and a widely cited modern analogue for palaeoglaciological reconstructions. A problem in the case of Ice Stream B was the fact that the till was too weak to support the basal shear stresses of the flowing ice. Therefore, it was proposed that the ice velocity is reduced by side drag at the ice stream walls and 'sticky spots' on the bed (MacAyeal 1992; Alley 1993; Anandakrishnan and Alley 1994; MacAyeal *et al.* 1995). This highlighted the fact that there is a large spatial variability in basal friction below ice streams due to till discontinuity, bed roughness and till inhomogeneity. Material extracted from the bed of Ice Stream B and subjected to shear experiments by Kamb (1991) failed along a single failure plane, contributing to the ongoing subglacial deformation debate (see the previous section). This was elaborated by Tulaczyk *et al.*

(2000a, 2000b), who reported that the failure surface in till below Ice Stream B migrates vertically on a diurnal cycle in response to changing water pressure and supply. Although some uncertainty remains about the linkages between deforming or shearing substrates, glacier flow and substrate streamlining, the recognition of extensive fields or flow sets of flutings both in Antarctic offshore surveys (e.g. Canals *et al.* 2000; Wellner *et al.* 2001; Ó Cofaigh *et al.* 2002) and in the terrestrial record (e.g. Clark and Stokes 2001a, 2001b, 2003; see also the following section) has resulted in the identification of palaeo-ice streams based largely upon the observations from the Antarctic ice streams. Additionally, Alley *et al.*'s concept (1987a) of a 'till delta' provided explanation for coarse-grained glacimarine depo-centres such as grounding line fans/wedges and more particularly trough mouth fans. The latter, located at the ends of vast offshore troughs, are now routinely interpreted as the products of palaeo-ice streams, especially when associated with flow sets of flutings (e.g. Boulton 1990; Powell and Domack 1995; Vorren and Laberg 1997; Vorren 2003).

Ice streams and overprinted ice flow signatures

A.S. Dyke and T.F. Morris (1988) Drumlin fields, dispersal trains, and ice streams in Arctic Canada. *Canadian Geographer* 32(1): 86–90

Ice sheet reconstruction remains central to understanding full glacial global climate conditions. Of specific interest are not only the extent and duration of ice sheet coverage but also the dynamics of ice sheets and their interaction with, and effects on, regional climate systems. Crucial in this respect has been the discovery that ice sheets were not single domed bodies that maintained the same internal flow geometries throughout glacial cycles but were instead composed of multiple dispersal centres that migrated through time in response to changing ice thicknesses and extents (Boulton and Clark 1990a, 1990b; Clark 1993, 1994). This gave rise to changes in flow patterns and the overprinting of geomorphic and sedimentary signatures in the landform record (see Benn and Evans 1998, section 12.4.4 for a review). This knowledge of palaeo-ice sheet dynamics has been accumulated through both the meticulous observations of field researchers and the application of modern glaciological concepts to interpretations of ice sheet landform assemblages. For example, drumlin fields have long been associated with the locations of former areas of sustained or fast flow in ice sheets and glaciers, but the discovery of fluted surfaces beneath the ice streams of Antarctica (see the previous section) prompted glacial geomorphologists to attempt identification of the imprints of palaeo-ice streams.

The paper by Dyke and Morris (1988) is significant in two respects. First, it combines various forms of evidence such as suites of subglacially streamlined landforms, regional till composition and erratic dispersal to identify ice streaming in the north sector of the former Laurentide Ice Sheet. Second, it clearly identifies cross-cutting patterns in the palaeo-ice flow directions, demonstrating that the ice streams changed location through the last glacial cycle. Although overprinted subglacial landforms had been reported previously (e.g. Fairchild 1929; Rose and Letzer 1977; Mollard 1984; Riley 1987), they had not been integrated into ice sheet dynamics at the scale proposed by Dyke and Morris (1988). Moreover, palaeo-ice streams were identified confidently for the first time, drawing upon intensive field mapping of glacial features over a vast region of arctic Canada (Dyke *et al.* 1982; Dyke 1983, 1984). The identification of overprinted ice stream signatures aided in the reconstruction of wet and cold based zones beneath the ice sheet and the location of the shifting M'Clintock ice divide. In addition to identifying the landform evidence for fast flow at the centre of the ice stream, Dyke and Morris (1988) also describe a lateral shear moraine formed at the boundary between warm, streaming ice and adjacent cold-based ice. The final stages of streaming in the region of their study were associated with localized draw-down of the ice into marine embayments flooded by glacio-isostatically controlled high sea-levels. Thus the field evidence was interpreted in the context of the physics of ice sheet behaviour (e.g. Hughes *et al.* 1985). Another remarkable feature described in Dyke and Morris's paper for the first time is the 'Boothia type' dispersal train, a feature that clearly records former ice stream flow within the till lithology of the region.

The operation and impacts of subglacial hydrology

M. Sharp, J.C. Gemmell and J.-L. Tison (1989) Structure and stability of the former subglacial drainage system of the Glacier de Tsanfleuron, Switzerland. *Earth Surface Processes and Landforms* 14: 119–34

The importance of the subglacial drainage system to glacier dynamics and sediment transfer in glacier systems is reflected in the large number of seminal papers on the subject over the last 30 years. Glaciologists now understand that meltwater moves subglacially either in discrete, channelized systems or in distributed systems. Discrete channels include Rothlisberger or 'R' channels cut into the ice and Nye or 'N' channels and tunnel valleys cut into the substrate (Rothlisberger 1972; Shreve 1972; Nye 1973; Lliboutry 1983). Distributed systems include water films, linked cavity networks, braided canal networks and porewater (Darcian) flow (Weertman 1964, 1969, 1972; Lliboutry 1968, 1976, 1979; Nye 1976; Walder and Hallet 1979; Walder 1982, 1986; Kamb 1987; Walder and Fowler 1994). Temporal changes in the type of subglacial drainage that exists beneath modern glaciers have been monitored in recent research (e.g. Kamb *et al.* 1985; Seaberg *et al.* 1988;

Willis *et al.* 1990; Fountain 1993, 1994; Hubbard *et al.* 1995). The identification of former subglacial meltwater systems in the geomorphological record has been the focus of several studies (e.g. Walder and Hallet 1979; Hallet and Anderson 1982; Sharp, Gemmell and Tison 1989). Additionally, the possible sedimentological signatures of subglacial drainage have been identified in ancient till sequences (e.g. Eyles *et al.* 1982; Brown *et al.* 1987; Clark and Walder 1994; Evans *et al.* 1995). Given the large number of important contributions on subglacial drainage networks it is difficult to identify the most significant critical concept paper on the subject. Of the papers that concentrate on the geomorphological impact of subglacial drainage, it is the contribution by Sharp, Gemmell and Tison (1989) that appears to have been most widely cited in modern research. Although it was not the first of its kind (see Walder and Hallet 1979; Hallet and Anderson 1982), the Sharp *et al.* (1989) paper reported on an intricate assemblage of subglacial meltwater forms cut in limestone bedrock and provided a clear linkage between glaciological theory, process measurements and geomorphology. As such it is a benchmark example of reconciling glaciology with glacial geomorphology.

A problem with melt-out

M.A. Paul and N. Eyles (1990) Constraints on the preservation of diamict facies (melt-out tills) at the margins of stagnant glaciers. *Quaternary Science Reviews* 9(1): 51–69

After the definition of melt-out till and its genesis by Boulton (1970b, 1972a; see pp. 11–12), a subglacial melt-out origin was proposed in a variety of glaciated basins for stratified diamictons (e.g. Shaw 1979, 1982, 1983b; Haldorsen and Shaw 1982). This was followed by research on environments of subglacial melt-out till production in which the optimum conditions for its accumulation were evaluated (e.g. Mickelson 1971, 1973, 1986; Lawson 1979a, 1979b; Ronnert and Mickelson 1992; Ham and Mickelson 1994). Melt-out till is still regarded as one of the prime subglacial till types by glacial geomorphologists but the geotechnical constraints on its preservation potential had not been systematically and objectively evaluated until Paul and Eyles (1990) undertook the task. The slow, passive release of sediment from melting debris-rich basal glacier ice was first proposed by Goodchild (1875) and the importance of geothermal heat provision to the subglacial melt-out process was modelled in a benchmark glaciological paper by Nobles and Weertman (1971). By applying thaw-consolidation theory, Paul and Eyles (1990) demonstrate that the optimum environmental conditions for the passive accumulation of melt-out till due to the geothermal heat flux to stagnant debris-rich glacier ice are unlikely to be widespread. Poor drainage and high ice contents will tend to produce high porewater pressures in the accumulating melt-out sequence thereby initiating failure, remobilization and dewatering. Where debris concentrations in the parent ice are high, the thaw consolidation ratio is low. In such situations delicate englacial structures may be preserved, especially where meltwater drainage is efficient. The degree of modification to former englacial foliation and structure is dictated by the initial debris concentration and the nature of the drainage conditions at the site of deposition. The implications for the interpretation of thick and laterally extensive till units are significant, suggesting specifically that the melt-out process is likely to produce subglacial melt-out tills that are discontinuous and patchy.

Quarrying explained

N.R. Iverson (1991) Potential effects of subglacial water-pressure fluctuations on quarrying. *Journal of Glaciology* 37(125): 27–36

When studied at the microscopic scale, striae (see p. 16) and subglacially quarried features, such as chattermarks and fractures, have much in common in that they are the products of rock fragment removal from the glacier bed by brittle failure. The removal of rock fragments ranging in size from microscopic chips within a striation to fragments larger than 1 cm on the quarried steps of a roches moutonnées is due to temporary stress concentrations below overriding clasts. These stresses enlarge existing weaknesses in the bed and lead to the removal of fragments due to discontinuous rock mass failure (Addison 1981). Although the products of quarrying or plucking had long been associated with the dragging of sharp tools, often in a jerky fashion, across rock surfaces, modelling of the stress patterns associated with the drag of individual clasts came only relatively recently (e.g. Ficker *et al.* 1980). Differential stresses in rock knobs were modelled by Morland and Boulton (1975), who explained the initiation of fractures in softer rocks but proposed that harder substrates needed to contain pre-existing joints if they were to fail. The removal of loosened fragments by regelation (*heat pump effect*) was elucidated by the seminal work of Robin (1976). Neil Iverson (1991) contributed a significant paper to the subglacial erosion debate when he provided an explanation for the fracture of intact hard rock beds. Fundamental to his model was the occurrence of low pressure cavities at the ice–bedrock interface. Iverson demonstrated that variability in water pressures was crucial to the quarrying of stepped rock beds, especially where supraglacial meltwater was accessing subglacial cavities. Specifically, falling water pressures led to a rise in principal stresses and vertical ice velocities, causing rock fragments to spall off the back of rock steps in the bed. The influence of cavities was later acknowledged as critical to the quarrying rate by Hallet (1996). He also proposed that the optimum conditions for quarrying were met during the 1982–3 surge of Variegated

Glacier in Alaska when effective pressures were reduced and an extensive linked cavity subglacial drainage system was in operation (Hallet *et al.* 1996).

Flutings and ploughing ice

S.M. Tulaczyk, R.P. Scherer and C.D. Clark (2001) A ploughing model for the origin of weak tills beneath ice streams: a qualitative treatment. *Quaternary International* 86: 59–70

Boulder ploughing and lodgement in soft subglacial sediment has long been recognized (e.g. Weertman 1964; Beget 1986; Brown *et al.* 1987; Clark and Hansel 1989) and is fundamental to some models of fluting genesis (e.g. Gordon *et al.* 1992; Benn 1994b). The potential for grooves in the ice base to plough through soft substrates has only recently been proposed by Tulaczyk *et al.* (2001) and has far reaching implications for subglacial till genesis and substrate streamlining. Tulaczyk *et al.* (2001) proposed ice ploughing as a till producing mechanism because previous models of subglacial deformation had been unable to explain the Coulomb-plastic till rheology reproduced in laboratory experiments and field measurements of basal sliding and/or shallow deformation. Additionally, the intra-till inhomogeneities in porosity, composition and microfossil content detected in tills recovered from beneath Ice Stream B (Scherer *et al.* 1998; Tulaczyk *et al.* 1998) are thought to be inconsistent with pervasive deformation to depths greater than 1.5 m. The ploughing model advocates disturbance and transportation of subglacial till up to a few metres thick by the keels in a sliding, bumpy glacier sole. Essentially, the till deforms around bumps in the ice sole as they are dragged through the substrate. At the base of the West Antarctic ice streams this model explains why streamlining of the substrate takes place even though the basal ice is debris-poor and the underlying till is fine-grained and clast-poor. Tulaczyk *et al.* (2001) propose an analogy with the fault gouge model of Eyles and Boyce (1998) in which the sliding ice base acts as the rigid upper fault plate. The asperities in this fault plate, whether clasts or bumps/keels in the ice sole, plough through the soft substrate, generating the deformable till (fault gouge layer) above the underlying, rigid strata (lower fault plate). Only large keels or bumps in the ice sole will generate new till, because they can protrude through the existing till layer into the underlying substrate. Smaller bumps act to transport the existing till.

Because the ploughing model requires no sediment input from the glacier base, it is particularly appropriate as an explanation of substrate deformation and streamlining of pre-existing sediments, avoiding inherent problems of till continuity (Alley 2000). It also explains many of the features observed in glacially streamlined terrains (e.g. Clark 1993, 1994; Canals *et al.* 2000; Clark and Stokes 2001a, 2001b; Wellner *et al.* 2001; Ó Cofaigh *et al.* 2002).

Sedimentologically, the ploughing model helps to explain extensive till units of relatively uniform thickness in that there is a 'stabilizing feedback' (Tulaczyk *et al.* 2001). Specifically, as a till thickens fewer ice bumps can penetrate through it to the underlying substrate, thereby arresting and ultimately terminating the thickening rate. Alternatively, if a till thins, more ice bumps will protrude into and erode the substrate, thereby generating more till. Time will tell whether or not the ploughing model holds up to intense scrutiny by the glacial geomorphological community.

References

Aber, J.S. (1982) Model for glaciotectonism. *Bulletin of the Geological Society of Denmark* 30: 79–90.

Aber, J.S., Croot, D.G. and Fenton, M.M. (1989) *Glaciotectonic Landforms and Structures*. Dordrecht: Kluwer, 200pp.

Addison, K. (1981) The contribution of discontinuous rock mass failure to glacier erosion. *Annals of Glaciology* 2: 3–10.

Agassiz, L. (1838) On the polished and striated surfaces of the rocks which form the beds of glaciers in the Alps. *Proceedings of the Geological Society of London* 3: 321–2.

Alley, R.B. (1989) Water pressure coupling of sliding and bed deformation. *Journal of Glaciology* 35: 108–39.

Alley, R.B. (1991) Deforming bed origin for southern Laurentide till sheets? *Journal of Glaciology* 37: 67–76.

Alley, R.B. (1992) How can low-pressure channels and deforming tills coexist subglacially? *Journal of Glaciology* 38: 200–7.

Alley, R.B. (1993) In search of ice stream sticky spots. *Journal of Glaciology* 39: 447–54.

Alley, R.B. (2000) Continuity comes first: recent progress in understanding subglacial deformation. In A.J. Maltman, B. Hubbard and M.J. Hambrey (eds) *Deformation of Glacial Materials*. Geological Society Special Publication no. 176, pp. 171–9.

Alley, R.B., Blankenship, D.D., Bentley, C.R. and Rooney, S.T. (1986) Deformation of till beneath Ice Stream B, West Antarctica. *Nature* 322: 57–9.

Alley, R.B., Blankenship, D.D., Bentley, C.R. and Rooney, S.T. (1987a) Till beneath ice stream B: 3. Till deformation: evidence and implications. *Journal of Geophysical Research* 92: 8921–9.

Alley, R.B., Blankenship, D.D., Bentley, C.R. and Rooney, S.T. (1987b) Till beneath Ice Stream B 4: A coupled ice-till flow model. *Journal of Geophysical Research* 92: 8931–40.

Alley, R.B., Blankenship, D.D., Rooney, S.T. and Bentley, C.R. (1987c) Continuous till deformation beneath ice sheets. In E.D. Waddington and J.S. Walder (eds) *The Physical Basis of Ice Sheet Modelling*. IAHS Publication 170, pp. 81–91.

Alley, R.B., Lawson, D.E., Evenson, E.B., Strasser, J.C. and Larson, G.J. (1998) Glaciohydraulic supercooling: a freeze-on mechanism to create stratified, debris-rich basal ice. II. Theory. *Journal of Glaciology* 44: 563–9.

Alley, R.B., Strasser, J.C., Lawson, D.E., Evenson, E.B. and Larson, G.J. (1999) Glaciological and geological implications of basal-ice accretion in overdeepening.

In D.M. Mickelson and J.W. Attig (eds) *Glacial Processes: Past and Present.* Geological Society of America, Special Paper 337, pp. 1–9.

Anandakrishnan, S. and Alley, R.B. (1994) Ice Stream C, Antarctica, sticky spots detected by microearthquake monitoring. *Annals of Glaciology* 20: 183–6.

Anderson, J.B., Kurtz, D.D., Domack, E.W. and Balshaw, K.M. (1980) Antarctic glacial marine sediments. *Journal of Geology* 88: 399–414.

Anderson, R.S. (2000) A model of ablation-dominated medial moraines and the generation of debris-mantled glacier snouts. *Journal of Glaciology* 46: 459–69.

Andrade, E.N. da C. and Fox, J.W. (1949) The mechanism of dilatancy. *Proceedings of the Physical Society, Sect. B*, 62: 483–500.

Andrews, J.T. (1963a) Cross-valley moraines of north-central Baffin Island: a quantitative analysis. *Geographical Bulletin* 20: 82–129.

Andrews, J.T. (1963b) Cross-valley moraines of the Rimrock and Isortoq River valleys, Baffin Island, North West Territories. *Geographical Bulletin* 19: 49–77.

Andrews, J.T. (1963c) The cross-valley moraines of north-central Baffin Island, NWT: a descriptive analysis. *Geographical Bulletin* 20: 82–129.

Andrews, J.T. (1965) Surface boulder orientation studies around the north-western margins of the Barnes Ice Cap, Baffin Island, Canada. *Journal of Sedimentary Petrology* 35: 753–8.

Andrews, J.T. and Matsch, C.L. (1983) *Glacial Marine Sediments and Sedimentation.* Norwich: Geobooks, 227pp.

Andrews, J.T. and Shimizu, K. (1966) Three-dimensional vector technique for analysing till fabrics: discussion and Fortran progam. *Geographical Bulletin* 8: 151–65.

Andrews, J.T. and Smithson, B.B. (1966) Till fabrics of the cross-valley moraines of north-central Baffin Island, NWT, Canada. *Bulletin of the Geological Society of America* 77: 271–90.

Armstrong, J.E. and Brown, W.J. (1954) Late Wisconsin marine drift and associated sediments at the lower Fraser Valley, British Columbia, Canada. *Bulletin of the Geological Society of America* 65: 349–64.

Ashley, G.M. (1995) Glaciolacustrine environments. In J. Menzies (ed.) *Modern Glacial Environments – Processes, Dynamics and Sediments.* Oxford: Butterworth-Heinemann, pp. 417–44.

Astakhov, V.I. and Isayeva, L.L. (1988) The 'ice hill': an example of 'retarded deglaciation' in Siberia. *Quaternary Science Reviews* 7: 29–40.

Baker, V.R., Benito, G. and Rudoy, A.N. (1993) Paleohydrology of late Pleistocene superflooding, Altay Mountains, Siberia. *Science* 259: 348–50.

Baker, V.R., Greeley, R., Komar, P.D., Swanson, D.A. and Waitt, R.B., Jr. (1987) Columbia and Snake River Plains. In W.L. Graf (ed.) *Geomorphic Systems of North America.* Boulder, Col.: Geological Society of America, Centennial Special Volume 2, pp. 403–68.

Ballantyne, C.K. (1997) Periglacial trimlines in the Scottish Highlands. *Quaternary International* 38–9: 119–36.

Ballantyne, C.K. and McCarroll, D. (1995) The vertical dimensions of Late Devensian glaciation on the mountains of Harris and southeast Lewis, Outer Hebrides, Scotland. *Journal of Quaternary Science* 10: 211–23.

Ballantyne, C.K., McCarroll, D., Nesje, A. and Dahl, S.O. (1997) Periglacial trimlines, former nunataks and the altitude of the last ice sheet in Wester Ross, northwest Scotland. *Journal of Quaternary Science* 12: 225–38.

Ballantyne, C.K., McCarroll, D., Nesje, A., Dahl, S.O. and Stone, J.O. (1998) The last ice sheet in north-west Scotland: Reconstruction and implications. *Quaternary Science Reviews* 17: 1149–84.

Banerjee, I. and McDonald, B.C. (1975) Nature of esker sedimentation. In A.V. Jopling and B.C. McDonald (eds) *Glaciofluvial and Glaciolacustrine Sedimentation.* Tulsa, Okla.: SEPM Special Publication 23, pp. 132–54.

Battle, W.R.B. and Lewis, W.V. (1951) Temperature observations in bergschrunds and their relationship to cirque erosion. *Journal of Geology* 59: 537–45.

Beaumont, P. (1971) Break of slope in particle size curves of glacial tills. *Sedimentology* 16: 125–8.

Bednarski, J. (1988) The geomorphology of glaciomarine sediments in a high arctic fiord. *Geographie Physique et Quaternaire* 42: 65–74.

Beget, J.E. (1986) Modelling the influence of till rheology on the flow and profile of the Lake Michigan lobe, southern Laurentide Ice Sheet. *Journal of Glaciology* 32: 235–41.

Benn, D.I. (1994a) Fabric shape and the interpretation of sedimentary fabric data. *Journal of Sedimentary Research* A64: 910–15.

Benn, D.I. (1994b) Fluted moraine formation and till genesis below a temperate glacier: Slettmarkbreen, Jotunheimen, Norway. *Sedimentology* 41: 279–92.

Benn, D.I. (1995) Fabric signature of subglacial till deformation, Breidamerkurjökull, Iceland. *Sedimentology* 42: 735–47.

Benn, D.I. and Ballantyne, C.K. (1994) Reconstructing the transport history of glacigenic sediments: a new approach based on the co-variance of clast shape indices. *Sedimentary Geology* 91: 215–27.

Benn, D.I. and Evans, D.J.A. (1996) The interpretation and classification of subglacially-deformed materials. *Quaternary Science Reviews* 15: 23–52.

Benn, D.I. and Evans, D.J.A. (1998) *Glaciers and Glaciation.* London: Arnold, 734pp.

Benn, D.I. and Ringrose, T.J. (2001) Random variation of fabric eigenvalues: implications for the use of A-axis fabric data to differentiate till facies. *Earth Surface Processes and Landforms* 26: 295–306.

Bennett, M.R., Hambrey, M.J., Huddart, D. and Glasser, N.F. (1998) Glacial thrusting and moraine-mound formation in Svalbard and Britain: the example of Coire a' Cheud-chnoic (Valley of Hundred Hills), Torridon, Scotland. *Quaternary Proceedings* 6: 17–34.

Bennett, M.R., Waller, R.I., Glasser, N.F., Hambrey, M.J. and Huddart, D. (1999) Glacigenic clast fabrics: genetic fingerprint or wishful thinking? *Journal of Quaternary Science* 14: 125–35.

Bishop, B.C. (1957) *Shear Moraines in the Thule Area, Northwest Greenland.* US Snow, Ice and Permafrost Research Establishment, Research Report 17.

Blankenship, D.D., Bentley, C.R., Rooney, S.T. and Alley, R.B. (1986) Seismic measurements reveal a saturated porous layer beneath an active Antarctic ice stream. *Nature* 322: 54–7.

Blankenship, D.D., Bentley, C.R., Rooney, S.T. and Alley, R.B. (1987) Till beneath ice stream B: 1. Properties derived from seismic travel times. *Journal of Geophysical Research* 92: 8903–11.

Bluck, B.J. (1974) Structure and directional properties of some valley sandur deposits in southern Iceland. *Sedimentology* 21: 533–44.

Boulton, G.S. (1967) The development of a complex supraglacial moraine at the margin of Sorbreen, Ny Friesland, Vestspitsbergen. *Journal of Glaciology* 6: 717–36.

Boulton, G.S. (1968) Flow tills and related deposits on some Vestspitsbergen glaciers. *Journal of Glaciology* 7: 391–412.

Boulton, G.S. (1970a) On the origin and transport of englacial debris in Svalbard glaciers. *Journal of Glaciology* 9: 213–29.

Boulton, G.S. (1970b) On the deposition of subglacial and melt-out tills at the margins of certain Svalbard glaciers. *Journal of Glaciology* 9: 231–45.

Boulton, G.S. (1971) Till genesis and fabric in Svalbard, Spitsbergen. In R.P. Goldthwait (ed.) *Till: A Symposium*. Columbus: Ohio State University Press, pp. 41–72.

Boulton, G.S. (1972a) Modern arctic glaciers as depositional models for former ice sheets. *Journal of the Geological Society of London* 128: 361–93.

Boulton, G.S. (1972b) The role of the thermal regime in glacial sedimentation. In R.J. Price and D.E. Sugden (eds) *Polar Geomorphology*. Institute of British Geographers, Special Publication 4, pp. 1–19.

Boulton, G.S. (1974) Processes and patterns of subglacial erosion. In D.R. Coates (ed.) *Glacial Geomorphology*. Binghamton: State University of New York, pp. 41–87.

Boulton, G.S. (1975) Processes and patterns of subglacial sedimentation: a theoretical approach. In A.E. Wright and F. Moseley (eds) *Ice Ages: Ancient and Modern*. Liverpool: Seel House Press, pp. 7–42.

Boulton, G.S. (1976) The origin of glacially fluted surfaces: observations and theory. *Journal of Glaciology* 17: 287–309.

Boulton, G.S. (1977) A multiple till sequence formed by a Late Devensian Welsh ice cap: Glanllynnau, Gwynedd. *Cambria* 4: 10–31.

Boulton, G.S. (1978) Boulder shapes and grain size distributions of debris as indicators of transport paths through a glacier and till genesis. *Sedimentology* 25: 773–99.

Boulton, G.S. (1979) Processes of glacier erosion on different substrata. *Journal of Glaciology* 23: 15–38.

Boulton, G.S. (1982) Subglacial processes and the development of glacial bedforms. In R. Davidson-Arnott, W. Nickling and B.D. Fahey (eds) *Research in Glacial, Glacio-fluvial and Glacio-lacustrine Systems*. Norwich: Geo Books, pp. 1–31.

Boulton, G.S. (1986) A paradigm shift in glaciology? *Nature* 322: 18.

Boulton, G.S. (1987) A theory of drumlin formation by subglacial sediment deformation. In J. Menzies and J. Rose (eds) *Drumlin Symposium*. Rotterdam: Balkema, pp. 25–80.

Boulton, G.S. (1990) Sedimentary and sea level changes during glacial cycles and their control on glacimarine facies architecture. In J.A. Dowdeswell and J.D. Scourse (eds) *Glacimarine Environments: Processes and Sediments*. Geological Society, Special Publication 53, pp. 15–52.

Boulton, G.S. and Clark, C.D. (1990a) A highly mobile Laurentide Ice Sheet revealed by satellite images of glacial lineations. *Nature* 346: 813–17.

Boulton, G.S. and Clark, C.D. (1990b) The Laurentide Ice Sheet through the last glacial cycle: drift lineations as a key to the dynamic behaviour of former ice sheets. *Transactions of the Royal Society of Edinburgh, Earth Sciences* 81: 327–47.

Boulton, G.S., Dent, D.L. and Morris, E.M. (1974) Subglacial shearing and crushing, and the role of water pressures in tills from south-east Iceland. *Geografiska Annaler* 56A: 135–45.

Boulton, G.S., Dobbie, K.E. and Zatsepin, S. (2001) Sediment deformation beneath glaciers and its coupling to the subglacial hydraulic system. *Quaternary International* 86: 3–28.

Boulton, G.S. and Eyles, N. (1979) Sedimentation by valley glaciers: a model and genetic classification. In C. Schluchter (ed.) *Moraines and Varves*. Rotterdam: Balkema, pp. 11–23.

Boulton, G.S. and Hindmarsh, R.C.A. (1987) Sediment deformation beneath glaciers: rheology and geological consequences. *Journal of Geophysical Research* 92(B9): 9059–82.

Boulton, G.S. and Jones, A.S. (1979) Stability of temperate ice caps and ice sheets resting on beds of deformable sediment. *Journal of Glaciology* 24: 29–43.

Bowen, D.Q. and Gregory, K.J. (1965) A glacial drainage system near Fishguard, Pembrokeshire. *Proceedings of the Geologists' Association* 76: 275–82.

Brennand, T.A. (1994) Macroforms, large bedforms and rhythmic sedimentary sequences in subglacial eskers, south-central Ontario: implications for esker genesis and meltwater regime. *Sedimentary Geology* 91: 9–55.

Bretz, J.H. (1923a) The channeled scablands of the Columbia Plateau. *Journal of Geology* 31: 617–49.

Bretz, J.H. (1923b) Glacial drainage on the Columbia Plateau. *Bulletin of the Geological Society of America* 34: 573–608.

Bretz, J.H. (1925) The Spokane Flood beyond the Channeled Scabland. *Journal of Geology* 33: 97–115, 232–59.

Bretz, J.H. (1927) Channeled Scabland and the Spokane flood. *Journal of the Washington Academy of Science* 17: 200–11.

Bretz, J.H. (1928a) Alternative hypotheses for Channeled Scabland. *Journal of Geology* 36: 193–223, 312–41.

Bretz, J.H. (1928b) Bars of the Channeled Scabland. *Bulletin of the Geological Society of America* 39: 643–702.

Bretz, J.H. (1928c) The Channeled Scabland of eastern Washington. *Geographical Review* 18: 446–77.

Bretz, J.H. (1932) *The Grand Coulee*. American Geographical Society, Special Publication 15, 89pp.

Bretz, J.H. (1959) *Washington's Channeled Scabland*. Washington Department of Conservation, Division of Mines and Geology, Bulletin 45, 57pp.

Bretz, J.H. (1969) The Lake Missoula floods and the Channeled Scabland. *Journal of Geology* 77: 505–43.

Bretz, J.H., Smith, H.T.U. and Neff, G.E. (1956) Channeled Scabland of Washington: new data and interpretations. *Bulletin of the Geological Society of America* 67: 957–1059.

Brown, N.E., Hallet, B. and Booth, D.B. (1987) Rapid soft-bed sliding of the Puget glacial lobe. *Journal of Geophysical Research* 92: 8985–97.

Canals, M., Urgeles, R. and Calafat, A.M. (2000) Deep sea-floor evidence of past ice streams off the Antarctic Peninsula. *Geology* 28: 31–4.

Carey, S.W. and Ahmad, N. (1961) Glacial marine sedimentation. In *Proceedings of the First International Symposium on Arctic Geology*. Vol. 2. Toronto: University of Toronto Press, pp. 865–94.

Chamberlin, T.C. (1895) Recent glacial studies in Greenland. *Bulletin of the Geological Society of America* 6: 199–220.

Charlesworth, J.K. (1929) The South Wales end moraine. *Quarterly Journal of the Geological Society of London* 85: 335–58.

Chorley, R.J. (1959) The shape of drumlins. *Journal of Glaciology* 3: 339–44.

Church, M. (1972) *Baffin Island Sandurs: A Study of Arctic Fluvial Processes.* Geological Survey of Canada Bulletin 216.

Church, M. and Gilbert R. (1975) Proglacial fluvial and lacustrine environments. In A.V. Jopling and B.C. McDonald (eds) *Glaciofluvial and Glaciolacustrine Sedimentation.* Tulsa, Okla.: SEPM Special Publication 23, pp. 22–100.

Clark, C.D. (1993) Mega-scale glacial lineations and cross-cutting ice flow landforms. *Earth Surface Processes and Landforms* 18: 1–29.

Clark, C.D. (1994) Large scale ice-moulded landforms and their glaciological significance. *Sedimentary Geology* 91: 253–68.

Clark, C.D. and Stokes, C.R. (2001a) Extent and basal characteristics of the M'Clintock Channel ice stream. *Quaternary International* 86: 81–101.

Clark, C.D. and Stokes, C. (2001b) Palaeo-ice streams. *Quaternary Science Reviews* 20: 1437–57.

Clark, C.D. and Stokes, C. (2003) Palaeo-ice stream landsystem. In D.J.A. Evans (ed.) *Glacial Landsystems.* London: Arnold, pp. 204–27.

Clark, P.U. and Hansel, A.K. (1989) Clast ploughing, lodgement and glacier sliding over a soft glacier bed. *Boreas* 18: 201–07.

Clark, P.U. and Walder, J.S. (1994) Subglacial drainage, eskers, and deforming beds beneath the Laurentide and Eurasian ice sheets. *Geological Society of America Bulletin* 106: 304–14.

Clarke, G.K.C. (1986) Professor Mathews, outburst floods, and other glaciological disasters. *Canadian Journal of Earth Sciences* 23: 859–68.

Clayton, K.M. (1965) Glacial erosion in the Finger Lakes region, New York State, USA. *Zeitschrift für Geomorphologie* 9: 50–62.

Croot, D.G. (1988a) Morphological, structural and mechanical analysis of neoglacial ice-pushed ridges in Iceland. in D.G. Croot (ed.) *Glaciotectonics: Forms and Processes.* Rotterdam: Balkema, pp. 33–47.

Croot, D.G. (1988b) Glaciotectonics and surging glaciers, a correlation based on Vestspitsbergen, Svalbard, Norway. In D.G. Croot (ed.) *Glaciotectonics: Forms and Processes.* Rotterdam: Balkema, pp. 49–61.

Czechowna, L. (1953) Zagadnienie drumlinow w swietle literatury. *Czasop. Geogr.* 23/24: 50–90. (English translation: 'The question of drumlins in literature' in D.J.A. Evans (ed.) (1994) *Cold Climate Landforms.* Chichester: Wiley, 269–91.)

Dahl, E. (1946) On different types of unglaciated areas during the ice ages and their significance to phytogeography. *The New Phytologist* 45: 225–42.

Dahl, E. (1955) Biogeographic and geological indications of unglaciated areas in Scandinavia during the glacial ages. *Bulletin of the Geological Society of America* 66: 1499–519.

Davis, W.M. (1926) The value of outrageous geological hypotheses. *Science* 63: 463–8.

Derbyshire, E. (1962) Fluvioglacial erosion near Knob Lake, central Quebec-Labrador, Canada. *Bulletin of the Geological Society of America* 73: 1111–26.

Domack, E.W. (1982) Sedimentology of glacial and glacial marine deposits on the George V Adelie continental shelf, East Antarctica. *Boreas* 11: 79–97.

Dreimanis, A. and Vagners, U.J. (1971) Bimodal distribution of rock and mineral fragments in basal tills. In R.P. Goldthwait (ed.) *Till: A Symposium*. Columbus: Ohio State University Press, pp. 237–50.

Dreimanis, A. and Vagners, U.J. (1972) The effect of lithology on the texture of till. In E. Yatsu and A. Falconer (eds) *Research Methods in Pleistocene Geomorphology*. Guelph: University of Guelph, pp. 66–82.

Drewry, D.J. and Cooper, A.P.R. (1981) Processes and models of Antarctic glaciomarine sedimentation. *Annals of Glaciology* 2: 117–22.

Dwerryhouse, A.R. (1902) Glaciation of Teesdale, Weardale, the Tyne Valley and their tributary valleys. *Quarterly Journal of the Geological Society of London* 58: 572–608.

Dyke, A.S. (1983) *Quaternary Geology of Somerset Island, District of Franklin*. Geological Survey of Canada, Memoir 404.

Dyke, A.S. (1984) *Quaternary Geology of Boothia Peninsula and Northern District of Keewatin, Central Canadian Arctic*. Geological Survey of Canada, Memoir 407.

Dyke, A.S. (1993) Landscapes of cold-centred Late Wisconsinan ice caps, arctic Canada. *Progress in Physical Geography* 17: 223–47.

Dyke, A.S., Andrews, J.T. and Miller, G.H. (1982) *Quaternary Geology of Cumberland Peninsula, Baffin Island, District of Franklin*. Geological Survey of Canada, Memoir 403.

Dyke, A.S., Dredge, L.A. and Vincent, J.-S. (1982) Configuration and dynamics of the Laurentide Ice Sheet during the Late Wisconsinan maximum. *Geographie Physique et Quaternaire* 36: 5–14.

Dyke, A.S. and Morris, T.F. (1988) Drumlin fields, dispersal trains and ice streams in arctic Canada. *Canadian Geographer* 32: 86–90.

Dyke, A.S. and Savelle, J.M. (2000) Major end moraines of Younger Dryas age on Wollaston Peninsula, Victoria Island, Canadian arctic: implications for paleoclimate and for formation of hummocky moraine. *Canadian Journal of Earth Sciences* 37: 601–19.

Elson, J.A. (1966) Early discoverers XXIII: Till stone orientation – Henry Youle Hind (1823–1908). *Journal of Glaciology* 6: 303–6.

Elverhoi, A., Liestol, O. and Nagy, J. (1980) Glacial erosion, sedimentation and microfauna in the inner part of Kongsfjorden, Spitsbergen. *Saertrykk Norsk Polarinstitutt Skrifter* 172: 33–61.

Embleton, C. (1961) The geomorphology of the Vale of Conway, North Wales, with particular reference to its deglaciation. *Transactions of the Institute of British Geographers* 29: 47–70.

Embleton, C. (1964a) Subglacial drainage and supposed ice-dammed lakes in north-east Wales. *Proceedings of the Geologists' Association* 75: 31–8.

Embleton, C. (1964b) The deglaciation of Arfon and southern Anglesey, and the origin of the Menai Straits. *Proceedings of the Geologists' Association* 75: 407–30.

Embleton, C. and King, C.A.M. (1975) *Glacial Geomorphology*. London: Edward Arnold, 573pp.

Engelhardt, H., Humphrey, N., Kamb, B. and Fahnestock, M. (1990) Physical conditions at the base of a fast moving Antarctic ice stream. *Science* 248: 57–9.

Evans, D.J.A. (1989) Apron entrainment at the margins of sub-polar glaciers, northwest Ellesmere Island, Canadian high arctic. *Journal of Glaciology* 35: 317–24.

Evans, D.J.A. (1999) Glacial debris transport and moraine deposition: a case study of the Jardalen cirque complex, Sogn-og-Fjordane, western Norway. *Zeitschrift für Geomorphologie* 43: 203–34.

Evans, D.J.A. (in press) Glacial depositional processes and forms. In T.P. Burt, R.J. Chorley, D. Brunsden, A.S. Goudie and N.J. Cox (eds) *The History of the Study of Landforms*.Vol. 4: *Quaternary and Recent Processes and Forms (1890–1965) and the Mid-Century Revolutions*. London: Routledge.

Evans, D.J.A. and England, J. (1991) Canadian landform examples 19: high arctic thrust block moraines. *Canadian Geographer* 35: 93–7.

Evans, D.J.A., Owen, L.A. and Roberts, D. (1995) Stratigraphy and sedimentology of Devensian (Dimlington Stadial) glacial deposits, east Yorkshire, England. *Journal of Quaternary Science* 10: 241–65.

Evans, D.J.A., Rea, B.R. and Benn, D.I. (1998) Subglacial deformation and bedrock plucking in areas of hard bedrock. *Glacial Geology and Geomorphology* rp04/1998 – http://ggg.qub.ac.uk/ggg/papers/full/1998/rp041998/rp04.html

Evans, D.J.A. and Twigg, D.R. (2002) The active temperate glacial landsystem: a model based on Breidamerkurjökull and Fjallsjökull, Iceland. *Quaternary Science Reviews* 21: 2143–77.

Evenson, E.B., Lawson, D.E., Strasser, J.C., Larson, G.J., Alley, R.B., Ensminger, S.L. and Stevenson, W.E. (1999) Field evidence for the recognition of glaciohydraulic supercooling. In D.M. Mickelson and J.W. Attig (eds) *Glacial Processes: Past and Present*. Geological Society of America, Special Paper 337, pp. 23–35.

Eyles, C.H., Eyles, N. and Miall, A.D. (1985) Models of glaciomarine sedimentation and their application to the interpretation of ancient glacial sequences. *Palaeogeography, Palaeoclimatology, Palaeoecology* 51: 15–84.

Eyles, N. and Boyce, J.I. (1998) Kinematics indicators in fault gouge: tectonic analog for soft-bedded ice sheets. *Sedimentary Geology* 116: 1–12.

Eyles, N. and Rogerson, R.J. (1978a) A framework for the investigation of medial moraine formation: Austerdalsbreen, Norway, and Berendon Glacier, British Columbia, Canada. *Journal of Glaciology* 20: 99–113.

Eyles, N. and Rogerson, R.J. (1978b) Sedimentology of medial moraines on Berendon Glacier, British Columbia, Canada: implications for debris transport in a glacierized basin. *Bulletin of the Geological Society of America* 89: 1688–93.

Eyles, N., Sladen, J.A. and Gilroy, S. (1982) A depositional model for stratigraphic complexes and facies superimposition in lodgement tills. *Boreas* 11: 317–33.

Fahnestock, R.K. (1963) Morphology and hydrology of a glacial stream – White River, Mount Rainier, Washington. *USGS Professional Paper* 422-A, 70pp.

Fairchild, H.L. (1929) New York drumlins. *Proceedings of the Rochester Academy of Sciences* 3: 1–37.

Fernald, M.L. (1925) Persistence of plants in unglaciated areas of boreal North America. *Amercian Academy of Arts and Science, Memoir* 15: 237–42.

Ficker, E., Sonntag, G. and Weber, E. (1980) Ansatzezur mechanischen Deutung der Rissentstehung bei Parabelrissen und Sichelbruchen auf glazialgeformten Felsoberflachen. *Zeitschrift für Gletscherkunde und Glazialgeologie* 16: 25–43.

Fountain, A.G. (1993) Geometry and flow conditions of subglacial water at South Cascade Glacier, Washington State, USA: an analysis of tracer injections. *Journal of Glaciology* 39: 143–56.

Fountain, A.G. (1994) Borehole water-level variations and implications for the subglacial hydraulics run-away: a mechanism for thermally regulated surges of ice sheets. *Journal of Glaciology* 40: 293–304.

French, H.M. and Harry, D.G. (1990) Observations on buried glacier ice and massive segregated ice, western arctic coast, Canada. *Permafrost and Periglacial Processes* 1: 31–43.

Gardner, J.S. (1987) Evidence for headwall weathering zones, Boundary Glacier, Canadian Rocky Mountains. *Journal of Glaciology* 33: 60–7.

Gaudin, A.M. (1926) An investigation of crushing phenomena. *Transactions of the American Institute of Mining and Metallurgical Engineers* 73: 253–316.

Geikie, J. (1894) *The Great Ice Age*. London: Edward Stanford.

Gilbert, G.K. (1896) The origin of hypotheses, illustrated by a discussion of a topographic problem. *Science* 3: 1–13.

Gilbert, R. (1971) Observations on ice-dammed Summit Lake, British Columbia, Canada. *Journal of Glaciology* 10: 351–6.

Gilbert, R. (1972) Observations on sedimentation at Lillooet Delta, British Columbia. In O. Slaymaker and H.J. McPherson (eds) *Mountain Geomorphology*. Vancouver: Tantalus Press, pp. 187–94.

Gjessing, J. (1960) Isvasmeltningstidens drenering dens forlop og Formdannende virkning i Nordre Atnedalen. *Ad Novas* 3.

Gjessing, J. (1965) On 'plastic scouring' and 'subglacial erosion'. *Norsk Geografisk Tidsskrift* 20: 1–37.

Gjessing, J. (1966) Some effects of ice erosion on the development of Norwegian valleys and fjords. *Norsk Geografisk Tidsskrift* 20: 273–99.

Glen, J.W., Donner, J.J. and West, R.G. (1957) On the mechanism by which stones in till become orientated. *American Journal of Science* 255: 194–205.

Goldthwait, R.P. (1951) Development of end moraines in east-central Baffin Island. *Journal of Geology* 59: 567–77.

Goldthwait, R.P. (1960) *Study of Ice Cliff in Nunatarssuaq, Greenland*. Snow, Ice, Permafrost Research Establishment, Technical Report 39: 1–103.

Goldthwait, R.P. (1961) Regimen of an ice cliff on land in Northwest Greenland. *Folia Geographica Danica* 9: 107–15.

Goldthwait, R.P. (1971) Introduction to till today. In R.P. Goldthwait (ed.) *Till: A Symposium*. Columbus: Ohio State University Press, pp. 3–26.

Goldthwait, R.P. (1979) Giant grooves made by concentrated basal ice streams. *Journal of Glaciology* 23: 297–307.

Goodchild, J.G. (1875) The glacial phenomena of the Eden Valley and the western part of the Yorkshire Dales district. *Quarterly Journal of the Geological Society of London* 31: 55–99.

Gordon, J.E., Whalley, W.B., Gellatly, A.F. and Vere, D.M. (1992) The formation of glacial flutes: assessment of models with evidence from Lyngsdalen, north Norway. *Quaternary Science Reviews* 11: 709–31.

Gorrell, G. and Shaw, J. (1991) Deposition in an esker, bead and fan complex, Lanark, Ontario, Canada. *Sedimentary Geology* 72: 285–314.

Gould, S.J. (1980) *The Panda's Thumb*. New York: Norton, 343pp.

Gravenor, C.P. (1953) The origin of drumlins. *American Journal of Science* 251: 674–81.

Gravenor, C.P. (1955) The origin and significance of prairie mounds. *American Journal of Science* 253: 715–28.

Gravenor, C.P. and Kupsch, W.O. (1959) Ice disintegration features in western Canada. *Journal of Geology* 67: 48–64.

Gravenor, C.P., von Brunn, V. and Dreimanis, A. (1984) Nature and classification of waterlain glaciogenic sediments, exemplified by Pliestocene, Late Palaeozoic and Late Precambrian deposits. *Earth Science Reviews* 20: 105–66.

Gregory, K.J. (1965) Proglacial Lake Eskdale after sixty years. *Transactions of the Institute of British Geographers* 36: 149–62.

Gripp, K. (1929) Glaciologische und geologische Ergebnisse der Hamburgischen Spitzbergen Expedition 1927. *Naturwissenschaften Verein in Hamburg Abhandlungen aus dem Gebiete der Naturwissenschaften* 22: 146–249.

Gripp, K. (1938) Endmoränen. International Geographical Congress Abstracts, Amsterdam 1938 TII, Section IIA, Géographie Physique, pp. 215–28. (English translation: 'End moraines' in D.J.A. Evans (ed.) (1994) *Cold Climate Landforms*. Chichester: Wiley, pp. 255–67.)

Gripp, K. and Todtmann, E.M. (1925) Die Endmorane des Green-Bay Gletschers auf Spitsbergen, eine studie zum Verstandnis norddeutscher Diluvial-Gebilde. *Mitt. geogr. Ges. Hamburg* 37: 45–75.

Gry, H. (1940) De istektoniske forhold i moleromraadet. *Meddelelser Dansk Geologisk Forening* 9: 586–627.

Gustavson, T.C. (1975) Sedimentation and physical limnology in proglacial Malaspina Lake, southeastern Alaska. In A.V. Jopling and B.C. McDonald (eds) *Glaciofluvial and Glaciolacustrine Sedimentation*. Tulsa, Okla.: SEPM Special Publication 23, pp. 249–63.

Haldorsen, S. (1981) Grain size distribution of subglacial till and its relation to subglacial crushing and abrasion. *Boreas* 10: 91–105.

Haldorsen, S. (1983) Mineralogy and geochemistry of basal till and its relationship to till-forming processes. *Norsk Geologisk Tidsskrift* 63: 15–25.

Haldorsen, S. and Shaw, J. (1982) The problem of recognizing melt-out till. *Boreas* 11: 261–77.

Hallet, B. (1979) A theoretical model of glacial abrasion. *Journal of Glaciology* 23: 39–50.

Hallet, B. (1981) Glacial abrasion and sliding: their dependence of the debris concentration in basal ice. *Annals of Glaciology* 2: 23–8.

Hallet, B. (1996) Glacial quarrying: a simple theoretical model. *Annals of Glaciology* 22: 1–8.

Hallet, B. and Anderson, R.S. (1982) Detailed glacial geomorphology of a proglacial bedrock area at Castleguard Glacier, Alberta, Canada. *Zeitschrift für Gletscherkunde und Glazialgeologie* 16: 171–84.

Hallet, B., Hunter, L. and Bogen, J. (1996) Rates of erosion and sediment evacuation by glaciers: a review of field data and their implications. *Global and Planetary Change* 12: 213–35.

Ham, N.R. and Mickelson, D.M. (1994) Basal till fabric and deposition at Burroughs Glacier, Glacier Bay, Alaska. *Geological Society of America Bulletin* 106: 1552–9.

Hambrey, M.J., Huddart, D., Bennett, M.R. and Glasser, N.F. (1997) Genesis of 'hummocky moraine' by thrusting in glacier ice: evidence from Svalbard and Britain. *Journal of the Geological Society of London* 154: 623–32.

Harmer, F.W. (1907) On the origin of certain canyon-like valleys associated with lake-like areas of depression. *Quarterly Journal of the Geological Society of London* 63: 470–514.

Harris, C., McCarroll, D. and Gray, J.M. (1990) Glanllynnau. In K. Addison, M.J. Edge and R. Watkins (eds) *North Wales – Field Guide*. Coventry: Quaternary Research Association, pp. 38–47.

Harrison, P.W. (1957) A clay till fabric: its characteristics and origin. *Journal of Geology* 65: 275–308.

Hartshorn, J.H. (1958) Flow-till in southeastern Massachusetts. *Bulletin of the Geological Society of America* 69: 477–82.

Hebrand, M. and Amark, M. (1989) Esker formation and glacier dynamics in eastern Skane and adjacent areas, southern Sweden. *Boreas* 18: 67–81.

Hicock, S.R., Goff, J.R., Lian, O.B. and Little, E.C. (1996) On the interpretation of subglacial till fabric. *Journal of Sedimentary Research* 66: 928–34.

Hind, H.Y. (1859) A preliminary and general report on the Assiniboine and Saskatchewan exploring expedition. *Canada Legislative Assembly Journal* 19, Appendix 36.

Hjulstrom, F. (1952) The geomorphology of the alluvial outwash plains (sandurs) of Iceland, and the mechanics of braided rivers. *Proceedings of the International Geographical Union, 17th Congress, Washington*, pp. 337–42.

Hjulstrom, F. (1955) The groundwater: The Hoffellssandur – a glacial outwash plain. *Geografiska Annaler* 37: 234–45.

Hollingworth, S.E. (1952) A note on the use of marginal drainage channels in the recognition of unglaciated enclaves. *Journal of Glaciology* 2: 107–8.

Holmes, C.D. (1941) Till fabric. *Bulletin of the Geological Society of America* 52: 1301–52.

Holmes, C.D. (1960) Evolution of till stone shapes, New York. *Bulletin of the Geological Society of America* 71: 1645–60.

Hooke, R. Le B. (1973a) Structure and flow at the margin of the Barnes Ice Cap, Baffin Island, NWT, Canada. *Journal of Glaciology* 12: 423–38.

Hooke, R. Le B. (1973b) Flow near the margin of the Barnes Ice Cap, and the development of ice-cored moraines. *Bulletin of the Geological Society of America* 84: 3929–48.

Hooke, R. Le B. and Hudleston, P.J. (1978) Origin of foliation in glaciers. *Journal of Glaciology* 20: 285–99.

Hooke, R. Le B. and Iverson, N.R. (1995) Grain-size distribution in deforming subglacial tills, role of grain fracture. *Geology* 23: 57–60.

Hoppe, G. (1948) Isrecessionen fran Norrbottens Kustland i belysning av de glaciala formelementen. *Geographica* 20: 112pp.

Hoppe, G. (1950) Nagra exempel pa glaci-fluvial dranering fran det Inre Norrbotten. *Geografiska Annaler* 32: 37–59.

Hoppe, G. (1952) Hummocky moraine regions, with special reference to the interior of Norrbotten. *Geografiska Annaler* 34: 1–72.

Hoppe, G. (1957) Problems of glacial morphology and the ice age. *Geografiska Annaler* 39: 1–18.

Hoppe, G. (1959) Glacial morphology and inland ice recession in northern Sweden. *Geografiska Annaler* 41: 193–212.

Hoppe, G. (1963) Subglacial sedimentation, with examples from northern Sweden. *Geografiska Annaler* 45: 41–51.

Hubbard, B. and Sharp, M. (1989) Basal ice formation and deformation: a review. *Progress in Physical Geography* 13: 529–58.

Hubbard, B., Sharp, M. and Lawson, W. (1996) On the sedimentological character of alpine basal ice facies. *Annals of Glaciology* 22: 187–93.

Hubbard, B., Sharp, M.J., Willis, I.C., Nielsen, M.K. and Smart, C.C. (1995) Borehole water-level variations and the structure of the subglacial hydrological system of Haut Glacier d'Arolla, Valais, Switzerland. *Journal of Glaciology* 41: 572–83.

Hughes, T.J., Denton, G.H. and Fastook, J.L. (1985) The Antarctic Ice Sheet: an analog for northern hemisphere paleo-ice sheets? In M.J. Woldenberg (ed.) *Models in Geomorphology*. Boston: Allen & Unwin, pp. 25–72.

Iverson, N.R. (1990) Laboratory simulations of glacial abrasion: comparison with theory. *Journal of Glaciology* 36: 304–14.

Iverson, N.R. (1991) Potential effects of subglacial water pressure fluctuations on quarrying. *Journal of Glaciology* 37: 27–36.

Iverson, N.R., Hooyer, T.S. and Baker, R.W. (1998) Ring-shear studies of till deformation: Coulomb-plastic behaviour and distributed strain in glacier beds. *Journal of Glaciology* 44: 634–42.

Iverson, N.R., Hooyer, T.S. and Hooke, R. Le B. (1996) A laboratory study of sediment deformation, stress heterogeneity and grain-size evolution. *Annals of Glaciology* 22: 167–75.

Iverson, N.R. and Iverson, R.M. (2001) Distributed shear of subglacial till due to Coulomb slip. *Journal of Glaciology* 47: 481–88.

Ives, J.D. (1966) Blockfields and associated weathering forms on mountain tops and the nunatak hypothesis. *Geografiska Annaler* 48A: 220–3.

Ives, J.D. (1978) The maximum extent of the Laurentide Ice Sheet along the eastern coast of North America during the last glaciation. *Arctic* 31: 24–53.

Johnson, W.D. (1904) Maturity in alpine glacial erosion. *Journal of Geology* 12: 571–8.

Jopling, A.V. (1975) Early studies on stratified drift. In A.V. Jopling and B.C. McDonald (eds) *Glaciofluvial and Glaciolacustrine Sedimentation*. Tulsa, Okla.: SEPM Special Publication 23, pp. 4–21.

Jowett, A. (1914) The glacial geology of East Lancashire. *Quarterly Journal of the Geological Society of London* 70: 199–231.

Kälin, M. (1971) *The Active Push Moraine of the Thompson Glacier, Axel Heiberg Island, Canadian Arctic Archipelago*. Axel Heiberg Island Research Reports, Glaciology no. 4, Montreal: McGill University, 68pp.

Kamb, B. (1987) Glacier surge mechanism based on linked cavity configuration of the basal water conduit system. *Journal of Geophysical Research* 92: 9083–100.

Kamb, B. (1991) Rheological non-linearity and flow instability in the deforming bed mechanism of ice stream motion. *Journal of Geophysical Research* 96: 585–95.

Kamb, B., Raymond, C.F., Harrison, W.D., Engelhardt, H., Echelmeyer, K.A., Humphrey, N., Brugman, M.M. and Pfeffer, T. (1985) Glacier surge mechanism: 1982–1983 surge of Variegated Glacier, Alaska. *Science* 227: 469–79.

Kehew, A.E. and Lord, M.L. (1986) Origin and large scale erosional features of glacial lake spillways in the northern Great Plains. *Bulletin of the Geological Society of America* 97: 162–77.

Kendall, P.F. (1902) A system of glacier lakes in the Cleveland Hills. *Quarterly Journal of the Geological Society of London* 58: 471–71.

Kirby, R.P. (1961) Deglaciation in central Labrador-Ungava as interpreted from glacial deposits. *Geographical Bulletin* 16: 4–39.

Klassen, R.A. (1982) Glaciotectonic thrust plates, Bylot Island, District of Franklin. *Geological Survey of Canada Paper* 82–1A: 369–73.

Kleman, J. (1994) Preservation of landforms under ice sheets and ice caps. *Geomorphology* 9: 19–32.

Kleman, J. and Bergström, I. (1994) Glacial landforms indicative of a partly frozen bed. *Journal of Glaciology* 40: 255–64.

Kozarski, S. (1959) O genezie chodzieskiej moreny czolowej. *Badania Fizjograficzne nad Polska Zachodnia* 5: 45–69. (English translation: 'On the origin of the Chodziez end moraine', in D.J.A. Evans (ed.) (1994) *Cold Climate Landforms.* Chichester: Wiley, pp. 293–312.)

Krigstrom, A. (1962) Geomorphological studies of sandur plains and their braided rivers in Iceland. *Geografiska Annaler* 44: 328–46.

Kuhn, T.S. (1962) *The Structure of Scientific Revolutions.* Chicago: University of Chicago Press, 172pp.

Kupsch, W.O. (1962) Ice-thrust ridges in western Canada. *Journal of Geology* 70: 582–94.

Lawrence, D.B. and Elson, J.A. (1953) Periodicity of deglaciation in North America. Part II. Late Wisconsin recession. *Geografiska Annaler* 35: 96.

Lawson, D.E. (1979a) *Sedimentological Analysis of the Western Terminus Region of the Matanuska Glacier, Alaska.* CRREL Report 79–9, Hanover, NH.

Lawson, D.E. (1979b) A comparison of the pebble orientations in ice and deposits of the Matanuska Glacier, Alaska. *Journal of Geology* 87: 629–45.

Lawson, D.E., Strasser, J.C., Evenson, E.B., Alley, R.B., Larson, G.J. and Arcone, S.A. (1998) Glaciohydraulic supercooling: a freeze-on mechanism to create stratified, debris-rich basal ice. I. Field evidence. *Journal of Glaciology* 44: 547–62.

Leighton, M.M. (1959) Stagnancy of the Illinoian glacial lobe east of the Illinois and Mississippi Rivers. *Journal of Geology* 67: 337–44.

Lewis, W.V. (1954) Obituary: Walter Ravenhill Brown Battle. *Journal of Glaciology* 2: 372–3.

Lewis, W.V. (1960) *Norwegian Cirque Glaciers.* London: Royal Geographical Society Research Series 4.

Linton, D.L. (1949) Unglaciated areas in Scandinavia and Great Britain. *Irish Geographer* 2: 25–33 and 77–9.

Linton, D.L. (1950) Unglaciated enclaves in glaciated regions. *Journal of Glaciology* 1: 451–3.

Linton, D.L. (1952) The significance of tors in glaciated lands. *Proceedings 17th International Geographical Congress, Washington 1947*, pp. 354–7.

Linton, D.L. (1963) The forms of glacial erosion. *Transactions of the Institute of British Geographers* 33: 1–28.

Lliboutry, L. (1968) General theory of subglacial cavitation and sliding of temperate glaciers. *Journal of Glaciology* 7: 21–58.

Lliboutry, L. (1976) Physical processes in temperate glaciers. *Journal of Glaciology* 16: 151–8.

Lliboutry, L. (1979) Local friction laws for glaciers: a critical review and new openings. *Journal of Glaciology* 23: 67–95.

Lliboutry, L. (1983) Modifications to the theory of intraglacial waterways for the case of subglacial ones. *Journal of Glaciology* 29: 216–26.

Lønne, I. (2001) Dynamics of marine glacier termini read from moraine architecture. *Geology* 29: 199–202.

Loomis, S.R. (1970) Morphology and ablation processes on glacier ice. *Proceedings of the Association of American Geographers* 2: 88–92.

Love, A. and Love, D. (eds) (1963) *North Atlantic Biota and their History*. Oxford: Pergamon, 442pp.

Maag, H. (1969) *Ice Dammed Lakes and Marginal Glacial Drainage on Axel Heiberg Island*. Axel Heiberg Island Research Report, Montreal: McGill University, 147pp.

MacAyeal, D.R. (1992) The basal stress distribution of Ice Stream E, Antarctica, inferred by control methods. *Journal of Geophysical Research* 97: 595–603.

MacAyeal, D.R., Bindschadler, R.A. and Scambos, T.A. (1995) Basal friction of Ice Stream E, Antarctica. *Journal of Glaciology* 41: 247–62.

McClintock, P. and Dreimanis, A. (1964) Reorientation of till fabric by overriding glacier in the St Lawrence Valley. *American Journal of Science* 262: 133–42.

McDonald, B.C. and Banerjee, I. (1971) Sediments and bed forms on a braided outwash plain. *Canadian Journal of Earth Sciences* 8: 1282–301.

Mackay, J.R. (1959) Glacier ice-thrust features of the Yukon coast. *Geographical Bulletin* 13: 5–21.

Mackay, J.R. (1960) Crevasse fillings and ablation slide moraines, Stopover Lake area, NWT. *Geographical Bulletin* 14: 89–99.

Mair, R. and Kuhn, M. (1994) Temperature and movement measurements at a bergschrund. *Journal of Glaciology* 40: 561–5.

Malde, H.E. (1968) The catastrophic late Pleistocene Bonneville flood in the Snake River Plain, Idaho. *USGS Professional Paper* 596, 52pp.

Mannerfelt, C.M. (1945) Nagra glacialmorfologiska formelement. *Geografiska Annaler* 27: 1–239.

Mannerfelt, C.M. (1949) Marginal drainage channels as indicators of the gradients of Quaternary ice caps. *Geografiska Annaler* 31: 194–9.

Mannerfelt, C.M. (1960) Oviksfjallen: a key glaciomorphological region. *Ymer* 80: 102–13.

Matthews, J.A. and Petch, J.R. (1982) Within-valley assymetry and related problems of Neoglacial lateral moraine development at certain Jotunheimen glaciers, southern Norway. *Boreas* 11: 225–47.

Mead, W.J. (1925) The geologic role of dilatancy. *Journal of Geology* 33: 685–98.

Menzies, J. (1979) A review of the literature on the formation and location of drumlins. *Earth Science Reviews* 14: 315–59.

Menzies, J. (1984) *Drumlins: A Bibliography*. Norwich: Geobooks, 117pp.

Mickelson, D.M. (1971) *Glacial Geology of the Burroughs Glacier Area, Southeast Alaska*. Institute of Polar Studies Report 40. Ohio State University, 149pp.

Mickelson, D.M. (1973) Nature and rate of basal till deposition in a stagnating ice mass, Burroughs Glacier, Alaska. *Arctic and Alpine Research* 5: 17–27.

Mickelson, D.M. (1986) Observed processes of glacial deposition in Glacier Bay. In P.J. Anderson, R.P. Goldthwait and G.D. McKenzie (eds) *Landform and Till Genesis in the Eastern Burroughs Glacier-Plateau Remnant Area, Glacier Bay, Alaska*. Institute of Polar Studies, Miscellaneous Publication 236, Ohio State University, pp. 47–61.

Miller, G.H. (1973) Late Quaternary glacial and climatic history of northern Cumberland Peninsula, Baffin Island, NWT, Canada. *Quaternary Research* 3: 561–83.

Miller, H. (1884) On boulder-glaciation. *Proceedings of the Royal Physics Society of Edinburgh* 8: 156–89.

Mollard, J.D. (1984) Extraordinary landscape patterns: a quest for their origin. *Canadian Journal of Remote Sensing* 10: 121–34.

Morland, L.W. and Boulton, G.S. (1975) Stress in an elastic hump: the effects of glacier flow over elastic bedrock. *Proceedings of the Royal Society of London* Series A 344: 157–73.

Murray, T. and Dowdeswell, J.A. (1992) Water throughflow and the physical effects of deformation of sedimentary glacier beds. *Journal of Geophysical Research* 97: 8993–9002.

Nakawo, M. and Young, G.J. (1981) Field experiments to determine the effect of a debris layer on ablation of glacier ice. *Annals of Glaciology* 2: 85–91.

Nakawo, M. and Young, G.J. (1982) Estimate of glacier ablation under a debris layer from surface temperature and meteorological variables. *Journal of Glaciology* 28: 29–34.

Nelson, A.R. (1981) Quaternary glacial and marine stratigraphy of the Qivitu Peninsula, northern Cumberland Peninsula, Baffin Island, Canada: summary. *Geological Society of America Bulletin* 92: 512–18.

Nesje, A., Anda, E., Rye, N., Lien, R., Hole, P.A. and Blikra, L.H. (1987) The vertical extent of the Late Weichselian ice sheet in the Nordfjord-Møre area, western Norway. *Norsk Geologisk Tidsskrift* 67: 125–41.

Nesje, A. and Dahl, S.-O. (1990) Autochthonous block fields in southern Norway: implications for the geometry, thickness, and isostatic loading of the Late Weichselian Scandinavian ice sheet. *Journal of Quaternary Science* 5: 225–34.

Nesje, A., Dahl, S.-O. and Rye, N. (1988) Block fields in southern Norway: significance for the Late Weichselian ice sheet. *Norsk Geologisk Tidsskrift* 68: 149–69.

Nesje, A. and Whillans, I.M. (1994) Erosion of Sognefjord, Norway. *Geomorphology* 9: 33–45.

Nobles, L.H. and Weertman, J. (1971) Influence of irregularities of the bed of an ice sheet on deposition rate of till. In R.P. Goldthwait (ed.) *Till: A Symposium.* Columbus: Ohio State University Press, pp. 117–26.

Nye, J.F. (1973) Water at the bed of a glacier. *Symposium on the Hydrology of Glaciers.* IASH Publication 95, Cambridge, pp. 189–94.

Nye, J.F. (1976) Water flow in glaciers: jokulhlaups, tunnels and veins. *Journal of Glaciology* 17: 181–207.

Ó Cofaigh, C., Pudsey, C.J., Dowdeswell, J.A. and Morris, P. (2002) Evolution of subglacial bedforms along a paleo-ice stream, Antarctic Peninsula continental shelf. *Geophysical Research Letters* 29, no. 8, 10.1029/2001Glo14488, 2002: 41/1–41/4.

Orheim, O. and Elverhoi, A. (1981) Model for submarine glacial deposition. *Annals of Glaciology* 2: 123–7.

Östrem, G. (1959) Ice melting under a thin layer of moraine and the existence of ice cores in moraine ridges. *Geografiska Annaler* 41: 228–30.

Östrem, G. (1962) Ice-cored moraines in the Kebnekajse area. *Biuletyn Peryglacjalny* 11: 271–8.

Östrem, G. (1963) Comparative crystallographic studies on ice from ice-cored moraine, snow banks and glaciers. *Geografiska Annaler* 45: 210–40.

Östrem, G. (1964) Ice-cored moraines in Scandinavia. *Geografiska Annaler* 46: 282–337.

Pardee, J.T. (1910) The glacial lake Missoula, Montana. *Journal of Geology* 18: 376–86.

Pardee, J.T. (1942) Unusual currents in glacial lake Missoula, Montana. *Bulletin of the Geological Society of America* 53: 1569–600.

Paul, M.A. and Eyles, N. (1990) Constraints on the preservation of diamict facies (melt-out tills) at the margins of stagnant glaciers. *Quaternary Science Reviews* 9: 51–69.

Petrie, G. and Price, R.J. (1966) Photogrammetric measurements of the ice wastage and morphological changes near the Casement Glacier, Alaska. *Canadian Journal of Earth Science* 3: 827–40.

Phillips, E.R., Evans, D.J.A., Hiemstra, J.F. and Auton, C.A. (in prep.) Subglacial till: formation, deformation and sedimentary characteristics. *Earth Science Reviews*.

Piotrowski, J. and Tulaczyk, S. (1999) Subglacial conditions under the last ice sheet in northwest Germany: ice-bed separation and enhanced basal sliding? *Quaternary Science Reviews* 18: 737–51.

Powell, R.D. (1981) A model for sedimentation by tidewater glaciers. *Annals of Glaciology* 2: 129–34.

Powell, R.D. (1984) Glacimarine processes and inductive lithofacies modelling of ice shelf and tidewater glacier sediments based on Quaternary examples. *Marine Geology* 57: 1–52.

Powell, R.D. (1990) Glacimarine processes and grounding line fans and their growth to ice-contact deltas. In J.A. Dowdeswell and J.D. Scourse (eds) *Glacimarine Environments: Processes and Sediments*. Geological Society Special Publication 53, pp. 53–73.

Powell, R.D. and Domack, E. (1995) Modern glaciomarine environments. In J. Menzies (ed.) *Modern Glacial Environments: Processes, Dynamics and Sediments*. Oxford: Butterworth-Heinemann, pp. 445–86.

Price, R.J. (1965) The changing proglacial environment of the Casement Glacier, Glacier Bay, Alaska. *Transactions of the Institute of British Geographers* 36: 107–16.

Price, R.J. (1966) Eskers near the Casement Glacier, Alaska. *Geografiska Annaler* 48: 111–25.

Price, R.J. (1969) Moraines, sandar, kames and eskers near Breidamerkurjökull, Iceland. *Transactions of the Institute of British Geographers* 46: 17–43.

Price, R.J. (1970) Moraines at Fjallsjökull, Iceland. *Arctic and Alpine Research* 2: 27–42.

Rea, B.R., Evans, D.J.A., Dixon, T.S. and Whalley, W.B. (2000) Contemporaneous, localized, basal ice-flow variations: implications for bedrock erosion and the origin of p-forms. *Journal of Glaciology* 46: 470–6.

Reynolds, O. (1885) On the dilatancy of media composed of rigid particles in contact. *Philosophical Magazine* 20: 469–81.

Richter, K. (1932) Die Bewegungsrichtung des Inlandeis reconstruiert aus den Kritzen und Langsachsen der Geschiebe. *Zeitschrift Geschiebeforschung* 8: 62–6.

Richter, K. (1933) Gefuge und Zusammensetzung des norddeutschen Jungmoranen-gebietes. *Abh. geol.-paleont. Inst. Greifswald* 11: 1–63.

Richter, K. (1936) Gefugestudien im Engebrae, Fondalsbrae und ihren Vorland-sedimenten. *Zeitschrift für Gletscherkunde* 24: 22–30.

Riley, J.M. (1987) Drumlins of the southern Vale of Eden, Cumbria, England. In J. Menzies and J. Rose (eds) *Drumlin Symposium*. Rotterdam: Balkema, pp. 323–33.

Ringrose, S. (1982) Depositional processes in the development of eskers in Manitoba. In R. Davidson-Arnott, W. Nickling and B.D. Fahey (eds) *Research in Glacial, Glaciofluvial and Glaciolacustrine Systems*. Norwich: Geobooks, pp. 117–38.

Robin, G. de Q. (1976) Is the basal ice of a temperate glacier at the pressure melting point? *Journal of Glaciology* 16: 183–96.

Ronnert, L. and Mickelson, D.M. (1992) High porosity of basal till at Burroughs Glacier, southeastern Alaska. *Geology* 20: 849–52.

Rooney, S.T., Blankenship, D.D., Alley, R.B. and Bentley, C.R. (1987) Till beneath Ice Stream B: 2. Structure and continuity. *Journal of Geophysical Research* 92B: 8913–20.

Rose, J. and Letzer, J.M. (1977) Superimposed drumlins. *Journal of Glaciology* 18: 471–80.

Rothlisberger, H. (1972) Water pressure in intra- and sub-glacial channels. *Journal of Glaciology* 11: 177–203.

Rotnicki, K. (1976) The theoretical basis for and a model of glaciotectonic deformations. *Quaestiones Geographicae* 3: 103–39.

Rudoy, A.N. and Baker, V.R. (1993) Sedimentary effects of cataclysmic late Pleistocene glacial outburst flooding, Altay Mountains, Siberia. *Sedimentary Geology* 85: 53–62.

Rust, B.R. and Romanelli, R. (1975) Late Quaternary subaqueous outwash deposits near Ottawa, Canada. In A.V. Jopling and B.C. McDonald (eds) *Glaciofluvial and Glaciolacustrine Sedimentation*. Tulsa, Okla.: SEPM Special Publication 23, pp. 177–92.

Salisbury, R.D. (1894) Superglacial drift. *Journal of Geology* 2: 613–32.

Salisbury, R.D. (1896) Salient points concerning the glacial geology of north Greenland. *Journal of Geology* 4: 769–810.

Sammis, C., King, G. and Biegel, R. (1987) The kinematics of gouge deformation. *Pure and Applied Geophysics* 125: 777–812.

Saunderson, H.C. (1975) Sedimentology of the Brampton esker and its associated deposits: an empirical test of theory. In A.V. Jopling and B.C. McDonald (eds) *Glaciofluvial and Glaciolacustrine Sedimentation*. Tulsa, Okla.: SEPM Special Publication 23, pp. 155–76.

Scherer, R.P., Aldaham, A., Tulaczyk, S., Kamb, B., Engelhardt, H. and Possnert, G. (1998) Pleistocene collapse of the West Antarctic ice sheet. *Science* 281: 82–5.

Schweizer, J. and Iken, A. (1992) The role of bed separation and friction in sliding over an undeformable bed. *Journal of Glaciology* 38: 77–92.

Schytt, V. (1956) Lateral drainage channels along the northern side of the Moltka Glacier, northwest Greenland. *Geografiska Annaler* 38: 64–77.

Seaberg, S.Z., Seaberg, J.Z., Hooke, R. Le B. and Wiberg, D.W. (1988) Character of the englacial and subglacial drainage system in the lower part of the ablation area of Storglaciaren, Sweden, as revealed by dye-trace studies. *Journal of Glaciology* 34: 217–27.

Sharp, M., Dowdeswell, J.A. and Gemmell, J.C. (1989) Reconstructing past glacier dynamics and erosion from glacial geomorphic evidence: Snowdon, North Wales. *Journal of Quaternary Science* 4: 115–30.

Sharp, M., Gemmell, J.C. and Tison, J.-L. (1989) Structure and stability of the former subglacial drainage system of the Glacier de Tsanfleuron, Switzerland. *Earth Surface Processes and Landforms* 14: 119–34.

Sharp, R.P. (1948) The constitution of valley glaciers. *Journal of Glaciology* 1: 182–9.

Sharp, R.P (1949) Studies of supraglacial debris on valley glaciers. *American Journal of Science* 247: 289–315.

Sharp, R.P. (1951) Glacial history of Wolf Creek, St Elias Range, Canada. *Journal of Geology* 59: 97–117.

Shaw, J. (1972) Sedimentation in the ice contact environment, with examples from Shropshire, England. *Sedimentology* 18: 23–62.

Shaw, J. (1979) Genesis of the Sveg tills and Rogen moraines of central Sweden: a model of basal melt-out. *Boreas* 8: 409–26.

Shaw, J. (1982) Melt-out till in the Edmonton area, Alberta, Canada. *Canadian Journal of Earth Sciences* 19: 1548–69.

Shaw, J. (1983a) Drumlin formation related to inverted meltwater erosional marks. *Journal of Glaciology* 29: 461–79.

Shaw, J. (1983b) Forms associated with boulders in melt-out till. In E.B. Evenson, C. Schluchter and J. Rabassa (eds) *Tills and Related Deposits*. Rotterdam: Balkema, pp. 3–12.

Shaw, J. (1996) A meltwater model for Laurentide subglacial landscapes. In S.B. McCann and D.C. Ford (eds) *Geomorphology Sans Frontières*. Chichester: Wiley, pp. 181–236.

Shaw, J. (2002) The meltwater hypothesis for subglacial bedforms. *Quaternary International* 90: 5–22.

Shreve, R.L. (1972) Movement of water in glaciers. *Journal of Glaciology* 11: 205–14.

Shreve, R.L. (1985a) Esker characteristics in terms of glacier physics, Katahdin esker system, Maine. *Geological Society of America Bulletin* 96: 639–46.

Shreve, R.L. (1985b) Late Wisconsin ice surface profile calculated from esker paths and types, Katahdin esker system, Maine. *Quaternary Research* 23: 27–37.

Sissons, J.B. (1958a) Supposed ice-dammed lakes in Britain, with particular reference to the Eddleston Valley, southern Scotland. *Geografiska Annaler* 40: 159–87.

Sissons, J.B. (1958b) The deglaciation of part of East Lothian. *Publications of the Institute of British Geographers* 25: 59–77.

Sissons, J.B. (1958c) Subglacial stream erosion in southern Northumberland. *Scottish Geographical Magazine* 74: 163–74.

Sissons, J.B. (1960a) Subglacial, marginal and other glacial drainage in the Syracuse-Oneida areas, New York. *Bulletin of the Geological Society of America* 71: 1575–88.

Sissons, J.B. (1960b) Some aspects of glacial drainage channels in Britain, Part I. *Scottish Geographical Magazine* 76: 131–46.

Sissons, J.B. (1961a) Some aspects of glacial drainage channels in Britain, Part II. *Scottish Geographical Magazine* 77: 15–36.

Sissons, J.B. (1961b) A subglacial drainage system by the Tinto Hills, Lanarkshire. *Transactions of the Edinburgh Geological Society* 18: 175–93.

Sissons, J.B. (1961c) The central and eastern parts of the Lammermuir-Stranraer moraine. *Geological Magazine* 98: 380–92.

Sissons, J.B. (1963) The glacial drainage system around Carlops, Peeblleshire. *Publications of the Institute of British Geographers* 32: 95–111.

Sissons, J.B. (1964) The glacial period. In J.W. Watson and J.B. Sissons (eds) *The British Isles, A Systematic Geography.* London: Nelson, pp. 131–51.

Slater, G. (1926) Glacial tectonics as reflected in disturbed drift deposits. *Proceedings of the Geologists' Association* 37: 392–400.

Slater, G. (1927a) The structure of the disturbed deposits in the lower part of the Gipping Valley near Ipswich. *Proceedings of the Geologists' Association* 38: 157–82.

Slater, G. (1927b) The structure of the disturbed deposits of the Hadleigh Road area, Ipswich. *Proceedings of the Geologists' Association* 38: 183–261.

Slater, G. (1927c) The structure of the disturbed deposits of Moens Klint, Denmark. *Transactions of the Royal Society of Edinburgh* 55: 289–302.

Slater, G. (1927d) The disturbed glacial deposits in the neighbourhood of Lonstrup, near Hjorring, north Denmark. *Transactions of the Royal Society of Edinburgh* 55: 303–15.

Slater, G. (1927e) Structure of the Mud Buttes and Tit Hills in Alberta. *Bulletin of the Geological Society of America* 38: 721–30.

Slater, G. (1931) The structure of the Bride Moraine, Isle of Man. *Proceedings of the Liverpool Geological Society* 14: 184–96.

Smalley, I.J. (1966) Drumlin formation: a rheological model. *Science* 151: 1379–80.

Smalley, I.J. and Unwin, D.J. (1968) The formation and shape of drumlins and their distribution and orientation in drumlin fields. *Journal of Glaciology* 7(51): 377–90.

Smith, N.D. (1970) The braided stream depositional environment: comparison of the Platte River with some Silurian clastic rocks, north-central Appalachians. *Bulletin of the Geological Society of America* 81: 2993–3014.

Smith, N.D. (1974) Sedimentology and bar formation in the upper Kicking Horse River, a braided meltwater stream. *Journal of Geology* 82: 205–23.

Smith, N.D. (1985) Proglacial fluvial environment. In G.M. Ashley, J. Shaw and N.D. Smith (eds) *Glacial Sedimentary Environments.* Tulsa, Okla.: SEPM Short Course 16, pp. 85–135.

Smith, N.D. and Ashley, G.M. (1985) Proglacial lacustrine environments. In G.M. Ashley, J. Shaw and N.D. Smith (eds) *Glacial Sedimentary Environments.* Tulsa, Okla.: SEPM Short Course 16, pp. 136–215.

Souchez, R.A. (1967) The formation of shear moraines: an example from south Victoria Land, Antarctica. *Journal of Glaciology* 6: 837–43.

Stalker, A. MacS. (1960) *Ice-pressed Drift Forms and Associated Deposits in Alberta.* Geological Survey of Canada, Bulletin 57, 38pp.

St Onge, D.A. and McMartin, I. (1995) *Quaternary Geology of the Inman River Area, Northwest Territories.* Geological Survey of Canada, Bulletin 446.

Sugden, D.E. (1968) The selectivity of glacial erosion in the Cairngorm Mountains, Scotland. *Transactions of the Institute of British Geographers* 45: 79–92.

Sugden, D.E. (1974) Landscapes of glacial erosion in Greenland and their relationship to ice, topographic and bedrock conditions. *Institute of British Geographers, Special Publication* 7: 177–95.

Sugden, D.E. (1976) A case against deep erosion of shields by ice sheets. *Geology* 4: 580–2.

Sugden, D.E. (1977) Reconstruction of the morphology, dynamics and thermal characteristics of the Laurentide ice sheet at its maximum. *Arctic and Alpine Research* 9: 27–47.

Sugden, D.E. (1978) Glacial erosion by the Laurentide ice sheet. *Journal of Glaciology* 20: 367–91.

Sugden, D.E., Marchant, D.R., Potter, N., Souchez, R.A., Denton, G.H., Swisher, C.C. and Tison, J.L. (1995) Preservation of Miocene glacier ice in East Antarctica. *Nature* 376: 412–14.

Swinzow, G.K. (1962) Investigation of shear zones in the ice sheet margin, Thule area, Greenland. *Journal of Glaciology* 4: 215–29.

Syverson, K.M., Gaffield, S.J. and Mickelson, D.M. (1994) Comparison of esker morphology and sedimentology with former ice-surface topography, Burroughs Glacier, Alaska. *Geological Society of America Bulletin* 106: 1130–42.

Thompson, H.R. and Bonnlander, B.H. (1956) Temperature measurements at a cirque bergschrund in Baffin Island: some results of W.R.B. Battle's work in 1953. *Journal of Glaciology* 2: 762–9.

Truffer, M., Harrison, W.D. and Echelmeyer, K.A. (2000) Glacier motion dominated by processes deep in underlying till. *Journal of Glaciology* 46: 213–21.

Tulaczyk, S., Kamb, B. and Engelhardt, H. (2000a) Basal mechanics of Ice Stream B: I. Till mechanics. *Journal of Geophysical Research* 105: 463–81.

Tulaczyk, S., Kamb, B. and Engelhardt, H. (2000b) Basal mechanics of Ice Stream B: II. Plastic-undrained-bed model. *Journal of Geophysical Research* 105: 483–94.

Tulaczyk, S., Kamb, B., Scherer, R.P. and Engelhardt, H.F. (1998) Sedimentary processes at the base of a West Antarctic ice stream: constraints from textural and compositional properties of subglacial debris. *Journal of Sedimentary Research* 68: 487–96.

Tulaczyk, S.M., Scherer, R.P. and Clark, C.D. (2001) A ploughing model for the origin of weak tills beneath ice streams: a qualitative treatment. *Quaternary International* 86: 59–70.

van der Wateren, F.M. (1985) A model of glacial tectonics, applied to the ice pushed ridges in the central Netherlands. *Bulletin of the Geological Society of Denmark* 34: 55–74.

Viellette, J.J. (1986) Former southwesterly ice flows in the Abitibi-Timiskaming region: implications for the configuration of the late Wisconsinan ice sheet. *Canadian Journal of Earth Sciences* 23: 1724–41.

Viellette, J.J., Dyke, A.S. and Roy, M. (1999) Ice flow evolution of the Labrador sector of the Laurentide Ice Sheet: a review, with new evidence from northern Quebec. *Quaternary Science Reviews* 18: 993–1019.

Vorren, T. (2003) Subaquatic landsystems: continental margins. In D.J.A. Evans (ed.) *Glacial Landsystems*. London: Arnold, pp. 289–312.

Vorren, T.O. and Laberg, J.S. (1997) Trough mouth fans – palaeoclimate and ice-sheet monitors. *Quaternary Science Reviews* 16: 865–81.

Walder, J.S. and Fowler, A. (1994) Channelized subglacial drainage over a deformable bed. *Journal of Glaciology* 40: 3–15.

Wakahama, K. and Tusima, M. (1981) Observations of inner moraines near the Terminus of McCall Glacier in Arctic Alaska and laboratory experiments on the mechanism of picking up moraines into a glacier body (abstract). *Annals of Glaciology* 2: 116.

Walder, J.S. (1982) Stability of sheet flow of water beneath temperate glaciers and implications for glacier surging. *Journal of Glaciology* 28: 273–93.

Walder, J.S. (1986) Hydraulics of subglacial cavities. *Journal of Glaciology* 32: 439–45.

Walder, J.S. and Fowler, A. (1994) Channelized subglacial drainage over a deformable bed. *Journal of Glaciology* 40: 3–15.

Walder, J.S. and Hallet, B. (1979) Geometry of former subglacial water channels and cavities. *Journal of Glaciology* 23: 335–46.

Warren, W.P. and Ashley, G.M. (1994) Origins of the ice-contact stratified ridges (eskers) of Ireland. *Journal of Sedimentary Research* A64: 433–49.

Weertman, J. (1957) On the sliding of glaciers. *Journal of Glaciology* 3: 33–8.

Weertman, J. (1961) Mechanism for the formation of inner moraines found near the edge of cold ice caps and ice sheets. *Journal of Glaciology* 3: 965–78.

Weertman, J. (1964) The theory of glacier sliding. *Journal of Glaciology* 5: 287–303.

Weertman, J. (1969) Water lubrication mechanism of glacier surges. *Canadian Journal of Earth Sciences* 6: 929–42.

Weertman, J. (1972) General theory of water flow at the base of a glacier or ice sheet. *Reviews of Geophysics and Space Physics* 10: 287–333.

Welch, R. and Howarth, P.J. (1968) Photogrammetric measurements of glacial landforms. *Photogrammetric Record* 6: 75–96.

Wellner, J.S., Lowe, A.L., Shipp, S.S. and Anderson, J.B. (2001) Distribution of glacial geomorphic features on the Antarctic continental shelf and correlation with substrate: implications for ice behaviour. *Journal of Glaciology* 47: 397–411.

Willis, I.C., Sharp, M. and Richards, K.S. (1990) Configuration of the drainage system of Mitdalsbreen, Norway, as indicated by dye-tracing experiments. *Journal of Glaciology* 36: 89–101.

Winters, W.A. (1961) Landforms associated with stagnant ice. *Professional Geographer* 13: 19–23.

Wright, H.E. (1957) Stone orientation in Wadena drumlin field, Minnesota. *Geografiska Annaler* 39: 19–31.

Wright, H.E. (1962) Role of the Wadena lobe in the Wisconsin glaciation of Minnesota. *Bulletin of the Geological Society of America* 73: 73–100.

THE CHANNELED SCABLANDS
OF THE COLUMBIA PLATEAU

J.H. Bretz

Source: *Journal of Geology* 31(8) (1923): 617–49.

Definition of "scabland"

The terms "scabland" and "scabrock" are used in the Pacific Northwest to describe areas where denudation has removed or prevented the accumulation of a mantle of soil, and the underlying rock is exposed or covered largely with its own coarse, angular débris. The largest areas of scabland are on the Columbia Plateau in Washington, north of Snake River. These scablands have a history which is believed to be unique. The prevailing feature of their topography is indicated in the term here used: channeled scablands.[1] They are scored by thousands of channels eroded in the underlying rock. The plateau in Washington, north of Snake River, has a total area of about 12,750 square miles, of which at least 2,000 square miles is channeled scabland. The scabland is widely distributed over the region in linear tracts among maturely dissected hills which bear the loessial soil (wheat lands) of the plateau.

Physiographic relations of the channeled scablands

The following features and relations of the scablands exist in all tracts. They must form the basis of any interpretation for the origin of channeled scabland. The map should be examined as this list is read.

1. Scabland tracts are developed invariably on or in the Columbia basalt formation.
2. Scabland tracts are invariably lower than the immediately adjacent soil-covered areas.
3. Scabland tracts are invariably elongate.

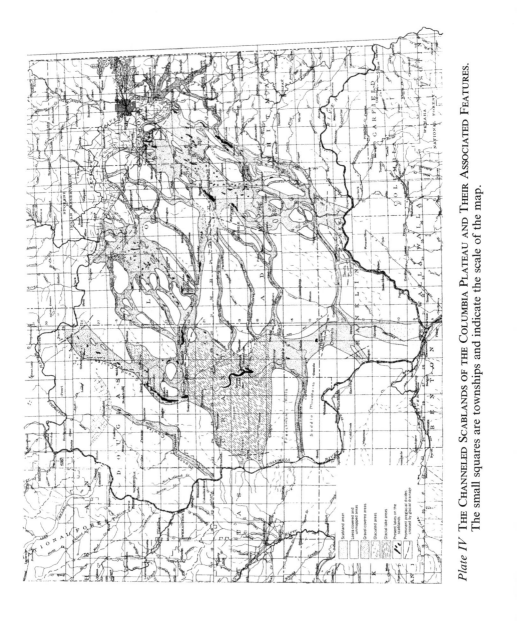

Plate IV The Channeled Scablands of the Columbia Plateau and Their Associated Features. The small squares are townships and indicate the scale of the map.

4. The elongation of scabland tracts is with the dip slope of the underlying basalt flows. There are eight known exceptions to this rule,[2] all minor affairs so far as length is concerned.
5. Scabland tracts, considered as units, invariably have continuous gradients.
6. Scabland tracts are invariably bounded by maturely eroded topography.
7. Scabland tracts are developed in pre-existing drainage lines of the mature topography.[3]
8. Scabland tracts are connected with each other.[4]
9. (a) The areas surrounded by scabland invariably have the dendritic drainage pattern, mature topography and loessial soil of the plateau. (b) They are almost invariably elongate with the scabland tracts. (c) They commonly have steep marginal slopes descending from 50 to 200 feet to the scabland. These slopes are almost invariably in loess. Slopes of 30° to 33° are not uncommon. They are much younger topographically than the slopes of the valleys among these mature hills.
10. Scabland tracts with steep gradient are narrow, while those with gentle gradient are wide.
11. The pattern of scabland tracts, where hills of the older topography are isolated in them, is anastomosing or "braided."
12. Scabland tracts invariably contain "channels." These are gorges or canyons or elongated basins eroded in the basalt. The channels are invariably elongate in parallelism with the tract as a whole and, in most cases, the channel pattern is anastomosing or braided.
13. (a) Scabland tracts invariably bear discontinuous deposits of basaltic stream gravel. (b) These deposits invariably contain a small proportion of pebbles and cobbles of rock foreign to the plateau. (c) These deposits invariably rest on an eroded, scabland surface of the basalt. (d) They commonly lie on the down-gradient side of eminences in the scabland.
14. (a) Scabland tracts invariably bear scattered bowlders of foreign rock. (b) The proportion of foreign débris, either the fragments in the gravel or the scattered bowlders, is invariably smaller with increasing distance down-gradient.
15. Scabland tracts are invariably without a mantle of residual soil.
16. Scabland tracts are traceable up-gradient to a narrow basalt plain bordering the south side of Spokane River in the northern part of the plateau.[5] This basalt plain bears many glaciated erratic bowlders and some patches of till, but no channeled scabland, no mature topography, and no loessial soil.
17. Only where the minor valleys of the mature topography adjacent to this basalt plain open northward on to the plain are any glacial erratics found in them.[6]
18. There are but ten scabland openings to this basalt plain to the north.
19. Scabland tracts are invariably traceable down-gradient to Snake River on the south or to Columbia River on the west. There are nine places

where scabland tracts enter these two streams. Only three of them were drainage ways before the scablands were formed.

20. There is no channeled scabland on the plateau in western Idaho or south of Snake River or west of Columbia River.

21. Nowhere in the scablands or the maturely dissected country, during ten weeks of field study, has a till been found, or any deposit of doubtful genesis which could be interpreted more satisfactorily as till than as non-glacial in origin.[7]

Generalized statement of the origin of the channeled scablands

This unique combination of topographic features of the Columbia Plateau in Washington has only one interpretation consistent with all the foregoing items. The channeled scablands are the erosive record of large, high-gradient, glacier-born streams. The basalt plain records the southern limit reached by the ice sheet from which these streams took origin. Before this glaciation occurred, the entire plateau of Washington was covered with a loessial soil, varying in depth from a few feet to 200 feet. This and the underlying basalt had been eroded to maturity and a network of drainage lines covered it. The major water courses of this mature topography were consequent on the warped surface of the plateau which descends in a general way from the northeast to the southwest.

The ice sheet approached and invaded the plateau from its northern high margin. It barely crossed to the headwaters of the consequent drainage. In places, it did not cross, but by blocking all other escapeways, its waters were forced to cross. By about a dozen different routes, at different altitudes and distributed along more than 150 miles of the ice front, water entered the mature drainage system. The capacity of the pre-existing valleys was wholly inadequate for the volume of most of these streams. Furthermore, gradients were high and the glacial waters eroded enormously, sweeping away the overlying loessial material, crossing low divides and isolating many groups of the maturely eroded hills to produce the anastomosing pattern of the scablands, biting deeply into the basalt to make the canyons and rock basins, and spilling into the Snake and Columbia in three times as many places as the pre-existing drainage had used.

This procedure of glacial streams was unique, so far as the writer is aware. It was unorthodox, at any rate, for no valley trains and but two outwash plains[8] were built on the plateau south of the basalt plain. The stream gravel of the scablands is almost wholly in separate bars.

The conception above outlined is amply sustained by every feature and relationship of the scablands. All other hypotheses meet fatal objections. Yet the reader of the following more detailed descriptions, if now accepting the writer's interpretation, is likely to pause repeatedly and question that interpretation. The magnitude of the erosive changes wrought by these glacial streams is nothing short of amazing. The writer confesses that during ten weeks' study of the region, each newly examined scabland tract reawakened a feeling of amazement that such huge streams could take origin from such small marginal tracts of an ice sheet, or that such an enormous amount of erosion, despite high gradients, could have resulted in the very brief time these streams existed. Not River Warren, nor the Chicago outlet, nor the Mohawk channel, nor even Niagara falls and gorge itself approach the proportions of some of these scabland tracts and their canyons. From one of these canyons alone[9] 10 cubic miles of basalt was eroded by its glacial stream.

The basalt plain, north of the scablands and mature topography

This physiographic feature extends westward from Spangle for 50 miles along the south side of Spokane River and varies from 3 to 12 miles in width. It is determined by the upper surface of the Columbia basalt formation. It is interrupted in places by short valleys tributary to Spokane River, and the different portions are known as prairies.[10] This plain is bounded on the south by channeled scabland and maturely eroded loessial hills. The differences between it and adjacent broader scablands are not marked, but the loessial hills are in striking contrast with it. These hills which, elsewhere on the plateau, come right to the edge of Snake and Columbia valleys,[11] nowhere overlook the Spokane Valley. They terminate abruptly on the southern margin of this plain. On the plain there is no mature topography and no channeled scabland. There is no area on the plateau like it. The nearest approach to it is the northern portion of Douglas County, back of the Wisconsin terminal moraine. This narrow plain must be the result of conditions which prevailed no farther south.

These conditions can be summed up in one word—glaciation. Deposits of till and many striated erratics have been found in every township examined on it. The till is patchy in distribution. No moraine margins the southern edge of the plain and no good moraine ridges occur on it, though here and there is morainic topography.

The genesis of the plain thus established, the questions of its character before glaciation and the method of its development arise. These are answered clearly when the adjoining mature loessial topography is studied. The larger valleys of this topography have been eroded to varying depths into the basalt. The bottoms of such as lead out across the basalt plain to the north are lower among the hills than the general surface of the plain. The profile of the plain, extended back among the hills, cuts their slopes somewhere between hill summits and valley bottoms. And this transection is at the contact of loess on basalt. The ice sheet which covered the plain, therefore, simply removed the upper, weaker formation, and only to a minor extent altered the surface of the basalt formation.

The mature topography

This is the dominant type of topography of the Columbia Plateau. The major drainage lines are structurally controlled, but the minor ones constitute a dendritic network. The pattern is eroded largely in a weak sedimentary deposit, chiefly loessial, which overlies the basalt. Maturity is expressed in the complete development of the drainage system, in the reduction of the original surface to valley slopes[12] and in the concavity of the lower part of many of these slopes. This maturity has been developed with reference to the underlying basalt as a base level, for progress of the cycle of erosion in the loess has been very much more rapid. Neglecting the loessial cover, the basalt plateau is in early youth, and will still be when the loess has been entirely removed. Nevertheless, the absence of deep trenches in basalt in the interior of the plateau, similar to Spokane, Columbia, and Snake River valleys about its margin, and the cutting through of the loess has allowed the development of shallow mature valleys in the upper part of the basalt.

The loessial deposit varies in thickness, in some places being only a soil, and in others being 200 feet or more in thickness. It is not all loess. There are places where it is chiefly a residual soil from the basalt, and others where it is a waterlaid sediment.[13] But many widely distributed sections show a succession of loessial deposits, with abundant root and rootlet casts throughout and with reddened upper surfaces of each deposit, aggregating 50 feet or more.[14]

The mature topography is older than the scablands and the basalt plain. Literally hundreds of isolated groups of maturely eroded hills of loess stand in the scablands. Their gentle interior slopes are identical with those far from the scabland tracts. But their marginal slopes, descending to the scablands, commonly are very steep, over large areas amounting to 30° and even 35° (Figs. 1, 2, and 3). These steep slopes are seldom even gullied, except where a drainage line leads out from the hill group to the scabland. Where the minor valleys transected by the steep slopes lead backward into the interior of a hill group they are simply hanging valleys.

There are few places where basalt occurs in these steepened slopes. Where present, it is always restricted to the lower part and shows itself in conspicuous ledges.

A very striking and significant feature of the steepened slopes is their convergence at the northern ends of the groups to form great prows, pointing up the scabland's gradient (Figs. 2 and 3). The nose of a prow may extend as a sharp ridge from the scabland to the very summit of the hill. It is impossible to study these prow-pointed loessial hills, surrounded by the scarred and channeled basalt scablands, without seeing in them the result of

Figure 1 One of a group of loessial hills in the scablands a few miles southwest of Rock Lake. One of the steepened slopes and its alignment are shown. Photo by O.C. Clifford.

Figure 2 An isolated loessial hill on the scabland south of Hooper. The prow of the hill is pointed at the observer. The hill is 180 feet high, more than half a mile long, has a very narrow crest, and sides which slope 35°. It is entirely surrounded by scrubbed basalt. Half a mile to the west is a canyon in the scabland 75 feet deep, with an abandoned waterfall at the head.

Figure 3 The same hill as shown in Figure 2. The prow is at the right. The corresponding steepened slope at the tail of the hill shows at the extreme left. Most of the apparent left slope of the crest is a matter of perspective. Photo by O.C. Clifford.

a powerful eroding agent which attacked them about their bases and most effectively from the scabland's up-gradient direction.

Details of a scabland surface

All scablands are channeled to a greater or lesser extent. These channels are eroded in basalt to depths varying from a few feet to hundreds of feet. Commonly there are many shallow channels on each tract. Most tracts also have a few deeper channels, of the proportions of canyons[15] (Figs. 4 and 5). All channels in a tract are intricately interlaced, resulting in a multitude of buttelike hills, knobs, and ridges among them. Few channels have accordant

Figure 4 Devils Canyon at mid-length, looking south. A double fall existed here when the canyon was eroded. The island and the eastern part still remain.

Figure 5 Devils Canyon near mid-length, looking north. Note the scrubbed basalt ledges above the canyon rim, and the profile of the loessial bluffs, still higher and farther back.

Figure 6 Longitudinal profile of the deepest of the Drumheller Channels across the nose of the anticline. Four rock basins are indicated, the largest of which is 75 feet deep.

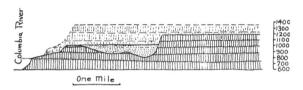

Figure 7 "The Potholes," longitudinal profile of the northern half, showing (1) cliff along northern side of the canyon, (2) amount of recession of the falls, (3) elongated rock basin below the falls, (4) great gravel bar along edge of the rock basin, and (5) approximate level of Columbia River when the cataract was formed.

grades where they unite or diverge, the bottoms of the shallower ones hanging above the floors of the deeper ones. Many canyoned channels have abandoned cataracts and cascades in them or at their heads.[16] Most canyoned channels have elongated rock basins (see Fig. 6). Even in the shallow channels, basins or pockets in the rock are common. Some of these rock basins clearly were produced by recession of a cataract whose scarp still exists.[17] Others were produced by plucking of the columnar basalt in the canyon floors where the gradient was high.[18]

These features of the channeled scablands on the Columbia basalt plateau do not closely resemble any other type of topography. The narrowness and elongation of channels and the continuous gradients of tracts as a whole suggest river valleys but these features are all inherited from pre-existing valleys. Furthermore, some tracts are nearly as broad as they are long. The pattern of channels on a tract, like the pattern of some tracts and their isolated loessial hills, is much like that of great braided streams (Fig. 8). But the scablands are erosional in origin, while the braided stream pattern is depositional.

The evidence for the origin of channeled scabland by stream erosion is overwhelming. The evidence of contemporaneity of action of all channels of a given tract, at least in its early history, is equally convincing. The only sequence indicated is that of development of the greater channels later in the epoch and consequent draining off of the shallower channels.[19] The scablands of the plateau in Washington are the beds of huge river courses in which the streams once spread completely from side to side and only later became concentrated in the deeper canyons.

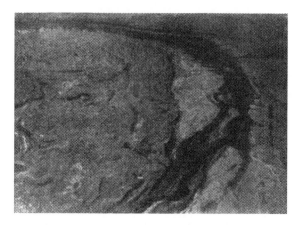

Figure 8 A part of Rock Creek (Lincoln County) and its associated scabland. The creek in the main channel is margined by vegetation. The scabland to the left is barren and the irregularity in lighting is due almost entirely to the cliffs which border the shallow anastomosing channels. (Aeroplane Photo by F.H. Frost.)

Altitudes and gradients of the scabland tracts

The heads of scabland tracts which are open to the basalt plain range in altitude between 2,350 and 2,500 feet above tide. There are six or eight of these, the number depending on just what is considered to constitute an individual scabland tract. At least the following should be recognized as unit tracts. The order in the list is from east to west.

1. Pine Creek channel. Altitude of head, 2,450 feet A.T.
   Gradient approximately 25 feet per mile
2. Cheney-Hooper tract
   Four heads:
   a) Marshall-Spangle, altitude, 2,350±
   b) Four Lakes, altitude, ?
   c) Medical Lake, altitude, 2,425
   d) Deep Creek, altitude, 2,350±
   Total width of the group is 22 miles
   Gradients:
      Spangle to Hooper, 21.9 feet per mile
      Cheney to Hooper, 22.6 feet per mile
      Medical Lake to Sprague, 24 feet per mile
      Medical Lake to Hooper, 22.5 feet per mile
3. Reardan channel. Altitude of head, 2,500+
   Gradient, Reardan to Odessa, 20 feet per mile
4. Davenport-Harrington channel. Altitude of head, 2,450±
   Gradient:

       Davenport to Harrington, 27 feet per mile
       Harrington to Odessa, 26 feet per mile

5. Telford tract
   a) Eastern head. Altitude, 2,500±
      Gradient:
        Rocklyn to Odessa, 30 feet per mile
        Rocklyn to Krupp (Marlin), 26 feet per mile
   b) Western head. Altitude, 2,500±
      Gradient:
        Near Creston to Krupp (Marlin), 32 feet per mile
        Near Creston to Wilson Creek, 30 feet per mile
      Total width of heads is 17 miles
   c) Wilbur branch
      Gradient:
        Creston to Wilson Creek, 32 feet per mile
        Wilbur to Wilson Creek, 25 feet per mile
        Almira to Wilson Creek, 26 feet per mile

The only scabland tracts which do not open on the basalt plain are Grand Coulee and Moses Coulee. For the head of Moses Coulee there are no altitude measurements. Grand Coulee has had a peculiar history, not yet fully deciphered, but the significant altitude at its head for present purposes is not the coulee floor (1,530 feet A.T.) but the scabland margining the brink of the canyon, about 2,500 feet A.T. The canyon has been cut subsequent to the first spilling over of glacial waters. The floor near Coulee City is 1,510 feet A.T. Most of this descent occurred within a few miles of Coulee City, the original slope being as steep as 20° in part and averaging perhaps 10° for 1,000 feet of descent. This is the chief reason for the great canyon across the divide. No other scabland head has been notably canyoned. None other had a gradient to exceed about 30 feet to the mile. All the canyons of the channeled scablands are located in places of exceptionally steep original gradients.

If the channeled scablands are the product of stream erosion, and if all tracts are of the same age, the fact that the scabland heads vary in altitude, though all but two are open to the same glaciated basalt plain, can have but one satisfactory explanation. Each glacial stream must have had a source of water which was unconnected with the others. This means that the ice sheet whose melting yielded these streams must have covered the basalt plain and in most places must have been in contact with the mature loessial hills which separate the scabland heads. It means, furthermore, that but a few miles of ice front supplied the water for streams so huge that they flooded over many preglacial divides of the mature topography even in the southern part of the plateau. It was this flooding across divides which produced the scabland plexus of the plateau.

About 40 miles of ice front in one case[20] yielded water sufficient to denude a non-elongated tract 250 square miles in area of a loessial cover about 100 feet in maximum thickness. This was done by lateral planation in the preglacial drainage lines of the tract. These lines were so shallowly intrenched in the basalt and the volume of the water was so great that, as the loessial hills were eroded away, the flood spread over the entire area, 13 miles wide. Gradients were low, however, and it did not develop canyoned channels. Steeper gradients farther from the edge of the ice, and greater capacity of the preglacial valleys, held the waters within the confines of these valleys, but six such[21] were necessary to contain the flood and they were all greatly eroded in the underlying basalt.

Another large scabland area[22] is 75 miles long and 15 miles in average width. Its total descent is 1,850 feet. Its altitude at the head is the lowest of all such tracts. It differs from the one above in its notable linear extent, in the possession of a large number of isolated groups of loessial hills of the older topography, in its greater gradient and in the development of canyons.[23] Though a much greater volume of water passed through this tract, the gradients were steep enough to draw off the flood and prevent a complete spreading over the area. Much water came to it from the ice margin to the east in Idaho and no estimate can now be made of the length of ice front which contributed to it. There were five or six places of distributary discharge where this flood crossed preglacial divides and one of them[24] eventually obtained most of the discharge. Along this distributary route, the glacial flood swept away the loessial hills for a width of 10 to 15 miles and eroded 500 feet into the basalt.

The glacial drainage route which possessed the highest crossing of the preglacial surface of the plateau is Grand Coulee. It also found the steepest gradient[25] and its volume was sufficient with this gradient to cause the deepest erosion in the basalt. Upper Grand Coulee, across the preglacial divide, is 1,000 feet deep. But its floor, after the epoch had closed, was lower than that of any other glacial drainage route at the head. In its early history it drew water from about 40 miles of ice front, but never spread widely. The gradient, determined here by exceptional warping of the basalt, prevented that. There was no noteworthy preglacial stream along its course. Grand Coulee is, therefore, the simplest but grandest case of canyon-cutting by glacial streams on the plateau.

Depth of glacial stream erosion in the scablands

Criteria

The courses of the larger valleys in the mature drainage system were determined by the warped surface of the basalt. The dominant feature of this warping is the southwestward dip from Spokane River and Columbia River on the north to Snake River on the south and Columbia River on the west. Many minor folds are superposed on this dip slope, so recent geologically that anticlines determine divides and synclines contain stream valleys.[26] Commonly, only the major valleys are intrenched in the basalt. The minor valleys, in general, are not eroded through the loessial mantle.

Except near the bounding canyons of Spokane, Columbia, and Snake Rivers, or in exceptionally upwarped parts of the plateau, the preglacial valleys in basalt which were unviolated by the glacial flood, have the same mature slopes as those in the loess. Where glacial streams found routes eroded but slightly in basalt, they commonly spread widely at first[27] and only later eroded canyons, if they did so at all. Such wide scablands commonly are bounded by steep bluffs of loess. Many of the mature valleys in basalt were sufficiently capacious to contain the glacial waters which entered them. In such cases[28] the scabland of the route lies on the sides and bottoms only, and unsteepened slopes of the bounding older topography may come to the edge of the scabland. Such a tract is narrow and instead of having a multitude of lateral shallow channels anastomosing with the main canyon, it consists of exceedingly roughened ledges of basalt outcropping on the slopes of the main valley. Shoulders in the curves of these valleys were treated with especial vigor, in some cases being wholly cut away to leave prominent cliffs on the valley walls.

By smoothing out these ledges, something of the cross-section of the preglacial valley may be obtained. Remnants of the old floors are recorded in isolated buttes of basalt on the present floor and in the lowest prominent rock terraces, below which is the canyoned channel eroded in mid-current by the huge glacial stream.

Instances

In Cow Creek, southwest of Ralston, the remarkable number of knobs and buttes in the lower part of the valley indicates clearly that the preglacial floor must have been at least 75 feet higher than the present. In Crab Creek Valley, between Krupp (Marlin) and Stratford, there are prominent buttes, isolated or partially isolated on the floor of the canyon. The tops of these are remnants of the preglacial floor and their height (100 feet in places) is a minimal measure of the depth of canyon-cutting by the glacial waters.

In Washtucna Coulee, numerous prominent rock terraces from 150 to 200 feet above the present floor are probably remnants of the earlier valley bottom. In Esquatzel Coulee, into which Washtucna opens, these terraces are 200 feet and more above the bottom.

In Lower Moses Coulee (Fig. 9), the mouths of preglacial tributary ravines hang approximately 400 feet above the rock floor. Some of this discordance of grade is due to widening of the preglacial coulee, which here

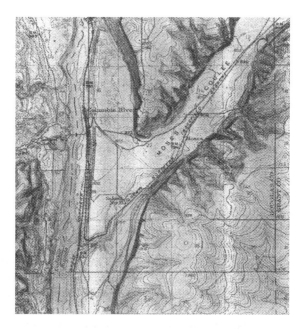

Figure 9 The lower part of Moses Coulee. Part of Malaga, Washington, topographic
map. Note truncated spurs, hanging ravines, alluvial fans, and the great
terrace in both Moses Coulee and Columbia Valley.

was a canyon, by the glacial stream but probably most of it is due to
deepening during the glacial epoch.

In addition to the deepening of valleys already existing, the glacial flood
actually made, *de novo*, drainage ways of greater width and depth than any
previously developed on the plateau. This happened where divides were
crossed and unusually high gradients down the farther slopes were found.
Five such places are especially noteworthy: Devils Canyon, Palouse Canyon
below Hooper, Drumheller Channels, Othello Channels, and Grand Coulee.

Preglacial Palouse River joined Snake River at Pasco, its subparallelism
with the larger stream for 150 miles being structurally determined. The
glacial flood from the north entered it in mid-length at several places
between Winona and Washtucna. The volume of this flood was more than
the valley could carry away. Two leaks across the divide to the Snake
developed, one near Kahlotus, and one near Hooper, and in both very great
gradients were encountered.

The Devils Canyon distributary, south of Kahlotus, cut 50 feet or so
through the loess and then by recession of waterfalls over ledges of basalt in
the north slope of Snake River Valley, it eroded a canyon 5 miles long, a
quarter of a mile wide and 500 feet deep (Fig. 10). Every fall but the lowest
retreated completely through the divide. The remaining ledge, like a dam

Figure 10 Devils Canyon, cross-section a mile and a half south of Kahlotus, showing (1) steepened loessial bluffs (33°), (2) narrow scabland above brink of basalt cliffs (3) canyon 450 feet deep and less than one-fourth mile wide, and (4) post-Spokane talus, three-fourths the height of the cliffs. Horizontal and vertical scale the same.

separating the two canyons, is less than 100 feet above the floor of Washtucna Coulee and not half a mile wide.[29] An abandoned half of a double cascade stands in mid-length of Devils Canyon (Fig. 4).

A much larger volume of water spilled across the preglacial divide south of Hooper. Before the great canyoned channel now transecting this divide had been eroded, the stream was 10 to 15 miles wide. Several channels of canyon proportion were initiated, but one outran the others in its deepening and finally secured all the discharge. Many of the shallower canyons enter the southern part of the main one over abandoned waterfalls. These channels could have carried water only when the wide scabland of the divide had no deep canyon completely across it. Their falls could have developed only after a deep main canyon existed *below* their junction. It follows, therefore, that Palouse Canyon was cut by retreating waterfalls, though these were destroyed later in the epoch. One only survived, now notched considerably by the post-glacial work of the Palouse River. The falls today are 198 feet high. Palouse Canyon is another Devils Canyon in all save its greater width and the fact that the preglacial divide was cut entirely in two.

Drumheller Channels and Othello Channels are two remarkable cases where the glacial flood crossed anticlinal ranges.[30] In both, the anticline is asymmetrical and the waters flowed down the gentle slope. In both, the flood at first was wide but became concentrated later in the deepening canyons. The maximum depth of erosion in Drumheller Channels was 400 feet, about 100 feet of which was in a weak sedimentary formation (probably the Ellensburg), and 300 feet in basalt (Fig. 11). The width of the Drumheller denuded tract is about 10 miles. This particular scabland area has a more striking and more complicated development of channels and rock basins than any other on the plateau. It is the only area of this type now topographically mapped. Below the Channels, along the northern flank of Saddle Mountains, the ancient river eroded 300 feet into the broader scabland.

But Drumheller Channels is not wholly the product of the glacial flood whose history we have been following. It and the main canyon of Grand Coulee carried drainage from a later ice sheet, and it has carried Crab Creek since the later glacial epoch. The amount of deepening during each Pleistocene epoch is difficult to determine. Othello Channels carried less water than

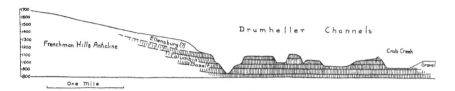

Figure 11 Cross-section of the head of Drumheller Channels, showing (1) structure
at eastern nose of Frenchman Hills anticline, (2) the pre-scabland topography, and (3) the scabland channels. Probably no preglacial drainage
crossed the anticline at this place. The lower 50 feet of the deepest channel
is a rock basin.

Drumheller Channels, consequently is a smaller tract. Furthermore, it received
water during only the earlier epoch. The degree of development of its canyons
and rock basins is comparable to that of the larger tract and makes it probable that most of Drumheller Channels were formed during the earlier epoch.

The features of Grand Coulee are of such magnitude and its history so
complicated by local conditions that an entire paper might well be devoted
to it. It affords the greatest example of canyon-cutting by glacial streams,
not alone for the Columbia Plateau, but for the world. The field evidence
indicates that no preglacial drainage route ever existed here. Scabland with
shallow channels margins the upper part of the Coulee, though 1,000 feet
higher than the adjacent coulee floor, and there are no tributaries in the mature
topography such as are possessed by Lower Moses Coulee and Washtucna
Coulee. A glacial river, 3 miles in minimum width, spilled southward here
over the divide and down a steep monoclinal slope. Judging by present
grades and altitudes of this structural slope, the stream descended nearly
1,000 feet on a grade of approximately 10°, a few miles north of Coulee
City. Such a situation is unparalleled, even in this region of huge, suddenly
initiated, high-gradient rivers. Across this monocline between Columbia River
and Coulee City, the canyon is 30 miles long and averages all of 2 miles in
width and 800 to 900 feet in depth. In the making, at least 10 cubic miles of
basalt were excavated and removed. Though a later flood of glacial waters
used this route,[31] it did but little to deepen it.[32] By far the greater part of the
erosion of Upper Grand Coulee was performed by the earlier glacial stream.
It is very probable that this immense task was performed by the stoping
of cascades and cataracts which retreated entirely through the monoclinal
uplift to the deeper valley of Columbia River and thus left the great notch.[33]

It also seems probable that when the retreat of the ice sheet began, the
plateau west of Grand Coulee was abandoned last. Earlier clearing of
the Spokane and Columbia valleys to the east allowed all the drainage
of the ice sheet to use the Grand Coulee route, which was then the lowest of
all. Grand Coulee's greatest flood and probably its greatest erosion thus
came after the other scabland routes had gone dry.

Figure 12 Cross-section of Washtucna Coulee and Snake River Valley, with longitudinal profile of Devils Canyon, showing (1) structure of the basalt, (2)
its loessial cover, (3) rock terraces flanking Washtucna (Kahlotus) Lake,
remnants of the preglacial valley floor, (4) the inner canyon eroded by the
glacial stream, (5) steepened slopes of the loess, facing the coulee, and (6)
approximate upper limit of the flood which spilled across to Snake River.

Volume of the glacial streams

If the channeled scablands of the Columbia Plateau are the erosive results
of glacial waters, certain statements as to the volume of the streams can be
made. Measurements are possible if remnants of the preglacial floor of the
main valley exist in places where the valley brimmed over with the glacial
flood to produce distributary courses. Should it appear that the amount of
canyoncutting by these glacial streams has been overestimated, the depth
of the stream to flood across the divide must be correspondingly increased.
This view promptly runs into an absurdity, for the less the canyon-cutting is
held to have been, the deeper and therefore more competent to erode the
stream must have been.

One of the best cases for measurement is Washtucna Coulee at the head
of Devils Canyon. Though there are but two small rock knobs out in the
coulee floor, the summit of neither indicating the original valley bottom,
there are good rock terraces to record it (Figs. 12 and 13). Near Kahlotus
they lie 1,000 to 1,200 feet above tide. The glacial stream here, at the beginning of its history, overflowed through the loessial hills to Snake River, at
an altitude of at least 1,350 feet. It was, therefore, from 150 to 350 feet deep.
Its width averaged at least a mile.

And this was no ponded condition, for Washtucna Coulee opened widely
into Esquatzel Coulee, an even more capacious valley, and Esquatzel in turn
into Columbia and Snake valleys, and the glacial waters cut deeply into the
bottom of both coulees. Figuring the preglacial floor as 1,000 feet A.T. at
Kahlotus and as 675 feet A.T. at Eltopia, 34 miles farther down the valley,
the great glacial stream had a gradient of about to feet 10 the mile.

Further evidence that glacial waters so filled Washtucna Coulee that the
former valley became simply a channel is found in the upper limit of glacial
stream gravels and of scablands. At the head of Devils Canyon, the highest
scabland is 1,250 feet A.T., 250 feet above the brink of the cliffs of the
canyoned channel. At Estes, a gravel bar deposited by the glacial stream lies
250 feet above the Coulee floor and about 125 feet above the rock terrace

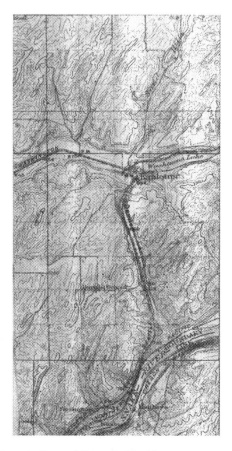

Figure 13 Devils Canyon. Part of Connell, Washington, topographic map.

which marks the old valley floor. Near Sulphur, the highest scabland surface, at the base of the steep loessial bluffs, is between 1,100 and 1,150 feet
A.T., 100 to 150 above the rock terrace. Northwest of Connell a terrace of
sand and fine gravel lies at 1,000 feet A.T. It marks the upper margin
of the scabland here and probably is a deposit of the glacial stream. It is
100 feet above the broad rock terrace. The canyoned channel here is cut
150 feet below the rock terrace.

Crab Creek Valley, below Odessa, received more water than it could carry
away, at least before its central canyon had been eroded. It overflowed
southwestward, by way of Black Rock Coulee and its associated scabland,
to the Quincy Basin which Crab Creek itself entered farther north. Measurements here are only approximate but they indicate the order of magnitude
of this glacial stream. Scabland and glacial stream gravel along the southern
edge of Crab Creek Valley lie 300 feet or more above the present stream and

extend a mile and a half back on the upland from the margin of the preglacial valley. This valley had been canyoned more than 100 feet by the glacial stream which, on the basis of these figures, was 200 feet deep at its inception.

The Telford scabland tract, 13 miles wide and 20 miles long, has been swept almost completely bare of the loessial deposit. The relief in a cross-section of the basalt surface now exposed, aside from the minor canyons, is about 50 feet. To have been so denuded, this tract must have had a sheet of running water of this depth completely over it.[34]

Further evidence of the depth of the glacial streams will be presented under the next subject.[35]

Deposits made by the glacial streams

The record of Pleistocene glacial streams almost everywhere is one of aggradation. Glacialists commonly think of the subject only in such terms, textbooks discuss it only in that light; it is the orthodox conception. But in the Columbia Plateau exceptional factors controlled. The preglacial valleys in general were small and of relatively high gradient, the volume of the glacial streams was very great, the amount of detritus from the ice was very small, and the rock crossed was either loess or closely and vertically jointed basalt, both of which yielded rapidly to the torrents. The result of these conditions was great deepening of the valleys, and deposits made by the streams are of minor importance. Their character, however, adds to the weight of evidence already presented for the origin of channeled scabland by glacial stream erosion.

The deposits are almost wholly of gravel. Sand is a minor constituent and clayey material is lacking. The gravel and sand are almost wholly of basalt, though all deposits contain fragments of rock foreign to the plateau. The basaltic gravel is not well rounded though most of it is sorted and stratified and indubitably of stream origin. Foreset bedding is common, the direction of dip according with the slope of the scabland tract. In some places, the deposits are composed of very coarse material, with abundant, sub-rounded, basaltic bowlders 3 and 4 feet in diameter. These were originally bowlders of decomposition and were derived from flows with particularly large columns, underlying or in the immediate proximity of the deposit. Where a few erratic bowlders are associated[36] the deposit itself might be misinterpreted as a bowldery till.

The gravel deposits rest on irregularly eroded basalt, essentially a buried scabland surface. Nowhere do they lie on or beneath the loess. Neither the gravels nor the underlying basalt are decayed.[37] Cementation with calcium carbonate has begun but has not advanced far.

Most isolated loessial hills in a scabland tract have a deposit of gravel depending from their down-gradient end. Many knobs and buttes of basalt have similarly situated gravel deposits.

In many cases, the gravel deposits constitute discontinuous terraces on the margins of the scabland tracts, suggesting remnants of former valley fills.[38] But the evidence seems conclusive that all gravel deposits of the scablands are bars, built in favorable situations in the great streams which eroded the channels.

The rounded profiles and ground plans of many gravel deposits in the scablands are in accord with this interpretation. The unfilled canyons and rock basins, in intimate association with the discontinuous gravel deposits, indicate clearly that both are products of the same episode. The only alternative hypothesis is that channeled scabland was formed, then buried in gravel, then in large part re-excavated by streams little short of the magnitude of those which eroded the scablands. This has no other field evidence to support it and requires a much more complicated history. Furthermore, such deep canyons were cut when the scablands were made, and such noteworthy divide crossings were made that a reoccupation of all the scablands by glacial drainage from a second ice sheet would be impossible. And the hypothesis of dissection by the postglacial streams of the scablands is quite inadequate. Lakes and pools still stand in the rock basins on the channel floors, almost as they were left by the glacial flood.

Gravel deposits in the deeply canyoned scablands occur on the broad upper scabland surfaces, on the roughened slopes of the preglacial valleys and down in the canyons. The interpretation of these deposits as bars requires no change in general conditions, as does the alternative hypothesis; it simply requires that gravel be deposited locally as conditions might favor, all through the epoch of erosion of the basalt. Deposits on the highest scablands[39] were made before the deepening canyons drew the waters into more restricted routes. Deposits in the deeper portions[40] were made during the latest stages of the episode. No gravel deposits were ever as thick as the rival hypothesis would require them to be.[41] That view demands that the canyons be eroded, then filled completely *with their own débris*, then re-excavated in large part.

However, there are two places on the plateau where the history of deposition by glacial streams has been somewhat different. One of these is Hartline gravel flat, the other is Quincy Basin. Both are structural depressions, not completely filled before the glacial floods arrived. The Hartline structural valley became filled with débris from the cutting of Upper Grand Coulee before Lower Grand Coulee had been eroded. The trenching of the lower coulee, and particularly the development and retreat of Grand Falls, incised the southern rim of the valley so that, by the close of the episode, the gravel fill, once the floor of the glacial stream, had been removed in its western part, and the remnant left 200 feet or so higher than the brink of Grand Falls.[42] The total fill in the Hartline structural valley is about 250 feet. It is composed of bowlders, cobbles, and gravel near Grand Coulee, and of sand 5 or 6 miles east, back from the main drainage line. An old channel from

Grand Coulee crosses its northern and eastern part and leads into Deadmans Gulch, a distributary which spilled across the southern rim into Spring Coulee before Lower Grand Coulee had developed into the dominant notch that drained and dissected the flat. The terrace form of Hartline flat is the result of erosion of a once complete gravel fill. Its scarp is not constructional, as are those of bars. But the fill and the subsequent erosion occurred because of special local conditions, not because general conditions changed.

Quincy Basin, like the Hartline Valley, does not have a scabland floor. Both lay too low to be eroded. But both belong to the glacial drainage plexus. Quincy Basin probably contains more gravel than all the scablands of the plateau together. It was an enormous settling basin for the glacial rivers from Grand Coulee and Crab Creek. The flood of waters entering it was so great that at first three discharge ways were simultaneously in operation.[43] The southern and larger one obtained all the discharge later, and by deep notching of the basin's rim, caused the glacial waters which traversed the fill to incise their deposits. Two great channels and one smaller one were thus formed. The two large channels are each about 3 miles wide. Each was eroded about 100 feet deep during the later part of the episode. The one which contains Rocky Ford Creek also carried the later Wisconsin discharge and was further modified then.[44]

Erratic bowlders, some of them striated, are widely distributed at all altitudes on the basalt plain and the scablands. They also occur in valleys of the mature topography which open northward on to the basalt plain, and in some which open on to scabland tracts. The size, angularity, and striated surfaces indicate that these erratic bowlders were not rolled to their positions by running water. In the scablands, they must have been carried by berg ice on the great rivers. In their peculiar and limited distribution in the valleys in loess is evidence of small glacial lakes, in which the driftbearing bergs floated.[45]

Depths of Snake and Columbia Valleys during the epoch

Evidence on this question may be obtained at the debouchure of glacial drainage routes into these master valleys. Five of the nine such debouchures will be examined.

The rock floor of Moses Coulee is fully as low as that of Columbia Valley at the junction of the two. Both contain a great gravel fill here. Columbia River has cut through it, a depth of more than 300 feet. There is no such trenching in Lower Moses Coulee, but a well at Appledale penetrates 300 feet of this gravel without encountering bedrock.

The two cataracts of The Potholes and Frenchman Springs, which operated in the early part of the epoch, descended nearly the full height of the present Columbia Valley walls there. At The Potholes the glacial cataract can be traced down to less than 200 feet above the present Columbia (Fig. 7).

Koontz Coulee, 20 miles north of Pasco, is cut in the weak Ringold formation. It is 250 feet deep and a mile wide. It is floored with basaltic stream gravel from the scablands farther upstream. Though the Ringold silts extend down to the level of the Columbia at this place, the mouth of the glacial river channel hangs 200 feet above. No cataract could have been maintained here, as was done at The Potholes and Frenchman Springs, and the level of the Columbia of this epoch at this place is thus clearly recorded.

At the mouth of Palouse River, there are two parallel canyons in the scabland, one containing the river, the other dry. A basaltic butte separates them. It stands nearly in the center of the valley and its summit is between 350 and 400 feet above Snake River. It is a part of the original north wall of Snake River Valley, over which the gigantic cascade tumbled when the glacial flood broke across the preglacial divide from the north. This "Goat Island" testifies to the existence of a Snake River Valley at this place as deep then as now.

The glaciation

Because the record of the ice sheet, from which came the streams that made the scablands, is best preserved on the basalt plain about the city of Spokane and along the south side of Spokane River, this has been named the Spokane glaciation.[46] It assuredly is not an early phase of the Wisconsin glaciation. That is recorded by pronounced moraines on the plateau west of Grand Coulee, in Columbia Valley north of the mouth of Spokane River, and in Colville Valley north of Spokane River. The Spokane ice left no terminal moraine and very little ground moraine. The reverse relation exists regarding the glacial waters of the two glaciations, for the Spokane glacial waters were of prodigious quantity and the Wisconsin waters of little consequence. Furthermore, a long time elapsed between the two glaciations as shown by the relative volume of talus accumulations in tracts swept by glacial streams of the two epochs. Post-Spokane talus in almost all places stands three-fourths to four-fifths of the total height of the basalt cliffs, post-Wisconsin talus stands about halfway up on the cliffs of Grand Coulee.[47]

For the absence of a terminal moraine along the southern edge of the area reached by the Spokane ice sheet, the writer has as yet no satisfactory explanation. It seems clear, however, that a moraine never was deposited, rather than that it was once built and subsequently removed. The functioning of some scabland tracts absolutely required glacial ice against the north slopes of the unglaciated hills at their heads. Floated granite erratics among some of these hills also require blocking of valleys by glacial ice. Yet there is no evidence of lateral drainage along the ice front in contact with the unglaciated hills, to which might be ascribed the removal of a terminal moraine.

The Spokane glaciation cannot be dated very far back in the Pleistocene, else the scablands should have a soil mantle of eolian sand and dust and disintegrated basalt, and the hundreds of lakes in the old channels should have been destroyed.

Leverett has suggested that a pre-Spokane till beneath loess at Cheney is of Kansan age.[48] If it is, and if the post-Spokane glaciation is correctly ascribed to the Wisconsin epoch, the Spokane glaciation should be either Iowan or Illinoisan in age. Farther than this, the writer does not care to go. Ordinary criteria in use east of the Rocky Mountains for differentiation of drift sheets cannot safely be used for the correlation of these glaciations in Washington. The only one relied on here is the moraine-building habit of the Wisconsin ice sheet, a character which seems to have been worldwide.[49]

There were no channeled scablands on the Columbia Plateau before the Spokane glaciation. A mantle of loess, with a mature topography, completely covered it. The evidence for this conclusion is found in the great and remarkably persistent width of the Cheney-Hooper scabland tract throughout a length of 70 miles, and the various distributary courses out of it, some of which never were eroded to the basalt. These features never could have been formed, had spillways like those of the present existed. But with early escape southward retarded by the loessial hills and their small drainage ways, a wide spreading among them necessarily occurred, and some distributaries were able to cross to Crab Creek drainage.

A puzzling situation regarding glacial drainage exists in the vicinity of the small Spangle lobe. There is no adequate drainage route around it for glacial waters which came from Idaho and western Montana and entered the Cheney-Hooper scabland tract. Two spillways exist north of Mica, between Lake Spokane[50] east of this lobe and Pine Creek channel. Both have erratics in them, the highest at 2,550 feet A.T., but neither carried much water. There is no error involved in mapping this lobe because an ice dam at Spangle is required for the operation of the Pine Creek channel. This channel carried far more water than the Mica spillways, water derived directly from the Spangle lobe. Yet it also is inadequate for the drainage in question. And much more water went down the Cheney-Hooper scabland tract, in proportion to the immediately tributary ice edge, than passed through any other scabland tract except Grand Coulee in its later stages. It seems necessary, therefore, to assume a prominent *subglacial* drainage line, across the area covered by the Spangle lobe. This is best located along the preglacial valley of Lake (Marshall) Creek and the rock basin of Farrington (Fish) Lake. Out of this rock basin, just beneath the edge of the ice, the waters from the east emerged and joined those coming directly from the ice.

If the battle between the diluvialists and the glacialists, out of which has emerged our conception of Pleistocene continental glaciation, had been staged in the Pacific Northwest instead of the Atlantic Northeast, it seems likely that the surrender of the idea of a debacle might have been delayed a decade or so. Fully 3,000 square miles of the Columbia plateau were swept by the glacial flood, and the loess and silt cover removed.[51] More than 2,000 square miles of this area were left as bare, eroded, rock-cut channel floors, now the scablands, and nearly 1,000 square miles carry gravel deposits derived from the eroded basalt. It *was* a debacle which swept the Columbia Plateau.

Notes

1 An earlier paper on this subject was published by the writer in the *Bulletin of the Geological Society of America*, Vol. XXXIV (1923), pp. 573–608. The study on which it was based involved about a 1,000-mile traverse of the plateau. Since then, the writer has studied the plateau more thoroughly, having added more than 2,000 miles to the previous total traverse. Much more detailed information and several modifications of the earlier interpretations justify the appearance of a second paper on the subject. The accompanying map (Plate III) is based on a field examination of every scabland there indicated. In a few places the boundaries are inferred (dashed lines) but future work will hardly do more than make minor changes or additions.

2 A part of Othello Channels, a part of Drumheller Channels, at Palisades and near Spencer in Moses Coulee, at Soap Lake and near Bacon in Grand Coulee, 6 miles south of Almira on Wilson Creek, and at Long Lake in Spring Coulee.

3 There are many exceptions to this rule, occurring where scabland tracts cross pre-existing divides, but the total length of such is only a small fraction of the aggregate length of all scabland tracts.

4 Moses Coulee is the only exception.

5 Grand Coulee and Moses Coulee are exceptions. Grand Coulee opens upgradient into Columbia River Valley and the upper end of Moses Coulee is obliterated by the drift of a later glaciation.

6 Exceptions due to a later episode in the history of the region are noted later.

7 One exception, noted later.

8 The Hartline gravel flat and the Quincy basin fill, both in structural depressions in the plateau.

9 Upper Grand Coulee.

10 Paradise Prairie, Sunset Prairie, Indian Prairie, Four Mound Prairie, etc. On the north side of Spokane River are Pleasant Prairie and Five Mile Prairie, also parts of this plain.

11 With the exception of northern Douglas County.

12 A few broad, undissected divides are still left. Michigan Prairie, south of Lind, is a good example.

13 Probably the Ellensburg formation. The Pleistocene Ringold formation, in Franklin County, is younger than the mature topography.

14 Near Harrington and near Kahlotus are two excellent cuts which show this very well.

15 Upper Grand Coulee (1,000 feet deep), Lower Moses Coulee (900 feet deep), Devils Canyon, Franklin County (500 feet deep), and Palouse Canyon (500 feet deep) are the most noteworthy examples.

16 Dry Falls (400 feet high) in Grand Coulee, The Potholes (350 feet high) south of Trinidad, Frenchman Springs (400 feet high) south of The Potholes, and The Three Devils (600 feet total descent) in Moses Coulee are especially noteworthy.

17 Deep Lake, below one of the Grand Coulee abandoned falls, has many associated huge potholes, drilled into the basalt at the foot of the falls as they retreated. Each of the two cataracts of "The Potholes," south of Trinidad, has a single elongated rock basin at the foot (Fig. 7).

18 Rock Lake in Whitman County, Goose Lake in Grant County, Washtucna Lake and Eagle Lake in Franklin County, Pacific Lake and Tule Lake in Lincoln County, Goose Lake in Grant County, Medical Lake, Silver Lake, and Farrington (Fish) Lake in Spokane County, and Big Swamp Lake and Cow Lake in Adams County are examples of hundreds of such basins.

19 Especially well shown in Lower Grand Coulee, in Moses Coulee between Palisades and Spencer, in The Potholes and Frenchman Springs south of Trinidad, and in Palouse Canyon south of Hooper, in all of which cataract recession in main channels cut off smaller channels alongside in the same scabland tract.

20 The Telford scabland and its dependencies.

21 Coal Creek, Duck Creek, Lake Creek, an unnamed creek, Connawai Creek and Wilson Creek.

22 The Cheney-Hooper area, extending from Spangle and Cheney to Snake River, south of Hooper.

23 Cow Creek, Rock Lake, and Creek and lower Palouse River now occupy the most striking of these canyons.

24 Palouse River Canyon below Hooper.

25 In one place, 1,000 feet in about a mile.

26 Examples are Moses Coulee east of Palisades, Wilson Creek above Almira, Crab Creek below Corfu, Washtucna Coulee, Lind Coulee below Lind, Snake River near Lewiston and Clarkston, Union Flat Creek, Rebel Flat Creek, and Palouse River above Winona.

27 As in the Telford and the Cheney-Hooper areas.

28 Washtucna Coulee, Esquatzel Coulee, Lower Moses Coulee, Lind Coulee, Pine Creek above Rock Lake, Rock Creek (Lincoln County) and Coal Creek are examples.

29 The Spokane, Portland and Seattle Railroad tunnels through this ledge. In 1916, Washtucna Lake was so high for a time that it overflowed through this tunnel into Devils Canyon.

30 Drumheller Channels crosses the eastern nose of Frenchman Hills anticline, Othello Channels crosses the eastern nose of Saddle Mountains anticline.

31 During the Wisconsin glaciation.

32 Grand Falls, below Coulee City, consists of Dry Falls, Deep Lake Falls and a smaller unnamed falls a mile east of Deep Lake. The lip of the smaller falls is 125 feet higher than the floor of the channel leading to Dry Falls. Yet all were made by the same glacial stream and only Dry Falls and Deep Lake Falls were used and modified by the later discharge.

Furthermore, the Wisconsin ice did not cross Spokane River or Columbia River east of Grand Coulee and its waters were free to use the lowest of the ten earlier routes. Only Grand Coulee was so used, showing that it had been eroded by the earlier discharge to a depth not far short of that which it now has.

33 This inference has no physiographic evidence in Grand Coulee to substantiate it, but is based on the known procedure of the glacial streams in similar situations, e.g., Lower Palouse Canyon, Devils Canyon, Frenchman Springs, and The Potholes.

34 That the ice sheet did not advance over the Telford denuded tract is shown by the presence of a few isolated loessial hills with characteristically steepened marginal slopes. One such group lies 7 or 8 miles north of Telford. It has a maturely eroded topography and a dark loessial soil without rock fragments of any kind. But it is cut by channels of glacial waters which eroded to the basalt. The fact that these waters went through the group, though the surface immediately north of them drops off into the deep canyons of Hawk Creek, a tributary of Spokane,

River, proves that glacial ice mast have crowded up against the northern side of the group.

35 If these enormous streams all came to the Columbia eventually, should not the great volume be recorded farther down the master stream? The writer has seen enough to convince him that it is so recorded, and hopes to publish on this subject in the future.

36 As west of Lantz, for example.

37 Local exceptions, as 1 mile southwest of Lamont, where ground water has been especially active.

38 So the terraces in Pine Creek channel were interpreted in the earlier paper by the writer. That view is here abandoned for one much more consistent with all other features of the scablands.

39 As about Gloyd along the Black Rock distributary from Crab Creek Valley.

40 As at the mouths of Duck Creek and Wilson Creek in Crab Creek Valley and at the junction of Crab and Coal creeks.

41 For example, gravel deposits north of the town of Washtucna lie on the slopes of the coulee, 350 feet above the present floor. They antedate the deeper canyoned portions of the coulee.

42 Estimated from the surviving eastern member of that complex waterfall, the only part which escaped modification by the Wisconsin glacial stream.

43 Frenchman Springs, The Potholes, and Drumheller Channels.

44 This interpretation is a modification of that published earlier by the writer. Further study of Grand Coulee, Quincy Basin, and Drumheller Channels has led to a magnification of the work of the earlier flood, and a minimizing of the results accomplished during the Wisconsin epoch. Grand Falls is now considered to be a pre-Wisconsin affair, none of the distributary canyons of Grand Coulee, except Dry Coulee, are thought to have functioned during the second flooding, the Adrian terrace is considered to be a part of the original fill and not of Wisconsin age, and all the deep canyons of Drumheller Channels are thought to date back to the earlier episode.

45 Below an altitude of about 1,250, erratic bowlders occur on every formation and type of topography on the plateau. But these are a younger deposit (see *Journal of Geology*, Vol. XXVII [1919], pp. 489–506) and do not much overlap the scablands. Where overlap does occur, however, it is impossible to distinguish bowlders of the two categories by any difference in the amount of decay.

46 J.H. Bretz, "Glacial Drainage on the Columbia Plateau," *Bulletin of the Geological Society of America*, Vol. XXXIV (1923), pp. 573–608.

47 This question of differentiating Pleistocene epochs by talus accumulation will be discussed more fully in a separate paper.

48 Frank Leverett, *Bulletin of the Geological Society of America*, Vol. XXVIII (1917), p. 143.

49 No till or other evidence of pre-Spokane glaciations has been found beneath the loess (save only the Cheney deposit). No till has been found in the scablands. The writer is unable to agree with J.T. Pardee who states (*Science*, Vol. LVI [December 15, 1922], pp. 680–87) that till occurs at "scores" of places on the plateau south of the limit of Spokane glaciation, as mapped in Plate III.

50 Thomas Large, *Science*, Vol. LVI (September 22, 1922), pp. 335–36.

51 Except in the Hartline and Quincy structural depressions.

MARGINAL DRAINAGE CHANNELS AS INDICATORS OF THE GRADIENTS OF QUATERNARY ICE CAPS

C.M. Mannerfelt

Source: *Geografiska Annaler* 31 (1949): 194–9.

Notwithstanding all the research work done in the Scandinavian and North American districts of Quaternary glaciation, our knowledge of glaciological and climatological conditions on such parts of the ice caps as are situated above the melting ice borders is still rather inadequate. Many apparently simple problems are, still unsolved. When did the first nunataks emerge on the ice divide? How far northwards had the continuous ice front melted when the progressing climatic improvement lifted the firn line above the highest peaks still ice-covered? What were the relations of the superficial annual thinning to the successive recession of the ice front? Was the surface of the melting ice flat and undulating, or did it slope rather sharply from its central to its marginal parts?

Reactions at the supra-aquatic ice front are as a rule functions of the glaciological changes in the accumulation area. In modern investigations the glaciologists are trying to deal with the whole ice mass as one three-dimensional unit, and the glacial-geologists ought similarly to extend their traditional field work—in which the melt formations at the ice front have so far been primarily considered—also to the central, higher portions of the ice. Few regions on earth are probably more suitable for this purpose than the isolated mountain massifs (1,000–1,700 m above sea level) of the late glacial ice divide between Storsjön in Jämtland (Sweden) and Lake Fæmund in Norway.

When the peaks of this district first emerged as nunataks from the inland ice, the firn line had already risen above the top level of the ice. Climatologically, the residual inland ice was dead. It had no permanent accumulation area, and its continued reduction was contingent on the gradual wasting effect of the ablation alone. To begin with, the melt water found its way through the highest mountain passes between the first nunataks. Overflow channels or col gullies were established, and these can now be seen as deep, dry valleys—evidence of very intensive ice-river drainage in altitudes above 1,000 m above sea level.

Large numbers of marginal drainage channels were formed between the gradually bared mountain sides and the remaining stagnant ice masses. When the melt water had undermined the dead ice, it was drained off into the subglacial chutes, in which the erosion abruptly changed at the ground water level of that time into accumulation, and formed chute eskers. These submarginally engorged eskers have since been covered by the moraine material from the surface of the stagnant ice.

The marginal drainage strakes may give some, even if uncertain, indications of the lowest level of the firn line and a chance of estimating the gradient and annual melting rhythm of the inland ice.

The fact that there were no local glaciers on the nunataks projecting round the ice divide proves the same thing, i.e. that during the melting of the inland ice in the continentally situated mountain massifs the firn line must have been considerably above about 1,500 m above sea level. In other words, the continuous thinning of the ice could not keep pace with the gradual rise of the glaciation limit—and accordingly also the firn line—to ever higher levels.

It should be possible to estimate the superficial gradients of the ice by a study of any occurring «Schliffgrenze» or nunatak limits, or by observing the gradients of the lateral moraines. But neither of these possibilities occur in the region of the ice divide. The marginal drainage gullies can, however, provide some clues of great fundamental value. Nowhere else do they occur in such splendid abundance as on Mount Sånfjället in Härjedalen. Dry valleys score the mountain sides everywhere. On Gråsidan in particular the erosion due to the late glacial melt water has given the details of the topography a peculiar character of its own. If looked down upon from the top of the slope, this remarkable erosion-landscape is reminiscent of a newly ploughed field with gigantic furrows following one another at almost regular intervals. Many of the gullies can be traced unbroken for several hundred metres as indicated on Fig. 1 and 2.

Before drawing any far-reaching conclusions regarding the relations between the slopes of the gullies and the surface gradient of the ice, the conditions under which the gullies were formed must be elucidated. The lateral drainage channels were formed mainly in the spring and early summer by the intensified melting of the snow. Sub-glacial drainage tunnels opened later in the summer. Through these the water was then drained off as the summer ablation lowered the surface of the ice, gradually leaving the lateral gullies of the spring flood dry. During next year's melting period another

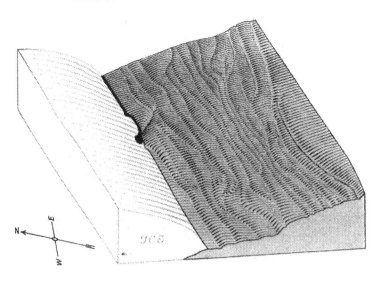

Figure 1 Diagram showing the genesis of lateral drainage channels and subglacial chutes at Gråsidan, Sånfjället, Sweden.

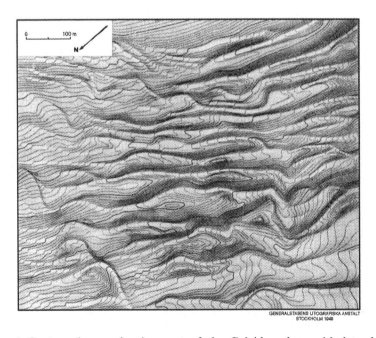

Figure 2 Contoured map showing part of the Gråsidan slope with lateral and submarginal channels.

Figure 3 Air photo towards Fulufjället, Sweden, with lateral drainage channels. Photo: G. Lundqvist.

gully was eroded at a lower level. Wherever these gullies occur in large numbers, they are invariably so regularly spaced that they must be assumed to reflect the annual ablation, seeing that hardly any other climatological periodicity is conceivable. If that assumption is correct, the annual thinning by ablation must have averaged 3.5–5 m, which seem likely values if compared with the recent observations in glaciated districts. Insolation was apparently of subordinate importance, as no tendency towards a dominant southerly orientation has been observed in any of the several hundred drainage channels examined.

If we want to estimate the superficial gradient of the ice, it is important to ascertain first whether the lateral gullies really reflect the shape of the free ice surface. There is always a natural tendency for the lateral water to undermine the margin of the ice and find its way into the subglacial chutes. Three different main types can be distinguished in the Scandinavian mountains:

1. *Strictly lateral terraces and channels*, mainly the results of glacio-fluvial accumulation or erosion between the bared mountain sides and the ice margin. These terraces are sometimes very persistent, and can in one or two cases be followed unbroken for more than 1,500 m. Their gradients have been determined, and indicate that the superficial slope of the damming ice mass was probably 10–20 m in every 1,000 m.
2. *Sublateral drainage gullies.* These are the commonest type. They are deeper and more marked, and generally scoured out of the moraines to a depth of several metres. They are not as persistent as the accumulation terraces, however, and as a rule one drainage gully suddenly merges

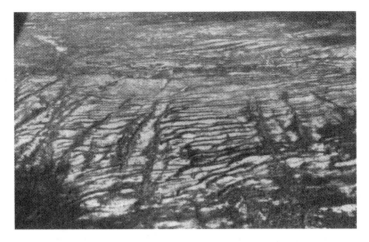

Figure 4 The subglacial chutes pierce the system of marginal drainage channels. Photo: G. Lundqvist.

    into the next below, thus creating an anastomosing system. The melt water has obviously found an oblique way down underneath the ice margin. Their average gradient of more than 3.5 m: 100 determined by numerous observations, can therefore hardly represent the superficial gradient of the ice mass itself, but rather the steeper gradient of the local erosion. These sublateral drainage gullies are mostly situated on the convex sides of the mountains.

3.   *Subglacial chutes.* These often originate from sublateral gullies' suddenly consistently turning round to plunge downhill in a serpentine course. The dry ditches left by them obviously have nothing to do with the superficial gradient of the ice.

As these different types are frequently combined, it is very difficult to obtain from them any reliable values of the superficial gradient of the ice. A critical analysis from case to case indicates, however, that an average value of 1–2 m: 100 is a likely figure. The slope was steeper close to the strongly melting margin, slightly less so on the free surface of the ice.

If, as a consequence of the above argument, we regard the ice melt as a three-dimensional problem, we come to the conclusion that the mountains in the neighbourhood of the ice divide must have emerged from the ice at a rather late stage. On the assumption that the ice gradient was as indicated above, the crest of Sånfjället cannot have been bared until the ice margin had approached to a distance from it of 50 or 100 km. The theoretical relation between the annual thinning of the ice from the surface and the recession of its supra-aquatic ice front can easily be computed trigonometrically if one of these values and the average slope of the ice surface and the underlying ground are known. Assuming the annual ablation of the ice to be, say, 4 m, and its average gradient 1.5 m: 100, the ice front must have receded 265 m per annum, provided of course that the underlying ground was plane and that the ice was no longer in motion. An ice front melting on an opposing slope would recede less, and much more on a downward slope. This relation of the surface ablation to frontal recession is much the same as that which may be deemed plausible on other grounds.

The one-sided evidence of the melting of the ice front provided by recessional moraines or warved clay does not always reflect the climatic changes satisfactorily. In sub-aquatic positions calving may distort the picture completely. In supra-aquatic positions the thin front is often hidden by a protective layer of ablation moraine, which favours the formation of irregular dead ices. It is therefore most important to try to obtain also some idea of how the free ice surface above the front has reacted. A more thorough study of the details of the highest marginal drainage channels should afford interesting contributions to our understanding of the glaciological conditions in the final stage of the Glacial Period.

Modern glaciology is an essential auxiliary to Quaternary, and particularly glacial, geology. I will conclude by appealing to Professor Ahlmann to supplement the glaciological program of the Norwegian–Swedish–British Expedition to Dronning Maud Land by a detailed study of the lateral and sublateral form elements, and of the conditions under which they are formed in nunatak regions in high altitudes. The classic monographs of O.D.v. Engeln and R.S. Tarr on the marginal parts of the Malaspina Glacier have, like P.G. Visser's from Himalaya and P. Woldstedt's from Vatnajökull, been of inestimable value to Quaternary geology. A corresponding detailed investigation from Antarctic nunatak regions would be a highly stimulating complement to them.

3

TEMPERATURE OBSERVATIONS IN BERGSCHRUNDS AND THEIR RELATIONSHIP TO CIRQUE EROSION

W.R.B. Battle and W.V. Lewis

Source: *Journal of Geology* 59 (1951): 537–45.

Abstract

Willard Johnson's bergschrund hypothesis assumes diurnal temperature changes at depth. Direct evidence from continuous temperature records in bergschrunds calls for a modification of this hypothesis. Evidence from northeastern Greenland and Norway shows that air temperatures in an open bergschrund during the summer vary between 30° and 32° F. In Norway diurnal changes of temperature were absent in spite of substantial changes in air temperature outside. New continuous recording instruments have been designed to measure temperatures and the number of times the temperature crosses the freezing point. Further direct evidence is necessary before adopting a thaw-freeze hypothesis.

Willard D. Johnson's (1904) justifiably renowned descent of a bergschrund and his careful account of what he saw opened a new era in the study of cirque erosion. It is scarcely an exaggeration to say that it heralded a change of emphasis from description of the characteristic features of glaciated mountain scenery to the investigation of the processes responsible for these features. The inferences he somewhat cautiously drew from his observations were taken up by his followers with an enthusiasm that embarrassed him. According to Isaiah Bowman (1920, p. 296), Johnson, before he died, had fully intended launching an attack on the so-called "bergschrund hypothesis."

Johnson's observations are important primarily because they pointed to thaw-freeze processes operating deep down in a bergschrund, but he also revealed the true nature of a bergschrund. The bergschrund he descended continued down until it reached rock on the up-glacier side. The *randkluft*, on the other hand, which often forms late in the melt season, is a crack wholly between the rock wall of the cirque head and the névé. The distinction, though useful, is somewhat academic, as what appear to be bergschrunds early in the melt season frequently turn into *randklufts* later on, especially at the end of seasons in which ablation has been excessive. Whether or not one should refer to a crevasse near the head wall as a bergschrund, as Bowman did, even though it does not reach bedrock, is as yet unsettled. It is probably wisest to keep the term "bergschrund" fairly wide, as, in fact, the crevasse or crevasse series to which the name was originally applied varies much from place to place and from time to time. It resembles a living organism with a life-cycle concerning which we are extremely ignorant.

The great advantage of the bergschrund hypothesis was that it provided a simple mechanism of attack by a glacier upon the rock wall at its head. The day-to-day changes of temperature across the freezing point which are so effective in shattering rocks lying high up on the mountain sides above the blanketing snow or ice have, nevertheless, still to be demonstrated deep down in such crevasses. Lewis (1940), having been largely initiated into the study of snow and ice in the exceedingly moist climate of southeastern Iceland, considered that meltwater must aid substantially the process envisaged by Johnson and his followers. He repeatedly observed meltwater descending into glaciers and névé fields, especially where any rock rose above the surface of the ice. This modification helped to overcome the rather serious objection to Johnson's theory that warm air cannot displace the heavier cold air occupying a crevasse, except under distinctly abnormal conditions. Equally, direct or indirect radiation from the sun cannot transfer much heat a hundred or more feet down a narrow crevasse in which air circulation and radiation alike tend to be seriously hindered by snow bridges. Furthermore, severe erosion of the rock wall was examined by Lewis in Switzerland *above* a bergschrund, where the tunnel from the Jungfraujoch railway station within the mountain emerges on to the Jungfraufirn. An examination of British cirques also suggested that erosion by glacial plucking and shattering—for which thaw-freeze seemed to offer the best explanation—occurred right to the base of 1,000-foot head walls. Thus it was argued that an erosive process similar to, if not identical with, thaw-freeze had once been active over zones of cirque head walls many hundreds of feet in vertical extent.

However, the meltwater hypothesis, like its parent-hypothesis, was based on somewhat casual observation and inference, and the search for definite evidence is overdue. Johnson, at least, was quite honest in admitting that he "was at the disadvantage for close observation of having to clamber over these [blocks] with a candle, in a dripping rain of meltwater." Battle therefore set himself the task of searching for more direct evidence of what really happened in bergschrunds by emulating Johnson's feat of penetrating deeply into a bergschrund, but armed with instruments for measuring temperature, preferably over long periods. This seems to be a new line of attack, perhaps owing to its arduous nature. Only a very few such attempts have been made previously, as far as we know. Von Klebelsberg (1920, 1948) examined bergschrunds encountered during tunneling operations on the Austrian-Italian frontier during the 1914–1918 war. He noted the absence of melting such as would be expected if warmth penetrated the bergschrunds according to the then widely held views. McCabe (1939, p. 463) recorded summer air temperatures in fairly shallow bergschrunds in Spitsbergen; he found them to be usually 0° C., even when the air temperatures outside were far higher. D.H. Chapman sent a letter in 1948, stating that his student, Robert Lange, took temperatures 50 feet down a bergschrund in Teton Glacier, Wyoming, with a thermograph, for most of a period of 3 weeks in the summer. These observations were, we understand, of a somewhat preliminary nature and have not yet been continued.

Battle (1949), as leader and glaciologist of the Leeds University Greenland Expedition in 1948, was able to examine the bergschrund of a corrie glacier[1] on the mainland of northeastern Greenland near Clavering Island (pl. 1, A). During the day the bergschrund, which was only barely visible on the surface as a narrow crack, had meltwater trickling down the névé and ice on the up-glacier side. This, however, seemed to be localized at one or two small streams. At the bottom, about 100 feet down, the bergschrund opened out into a huge cavern with a roof of snow blocks. Even at night, when air temperatures in the open were as low as 23° F., it was comparatively warm beneath, where some of the icicles continued dripping. The rock wall appeared in only one or two places as a smooth black slab, mostly covered with a coating of transparent ice, 2 inches or more thick. It was quite impossible to tell whether the floor reached bedrock, as it was covered by a confusion of large blocks of ice fallen from the roof nearly 100 feet above.

The almost uniform cover of ice which remained day and night during the 10 days of observation at the height of summer was surprising and suggests that relatively little damage was being done to the rock face by thaw-freeze or any other process. There were no signs of diurnal changes of temperature. Conditions of melting and of the unfrozen state of one's outer clothing remained unchanged day and night, as far as one could judge without thermometers. If these conditions—seemingly harmless to hard rock—are in any way typical of summertime, it is interesting to speculate as to when the cirque was formed. Lower temperatures, such as occur in winter, would

A

B

C

Plate 1 Cirque features in Greenland and Norway.

A, Grif Gletscher, the corrie glacier above the Pasterze Glucier, northeastern Greenland, August, 1948. The bergschrund is just visible as a thin line below the head of the glacier. Photograph by W.R.B. Battle.

B, The interior of a 30-foot-deep bergschrund in Hellstiggutind, Jotunheim, August, 1950. The transition from névé to ice is plainly visible. The dirt band at the base of the névé marks this transition. The irregular surface below consists of ice masking the rock wall. Photograph by W.R.B. Battle.

C, Large blocks of gabbro broken loose from the head wall of Juvass Kjedlen, August, 1947. In a normal year this whole area is covered by the névé of the cirque glacier. Photograph by W.V. Lewis.

produce little or no meltwater; only relatively hot or wet spells in summer would presumably produce sufficient water to melt through to bedrock. The summer of 1948 was a relatively warm one, and when the party left on August 28 the winter freeze-up had begun. One was left with the strong impression that erosion was not active under present climatic conditions. It would have been even less so half a century earlier, when Koldeway (1873) records that the ice covered everything except one or two nunataks. Little is known about the previous extent of the ice, but Ahlmann (1948, p. 166) and Brooks (1949, p. 357) suggest that it was much less extensive during the period of the Viking discoveries from 900 to 1,400. Cirque erosion may have been more active in Greenland then.

These indeterminate but thought-provoking observations in Greenland stimulated the co-author to continue his work in the more accessible Jotunheim, where other glaciological field work was being undertaken by the Cambridge group. The instruments available for this work in the summer of 1950 were a mercury-insteel Negretti and Zambra thermograph with a 10-inch bulb, very kindly loaned by Professor Gordon Manley, of Bedford College, London, and six small radioactive temperature recorders designed and loaned by Dr. Eilif Dahl, of the botany department of the University of Oslo. The thermograph was a standard instrument driven by an 8-day clock. Its accuracy to about 0.3° F. satisfied our requirements, but it was distinctly cumbersome, weighing 30 pounds. Also, it needed visiting each week—a wise precaution for such work, as any minor mishaps to the instruments can then be promptly remedied.

The Dahl (1949) recorders are extremely ingenious and stoutly constructed devices, measuring only $6 \times 3 \times 1^1/_2$ inches and weighing only a pound or so. A bimetallic spring moves a radioactive source, which emits a sheet of alpha particles on to a small photographic plate. The length of time for which the temperature remains at any particular value is estimated from the intensity of the record on the photographic plate. These recorders were designed for botanical work, such as measuring the duration of the growing season in the distant coniferous forests of northern Norway. They possess the great advantage of being able to be left out for a period of up to 3 months. This made them most convenient for use in bergschrunds. One was placed alongside the thermograph for standardization, and the others were placed in distant bergschrunds, such as that in one of the left-hand tributaries to Heillstuggubreein (pl. 1, *B*). The chief limitation for our purpose is that these recorders do not indicate the number of times the temperature crosses the freezing point but only give the approximate length of time at various temperatures.

The thermograph record was made at the bottom of a 30-foot bergschrund at the head of Tverrabreein, Midt-Jotunheim (6,875 feet.).[2] The bergschrund was narrowly open to the air, the gap widening to 3 feet at the surface by the middle of August. The bulb was fixed within 3 inches of the rock wall,

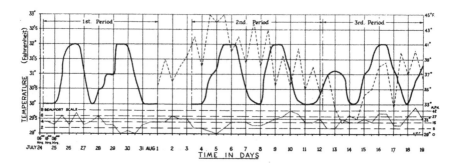

Figure 1 The continuous line with scale on the left-hand side is the record of air
temperatures at the bottom of the 30-foot bergschrund. The broken line
indicates air temperatures at these times at Fannaraaken, the nearest
meteorological station; the scale is on the right-hand side. The wind speeds
at Fannaraaken are shown by the dotted line at the foot of the graph.

which had a slight covering of glazed frost on it and was remarkably dry.
The chart was changed three times, but, owing to other demands on the
personnel for the varied glaciological program, the changing of the chart
was unavoidably delayed the first week and the pen ran dry. With this 2-day
break, the record ran for 26 days. In figure 1 the maximum and minimum
temperatures and wind speeds (Beaufort scale) at the meteorological station
on Fannaraaken, less than 14 miles to the west and only 100 feet lower, are
added. Both temperature scales start at 29° F., but the Fannaraaken record,
on the right-hand side, is reduced four times compared with the bergschrund
temperatures, given on the left of the figure. Comparisons of temperature on
the meteorological station at the Jungfraujoch, Switzerland, which, like the
Fannaraaken station, is situated on a peak, with those taken on the glacier
below by Seligman's party in 1938, were found to be somewhat erratic.
Therefore, too much confidence must not be placed in the Fanaraaken records
as an indication of air temperatures on the surface above the Tverrabreein
bergschrund. Also the standard screen which records temperatures four feet
above the ground is not well suited to measuring the temperature of what
may be no more than a shallow downdraught of air spilling over the rock
wall of the cirque. This cirque culminates in the peak of Bukkehöe, 800 feet
higher than Fannaraaken. However, lacking meteorological records taken
at the bergschrund, we are fortunate in having them in the same mountain
mass and at a similar altitude.

The truncation of the curves at the freezing point is a striking and significant feature of the graph of bergschrund temperatures. This occurred
in spite of the fact that the outside air temperature rose to 45° F. at
Fannaraaken. It seems that the peak temperature in the bergschrund is
determined by that of the surrounding ice, which, by late July, would be
at 32° F. This fact is extremely important, and it is supported by the

preliminary results of the Dahl recorders placed in other bergschrunds and which Dr. Dahl has kindly just forwarded to us. This fact alone must go a long way toward disposing of the oversimple interpretation of the bergschrund hypothesis of Willard Johnson, which requires temperatures at the bottom of much deeper bergschrunds than this one to rise repeatedly above the freezing point.

A second feature, the significance of which is not yet apparent, is the uniform lower limit of 30° F. Close examination of the instrument trace showed that the uniformity of this lower limit was not quite so marked as that of the upper one. There were two falls of temperature to nearly 29.5° F., lasting for an hour or more. Why the indraught of cold air which presumably lowers the bergschrund temperature below that of the enclosing walls of ice did not produce a more variable fall of temperature is puzzling.

Perhaps the most striking feature of all is the complete absence of any diurnal change of temperature in the bergschrund, in spite of the very substantial changes at Fannaraaken of nearly 8° F. This, again, is consistent with the impressions which Battle has gained from all the bergschrunds he has worked in. This lack of direct correlation with air temperatures suggests that other factors may operate, such as calm conditions or falls of pressure. The former may allow a stream of cold air to pour down the back wall into the bergschrund, whereas the latter may encourage this cold air to be sucked in.

The correlation between drops in temperature and the prevalence of relatively still air conditions is at least suggestive. Fannaraaken is an exposed peak. Let us assume, therefore, that a wind of 15 miles per hour at Fannaraaken would correspond with relatively calm conditions in the more sheltered vicinity of the bergschrund. Then it would seem that each fall in temperature occurred during calm spells, with the notable exception of August 10. This still leaves unexplained the undoubted 3^1/$_2$-day cycle, which may be connected with the time it takes for the volume of cold air filling the bergschrund to be warmed to the temperature of the surrounding ice. This cyclical temperature change in the bergschrund seems to be modified by distinct cold spells, such as that from August 11 to August 15. During the whole of this period the temperature in the bergschrund rose barely above 31° F. Even this short record suggests that cold spells have more effect in lowering the bergschrund temperature than warm spells have in raising it. The marked rise in air temperature between August 4 and August 6 produced only a slightly more marked "peak" than the others shown. There is patently need for further data before any firm conclusions can be drawn.

Battle, in his descents, nearly always found the bergschrund walls frozen and dry. A general impression is shown in plate 1, B, where the rock wall may be present on the right beneath a layer of transparent ice a foot or more thick. The steeply sloping névé layers shown in the center of the photograph are also underlain by a foot or more of hard ice actually in contact with the ice covering the rock wall. The thermograph record further confirms this impression of dry, frozen conditions, in which all is inert except for the periodic bombardment of névé and ice blocks falling from the roof. In fact, there is little or no direct evidence of thaw-freeze processes operating. Lewis is, as yet, mildly unrepentant and points out that the summer of 1950 was one of the coldest and snowiest for years and that therefore melting must have been far less than usual at that season and the bergschrunds very much less open. In fact, the meltwater variant of the bergschrund hypothesis requires that temperatures in bergschrunds should be below the freezing point. Battle's experiences in silent bergschrunds, where no dripping or trickling of water could be heard, are contrary to Lewis' experiences in shallower bergschrunds and on the glacier surfaces in névé regions. The whole may well help to build up a truer picture of the variations that can occur from locality to locality and from season to season. It is abundantly clear, however, that no preconceived views must be held overtenaciously in the face of new and important data.

Battle reports that his Easter, 1951, observations in the bergschrunds of the Jungfraufirn, Switzerland, confirm his impression of the low temperatures that generally prevail in Jotunheim bergschrunds. On this occasion the thermograph was placed at the bottom of a 100-foot bergschrund which was almost totally enclosed at the top. In this the temperature wavered between 26° and 28° F., and again all seemed frozen and still. We await with increased interest the complementary observations which he proposes to make at the same localities in the autumn of this year. He has buried several recorders, designed by J.W. Glen, of the Cavendish Laboratory, Cambridge, and himself, which indicate on post-office counters the number of times the temperature crosses the freezing point or any selected point a few degrees lower, according to the freezing mixture used.

A trial test is also being made with a new thermograph designed by J.E. Jackson, a lecturer in the department of geodesy and geophysics, Cambridge, and Lewis. It is based on Dahl's design, in that temperature changes affect a bimetallic spring which carries a needle point. This scratches a thin mark on a circular glass plate covered with lamp black. The plate is turned slightly each time the temperature rises and is held stationary each time it falls. So the trace should be recognizable even if the instrument is left unattended for a year or more. Unfortunately, we shall not know the date of any temperature changes recorded. An improvement which will give a finer record, with no friction on the moving parts, and a rough time scale is being developed with the kind assistance of Dr. D.H. Wilkinson, of the Cavendish Laboratory. This will replace the needle by a fine jet of alpha particles from a radioactive source attached to the free end of the bimetallic spring, and the lampblack by a photographic plate. We hope to arrange for the alpha particles to emerge at an average frequency of one per hour. This will give an approximate time scale, either by counting the number of marks on the photographic plate or, at worst, by estimating the intensity of any blurs produced while the plate was stationary.

These comparatively low temperatures recorded in the bergschrunds form an instructive comparison with those which the Swiss engineers (Haefeli, 1951, and personal communication) experienced at the rock-wall end of the tunnel through the foot of the Mount Collon ice-fall of the Lower Arolla Glacier. There, far below the névé field, the winter temperatures were always near the freezing point, and the workmen were loath to emerge into the far colder air outside the tunnel. In such circumstances it is difficult to imagine how the water, pouring down the rock step beneath the glacier, could ever freeze. Battle's experience in bergschrunds with temperatures slightly below the freezing point also leads him to doubt whether freezing readily occurs. So the best evidence of thaw-freeze is still the shattered nature of rocks newly exposed.

Plate 1, C, shows the destruction caused by thaw-freeze in the strong gabbro in the head wall of Juvashytte cirque, Mid-Jotunheim. Great blocks of rock, many weighing tons, were wedged loose—and pushed slightly away —from the rock wall. The site is normally covered by névé, but the photograph was taken in the summer of 1947, when phenomenal ablation followed a winter in which little snow had fallen. Professor Faegri, of Bergen, reported that the snouts of many Norwegian glaciers receded 30 meters a month during that summer. Thus we were given a glimpse of bedrock which had probably not been exposed for a thousand years or more.

Figure 2 records the mean monthly maximum and minimum temperatures at Fannaraaken (1933–1946) and at Ben Nevis (1884–1903). The minimum at Fannaraaken is below the freezing point except in July, so this is almost always low enough to allow freezing to occur. On the other hand, the maximum is above the freezing point only in June, July, and August and is very

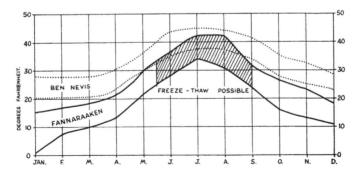

Figure 2 Mean monthly maximum and minimum temperatures at Fannaraaken, 6,763 feet (1933–1946) and Ben Nevis, 4,406 feet (1884–1903), showing approximate period when freeze-thaw could take place.

slightly below in May and September. Does this mean that at Fannaraaken it is possible ordinarily for thaw-freeze to operate only in the three summer months of the year? Priestley (1923, p. 33) records that thawing occurred in the Antarctic when the air temperature was below the freezing point, provided that there was much black surface rock or debris about. This may mean that limited thaw-freeze can occur in May and September and occasionally in colder months in the Jotunheim.

The Ben Nevis results form an interesting comparison. Ahlmann (1948) and Manley (1949, p. 190) have both estimated that this mountain is only a few hundred feet too low to be glaciated today. The Ben Nevis temperatures suggest that thaw-freeze might operate in April and May and again in September and October.

These two examples give no more than a rough indication of when thaw-freeze may operate and when it cannot. In conclusion, it should be stressed that the range of air temperatures, which must influence in a subdued form the changes of temperature in bergschrunds, varies much from place to place today. In summer the diurnal range is small in Greenland, moderate in the Jotunheim, and considerable in the Alps at the respective altitudes of the névé fields.[3] It would thus appear that the opportunity for freeze-thaw is greater in lower latitudes than nearer the poles. Furthermore, it is clearly greater in shallow, open bergschrunds than in deep, confined ones. An increasing number of fluctuations of climate is being revealed as our knowledge of the Pleistocene expands. We must therefore allow for considerable changes occurring over the centuries in the conditions associated with any given bergschrund locality. The absence of frost-shattering today in a particular bergschrund does not necessarily prove that it has not occurred previously. It certainly induces caution in advocating thaw-freeze, but our real lesson should be to continue with our researches.

Acknowledgments

The authors are indebted to the following bodies, which have met the cost of this research work: the Royal Society, the Royal Geographical Society, the University of Leeds, the Worts and Tennant Funds of the University of Cambridge, and the Board of Education (F.E.T.S.).

Notes

1 The Danish State Department Committee on Greenland Names has just authorized the suggested name for this glacier, "Gryphon Gletscher," in the Danish form. "Grif Gletscher."

2 See 1:50,000 map, Midt-Jotunheimen, 1935 Norges Geografiske, Opmåling.

3 The following absolute maximum and minimum temperatures were recorded on the Jungfraufirn in 1938 by Seligman's party (unpublished monograph by E.A. Ferguson):

Month	Max. (° F.)	Min. (° F.)
May	60.8	7.7
June	72.0	14.0
July	66.2	14.0
August	63.5	19.4

References cited

AHLMANN, H.W. (1948) The present climatic fluctuation: Geog. Jour., vol. 62, pp. 165–193.

BATTLE, W.R.B. (1949) Leeds University expedition to East Greenland, 1948: Polar Record, vol. 5, nos. 37, 38, 339 pp.

BOWMAN, I. (1920) The Andes of southern Peru, London, Constable & Co., Ltd.

BROOKS, C.E.P. (1949) Climate through the ages, London, Ernest Benn, Ltd.

DAHL, E. (1949) A new apparatus for recording ecological and climatological factors, especially temperatures over long periods: Physiologia Plantarum, vol. 2, pp. 272–286.

HAEFELI, R. (1951) Some observations on glacier flow: Jour. Glaciology, vol. 1, pp. 496–498.

JOHNSON, W.D. (1904) Maturity in Alpine glacial erosion: Jour. Geology, vol. 12, pp. 571–578.

KLEBELSBERG, R. VON (1920) Glazialgeologische Erfahrungen aus Gletscherstollen: Zeitschr. Gletscherkunde, vol. 11.

—— (1948) Handbuch der Gletscherkunde und Glazialgeologie, vol. 1, Vienna, Springer Verlag.

KOLDEWAY, K. (1873) Die zweite deutsche Nordpolarfahrt in den Jahren 1869 und 1870 unter der Führung des Capt. K. Koldeway, Leipzig, F.A. Brockhaus.

LEWIS, W.V. (1940) The function of meltwater in cirque formation: Geog. Rev., vol. 30, pp. 64–83.

McCABE, L.H. (1939) Nivation and corrie erosion in west Spitsbergen: Geog. Jour., vol. 94, pp. 447–465.

MANLEY, G. (1949) The snowline in Britain: Geog. Annaler, nos. 1–2, pp. 179–193.

PRIESTLEY, R.E. (1923) Physiography (Robertson Bay and Terra Nova Bay Region), London, Harrison and Sons.

4

ON THE MECHANISM BY WHICH STONES IN TILL BECOME ORIENTED

J.W. Glen, J.J. Donner and R.G. West

Source: *American Journal of Science* 255 (1957): 194–265.

Abstract

In most till deposits the long axes of the stones tend to be aligned parallel to the direction in which the ice flowed. Sometimes there is also a tendency for long axes to be oriented transverse to the flow direction. This transverse peak is particularly noticeable in narrow band tills. Laboratory experiments and theoretical studies by various authors have shown that an initially random collection of elongated objects immersed in a flowing liquid will develop a long axis distribution with a peak parallel to the direction of flow in a very short time, but that if flow is continued for a long time a transverse peak will also develop. The observed orientations could thus be entirely the result of flow in moving ice. Collisions between particles and the mode of deposition could change the relative distribution of long axes between the two peaks. The theoretical results are compared with distributions of long axes found in various tills.

Introduction

The purpose of this paper is to discuss the mechanisms by which stones in glacial deposits can acquire a preferred orientation with respect to the direction of flow of the ice which deposited them. In order to refer to the various directions relative to the flow of ice, let us consider a block of ice forming part of the bottom of a glacier flowing over a plane bed as shown in figure 1. The direction parallel to the direction of flow will be referred to as the *parallel* direction, the direction perpendicular to the direction of flow but parallel to the bed will be referred to as the *transverse* direction, and the

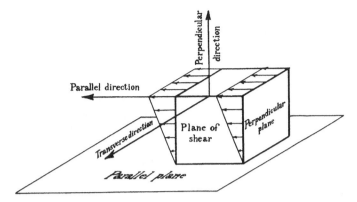

Figure 1 Diagram to show the terms used for the different planes and directions in a material flowing by simple shear.

direction perpendicular to both the direction of flow and the bed will be referred to as the *perpendicular* direction. The plane containing the parallel and perpendicular directions will be termed the *plane of shear*, the plane containing the parallel and transverse direction will be termed the *parallel plane*, the plane perpendicular to the direction of flow will be termed the *perpendicular plane*. All of these terms are shown in figure 1.

It has long been known that the stones or rock fragments in till are not oriented at random, but that the long axes of stones tend to have a preferred orientation in the parallel direction. It is sometimes also found that a subsidiary peak of long axis orientations lies in the transverse direction. Holmes (1941) made a detailed study of the preferred orientations of stones in till in central New York State, determining not only the long axis directions but also the directions of the intermediate axes, the form and the roundness of each of over 1,000 stones. He noted various systematic tendencies between various characteristics of the stones, and also discussed the mode of formation of the peaks in long axis distribution. He attributed the prominent parallel peak to orientation by gliding of stones in contact with the glacier floor (or possibly along well-defined shear planes in the ice), while he attributed the transverse peak to the tendency of stones totally immersed in the ice to assume an orientation in which they rotate about their long axes.

If these were the only mechanisms which could produce preferred orientations of stones, then conclusions about the nature of ice flow could be deduced from observations of the number of stones near the two peaks in the long axis distributions. Thus a till built up beneath active ice would be expected to give a strong parallel peak, while a till formed by deposition from dead ice might be expected to show a more prominent transverse peak, inherited from the orientations assumed in the flowing ice. However there are physical reasons for suggesting that the parallel orientation can also

91

be produced in shearing ice, without the necessity for any dragging or sliding. This being so, it seems opportune to review all the possible methods by which preferred orientations may have developed, to see what deductions, if any, can be drawn from observations of stone orientations. These results can be compared with measurements of stone orientations taken on tills still surrounded by ice in Nordaustlandet (Donner and West, 1956), where large transverse orientations appeared to be formed only in narrow band tills, and with orientation measurements on deposits in eastern England (West and Donner, 1956).

Although the considerations brought forward in this paper are related to the orientation of stones, similar considerations will apply to the orientation of the microscopic particles in till, although the importance of the various factors will be rather different. Seifert (1954) has studied the preferred orientations in this case, and his results also can be considered in the light of the physical arguments here suggested.

Possible causes for preferred orientation of stones in till

The causes which could be responsible for preferred orientations can be divided into three main groups, those stemming from the method of entrainment of the stones in the ice, those inherent in the flow process itself, and those resulting from deposition or subsequent movement of the deposit.

Entrainment of debris in the ice

The debris which forms the stones in till may enter the ice in several ways. It may be plucked from rocks which are being shattered by freezing, it may be washed along a melt water stream, or it may fall onto the surface of the glacier from rock walls. All of these processes will place stones in the ice in initial orientations that are probably not random. For example, stones are unlikely to enter the ice with their long axes vertical. It is difficult, however, to estimate exactly what distribution is to be expected, and probably the best assumption to make in computing the effects produced in the subsequent stages is to assume that the stones do in fact enter the ice quite randomly oriented. This assumption is unlikely to lead to spurious preferred orientations.

Flow in the ice

A stone completely immersed in ice which is flowing by shear will be rotated by the ice as well as being carried forward. The rotation rate depends on the orientation of the stone, and does not remain constant as the stone is moved. The theory of the movement of non-spherical objects suspended in a flowing liquid has been studied for the case of ellipsoids in a viscous liquid by Jeffery

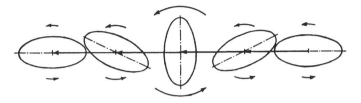

Figure 2 Diagram to demonstrate the movement and rotation of an ellipsoidal object in a shearing material.

(1922). His calculations show that, provided the velocity of flow is small, the ellipsoids, as well as being carried forward, rotate in orbits about their centers, their long axes returning after a time to their original orientation. The rate of rotation varies, being least when the long axis of the stone is most nearly parallel to the direction of flow. For example, for the simple case of a prolate spheroid (i.e. an object shaped like a rugby football) whose long axis is initially in the plane of shear (fig. 2), the long axis rotates so that it remains in this plane, but the angular velocity has a maximum when the long axis is in the perpendicular direction (the position in the center of fig. 2) and a minimum when it is in the parallel direction (at the left and right in fig. 2). The ratio of the maximum to the minimum angular velocity is the square of the axial ratio of the spheroid. The spheroid therefore spends much more time with its long axis near to the parallel direction than it does perpendicular to this. This result is very reasonable, as, when the long axis is perpendicular the top and bottom of the spheroid extend into regions of the fluid flowing relative to the centre of the spheroid at a faster rate than the regions reached when the long axis is parallel in direction.

For a prolate spheroid in a more general orientation the result is similar but more complicated. Jeffery's calculation shows that the angular velocity of the long axis about an axis in the transverse direction still has a minimum when the long axis is in the parallel plane, but there is a further effect, for the angle between the long axis and the plane of shear also varies cyclically and is least when the long axis lies in the parallel plane, and greatest when the axis is in the perpendicular plane. This effect crowds the long axes together towards the parallel direction, and so increases the effect due to the angular velocity variations by which, at any given time, there is a greater chance of finding a given spheroid near to the parallel direction. The preferred orientation resulting from this flow can be calculated from Jeffery's formulae if the initial positions of the spheroids are known. A completely random distribution of spheroids is not stable, i.e. if it existed at one time, then a short time later the flow would have produced a preferred orientation. If it is assumed that the stones are placed in the liquid in random orientations, then the number of stones in any given orbit will be proportional to the angle subtended by that orbit at the centre of the spheroid, i.e.

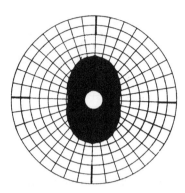

Figure 3 Long-axis rose diagram computed from Jeffrey's theory for prolate spheroids
with an axial ratio of 2:1 in a linear viscous liquid assuming the spheroids initially to have been placed in the liquid at random. The top and bottom of
the diagram represent the orientation parallel to the direction of flow.

to the number of different orientations that lead to that orbit. However
within any one orbit, the chance of finding a spheroid in any given part of
the orbit will be inversely proportional to the angular velocity at that part.
In this way the probability of any given long axis orientation can be
deduced, and from this a rose diagram can be constructed to show the
distribution of the projection of long axes on the parallel plane. This
computation has been carried out graphically for the case of an axial ratio
of 2:1, and the resulting rose diagram is shown in figure 3. This diagram
has been drawn so as to be directly comparable with the rose diagrams
constructed from measurements on tills, the radii vectores are measured
from the circumference of the small white center circle, and are proportional
to the number of spheroids which would have their long axes in the plane
perpendicular to the parallel plane and containing the direction of the radius
vector. The parallel direction is at the top and bottom of the diagram, the
transverse direction at the sides. It will be seen that for this axial ratio, there
are just over twice as many stones lying near the parallel direction as near
the transverse direction.

In the solution of the viscous problem given by Jeffery there is no
tendency of the spheroids to move from orbit to orbit. Thus after all the
spheroids have rotated once or twice, the preferred orientation pattern should
not change with further flow. In the transverse direction there is a minimum
in the rose diagram, and there is no obvious reason why this situation
should change. However Jeffery himself pointed out that the energy dissipated by the spheroid is a minimum if the long axis is in the transverse
direction. A similar result is true for a more general ellipsoid, while for
an oblate spheroid the minimum energy is dissipated if it rotates about a
diameter (in this case the discus-shaped object has no definite longest axis,

and a diameter is just one of the infinite number of longest axes). Jeffery suggested that higher order terms in the equation, which he had neglected, might be expected to favor the orientation which dissipates minimum energy, though it is to be expected that such a reorientation might take a very long time, and as the viscosity enters into the terms neglected, the higher the viscosity, the longer one might expect this time to be.

Various experiments have been performed to check the results of Jeffery's calculation and his surmise about the transverse orientation. Taylor (1923) immersed both prolate and oblate spheroids in water-glass and studied their behavior during shear between concentric cylinders. He found that the ellipsoids did rotate in orbits of the kind predicted by Jeffery, and he also found that after a long time the particles tended to assume the orientation predicted by the minimum energy hypothesis. Thus the prolate spheroids slowly shifted until their long axes were transverse. However the time taken to establish this reorientation was quite large, the spheroids having rotated about 370 times before their long axes were transverse. Taylor remarks how difficult it is to understand the mechanism by which this reorientation takes place, and Saffman (private communication) has suggested that the cause may not be the terms neglected by Jeffery in his calculation of the solution for the perfect viscous liquid, but may be due to the existence in the liquid of thixotropy, in particular of cross viscosity (i.e. the expansion of the liquid in the transverse direction when sheared). Saffman has found that reorientation occurred quite rapidly in a thixotropic liquid, whereas in a Newtonian viscous liquid no reorientation occurred. The experiments of Taylor and of Saffman were made with spheroids of axial ratio about 2:1, and these results are probably most comparable to the behavior of stones in ice. Most other experiments have been made on very long thin specimens such as hair, paper fibres or fine glass rods (Eirich, Margaretha and Bunzl, 1936; Binder, 1939; Trevelyan and Mason, 1951). All of these workers find that Jeffery's equations correctly predict the behavior of the particles for a short time, but with a long time reorientation occurs. For particles with a small axial ratio (less than 15:1) Binder found that the reorientation was towards the transverse direction, while for greater axial ratios he found that, despite Jeffery's hypothesis, the parallel direction was that to which all particles tended. Eirich, Margaretha and Bunzl also observed that the particles with smaller axial ratios were more inclined to take up the transverse orientation.

To sum up this discussion, objects placed at random in a flowing liquid soon develop a parallel orientation of their long axes, and then, as time proceeds, they slowly develop a transverse orientation if their axial ratios are less than 15:1. Ice is not a linear viscous liquid, though the way in which it differs is not necessarily that assumed by Saffman. For ice the relation between shear stress and shear strain rate is not linear (Glen, 1955), and it is not yet known whether ice possesses cross-viscosity to any marked extent.

Jeffery's theory will not therefore apply quantitatively to ice, but it is to be expected that the qualitative deductions will still be quite accurate, as they are based on rather general properties of a flowing material.

Collisions between stones

Interactions between stones may influence the distribution of long axes. It is not simple to calculate what effect such interactions will have, as the effect of two stones interacting with each other will depend rather critically on the shape of the stones and the exact nature of the collision. Collisions may aid the stones to move from orbit to orbit towards the orbit of minimum energy, and in this way could favor the transverse orientation. In support of this suggestion is the observation of Manley, Arlov and Mason (1955) that the orbit of a single rod remained constant for a long time, whereas when the density of particles was increased the transverse orientation developed. This has led Manley, Arlov and Mason (1955, see also Manley and Mason, paper to be published) to suggest that the reorientation is primarily a collision effect.

A reorientation due to collision is to be expected on grounds other than those connected with Jeffery's minimum energy hypothesis. A stone which is rotating in a given orbit has a chance of colliding with another stone which is larger the larger the volume swept out by the stone in rotating in the orbit. Stones have, therefore, less chance of colliding, and so stay longer in an orbit, if the volume swept out in that orbit is small. Thus, if the main effect of collisions is to move stones from orbit to orbit at random, the effect of collisions will be to favor those orbits which sweep out the least volume. This argument is rather similar to that employed to demonstrate the development of the parallel peak during free shear; the stone spends the longest time in the orbit which sweeps out the least volume. For a prolate stone the orbit which sweeps out the least volume is that in which the long axis is transverse, so that the stone rotates about it. Thus for prolate stones collisions favor the transverse orientation for the long axis, and this happens to be the orientation favored also by the minimum energy hypothesis. For oblate stones, however, the orbit sweeping out least volume is that in which the *smallest* axis is transverse so that the stone rotates about that. This orbit is that of maximum energy dissipation for oblate spheroids, so that for oblate stones the minimum energy hypothesis still favors the transverse peak, while the collision probability favors the parallel peak.

For a stone which is neither prolate nor oblate (i.e. whose intermediate axis is not equal either to the shortest or the longest axis) the orientation favored by collisions will be transverse if the intermediate axis is more nearly equal to the shorter axis (prolate type stone) and will be parallel if the intermediate axis is more nearly equal to the longest axis (oblate type stone). The former case follows directly from the argument above, while for the latter case, the orbit sweeping minimum volume will be that in which

96

the shortest axis is transverse, i.e. in which both longest and intermediate axes are in the plane of shear, and for this orbit the rate of rotation will vary, being least when the long axis is in the parallel direction, which is thus the most probable orientation.

The argument we have used here suggests that due to collisions the stones showing the transverse peak should statistically be more prolate in form than those in the parallel peak. It is perhaps significant that Holmes (1941) found "a slight, though gradual, increase in preference for parallel orientation with decreasing difference in axial lengths". Holmes is here referring to the difference in lengths between the longest and the intermediate axes so that a small difference in length corresponds to an oblate type stone.

Interaction with the bed or across a shear plane

The next type of interaction that we must consider is that between particles embedded in the moving ice and the stationary layer beneath. This layer may be bed rock, but it is more likely to be deposited till or perhaps even another ice layer, if the moving ice is overriding dead or slower moving ice along a shear plane. As Holmes has pointed out, a stone which projects to the bottom of the moving layer will be dragged along the layer underneath. This will tend to orient the stones in the parallel direction and may also reorient in this direction stones in the underlying layer. Holmes discussed this process in some detail and considered it the primary mechanism by which the parallel orientation is produced; for that reason we will not enter into great detail here. It is obviously a powerful method of orientation, and will, as Holmes has pointed out, tend to orient the long edges of stones rather than their long axes; thus for wedge-shaped stones one of the two sides of the wedge will tend to be parallel to the flow rather than the bisector of the angle. This is confirmed by the observation that wedgeshaped stones tend to have striae parallel to the sides of the wedge.

In confirmation of this type of orientation we can note the observation of Eirich, Margaretha and Bunzl that when their suspension of cylinders in a liquid was forced down a tube, the cylinders near the wall soon assumed a parallel orientation, whereas near the axis of the tube (where there is no strong shearing) the cylinders remained in apparently random orientations.

Deposition of till from the ice

The process of deposition of the till may well affect the preferred orientation developed during flow. If till is deposited from active ice, so that material is added onto the top of a growing layer, the principal effects will be those due to interaction between the moving ice and the stationary till layer. As we have just seen, this tends to emphasize the parallel orientation. Holmes has pointed out another effect that may enter here. If a stone is rotating about an axis which is not one of circular symmetry, then the theory of Jeffery shows that the longest direction in the plane of shear spends comparatively little time near the perpendicular direction, and so has a preferred orientation which is horizontal. However, the chance of a collision with a stone above is highest when the axis is near to perpendicular, for it is then projecting into a region of much faster moving ice. If while it is rising into the perpendicular direction, a stone above, and thus moving faster than our stone, interacts with it from above, it will tend to force our stone downwards, and thus may push it into the till layer and bring it to rest. Thus in active ice the perpendicular orientation is more favored than would be predicted from the flow orbits alone. When the long axis is transverse, the same argument can be used to predict a greater number of perpendicular intermediate axes than are expected during flow, and Holmes explained in this way his observation that the class of stones with steeply inclined intermediate axes had a larger percentage of long axes in the transverse direction than had the class with low intermediate axis dip.

When till is formed from dead ice the effects will be rather different. Apart from bodily movements of the dead ice mass, the principal effect will be that, as the ice melts, the stones will fall down onto the material below. This will tend to drop the stones down so that their long axes are parallel to the bed (usually approximately horizontal), but will not affect the orientation distribution of long axes projected onto the parallel plane. It is for this reason that the rose diagram is probably a more satisfactory indicator of processes connected with ice flow than the contoured equal area diagram. In the rose diagram the number of stones whose long axes project onto the horizontal within a given angular range is plotted on polar co-ordinates against the angle, and this will be affected by deposition only to the extent that stones twist about a vertical axis as they are released from the melting ice. If the dip of the till stratum is large, then it is preferable to plot rose diagrams for the plane of the stratum rather than for the horizontal, but unless the angle is greater than about 30° no great errors will enter if this refinement is ignored. Contour diagrams, in which the dips as well as the horizontal angles of the long axes are recorded, will be much influenced by the way in which a stone drops onto the material beneath it.

Disturbance of till after deposition

After till has been deposited from the ice further influences may upset its preferred orientations. Slumping of the whole deposit due to the melting of dead ice beneath may cause complications in the pattern, especially if the deposit itself flows during such slumping. For example a deposit on the edge of a kettle hole may flow into the hole and so obtain a preferred orientation in the direction of flow into the hole. Another process of this sort has been suggested by Hoppe (1952) to account for the observation that in hummocky moraine regions the preferred orientation of the stones is frequently perpendicular to the direction in which the moraine ridges run, even when these form curved loops. He suggested that till is pushed by the weight of overlying ice up into subglacial crevasses and in this way acquires its preferred orientation.

Solifluxion after the retreat of all the ice from the neighbourhood of the till may also be expected to change the preferred orientation, particularly if it is continued for any time, and if a mass movement of the deposit results. None of these effects will in general be related to the old direction of ice flow, and in any case in which stone orientation measurements are used to determine directions of ice flow, care must be taken to select deposits that do not appear to have been seriously disturbed since deposition.

Summary of effects producing preferred orientations

We can now summarize the processes which have been discussed in the last section.

Preferred orientation parallel to the direction of flow: is produced by
free flow in the ice,
stone collisions (oblate stones only),
dragging on a stationary layer or over a plane of discontinuous shear (particularly during deposition from active ice).

Preferred orientation in the transverse direction: is produced by
protracted free flow in the ice,
stone collisions (prolate stones only),

Preferred orientation in the perpendicular direction: is produced by
stones being forced into a stationary layer by faster moving material above.

Preferred orientations unrelated to the direction of ice flow: is produced by
slumping of stones during deposition from dead ice (tending to eliminate large angles of dip),
subsequent flow of the till due to the melting of dead ice or extrusion from under masses of dead ice,
solifluxion.

Comparison of predictions with observed stone orientations and deductions which can be drawn from orientation measurements

We have already mentioned several ways in which the theory of stone orientations in till deposits is confirmed in practice. The majority of undisturbed deposits show either one or two directions in which the orientations have a maximum on a rose diagram. In most the largest peak is parallel to the direction of ice flow. Where there is a second peak it is usually at right angles to the first, i.e. transverse to the direction of ice flow. The second peak may exceed the first in magnitude. In no case does the transverse peak exist by itself. These observations alone show how well the stone orientations agree with predictions. There are probably significant correlations between prolateness and the tendency to have a transverse peak, which we would account for by the effect of collisions between stones. What remains to be explained, however, is why some rose diagrams lack the transverse peak. Two possibilities can be considered. Either the stones have not been in the ice long enough to develop the transverse peak, or they may have had it and lost it during the deposition process. In the former case the stones must have travelled only a short distance actually entrained in the ice, though they may have travelled further while being dragged beneath the glacier proper, while in the second case the deposition process must have involved a protracted period of dragging of the material.

Some confirmation for this approach is obtained from observations on tills in process of formation at Brageneset, Nordaustlandet (Donner and West, 1956). Here it was possible to measure stone orientations in tills which, although free of ice themselves, had ice both above and below them. Large transverse peaks were found only in narrow band tills (cf. diagram 3 in fig. 4), whereas in the thicker deposits only a single major peak was observed (diagrams 1 and 2 in fig. 4 show the extreme cases obtained). If the explanations suggested above are accepted, this would imply that the thick band tills were formed by a plastering-on process of lodgement which effectively eliminated the transverse peak, or that the stones in this till had never been completely immersed in flowing ice for a long time, while the narrow band tills were probably freer from the sliding process in formation, and, being relatively freer to flow with the ice, developed the transverse peak by free flow or as a result of collisions during free flow. It would be dangerous to generalize from this one place and to state that transverse peaks are always associated with narrow band tills, but this experience does show that in the case of the Brageneset deposits there was a qualitative difference in morphology between the deposits with large transverse peaks and those without. The principle may have validity in other locations, however, as observations of stone orientations from a narrow band till in the Cromer deposits (about 20 cm wide) also showed two peaks at right angles of similar magnitude to each other (diagram 4 of fig. 4), a result which was not obtained from any of a large number of thicker lodgement tills in East Anglia. In the thicker tills the predominance of the parallel peak in the preferred orientation rose diagram is usually quite marked, and, provided care is taken to select only sites in which no subsequent movement or disturbance of the till has occurred, this method can be most useful in determining the direction of ice flow. In eastern England it has proved possible to differentiate between different ice advances using this method, as the direction of ice flow differed (West and Donner, 1956). Diagrams 5 to 8 of figure 4 are from these till deposits, and show the kind of preferred orientation found in practice.

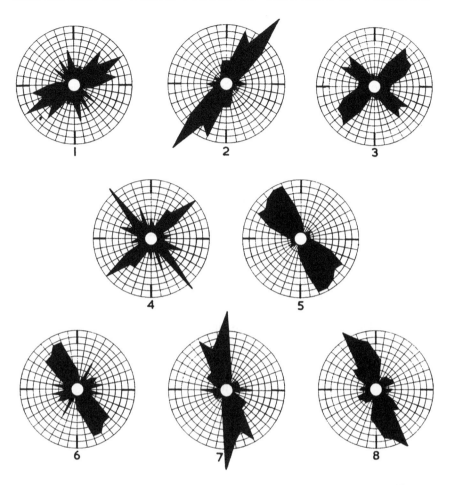

Figure 4 Rose diagrams showing long-axis orientation of stones in various tills. The four cardinal points of the compass are shown on the diagrams, magnetic north at the top. The diagrams are from the following tills:
1. and 2. thick lodgement tills, Brageneset, Nordaustlandet, Spitsbergen.
3. narrow band till in ice, Brageneset, Nordaustlandet, Spitsbergen.
4. narrow band till from the Contorted Drift on the Cromer coast near Sheringham, Norfolk, England.
5. to 8. various lodgement tills in eastern England.

The predominance of the parallel orientation has also been confirmed by the work of Holmes (1941) and Lundqvist (1948) and others. Seifert (1954) has used a similar method to determine the direction of ice flow, and hence to deduce the glaciation sequence on the north German coast. He studied the orientation of the microscopical particles in the till. He found that in a thick till layer the parallel orientation was favored near the edge, while the transverse orientation predominated near the centre. This result is different from that found for the larger particles, but this is not very surprising as the relative importance of the various processes influencing orientation must be very different in the two cases. For example, collisions are probably much more important for the small clay particles than for the larger stones. Unfortunately Seifert's microscopical technique did not allow him to distinguish between oblate and prolate particles, but if the particles are mostly prolate, it might be expected that, near the center of a band, the collisions would assist the production of the transverse peak, while near the edge, where the density of particles is less, the parallel peak would predominate. It is difficult to imagine that the interaction over a shear plane is of great importance for these very small particles.

The interpreation of rose diagrams showing the preferred orientations of long axes in terms of the direction of flow of the ice has been shown to be reasonable on physical as well as experimental grounds. Any further deductions made from the nature of the diagrams such as the relative magnitude of the transverse peak, are somewhat more dubious, owing to the large number of processes that could have acted to give the peaks their relative magnitudes. It would seem, therefore, that stone orientation measurements are most valuable for determining the direction of ice flow, but less significant for other purposes.

Acknowledgment

We would like to thank Mr. P.G. Saffman for permission to quote from his unpublished work, and for his help in finding the literature references on the behavior of suspensions of objects in liquids

References

Binder, R.C., 1939, The motion of cylindrical particles in viscous flow: Jour. Applied Physics, v. 10, no. 10, p. 711–713.

Donner, J.J., and West, R.G., 1956, The Quaternary geology of Brageneset, Nordaustlandet, Spitsbergen: Norsk Polarinstitutt Skrifter, in press.

Eirich, F., Margaretha, H., and Bunzl, M., 1936, Untersuchungen über die Viskosität von Suspensionen und Lösungen. 6. Über die Viskosität von Stäbchensuspensionen: Kolloid Zeitschrift, Band 75, Heft 1, p. 20–37.

Glen, J.W., 1955, The creep of polycrystalline ice: Royal Soc. London Proc., ser. A, v. 228, no. 1175, p. 519–538.

Holmes, C.D., 1941, Till fabric: Geol. Soc. America Bull., v. 52, no. 9, p. 1299–1354.

Hoppe, G., 1952, Hummocky moraine regions with special reference to the interior of Norrbotten: Geografiska Annaler, Årg. 34, Heft. 1–2, p. 1–72.

Jeffery, G.B., 1922, The motion of ellipsoidal particles immersed in a viscous fluid: Royal Soc. London Proc., ser. A, v. 102, no. 715, p. 161–179.

Lundqvist, G., 1948, Blockens orientering i olika jordarter: Sveriges Geologiska Undersökning, no. 492, (Årsbok 42, no. 6,) 29, p.

Manley, R. St. J., Arlov, A.P., and Mason, S.G., 1955, Rotations, orientations and collisions of suspended particles in velocity gradients: Nature, v. 175, no. 4459, p. 682–683.

Seifert, G., 1954, Das mikroskopische Korngefüge des Geschiebemergels als Abbild der Eisbewegung, zugleich Geschichte des Eisabbaues in Fehmarn, Ost-Wagrien und dem Dänischen Wohld: Meyiniana, Band 2, p. 124–184.

Taylor, G.I., 1923, The motion of ellipsoidal particles in a viscous fluid: Royal Soc. London Proc., ser. A, v. 103, no. 720, p. 58–61.

Trevelyan, B.J., and Mason, S.G., 1951, Particle motions in sheared suspensions. I. Rotations: Jour. Colloid Science, v. 6, no. 4, p. 354–367.

West, R.G., and Donner, J.J., 1956, The glaciations of East Anglia and the East Midlands. A differentiation based on stone orientation measurements of the tills: Quart. Jour. Geol. Soc. London, v. 112, p. 69–91.

5

ICE-DISINTEGRATION FEATURES IN WESTERN CANADA

C.P. Gravenor and W.O. Kupsch

Source: *Journal of Geology* 67 (1959): 48–64.

Abstract

The wasting Wisconsin glacier left predominantly till deposits in western Canada and only subordinate amounts of stratified drift. In the final phases of wasting, the ice separated in places into a large number of small, dead, ice blocks: it disintegrated. This disintegration caused the preservation of many different land forms, some of which were initiated during the time of ice flow, others originated after flow ceased. Those features that show the influence of the previous live ice are said to be "controlled." Such control may be exerted by crevasses and thrust planes which are the response to stresses operative in a living glacier. Uncontrolled deposits do not reveal the influence of former flow. All gradations between controlled and uncontrolled disintegation can be observed.

The depositional disintegration features include hummocks, moraine plateaus, round and irregularly shaped closed ridges, linear ridges, and washboard moraines. Ice-walled channels are an erosional form. The hummocky terrain and the closed ridges are regarded as the dominant product of uncontrolled disintegration. The linear and washboard ridges developed along inherited lines of weakness in the disintegrating ice and are regarded as controlled disintegration features.

Both uncontrolled and controlled deposits resulted from the sloughing of ablation material into cracks and cavities in the ice and from the squeezing of till upward into openings at the base of the ice.

Introduction

Although glacial features associated with stagnant ice of late Wisconsin age have been described from other regions, the writers feel justified in calling attention to those of western Canada for the following reasons:

1. Gradations have been observed between several glacial features generally believed to be of different origin but which are all preserved because of final stagnation, ice wastage, and consequent disintegration.
2. Unlike eastern North America, where stagnant ice features are largely glaciofluvial in origin, those in western Canada are composed mainly of till.
3. They are well preserved as a result of the dry climate and consequent scant postglacial erosion. They are also well displayed both on the ground and on air photos because of the lack of trees and, furthermore, have been little altered by human activities.

In view of the fact that the terminology of till features associated with stagnant ice is at present not clearly defined or universally accepted, the writers take this opportunity to propose a new descriptive classification of these features, supported by a discussion of their characteristics and possible origin.

Definition of ice disintegration

The term "wastage" applied to a glacier is usually considered as including melting, wind erosion, evaporation, and calving (Howell, 1957, p. 318). A wasting glacier may maintain a distinct receding terminus while flow continues actively, or it may, because of widespread thinning, stagnate and separate (Flint, 1957, p. 33). Along some parts of an ice margin, stagnation extended for many miles back from the front, whereas elsewhere along the same margin the ice was actively flowing and deposited or pushed up end moraines. In still other places the ice was receding at a high rate, thus leaving no separate blocks of ice. As a result, only a sheet of ground moraine without any patterned relief remained.

Although the writers use the term "separation," they prefer *ice disintegration* to describe the process of breaking up into numerous small blocks, which finally comes about in a stagnant and thus wasting glacier. It best describes the falling-to-pieces of the ice as a result of physical and chemical action and thus conforms to the definition of the geological meaning of "disintegration" as given in Webster's *Dictionary* (Neilson, 1956): "the falling to pieces of rocks as a result of chemical or physical action, weathering, frost, etc."

Disintegration may give rise to a variety of land forms, depending upon a great number of factors, such as the amount of debris carried by the ice, the position of the debris on, in, or under the ice, the amount of meltwater, and the resultant erosion and deposition. They all have in common, however, that they are preserved because they represent the last phase of glacial deposition and were not destroyed by later glacier advances.

When the forces that operate to break up an ice sheet are equal in all directions, the disintegration may be said to be *uncontrolled*, and the result is a field of round, oval, rudely hexagonal or polygonal features, and a general lack of dominant linear elements. Where the ice separated along fractures or other lines of weakness, the disintegration may be said to be *controlled*, and the result is a field of linear or lobate land forms. In places the ice broke along open crevasses or along thrust planes, both of which formed when the ice was still flowing, and the disintegration thus shows *inherited flow control* (pl. 3). The linear elements then bear a direct and understandable relationship to the preceding flow directions and are usually parallel, perpendicular, or at 45° to the direction of flow. In some areas, however, the fractures may have resulted from causes other than earlier ice flow, such as the settling of ice into bedrock joints. Gradations between controlled and uncontrolled disintegration features are to be expected and can be found in many places.

In places disintegration features are *superimposed* on previously formed live-ice features, such as drumlins and flutings parallel to the direction of ice movement or end moraines perpendicular to it.

Relief of ice-disintegration deposits may vary from pronounced to hardly noticeable in the field. In the latter case it may be evident only from the study of air photographs. *High-relief deposits* have local relief greater than 25 feet; *intermediate-relief deposits* from 10 to 25 feet; *low-relief deposits* less than 10 feet.

Ice-contact stratified drift deposits are generally believed to be associated with the thinning of an ice sheet. Deposits of this type in western Canada are subordinate to till features that resulted from stagnation. Because they have been described as stagnation features from elsewhere and because they can be independent of disintegration, they will not be considered here.

Recognition of ice-disintegration features

The distinction between end moraines and ground moraines on some of the older glacial maps of western Canada (Johnston *et al.*, 1948) is based largely on relief. Areas of drift having a local relief of more than 15 feet and consisting predominantly of till were regarded as end moraines. Areas of low relief were mapped as ground moraine. The following characteristics have now been found:

1. Some uplands mapped as end moraines have only a very thin till cover, and both their relief and their trend are due primarily to topographically high underlying bedrock.

2. Till ridges representing true end moraines cut across the trend of the high-relief uplands formerly mapped as end moraines with a different trend.
3. End moraines of low relief occur in areas formerly mapped as ground moraine.
4. Some areas mapped as end moraine appear as "blobs" on maps and show no trend, either as a whole or in detail.

These findings necessitate criteria for the recognition of areas of predominantly stagnant-ice deposition. The distinction between end moraine and ground moraine should be based on observable *trends*, in addition to *relief*. The writers regard a true end moraine as a ridge or narrow elongated area consisting of glacial material produced at the front of actively flowing ice or as a ridge of pre-existing material structurally disturbed by ice-push, with or without a covering of glacial material. This definition attempts to embody the concepts generally accepted by European geologists (Woldstedt, 1954, p. 101–110).

Some of the high-relief areas formerly regarded as end moraines and believed to have been deposited by live ice are interpreted by the writers as the result of ice disintegration; others appear to be complexes of end-moraine ridges. In some places low-relief ground moraine may show the effects of wasting of unoriented dead-ice blocks. In others the effects of the living ice are still visible, and the position of the former ice front can be reconstructed from the controlled disintegration pattern.

Uncontrolled disintegration features

Hummocky disintegration moraine

Description

Broad tracts of rough morainal topography are common features over much of the western plains. Many of these areas are irregular in outline and show no pronounced elongation. They consist of a nondescript jumble of knolls and mounds of glacial debris separated by irregular depressions. The knolls do not align into ridges, and no dominant trends are discernible. These areas have the characteristic "knob-and-kettle" topography. They are tracts of high local relief and were mapped as end moraines by early workers, even though clearly defined ridges may be absent and the area itself lacks any definite elongation. Thus the Moose Mountain upland in southeastern Saskatchewan was regarded as a northwest-trending end moraine (Johnston and Wickenden, 1931, p. 40). Recently, however, Christiansen (1956, p. 31) found that this area is devoid of any trends and shows no visible traces of live-ice deposition. It is therefore considered to be deposited by stagnant ice.

Washboard moraines, indicating the former position of the ice front, are found encircling the Moose Mountain upland but not on it.

The high-relief disintegration moraine may cover a preglacial upland, a condition which appears to hold true for the Max Moraine of the Missouri Coteau (Townsend and Jenke, 1951, p. 842), for the Turtle Mountain area in Manitoba, and for the Moose Mountain area in Saskatchewan. Dead-ice moraine of pronounced relief may also occur in areas that were topographically low in preglacial times. Limited drilling carried out on the Viking moraine (which the writers believe is a disintegration moraine and not a true end moraine) of east-central Alberta suggests that the depth to bedrock is greater there than in the surrounding ground moraine.

A careful analysis of a hummocky moraine generally reveals the presence of several distinct elements besides the dominant knobs and kettles. Among these are (1) moraine plateaus, (2) closed and linear ridges, (3) ice-walled channels (pls. 1 and 6). The last two features also appear outside the high-relief areas and will therefore be treated separately.

Moraine plateaus are relatively flat areas in the hummocky moraine (pl. 1, *A*). Their level surface is generally at the same elevation or slightly higher than the summits of the surrounding knobs. In east-central Alberta drilling operations show that the moraine plateaus are composed of clayey till. In some places there is a thin cover—2–10 feet thick—of lacustrine silts and clays at the surface overlying the till. The moraine plateau are roughly circular, elongate, or irregular in plan. They may be pitted with kettles, which in rare instances have poorly developed rims. The edges of the moraine plateaus may be defined by minor till ridges, referred to as "rim ridges," but in the plateaus of east-central Alberta such rim ridges are the exception rather than the rule. In some parts of the Missouri Coteau in Saskatchewan, on the other hand, where the plateaus are distinctly "pock marked" with kettles, rim ridges are common. In places meltwater channels may lead away from the plateaus (pl. 1, *E*). Similar situations are also recorded from till moraine plateaus in Sweden (Hoppe, 1952, p. 9).

Moraine plateaus composed chiefly of stratified silt and clay, of probable lacustrine origin, have been noted on Moose Mountain, Saskatchewan (Christiansen, 1956, p. 11). They appear to be similar to the plateau hills described from Denmark by Schou (1949, p. 10).

The knobs of hummocky dead-ice moraine in western Canada consist mainly of till, which in places is quite compact. Sand and gravel knobs, however, are common and generally show the characteristic "collapse" structures of ice-contact deposits. Some excavations revealed knobs composed of crudely stratified drift overlain by a layer of loose "washed" drift. Some of the knobs formerly believed to be composed of till may therefore have sand and gravel in their cores. In others, layers of sorted sand and gravel cover till cores. The knobs may occur as isolated mounds, in clusters, or connected by ridges.

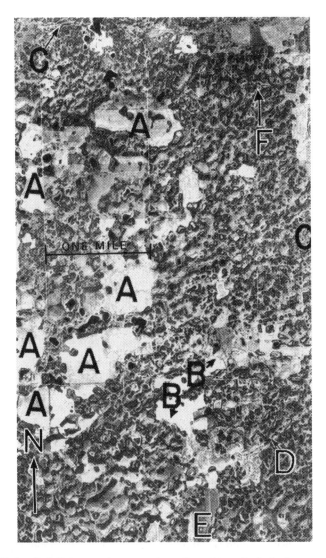

Plate 1 High-relief disintegration moraine. (Location of pl. 1 shown on fig. 4.)
    Location: T. 46, R. 12, W. 4th mer., Alberta. (Air photograph reproduced
    by permission of government of Alberta, Department of Lands and Forests.)
A, Moraine plateau.
B, Closed till disintegration ridge on moraine plateau (rim ridge).
C, Linear till disintegration ridge, sinuous.
D, Linear ridge, beaded.
E, Moraine plateau which forms the headwater region of an ice-walled channel
leading off to the south.
F, Closed till disintegration ridges, surrounding depressions (rimmed kettles).

109

Plate 2 Closed disintegration ridges in low-relief till area. (Location of pl. 2 shown on fig. 4.) Location: Sec. 33, T. 3, R. 3, W. 2d mer., Saskatchewan. (Air photo reproduced by permission of Royal Canadian Air Force.)

Plate 3 Linear disintegration ridges. (Location of pl. 3 shown on fig. 4.) Location: T. 48, R. 1, W. 4th mer., Alberta. (Air photograph reproduced by permission of government of Alberta, Department of Lands and Forests.) The ice movement in this area was slightly east of south, which is at right angles to the curvilinear or lobate ridges. On the eastern side of the photograph the intersecting ridges form a box pattern.

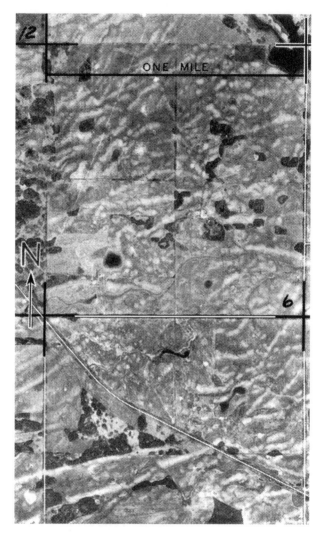

Plate 4 Linear disintegration ridges. (Location of pl. 4 shown on fig. 4.) Location: T. 50, R. 27, W. 3d mer., Saskatchewan. (Air photograph reproduced by permission of Royal Canadian Air Force.)

Circular undrained depressions are present at the tops of many mounds. They also occur between the knobs, but irregular hollows are more common in that position.

Distinctive features

It is likely that the hummocky, knob-and-kettle topography has various origins (Woldstedt, 1954, p. 96). An end-moraine complex of hummocky topography deposited by the live-ice front may appear very similar to a hummocky moraine deposited by dead ice. Inasmuch as blocks of ice may be left in the end moraine on retreat, some dead-ice forms can occur in the end moraine. Therefore, it becomes necessary to distinguish between a hummocky moraine which still shows the former position of the ice front from a hummocky moraine which does not. Detailed field work supported by study of air photographs and large-scale topographic maps generally provides the answer. If no trends, such as aligned knobs or hollows, are apparent from the air photographs and if the moraine itself is not elongated and narrow but round and broad, the moraine is regarded as deposited by stagnant ice. The hummocky disintegration moraine generally has a rounded outline, its borders are indistinct, and they may grade almost imperceptibly into the surrounding low-relief ground moraine (pl. 5).

In hummocky moraines deposited as true end moraines, the knobs are aligned as more or less clearly defined ridges, which are emphasized by intervening aligned depressions. These trends define successive positions of the ice front. Generally they are gently curved in outline and are regularly spaced.

Some hummocky moraines show alignment of knobs or ridges in two directions, which is indicative of ice-fracture control. These fractures were developed at the time when the ice was still active, and their trends do not necessarily bear a simple relationship to the trend of the ice front.

Terminology

The purely descriptive term *hummocky moraine* is recommended to designate areas of knob-and-kettle topography. Where an origin along a live-ice front is indicated, the term *hummocky end moraine* is recommended. Where stagnant ice was responsible for deposition, the deposit may be referred to as *hummocky disintegration moraine* or *hummocky dead-ice moraine.* German authors often use *kuppige Grundmoräne* ("hummocky ground moraine") to designate such areas (Woldstedt, 1954, p. 95–96). Hoppe (1952) extended the term *hummocky moraine* to low-relief features, including some long ridges that resulted from dead-ice deposition. Such an extension does not appear desirable because the term *hummock* to most geologists means a hillock of more or less equidimensional shape and not a ridge.

Closed disintegration ridges

Description

At many localities on the Western Plains, circular, oval, or irregular closed ridges of glacial material are present. They are most noticeable on air photographs, but some larger, more regularly shaped ones are also obvious in the field. The closed ridges may be one of the elements of a hummocky moraine (pl. 1, *F*), or they may occur in low-relief areas where they occur in large numbers alone or together with linear ridges. Gradations between closed and linear ridges are not uncommon.

The most symmetrically shaped type of closed ridge resembles a giant doughnut from the air. Perfectly circular ridges range in height from a few to 20 feet and in diameter from 20 to 1,000 feet. The ridge, which surrounds a central depression, is commonly referred to as a "rim." It is generally unbroken and of the same height around the depression. Concentric ridges have been noted, and in places a low circular knoll takes the place of the more common central depression. The circular ridges are similar to one type of "prairie mounds" described by Gravenor (1955), but in "prairie mounds" the base of the central depression lies well above the general ground level, whereas the base of the central depression in the circular ridges lies at or below that level. Intermediate forms, however, exist. In general, the low circular ridges surround a depression that is shallow, but the floor is somewhat below the general level of the till plain (pl. 2).

Locally the circular ridges occur with ridges that are oval or irregular in plan. Such forms are also described by Deane (1950, p. 14). They surround depressions which may vary in shape from circular to irregular. All such ridges are referred to as *closed ridges* by the authors, even though they may show some minor breaks resulting from irregular deposition or later erosion.

Some closed ridges do not surround depressions but mounds of glacial material or moraine plateaus in hummocky moraine (pl. 1, *B*). They occur on the outside edge or within the confines of the plateau. Hoppe (1952, p. 5) has described similar ridges from moraine plateaus in Sweden. In western Canada they are not common, and many moraine plateaus are devoid of ridges of any type.

Although most closed ridges in the areas studied are composed of till, some are of stratified materials and are usually associated with eskers, kames, and crevasse fillings. The sand and gravel ridges vary in diameter from 50 to 800 feet and from 5 to 30 feet in height. In general, these ridges are higher and have steeper sides than the till ridges, and the base of the central depression lies at or above the level of the surrounding stratified drift.

Terminology

Well-developed circular ridges of glacial material are colloquially referred to as "doughnuts"; mounds with round depressions are often referred to as "humpies." The depressions and the surrounding rim have been more formally designated as "rimmed kettles" (Christiansen, 1956, p. 11), the ridges alone as "ice-block ridges" (Deane, 1950, p. 14). More irregular ridges may produce a "brainlike" or "vermicelli" pattern on air photographs. Ridges on the edges of moraine plateaus are called "rim ridges" by Hoppe (1952, p. 5).

Closed ridge is proposed as a purely descriptive term to designate both the regular and the irregular forms. The adjective *circular, elliptical,* or *oval* can be applied to the more regular forms of closed ridges. Most authors seem to agree that the closed ridges resulted from stagnant ice that separated into individual dead-ice blocks, even though they do not agree as to the mechanics of deposition. If this origin is accepted, the term *closed disintegration ridge* may be used. *Ice-block ridges* and *rimmed kettles* are alternatives, provided that they do not imply construction of the ridge from material on the ice alone but leave room for the concept that till was squeezed upward from beneath the block. Generally, fields of closed ridges show no evidence of depositional control, even on a regional scale. In places, however, distinct end moraines are composed of hummocky moraine in which closed ridges are noticeable. The trends indicate a depositional control oriented transverse to the direction of ice flow. In other places the closed disintegration ridges are aligned because they are superimposed on previously formed trends (pl. 6). This is the case in some parts of the Missouri Coteau in Alberta and Saskatchewan, where structural end moraines of ice-shoved bedrock form ridges which caused the alignment of rimmed kettles in the overlying moraine.

Controlled disintegration features

Linear disintegration ridges

Description

Linear ridges resulting from ice disintegration are common features over much of the Plains region of western Canada and the United States. They have been described by many workers (e.g., Colton, 1955; Gravenor, 1956) and have been observed over large areas as undescribed.

Linear disintegration ridges are composed chiefly of till and may or may not have included pockets of stratified materials. A thin layer of gravel commonly lies on top and on the flanks of the ridges. In other places the ridges may consist predominantly of stratified material. They vary in height from 3 to 35 feet, in width from 25 to 300 feet, and in length from a few yards to 8 or more miles. They are straight or slightly arcuate. An important characteristic is that, in general, two sets intersect at acute or right angles. Such intersection of ridges forms a "waffle," "diamond," or "box pattern." These patterns indicate controlled deposition; orientation with regard to the direction of ice movement, however, may vary in different localities. In the Wolf Point area of Montana (Colton, personal communication) the majority of the longer and better-developed ridges lie at 45° to the direction of ice movement. In the Dollard district of Saskatchewan, on the other hand, the more distinctive ridges lie parallel to the ice flow (Kupsch, 1955, p. 329). In the Vermilion area of Alberta, the ridges show distinct parallelism to the lobate ice front in the center of the lobe, but with a more confused box pattern along its margin (pl. 3). Therefore, although the ridges are not of themselves reliable indicators of ice-movement direction at all localities, it is nevertheless significant that the most prominent ridges commonly lie normal, parallel, or at 45° to the direction of flow and hence show flow control.

In some areas three or more sets of linear disintegration ridges are present (pl. 4), and a complex pattern of linear elements results. Where two or more ridges meet, they produce forms resembling hairpins, wish-bones, crosses, and many other angular shapes. Especially interesting are those which represent gradations between closed and linear disintegration ridges, with the resultant shape of a shepherd's crook.

The junction of two ridges may be at the same level, or one ridge may be superimposed on the other (fig. 1). Superimposition is not common but has been observed (Kupsch, 1956, fig. 1), and no distortion of the lower ridge is apparent. The material of the upper ridge was apparently let down without disturbing the lower ridge.

The most striking, easily recognizable, linear disintegration ridges occur in till areas of low to intermediate relief. They are distinct because of their straightness and field pattern. In high-relief moraine they may be straight but are more commonly irregular or slightly sinuous (pl. 1, *C*). In some

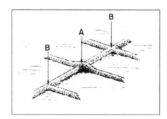

Figure 1 Diagram to illustrate different ridge junctions. *A*, superposed junction; *B*, confluent junction.

places they connect a series of closed till ridges, thus forming a feature like a string of beads (pl. 1, *D*). Linear ridges in high-relief areas may have parallel trends, as in the Viking disintegration moraine of east-central Alberta, where they are oriented northeast at right angles to the direction of ice movement.

Although linear ridges composed of till predominate in the areas studied, some consist of sand and gravel and thus suggest ice-contact deposition; the proportion of till and stratified materials varies between wide limits. Gravenor (1956, p. 14) describes such a ridge of stratified drift from east-central Alberta, where it terminates in a broad kame.

Linear disintegration ridges of "sandy moraine" are described from the Cree Lake area, Saskatchewan, by Sproule (1939, p. 104) as "ice-crack moraine." The material composing these ridges is a very sandy till and not a stratified drift. The area lies south and west of a large exposure of the pre-Cambrian Athabasca sandstone that lay in the path of glacier advance. It is interesting to note that some of the linear disintegration ridges of the Cree Lake area occur on bare bedrock.

Terminology

Deane (1950, p. 14) used "ice-block ridge" for both closed and linear till ridges. The most widely used term for the ridges is perhaps "crevasse filling." It was introduced by Flint (1928, p. 415) and is applicable to the till ridges of western Canada, which are thought to have originated in crevasses in the ice. The crevasse fillings described by Flint, however, are composed of stratified material, and, to indicate the different composition of the ridges in the plains region, Gravenor (1956, p. 10) has suggested the term "till crevasse filling." A more general term is *linear disintegration ridge*, which implies less about the origin than any other term. It merely indicates that the ridge originated during stagnation or near-stagnation of the ice. Most geologists apparently agree to this general origin but differ as to the specific mode of formation.

Washboard ridges

Gwynne (1942) described minor transverse till ridges from Iowa, where they form what he referred to as a "swell and swale pattern." Such patterns are common in many parts of the northern Great Plains and are now generally referred to as "washboard moraines." Gwynne (1942, p. 206) visualized them as being formed by periodic retreat and readvance of the live glacier front pushing previously deposited ground moraine into ridges. They would thus be minor end moraines or recessional ridges. The washboard moraines of Alberta and Saskatchewan were deposited on land; indications of a subaqueous origin are lacking. They are therefore different from those other minor end moraines, the "winter moraines" described from Sweden which apparently were formed under water (Woldstedt, 1954, p. 149).

Elson (1957, p. 1721) examined washboard moraines in Manitoba, where he found them to be composed of parallel, discontinuous, sandy-silty till ridges as high as 15 feet and as long as a mile, spaced 300–500 feet apart and forming lobate patterns. He concluded that the washboard ridges were deposited subglacially at the base of thrust planes in the ice.

Although it is realized that geologists are not in agreement as to the origin of washboard moraine, the writers tend to favor Elson's hypothesis and regard the preservation of the ridges as the result of ultimate stagnation. The washboard ridges were preserved because, finally, the glacier separated along the thrust planes. They are therefore to be regarded as disintegration features exhibiting inherited flow control. Of all the various disintegration features, they show the most obvious flow control, and they are thus helpful in reconstructing former ice-front positions if it is supposed that the thrust planes are everywhere parallel to the margin.

The linear disintegration ridges are in places different from washboard ridges, in that the linear ridges may be even-crested and have a fairly consistent width (pl. 3), whereas the washboard ridges are almost everywhere discontinuous and of variable height and width, forming a pattern of light "dapples" on air photographs (Gwynne, 1942, p. 202), like some of the disintegration ridges shown on plate 4. This difference may be due to the general open nature of crevasses into which till was squeezed to form continuous linear ridges and the narrow thrust planes along which the discontinuous till masses of the washboard ridges accumulated.

The washboard pattern is characteristically lobate, outlining the position of the thrust planes in the live ice, locally modified by small to large re-entrants formerly occupied by meltwater streams, which may or may not have deposited eskers. Irregular or branching ridges occur in places, but, in general, the washboard "waves" or parallel troughs and ridges are well developed. Gradations from washboard moraines to patterns indicating linear disintegration along crevasses have been observed. Lawrence and Elson (1953) noted a second direction of trends of acute or right angles to some washboard moraines. The presence of two or three directions is an outstanding characteristic of the field pattern of linear disintegration ridges, indicating a system of fractures composed of several sets. One of these directions may be parallel to the former ice front and show a lobate outline (pl. 3). The ridges are then similar to washboard ridges except that they were probably deposited in continuous open crevasses. Locally the development of a pattern controlled by several sets of fractures (as along the margins of the lobe shown in pl. 3) reveals the linear disintegration origin and distinguishes the fillings of transverse crevasses from washboard ridges caused by squeezing along transverse thrust planes.

Erosional disintegration features

Ice-walled channels

Description

Meltwater channels in areas of stagnant ice are among the more spectacular glacial features of the Plains region (pl. 5, *A, B, C*). They are especially well developed in east-central Alberta, where they can be divided into two types: (1) those filled with till, which are recognizable on air photographs by a chain of kettles in the bottom of the valley, and (2) those which are not filled with till but are broad, open troughs with some sand and gravel on their floors. Both types are generally marked by a topographic depression which outlines the channel, but the first type may be completely filled and thus blend in with the surrounding moraine. Gradations between the two extreme types are encountered in many places. The first type is most common in high-relief disintegration moraine, and the bedrock is at a considerable depth below the base of the channel. The second type is most common in areas of thin, low-relief ground moraine, and in many places the bedrock is exposed in the sides of the channel. Some of the channels have been observed to feed into esker systems.

In map view, the channels present a complex pattern composed of parallel and intersecting elements. In east-central Alberta the major channels are roughly parallel to the ice-movement direction, and the intersecting ones are at right angles. That the two directions were not everywhere used at the same time is indicated by the fact that the floor of one channel commonly truncates the deposits of another. The pattern is only roughly controlled, and over small areas control is not very obvious.

Terminology

Erosional channels occupied by subglacial meltwater streams are referred to as *Rinnentaler* in the German literature. It may be that some of these channels were not eroded by meltwater alone but that they were initiated by glacial erosion and later occupied by meltwater (Woldstedt, 1954, p. 38). The term *Rinnentaler* ("channel valleys") would seem suitable for the ice-walled channels of western Canada, but it is difficult to translate with retention of any descriptive meaning. In Denmark the trenches cut by streams flowing beneath the ice are called *tunneldale* ("tunnel valleys"), and an origin in closed tunnels is implied (Schou, 1949, p. 20). This term is applicable where such an origin is indicated, but the writers prefer *ice-walled channels* to include streams in open trenches as well as in closed tunnels.

119

Origin of disintegration features

General remarks

All studies made by the writers on ice-disintegration features suggest that the features were formed late in the existence of a glacier. The features resulted from the letting-down of till due to ablation, from the squeezing-up of till into openings at the base of the ice, or from a combination of both causes.

Vertical air photographs of the Cree Lake area reveal that some of the linear disintegration ridges cross over drumlins without any apparent disturbance of the older deposits. Near North Battleford, Saskatchewan, linear ridges cross such streamlined live-ice features as elongated drumlins and intervening grooves, collectively referred to as "fluting." In both areas the disintegration features are clearly younger than the live-ice features.

Closed ridges that are aligned into end moraines may owe their alignment to *inherited flow control* because the hollows and surrounding rims were formed by blocks of dead ice left in the constructional end moraine after retreat of the ice. The possibility that the development of trends in a hummocky moraine is later than ice disintegration has to be considered if the origin of dead-ice thrust moraines, as described by Seifert (1954, p. 132–133), is accepted. He visualizes the thinning of ice along the margin and the accumulation of morainal material on it as being followed by a thickening of the glacier. The regenerated ice would then override the stagnant ice with its moraine and push it into thrust ridges. If this process caused trends in hummocky disintegration moraine, it could be said to exhibit *regenerated flow control.*

Origin by ablation

In a wasting glacier, debris which was originally in the ice gradually accumulates on the surface of the ice. As ablation progresses, this surface debris is let down on the basal-ice deposits. While the material is still on the ice, it forms chaotic ridges, knolls, and depressions which reflect mainly relief of the ice surface itself. The drift let down irregularly from the surface of the glacier to the ground gives rise to ablation moraine (Flint, 1957, p. 120). This letting-down process can be used to explain ice-disintegration features (fig. 2). Hummocky moraine may be a reflection of irregular accumulations of debris on the ice (Gravenor, 1955, p. 477). Ablation material is moved by mass wasting into openings in the ice. Although the circular (or polygonal) shape is common for a dead-ice mass and the ablation material thus sloughed off into round closed ridges, in many places the blocks have an angular outline and represent remnants of a glacier that disintegrated along pre-existing fracture systems. When final melting of the ice blocks occurs,

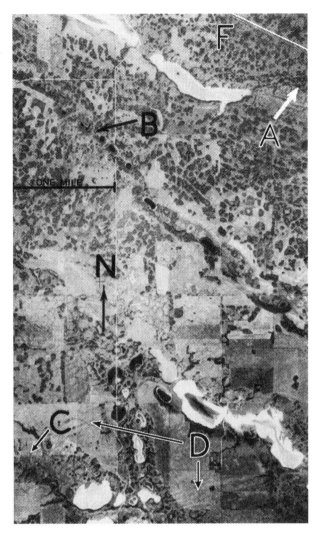

Plate 5 Ice-walled channels. (Location of pl. 5 shown on fig. 4.) Location: T. 45, R. 10, W. 4th mer., Alberta. (Air photograph reproduced by permission of government of Alberta, Department of Lands and Forests.)

A, Terrace of the Kinsella ice-walled channel, now covered by dead-ice moraine. To the west this channel is cut into bedrock.

B, Ice-walled channel, almost completely filled with drift.

C, Broad, open type of ice-walled channel, filled with sand, gravel, and some till, showing circular disintegration ridges.

D, Poorly developed flutings (?) in lodgment ground moraine, possibly indicating live-ice deposition, covered by thin veneer of dead-ice (ablation) moraine.

F, Dead-ice moraine, intermediate relief.

121

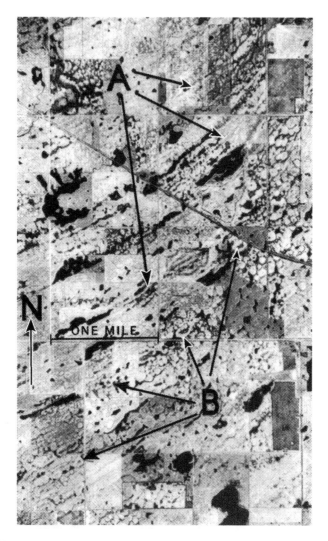

Plate 6 Closed disintegration ridges and hummocky dead-ice knobs superimposed on morainal ridges. (Location of pl. 6 shown on fig. 4.) Location: T. 19, R. 26, W. 4th mer. This photograph shows a series of parallel, curved end-morainal ridges (*A*) (much like "washboard" moraine) on which has bun superimposed depositional features of ice disintegration (*B*). Note that the ice-disintegration features show definite alignment which is interpreted m control derived from lines of weakness in the active ice. This photograph suggests that the end-moraine ridges were developed under the ice and at a later date—but along the same lines of weakness—than the dead-ice features were developed.

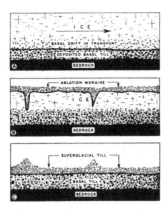

Figure 2 Disintegration features formed by ablation (modified after Flint, 1947, p. 112).

round or linear disintegration ridges remain, depending on the shape of the ice block (fig. 2).

It seems obvious from several considerations that many depositional disintegration features owe their origin, at least in part, to material that fell into crevasses, subglacial channels, and irregularly shaped hollows under and in the ice.

Observations on the Kinsella ice-walled channel in Alberta by Gravenor and Bayrock (1956, p. 6–8) show that at one point gravel outwash was deposited which lies *above* the level of the surrounding high-relief disintegration moraine. This outwash is not covered by till and shows no evidence of having been overridden by ice. Downstream from the outwash the channel is cut into bedrock and is floored with till in the form of a high-relief disintegration moraine (this part of the channel is shown on pl. 5). The fact that the outwash lies above the surrounding high-relief disintegration moraine and has not been overridden suggests that the outwash was deposited between ice walls and by the last ice to cover the area. The downstream portion of the ice-walled channel must have been infilled by debris which slumped in from the ice walls. That the channel was not used to any great extent by water after deposition of the ablation moraine is evident from the absence of any notable erosion of the ablation deposits by running water.

Whereas many smaller ice-walled channels may have been true ice tunnels, the width of the larger trenches—over 1 mile—and the fact that they contain little till (which may have slumped in from the ice walls along the sides) would suggest that the larger channels were open to the atmosphere at the time of final ice disintegration.

The superposed junction of linear disintegration ridges (fig. 1) also reveals that some material was let down from the stagnant ice. This type of junction suggests that first one crevasse was open, that material was dumped in it, that another crevasse opened up crossing the first, and that drift was deposited in this second crevasse without disturbing the older deposits.

In hummocky moraine the moraine plateaus of stratified material are generally explained as meltwater or lake sediments in a depression surrounded by ice. On final melting, the fill may then be higher than the surrounding ablation moraine (Schou, 1949, p. 10). If deposition of the water-laid sediments takes place on ice, the melting of this floor will cause "collapse structures" in the sediments. Such "collapse structures" are indeed common in the hummocky dead-ice moraine of western Canada (Christiansen, 1956, p. 12), and Flint (1955, p. 114) noticed them in eastern South Dakota.

Although there are many indications that ablation moraine was being deposited during disintegration of an ice sheet, the hypothesis that all material of a dead-ice origin was let down from above fails to explain certain observed relationships. Ablation material is characteristically loose, non-compact, and non-fissile and contains abundant gravel and larger stones (Flint, 1957, p. 120). Melting is active as ablation progresses, and trickles of water cause crude sorting and a general "washed" appearance. Meltwater deposits such as kames, eskers, and crevasse fillings of stratified drift, form during the disintegration phase. Such deposits are common in many areas in the eastern part of North America, where they have been interpreted as stagnation features, as, for example, the crevasse fillings described by Flint (1928, p. 411–416). In western Canada, however many features that apparently resulted from disintegration are composed of till which may be very compact. This material may be the only constituent of the disintegration feature, or it may be overlain by a layer of non-compact till of varying thickness, or the latter deposit may make up the whole land form. The presence of the compact till is difficult to account for if all material is thought to have been deposited by ablation.

Another objection is raised by many workers, who point out that even in Alaska, where thick ablation moraine is common on the lower reaches of living valley glaciers, clear glacier ice is the normal condition and moraine-covered ice the exception. It is held that in ice sheets a complete covering by ablation moraine is impossible (Hoppe, 1952, p. 26). If this is so and if it is believed that dead-ice features result from drift that was let down, it becomes necessary to find a mechanism to get large quantities of drift on top of the ice. Flint (1955, p. 114) therefore assumes a readvance of the glacier which overrode thin residual ice and deposited basal till on top of it. With renewed deglaciation the buried ice masses melted out, and the overlying till was let down. Such conditions are known to have occurred, but it appears that in many cases they are assumed without any corroborative evidence for the assumption of a readvance. Drift can also be brought up from the basal portion of the ice to its surface along thrust planes which curve obliquely upward in the terminal zone of the glacier. The drift melts out onto the ice surface along the thrust line and forms ridges on the ice. This process occurs in live ice, but, when the ice subsequently becomes stagnant, the till ridges on the ice may be let down onto the underlying surface when the supporting ice melts. Investigations of glaciers in Spitzbergen (Hoppe, 1952, p. 28–29; Woldstedt, 1954, p. 54) revealed the presence of ridges parallel to the ice margin and, in addition, others transverse to it. Basal drift is apparently also brought to the surface of the ice along open crevasses oriented transversely to the ice margin probably by squeezing due to ice flow or pressure from the weight of ice blocks. The resulting complex of till ridges (*Lehmmauern*, or "till walls") is strikingly similar to the disintegration pattern of linear ridges encountered in many places in western Canada.

Origin by squeezing

Although it has been demonstrated that ablation moraine may accumulate on the ice during wasting and that it may become concentrated in ridges and heaps as a result of many different processes, some investigators feel that the quantity of ablation material is not sufficient to account for the disintegration features which can be observed. Hoppe (1952) is a proponent of this school of thought. He believes that the dead-ice features were formed underneath the ice through squeezing of debris into basal cavities such as crevasses and irregular cavities caused by meltwater (fig. 3). The drift could be squeezed up into these cavities because it was soaked with water and therefore in a plastic state. The weight of the ice itself exerted the necessary pressure, but some slight ice-flow movement may have contributed in certain places. Hoppe recognizes that some drift may have been squeezed all the way to the surface of the ice along crevasses (*Lehmmauern*), but he believes that it is quantitatively insufficient to account for all dead-ice moraines features. Hoppe's arguments for assuming that dead-ice features

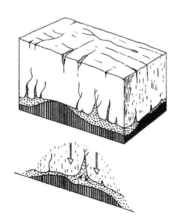

Figure 3 Disintegration features formed by squeezing (after Hoppe, 1952, p. 55).

resulted from basal till squeezed into openings, closed to the sky, on the underside of the ice, may be summarized as follows:

1. The till does not show evidence of washing, as would be the case with superglacial material that had fallen from the ice surface or from the side walls into an open crevasse.
2. The till is compact and has all the characteristics of basal till.
3. The till contains pebbles and cobbles that show a distinctive fabric with their long axes oriented at right angles to the long dimension of the ridge. This fabric is regarded as a primary characteristic of the till induced by the lateral pressure of the ice blocks which squeezed the till up into the crevasse.

Although Hoppe stresses the third observation and presents a great number of orientation studies on various types of disintegration ridges in Sweden, it is not clear why a similar orientation could not result from lateral outward flow of the plastic material by soil creep or solifluction when the debris moved slowly down a stagnant ice-block surface. Hoppe's investigations demonstrate, however, the important relationship between minor, or even micro-, relief features and till fabric. Many fabric studies in high-relief disintegration moraine previously considered as "end moraine" may have yielded incomprehensible or conflicting results because due account was not taken of this relationship.

The process envisaged by Hoppe hinges essentially on three assumptions: (1) the ice was relatively clear, and most of the debris was underneath the ice; (2) the debris under the ice was not frozen but in a water-saturated state and hence susceptible to plastic deformation; and (3) the fabric of the till in the ridges is the result of squeeze. The first assumption is made by many students of glacial geology. However, in view of the various processes that can bring debris to the top of the ice, it is doubtful that this condition held over the entire continental glacier. It seems likely that locally the margin of the glacier was covered by large quantities of ablation material. The second assumption appears valid under certain conditions, inasmuch as permafrost is generally absent under thick masses of "warm" ice (Woldstedt, 1954, p. 60). In "cold" glaciers, however, it is common for the ground under the ice to be in a permanently frozen state. The third assumption appears to be unproved, as the fabric may be produced during the final melting of the supporting ice, regardless of whether the till ridge was formed by "squeezing-up" or "slumping-in."

Perhaps the main merit of the squeezing theory is that it accounts for the character of the till in the ridges. It should be pointed out, however, that, once the ice had a thin layer of ablation debris on its surface, further down-melting would be extremely slow, and the clayey nature of the till might well be preserved. Recently, Harrison (1957) has argued that the bulk of till in ground moraine originated as englacial material and was deposited by the slow melting-out of debris from the basal zones of the ice. The compact nature or "toughness" of till may, therefore, not be a result of "plastering-on" or squeezing but may simply be due to the till's original texture, structure, and fabric. Hence the character of till in ridges or in hummocky moraine cannot be used as strong evidence in the determination of the origin of the feature.

Arguments against the squeezing theory (such as the superimposed crossing of two ridges, the presence of linear disintegration ridges on bare bedrock, observable "collapse structures" which indicate deposition on ice masses and subsequent lowering, and the characteristics of ice-walled channel fills) have already been presented as favoring the ablation theory.

Combined processes

The hypotheses of ablation and squeezing for the origin of disintegration features are not mutually exclusive, and the writers hold that, in the formation of many of the features observed, both processes may have been operative to varying degrees. In the marginal zone of the glacier, till may be squeezed upward in fractures to the surface of the ice and subsequently let down to the ground by ablation. The ridges thus formed on the ice may lose their identity during ablation and by slump action, eventually becoming disintegration moraine (Woldstedt, 1954, p. 54). Some fields of clearly recognizable linear till ridges may owe their origin to these combined processes, but with the preservation of the till ridges more or less as they existed on the ice.

Material squeezed into subglacial crevasses and other openings may, in addition, receive material from above during ablation. Some disintegration features appear to be composed of cores of clayey till squeezed upward and overlain by a covering of loose material dropped from above. Such cores may have been squeezed upward, the covering dropped from above. The amount of meltwater involved in ablation varies between wide limits, and all gradations between till disintegration features and those composed of stratified drift are to be expected. In places, meltwater erosion, instead of deposition, took place, and subglacial channels were cut in tunnels or in trenches open to the sky. Such channels are therefore the erosional counterparts of eskers.

Because combinations of several depositional processes may be responsible for dead-ice land forms, the writers prefer *disintegration features* to the term "ablation moraine," as the latter implies not only an origin from stagnant ice but also a certain process of deposition.

Terminology of moraines

Flint (1955, p. 111–120) reviewed the history of the terms "moraine," "ground moraine," and "end moraine," whose definitions vary widely with different authors. The recognition of ice-disintegration moraines further complicates any classification adopted for mapping. Dead-ice land forms may occur as high-relief features in "moraines" or as low-relief forms in "ground moraine." The term "ground moraine" has always implied deposition of material from the base of the ice, back from its margin (Flint, 1955, p. 111). If, however, the high-relief disintegration features owe their origin to squeezing-up of basal material, as Hoppe (1954, p. 8) and other investigators state, then some areas previously considered "moraine" or even "end moraine" are more properly "high-relief ground moraine." Townsend and Jenke (1951, p. 857) suggested this origin for the Max Moraine in North Dakota when they wrote: "The Max Moraine may be a special type more nearly related in extent and mode of deposition to ground moraine than to end moraine."

The German term *kuppige Grundmoräne* also implies a ground-, not end-, moraine origin for hummocky moraines. Such moraine may be deposited along the ice front at localities where no true end-moraine ridges are formed.

In reconnaissance mapping of glacial deposits in western Canada, where little is known about the direction of ice movement, it is of importance to distinguish between areas of predominantly live-ice deposition and those of stagnation. Features that indicate live-ice deposition are either parallel to the direction of ice movement or transverse to it. Parallel features are streamlined and range from striae to drumlins. Transverse features consist of lobate systems of true end moraines. Dead-ice features may show control inherited from live ice, or they may show no control whatever.

Not only do live-ice features grade into those of stagnant ice, but the superimposition of disintegration features on live-ice features can be noted in many places. The linear disintegration ridges that cross flutings near North Battleford, Saskatachewan, are but one example. Commonly an almost continuous blanket of disintegration moraine hides any underlying streamline features of the live-ice phase of the glacier. This blanketing layer may show dead-ice features that would be scarcely recognizable, were it not for the gradation into areas with higher relief (pl. 5, gradation from *E* to *F*). Areas generally considered ground moraine because of their low relief therefore do not necessarily show basal till at the surface but may include some ablation till if the disintegration originated by the combined processes of squeezing-up and lowering. Locally the disintegration blanket of the stagnant phase may be so thin that it does not effectively hide the streamlined "megafabric" of the underlying basal till deposited by lodgment from live ice (pl. 5, *D*). In other places the disintegration cover may be removed by erosion after final retreat of the ice, and the "megafabric" of the basal till may be secondarily brought out by that erosion. This may be one reason why some flutings are especially well developed in areas that at one time were occupied by streams (Christiansen, 1956, p. 16).

Other similar features

General remarks

Some land forms may appear morphologically similar to disintegration features but may have originated in a different way. Hummocky moraine may result from live-ice deposition as well as from dead-ice disintegration and is thus polygenetic. Patterns indicating deposition from live ice, such as alignment of the hummocks into lobate trends, may be hardly discernible, or the disintegration moraine may still vaguely show some inherited flow control. Thus it may be difficult or impossible to assign a definite origin to a particular feature. On the other hand, the end members of the live-ice–dead-ice sequences are recognizable and should be differentiated in mapping. If there is any doubt, the feature should be considered with regard to the following three main characteristics:

1. *Individual characteristics.*—What differences exist between the individual feature and others similar in form but known to be of different origin? These characteristics are best studied on the ground with the aid of large-scale vertical air photographs.
2. *Pattern characteristics.*—What differences exist in the pattern of a group of individual features with respect to the pattern of other groups of features of different origin? The pattern characteristics are most clearly seen on air-photograph mosaics. They are generally more revealing than the individual characteristics.
3. *Positional characteristics.*—Is the postulated origin compatible with other glacial features in the region? In other words, do the supposed live-ice features, disintegration moraines, or frozen-ground features occur where they are understandable with respect to the particular geological setting? These characteristics will generally be clear only after a large area or region has been mapped. Accordingly, subsequent reassignment of some features to a different origin may be necessary as mapping progresses. In general, the positional characteristics provide the strongest evidence for a particular origin.

Live-ice features

Live-ice features oriented in the direction of ice flow, such as ice-abraded bosses, crag and tail, and elliptical drumlins, are not readily confused with disintegration features. Extremely elongated drumlin ridges may, however, be similar to long crevasse fillings. The long linear ridges near Velva, North Dakota, were first considered to be crevasse fillings but are now reinterpreted as drumlins (Lemke, 1958). The streamlined shape of these elongated drumlins with a steeper stoss side and a long, gently tapering tail, the field pattern of the strikingly straight and parallel ridges with intervening grooves giving the country a "scratched" or "fluted" appearance, and their orientation in the direction of ice movement, as indicated by other evidence, will generally help to identify the ridges as live-ice features.

Large constructional or structural (push) end-moraine ridges with a distinctive lobate pattern are generally clearly evident as live-ice features. Similar, but smaller, transverse ridges formed by live ice may not be so easily distinguishable from some disintegration ridges.

Frozen-ground features

Land forms created by thawing ground ice may be very similar to those formed by melting and evaporation of dead glacier ice buried by debris. Some features may actually have the same shape and composition, so that only the relationship to other land forms of known origin can provide some hints as to their genesis. The writers prefer the inclusive term *frozen-ground features* (Flint, 1957, p. 195–196) for all features that result from the presence and final thawing of ground ice.

Rimmed depressions in the northern Netherlands previously regarded as kettles have recently been reinterpreted as pingo-remnants (Maarleveld and Van Den Toorn, 1955). Their similarity to true kettles is so great that they are referred to as *pseudo-sölle* ("pseudo-kettles"). They are distinguishable from kettles only because stratigraphic relationships indicate formation during a cold climate long after the final melting of the glacier ice. Such detailed

Figure 4 Index map showing location of plates.

stratigraphic information is not available in western Canada. It appears unlikely, however, that the round disintegration ridges surrounding depressions in low-relief areas (pl. 4) are formed by pingo-collapse. As far as the writers are aware, pingos do not occur in large fields of many square miles. Nor are they known to be so closely spaced that they touch each other and form a continuous pattern of round or modified polygonal features.

Elson (1955, p. 140) regarded features such as those shown in plate 4 as possible thaw depressions which resulted from subsidence following the thawing of perennially frozen ground (Hopkins, 1949). Thaw depressions, like pingos, occur in predominantly silty soils, whereas the disintegration features of western Canada are notably well developed in rather impervious till which has a high clay content.

Any alternative explanation for the origin of features which are here interpreted as ice-disintegration land forms will have to take into account the regional setting in relation to other glacial evidence. Development of frozen-ground features is a function of sediment characteristics, not of glacier motion. The regional setting of disintegration features, on the other hand, shows their dependence on glaciation. Disintegration is a marginal phenomenon of an ice sheet and operates irrespective of the sediment type.

Summary of classification

The various depositional ice-disintegration features composed predominately of till and their classification can be briefly summarized as follows:

A. Uncontrolled disintegration

Generally uncontrolled, but gradations to controlled disintegration exist, which is most commonly *inherited flow control*. In places *regenerated flow control* may have been operative.

Features: *Hummocks* (knob and kettle), *moraine plateaus, closed ridges* (rimmed kettles; ice-block ridges).

B. Controlled disintegration

1. Disintegration along open crevasses
Generally well controlled, but gradations to uncontrolled disintegration can be observed. Control commonly inherited from previous ice flow but could be caused by diastrophic movements, jointing in bedrock, and other causes.

Features: *Linear ridges* (crevasse fillings; ice-crack ridges; ice-block ridges).
2. Disintegration along thrust planes
Distinctive inherited flow control evident from lobate pattern.

Features: *Washboard ridges* (swell and swale topography; minor recessional ridges).

C. Superimposed disintegration

1. Superimposition on live-ice features parallel to ice flow, such as drumlins, flutings.
2. Superimposition on live-ice features transverse to ice flow, such as depositional and structural end moraines.

In addition to the above-mentioned till features, land forms composed of stratified drift are part of the disintegration moraine. Of the erosional features, *ice-walled channels* appear to be the most prominent.

Acknowledgments

The writers are indebted to V.K. Prest, R. Lemke, and J.A. Elson, who read the manuscript and offered valuable comments and criticisms, many of which have been incorporated in this paper. They would also like to thank the Research Councils of Alberta and Saskatchewan for financial support which made this study possible.

References cited

CHRISTIANSEN, E.A., 1956, Glacial geology of the Moose Mountain area, Saskatchewan: Sask. Dept. Mineral Resources, Rept. 21.

COLTON, R.B., 1955, Geology of the Wolf Point quadrangle, Montana: U.S. Geol. Survey, Geol. Quadrangle Map.

DEANE, R.E., 1950, Pleistocene geology of the Lake Simcoe district, Ontario: Geol. Survey Canada Mem. 256.

ELSON, J.A., 1955, Unpublished Ph.D., thesis: Yale University.

—— 1957, Origin of washboard moraines: Geol. Soc. America Bull., v. 68, p. 1721.

FLINT, R.F., 1928, Eskers and crevasse fillings: Am. Jour. Sci., v. 15, p. 410–416.

—— 1955, Pleistocene geology of eastern South Dakota: U.S. Geol. Survey Prof. Paper 262.

—— 1957, Glacial and Pleistocene geology: New York, John Wiley & Sons.

GRAVENOR, C.P., 1955, The origin and significance of prairie mounds: Am. Jour. Sci., v. 253, p. 475–481.

—— 1956, Air photographs of the plains region of Alberta: Research Council of Alberta, Prelim. Rept. 56–5.

—— and BAYROCK, L.A., 1956, Stream-trench systems in east-central Alberta: Research Council of Alberta, Prelim. Rept. 56–4.

GWYNNE, C.S., 1942, Swell and swale pattern of the Mankato lobe of the Wisconsin drift plain in Iowa: Jour. Geology, v. 50, p. 200–208.

HARRISON, P.W., 1957, A clay-till fabric: its character and origin: Jour. Geology, v. 65, p. 275–308.

HOPKINS, D.M., 1949, Thaw lakes and thaw sinks in the Imbruk Lake area, Seward Penninsula, Alaska: Jour. Geology, v. 57, p. 119–131.

HOPPE, GUNNAR, 1952, Hummocky moraine regions, with special reference to the interior of Norbotton: Geog. Annaler, v. 34, p. 1–71.

HOWELL, J. V. (ed.), 1957, Glossary of geology and related sciences: Washington, D.C., N.A.S.N.R.C. Pub. 501.

JOHNSTON, W.A., et al.., 1948, Surface deposits, southern Saskatchewan: Geol. Survey Canada, Paper 48-18, map.

—— and WICKENDEN, R.T.D., 1931, Moraines and glacial lakes in southern Saskatchewan and southern Alberta, Canada: Royal Soc. Canada Trans., 3d ser., v. 25, sec. 4, p. 29-44.

KUPSCH, W.O., 1955, Drumlins with jointed boulders near Dullard, Saskatchewan: Geol. Soc. America Bull., v. 66, p. 327-338.

—— 1956, Crevasse fillings in southwestern Saskatchewan, Canada: Verh. K. Nederlandsch Geologisch-Mijabouwkundig Genoot., geol. ser., v. 16, p. 236-241.

LAWRENCE, D.B., and ELSON, J.A., 1953, Periodicity of deglaciation in North America. Pt. II. Lake Wisconsin recession: Geog. Annaler, v. 35, p. 96.

LEMKE, R.W., 1958, Narrow linear drumlins near Velva, North Dakota: Am. Jour. Sci., v. 256, p. 270-283.

MAARLEVELD, G.C., and VAN DEN TOORN, J.C., 1955, Pseudo-sölle in Noord-Nederland (Dutch, with English summary): Tijdschr. K. Nederlandsch Aardrijksk. Genoot., v. 72, no. 4, p. 344-360.

NEILSON, W.A. (ed.), 1956, Webster's new international dictionary of the English language: 3d ed., unabridged, Springfield, Mass.

SCHOU, AXEL, 1949, Atlas of Denmark: Copenhagen, Hagerup.

SEIFERT, GERHARD, 1954, Das mikroskopiche Korngefüge des Geschiebemergels: Meyniana, v. 2, p. 124-184.

SPROULE, J.C., 1939, The Pleistocene geology of the Cree Lake region, Saskatchewan: Royal Soc. Canada Trans., 3d ser., v. 33, sec. 4, p. 101-109.

TOWNSEND, R.C., and JENKE, A.L., 1951, The problem of the origin of the Max Moraine of North Dakota and Canada: Am. Jour. Sci., v. 249, p. 843-858.

WOLDSTEDT, PAUL, 1954, Das Eiszeitalter: Stuttgart, Enke.

6

ICE MELTING UNDER A THIN LAYER OF MORAINE, AND THE EXISTENCE OF ICE CORES IN MORAINE RIDGES

G. Östrem

Source: *Geografiska Annaler* 41 (1959): 228–30.

Introduction

It has been known for a long time that some of the moraine ridges in the Tarfala valley are ice-cored. As this ice melts away, the moraine material which lies on the ice slides down and protects the lower part of the ice core from further melting. Meanwhile in the upper part, the process will continue, especially during warm, sunny days.

In order to get information about a moraine layer's influence on ice melting, some ice melting experiments under a thin layer of moraine cover were carried out during the summer of 1956.

Experiments with ice melting

On Isfallsglaciären, sand and gravel was placed on the clear glacier ice in test fields about 2 m² in area. The melting of the ice under the sand and gravel cover was measured by means of bamboo stakes. Control measurements on the uncovered glacier ice were also carried out by means of bamboo stakes.

By using moraine material of different grain size, and by making the test covers of different thicknesses, the experimental conditions could be varied.

Meteorological factors were observed at the nearby research station in Tarfala.

The temperature distribution between the upper and the lower level of the sand layers could be observed by means of thermistors. By extrapolation from the measured temperatures at different depths, the mean temperature on the upper sand surface during the measuring period (10/7–5/8 1956) could be computed (it was +9° C). As the mean air temperature during the same period was only +5.4° C, the warming of the sand and gravel layers must come to a great extent from the insolation. The absorbed amount of energy will be distributed among the following:

1. Outgoing (longwave) radiation.
2. Energy loss to the air by convection and conduction.
3. Energy loss through evaporation of melt water which has risen in the sand and gravel by capillary action.
4. Melting of glacier ice.

It was evident that the distribution of energy among these four factors varied with grain size and thickness of dirt layer.

To get more precise data on ice cores in moraine ridges, special attention was given to the effects of grain sizes and layer thicknesses upon the rate of melting. It was evident from direct measurement, that the uncovered ice melted with a mean rate = 4.5 cm/day. This rate decreased when the cover was more than about 1/2 cm thick. For example, under a 6 cm thick cover the mean melting speed was 3 cm/day; under 20 cm cover it was less than 1 cm/day. (Fig. 1).

[Not only will the melting be slower under a moraine cover, but also the ablation period will be shorter for the covered ice.]

Under really thin layers, the melting speed will accelerate, but as exact measurements are difficult to obtain because of melt water erosion, this part of the graph has been dotted only. The maximum point of the curve has been computed from the known value of incoming radiation and by setting

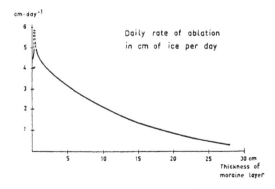

Figure 1 The daily rate of ablation during the measuring period 10/7–5/8 1956 when the normal ablation of exposed glacier ice was 4.5 cm/day. Computed from measured results at the test sites on Isfallsglaciären.

the different losses (to evaporation, etc.) as great as the losses would be from an uncovered ice surface.

Since, as seen from fig. 1, the curve appears to approach the X-axis as an asymptote, it is difficult to determine the thickness of moraine cover which should be great enough to permanently preserve ice from all melting. (It will be the point where the curve meets the X-axis). To get information about this limit value, it is necessary to measure the thickness of moraine cover on old ice deposits. This means principally on old moraine ridges.

Two methods have been used in these investigations: the seismic and the electric resistivity methods.

Localisation of ice cores in moraine ridges

In March, 1959, measurements were made at different places in the Tarfala valley, as well as on the lateral moraine at Isfallsglaciären. These measurements were made with a 12-channel seismic refraction instrument.

The thickness of the ice core was measured as 8–10 meters, but since the moraine cover was quite thin, the seismic method did not give information about this layer's thickness. Consequently, it was necessary to use other methods.

Efforts have been made to dig in the unconsolidated material on the moraine ridge, but without results.

Therefore, in August, 1959, several electrical earth resistivity measurements were made in the valley. Both the so-called Wenner (See Dobrin, 1952, p. 295) and the Schlumberger electrode configurations (See Lasfargues, 1957, p. 66) were used. At the same point on the lateral moraine of Isfallsglaciären where seismic measurements had been made, it could be shown that the moraine cover was only 1.8–2.6 metres. (The difference is the result of different computing methods.) Under this layer was an insulating body, the resistivity of which indicated that it must be ice. The thickness of this ice core could not be computed from the electric resistivity measurements.

Based upon these parallel experiments in the Tarfala valley, more investigations were made in the Jotunheimen area in Norway during October, 1959, and an additional survey in the Kebnekajse massif is planned for the summer of 1960.

As examples of the use of the two methods in localisation of ice cores in moraine ridges, the results of two parallel investigations from the Jotunheimen-measurements of October, 1959, will be shown here.

The first was a lateral moraine at Svellnosbreen, 30–50 meters high, which was thought to have an ice core. As can be seen from the seismic travel-time graph (fig. 2), the ridge only consisted of morainic material (sound velocity 710 m/sec.). Below this material the bed rock gives a velocity of 5,550 m/sec.

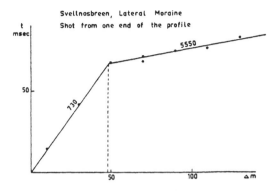

Figure 2 Travel-time graph made from seismic shots on the lateral moraine ridge at Svellnosbreen. No ice core can be determined. The moraine material lies on bed rock.

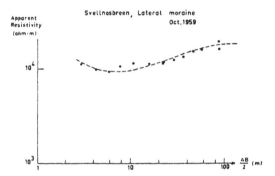

Figure 3 Earth resistivity curve obtained on the lateral moraine ridges at Svelinosbreen. (The same locality as fig. 2, see text). Horizontal scale refers to the distance between current electrodes A and B.

The earth resistivity curve from the same ridge (fig. 3) shows first a decrease of the earth resistivity, presumably because of increased humidity at greater depth. Thereafter it increases again, but not so rapidly as to indicate the existence of an ice core.

The second case, Veslegjuvbreen, has a terminal moraine 1,840 m above sea level. The seismic travel-time graph (fig. 4) shows two different velocities, 3,300 m/sec. which is a velocity for ice, and 5,000 m/sec. which comes from the bed rock. The thickness of the uppermost moraine layer could not be measured directly, but by means of measurements on similar material, it was assumed that the velocity would be approximately 740 m/sec. From this a moraine cover thickness of about 2 metres could be computed. The lower

137

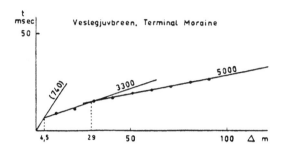

Figure 4 Travel-time graph obtained by seismic shooting on the terminal moraine at Veslegjuvbreen. The morainic velocity (740 m/sec.) has been computed from measurements in similar materials, and three layers can be recognized, the middle of which is obviously an ice core.

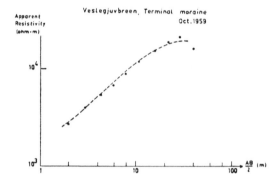

Figure 5 Earth resistivity graph from the same locality as fig. 4. (See text).

border between the ice and the bed rock was about 8 metres below the upper surface. (See fig. 4).

All the seismic work has been done by Stig Rune Ekman, Stockholm.

The earth resistivity curve is shown in fig. 5. When computing from this curve, it is obvious that an upper discontinuity lies 1.2 metres below the surface. A deeper boundary is difficult to compute exactly, but it is possible to see another discontinuity at about 12 metres below the surface; this figure is not exact.

Since the earth resistivity does not increase as rapidly, as would be the case if there was a clear ice core, it is possible to state that the ice is more or less mixed with morainic material.

As can be seen from these examples, the two methods complement each other. By using the resistivity method, dead ice masses can be easily located and the depth to the upper ice surface determined. Only the seismic method, however, can be used for measuring the thickness of the dead ice. The plan is to continue the investigations with earth resistivity measurements on different moraine ridges, and it is hoped that a technique can be developed to use the equipment also to indicate the presence of permafrost.

References

DOBRIN, MILTON B, 1952: Introduction to Geophysical Prospecting, New York.
LASFARGUES, PIERRE, 1957: Prospection Électrique, Paris.

7

MECHANISM FOR THE FORMATION OF INNER MORAINES FOUND NEAR THE EDGE OF COLD ICE CAPS AND ICE SHEETS

J. Weertman

Source: *Journal of Glaciology* 3 (1961): 965–78.

Abstract

A new mechanism is described which explains the formation of moraines in the ablation areas of cold ice sheets. The mechanism involves the freezing of water onto the bottom surface of an ice sheet. This water comes from regions of the bottom surface where the combination of the geothermal heat and the heat produced by the sliding of ice over the bed is sufficient to melt ice. A number of criticisms are made of the shear hypothesis, which has been advanced to explain moraines occurring on Baffin Island and near Thule, Greenland. It is concluded that this older hypothesis may be inadequate to account for these moraines.

Although in theory the mechanism proposed in this paper undoubtedly will lead to the formation of moraines, the existing field data are insufficient to prove conclusively that actual moraines have originated by means of this mechanism.

Introduction

In this paper a new explanation is offered for the formation of a special kind of moraine which has been studied in great detail in the Thule area of Greenland[1,2,3] and on Baffin Island.[4,5] This type of moraine also occurs in the Antarctic.[6]

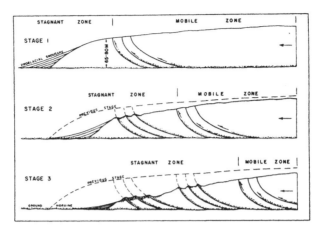

Figure 1 Diagram showing formation of successive moraines by the shear hypothesis. Figure is taken from Bishop's paper[1].

The structure of the Thule-Baffin type of moraine* has been known for a long time and is discussed in detail by Chamberlin and Salisbury.[7] They proposed the shear mechanism hypothesis to explain the origin of this phenomenon. Figure 1, which is taken from Bishop's article[1] (a similar picture is given in Ward's paper[4]), serves to illustrate the mechanism as well as to describe the Thule-Baffin moraine. Figure 1 shows a cross-section of the edge of a cold ice sheet which contains a number of Thule-Baffin moraines. It is assumed that "active" ice from the interior of the ice sheet moves outward to the edge, where it is blocked by a zone of "dead" ice. The active ice rides over the dead ice by slippage over active shear planes. It is further assumed that these active shear planes extend to the bottom of the ice sheet and that debris from the bed can be scraped up, carried into the ice to form debris layers, and ultimately be transported to the upper surface along these planes. Of course, the edge of the ice sheet must be in an ablation zone if the debris is to be exposed at the upper surface. In time, a layer of dirt, rock, sand, etc. will accumulate at the upper surface. This layer then protects the ice underneath from further extensive ablation. A debris layer whose thickness is only of the order of 0.5–1.0 m. will furnish such protection and can lead to the formation of ice hills approximately 15 m. in height. These debris-veneered ice hills are the Thule-Baffin moraines which

---

* Since this type of moraine has been investigated so thoroughly on Baffin Island and near Thule, Greenland, I prefer the name "Thule-Baffin moraine" to the commonly-used designation, "shear moraine." The latter name implies that the mechanism of formation is well established, which is not the case.

are considered in this paper. They are often called shear moraines, because it has been taken for granted that the mechanism advanced in Figure 1 for their formation is the correct explanation.

It is highly plausible to use the shear mechanism to explain the Thule-Baffin moraine. Yet there are serious difficulties associated with the theory which have not been pointed out in the literature and which will be discussed in the next section. Because of these difficulties, it is believed that another mechanism must be considered in order to give a satisfactory explanation of the phenomenon. In this paper it is proposed that the debris-carrying layers of ice are not formed by a scraping action of cold ice over the bed of an ice sheet, but rather are frozen into the ice by the freezing of water onto the bottom of the ice sheet.* The source of this water is ice which has been melted in the interior of the ice sheet (at the bottom surface) by geothermal heat and heat produced by the sliding of ice over the bed of the ice sheet. This water is forced by the pressure head to flow to the edges of the ice sheet, where it is refrozen to the bottom.

A third explanation for the formation of the Thule-Baffin moraine can be ruled out as highly unlikely. This hypothesis proposes that the debris-carrying ice layers originally were formed farther inland at the upper surface of the ice cap from rock, dirt, etc., which was derived from a protruding nunatak. This debris was buried later by snowfalls and eventually reappeared at the surface because of the ablation at the edge of the ice sheet. If this explanation were correct, one would not expect to find the Thule lobe of the Greenland Ice Sheet so completely fringed with Thule-Baffin type moraines. Instead one would see only an occasional moraine, each traceable to some local nunatak. It appears extremely probable that the debris which causes the Thule-Baffin moraines comes from the bottom of the ice sheet and not from the top surface. The fundamental question to be answered is: How does this debris become incorporated at the bottom surface into the ice mass?

---

* Mr. J. Hollin (private communication) has informed me that he also has been considering the possibility that ice may be formed at the bottom surface of a polar ice sheet.

After writing this paper, it came to my attention that E. von Drygalski[22] has mentioned the possibility of debris layers being frozen onto the bottom surface of an ice sheet. His idea (he bases it on some work of A. Blümcke and S. Finsterwalder[23] on glacial erosion) is that the pressure at the bottom of an ice sheet could cause melting and the water from this melting will flow into the sand or dirt underneath the ice and refreeze. It is not clear how this mechanism would lead to the formation of alternate layers of clear and debris-carrying ice. A criticism that can be made of it is that, although pressure will depress the freezing point of ice, it is still necessary to supply heat, if appreciable melting is to take place. Pressure by itself will not cause the melting envisaged in Drygalski's mechanism.

Comments on the shear hypothesis

Appearance of the debris layers

There are a number of criticisms that can be made of the shear hypothesis. Perhaps the most serious of these is the appearance of the debris layers themselves. In the Thule area, tunnels have been dug into the ice sheet, through these layers. As a consequence, the layers are easily examined. Their appearance varies greatly. They can occur as solid layers of stone, sand, etc. up to about 0.5 m. in thickness. At the other extreme they may consist of a layer of slightly dirty ice containing a very fine dispersion of sand or dirt particles. These slightly dirty layers, which may be up to 1–2 m. in thickness, are quite common in occurrence.

Now it may be plausible that ice can scrape layers of solid debris into a shear plane. But it is highly unlikely that a meter-thick layer of very slightly dirty ice would be formed by such a scraping action. What would be the mechanism for dispersing the particles picked off the bed through such a thickness of ice? Yet, if the shear hypothesis is to be accepted, it must account for layers of ice containing a fine dispersion of particles. On the other hand, if water were being frozen to the bottom surface of an ice sheet, one would not be surprised to find that thick layers of this refrozen ice were sometimes slightly dirty.

The possibility of cold ice scraping up debris

A question that the shear hypothesis must answer satisfactorily is: Can debris be scraped from the bed? The present answer to this question is that it cannot if the bottom of the ice sheet is below the freezing point. Direct observation has shown that cold ice of an ice sheet does not slide over a boulder-strewn bed. Goldthwait[2] has made this observation from the bottom of a vertical shaft dug from a horizontal tunnel which was excavated into the edge of the ice sheet in the Thule area. Although he found appreciable differential flow in the cold ice just above the ice sheet bed, there was no sliding over the bed. Even if an immeasurably small amount of sliding does occur, it is very ineffective in picking up debris. If sliding does occur, it is unable to remove even the moss covering the boulders at the bed. This moss was carbon dated as about 200 yr. old.[2]

If, in the interior of the ice sheet, the bottom is at the melting point, it could be possible for debris to be picked up. Direct observation by McCall[8] has shown that sliding does occur at the bed of a temperate glacier. McCall's study was made at the end of a tunnel dug to bedrock in a small Norwegian glacier. However, if one assumes that the debris is being scraped up in the interior of an ice sheet by temperate rather than cold ice, then one also must consider the mechanism that will he proposed for the formation of the debris layers.* This theory is based on the assumption that temperate ice does exist somewhere at the bottom of ice sheets which contain Thule-Baffin moraines. It might be mentioned that, from theoretical considerations[9] as well as field observations, one expects that cold ice will not slide over a rock interface, but that ice at its melting point will.

Since direct observation has shown that cold ice does not slide over a rock when subjected to a stress which is sufficient to produce appreciable differential flow in the ice itself, it would appear highly unlikely that the shear hypothesis can work in an ice sheet which is cold everywhere. Against the field observation just mentioned, one might argue that cold ice *will* slide over a rock–ice interface if the shear stress in a particular area is sufficiently high. However, experiments of Raraty and Tabor[10] have shown that the shear strength of an ice–solid interface is usually of the order of 10 bars if the solid is stronger than cold ice and if the solid is wetable. This shear strength of the interface is an order of magnitude larger than the shear stresses which occur at the bottoms of glaciers and ice sheets (0.4 to 1.0 bar).

The shear across a debris layer

It is a requirement of the shear hypothesis (although not sufficient proof of the validity of this mechanism) that appreciable discrete shear displacements occur across the debris layers (or at least that a very large amount of differential flow occurs in a layer of finite thickness). The evidence for the existence of these shear displacements is conflicting. Butkovich and Landauer[11] observed no discrete shearing motion across a dirt layer exposed in a tunnel in the Thule area, although there was differential flow in the ice. Hilty[12] made a similar observation in another tunnel in the Thule area. These measurements are evidence against the shear hypothesis. On the other hand, Ward[4] has published a photograph of the surface of the Barnes Ice Cap which shows a discrete shear across a dirt band. It can be argued, however, that Ward's observation was made in the summer time and was only a surface observation. During the Summer the surface layer is at the melting point. According to the sliding mechanism here proposed,[9] a discrete shear displacement of ice across a solid debris layer can occur when the ice is at

---

* Dr. H. Roethlisberger (private communication) has pointed out that debris might be brought into the ice mass in the boundary region at the bottom surface between temperate and cold ice. The thickness of dirt, sand, etc. which is frozen to the bottom, could be expected to be small in the region where the freezing point isotherm just begins to descend into the ground. If the shear strength of the interface between frozen and unfrozen dirt is small, it might be possible, if the shear hypothesis mechanism can operate, to have this thin layer of dirt carried into the ice mass along a shear plane. If a shear plane did become active in this boundary region, it could also carry debris picked up by sliding temperate ice and transported to this region.

the melting point. The shear seen by Ward may have occurred, therefore, only to those depths for which the ice is at the melting point.*

Even if further observations do show that appreciable shear does occur across debris layers, this result in itself would not be sufficient to prove the shear hypothesis. A condition necessary to the proof of the shear hypothesis would indeed have been met, but not a sufficient condition. Suppose that the debris layers were formed by a mechanism other than scraping. Then, let a layer be subjected to a stress which has a shear component such as is always found within a glacier or ice sheet. If the debris layer has the property that discrete shear takes place within it when such a stress is applied, this shear will occur regardless of how the debris layer was formed originally. A measurement of the shear across a debris layer thus yields information on a property of debris layers, but reveals nothing of their formation.

The geometry of the debris layers

Another criticism of the shear hypothesis is that it suffers from an "embarrassment of riches." If one accepts the shear hypothesis, one can account for a few debris layers. But can one really explain the very large number of debris layers that actually are observed? For example, in the ice tunnel constructed in 1959 next to Camp TUTO in the Thule area, one can see countless numbers of distinct dirt and debris layers within a distance of about 20 m. These layers often are separated from each other by distances of the order of millimeters. How could such a fine spacing be achieved by the shear model of Figure 1? One way might be for one "active" shear plane to start operation and carry debris to the surface. It then becomes "inactive" and another plane very close to it becomes active and carries a load of debris to the surface. In turn, the second plane stops and another nearby plane becomes active. An explanation along these lines would be difficult to accept since one must explain why one shear plane is active and another inactive when the planes are separated by distances of the order of only 0.01 to 1.00 m.

On the other hand, one might propose that many closely spaced shear planes commence at about the same time to carry debris to the surface. But again one is in difficulties. The sequence that would be followed in this case is shown in Figure 2. Since it must be assumed that the spacing of the planes is close, one shear band will catch up with another, as shown in Figure 2. Thus, the slope of the shear planes will increase the farther one goes back into an ice sheet. In the tunnel dug near Camp TUTO the slope of the "shear" planes actually decreases the farther one goes into the tunnel, the opposite of the behavior predicted by this explanation.

---

* McCall, Nye and Grove[24] have shown that differential ablation can lead to an apparent thrust plane such as was observed by Ward.

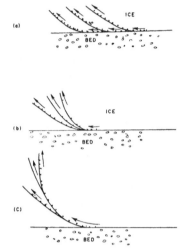

Figure 2 Sequence of events after the start of several closely spaced shear planes: (a) Planes just starting to bring up debris from the bottom; (b) Intermediate stage, the shear layers have just caught up to each other by shearing over the bed; and (c) A late stage, in which the layers have joined each other and only a single layer can exist at the bottom.

Although the shear hypothesis might account for debris planes which are separated by distances of the order of the thickness of an ice sheet (near the edge of the Thule ice lobe the separations would be of the order of 50 to 100 m.), it appears to be difficult to explain* with this mechanism the small scale separations which are actually observed. There really are too many "shear" planes present to be accounted for by the shear hypothesis.

The actual upward-curving form of an individual debris layer, which is approximately that shown in Figure 1, can be explained without the assumption that discrete shear occurs across a debris layer. It has been known

---

* Dr. H. Roethlisberger has suggested to me that the small separations might be explained if a single debris layer can become highly folded over a period of time. A multilayered debris layer then could be produced from a single debris layer. Of course, such a possibility cannot be ruled out. If this is the explanation of the field observations, the shear hypothesis must be a much more complex process than it has been pictured hitherto. It also would lose that simplicity of conception which is such a strong point in its favor. It is difficult to imagine the processes that would lead to a high degree of folding and at the same time cause the folded layers to assume the overall orientation pictured in Figure 1. Since moraines are general features of an ice sheet, this difficulty cannot be explained by local peculiarities in the shape of the ice sheet bed. A general rather than special cause will have to be invoked to explain the multilayered debris bands.

for a long time from arguments based on the principle of conservation of mass that the flow lines of ice in an ablation area must go upwards. (The shape of flow lines in glaciers and ice sheets has been discussed recently by Nielsen and Stockton.[13]) Thus, even without the shear hypothesis, it is possible to understand the general shape of the debris layers.

Freezing model

As an alternative to the shear hypothesis the mechanism illustrated in Figure 3 is proposed to explain the formation of Thule-Baffin moraines. Figure 3 shows a cross-section of an ice sheet whose edge is frozen to its bed, the actual situation existing at the Thule ice lobe and at Baffin Island. It is assumed that farther inland from the edge the bottom is at the melting point. This second region can be divided into two parts. In the furthermost inland part, the combination of geothermal heat and heat produced from sliding is greater than can be conducted down a temperature gradient in the ice. As a consequence ice is melted to water. This water is forced by the pressure gradient outward to the edge of the ice sheet. As it moves outward, it enters into a region where the temperature gradient in the ice can conduct away more heat than is produced by any sliding or comes from the geothermal heat. In this region the water refreezes to ice and rejoins the ice sheet. The bottom of the ice sheet will still be at the melting point in this region; the extra heat required to keep it at this temperature comes from the latent heat of freezing which is given up as the water freezes.

The scheme presented in Figure 3 should always occur in any ice sheet whose edge is frozen to the bed and whose bottom surface at some positions in the interior is at the melting point. Thus, if direct measurements show that in the interior of a cold ice sheet the bottom surface is at the melting

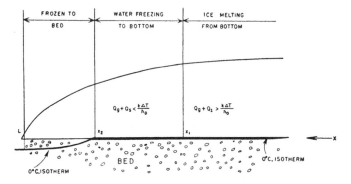

Figure 3 The freezing model. A cross-section of an ice sheet is shown. The edge of the ice sheet is frozen to the bed and the 0° C. isotherm reaches the bottom of the ice sheet inland from the edge.

point, and that no water is escaping underneath the ice sheet, one can be sure that ice is being formed at certain places at the bottom surface.

Under steady-state conditions, the model of Figure 3 probably would not lead to the inclusion of solid dirt layers into the ice sheet, since ice would merely form at the ice surface. However, no glacier or ice sheet is ever in a completely steady-state condition. The thickness, among other things, varies as a function of time. A change in thickness changes the stress acting on the bed and, hence, affects the speed of sliding and, thus, the amount of heat available to melt ice or retard the freezing of water. This change will cause a shift in the position of the border between the region where the ice sheet is frozen to the bottom and where it is at the melting point. Figure 4 shows how a cyclic shift in the amount of heat* produced by sliding will lead to a freezing-in of debris. Since the border being considered is simply the point at which the 0° C. isotherm ceases to coincide with the bottom of the ice sheet, a shift of this border inland simply causes the zero point isotherm to descend into the material upon which the ice is resting. Debris can thus be frozen onto the bottom of the ice sheet. If the border shifts towards the edge of an ice sheet, the region in which water is being frozen to the bottom of the ice sheet is extended. If debris has already been frozen to the bottom, the new ice will have to start forming underneath this frozen-on debris, for it is here that the temperature is at the melting point. The debris will therefore be incorporated into the ice sheet with ice surrounding it, as shown in Figure 4. Through numerous repetitions of the cycle, a larger number of debris layers can be incorporated into the ice of an ice sheet lying on unconsolidated material. The thickness of these layers depends on the time frequency of the cyclic change and can have any value. The length of the debris layers depends on the magnitude of the change in slope or thickness of the ice sheet, as this change controls the distance the border in Figure 4 will shift.

Once horizontal layers of debris are incorporated into the ice, they flow with the ice. If the edge of an ice sheet is an ablation zone, as is the case in the Thule area and on Baffin Island, the flow lines of ice have to come up to the surface in this region. Hence the debris layers will become exposed in the ablation area and can form Thule-Baffin moraines.

Theory

In this section are set down some formulae which will indicate the conditions under which the freezing mechanism will operate. Of course, it would

---

* Temperature changes at the upper surface also affect the amount of water being frozen or melted at the bottom surface. However, because of the extremely great length of time involved for temperature changes at the top surface to penetrate to the bottom surface, this variation would be unimportant compared to that produced by changes in the thickness or slope of an ice sheet.

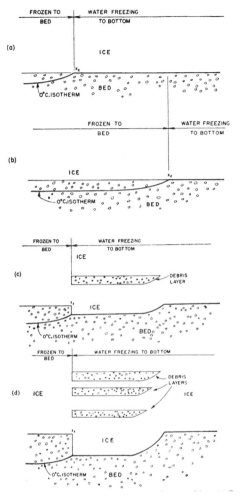

Figure 4 Illustrating how debris is incorporated into the ice sheet under non-steady state conditions: (a) The starting point. To the left of point x_2, the 0° C. isotherm descends into the debris which makes up the bed. To the right of point x_2, the bottom is at the melting point and water is being frozen to the bottom surface. Water flowing from the right permits the bottom to be kept at 0° C. but the water supply is exhausted at point x_2; (b) Less water is flowing from the right and the point of descent of the 0° C. isotherm has shifted to the right; (c) A greater supply of water now flows from the right. Water now freezes to the 0° C. isotherm position given in (b) and pushes up into the ice the debris frozen on the bottom shown in (b). Point x_2, which marks the limit of the flow of water, has moved to the left; and (d) The cycle repeated after many times with a number of debris layers fixed into the ice.

be desirable to be able to take known field data and prove conclusively that this mechanism has led to the formation of existing moraines. That hope, however, cannot be satisfied with the measurements at present available.

Consider an ice sheet whose thickness across a particular cross-section is given by $h(x)$, where x is horizontal distance. Let $a(x)$ be the accumulation or ablation at the position x; this function has a positive value for accumulation and a negative value when there is ablation.

Under steady-state conditions, the heat flowing down the temperature gradient at the bottom surface is exactly equal to the sum of the geothermal heat, Q_g, the heat produced by any sliding, Q_s, and any heat given off or absorbed by the freezing of water or melting of ice at the bottom surface. Values of the geothermal heat vary between different places on the Earth.[14] An average value listed by Bullard[14] is 39 cal. cm.$^{-2}$ yr.$^{-1}$. The heat of sliding is given by the equation

$$Q_s = \frac{V_\tau}{J}, \qquad (1)$$

where J is the mechanical equivalent of heat ($J = 4.185 \times 10^7$ erg cal.$^{-1}$), V is the velocity of sliding, and τ is the shear stress acting at the bottom surface which produces the sliding. To a good approximation, this shear stress is equal to $\rho g h\,\alpha$, where ρ is the average density of ice and α is the slope of the upper surface of the ice sheet. The velocity, V, itself also depends on shear stress. A theoretical expression[9] for the velocity is

$$V = B\tau^m, \qquad (2)$$

where m and B are constants. Reasonable values of these constants[15] are $m = 2$ and $B = 81$ m. yr.$^{-1}$ bar^{-2}. Since typical values[16, 17] of the shear stress on the bed of glaciers and ice sheets are in the range of 0.4 to 1.0 bar, equation (2) would give velocities of the order of 13 to 81 m. yr.$^{-1}$. From equation (1), therefore, one would expect the heats of sliding to be of the order of 13 to 190 cal. cm.$^{-2}$ yr.$^{-1}$. These values are about the same order of magnitude as the geothermal heat.

Under steady-state conditions, the heat flowing through the bottom interface of an ice sheet must balance the heat from the geothermal heat flow, the heat produced by sliding, and the heat given up or absorbed by any melting or freezing at the bottom. If $A(x)$ represents the thickness of ice being frozen to the bottom surface (it would be negative if the ice is melting) and $\gamma(x)$ represents the vertical temperature gradient at the bottom surface, the condition for steady-state heat flow at the bottom interface is given by the following equation:

$$k\gamma = Q_g + Q_s + SA, \qquad (3)$$

where S is the heat of fusion ($S = 72$ cal. cm.$^{-3}$ ice) and k is the coefficient of heat conductivity of ice ($k = 1.7 \times 10^5$ cal. cm.$^{-1}$ yr.$^{-1}$ ° C.$^{-1}$).

The temperature gradient at the bottom surface can be calculated if the temperatures at the top and bottom surfaces are known. This calculation has been done by Robin[18] for the case where heat flow in the horizontal direction (the x direction) can be neglected and where the longitudinal strain rate of the creep flow of ice is not a function of vertical distance. Since this longitudinal strain rate is approximately a/h, Robin was able to obtain the following equation for γ:

$$\gamma = \frac{\Delta T}{h^*}, \tag{4}$$

where ΔT is the temperature difference between the top and bottom surfaces and is considered to be positive if the bottom is warmer than the top, and h^* is a "compensated" thickness given by†

$$h^* = \int_0^h \exp\left(-\frac{cay^2}{2kh}\right) dy, \tag{5}$$

where c is the specific heat of ice ($c = 0 \cdot 45$ cal. cm.$^{-3}$). When the longitudinal strain rate is zero (and there is no melting or freezing at the bottom surface) this "compensated" thickness is equal to the actual thickness of the ice sheet.

If equation (4) is substituted into equation (3) one obtains

$$\frac{k\Delta T}{h^*} = Q_g + Q_s + SA. \tag{6}$$

If the bottom surface is below the freezing point, so that $A = Q_s = 0$, this equation will determine the bottom temperature, if the values of the temperature at the upper surface and the geothermal heat are known. On the

---

† This equation of Robin neglects any melting of ice or freezing of water at the bottom surface. If this factor is taken into account, the term inside the integral becomes

$$\exp\left\{-\frac{c}{k}\left(\frac{d}{2h}y^2 - Ay\right)\right\}.$$

The additional term in the exponent usually can be neglected if $|a|$ is much greater than $|A|$.

other hand, if the bottom surface is at the melting point, so that $A \neq 0$, equation (6) will determine the amount of melting or freezing once the temperature of the top surface and the amount of sliding are fixed.

From equations (5) and (6) the conditions can be determined under which the bottom of an ice sheet can be at the melting point. As a very simple example, suppose that the accumulation rate is zero. It can be seen from equation (6) that the bottom surface will be at the melting point when the ice thickness is equal to h_0, where

$$h_0 = k \frac{\Delta T}{Q_g + Q_s} \tag{7}$$

and ΔT is taken to be the difference in temperature between the melting point of ice and the upper surface temperature. When equation (7) is valid, the heat conducted through the ice is exactly equal to the geothermal heat plus the heat of sliding. The ice at the bottom is at the melting point, but neither ice is melting nor is water freezing there. If the ice thickness is greater than h_0, less heat can be conducted through the ice than is being supplied by the geothermal heat and the sliding heat. In this situation, ice melts off the bottom surface to make up the heat balance. If the ice thickness is less than h_0, more heat is conducted away than is being supplied. The bottom can remain at the melting point only if water is being frozen onto the bottom surface. However, if no water is available to give up its latent heat by freezing, the bottom surface has to be at a temperature lower than the melting point of ice and the ice sheet is frozen onto its bed. Thus, when the accumulation is zero, the thickness h_0 separates the condition of freezing from that of melting.

Now consider the more general case when the accumulation or ablation is not zero. Corresponding to any particular set of values for ΔT, Q_g, and Q_s, there will be a curve on a plot of a versus h which divides freezing conditions from melting conditions. This curve is determined by setting A equal to zero in equation (6). One obtains the equation

$$h_0 = \int_0^h e^{\beta y^2} dy, \tag{8}$$

where $\beta = -ca/2kh$ and h_0 is given again by equation (7). This equation can be written as

$$h_0 = h e^{\beta h^2} \left\{ 1 - \frac{(2\beta h^2)}{3} + \frac{(2\beta h^2)^2}{5 \cdot 3} - \frac{(2\beta h^2)^3}{7 \cdot 5 \cdot 3} + \frac{(2\beta h^2)^4}{9 \cdot 7 \cdot 5 \cdot 3} \cdots \right\}. \tag{9}$$

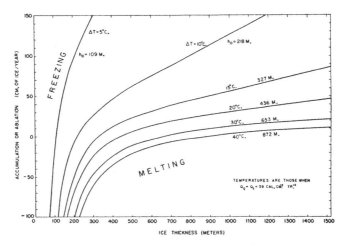

Figure 5 A plot showing freezing and melting conditions at the bottom of ice sheets. Each curve is calculated for a different value of $h_0 = k\Delta T/(Q_g + Q_s)$, where k is the thermal conductivity of ice, ΔT the difference between the melting point of ice and the upper surface temperature of an ice sheet, and Q_g and Q_s are the geothermal heat and the heat produced by the sliding of ice over the ice sheet bed. The region of the figure to the right of a particular curve corresponds to conditions where melting occurs, and the region to the left where freezing can take place.

For large values of h and a, equation (8) reduces to

$$a = \frac{h\pi k}{2ch_0^2}. \tag{10}$$

Once h_0 has been found from equation (7), equations (9) and (10) determine a curve of a versus h. Figure 5 shows such curves for various values of h_0. Also indicated in Figure 5 are the temperature differences, ΔT, which correspond to these values of h_0 when Q_g is the average geothermal heat of 39 cal. cm.$^{-2}$ yr.$^{-1}$ and Q_s is taken to be equal to Q_g. This value of the sliding heat corresponds to a shear stress of 0.58 bar and a sliding velocity of 27 m. yr.$^{-1}$ (or any other combination of stress and sliding velocity whose product is the same).

If the values of the ice thickness and accumulation or ablation in a region of an ice sheet are such that they correspond to a point lying to the *right* of the appropriate curve in Figure 5, the bottom of the ice sheet is *melting* away. If the thickness and accumulation are such that the point lies to the *left* the bottom is either frozen to the bed or water is *freezing* to the bottom. Hence, if the temperature at the upper surface and the geothermal heat and heat of sliding are known, it is possible to tell from the measurements of ice

153

thickness and rate of accumulation or ablation whether or not it is possible for ice at the bottom to be melting or freezing.

The bottom surface of an ice sheet which is located in a region corresponding to a point in the freezing zone can remain at the melting point only if there is a water supply available. The melting zone can supply this water if a hydrostatic pressure head exists to push water from one region into the other. From Figure 5 it can be seen that, other things being equal, the thicker parts of an ice sheet are more likely to be in a melting zone, if a melting zone does exist. The greater weight of ice in the melting zone will supply the necessary pressure head to move water into the freezing zone.

The principle of the conservation of mass enables one to calculate the extent of the region in the freezing zone where the bottom of an ice sheet is at the melting point of ice. That is, all the water created in the melting zone must be turned back to ice in the freezing zone if no water is to escape from underneath a cold ice sheet. The amount of ice, $A(x)$, which is either being melted from, or frozen to, the ice sheet in the regions where the ice is at the melting point is given by equation (6):

$$A(x) = \frac{1}{S}\left\{\frac{k\Delta T}{h^*} - (Q_g + Q_s)\right\}. \tag{11}$$

Consider how equation (11) can be applied to an actual problem. Suppose a two-dimensional ice sheet (center at $x = 0$ and edges at $x = +L$) has a melting zone extending over the distance from x_0 to x_1, where $x_0 < x_1 < L$. What would be the extent of the freezing zone in which the ice at the bottom of the sheet is at the melting point? In order to satisfy the principle of conservation of mass, when no water escapes from underneath the ice sheet, the following equation must be obeyed:

$$\int_{x_0}^{x_1} A(x)dx = -\int_{x_1}^{x_2} A(x)dx, \tag{12}$$

where the first integral covers the melting zone and the second the freezing zone. This equation fixes the value of x_2 ($x_2 < L$). Since ΔT (the difference between the melting point of ice and the surface temperature) and h^* (determined from the thickness of ice and the rate of accumulation) can be determined as functions of distance, once the heat of sliding and the geothermal heat are known, equations (11) and (12) determine uniquely the extent of the freezing zone at the melting point. At distances greater than x_2 the bottom of an ice sheet is below the freezing point of ice and the sheet is frozen to its bed. In this region, there is no water available to freeze onto the bottom and hence raise the ice temperature to the melting point.

Discussion

In the previous section, it has been shown that under certain conditions a part of the bottom surface of a cold ice sheet may be at the melting temperature of ice and that, in this part, ice can be melting away from the bottom in some parts and water can be freezing onto the bottom in the other parts. Of course, one would like to know if the ice sheets which contain Thule-Baffin moraines meet the special conditions that lead to the ice at the bottom surface being melted in one place and refrozen in another. Since the edges of the Barnes Ice Cap on Baffin Island and of the Thule lobe of the Greenland Ice Sheet are known to be frozen to their beds, it is only necessary to show that in some inland regions the conditions are such that the bottom is at the melting point (or more exactly, *was* at the melting point during the time the debris which forms the moraines on the edges of these ice sheets was brought into the ice). Unfortunately, from the field evidence at hand, it is not possible to prove conclusively whether or not there exist regions on either of these ice sheets where the bottom surface is at the melting point of ice.

Consider first the Barnes Ice Cap on Baffin Island. The upper surface of the Barnes Ice Cap appears to be at $-10.7°$ C. from one measurement made by Ward;[4] its thickness goes from 0 to about 450 m.;[16] its accumulation ranges from -170 to $+20$ cm. of ice yr.$^{-1}$, the value depending on the elevation of the upper surface. From these data, Figure 5 would predict that the bottom surface in the interior of the ice cap definitely is at the melting point. In order to arrive at this conclusion, consider the curve of Figure 5 marked $\Delta T = 10°$ C. This curve represents a temperature close to the actual temperature of $-10.7°$ C. Any thickness greater than 280 m. corresponds to a point in the melting region, even for the maximum rate of accumulation of 20 cm. of ice yr.$^{-1}$. One must remember, however, that the curves in Figure 5 were calculated on the assumption that the heat of sliding is equal to the geothermal heat, and it was assumed further that the geothermal heat is equal to 39 cal. cm.$^{-2}$ yr.$^{-1}$. Suppose the estimate of the heat passing through the bottom is too great by a factor of 2. Then the curve to be considered is not the one marked 10° C., but rather the curve marked 20° C. In this case, it is open to question whether any part of the Barnes Ice Cap lies in the melting region and it may well be that the entire ice cap is frozen to the bottom. The only way one could really be sure is to sink bore holes to the bottom of the ice cap and actually measure the temperature.

The same uncertain situation occurs in the Thule ice lobe. The thickness of this ice has been only partially measured. Roethlisberger[19] found a thickness of 260 m. at a distance of 5 km. inland from the edge. (The ice sheet is about 33 km. wide and a somewhat greater thickness can be expected farther inland.) Accumulation and ablation rates[20,21] depend on position and range from about -100 cm. of ice yr.$^{-1}$ close to the edge to 70 cm. yr.$^{-1}$ in the central region of the ice sheet. The temperature in the ablation zone[20] near the edge of the ice sheet is about −12° C. Temperatures have been measured[20] in the inland region at two points down to a depth of 9 m. The temperatures at 9 m. seem to be of the order of −3° to −6° C.* If these inland temperatures are representative values and if the curves of Figure 5 are valid, there is no question but that the bottom surface in the interior of the Thule ice lobe is at the melting point. On the other hand, if these temperatures are not reasonable and if other values of the heat of sliding and the geothermal heat are used, it is quite possible that the ice sheet is frozen everywhere to its bed.

Another problem connected with the determination of the temperature at the lower surface of a cold ice sheet concerns the question as to whether or not the ice is actually sliding over the bottom bed. If no sliding occurs, no heat will be produced by this mechanism. It is thus conceivable to have an ice sheet which would be frozen to its bottom surface, if the heat of sliding were not available, but whose lower surface would be at the melting point if this heat were present. Once the ice sheet started to slide, the heat produced would keep the bottom surface at the melting point and permit the sliding to continue. If the sliding were stopped, the ice sheet would refreeze to its bed and no further sliding would occur. In the absence of direct measurements, a knowledge of the past history of such an ice sheet would be necessary in order to determine whether or not the bottom is at the melting point.

Conclusion

It is concluded that there are a number of objections that can be raised against the shear hypothesis for the formation of a type of moraine found on Baffin Island and in Greenland. It is further concluded that in certain situations it is possible for ice to be formed at the bottom of an ice sheet and that this accretion of new ice can lead to a freezing-in of loose debris lying on the bed of the ice sheet. In turn, this frozen-in debris can result in the formation of moraines in the ablation areas of an ice sheet.

Although one can demonstrate the possibility that the moraines found on the Thule ice lobe in Greenland and on the Barnes Ice Cap of Baffin Island are formed by the mechanism proposed here, it is not possible from the field data at hand to prove conclusively that these moraines have been so formed.

---

* The inland ice appears to be warmer than the ice at the edge because in the ablation region the melt water produced in the Summer simply runs off the upper surface and does not soak into the solid ice surface. On the other hand, in the inland region, the melt water produced in the Summer does not run off the surface, but trickles into the firn layers below, where it refreezes and gives up its latent heat. Thus, in the Summer, the accumulation area is able to warm up more than the ablation area.

Acknowledgements

I wish to thank Mr. James Bender for the opportunity of seeing at first hand the moraines near Thule, Greenland. I am indebted to him and to Dr. Hans Roethlisberger and Dr. George Swinzow for informative discussions and arguments on the formation of these moraines.

References

1. Bishop, B.C. Shear moraines in the Thule area, northwest Greenland. *U.S. Snow, Ice and Permafrost Research Establishment. Research Report* 17, 1957.
2. Goldthwait, R.P. Formation of ice cliffs. (*In* Study of ice cliff in Nunatarssuaq, Greenland. *U.S. Snow, Ice and Permafrost Research Establishment. Technical Report* 39, 1956, p. 139–50.)
3. Nobles, L. Investigations of structures and movement of the steep ice ramp near Red Rock Lake, Nunatarssuaq, Greenland. *U.S. Snow, Ice and Permafrost Research Establishment. Report on Contract No. DA-11-190-ENG-12, 1960.*
4. Ward, W.H. The glaciological studies of the Baffin Island Expedition, 1950. Part II: The physics of deglaciation in central Baffin Island. *Journal of Glaciology*, Vol. 2, No. 11, 1952, p. 9–23.
5. Goldthwait, R.P. Development of end moraines in east-central Baffin Island. *Journal of Geology*, Vol. 59, No. 6, 1951, p. 567–77.
6. Hollin, J.T., *and* Cameron, R.L. I.G.Y. glaciological work at Wilkes Station, Antarctica. *Journal of Glaciology*, Vol. 3, No. 29, 1961, p. 833–42.
7. Chamberlin, T.C., *and* Salisbury. R.D. *Geology. Vol. I. Second edition, revised.* New York, Henry Holt and Co., 1904, ch. 5.
8. McCall, J.G. The internal structure of a cirque glacier. *Journal of Glaciology*, Vol. 2, No. 12, 1952, p. 122–30.
9. Weertman, J. On the sliding of glaciers. *Journal of Glaciology*, Vol. 3, No. 21, 1957, p. 33–38.
10. Raraty, L.E., *and* Tabor, D. The adhesion and strength properties of ice. *Proceedings of the Royal Society*, Ser. A, Vol. 245, No. 1241, 1958, p. 184–201.
11. Butkovich, T.R., *and* Landauer, J.K. A grid technique for measuring ice tunnel deformation. *Journal of Glaciology*, Vol. 3, No. 26, 1959, p. 508–11.
12. Hilty, R.E. Measurements of ice tunnel deformation, Camp Red Rock, Greenland. *U.S. Snow, Ice and Permafrost Research Establishment. Special Report* 28, 1959.
13. Nielsen, L.E., *and* Stockton, F.D. Flow patterns in glacier ice. *Journal of Applied Physics*, Vol. 27, No. 5, 1956, p. 448–53.
14. Bullard, E. The interior of the earth. (*In* Kuiper, G.P., *ed. The Earth as a planet.* Chicago, University of Chicago Press, 1954, p. 57–137.)
15. Weertman, J. Equilibrium profile of ice caps. *Journal of Glaciology*, Vol. 3, No. 30, 1961, p. 953–64.
16. Orvig, S. The glaciological studies of the Baffin Island Expedition, 1950. Part V: On the variation of the shear stress on the bed of an ice cap. *Journal of Glaciology*, Vol. 2, No. 14, 1953, p. 242–47.
17. Ward, W.H. Studies in glacier physics on the Penny Ice Cap, Baffin Island, 1953. Part IV: The flow of Highway Glacier. *Journal of Glaciology*, Vol. 2, No. 18, 1955, p. 592–98.

18. Robin, G. de Q. Ice movement and temperature distribution in glaciers and ice sheets. *Journal of Glaciology*, Vol. 2, No. 18, 1955, p. 523–32.
19. Roethlisberger, H. Seismic survey 1957, Thule area, Greenland. *U.S. Snow, Ice and Permafrost Research Establishment. Technical Report* 64, 1959.
20. Schytt, V. Glaciological investigations in the Thule Ramp area. *U.S. Snow, Ice and Permafrost Research Establishment. Report* 28, 1955.
21. Griffiths, T.M. Glaciological investigations in the TUTO area of Greenland. *U.S. Snow, Ice and Permafrost Research Establishment. Technical Report* 47, 1960.
22. Drygalski, E. von. *Grönland-Expedition der Gesellschaft für Erdkunde zu Berlin, 1891–1893.* Bd. 1. Grönlands Eis und sein Vorland. Berlin, W.H. Kuhl, 1897, p. 109.
23. Blümcke, A., *and* Finsterwalder, S. Zur Frage der Gletschererosion. *Sitzungsberichte der Kgl. Bayerischen Akademie der Wissenschaften zu München*, Math. -phys. Klasse, Bd. 20, 1890, p. 435–44.
24. Lewis, W.V., *ed. Investigations on Norwegian cirque glaciers.* London, Royal Geographical Society, 1960, p. 49. (R.G.S. Research Series, No. 4.)

8

THE FORMATION AND
SHAPE OF DRUMLINS AND
THEIR DISTRIBUTION
AND ORIENTATION IN
DRUMLIN FIELDS

I.J. Smalley and D.J. Unwin

Source: *Journal of Glaciology* 7(51) (1968): 377–90.

Abstract

If glacial till contains more than a certain minimum boulder content, it is dilatant and requires a much larger stress to initiate shear deformation than to sustain it. If the stress level at the glacier–terrain interface drops below a certain critical level, or the till reaches its critical boulder-content density, then the till beneath the glacier packs into stable obstructions. These are shaped into streamlined forms by the glacier and are found distributed at random in drumlin fields. Due to drumlin coalescence there is a normal distribution of drumlin axes about the direction of ice movement.

Introduction

The Pleistocene glaciers left many signs of their passage; some of the most remarkable are the low, smooth hills known as drumlins. Drumlins have been observed and investigated for a considerable time but to date no completely satisfactory theory has been evolved to explain their formation. Several factors appear to be involved and the problem of choosing the significant and avoiding the irrelevant is very difficult. This paper represents the development of the theory, already published in outline (Smalley, 1966[b]), that the formation of glacial-till drumlins is a consequence of the dilatancy of the material of which they are composed. Most drumlins are formed from till, a material with very complex rheological properties and this paper

159

is mainly concerned with these, but any other formation which may be called a drumlin (e.g. a rock drumlin) is also considered.

The drumlins we see now were formed in the Pleistocene and like most aspects of Pleistocene geology there is a vast literature relating to them. Literature up to the mid-1950s has been surveyed by Charlesworth (1957) in his great compendium. He does, however, tend to concentrate on an assessment of theory and opinion rather than on the collection of available drumlin facts. The best collection of drumlin-shape and size data is probably still that by Ebers (1926). There have been several notable papers since the Charlesworth survey and these are discussed in later sections of this paper.

Two questions need to be answered to provide an explanation of the observed nature of drumlins. These concern (1) the nature of the geomorphic force which shaped the drumlins, and (2) the way in which the force accomplished the shaping. The necessary geomorphic force is generally ascribed to the action of glaciers, so only one problem remains, that of the mode of interaction of the glacier and the terrain which leads to the formation of drumlins. The consequences of this interaction are low streamlined hills which tend to occur in groups or fields and within these fields they have a certain distribution of orientations and certain positional relationships.

In this paper the formation of the drumlins is explained by invoking the dilatancy mechanism, the shape by the requirements for streamline flow, and the orientation is simulated by a random model. The distribution of drumlins in a drumlin field has been investigated in two ways. Random models have been produced and the spacings of drumlins in these compared with the measured spacings of real drumlins, and drumlin-spacing data derived from maps have been analysed for indications of random spacing. Both approaches indicate that drumlins do occur at random within drumlin fields.

Formation

The relative scarcity of drumlin forms suggests that the conditions necessary for their formation were rigorous, and infrequently achieved. It is proposed that the basic conditions for the formation of glacial-till drumlins were:

(i) The glacier–terrain relationship was such that at the base of the glacier the terrain material was being continuously deformed. Some of this terrain material was carried along by the glacier so that shear deformation occurred within the terrain material.

(ii) The deformed layer was composed of a concentrated dispersion of boulders and large rock particles in a dense clay–water system, the material usually called boulder clay or glacial till. For drumlins to form, the large particles in this till layer had to form a dilatant system.

160

Dilatancy is a property of granular masses. When a granular mass, for example some dry sand, is at rest, it forms a stable heap and the particles in the heap are relatively closely packed together. When the granular material is deformed, it expands; this is the phenomenon of dilatancy, first observed by Reynolds (1885), related to geology by Mead (1925), and fully investigated by Andrade and Fox (1949). There is no completely satisfactory definition of dilatancy. Boswell ([1961], p. 73) stated that dilatant systems are those in which the anomalous viscosity increases with increase of shear. This is just an elaborate way of saying that dilatant materials are more resistant to shear stresses than might be expected. It is the high resistance to initial deformation of till which leads to the formation of drumlins.

The dilatancy of granular masses under compressive and shear stresses has been demonstrated using the simple apparatus illustrated in Figure 1a.

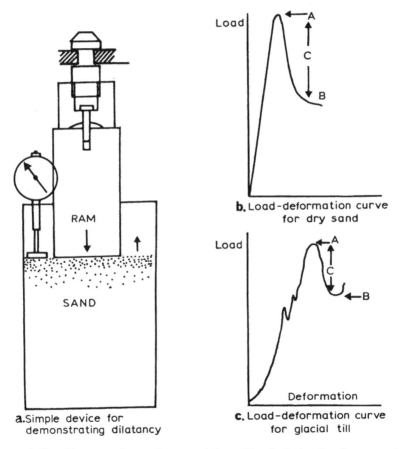

a. Simple device for demonstrating dilatancy

b. Load-deformation curve for dry sand

c. Load-deformation curve for glacial till

Figure 1 The dilatancy of granular materials. a. Simple device for demonstrating dilatancy; b. Load-deformation curve for dry sand; c. Load-deformation curve for a glacial till.

A cylindrical ram of diameter 7.5 cm is forced down at a constant rate by a hydraulic press into a cylindrical container of diameter 15 cm which contains dry sand of particle size about 0.5 mm. A graph showing load vs deformation (i.e. ram travel) is produced automatically by the machine recorder as shown in Figure 1b. When the load is first applied, the sand aggregate starts to expand because its natural close packing is being disturbed and a more open packing is developing. The material continues to expand and to resist deformation until it reaches the state of maximum expansion, at point A in Figure 1b. Further deformation causes a collapse and the required deforming load drops; point B on the graph is reached and further deformation only requires loads of about this magnitude.

The curve shown in Figure 1b is an *ideal* example produced with dry sand as the granular medium. A curve of similar form is produced when dilatant glacial till is deformed from rest. The constraints produced by the container are more noticeable when the till is deformed and because of these constraints quantitative measurements were not attempted; unfortunately an uncomfortably large container would be required to eliminate the constraints. The very large particles in the till cause irregular deformation but the general shape of the load-deformation curve (Fig. 1c) is similar to the ideal model case shown in Figure 1b.

In the suggested drumlin-forming mechanism the glacial till is being continuously deformed by the movement of the glacier and a stress level in the general range indicated by C in Figure 1b and c is involved. Within the thin deformed layer of till there is a certain variation in stress level. If the stresses drop below level B then the expanded material collapses into the static stable form and there are no stresses of magnitude A available to cause sufficient dilation to get the compacted material moving again so the flowing till flows around it, shaping it so that it causes the minimum of disturbance in the flowing stream of till.

If the glacier–terrain relationship is such that the general stress level is greater than A, then no drumlins can form; the glacier sweeps all before it. If the general stress level is below B, no drumlins can form because continuous deformation of the till is impossible. In most large continental glacier systems one would expect the mean stress level to be greater than A so that most tracts of glaciated land have no drumlins. Towards the periphery of the ice sheets the stress levels drop and there is a region in which the stress levels are such that drumlins form and farther towards the periphery the stress levels drop too low to allow drumlin formation and end-moraine structures form. The drumlins form when the general stress level beneath the ice sheet is in the region represented by range C in Figure 1b and c, i.e. below A but above B; this is illustrated diagrammatically in the ice-sheet section shown in Figure 2. Fairly rapid thinning might be expected to occur at the edges of large ice sheets, giving rise to a relatively narrow drumlin belt. Small local glaciers, such as covered the northern half of Ireland during

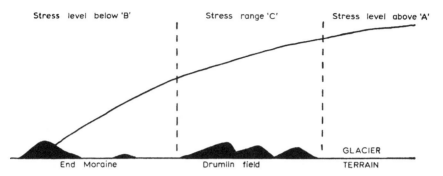

Figure 2 Cross-section at the edge of the ice sheet with critical stress regions indicated.

the late Weichsel glaciation, mostly produced stress levels in the C range and thus very large and extensive drumlin fields exist in Ireland.

It is the large rock fragments in glacial till which make it dilatant; it may be that the clay part of the material is thixotropic and thus aids the shaping process. When part of the till layer packs into an obstruction, a local high-pressure zone is formed as the rest of the till layer flows past the obstruction. Under the influence of this local high pressure the clay part of the till becomes more fluid. The more fluid clay flows more easily around the obstruction carrying the large particles with it and giving the obstruction a smooth streamlined shape. The phenomenon of thixotropy is even more difficult to define satisfactorily than that of dilatancy. The two phenomena are effectively opposites in that thixotropic materials are less resistant to shear stresses than might be expected. A well-known example of a thixotropic material is non-drip paint; in the can this appears almost solid but under the pressure of the brush it flows quite smoothly and easily. In the same way the till clay flows more easily at the drumlin–till flow interface and thus facilitates the formation of a smooth streamlined shape.

Thus the model for the drumlin-forming mechanism depends on two postulates: that the glacier is separated from the terrain by a continuously deformed layer of till, and that within this till layer there is a range of stress levels. The drumlins formed are basically accretional forms, although the material is gathered by local erosion; this falls in fairly neatly with the views of Thornbury ([1954], p. 391). He described drumlins as characteristically lying several miles to the back of end moraines and suggested that their streamlined forms indicated lodgement of till, with subsequent over-riding and reshaping by the ice. He also suggested that drumlin development just behind end moraines may possibly be related to rapid thinning of the ice in this zone; this is the state of affairs illustrated in Figure 2.

If the continuously deformed till-layer postulate is discarded (the action being justified by reference to Occam's razor), a related erosional mechanism suggests itself. It can be imagined operating in Ireland where the late

Weichselian glaciers advanced over morainic material resulting from an earlier glacial phase. Within this earlier till there was a certain distribution of relatively large rock fragments, in some places more tightly packed together than in others. The glacier advanced with stress level c in operation at the working interface and, while the till with the loose packing of boulders was smeared easily across the landscape, the till patches with the high boulder content were obstinately dilatant. For the closely packed parts to be eroded they had first to expand but the weight of the glacier pressing down prevented them from doing so. The glacier provides the eroding force but, paradoxically, this very force prevents erosion. The obstruction, *very* closely packed by the passage of some of the ice, was sculptured into the most convenient shape as the glacier forged on. The most convenient shape was the one which caused the least interference with the flow pattern; this was a streamlined shape (Chorley, 1959) and thus drumlins achieved their characteristic form.

Drumlins formed from older, stratified drift material are presumably formed by this simple mechanism and drumlins formed from the latest drift are formed by the more complex method. Actually, both can operate at the same time, and must in fact necessarily do so. There is, as Gravenor (1953) noted, both an erosional and a depositional aspect to the formation of drumlins. Aronow (1959) suggested that to explain the phenomenon of drumlin formation recourse must be had to "something" in the now vanished ice. He was only very slightly off target, and it is suggested that this "something" was the dilatancy of the till.

Flint ([1957], p. 68) stated that ". . . there is a complete gradation, independent of outward form and within a single field, from rock to drift. This suggests that any one group was molded contemporaneously under a single set of conditions". Within the various forms there will be basic material requirements; solid obstructions will require stresses far above the A level in order to be eroded away completely, these gain an accretionary smoothing of clay to give streamline flow contours. The drumlins with high boulder contents also represent zones of resistance because the material was dilatant and required stresses consistently at the A level for complete erosion. As the rock content decreases a minimum point is reached where the mean c level stresses at the glacier–terrain interface cause erosion so effectively that no drumlins can form. Thus drumlins can have assorted contents, provided the minimum rock or boulder content is reached.

It will be seen from Figure 2 that due to the rapid thinning at the periphery of the ice sheet the critical stress range at the A level is passed through much more slowly than the critical stress at the B level. In other words, the area where the drumlin-forming mean stress level c is operating is bounded at the up-stream side by a fairly diffuse boundary and on the downstream side by a more precise boundary. If a drumlin field could be examined in its entirety, it should exhibit a concentration of drumlins near the fringe of the ice sheet, behind the end moraine, and the frequency of drumlins should decrease in the parts of the field which were farther under the ice. When the general stress level drops below the A range, drumlins may begin to form but the general stress level is too high for large-scale formation; an occasional drumlin forms, possibly with an elongated shape due to the relatively high ice pressure. The stress level drops as the ice thins and more drumlins form. As the edge of the ice sheet is approached the ice thins rapidly and the stress level drops rapidly below the critical B range. At this level the glacier is still an efficient debris transporter and the carried till load is eventually dumped as end moraine.

Shape

Two methods have been advocated to describe the shape of drumlins. Chorley (1959) has proposed that the plan form of the drumlin can be described by a polar equation of the form $\rho = l \cos k\theta$. In this equation l is the length of the drumlin, k is a dimensionless number which effectively indicates the width of the drumlin, ρ and θ are the two variables. Reed and others (1962) preferred to use an equation of the form

$$(x^2/a) + (y^2/b) + (z^2/c) = 1.$$

This produces an ellipsoid with its centre at the origin; a, b and c are the semi-axes, and x, y and z are the rectangular coordinate directions.

Each method has advantages and disadvantages, but the Chorley method appears to be the more useful. The disadvantages of the Chorley method are that it only operates in two dimensions and the equation is in polar coordinate form which is perhaps more difficult to manipulate than the more common rectangular coordinate system which is used by Reed and others. The great advantage of the Chorley method is that the value k serves to classify the shapes of drumlins; if the value of k is known then the shape of the drumlin is known. The other desirable thing about the Chorley equation is that it is obviously related to the mode of formation of the body it describes. It is accepted that drumlins are a consequence of glacier flow and flowing systems flow most easily around a streamlined object. Chorley's equation is that for a streamlined form; actually it is a slight simplification but in terms of goodness of fit and simplicity it is the best available.

Reed and others claimed no particular advantage for their representational method but it appears to have two: it gives a three-dimensional model and it works in rectangular coordinates. Actually, these advantages are relatively trivial and are completely overshadowed by the method's overwhelming disadvantage—that it lacks the physical meaning which Chorley's model has. Reed and others decided that drumlins have ellipsoidal shapes and stated the equation for an ellipsoid without really considering the consequences or implications of either the observation or the equation.

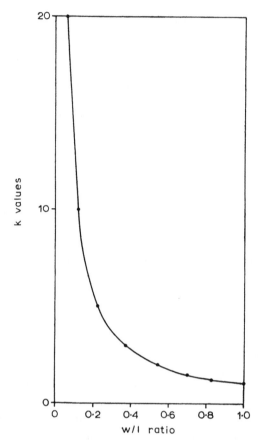

Figure 3 Variation of Chorley value *k* with width/length ratio for lemniscate-type curve.

The value *k* has been calculated by Chorley (1959) for some drumlins described by Alden (1905). He gave values for 23 drumlins; the mean value was just under 6 and the mode value was between 3 and 4. If comparisons are to be made with these pioneering values of Chorley, then it is necessary to facilitate the production of *k* values for observed drumlins. Figure 3 is a nomogram devised to convert the width–length ratio (*w/l*) into the *k* value. The curve of $\rho = l \cos k\theta$ was drawn for different values of *k* and the widths were measured, the length *l* remained constant. The *w/l* ratio was then plotted against *k* to produce the graph shown in Figure 3. The *w/l* ratio is preferred to the *l/w* ratio because the limits are 0–1 (real) rather than 1–∞ (unreal).

Charlesworth (1957, p. 389) has given some data for average dimensions of drumlins, indicating that the mean *w/l* ratio is of the order of 0.40; this

166

gives a k value of just under 3. Some more recent observations by Heidenreich (1964) include values of l and w for Canadian drumlins. He found that in the four drumlin fields investigated w/l ratios decreased fairly uniformly until a critical width was reached, after which the drumlins could increase in length while width remained constant. Drumlins appeared to be divisible into two distinct populations, those of k value around 3–4 and those of high k values which usually had reasonably constant width within the same field. Heidenreich stated that until constant width was reached, width and length increased fairly constantly at a ratio of 0.37 which gives a k value of 3. It appears that a k value of about 3 is the most usual.

Distribution

If either the accretionary or the erosional or, as seems most likely, both, mechanisms are working to produce drumlins their distribution might be expected to be completely random, or in other words there is no reason to expect a non-random distribution. Within the boundary conditions for drumlin formation the most important variable is probably the variation of properties in the available glacial till and these can be expected to vary randomly. The distribution of drumlins has been investigated in two ways. A random model has been devised and the distribution of drumlins in the model field has been compared with the observations of Reed and others (1962) and Vernon (1966), and drumlins in real fields have been subjected to nearest-neighbour analysis.

Random-placement model

The random process used to produce the model was devised initially to give one-dimensional random packings (Smalley, 1962) for subsequent comparison with three-dimensional packings of sedimentary particles (Smalley, 1964). To produce a model of a drumlin field, the method must function adequately in two dimensions; it has been applied successfully in two dimensions in producing models of crack systems in lava flows (Smalley, 1966[a]).

The terrain is represented by a square 100×100 frame (the numbers have arbitrary units). The glacier flows over this chosen piece of terrain and drumlins are formed within the square demarcation. The drumlins are placed at random, the points at which they occur being indicated by coordinates taken from a set of random-number tables. A very suitable set of tables is that by Kendall and Smith (1951); this gives lists of two-digit random numbers between 00 and 99 which are ideal when taken in pairs to represent positions within the square field. Two adjacent sides of the square frame are considered as axes, the first two-digit number of a pair represents the x coordinate, the second number the y coordinate. The plotted point marks the stoss end of the drumlin. The number of points plotted depends on the density of drumlins in the field; the mode density according to Charlesworth (1957, p. 389) is 3 per square mile. Thus if the side of the square frame represents 2 miles, 12 random points are needed.

In this model some overlaps will probably occur, giving rise to some rather odd shapes and this also appears to occur in nature. It is possible, for comparison, to produce a model in which overlaps do not occur. Each drumlin is drawn in as soon as its position is known and any subsequent drumlins which would overlap are rejected.

To produce a model field containing 12 drumlins the following procedure would be adopted: 12 random points are plotted in the square field and a drumlin is drawn, using a template, for each point. Some of these drumlin shapes will overlap giving several larger, more complex drumlins. Each of these larger drumlins counts as one so more coordinates are plotted until 12 distinguishably separate drumlins are formed; this is the model distribution and spacing measurements can be made and compared with real fields. Reed and others (1962) made perpendicular and parallel measurements which are difficult to define and make. The Vernon (1966) direct measurements are capable of a more rigorous definition; every drumlin has two measurements associated with it. These are the distances of the two nearest drumlins in an up-stream direction, one on each side of the long axis of the reference drumlin.

In order to produce a comprehensive set of measurements of the drumlin in the random-model field special boundary conditions have to be introduced for the field otherwise the drumlins near the edge have no adjacent drumlin. The edge effect is eliminated by introducing edge drumlins into special adjacent fields. If a drumlin is placed at 05 51 by the random number coordinate then one is also placed at 105 51; similarly, one at 62 02 has a corresponding placement at 62 102. This has the effect of producing a closed field so that every drumlin has the required neighbours. Overlaps at the top are introduced at the bottom and so on. A sample field is shown in Figure 4; the measurements taken are indicated. Four closed fields were produced, each containing 24 drumlins. Each field yielded 48 measurements and these were amalgamated to produce the histogram shown in Figure 5; this should be compared with the histogram produced by Vernon (1966) from similar measurements on the Ards Peninsula drumlins, a distinct similarity is apparent.

Nearest-neighbour analysis

It is possible to test the fit between a natural field and a random model without recourse to direct simulation since the "nearest-neighbour" statistic tests the manner and degree to which the distribution of individuals in a population in a given area departs from that of a random distribution. The

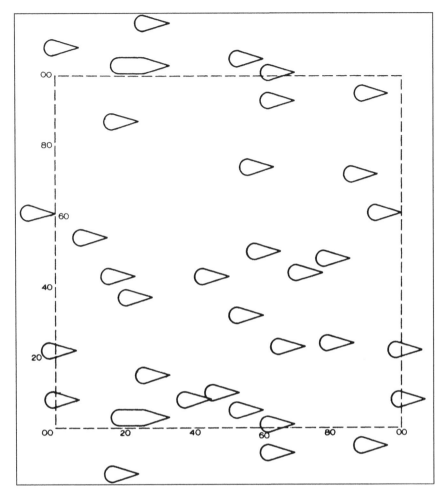

Figure 4 Drumlin-field model produced by random placement of drumlins in a square field; edge of field represents a length of 4,572 m, each drumlin is 457.2 m long.

statistic, which was devised initially for use in the biological sciences by Clark and Evans (1954) and has since been used in settlement geography (Hagget, [1965]), is defined as:

$$R = D_{obs}/0.5(A/N)^{-1/2}$$

where D_{obs} is the linear distance between any one point in a specified area A and the nearest neighbouring point, and N is the number of points within the area. The statistic is such that values of R range from zero for maximum

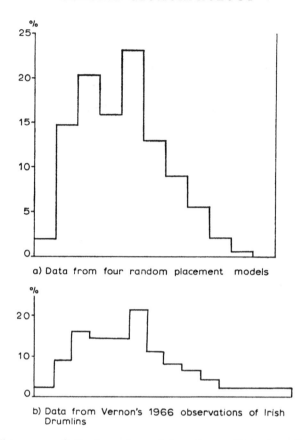

a) Data from four random placement models

b) Data from Vernon's 1966 observations of Irish Drumlins

Figure 5 Histograms of direct spacings of drumlins. a. Data from four random-placement models; b. Data from Vernon (1966), observations of Irish drumlins.

aggregation to 2.1491 for maximum (hexagonal) spacing. Random distributions give values of unity. However, it is important to note, as Clark and Evans pointed out, that a random distribution in this sense is defined as one in which any point has the same chance of occurring on any sub-area as any other point, and that any sub-area of specified size has the same chance of receiving a point as any other sub-area of that size. So defined, randomness is a spatial concept, depending upon the boundaries chosen. Thus a distribution may be random with respect to one area but non-random with respect to a larger area which includes it.

The method can be easily applied to a drumlin field. Natural patterns are taken off either aerial photographs or maps of any scale consistent with identification of the features and the nearest-neighbour distances are found for a specified number of individuals within a sub-area. This is to eliminate

Table 1 Nearest-neighbour Analyses for Four Natural Drumlin Patterns.

Number	Location	Area km²	Density number/km²	Number	D_{obs} m	R
1	Co. Clare, Ireland	24.08	1.33	32	563.9	1.3096
2	Co. Clare, Ireland	24.08	1.53	37	554.2	1.3842
3	Co. Clare, Ireland	56.16	1.62	91	497.9	1.2784
4	Vale of Eden, England	40.04	0.92	37	583.7	1.1295

side effects; a drumlin's nearest neighbour may fall outside the area *A* chosen. The distances are summed and a mean value D_{obs} calculated. Knowing the area, values of *R* can be calculated directly.

Table I presents the results of such an analysis performed using 1 : 25,000 topographic maps for small areas in Ireland, and one in England. Recognition of drumlins on the maps was done on the basis of contour pattern, drumlins being recognized where ellipsoidal patterns occur. This may result in the recognition of some forms which are not drumlins, but it is thought that the bias introduced is not great. Figure 6 shows the field which was used in England.

As the table shows, the *R* values derived for natural patterns range from 1.1295 to 1.3842, indicating that the distribution lies somewhere between "uniformly spaced" and "random", as the figures are usually interpreted. It should be noted, however, that the random-placement models gave *R* values of 1.1500, 1.0441, 1.2766 and 1.1367, indicating that the real distributions are actually truly random.

In general, both methods used indicate that the drumlin fields examined contain a random distribution of drumlins.

Orientation

If a closely packed drumlin field is produced by the random-placement method, there will be considerable overlapping. Figure 7A shows a close packing of drumlins in which the outlines have been slightly modified to produce coalescence and a smooth outline for the complex drumlins. This produces some drumlins of a more complicated shape and having orientations which do not lie exactly along the line of glacier travel. The direction of glacier travel is chosen at random after a suitable number of location points have been plotted. In Figure 7A the apparent drumlin axes are marked; the orientations have been measured and the results are shown as a histogram in Figure 7B. A normal symmetrical distribution of orientations is produced; this agrees exactly with the orientation measurements on real drumlins reported by Reed and others (1962).

171

Figure 6 Drumlin distribution in the Vale of Eden.

Discussion

Initially, the rigorous nature of the postulated boundary conditions suggests that drumlins should only rarely be observed in nature. Most authors who have written on the subject have agreed that this is the case. Fairchild (see Alden, 1911, p. 734) spoke of "The combination of several factors which do

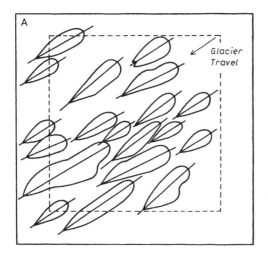
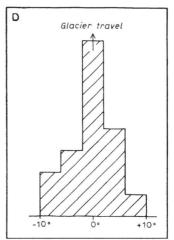

Figure 7 Orientation of drumlins. A. Random-placement orientation model; B. Histogram showing normal distribution of orientations about chosen glacier-flow direction.

not commonly occur in nature", whilst Aronow (1959) considered drumlins to be related to "unknown conditions" which are not simply related to either the terrain or to the nature of the available materials. The dilatancy mechanism suggested requires both that the material be dilatant and that the stresses within the glacier system are within a certain critical range, and it is suggested that these are the "unknown conditions".

Available evidence indicates that drumlin fields are to be found in bands paralleling, and some distance to the rear of, end moraines. Vernon (1966) has noted that the drumlin fields of eastern Ireland are bounded to the south by the Carlingford Re-advance moraine, whilst Alden (1911) showed that in eastern Wisconsin the drumlin fields are confined to a zone within 30 to 50 miles (48 to 80 km) of the limits of an ice lobe where the ice was radiating towards its margin. He suggested that the height of ice cover over the drumlin field varied from 1,450 ft (442 m) over the initial part to 450–830 ft (137–253 m) where drumlins cease to be found, some 5 miles (8 km) from the limit of the advance. Taylor (1931) noted that drumlins occur in wide basins or lowland areas such as some of the Great Lakes basins and the lowlands of Ireland, Scotland and Scandinavia, suggesting that either dying or slowly moving ice is an essential pre-requisite to their formatiom. Similarly, Aronow (1959) has noted that there is a continuous gradation from perfect drumlin forms through drift patches to true end moraine; this is a critical observation since this is exactly what would be expected if the dilatancy mechanism were responsible for drumlin formation. As the stress level falls below the lower critical level B, drumlins cease to form and end-moraine material may begin to be deposited.

The observed distribution of drumlins relative to each other is as would be expected from the dilatancy mechanism. As Flint ([1957]) pointed out, single isolated drumlins are rare, possibly indicating that certain boundary conditions are involved and that a definable stress range is required for drumlin formation. This tends to occur over a relatively wide area when it does occur, although within a large ice sheet covering a large area of suitable terrain the occasional anomaly may occur to produce an isolated drumlin.

It has been the usual practice to divide theories of drumlin formation somewhat arbitrarily into erosional and accretional ones. Aronow (1959) placed in the erosional group suggestions by Gravenor (1953) and Flint ([1957]), and in the accretional group theories by Thwaites (1957), Thornbury ([1954]) and Charlesworth (1957). Reed and others (1962) observed the traditional dichotomy but also suggested that both erosion and deposition must be involved.

Few, if any, of the so-called theories are actually theories, if a theory is required to account for the formation of the streamlined hill of till or rock. Most "theories" are really suggestions for boundary conditions within which some unspecified process operates, and the stipulated conditions may be remarkably imprecise. They are perhaps none the worse for that and it is dangerous to be too precise with very little data to base judgements on and, although much has been written about drumlins, very little hard fact has emerged. Only recently, in accordance with the general trend in geomorphology, has the quantitative investigation of drumlins and drumlin fields been undertaken. Chorley (1959) has given a meaningful interpretation of the shape of drumlins; Reed and others (1962) have measured distributions and orientations, and Vernon (1966) has measured spacings and distribution; these three papers represent the basis of the new approach to the problem of drumlin formation.

Vernon (1966) has suggested that drumlins form when ice flow induces a pressure differential between the front and back of an obstacle. Pressure melting at the front of the obstacle creates a zone of greater mobility in the ice which moves to the zone of low pressure behind the obstacle, leaving the debris behind as till; presumably in front of the obstacle. The idea of pressure melting is attractive and obviously feasible but there are objections to incorporating it into the drumlin-forming mechanism. If the glacier is moving enough till to subsequently form a drumlin or an end-moraine deposit, the bedrock obstacle, unless it is drumlin-sized itself, will tend to be insulated from the actual ice by the till load. Also, the fact that pressure melting may be involved does not throw any light on the problem of drumlin distribution.

Gravenor (1953) has listed ten conditions which must be satisfied by a theory of drumlin formation and these can be summarized as follows:

1. Drumlins may consist of a variety of materials.
2. They may have layers of stratified materials which may be faulted or folded.
3. Rock and till drumlins have the same shapes and occur in the same fields.
4. Many glaciated areas do not have drumlins.
5. They occur in fields which are wider than most moraines and they rarely occur singly.
6. They have a streamlined shape with the blunt end pointing up-stream.
7. Lamination may be present.
8. Some drumlins may have cores but most do not.
9. Drumlins are found behind end moraines.
10. They are aligned parallel to the ice-flow direction.

It is suggested that the dilatancy theory provides a mechanism which satisfies these conditions, and especially those relating to distribution (numbers 4, 5 and 9). Perhaps one further condition could be added to Gravenor's ten:

11. They are formed beneath temperate glaciers.

The suggestion by Nye (1965) that temperate glaciers slip on their bed is to some extent supported by the observation of MacNeill (1965) that there appeared to be free water present when some of the drumlins in south-western Nova Scotia were formed. A glacier whose bottom ice is below the melting point probably does not slip on its bed and thus cannot form drumlins.

Conclusions

A comprehensive theory of drumlin formation can be derived from two basic precepts: (a) glacial-till drumlins are formed from dilatant material, and (b) they were formed when the stresses in the till–glacier interface zone were within certain critical limits. There must be deformation of the actual till, which is the dilatant material. The bulk of transported till may be carried within the ice sheet; for drumlins to form there must be shear deformation in till at the glacier–terrain interface. It may be that this requires the ice to be advancing over already deposited till.

References

Alden, W.C. 1905. The drumlins of south-eastern Wisconsin. *U.S. Geological Survey. Bulletin* No. 273, p. 9–46.

Alden, W.C. 1911. Radiation of glacial flow as a factor in drumlin formation. *Bulletin of the Geological Society of America*, Vol. 22, p. 733–34. [Abstract. Discussion by H. Fairchild, p. 734.]

Andrade, E.N. da C., *and* Fox, J.W. 1949. The mechanism of dilatancy. *Proceedings of the Physical Society*, Sect. B, Vol. 62, Pt. 8, p. 483–500.

Aronow, S. 1959. Drumlins and related streamline features in the Warwick–Tokio area, North Dakota. *American Journal of Science*, Vol. 257, No. 3, p. 191–203.

Boswell, P.G.H. [1961.] *Muddy sediments.* Cambridge, W. Heffer and Sons, Ltd.

Charlesworth, J.K. 1957. *The Quaternary era, with special reference to its glaciation.* London, Edward Arnold. 2 vols.

Chorley, R.J. 1959. The shape of drumlins. *Journal of Glaciology*, Vol. 3, No. 25, p. 339–44.

Clark, P.J., *and* Evans, F.C. 1954. Distance to nearest neighbour as a measure of spatial relationships in populations. *Ecology*, Vol. 35, No. 4, p. 445–53.

Ebers, E. 1926. Die bisherigen Ergebnisse der Drumlinforschung. Eine Monographie der Drumlins. *Neues Jahrbuch für Mineralogie, Geologie und Paläontologie. Beilagebände*, 53, Abt. B, p. 153–270.

Fairchild, H. 1911. Discussion. [See Alden, 1911, p. 734.]

Flint, R.F. [1957.] *Glacial and Pleistocene geology.* New York, John Wiley and Sons, Inc.

Gravenor, C.P. 1953. The origin of drumlins. *American Journal of Science*, Vol. 251, No. 9, p. 670–81.

Hagget, P. [1965.] *Locational analysis in human geography.* London, Edward Arnold.

Heidenreich, C. 1964. Some observations on the shape of drumlins. *Canadian Geographer*, Vol. 8, No. 2, p. 101–07.

Kendall, M.G., *and* Smith, B.B. 1951. *Tables of random sampling numbers.* Cambridge, University Press. (Tracts for Computers, No. 24.)

MacNeill, R.H. 1965. Variation in content of some drumlins and tills in southwestern Nova Scotia. *Maritime Sediments*, Vol. 1, No. 3, p. 16–19.

Mead, W.J. 1925. The geologic rôle of dilatancy. *Journal of Geology*, Vol. 33, No. 7, p. 685–98.

Nye, J.F. 1965. The flow of a glacier in a channel of rectangular, elliptic or parabolic cross-section. *Journal of Glacioloy*, Vol. 5, No. 41, p. 661–90.

Reed, B., *and others.* 1962. Some aspects of drumlin geometry, by B. Reed, C.J. Galvin, Jr., and J.P. Miller. *American Journal of Science*, Vol. 260, No. 3, p. 200–10.

Reynolds, O. 1885. On the dilatancy of media composed of rigid particles in contact. *Philosophical Magazine*, Fifth Ser., Vol. 20, No. 127, p. 469–81.

Smalley, I.J. 1962. Packing of equal O-spheres. *Nature*, Vol. 194, No. 4835, p. 1271.

Smalley, I.J. 1964. Representation of packing in a clastic sediment. *American Journal of Science*, Vol. 262, No. 2, p. 242–48.

Smalley, I.J. 1966[a]. Contraction crack networks in basalt flows. *Geological Magazine*, Vol. 103, No. 2, p. 110–14.

Smalley, I.J. 1966[b]. Drumlin formation: a rheological model. *Science*, Vol. 151, No. 3716, p. 1379–80.

Taylor, F.B. 1931. Distribution of drumlins and its bearing on their origin. *Bulletin of the Geological Society of America*, Vol. 42, No. 1, p. 201. [Abstract.]

Thornbury, W.D. [1954.] *Principles of geomorphology.* New York, John Wiley and Sons, Inc.

Thwaites, F.T. 1957. *Outline of glacial geology. Revised edition.* Ann Arbor, Edwards Brothers Inc.

Vernon, P. 1966. Drumlins and Pleistocene ice flow over the Ards Peninsula/ Strangford Lough area, County Down, Ireland. *Journal of Glaciology*, Vol. 6, No. 45, p. 401–09.

176

9

MORAINES, SANDAR, KAMES AND ESKERS NEAR BREIDAMERKURJÖKULL, ICELAND

R.J. Price

Source: *Transactions of the Institute of British Geographers* 46 (1969): 17–43.

The area of study is located in south-east Iceland (Fig. 1) approximately 80 km west of Hofn. On the southern side of Vatnajökull, glaciers descend from the plateau ice almost to sea level. One of these glaciers, Breiðamerkurjökull, and its proglacial area has been the subject of detailed study by members of the Department of Geography, University of Glasgow, during the period 1964 to 1968. This paper is concerned with a relatively small part of the proglacial area of Breiðamerkurjökull (3.5 × 2 km) which exhibits a very complex sequence of landforms. These landforms have all developed since 1890 and are associated with the very rapid wastage of Breiðamerkurjökull.

Breiðamerkurjökull can be regarded as a valley glacier some 20 km in length which is 10 km wide at a distance of 1 km from the ice front (in 1965) but which broadens out into a lobe towards its terminus (Fig. 1). This glacier originates in the extensive accumulation area of Vatnajökull and is fed by several distinct tributaries. The confining walls of Breiðamerkurjökull are high ridges (600–800 m) of extrusive volcanic rocks, mainly basaltic lavas.

The proglacial area of Breiðamerkurjökull consists of two parts. The area beyond the limit of the last readvance of the glacier (that is, beyond the limit of the outermost, 1890, moraine) consists of sandar (outwash plains)[1] which descend from 30 m at their proximal margins to sea-level. The surfaces of the sandar are crossed by thousands of abandoned stream channels providing a local relief of 0.3 to 5 m. Inside the 1890 moraine, the area consists of till plains, moraine ridges, small sandar, eskers, kame and kettle

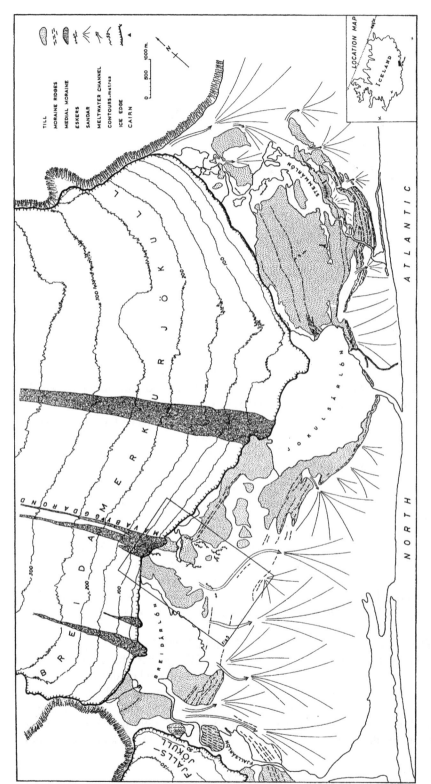

Figure 1 Breiðamerkurjökull and its proglacial area (based on photogrammetric maps produced by R. Welch and P. Howarth).

topography, lake basins and meltwater channels. This paper is concerned with only one small part of the area inside the 1890 moraine.

The area was first studied by examining, stereoscopically, pairs of aerial photographs (1:15,000). Parts of these aerial photographs were then enlarged to 1:7,500 and were used in the field as base maps. The information obtained in the field was eventually transferred to a contour base map, constructed photogrammetrically by R. Welch. Pits were dug at several localities and samples of the sediments constituting the various landforms collected and examined. The orientation of stones in both the ground moraine and in moraine ridges was also measured. Measurements of ablation and horizontal retreat of the ice edge were also obtained during the summer of 1966. A large-scale map (1:4,000) of an esker was produced by tachyometric methods.

History of deglacierization since 1890

It is believed that, for at least a century prior to 1890, Breiðamerkurjökull was advancing. Since 1890 there has been rapid and continuous retreat (Fig. 2). The positions of the ice margin in 1890 and 1937 are based on information provided by F. Björnsson of Kvisker. The 1937 position is verified by 1:100,000 maps produced by the Geodetic Institute of Copenhagen based on a survey carried out in 1935. The positions of the ice margins in 1946, 1961 and 1965 were obtained from aerial photography. The 1951 ice-margin position was obtained from a map produced by the Durham University Exploration Society.

The average rate of retreat of the ice front in the area studied in this paper, between 1890 and 1937, was 12.8 m/year. Between 1937 and 1945 the ice front retreated at a rate of 93.7 m/year. The record of frontal retreat for the period 1945–65 is much more accurate than the record for the period 1890–1945 because of the availability of aerial photographs. The annual rate of frontal retreat for the latter period ranged between 60 and 75 m depending on the location. From these figures it can be seen that the rate of frontal retreat was very slow between 1890 and 1937, very rapid between 1937 and 1945 and continuous but steady from 1945 to 1965. The frontal retreat was of course accompanied by a general thinning of the glacier. The rate of down-wastage on the ice surface in the frontal zone for the period 1945 to 1965 ranged between 5.0 and 7.5 m/year (Welch, 1967).

Both the rates of frontal retreat and down wastage have been affected by the slope of the ice surface and the slope of the underlying ground surface. Since the late 1950s, the front of Breiðamerkurjökull, in the area of study, has been retreating across a reverse slope. This means that the horizontal equivalent of the distance through which the ice front has retreated is much less (Fig. 3). Ablation rates and frontal retreat were measured during the summer of 1966. Four stakes were drilled into the ice west of Mavabyggdarond (Fig. 4). The ice stakes were placed approximately 100 m apart, the lowest

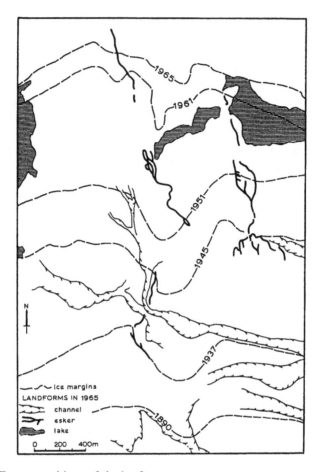

Figure 2 Former positions of the ice-front.

stake being 100 m from the ice edge on 3 July 1966. The total ablation recorded at each stake over the period 3 July–21 August was as follows:

Stake 1 (top)	3.98 m
2	4.05 m
3	4.21 m
4	4.12 m

These measurements indicate an average daily rate of ablation of 8 cm/day.

The total weekly ablation at all four stakes fell continuously from the first week (3–10 July) to the fourth week (ending 31 July). During the fifth week, ablation amounts increased to a total greater than that recorded during the first week. During the sixth week, the amount of ablation decreased, only to

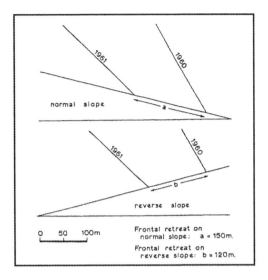

Figure 3 The retreat of an ice margin across normal and reverse slopes.

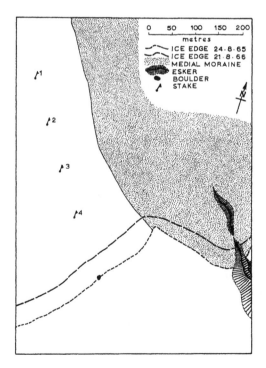

Figure 4 The location of stakes drilled into the ice for the purpose of measuring ice movement and ablation.

rise slightly during the last week of observations. The summer of 1966 was a very dry one. Rain was only recorded on 15 days out of 45. Apart from two periods (3–6 August and 17–20 August), the periods of rain were of short duration. During both the periods of heavy precipitation, ablation amounts increased.

Measurements of frontal retreat were made from a large boulder (Fig. 4). This relatively simple exercise had its difficulties as it was often not easy to define the margin of the ice. The ice edge was defined as that point at which there was a distinct break in slope between the ice surface and the ground moraine. On some occasions the ice was observed to extend beneath this ground moraine for an unknown distance. During the period 3 July–21 August 1966, the total amount of frontal retreat was 17.7 m and the average weekly rate of retreat was 2.5 m. If these figures are typical, there can be very little frontal retreat during the winter, the annual rate of frontal retreat over the period 1945–65 being between 60 and 75 m.

Further measurements were made by F. Björnsson of Kvisker between the boulder and the ice front on the following dates:

Date	Distance between boulder and ice front
20 March 1967	40.60 m
2 June 1967	44.00 m
18 October 1967	93.00 m

The above information was generously provided by Mr. Björnsson and indicates that, through the winter of 1966–67, the ice front retreated at an average weekly rate of 0.22 m. However, during the summer of 1967 the rate of retreat increased to 2.9 m/week.

It is evident, therefore, that since 1890 the Breiðamerkurjökull ice front has been retreating and the ice surface down-wasting at a rapid rate at least during the summers. There is no evidence that, at any time since 1890, the ice front has advanced even for short periods.

The measurement of frontal retreat and down-wastage only provides information on one side of the glacier's budget. As the ice front is retreating and the ice surface is down-wasting, new ice is being brought to the frontal zone. Only a limited amount of information is available about rates of ice movement on Breiðamerkurjökull west of the Jökulsá. A cairn built on one of the medial moraines, 2 km from the ice front, was accurately located by triangulation on 24 August 1965 (Fig. 1). The same cairn was rebuilt during the summer of 1966 and accurately located on 14 August by triangulation. The cairn had moved 24 m during the period between the observations, giving an approximate average daily rate of movement of 6 cm.

The positions of the four stakes used for ablation measurements (Fig. 4) were also accurately located by triangulation on 5 July. Throughout the summer of 1966 the stakes were re-drilled for the purpose of measuring ablation and, on 20 August, their positions were again accurately determined by triangulation methods. The highest stake moved at an average rate of 2 cm/day, and the lowest stake at 8 mm/day, there being a steady decrease in the rate of movement between the highest and lowest stakes. The limited amount of information available makes it dangerous to draw any sweeping conclusions. If the decrease in rate of movement from a point 2 km up the ice (6 cm/day) through a point 500 m up the ice (2 cm/day) to a point 100 m up the ice (8 mm/day) is a meaningful relationship, then it could be explained by the large number of steeply dipping shear planes which were observed near the ice front. It is possible that the horizontal component of movement recorded at 2 km up the ice is transformed into an almost vertical component near the ice front so that the measured horizontal displacement of the stakes is less than would be expected if no shearing was taking place.

Even though the ice is moving forward in the frontal zone, ablation is sufficient to produce a net lowering of the ice surface and a retreat of the ice front. It is the landforms produced by this ice wastage during the period 1890–1965 that are the main concern of this paper.

Moraine ridges

Breiðamerkurjökull has produced a very impressive suite of moraine ridges during its retreat from its recent maximum extent in the late nineteenth century. The outer moraines are usually the biggest and are between 5 and 10 m high (Plate I). The younger moraines are usually 1 to 3 m high. In the area of study, moraine ridges only cover a very small part of the surface area. There are four moraine systems (M1–4, Fig. 5). All of the moraine ridges consist of rounded volcanic fragments ranging in size from clay to boulders. In most of the moraines, less than 30 per cent by weight of the material is finer than 0.06 mm. The material forming most of the moraine ridges is therefore mainly gravel with some fine material acting as a rather weak matrix.

Apart from the 1890 moraine, which is a complex feature consisting of a major ridge-form on top of which smaller ridges have been deposited, the moraine ridges in this area are simple linear features. In some localities, particularly at M4 (Fig. 5), the linear ridges are made up of individual crescentic ridges which are often linked together. On first inspection it is only the alignment of these ridges (parallel to former ice fronts) which suggests that they are moraines. The material forming the ridges is so well-rounded that it could have been waterlaid. However, the form of these ridges, their position, and the fabric of the sediments constituting the ridges leave no doubt that they are moraines formed in association with an ice margin.

Several attempts were made to measure the orientation of the long axes of pebbles in the moraine ridges. However, the general lack of a stiff matrix

Figure 5 The location of features and sites discussed in the text.

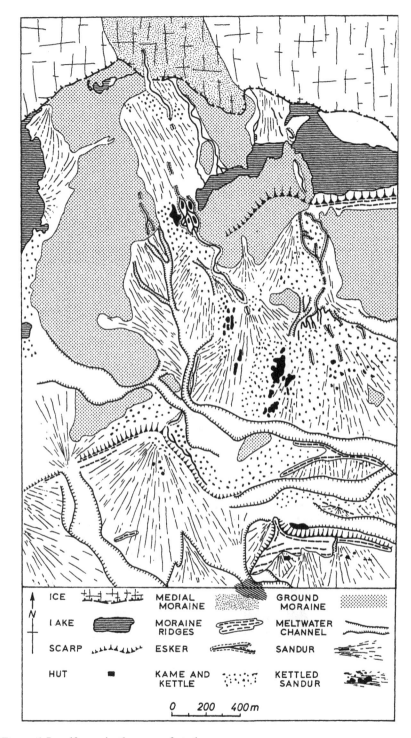

ICE

N

LAKE

SCARP

HUT

MEDIAL
MORAINE

MORAINE
RIDGES

ESKER

KAME AND
KETTLE

GROUND
MORAINE

MELTWATER
CHANNEL

SANDUR

KETTLED
SANDUR

0   200   400m

Figure 6 Landforms in the area of study.

and the fact that the walls of pits dug into the ridges frequently collapsed, made this exercise very difficult. It was possible to collect some data from two locations (Fig. 5), two pits in moraine 1 (a) and three pits in moraine 4 (j). Fifty pebbles were measured in each pit. At site a (Moraine 1), one pit was dug on the proximal side of a moraine ridge and one pit on the distal side. The till fabric for each pit (Fig. 7, A,B) clearly indicates a strong orientation of pebbles at right angles to the ridge crest. At site j (Moraine 4), three pits were dug in the crest of a 2 m-high crescentic ridge. From the till fabric diagrams (Fig. 7, C,D,E) it can be clearly seen that there is a strong orientation of pebbles at right angles to the ridge crest and that a majority of pebbles dip towards the proximal side of the ridge. It is interesting to

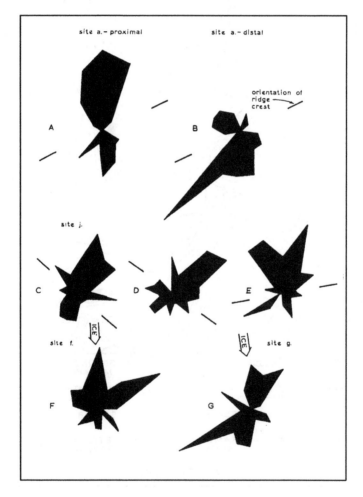

Figure 7 Till fabrics at sites a, f, g and j (Fig. 5).

note that the dominant orientation of pebbles changes from fabric c to D to E as the orientation of the ridge crest changes.

From this limited information, it is only possible to suggest a possible mode of origin for these moraines. The strong orientation of the particles in these moraines clearly indicates that they were not dumped or bulldozed by the ice. One possible mechanism of formation leading to these strong fabrics would be the squeezing of water-saturated till from beneath the ice to form ridges, the crest lines of which would be parallel to the ice front. This mechanism of formation would not only explain the strong fabrics at right angles to the ridge crests but also the lobate pattern of many of the moraine ridges.

Ground moraine

About one-third of the area discussed in this paper is covered by ground moraine. It consists of till in the form of a gently undulating surface of low relief (1–2 m) and in certain areas it is distinctly fluted.

The distribution of ground moraine is indicated on Figure 6. It only occupies those areas not affected by fluvioglacial erosion and deposition and was presumably much more extensive prior to the modification of the area by numerous meltwater streams which developed as the glacier retreated. The surface expression of the material is dull and monotonous. Frequently the surface is scattered with large blocks of rock up to 5 m across which were just dumped by the glacier as it wasted away. The gently undulating surface of the ground moraine generally increases in altitude from south to north reaching altitudes of between 40 and 50 m in the northern part of the area. The surface material appears to be very similar to fluvioglacial material except that it shows no signs of size sorting. There is no reason to doubt that, prior to the last advance of the glacier, the proglacial area was similar to the present proglacial area which is dominated by sandar consisting of fluvioglacial sands and gravels. As the glacier advanced, it picked up these fluvioglacial deposits and incorporated them in the ground moraine. This explains the high frequency of rounded and sub-rounded fragments in the ground moraine and the similarity in its surface appearance to that of a fluvioglacial deposit.

The material beneath the surface of the ground moraine is compact, consisting of pebbles and cobbles embedded in a silty clayey matrix. Once again there is a dominance of coarse material and in some locations the finer particles do not form a very good matrix. Two pits were dug in the ground moraine (Fig. 5, F and H), which at each location was 1 m thick. Fifty stones between 1 and 6 cm long and having a distinct long axis were measured. These measurements were then used for the construction of rose diagrams (Fig. 7, F and G). Both the fabric diagrams show a wide dispersion of values and neither of them shows a really dominant orientation parallel to the direction of ice movement.

At numerous locations along the margin of Breiðamerkurjökull in 1965 and 1966, extensive areas of semi-liquid till were observed. It is not often possible to cross this material on foot without sinking into it to considerable depths. In several places it was possible to insert an ice-axe handle into the semi-liquid material to a depth of at least 1 m. The presence of such areas of semi-liquid till in front of the retreating ice margin suggests that all fabrics in the upper 2 m, and possibly at greater depths, may be at least modified if not solely created by the 'flowage' of the material subsequent to deposition, rather than being the product of ice movement.

Sandar and kames

Approximately two-thirds of the surface of the proglacial area of Breiðamer-kurjökull consists of fluvioglacial deposits (Fig. 1). Even the remainder of the area consisting of ground moraine, moraine ridges and lakes, is underlain by fluvioglacial deposits. All these deposits consist of sand, gravel, cobbles and boulders almost entirely derived from extrusive igneous rocks. The deposits have been carried relatively short distances by melt-water streams. The high degree of rounding exhibited by the constituent rock particles may well be caused by the fact that, although the distance travelled by individual particles may not have been great (probably less than 5 km) during the most recent cycle of erosion, transport and deposi-tion, these same rock fragments have probably been subjected to several cycles of transport by water. The large area of fluvioglacial deposits which can be seen in front of Breiðamerkurjökull at present presumably repres-ents several advances and retreats of the glacier and the morphology of the fluvioglacial deposits to be discussed in this and the next section only represents the relatively recent developments in the accumulation of a very large body of material. The thickness of fluvioglacial deposits is unknown because the depth to the solid rock floor of the Breiðá valley has never been determined. Since the greatest depth of the Breiðárlón (Fig. 1) is 57 m below sea-level and fluvioglacial deposits were observed at 40 m above sea-level in the same vicinity, it is possible to state that the combined thickness of glacial and fluvioglacial deposits in the area of study is at least 97 m. The age of the lower layers of these deposits is unknown. It has been suggested by H.W. Ahlmann (1938, p. 215) that great accumulations of sand and gravel were laid down during a very short period when large quantities of glacier ice and snow were melted on the occasion of the eruption of Öræfa Jökull volcano in 1362: 'The activity of the volcano must then have been so great that not only one of the glaciers which covered it slid down, but the majority of them hurled themselves with immeasurable masses of gravel, sand and water over the surrounding country. It is more than probable that a large number of the sandy plains at the foot of Öræfa Jökull obtained their present-day characteristics at that time. The story is told that off

Plate I The '1890' moraine with sandur surface in foreground.

Plate II Extensively pitted sandur surface (kame and kettle).

Plate III Ice underlying sandur surface.

Plate IV Kettle hole developing in sandur surface (Plate III). Ice can be seen underlying gravels forming the wall of the kettle hole.

Plate V Esker E5 viewed from medial moraine on the ice surface.

Plate VI Stratified sand and gravel exposed in the side of esker E5.

Plate VII Ice exposed in the side of esker E5.

Plate VIII Esker E6.

Plate IX Channel C4 looking east. At the top right of the plate can be seen the ridges of moraine M3.

Plate X Aerial photograph taken in 1945 of the area covered by Figures 5 and 6. Approximate scale 1:18,000.

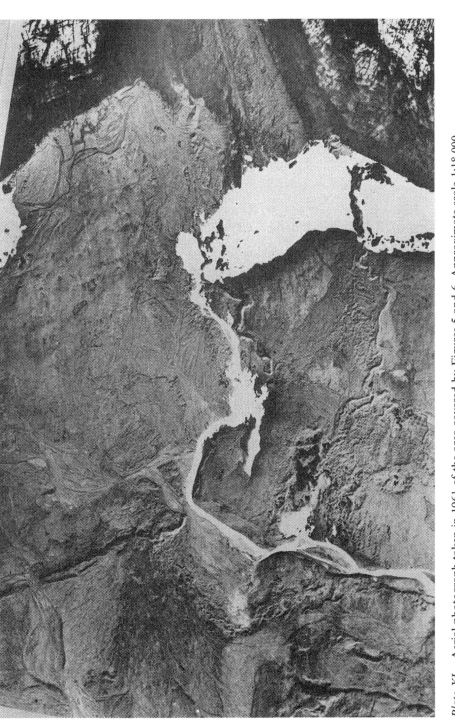

Plate XI   Aerial photograph taken in 1961 of the area covered by Figures 5 and 6. Approximate scale 1:18,000.

Plate XII Aerial photograph taken in 1965 of the area covered by Figures 5 and 6. Approximate scale 1:18,000.

the coast where the sea used to be 60 m deep, a sandur grew up almost over-night'.

Large amounts of fluvioglacial material were presumably laid down during the advance of Breiðamerkurjökull during the nineteenth century. These deposits have subsequently been partly eroded beyond the limit of the 1890 moraine, and both eroded and added to inside the 1890 moraine.

The large outwash plain, usually referred to in Iceland as a sandur, which occupies so much of the area beyond the 1890 moraine was not studied in detail and it will not be discussed in this paper. Detailed information about the smaller sandar inside the 1890 moraine will be presented. The modification of these sandur plains by the melting-out of buried ice subsequent to their formation will also be discussed.

The sandar inside the 1890 moraine (Figs. 5 and 6) are not so large as the ones outside the moraine. Most of them were built up over relatively short periods of time and in some cases were partly destroyed by subsequent fluvioglacial erosion. However, many of the sandar are interesting because of their morphology and their relationships with moraines and eskers. It must be stressed that parts of these sandar have been destroyed by the development of kettle holes as a result of the melting-out of ice buried beneath the sandar (Plate II). The process can be interpreted from Figure 6 on which the symbols for sandar, sandar with kettle holes, and kame and kettle topography, merge one into the other. In the field it can be clearly seen that, even in the most severely 'kettled' areas, the deposits were once a part of a sandur surface because the anastomosing channels of the sandur surface can be traced between the kettle holes.

Although the sandar inside the 1890 moraine are relatively small, most of them have quite steep gradients (1:30 to 1:45). They were deposited by meltwaters which were moving away from the ice as proglacial streams. Some of these streams had to cut through moraine ridges to establish their routes. Other streams originated from supraglacial, englacial or subglacial channels because in some instances the sandar are joined to eskers or esker systems (Fig. 6).

The surfaces of the sandar are dominated by abandoned anastomosing channels. Most of these channels are between 0.3 and 2 m deep, and 1 to 5 m wide. The kettle holes which occur in the sandar do so in their largest numbers at the proximal end of the sandar. Some very large kettle holes occur just beyond the 1890 moraine (up to 10 m deep and 15 m in diameter) but it is in the sandar inside the moraine that the largest number of kettle holes occurs. The kettle holes become more numerous, larger and deeper towards the proximal part of any sandur. Most of the kettles in the sandar are 1–3 m deep and 3–10 m in diameter, but some of the kettles to the south of E4 (Fig. 5) are up to 15 m deep and 100 m in diameter. Some of the kettle holes contain water. Between kettles holes there are a few areas of the original sandur surface still surviving and anastomosing channels can be

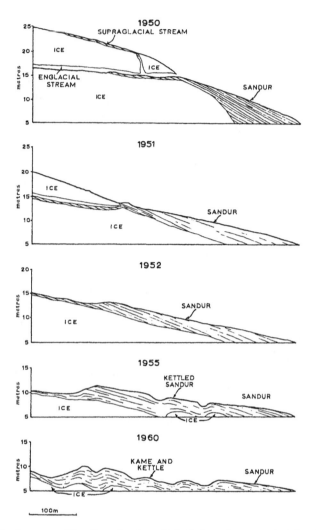

Figure 8 The forms resulting from the deposits laid down by supraglacial and englacial streams on top of stagnant ice (schematic diagram).

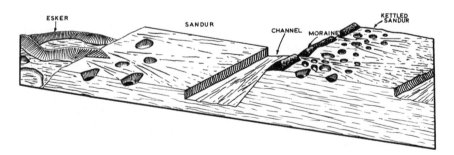

Figure 9 An idealized block-diagram of the relationship between sandar, kettled-sandar, moraines and eskers.

seen on these surfaces. The origin of these kettle holes is related to the development of sandar across areas of stagnant ice, either in the form of a continuous sheet or isolated blocks. The fact that sandar developed on top of a sheet of ice is supported by the fact that, in some areas, little or nothing of the original sandur surface survived after the melting-out of the buried ice. The development of these extensively pitted areas must have been related to the emergence near the ice front of either englacial or supraglacial streams which carried sediments on to either large blocks of ice or a continuous sheet of detached stagnant ice (Fig. 8).

Throughout the above discussion of the development of kettle holes in sandar the term *kame* has been avoided. When the pitting of a sandur surface is very extensive, it is possible that no part of the original sandur surface will remain. The chaotic assemblages of ridges and mounds, separated by kettle holes, so produced could be described as kame and kettle topography. However, it must be remembered that the kames are the result of the development of a large number of kettle holes. It can be seen from Figure 9 that the development of kame and kettle topography could be regarded as the end-product of the destruction of a part of a sandur as a result of the melting-out of buried ice from beneath the sandur. The difference between a kettled (or pitted) sandur and kame and kettle topography is simply the number of kettle holes. If the kettle holes are very numerous, little of the original sandur surface survives because of the slumping of material into the kettle holes.

There were areas of buried ice in front of Breiðamerkurjökull in 1966. The large meltwater stream issuing from a subglacial tunnel and flowing into the north-western corner of the Breiðárlón (Fig. 1) has cut deep banks into the proglacial fluvioglacial deposits (Plate III). In the banks of this meltwater stream, ice could be seen under 4 m of sand and gravel. In the bottom of kettle holes to the south-west of this stream, ice could be seen forming the walls of the kettle holes and underlying 3–4 m of sand and gravel (Plate IV). In the area between the ice edge and the north-eastern part of the Stemmárlón, kettles were observed developing in a recently formed sandur surface. In some of the kettle holes, water could be seen bubbling up under hydrostatic pressure.

Both in the past and at present, meltwater streams have deposited fluvioglacial deposits on top of ice in front of Breiðamerkurjökull. These sandar, which were partly destroyed by the subsequent development of kettle holes, have some interesting relationships with eskers and moraines which will be discussed in a subsequent section.

Eskers

Apart from a few eskers, all the major eskers and esker systems in front of Breiðamerkurjökull are located in the area being dealt with in this paper

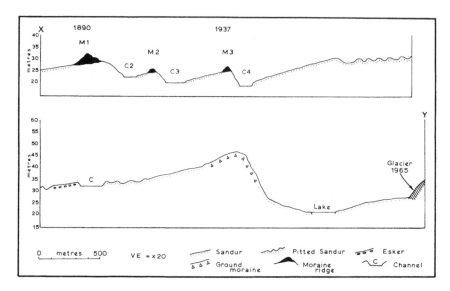

Figure 10 Profile across the proglacial area (x-y, Fig. 5).

(Figs. 5 and 6). There are over 5 km of sharp-crested, steep-sided gravel ridges within this area. These ridges consist of sand, gravel and cobbles and in some localities these deposits can be seen to be stratified. These ridges of fluvioglacial deposits are eskers, and represent the former courses of melt-water streams either in tunnels under or in the ice or in channels on the ice surface (Price, 1966).

The location of eskers 1–10 (Figs. 5 and 6) is interesting for two reasons. First, the eskers are confined to a relatively narrow area, less than 1.5 km wide, which is directly in line with three medial moraines on the surface of Breiðamerkurjökull. Secondly, the eskers occur in an area in which the topography is arranged in a series of steps between the ice margin and the 1890 moraine (Figs. 5 and 10). Apart from eskers E1 and E2 (Fig. 5), all the eskers occur on ground which has been deglacierized since 1945. The two largest esker systems, E4–5 and E6–7–8 occur both on the north- and south-facing slopes of the escarpment which overlooks the ice front.

Esker 1 (Figs. 5 and 6) has a most unusual position. It is located at the base of a scarp 10–15 m high which forms the proximal boundary of a major sandur. The esker is 4 to 5 m high and consists of a series of ridges which bifurcate and rejoin. The crest lines are irregular in long profile and sinuous in plan.

Esker 2 is 10–12 m high and is joined, at its distal end, to a kame and kettle area which forms the proximal part of a major sandur. Esker 2 is larger than esker 1 and, prior to the cutting of channel C4–5, was probably an extension of esker 3. Esker 3 is a 6–8 m high sinuous ridge with an irregular long profile and steep sides. Several angular boulders up to 2 m across occur on the crest and sides of this ridge. Two pits were dug at site d (Fig. 5) and no distinct stratification was observed. All the pebbles were well rounded and the matrix was a 'pea-gravel'.

Esker 4 is a very large esker with a sharp, undulating crest line. This esker has two distinct parts (a/b) which appear to be related and are in fact connected by low ridges (1–2 m high). Esker 4a is generally between 5 and 10 m high and is very sinuous. At its distal end the main ridge bifurcates to form a loop. A pit dug near the almost right-angle bend (i, Fig. 5) and some 3 m below the crest, revealed good stratification. There were strata of fine sand grading up into pea gravel and cobbles. All the rock particles are very well rounded except for a surface covering of brown, angular fragments (ablation moraine similar to that seen in the present medial moraines on the glacier) in a few localities. The base of this esker (4a) generally falls in altitude from the proximal to the distal part (a drop in altitude of 7 m) but there are irregularities in the sub-esker surface which are reflected in the crest profile (Fig. 11). The proximal part of the base of esker 4a is some 10 m higher than the distal part of 4b.

Esker system 4b is a complex one consisting of a series of sub-parallel ridges 3–5 m high often separated by deep circular or oval kettle holes. In the wall of one of these kettle holes, buried ice was observed in the summer of 1966. The proximal end of the system merges into an area of meltwater channels which contains several kettle holes (Fig. 11). There is a slight increase (5–10 m) in the altitude of this system towards the proximal end.

Esker 5 (Plate V) is the biggest esker, in terms of height (up to 20 m) and cross-sectional area, within the area of study. It is a sinuous, sharp-crested ridge with steep sides (30°). In several places along this ridge, recent slumps revealed good stratification in the fluvioglacial deposits (Plate VI). The esker continues on to the glacier and is definitely underlain by ice at its proximal end.

On the 1965 air photograph (Plate XII), this esker is clearly seen as a continuous ridge extending on to the ice surface. By the beginning of the summer of 1966, a small stream flowing from west to east had cut a tunnel underneath the esker (Plate VII) and through the ice occurring in the core of the esker. By the end of the summer of 1966, this ice tunnel had collapsed and the esker beyond the ice margin was detached from the esker on the ice surface.

A detailed map of esker 5 was made by tachyometric methods in July 1966 (Fig. 12b). Although the detail and accuracy of the map is considerably greater than that produced by photogrammetric methods (Fig. 12a), it is interesting to compare the long profiles of the eskers based on these maps (Fig. 13). The greater smoothness of the 1965 profile is largely the function of the fewer number of spot heights available from which to construct the profile. However, in the proximal part of this esker the altitude of the crest was generally 3–7 m lower in 1966 than it was in 1967. This lowering may result from at least three processes: (i) the general slumping of steep, unstable

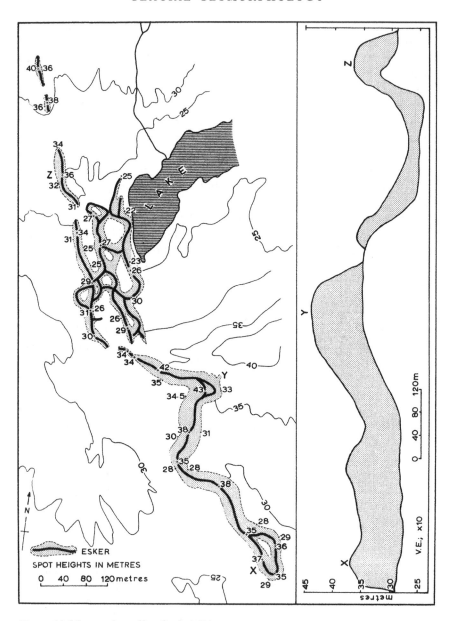

Figure 11 Map and profile of esker E4.

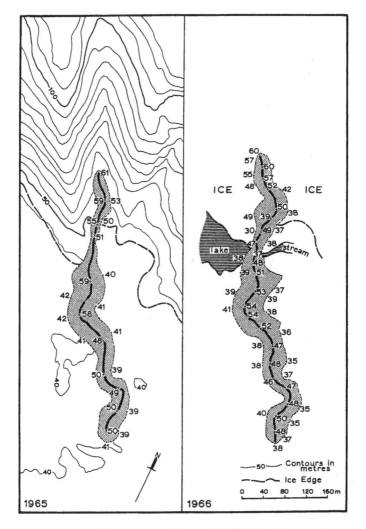

Figure 12 Esker E5: (a) photogrammetric map (1965); (b) tachyometric map (1966).

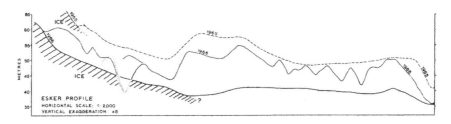

Figure 13 Long-profiles of esker E5, based on Figure 12a and b.

gravel slopes; (ii) wind erosion; (iii) the melting of buried ice from beneath the esker.

Both slumping and wind erosion were observed in the field in 1966. The occurrence of buried ice beneath the proximal part of the esker was proven by pits dug into the esker in which ice was encountered at a depth of 4–5 m from the ridge crest. The significance of the melting of buried ice from beneath the esker as a means of lowering the altitudes of the crest of the esker is difficult to estimate because of the short period of time between the two sets of measurements. However, there can be little doubt that at least the proximal half of this esker has been lowered to some extent by the wastage of buried ice.

Eskers 6, 7, 8 and 9 (Fig. 5) can be regarded as forming one system which is sub-parallel to the system made up of eskers 4a, 4b and 5. In plan, eskers 6, 7, 8 and 9 clearly resemble an anastomosing stream pattern. The ridges making up the pattern are larger (10–15 m high) and more simple in plan, at the proximal end of the system, and smaller and more complicated in plan at the distal end of the system where they merge into a sandur surface with numerous kettle holes.

Esker 6 (Plate VIII) is the largest one in the system. It is over 15 m high in places (Fig. 14), has a sharp crest and is steep sided. There are a few kettle holes actually in the ridge and on either side of it, indicating that the ridge certainly contained ice in the past and may have had ice still beneath it in 1966, although none was actually observed. A part of esker 6 has been destroyed as a result of the lowering of the level of the large lake to the east and west of it. As the surface of this lake was lowered, currents developed between the smaller lake to the west and the large lake to the east and these currents destroyed part of the esker ridge. The base of esker 6 rises in altitude from north to south and actually climbs up a north-facing scarp (Fig. 15). Just to the south of the crest of this scarp, esker 6 bifurcates to form eskers 7 and 8. Esker 7 consists of a series of ridges (2–5 m high) linked together by almost right-angled bends. Esker 8 is a single, sinuous ridge, 4–7 m high. Irregularities in the sub-esker surface appear to be reflected in the profile of the crest of the esker. Eskers 7 and 8 eventually rejoin and, apart from a short break, continue towards a complex system of ridges and kettle holes (Figs. 5 and 14). This system consists of sinuous ridges 5–10 m high which eventually merge into a sandur surface in which there are many kettle holes. Within this ridge and kettle-hole topography there are a few flat areas upon which remnants of the former channel systems can be seen. Many of the channel systems are now truncated by deep kettle-holes (up to 15 m deep). There can be little doubt that the meltwaters which deposited the gravels forming eskers 6, 7 and 8 in definite channels or tunnels in the ice, were more widely dispersed in numerous interconnected channels at the time when esker system 9 was deposited. The sandur to which esker system 9 is linked was deposited in a proglacial environment but the sandur

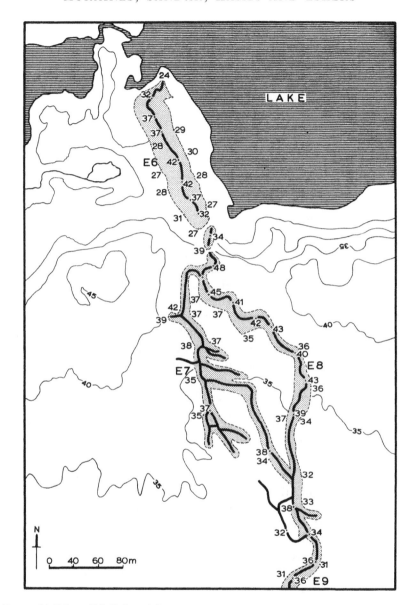

Figure 14 Eskers E6, 7, 8 and 9.

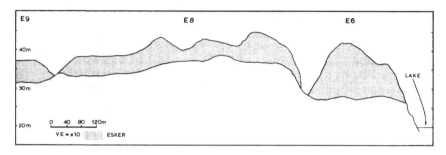

Figure 15 Long-profile of eskers E6, 8 and 9, based on Figure 14.

was underlain by buried ice which subsequently melted out to produce numerous kettles.

It is probable that the ridges of fluvioglacial deposits described above were laid down by meltwater streams, the courses of which were determined by the presence of ice. The material which makes up these ridges was probably derived from the very large medial moraines which occur on the surface of Breiðamerkurjökull at the proximal ends of the esker systems. This material was carried by supraglacial, englacial or subglacial streams and deposited in these channels or tunnels. In which of these three environments the esker ridges were actually built it is impossible to say. In 1945, only eskers 1, 2, 3 and 9 (Fig. 5 and Plate X) were exposed. By 1961, all the eskers 4a, 4b, 6, 7, 8 and 10 (Fig. 10 and Plate XI) were visible. Photogrammetric measurements with accuracies of ±2 m, made by R. Welch (Welch and Howarth, 1968) on photography taken in 1945, 1960, 1961 and 1965, indicate that eskers 4b, 5, 6, 7, 8 and 9 have all been lowered with the passage of time. The range of altitudinal changes is between 2 and 13 m, the greatest amount of lowering (13 m) being recorded at esker 5 between 1960 and 1961, and during the same period the proximal end of esker 6 was lowered by 10 m. Such rapid lowering of the esker ridge without its destruction is quite remarkable and is much more rapid than the rates measured at the Casement Glacier by the writer (Price, 1966; G. Petrie and Price, 1966).

These measurements, together with the profiles of esker 5 (Fig. 13) and the fact that ice was actually observed beneath eskers 5 and 4b, and that esker 5 continued on to the glacier in 1966 strongly suggests that at least parts of the eskers in this area were deposited by supraglacial or englacial streams. The fact that irregularities in the sub-esker surface are often reflected in the profiles of the ridge crests supports the hypothesis that at least parts of these eskers were superimposed from an ice surface on to the subglacial topography. If this hypothesis is accepted, then the fact that these esker systems descend a south-facing slope, cross a depression and rise up over a north-facing slope (Figs. 11 and 15) need not be explained by large volumes of water flowing uphill under hydrostatic pressure. It may be that some parts of an esker ridge were deposited subglacially while others were subsequently let down by the meltingout of ice which once formed the floors and sides of englacial tunnels or supraglacial channels.

In parts of eskers 4b, 7 and 9, the juxtaposition of kettle holes and gravel ridges raises the question whether esker-like ridges can in fact be produced by the melting-out of buried ice from beneath a spread of fluvioglacial deposits. Particularly in the case of esker system 9, which merges into a sandur containing many kettle holes, it is difficult to decide whether a ridge separating a series of kettles is the product of the formation of the kettles or is a deposit of a meltwater stream confined between walls of ice. This distinction is really irrelevant in this particular instance because it is the relationship between the former meltwater stream which deposited eskers 6, 7, 8 and 9 and the sandur at the distal end of the system which is most significant.

The development of landforms and drainage systems, 1890–1966

There is a repeated pattern observable in the relationship which exists between sandar, kame and kettle complexes (kettled sandar), eskers and moraines. This pattern is demonstrated by a profile drawn from south to north (Fig. 10). In a horizontal distance of nearly 3 km from the 1890 moraine to the position of the ice margin in 1965, the following sequence of landforms occurs.

The 1890 moraine (M1, Fig. 5) consists of a series of sub-parallel mounds attaining heights of 15 m. To the south and to the north of this moraine there are kettle holes up to 6 m in depth. When the ice front stood at the 1890 moraine, numerous streams issued from the ice front and crossed the wide sandur towards the sea. Running parallel to the 1890 moraine, and to the north of it, there is a meltwater channel (C2, Fig. 5) which carried meltwater in an easterly direction between the ice edge and the 1890 moraine. As the ice front retreated northward, anastomosing streams deposited a small sandur in front of a small moraine (M2, Fig. 5), and a somewhat larger sandur to the east of M2 contains numerous kettle holes in its proximal part.

North of channel 3 there is a sandur which, once again, contains kettle holes in its proximal part. These kettles holes, which are 1–5 m deep, are best developed in front of the 1937 moraine. In 1937 the ice front probably had a shape in plan very similar to that of the 1965 ice front. The embayment near E1 (Fig. 5) was probably occupied by that part of the ice front covered by the medial moraines. The moraine ridges (M3, Fig. 5) deposited around 1937 consist of a series of sub-parallel mounds 1–4 m high. On the north side of this moraine system there is a steep slope down to the floor of channel 4 which is some 10 m below the crest of the 1937 moraine (Plate IX).

Channel 4 was initiated some time between 1937 and 1945 because, on the aerial photograph taken in 1945 (Plate X), it is seen to be occupied by a larger meltwater stream. This stream originated from one large and several small outlets at the ice front. To the south of the ice front, east of the medial moraines, a large sandur (500 m north-south) with a few kettle holes, can be clearly seen. Esker E9 is seen to be connected to this sandur and emerges from the ice.

The area was photographed again from the air on 4 July 1961 (Plate XI). The contrast between the 1945 and the 1961 photographs is quite remarkable. Channel 5 (Fig. 5) had been abandoned and channel 4 was now carrying water overflowing from a large lake at the ice front. By this date nearly all the major eskers had been formed. One of the most significant differences in this area between 1945 and 1961 was the development of kettle holes in the sandur to the south of E9. By 1961 this sandur had been, in its northern and western parts, largely destroyed by the development of many kettle holes. From photogrammetric measurements made by Welch (Welch and Howarth, 1968) it is known that parts of this sandur surface had been lowered by as much as 3 m over this period, presumably by the melting-out of buried ice.

The large proglacial lake seen on the 1961 aerial photograph was dammed up between a north-facing scarp (25 m high) and the ice front. This lake partly submerged eskers 4b and 6. The surface of this lake stood at an altitude of approximately 30 m.

By 1965 (Plate XII), the ice front had retreated sufficiently farther north to permit an eastward escape of meltwater towards the Jokulsá (Fig. 1) and the proglacial lake was lowered to an altitude of approximately 20 m. The lowering of this water-level resulted in the abandonment of channels 6 and 4, the division of the proglacial lake into two parts and emergence of eskers 4b and 6.

The sequence of events described above, so clearly illustrated by the aerial photographs (Plates X, XI), and XII) has provided the area with a most complex topography. However, the occasional standstill in the retreat of the ice front allowed the formation of a moraine. On the distal side of this moraine, meltwater streams often developed sandar and sometimes these sandar were deposited on top of stagnant ice. When the ice beneath the surface of the sandur eventually melted out, part of the sandur surface was destroyed and, depending on the extent of this destruction, either a sandur with kettle holes or an area of kame and kettle topography was the resultant form. Some of the streams which produced the sandar originated on, in or under the ice as major drainage routes. The deposits laid down in these channels or tunnels in the ice survived the process of wastage of the ice from around and beneath them and remained in the proglacial area as eskers.

Even in an area such as this, in which a few of the stages in the development of the glacial and fluvioglacial deposits and landforms since 1945 are recorded on aerial photographs, the origins of the deposits and landforms seen in the field in 1965 and 1966 are difficult to interpret. When the geomorphologist is required to interpret the origins of deposits and landforms many thousands of years after their formation and after they have been subjected to modifications by weathering and mass movement it is like asking someone to work out the plot of a 1000-page detective novel from the last five pages. At least at Breiðamerkurjökull the geomorphologist's task is made a little easier by the existence of a few of the earlier pages of the story that have been recorded on the aerial photographs.

Acknowledgements

The field-work, on which this paper is based, was carried out as part of the Breiðamerkurjökull Project (1964–67) organized by the Department of Geography, University of Glasgow. Financial support for this project was obtained from the Natural Environment Research Council, the Carnegie Trust for the Universities of Scotland, the Court of the University of Glasgow, the Royal Geographical Society and the Gino Watkins Memorial Fund.

The author is pleased to acknowledge assistance in the field and laboratory from the following: P. Howarth, A. Kelly, D. Nicol, R. Ryder, J. Senior, J. Shearer, R. Welch and H. Williams. The 1:15,000 photogrammetric map of this area produced by Dr. R. Welch has been used as a base map. The author is particularly indebted to his colleague Mr. G. Petrie for his co-operation throughout the project and to Landmælingar Islands (Icelandic Survey Department), Vegamalastjorinn (Icelandic Highways Department), Dr. S. Thorarinsson (Icelandic Museum of Natural History) and the Björnsson brothers of Kvisker for their assistance.

A grant was made by the Carnegie Trust for the Universities of Scotland to cover part of the cost of the illustrations.

Note

1 Sandur is an Icelandic word meaning 'outwash plain' (Plural: sandar).

References

AHLMANN, H.W. (1938) *Land of ice and fire*

PETRIE, G. and R.J. PRICE (1966) 'Photogrammetric measurements of the ice wastage and morphological changes near the Casement Glacier, Alaska', *Can. J. Earth Sci.* 3(6), 827–40

PRICE, R.J. (1966) 'Eskers near the Casement Glacier, Alaska', *Geogr. Annlr*, Ser. A, 48(3), 111–25

WELCH, R. (1967) 'The application of aerial photography to the study of a glacial area: Breiðamerkur, Iceland' (Unpubl. Ph.D. thesis, Univ. of Glasgow)

WELCH, R. and P.J. HOWARTH (1968) 'Photogrammetric measurements of glacial landforms', *Photogramm. Rec.* 6(31), 75–96

10

BIMODAL DISTRIBUTION OF ROCK AND MINERAL FRAGMENTS IN BASAL TILLS

A. Dreimanis and U.J. Vagners

Source: R.P. Goldthwait (ed.) *Till: A Symposium*, Columbus: Ohio State University Press, 1971, pp. 237–50.

Abstract

Every lithologic component of basal till, fragments of both rocks and their constituent materials, has a bimodal particle-size distribution, if the rock is monomineralic or consists of minerals of similar physical properties. One of the modes is in the clast-size group, the other reflects the mineral fragments in the till matrix. Several such modes develop in the till matrix, if the rock comminuted by glacial transport consists of minerals with differing physical properties.

The clast-size mode is relatively larger than is that of the matrix mode near the source, where the glacier picked up the rock fragments. With increasing distance of glacial transport from the source, the matrix modes become larger and the clast-size modes are reduced or may even disappear.

The matrix modes are restricted to certain particle-size grades, which are typical for each mineral. These mineral grades are called "terminal grades," because they are the final product of the glacial comminution. The particle-size grade produced depends upon the original sizes and shapes of the mineral grains in the rocks, and upon the resistance of each mineral to comminution during glacial transport. Predominance of "terminal grades" over clast-size modes indicates a higher degree of maturity of till, as compared with shorter-transported, less mature tills which consist mainly of clast-size particles.

Introduction

The conclusions on rules which govern the textural and the lithologic composition of till, as presented in this paper, are based mainly upon investigations of Wisconsin-age tills in Ontario (Fig. 1). These tills were deposited by a continental ice sheet; therefore all their clastic components were derived from drift material which was incorporated in the ice probably exclusively at the base of the glacier. Absence of any evidence of nunataks rising above the Wisconsin ice sheet precluded addition of true superglacial drift. After incorporation of the drift material (fragments of bedrock and of older Quaternary sediments), the drift was transported in the ice either near the base of the glacier ("basal transport") or higher up within the glacial ice ("englacial transport"). Both these modes of transport caused comminution of the rock fragments and of the lumps of sediments carried by glacier. The most probable mechanism by which comminution was accomplished was by crushing and abrasion (Dreimanis and Vagners, 1965), though the disruptive effect of the growth of ice crystals and solution by water should also be considered.

Some of the englacially transported drift material became superglacial during surface melting of the ice sheet, thus producing ablation till. The effect of the superglacial environment upon the initially englacially transported material will not be discussed here, as all the samples investigated were taken from dense basal till, deposited most probably from englacially or basally transported drift (Fig. 2). The composition of the deformed till which has been subjected to hardly any transport, and the waterlaid till, which has resulted from a combination of glacial and lacustrine or marine sedimentation (Fig. 2), will also not be discussed here. It is only the basal till which will be considered.

Various investigations of basal till (for a summary, see Flint, 1957, p. 122–29) point out that its composition depends mainly upon the following two factors: (1) the multilithologic source material, consisting of bedrock and of older Quaternary sediments, and (2) comminution of this material during glacial erosion and transport, which in turn depends upon the durability of the rocks, mode and position of transport (basal or englacial), and distance of transport.

Investigations of the compositions of tills in Ontario were begun by the senior author in 1950, with their principal objective to develop quantitative criteria for (a) stratigraphic correlations, (b) differentiation of various till units, and (c) indicator tracing in a search for ore deposits. The first phase of these investigations dealt mainly with application of various quantitative laboratory and field methods and modification of them (Dreimanis and Reavely, 1953; Dreimanis and others, 1957; Dreimanis, 1961 and 1962).

Following a standard practice of most till investigators, arbitrary particle-size boundaries were used at the beginning, for both the textural and the

209

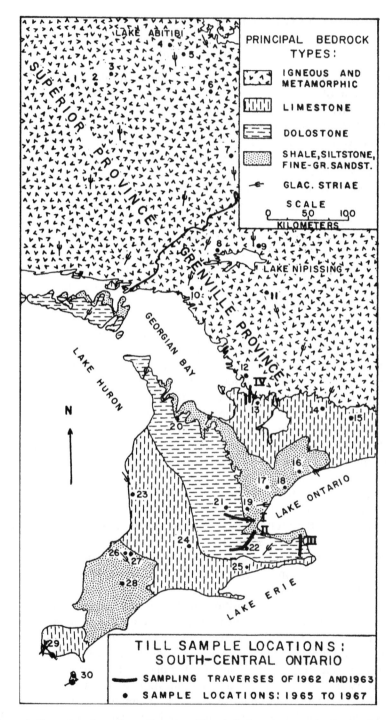

Figure 1 Central and southern Ontario: till sample locations, major bedrock types, and directions of glacial movements as indicated in striae.

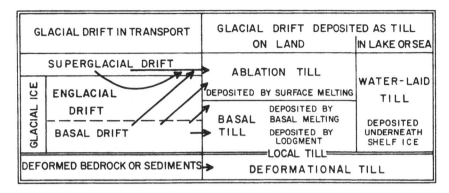

Figure 2 Classification of tills and their relationship to glacial drift in transport. (After Dreimanis, 1967a.)

lithologic studies. The first step toward selecting a natural particle-size range for quantitative determination of carbonates in the till matrix was stimulated by finding that carbonates had a bimodal distribution in tills and that the carbonate mineral modes were always in the silt- and clay-size fractions (Dreimanis and Reavely, 1953). In order to investigate the distribution of noncarbonate minerals and rocks in tills, and to select the particle-size grades most suitable for their investigation, the next step was to study the factors which determine the composition of till. These problems have been investigated since 1962, particularly by the junior author. By analysing large till samples, an effort was to relate their lithologic composition to (1) the distance from the most probable bedrock source, (2) the mode of transport, and (3) the mode of deposition. Several progress reports have resulted from these studies (Dreimanis and Vagners, 1965; Vagners, 1966; Dreimanis, 1969; Vagners, 1969). As both authors are still continuing these till investigations – one of them on Ontario, and the other in the Atlantic Provinces of Canada – this paper is essentially another progress report.

Methods of investigation

In the initial study, twenty-five till samples, 0.5–1.5 cubic meters large, collected over limestone and dolostone bedrock areas, were investigated (Vagners, 1966). Main objective of this study was to determine the mode of incorporation of these two rock types in the associated, overlying till and the distribution of their fragments throughout the particle-size range, from cobbles to clay-size, along four selected transects parallel to glacial movement (I through IV in Fig. 1). The samples were split into 18 size fractions according to Wentworth's (1922) scale. The quantitative lithologic composition of each fraction was then investigated separately. In the 0.062-mm-to–256-mm grades, all the rock and mineral grains were identified visually, but in the less-than-0.062 mm grades, only the percentages of calcite and dolomite were determined, using the gasometric method of Dreimanis (1962) (for details, see Vagners, 1966).

After the rule of the bimodal distribution for several rock groups investigated in basal tills was etablished (Dreimanis and Vagners, 1965), more detailed studies were conducted, dealing with the distribution of individual common minerals throughout various particle-size fractions in tills. In this study, begun in 1965, the junior author investigated 30 till samples, randomly collected over various bedrock terrains – Precambrian igneous and metamorphic bedrock, and Ordovician, Silurian, and Devonian limestones, dolostones, and fine-grained clastics in central and southern Ontario (Fig. 1). The samples were taken along transects 300 to 750 km long and parallel to the predominantly north-south to northeast-southwest-oriented glacial movements. The main objective of the study was to determine the results of comminution produced by glacial transport on such common minerals as quartz, the plagioclases, the potassium feldspars, amphiboles and pyroxenes as a group, garnets, heavy minerals, calcite, and dolomite. As in the previous study, each till sample was split according to the Wentworth scale, but this time in the size range from 8 to 0.001 mm. Staining methods were used for identification of several mineral groups in the particle-size fractions coarser than 0.062 mm, but X-ray-diffraction techniques were applied for quantitative determinations in the less-than-0.062-mm particle-size fractions (for details, see Vagners, 1969).

Results and conclusions

Some results of the above investigations have been reported by Dreimanis and Vagners before (1965, 1969). Only the main conclusions will be given here, each of them illustrated by typical examples (Figs. 3–8).

Conclusion I

When, during glacial erosion and transport, rocks become comminuted to their constituent minerals, abrupt transition from rocks to their minerals takes place (Figs. 3 and 4). As a result, at least two modes or two groups of modes develop for each lithic component of the till; one is in the clast size, consisting predominantly of rock fragments, and the other is in the till matrix, consisting mainly of mineral fragments. Several mineral modes may develop in the till matrix, depending on the different physical properties of the minerals present (see Conclusion III for elaboration on the mineral modes).

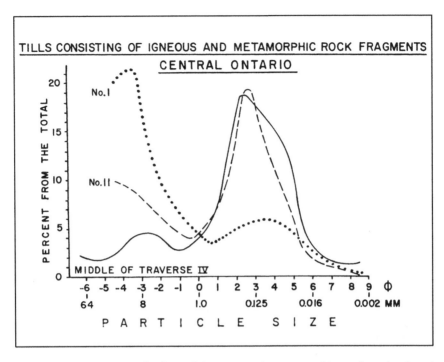

Figure 3 Frequency distribution of igneous and metamorphic rock and mineral fragments in three basal tills from the Canadian Shield, central Ontario. Maximum transport distances of igneous and metamorphic rock fragments prior to the deposition of the tills: No. I: 220 km; No. II and the Middle of Traverse IV: 1,200 km (probably less); see Figure 1 for location of samples. (After Dreimanis, 1967a.)

Conclusion II

Near the source, where the glacier picked up the rock fragments, the clast-size mode is always relatively larger than the matrix mode. For instance, in the local till at Niagara (0–3 km from bedrock source in Fig. 4), the fragments consist mostly of dolostone clasts.

With increasing distance of glacial transport away from the source, the matrix modes grow larger, recording increasing comminution of clast-size particles (Figs. 3 and 4), while the clast-size mode becomes relatively smaller. For instance (Fig. 4), the clast-size dolostone mode has nearly disappeared after the 300-to-500-km transport from the Beekmantown Dolostone bedrock area in the St. Lawrence Lowland down to the Hamilton area at the northeastern end of traverse II in Figure 1.

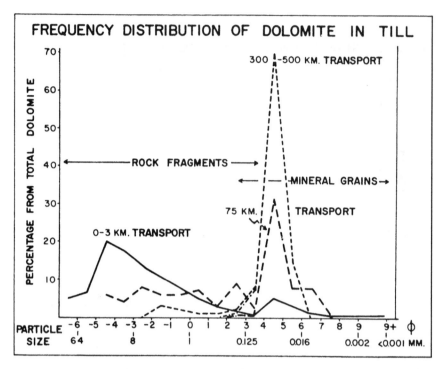

Figure 4 Frequency distribution of dolostone-dolomite in three selected till samples from the Hamilton-Niagara area (Traverses II and III, see Figure 1).

Conclusion III

The matrix mode, or modes, resulting from multi-mineral rocks, is restricted to certain particle-size grades, typical for each mineral. These mineral grades are called the "terminal grades," because they are the final product of the glacial comminution. They depend upon the original sizes and shapes of the mineral grains in the rocks, and upon the resistance of each mineral to comminution during glacial transport.

The "terminal grades" of some minerals common in tills are shown in Figure 5. Their gradual development and their final restriction to a certain particle-size range typical for each mineral is demonstrated by two examples: dolomite (Fig. 4), representing a medium-hard mineral with excellent cleavage, and garnet (Fig. 6), a much harder mineral with very poor cleavage.

For each mineral investigated (Fig. 5), the particle-size range of the "terminal grade" appears to be relatively constant, as concluded already by Dreimanis and Vagners (1965). Thus this "terminal grade" is between 0.031 and 0.062 mm for dolomite in Ontario, independent of the distance

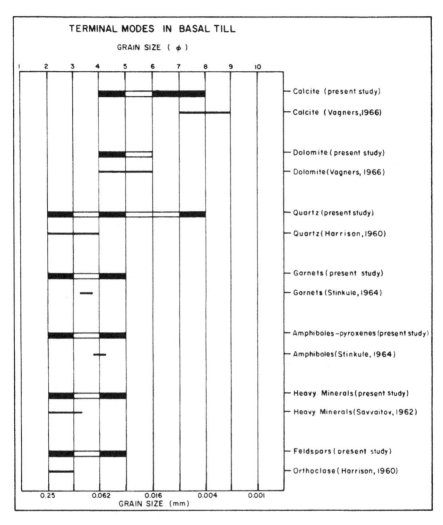

Figure 5 "Terminal grades" (both white and black bars) and major modes in them (black bars) of selected minerals and their groups in basal tills. (After Vagners, 1969.)

of its transport. The "terminal grade" of dolomite apparently represents its equilibrium size, beyond which the dolomite grains do not become crushed or abraded any more by lengthy glacial transport.

Several minerals (Fig. 5) have bi- or tri-modal "terminal grades." These may be explained in several ways. For chemically and mechanically very resistant minerals, such as quartz and garnet, the main cause for the several modes in their "terminal grades" may be the different original sizes of these

215

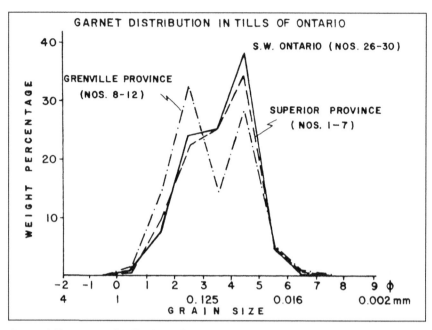

Figure 6 Frequency distribution of garnets in three groups of tills in Ontario. The bedrock source areas of garnets are the Grenville and the Superior provinces of the Canadian Shield. See Figure 1 for Iocations. (After Vagners, 1969.)

mineral grains in their source rocks. Thus Figure 6 suggests predominance of finer grained garnet in the Precambrian rocks of the Superior province, while garnets from the Grenville province, which is the main source of garnets in Ontario, show a bimodal distribution, containing both coarse and fine grains. Though the tills of southern Ontario contain greater quantities of garnets from the Grenville than from the Superior province bedrock (Dreimanis and others, 1957), the finer grained mode (maximum between 0.031 and 0.062 mm.) predominates in the tills 300–600 km down-glacier from the Grenville province. Apparently, some of the coarse-grained garnets (0.125-to-0.25-mm mode) have become comminuted during a lengthy transport of hundreds of kilometers from the Grenville province to southwestern Ontario. Therefore the 0.031-to-0.062-mm grade should be considered the true terminal modal class for garnets.

Quartz has a tri-modal distribution in its terminal grade, because it has derived from rocks containing quartz in particle sizes ranging from pebbles to silt size. As quartz has been picked up by the glacier not only from Precambrian rocks but also from belts of the various Paleozoic clastic sediments, it is difficult to determine how much comminution of quartz grains occurred during glacial transport.

Besides the original particle size of minerals and the effect of crushing or abrasion, solution during or after glacial transport must be also considered as one of the factors affecting the "terminal grade." This would apply particularly to the more soluble minerals, such as calcite. This is especially true if the content of the mineral in the till is low, and if the waters circulating through the till are not saturated with calcium bicarbonate. Vagners (1969) noticed that, in those tills which were low in calcite, the fine-grained side of its "terminal grade" is reduced.

The "terminal grades" of all the minerals listed in Figure 5 have been discussed in more detail by Vagners (1969). Because of the variety of factors involved in producing the "terminal grades," these investigations are still far from complete.

Conclusion IV

Predominance of "terminal grades" over clast-size modes indicates a higher degree of maturity of such tills, as compared with short-transported tills, for instance in local moraines, which consist mainly of clast-size particles. The highest degree of maturity of a till is attained when all of its rocks have become comminuted to the terminal grades of all of their constituent minerals. In Ontario, none of the tills have reached this degree of maturity.

Similar investigations in other areas

The bimodal distribution of rocks and their constituent mineral fragments in tills have been overlooked by most till investigators. However, some have noted either both the rock and the mineral modes, or merely the concentration of minerals in certain particle-size ranges in till matrix. Examples include Harrison (1960), who related till composition in Indiana to its probable source; various workers in the Baltic area, for instance Raukas (1961) in Estonia, Savvaitov (1962) and Stinkule (1964) in Latvia, and Gaigalas (1964) in Lithuania; and in Ohio, Smeck and others (1968) and Wilding and others (1971). Their accounts (see summary in Fig. 6) are merely descriptive; they do not discuss the dynamic aspects of the transition from the clast-size modes to the mineral modes during glacial transport. Nevertheless, their findings on the modes of various minerals in tills are in agreement with those discussed in this paper. Apparently the "terminal grades" which have been found in Ontario are not merely local phenomena. Nevertheless, many other areas should be investigated in order to establish possible variances in the "terminal grades" for each mineral or its varieties, for instance quartz derived from sandstone, siltstones, shales, intrusives, extrusives, metamorphics, etc. Kalinko (1948) and Sindowski (1949) have noticed that also in non-glacial clastic sediments each mineral has its maximum abundance in a specific particle-size range. Apparently "terminal grades" develop not only in tills, but also in other clastic sediments.

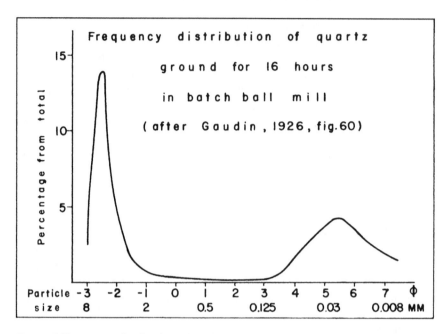

Figure 7 Frequency distribution of artificially ground quartz. (After Gaudin, 1926.)

Experimental simulation of glacial comminution of rocks and their minerals is difficult, because most of the artificial grinding, milling, or crushing equipment does not work in the same manner as does comminution by a glacier. However, it is interesting to note that artificial grinding in ball mills and pebble mills (Gaudin, 1926) also produced bimodal distribution of rock and mineral fragments in most cases (as in Fig. 7). However, by increasing the time of grinding, the fine-grained mode also becomes progressively finer. The cumulative curve of artificially crushed gneiss (Elson, 1961) also suggests a bimodal distribution. Crushing of hard shale (Elson, 1961), however, did not produce a bimodal distribution of its fragments, because the shale had been comminuted only to approximately 0.05 mm, which is considerably coarser than the individual mineral grains usually present in shale.

The straight-line cumulative curves of the granulometric composition of tills, if plotted on Rosin and Rammler's "law of crushing" paper, mentioned by Elson (1961) as being characteristic for tills, may develop because of the multi-lithologic composition of most tills. If several bimodal distribution curves are superimposed one upon the other, the resulting curve becomes more or less straight. However, if a monomineralic rock type predominates in till, the cumulative curve of its granulometric composition does not approach a straight line (Fig. 8; see also Figs. 5 and 6 in Dreimanis and Vagners, 1969).

DOLOMITIC TILLS OF S.W. ONTARIO, CANADA

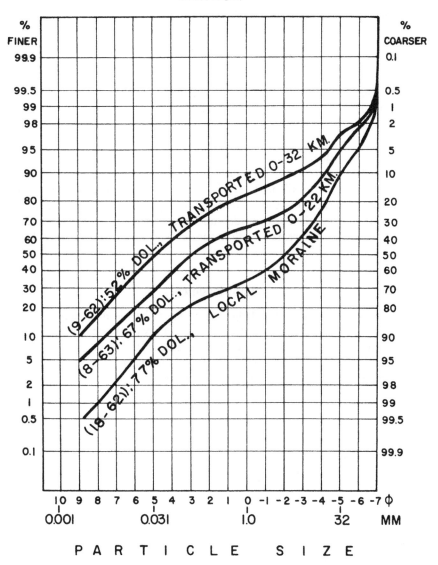

Figure 8 Cumulative granulometric curves of three dolostone-dolomite-rich tills from the traverses I (8–63), II (9–62), and III (18–62). See Figure 1 for locations. (After Dreimanis, 1967b.)

Implications of the bimodal distribution
of the rock-mineral components of till

The bimodal or even multimodal distribution of fragment sizes of each rock and its constituent minerals in till requires that at least two particle-size groups should be investigated for description of lithologic composition of a till (Dreimanis, 1969):

A. the clasts consisting predominantly of rock fragments;
B. the till matrix, consisting predominantly of minerals. Preference should be given to their "terminal grades."

The bimodal distribution should also be considered when separating the till matrix from clasts, for instance when determining the granulometric composition of till matrix. The granulometric composition of tills is strongly governed by their lithologic composition and by the interrelationship of the clast-size modes and the "terminal modes." Another factor, not discussed here, is the admixture of incorporated nonconsolidated sediments.

Acknowledgments

The authors gratefully acknowledge support for the till investigations provided by the Geology Department of the University of Western Ontario, the Ontario Research Foundation, the Department of the University Affairs of Ontario, and the National Research Council of Canada, and are thankful to J.L. Forsyth for critical reading of the manuscript.

References

Dreimanis, A., 1961, Tills of Southern Ontario: *in* Legget, R.F. (ed.), Soils in Canada, Royal Soc. Canada Spec. Public. No. 3, p. 80–94.

——, 1962, Quantitative gasometric determination of calcite and dolomite by using Chittick apparatus: Jour. Sed. Petrology, v. 32, p. 520–29.

——, 1969, Selection of genetically significant parameters for investigation of tills: Zesz. Nauk. Univ. A. Mickiewicza, Geografia No. 8, Poznan, Poland, p. 15–29.

Dreimanis, A., and Reavely, G.H., 1953, Differentiation of the lower and the upper till along the north shore of Lake Erie: Jour. Sed. Petrology, v. 23, p. 238–59.

Dreimanis, A., Reavely, G.H., Cook, R.J.B., Knox, K.S., and Moretti, F.J., 1957, Heavy mineral studies in tills of Ontario and adjacent areas: Jour. Sed. Petrology, v. 27, p. 148–61.

Dreimanis, A., and Vagners, U.J., 1965, Till-bedrock lithologic relationship: Abstracts, INQUA VII Internat. Congr. Gener. Sess., p. 110–11.

——, 1969, Lithologic relation of till to bedrock, *in* Wright, H.E., Jr. (ed.), Quaternary geology and climate: Washington, D. C., Nat. Acad. Sci., p. 93–98.

Elson, J.A., 1961, The geology of tills: Proceed. 14th Can. Soil Mech. Confer., Nat. Res. Counc. Can. Assoc. Com. Soil and Snow Mech. Techn. Mem. no. 69, p. 5–36.

Flint, R.F., 1957, Glacial and Pleistocene geology: New York, J. Wiley and Sons, 553 p.

Gaigalas, A.I., 1964, Mineralo-petrograficheskii sostav moren pleistocena iugovostochnoi Litvy: Lietuvos TSR Mokslu Akad. Darbai, ser. B, v. 4 (39), p. 185–211.

Gaudin, A.M., 1926, An investigation of crushing phenomena: Transact. Amer. Inst. Min. and Metallurg. Engineers, v. 73, p. 253–316.

Harrison, P.W., 1960, Original bedrock composition of Wisconsin till in central Indiana: Jour. Sed. Petrology, v. 30, p. 432–446.

Raukas, A., 1961, Mineralogiia moren Estonii: Eesti NSV Teaduste Akad. Toimetised X Koide, Füüsik.-Matemat. ja Tehn. Teaduste Ser. Nr. 3, p. 244–58.

Savvaitov, A.S., 1962a, O sostave melkozema morennykh otlozhenii v baseine r. Salaca, *in* I. Danilans (ed.), Questions on Quaternary geology, I: Akad. Sci. Latvian SSR, Instit. geol., Riga, p. 115–22.

——, 1962b, O soderzhanii tiazhelykh mineralov v morennykh suglinkakh, *in* I. Danilans (ed.), Questions on Quaternary geology, I: Akad. Sci. Latvian SSR, Inst. geol., Riga, p. 123–28.

Sindowski, F.K.H., 1949, Results and problems of heavy mineral analysis in Germany: a review of sedimentary-petrological papers, 1936–1948: Jour. Sed. Petrology, v. 19, p. 3–25.

Smeck, N.E., Wilding, L.P., and Holowaychuk, N. 1968, Properties of argillic horizons derived from Wisconsin-age till deposits of varying physical and chemical composition: Proc. Soil Sci. Soc. Amer., v. 32, p. 550–56.

Stinkule, A.V., 1964, O raspredelenii khimicheskikh elementov v melkozeme moreny, *in* I. Danilans (ed.), Questions on Quaternary geology, III: State Com. Geol. USSR, Instit. geol., Riga, p. 311–20.

Vagners, U.J., 1966, Lithologic relationships of till to carbonate bedrock in southern Ontario: M.Sc. thesis, Univ. West. Ontario, London, Canada. 154 pp.

——, 1969, Mineral distribution in tills, south-central Ontario: Ph.D. dissertation, Univ. West. Ontario, London, Canada, 270 p.

Wentworth, C.K., 1922, A scale of grade and class terms for clastic sediments: Jour. Geology, v. 30, p. 377–92.

Wilding, L.P., Dress, L.R., Smeck, N.E., and Hall, G.F., 1971, Mineral and elemental composition of Wisconsin-age till deposits in west-central Ohio, *in* this volume.

11

THE ACTIVE PUSH MORAINE OF THE THOMPSON GLACIER

Axel Heiberg Island, Canadian Arctic Archipelago

M. Kälin

Source: Axel Heiberg Island Research Reports, Glaciology, No. 4, Montreal: McGill University, 1971, 68 pp.

Chapter 1
Introduction

Definition

The definition of a push moraine is given by Chamberlin (1890, p. 28): "A glacier pushes matter forward mechanically, ridging it at its edge, forming what may be termed push moraine".

The definition implies that the glacier, considered to be the cause, is acting on predeposited material and is producing a feature resembling a terminal moraine, the push moraine. Geologically this is a descriptive as well as a genetic definition of a feature resulting from glacial erosion and redeposition. Mechanically this means that the glacier is superimposing a variable stress field on the underlying material. The stresses exceed the strength properties of the material involved. Reiner (1960, p. 95) refers to two independent types of strength: resistance against rupture and resistance against plastic deformation. Thus we can understand the push moraine as a failure zone. This phenomenon is surely not restricted to the observable part; it is bound to extend underneath the glacier.

For the purpose of a general description the push moraine system consists of three parts: the glacier, the underlying material and the push moraine, the latter being the result of an interaction between the two former elements.

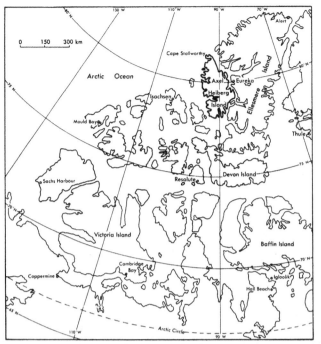

Canadian Arctic Archipelago

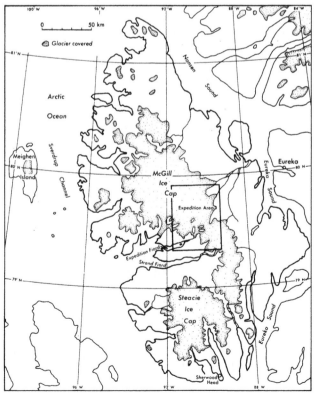

Axel Heiberg Island

Previous work

There are three groups of records: 1) observations of recent push moraines from alpine regions in moderate latitudes, 2) observations of recent push moraines in the Arctic, and 3) observations of fossil push moraines.

De Charpentier (1841, p. 41) describes the first type of push moraines.

> En 1818, le glacier de Schwarzberg, dans la vallée de Saas, déplaça un bloc de serpentine qui, par ses dimensions gigantesques, avait attiré l'attention des habitants de la vallée, qui l'appellent le Blaustein. Ce bloc a, d'après les mesures de Mr. Venetz, 244,000 pied cubes de volumes (approx. 9,000 m³). Cette même année, où les glaciers acquirent un developpement extrêmement considérable, nous avons vu le glacier de Trient détruire une portion de forêt en s'insinuant entre le roc vif et la terre, et renverser sur lui-même le terrain dans lequel les arbres étaient enracinées.
>
> En 1818, le glacier de Tour dans la vallée de Chamonix avait avancé, sans creuser, d'environs 80 pieds (approx. 24 m) sur un terrain graveleux et dégarni de terre; mais au bout de cet espace il rencontra des prairies, dont le sol était de la terre un peu maré-cageuse, fut entièrement soulevé et bouleversé.

Already in the 19th century his observations were confirmed by Heim (1885, p. 377). Instructive further examples are given, for the Alps by de Quervain (1919, p. 336), and, more recently, for a push moraine associated with a surging glacier in the Rocky Mountains, Canada, by Bayrock (1967, p. 16). The glaciers involved are advancing and of the alpine type. The push moraines consist of loose, soaked drift and soil in an unfrozen state. Height and width seldom exceed some metres while the extent along the glacier edge depends on the type and distribution of the material in front of the push-ing glacier.

The only arctic region extensively studied for push moraines is that of Spitzbergen (Gripp and Todtmann, 1926, p. 43; Gripp, 1927, p. 145). Gripp enumerates 15 push moraines including three uncertain cases. The glaciers, usually outlets of ice caps, are either advancing or showing signs of sub-recent advance. The tongues are often several kilometres wide. The push moraines consist of clastic rocks which are mostly permanently frozen. They are generally half-moon shaped and up to about 5 km long and 1 km wide. These push moraines differ from those of the moderate latitudes mainly by the frozen state of the material and the size which is greater by about two orders of magnitude. Isolated push moraines are known from Novaya Zemlya (Gripp, 1927, p. 226) and Greenland (M. Allen, ETH, Zürich, personal communication). From the Canadian High Arctic, Robitaille and Greffard (1962, p. 85) describe the push moraine of the Thompson Glacier, which is the subject of the present, more detailed report, and mention some 30 additional push moraines in this region.

Johnstrup (1874) first interpreted fossil disturbances as resulting from ice thrust. He noticed that the deformations of Cretaceous and Tertiary sediments which have been discovered by Puggaard (1851) on Møen Island, Denmark, do not extend to depth and concluded an exogenic cause, namely the movement of Pleistocene glaciers. Numerous observations from the marginal zones of the Pleistocene glaciations in northern Europe, the Alps and North America support Johnstrup's interpretation (Berger, 1937, p. 417; Bersier, 1954; Brinkmann, 1953, p. 231; Bubnoff, 1956, p. 557; Carlé, 1938, p. 27; Credner, 1880, p. 75; Fries, 1933; Fuller, 1914; Hopkins, 1923, p. 419; Jessen, 1928; Kupsch, 1962, p. 582, Lamerson and Dellwig, 1957, p. 546; Lang, 1964, p. 207; Leffingwell, 1919; Mackay, 1956, p. 218, 1959, p. 5; Roethe, 1930, p. 498; Schott, 1933, p. 54; Schulz, 1965, p. 564, 1966, p. 174; Wahnschaffe, 1882, p. 562). The zones of disturbed material often form elongated hills and clearly reflect in vertical view the shape of the ice margins. These zones can be tens of kilometres long and some kilometres in width. The deformations described occur usually to a depth of some 10 m, however, in exceptional cases, to some 100 m. The strata involved are normally of Pleistocene to Upper Cretaceous age. The deformations observed contain folds of all types and sizes with amplitudes that may reach hundreds of metres and wave-lengths of some kilometres. Overthrusts, faults and rock fracturing are some of the frequently mentioned characteristics.

Scope of the study

The present report gives a field description of the high arctic push moraine of the Thompson Glacier, emphasizing measurements of the mechanical behaviour. A simple theoretical model is developed. Some further push moraines on Axel Heiberg Island and Ellesmere Island were studied on air photos.

Chapter 2
The active push moraine of the Thompson Glacier

General description

Axel Heiberg Island is the second northernmost island of the Canadian Arctic Archipelago, situated at the border of the Arctic Ocean and to the west of Ellesmere Island (Frontispiece maps). It measures roughly 400 km by 200 km with its longitudinal axis running NS. Two ice caps in the interior cover about one third of the area; their outlet glaciers usually do not reach sea level. There are push moraines in front of several of the glaciers that reach low level. In the Expedition Area a well developed push moraine

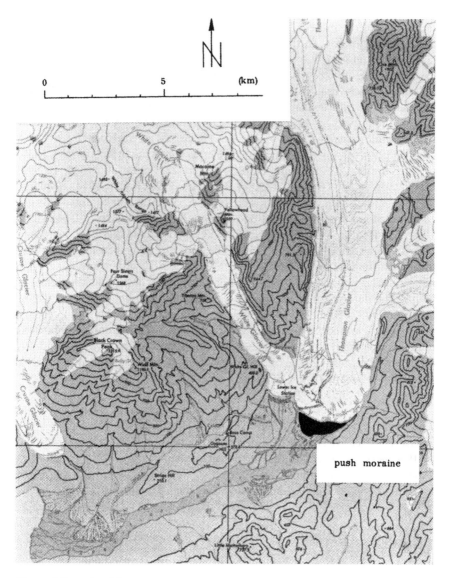

Figure 1 Location Map.

is located in front of the Thompson Glacier (Fig. 1), which is advancing in the Expedition River valley. The valley filling consists, at least on the surface, of permanently frozen fluvioglacial sediments which form a gently dipping outwash plain. The snout of the Thompson glacier is bulldozing the frozen detritus to a push moraine. The more or less stationery White Glacier is partly coalescing with the Thompson Glacier snout and accumulating a terminal moraine. The fundamental difference between the two types of

226

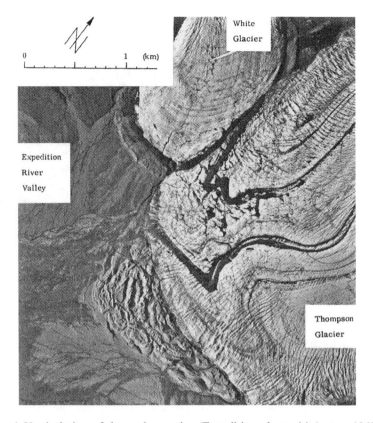

Figure 2 Vertical view of the push moraine (Expedition photo, 14 August 1967).

moraine is readily visible on the aerial photograph (Fig. 2). The push moraine appears as a half moon-shaped bulge, subdivided into ridges running roughly transversely to the glacier. In oblique view, however, it appears as an agglomeration of frozen blocks (Fig. 3). The geographical co-ordinates of the centre of the push moraine are 79° 24′ N and 90° 34′ W.

The Thompson Glacier

The Thompson Glacier is an advancing outlet of the McGill Ice Cap. The accumulation and the ablation areas of the main stream face south. The mean width of the main stream measures some 3 km and the down-valley dip of the upper surface in the ablation area some 1.5°. The accumulation area measures about 180 km^2; it consists of compound basins integrating over several sub-basins of different size, height above sea level, exposition and relation to the main stream. The complex composition of the accumulation area projects itself in the ablation area in the form of an equally complex

227

Figure 3 Oblique aerial photograph of the push moraine (Photo F. Müller, August 1961).

picture of medial moraines (Fig. 4). These are intersections of intraglacial moraines with the topographical surface. The most striking of these surface structures are folds along the west side of the lower main stream. An S-fold with an axial plane, probably close to the horizontal, is exposed at the frontal cliff of the glacier (Fig. 5). These structures may be explained as the result of intermittent advances of the tributary glaciers. The front of the glacier is rimmed in the centre and on the east side over a distance of about 2.0 km by the push moraine and on the west side by an ice cliff of 30 to 50 m high and about 0.8 km long.

Ablation occurs through melting and calving. Breaking-off of ice blocks along the front takes place all year round. The ice debris zone extends as far as 80 m ahead of the glacier. A single collapse may contain several 1,000 m³ of ice. Ablation through melting was recorded in a transverse profile about 8.5 km up-glacier from the front at the Eureka Profile (Müller, 1963a, p. 66). Twelve ablation stakes, situated 425 to 475 m a.s.l., yielded a mean ablation of 133 cm/yr over a period of 743 days (1960–62).

In the same profile, surface movement rates were determined (Müller, 1963a, p. 69). The velocity profile is U-shaped and fairly symmetrical. In the centre the long term velocity was found to be 19.0 cm/day increasing to 21.8 cm/day during the summer (43 days). The profile measures 3,170 m across.

Despite the push moraine, the advance of the glacier front is fairly uniform over the entire width (Fig. 6). The maximum of advance during the 1960–67 period amounts to 180 m, that is 26 m/yr or 7.1 cm/day, the average for the same time lapse being 160 m, that is 23 m/yr or 6.3 cm/day.

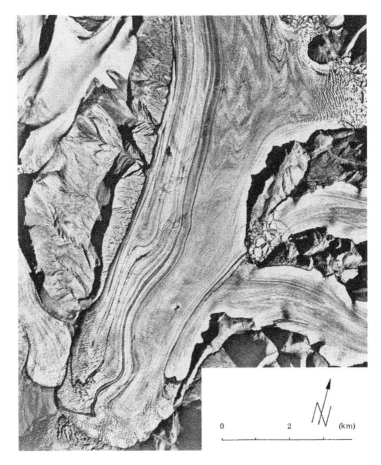

Figure 4 Lower ablation area of the Thompson Glacier (part of No. A 16,755–118, August 1959, courtesy RCAF).

The mean direction of the advance has an azimuth of 210°. The year-to-year displacement of the cliff was recorded for the interval 1960–68 (Fig. 7). From two base points which allowed different tangential views of the cliff, a point has been intersected ahead of the outward-convex cliff. The point is projected in a direction parallel to that of the estimated flow (azimuth 230°). Redetermining this point yields the displacement of the cliff. The displacement refers to the upper edge of the cliff which usually forms an overhang. Figure 7 yields values for the velocity of the ice at the upper front edge, reduced by the cliff ablation which is estimated to be some 6 m/yr. Over the eight years of observation the advance is fairly constant.

   Glacial thickness along the Eureka Profile was measured by the gravity method (Becker, 1963b, p. 98). Becker's measurements have been verified at

Figure 5 S-fold exposed on the Thompson Glacier cliff. The cliff is some 40 m high. (Photo: F. Müller, 1961).

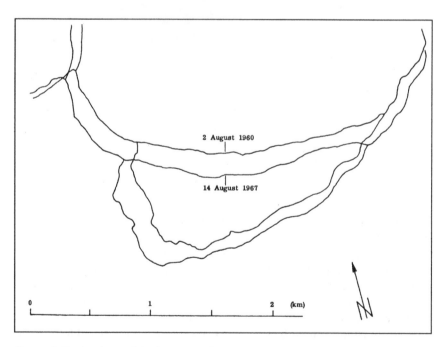

Figure 6 Comparison of the frontal positions of the Thompson Glacier and the push moraine based on the topographical map (1960) and on the orthophoto map (1967).

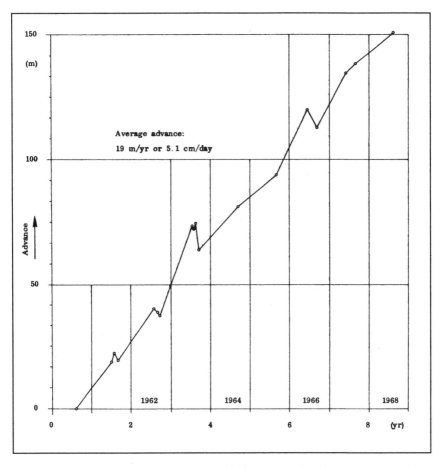

Figure 7 Displacement of the Thompson Glacier cliff 1960–68 (partly after Müller, 1963a, p. 76).

one point by seismic methods (Redpath, 1965, p. 20). The cross-section looks roughly like a segment; the maximum depth is 500 m and reaches up to about 50 m below sea level in a central section of 700 m width.

Despite the heterogeneous and asymmetrical appearance of the lower ablation area these kinematic and geometric observations indicate a fairly symmetrical dynamic behaviour of the snout region with respect to the longitudinal vertical axial plane.

The situation in the terminal region of the Thompson Glacier is complicated by the White Glacier joining it from the northwest. The mass of the White Glacier in the tongue region is approximately one order of magnitude smaller than that of the Thompson Glacier and is therefore forced to change its course. The Thompson Glacier does not seem to be much influenced by this contact.

The Expedition River valley

The geological structure of the valley is characterized by faults at both sides trending in a north-easterly direction (Fricker, 1963, p. 138). The tectonics seem to have predetermined the course of the Thompson Glacier either by a depression already present before the glaciations, or by easing the formation of a valley during the glacial fluctuations.

After more than 10 km of an almost straight course (azimuth 7°), the Thompson Glacier enters a bend of the Expedition River valley changing direction by 58° (azimuth 65°) (Fig. 1). In this area the width of the outwash plain narrows from 2.0 km to 0.5 km. The character of the valley changes at this bend: leaving a structually well-marked path, it cuts the structural trend and crosses an evaporite diapir, the Expedition Diapir (Hoen, 1964, p. 37) (Fig. 8). This diapir has influenced the morphology remarkably. The anhydrite and gypsum are more resistant to the erosion than other sediments in the area (limestone, shales, etc.). The mean slope of the Expedition River from the terminus of the Thompson Glacier to the shoreline of the Expedition Fiord, a distance of 13 km, is less than 0.15°. On both sides of the push moraine large outwash fans form with surface slopes of 1°, steepening at their apexes to 1.5°. They are about 1.5 km in length and 0.8 km in width (Fig. 8). In down-valley direction the sediment gradually changes from a coarse gravel near the glacier to a fine sand or silt on the beach.

Müller (1963b, p. 169) has clarified some of the valley's recent history. Two important findings are: evidence of post-glacial emergence of this valley, and 5,000 to 6,000 years ago the Thompson Glacier terminated at least 1.5 km further up-valley. It can be assumed that marine deposits,

Table 1 The Thompson Glacier.

Highest glacier elevation	(1750 ± 30) m a.s.l.
Lowest glacier elevation	(40 ± 10) m a.s.l.
Elevation of snow line	(800 – 1200) m a.s.l.
(Variation due to exposition, data valid for end of July 1959)	
Mean accumulation area elevation	(1350 ± 50) m a.s.l.
Mean ablation area elevation	(600 ± 50) m a.s.l.
Maximum length of ablation area	(30.1 ± 0.1) km
Total glacier length	(35.3 ± 0.1) km
Mean width of main stream	approx. 2.9 km
Slope of accumulation area	(4 – 7)°
Slope of ablation area	approx. 1.5°
Total surface area	(310 ± 30) km^2
Area of accumulation	(180 ± 20) km^2
Area of ablation	(130 ± 20) km^2
Accumulation area ratio	59%
Mean depth	(240 ± 30) m
Volume	(70 ± 15) km^3

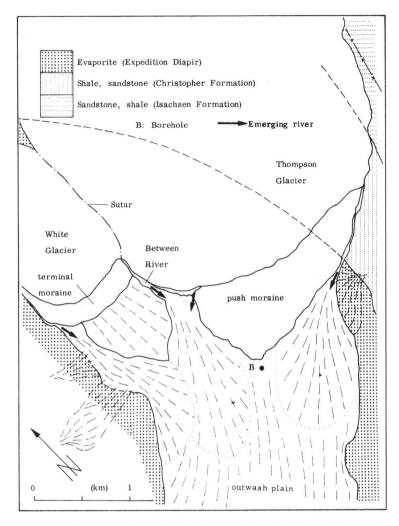

Figure 8 General geomorphic and geologic map (partly after Fricker, 1963).

buried underneath the alluvials, extend up-valley beyond the present glacier terminus. The outwash fan in front of the White Glacier is about 2200 years old and covered by pioneering vegetation, contrary to the completely barren outwash plain and the recent fans.

The development of the outwash complex is controlled by the cumulative action of the meltwater from the glaciers and the peculiarities of the temperature regime in the area. The geological cycle is characterized by approximately 8 months of subfreezing temperatures. During this time the surface activities are restricted to snow accumulation and snow drift.

Melting usually starts in May. Before run-off sets in, much of the snow disappears through sublimation and evaporation especially where the snow cover is discontinuous. In 1968 stream channels and hollows were filled up to 1 m with snow whereas the upward-convex areas were mainly bare. The first melt was recorded in front of the push moraine on 1 May. Later, slush formed still retaining the meltwater. During the night or periods of subfreezing temperature the slush was covered with a lid of ice. The melt in the push moraine, the White Glacier moraine and on the surrounding slopes proceeded faster than in the outwash plain. Gradually water began to flow in selected channels, favouring the largest ones. At the end of May and in early June the run-off built up to its maximum within about one week. Four main rivers emerged: two lateral ones, one to the west of the White Glacier and one to the east of the Thompson Glacier; one interglacial river between the White and the Thompson glaciers, the Between River; since 1967 a fourth river has been observed to originate intra- or subglacially in the corner between the Thompson Glacier and the push moraine. Especially the early floods transport considerable ice masses that later on form kettle holes in the active layer if embedded in the sediment. Upon entering the outwash plain, the rivers branch in a network of smaller and smaller streams. The largest channels near the glacier front measure up to some 300 m^2 in cross-section and some 10 m in depth (Maag, 1969, p. 81). The rivers are able to transport boulders weighing up to more than one ton. In the autumn the amount of running water decreases gradually and slowly. The main channels still carried water at the end of the observation period (end of August or early September).

In the Expedition River valley the processes of accumulation, erosion and differentiation of the material by proglacial rivers were studied by Maag (1969, p. 71). From his findings it can be deduced that the sediment varies in horizontal and vertical direction considerably in grading, maximum grain size and ice content. The degree of heterogeneity is increased by the occurrence of embedded fluvial ice and aufeis which form layers up to 1 m thick.

In 1968 a borehole was drilled in front of the push moraine and in the middle of the valley (Fig. 8, local coordinates 60244/32727) to obtain the depth to the bedrock, the stratigraphical profile of the detritus and the geothermal profile. Due to the technical difficulties of drilling in permanently frozen detritus the borehole did not reach the bedrock. The results are summarized in Figure 9. The lithological profile is based on a few core barrel samples, backwash samples and the drillers' experience. The sediment varies between a coarse gravel and sand, cemented by ice. Near the surface (approx. 23 m a.s.l.) there is outwash material. In the first few metres the grain size gradually decreases downwards. This observation cannot be confirmed in the push moraine, where some 5 m of the outwash is exposed. The change from alluvial outwash to marine delta sediment can be expected at a depth of some 20 m, but was not observed. At a depth of 33.7 m the

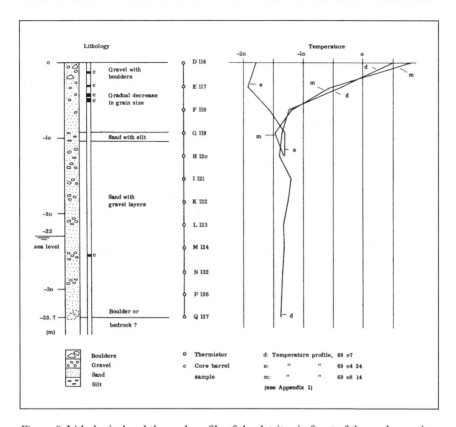

Figure 9 Lithological and thermal profile of the detritus in front of the push moraine.

drillers struck large boulders (diorite, sandstone) or possibly the bedrock. The drilling had to be abandoned, the hole was rinsed with water and a thermocable inserted, containing thermistors every 3 m. By extrapolation from the cooling rates the predrilling temperatures could be established (curve d in Fig. 9). Some 50 hours after termination of the refreezing, the thermocable ruptured between 12.4 and 15.4 m in depth. The geothermal data are listed in Appendix 1.

In late August 1968 the active layer of outwash was investigated in two pits, one situated 35 m to the south of the borehole, the other equidistant in the opposite direction. Water was found at 0.7 to 0.9 m below the surface. The permafrost table could not be reached. It was deeper than 1.1 to 1.3 m below the surface.

The geothermal data allow an estimation of the depth of the permafrost. The predrilling temperature at the lowest thermistor at a depth of 33.7 m was recorded to be $-13.8 \pm 0.3°C$. At this depth the seasonal temperature fluctuations can be neglected. The geothermal heat flow of the region, geometry and thermal conductivity of the underlying material are not known. As a first approximation the world mean geothermal gradient of $3.3 \cdot 10^{-4}$ °C/cm is assumed yielding the depth of the permafrost to be about 450 m. The possible positive geothermal anomaly caused by the Expedition Diapir (indicated by sulphur springs; Beschel, 1963), the ocean and the glaciers all of which are heat sources, suggests that the estimated depth of the permafrost is an upper limit.

In early May 1967 a fissure was observed approximately 1 km in front of the push moraine in the centre of the valley. Vertically intersecting the surface, it was some 10 m long and a few millimetres wide and was the only crack of this kind.

The push moraine

Geologic-hydrologic, structural and tectonic maps

Viewed from the air the push moraine is approximately sickle-shaped (Fig. 10). It is, however, noticeably asymmetrical with the eastern end gradually narrowing and the western outer edge abruptly turning towards the glacier. The "missing" portion is marked on Figure 10.

Three structural trends are recognizable in the push moraine. Two of them are straight and intersect at approximately right angles (I and II in Fig. 10). One bisector is approximately parallel to the axis of the Expedition River valley (azimuth 50°), the other normal to it. These two structures are most strongly developed near the outer edge of the push moraine. Thus, it seems more appropriate to describe the shape of the push moraine as a right-angled triangle with the sides enclosing the right angle coinciding with the structural directions I and II. The third structural trend is circular-concentrical (III in Fig. 10), the centre of the circle being some distance up-glacier. This trend is particularly well developed near the glacier and in the centre of the push moraine. Morphologically, the three structural trends express themselves as valleys and ridges.

Four maps (Figs 11–14) are used to present the main aspects of the moraine as an entity. Low-level aerial photographs (flying height 2,000 to 4,000 m above ground), taken on 14 August 1967, formed the basis of an orthophoto map at a scale of 1:5,000 and a 5 m contour interval map, at the same scale. These maps, prepared by the Photogrammetric Research Section of the National Research Council of Canada in cooperation with the expedition, were used for the field mapping program in 1968. Changes in the topography of the push moraine between August 1967 and the time of mapping were taken into consideration.

The geologic-hydrologic map shows that the delineation of the push moraine is uncertain in two areas: 1) along the glacier edge where ice blocks fall and subsequently melt or become incorporated again in the advancing

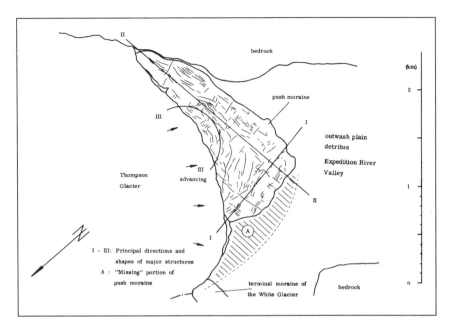

Figure 10 Situation 1967: Plan view of glacier snout, push moraine and outwash plain (from air photos).

glacier, and 2) along the front of the push moraine where fans obscure the boundary. The mapping of the geologic and hydrographic features yields the following main points of observation. 1) The push moraine consists mainly of deformed outwash which is locally modified by the activities of water at the surface. 2) Along the glacier edge is a narrow strip of some 50 m width of till and ice blocks. 3) Scattered throughout the push moraine are lakes and lacustrine sediments. They are less frequent in a belt of some 200 m along the outer edge of the push moraine. This belt does not extend to the glacier on the western side. 4) A very complex drainage system with truncated valleys, many river sources and sinks dissects the push moraine. 5) A very large river emerges at the western intersection of the glacier with the push moraine.

The structural map 1 (Fig. 12) shows S- and thrust surfaces. The S-surfaces are sedimentary upper surfaces of outwash blocks often with a pattern of river channels. They are almost flat. Each block of outwash with a surface larger than some 10 by 10 m was measured and represented by a point. The azimuth of the dip is given by the direction of a line; the length of the line, subdivided into 5 classes, is proportional to the dip. Since each dot represents a block, this graph readily shows the frequency distribution of the blocks. Due to the unreliability of the magnetic compass on Axel Heiberg Island, the azimuth had to be taken from the 1967 map whereas the dip was measured in the field in 1968 to an accuracy of 2°.

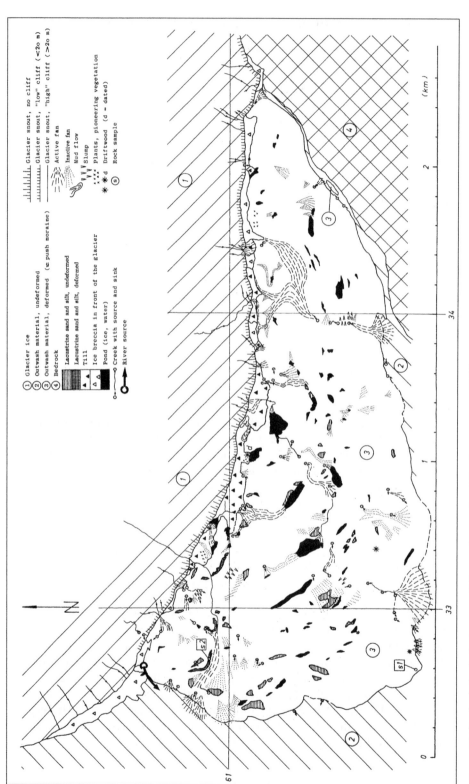

Figure 11 Geologic-hydrologic map of the push moraine (situation summer 1968).

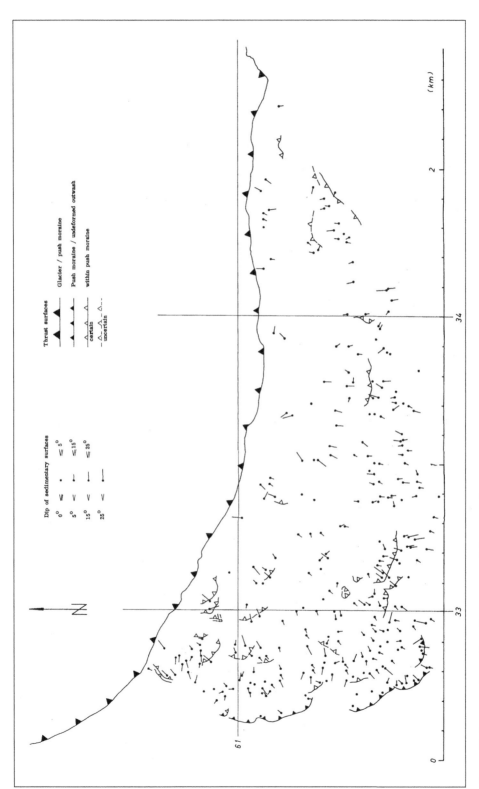

Figure 12 Structural map 1, sedimentary and thrust surfaces.

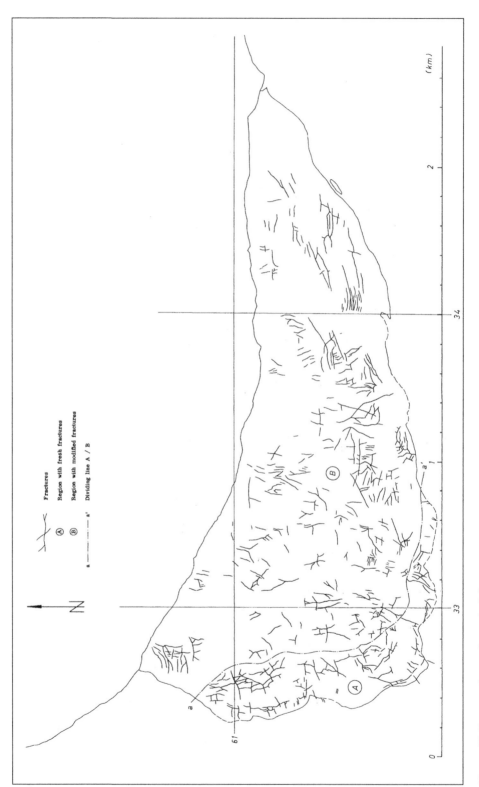

Figure 13 Structural map 2, vertical fractures.

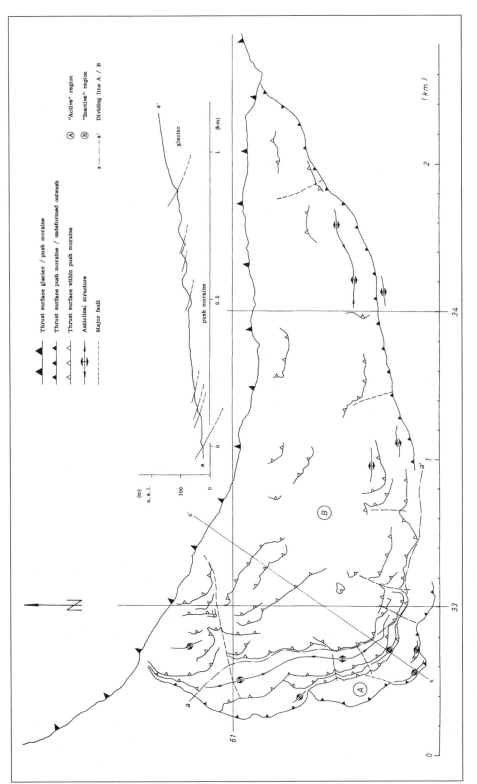

Figure 14 Tectonic map with cross-section.

Three types of thrust surfaces can be distinguished: those of the glacier over the push moraine, those of the push moraine over the undeformed outwash and those within the push moraine.

In summary the map shows: 1) The individual blocks are most frequent in the western portion of the push moraine. As the blocks decay rather rapidly, it could be concluded that the push moraine is younger and more active in its western and southern portions than in the interior and on the eastern side. The uneven density of blocks indicates differences in the development of the push moraine. In fact, a comparison of the 1960 and 1967 positions of the outer-convex boundary (Fig. 6) shows that the push moraine is shifting toward the centre of the valley. 2) The general trend of the dip directions is fan-shaped and runs approximately parallel to the dip of the thrust surfaces.

In Figure 13 a qualitative distinction is made between fresh and modified fractures. The former show little decay, are usually vertical and are clearly restricted to a belt in the outermost part of the push moraine. They can readily be distinguished from the older fractures covering the rest of the push moraine (see line a – a' on Fig. 13). The older fractures are modified by the action of meltwater and often filled with gravel. The dip of the fractures seems generally to be vertical or close to vertical.

Figure 13 was drawn on the basis of field observations. The three principal directions visible on Figure 10, recognized from great height, are no longer obvious on the ground.

The tectonic map (Fig. 14) is to some extent a summary of Figures 11–13. A distinction between an "active" (A) and an "inactive" (B) region is made. Region (A) contains more detail owing to its well preserved state. Eluvial and fluvial erosion effects are minimal. This means that the region was the last to be integrated into the push moraine. At present it is subjected to more modification than region (B).

Viewed as a whole the push moraine appears as a pile of layers separated by thrust surfaces and by faults of roughly radial trend. The base of the push moraine may be of upward-concave shape.

Petrology

In Table 2 the various rock types of the push moraine are synoptically presented in a genetic classification. The outwash is a grey, coarse and well-graded gravel cemented by interstitial ice (Fig. 15). The sedimentological data for this material are presented in Appendix 2 and the granulometric results are shown in Figure 16. The logarithmic grain size distribution for grains with diameters greater than about 6 mm closely approximates a Gaussian distribution (Appendix 3). Thus the mean grain size of 3.7 cm was found with a standard deviation range of 0.58 cm to 24 cm. On the basis of this approximation, it would seem that about 5% of the particles exceed a diameter of 80 cm. Particles larger than 1 cm (about 75% of the total volume)

Table 2 The rocks of the push moraine.

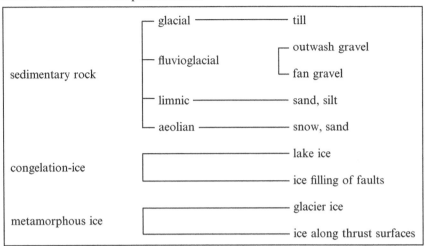

sedimentary rock	glacial	till
	fluvioglacial	outwash gravel
		fan gravel
	limnic	sand, silt
	aeolian	snow, sand
congelation-ice		lake ice
		ice filling of faults
metamorphous ice		glacier ice
		ice along thrust surfaces

Figure 15 Outwash gravel. The rod measures 3.2 m. The photograph was taken near the front of the push moraine and shows the side of a crack formed in the second outermost rim.

are mainly composed of sandstone and basalt and are mostly of subangular shape.

The surface slope of the gravel in the unfrozen state dips at an angle of 29° to 35°, averaging 32°. The interstitial ice coats the particles with a film up to some millimetres in thickness.

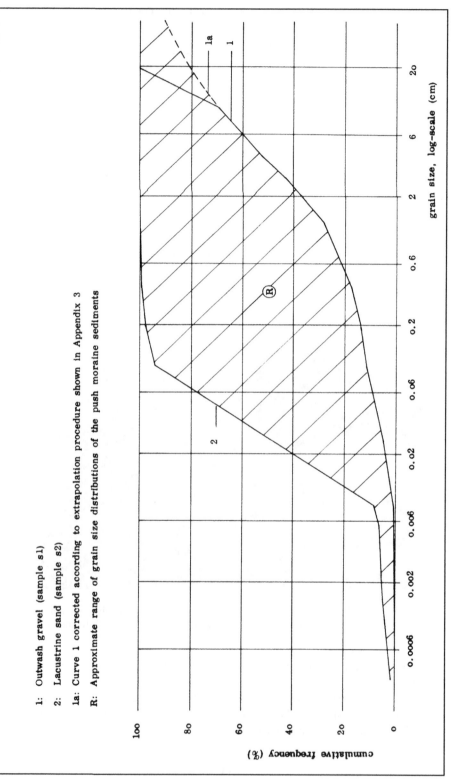

1: Outwash gravel (sample s1)

2: Lacustrine sand (sample s2)

1a: Curve 1 corrected according to extrapolation procedure shown in Appendix 3

R: Approximate range of grain size distributions of the push moraine sediments

Figure 16 Cumulative frequency curves of grain size distributions of outwash gravel and lacustrine sand.

The outwash gravel as a whole appears fairly homogeneous, but if observed in bulk of several cubic metres it proves to be quite heterogeneous. This is illustrated by Figure 15 where the finer grained band to the right of the rod results from a channel filling.

Fan gravel is a grey, coarse and well-graded sediment very similar to the outwash gravel. It differs mainly from the outwash by its smaller maximum grain size of about 20 to 50 cm compared to about 150 cm for the latter. This corresponds to the smaller transportation capacity of the push moraine rivers. Probably the fan gravel is less well graded then the outwash gravel.

Lacustrine sediments are grey sands in layers of up to about 1 m thickness. They are stratified in the order of centimetres to decimetres; the strata vary between fine sand with silt and coarse sand with pebbles. The profiles barely correlate over more than a few decametres. The sedimentological data are given in Appendix 2 and the granulometric results in Figure 16. These sediments mainly originate from outwash. An approximately Gaussian distribution is maintained for grain diameter greater than 0.007 cm (Appendix 3). Thus a mean diameter of 0.027 cm was found with the standard deviation limits at 0.010 cm and 0.068 cm. The top layer often consists of some centimetres of stratified silt with a surface crust.

Based on the data given in Appendix 2 the density of outwash gravel and lacustrine sand is estimated. The limits of density, assumed to correspond to an ice-saturation of 80% and 100%, are 2.1 g/cm^3 and 2.3 g/cm^3 for the outwash gravel and 1.7 g/cm^3 and 1.9 g/cm^3 for the lacustrine sand.

Till is a grey to dark grey, poorly-graded sediment lacking any stratification. The particles reach up to several metres in diameter. They vary in shape from round to angular and are occasionally striated. The bulk of the sediment consists of glacier ice. The active layer of the till complex occasionally forms mud-flows. Till is restricted to a belt of some 50 m in width along the glacier front (Fig. 11) and is interspersed with ice breccia originating from ice falls.

It is possible to interpret this belt as the remainder of the frontal glacial ablation. If the glacier's advance on to the push moraine is less than the ablation in front, the accumulation of till and ice blocks is inevitable.

Aeolian sediments occur as snow and sand. As a result of the semi-arid climate the uppermost sediment layer is usually dry which eases the corrosion. During storms, sand and smaller particles are transported in dust clouds.

The mostly clear lake ice is covered with a layer of stratified white ice. The maximum thickness was not recorded, but is estimated not to exceed 2.5 to 3 m. The white ice reaches several decimetres in thickness. The lakes stay ice-covered for about ten months (September to June). The melting of the ice begins at the edges and at the top.

The ice fillings of vertical faults are seldom exposed. Figure 17 shows a characteristic example. Along the crack edge there is a layer of clear ice with

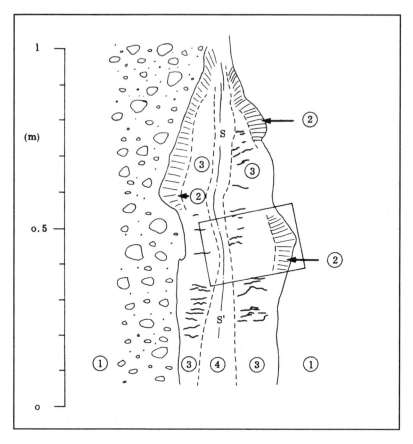

1: Outwash gravel.
2: Clear ice with isometric grains which become visible
   through air tubes running along the grain edges.
   The average grain diameter is approximately 1 cm.
   Elongated air bubbles, straight to slightly curved,
   intersect roughly at right angles with the edge of the
   fault.
3: Clear ice with isometric grains of 0.5 to 1.0 cm in diameter and much less air inclusions than in 2. The
   bubbles are elongated and knobby.
4: White ice with grains of 0.1 to 0.2 cm in diameter and
   isometric air bubbles.
S - S': Suture marked by dirt and a change in the air
   bubble content.

Figure 17 Ice filling of a vertical fault (cross-section).

Figure 17a Detail from Figure 17.

tubular air bubbles aligned vertically to the boundary between ice and outwash. The central slit is filled with white ice containing isometrical air bubbles, often subdivided by a vertical suture. Sometimes the ice filling is subdivided into more or less horizontal layers (Fig. 18) which vary in thickness from a few centimetres to a few decimetres.

These observations lead to the conclusion that the ice of vertical faults is congelation ice (Shumskii, 1964, p. 175). The ice in Figure 18 would be a one-time crevasse filling whereas the horizontal structure of the ice in Figure 19 may originate from an intermittant water supply.

Slickensides of deformed ice cover about half the exposed surfaces of the push moraine's outermost thrusts (Fig. 12). The slickenside ice measures up to several decimetres in thickness and is restricted to the surface of the upper, i.e., overthrusting limb. It is a milky ice with mosaic-shaped grains of 0.5 to 1.0 cm in diameter. The air bubbles are isometric or elongated reaching up to 0. 1 cm in diameter. Dirt particles contribute up to 2% of the mass and are distributed either uniformly or, more frequently, in layers. Dirt particles and air bubbles mark a lineation which is roughly parallel to the thrust surface and to the direction of the displacement. The orientation of the crystallographic c-axis of the ice was determined by a universal stage in one thin section. The fabric revealed a monoclinic symmetry. The c-axis lay normal to the thrust plane and parallel to the direction of displacement, with a weak frequency maximum normal to the thrust plane.

Figure 18 Ice filling of a vertical fault (longitudinal section). The exposed shield measures about 2 × 2 m.

Figure 19 shows the genetic relation of the examined slickenside ice to that of vertical faults because this ice layer seems to have been protected against large shear strains during the overthrust process. Next to the contact area of ice and outwash gravel, clear ice of several centimetres thickness containing tubular air bubbles was found. The elongations of the tubular air bubbles' axes intersect the contact area on a slant or vertically. The protruding part to the left was deformed and bent back, still leaving an opening and a suture.

It is concluded that the slickensides consist of congelation ice differing from that of the vertical cracks only by its dynamometamorphous state.

Based on the numerous surface exposures it is estimated that the push moraine consists of up to 90% by volume of outwash gravel. Till and fan gravel each cover some 5% of the surface. It is uncertain whether they are restricted to the surface or extend to greater depth. Lacustrine sand covers some 1% of the surface. The volume of the other rock types is negligible. But the findings of Maag (1969, p. 110) indicate that considerable masses of ice in the ground may occur, i.e. the judgement given could be misleading because of this unknown quantity.

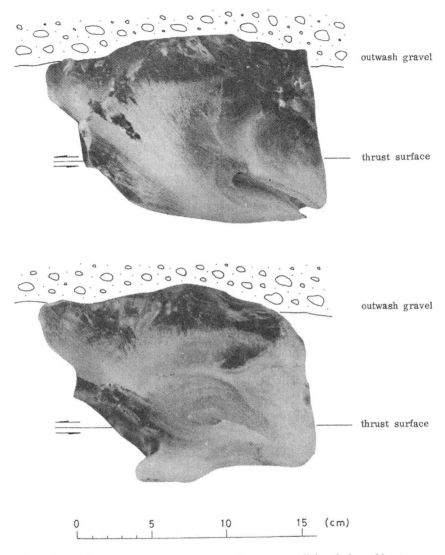

outwash gravel

thrust surface

outwash gravel

thrust surface

0   5   10   15   (cm)

Figure 19 Slickenside ice. The two cross-sections are parallel and about 20 cm apart.
The thrust surface is approximately a plane of simple shear deformation.

Hydrology

Water occurs in the push moraine during about three to four months of the
year (end of May to late August). It originates from the glacier or from
the push moraine itself. The glacier contributes water from a catchment area
which covers about 2 km^2 and ranges in elevation from about 100 to 300 m
a.s.l. Based on the ablation measurements from the White Glacier (Müller,

249

1963c, p. 38) the year to year variation of the average net ablation in the catchment area is estimated to vary between about 1 and 3 m of water equivalent. According to Andrews (1964, p. 18) and Müller and Roskin-Sharlin (1967, p. 21) the annual amount of rain can be estimated to be less than 0.1 m for the same area, i.e., the glacier contributes between about $2 \cdot 10^6$ and $6 \cdot 10^6$ m³/yr of water, which is mainly glacial melt water. Within the push moraine water is obtained from the melting of the seasonal snow cover, meteoric water and from the ice in the active layer. The amount of glacial melt water is estimated to exceed these water resources by an order of magnitude. The discharge of the largest rivers crossing the push moraine, though not measured, approximately reaches peak values of some 10 m³/sec.

The glacial run-off flows mainly in supraglacial channels and cascades over the front edge on to the push moraine. These streams either cross the push moraine directly or pass through lakes. The stream volume fluctuates according to the glacial run-off. Some 70 lake cover about 2% of the push moraine area, the largest ones having a surface area of about 3,000 m². Most of the lakes are cut off from water supply other than meteoric water, melted snow from the adjacent slopes and possibly, sublacustrine water. Figure 20 shows an active lake in the centre of the push moraine, whose level fluctuations are marked by remnant lacustrine sediments, the highest one being about 2 m above the present one. The course of the streams and the occurrence of sources and sinks depend heavily upon the structure of the push moraine. Thrust surfaces near the outer boundary often contain sources (Fig. 21). Dried-up sources are often recognized because of small gravel fans; such fans are thus indicative of hidden thrust planes. No observational

Figure 20 Lake in the centre of the push moraine. There are level fluctuations of several metres.

Figure 21 Sources emerging from a thrust plane. The source to the left emerges at the intersection of the thrust plane with an eroded vertical fracture. The water penetrates a sink in the same thrusts some 50 m to the right.

evidence could be found that the water penetrates deeper than a few metres into the push moraine; the sources could always be linked to some nearby sinks. About 51 sources were found in the push moraine. They occur four times more frequently than sinks (Fig. 11). These findings would indicate that the water in the push moraine flows close to the surface in veins which have the tendency to branch out.

Water is the dominant agent modifying the push moraine morphology. The prevailing processes of alluvial erosion are the lateral undercutting of the river bank (thermo-erosional niche) followed by its collapsing and disintegrating due to thawing and the removal of the unfrozen gravel from the river bed. The river erosion in frozen gravel was studied by Maag (1969, p. 99); his findings well apply to the push moraine. Melting of interstitial ice causes eluvial erosion, i.e., decay of the frozen detritus as a result of loss of cohesion.

In 1967 a river began to emerge subglacially in the eastern corner between the Thompson Glacier and the push moraine. Since the beginning of the Expedition in 1959 this was the first observation of such a river source in front of the Thompson Glacier. At the end of June in 1968 it began to flow again. The discharge increased rapidly and exceeded that of the Between River during July and August (the peak discharge of the Between River was observed to be 40 m³/sec in 1961 and 140 m³/sec in 1963, (Maag, 1969, p. 91)). The river continued to exist in 1969 and 1970. Since 1968 the discharge of the eastern marginal river of the Thompson Glacier remarkably decreased compared with that of the previous years (F. Müller, personal communication), i.e., the marginal river possibly feeds this sub- or intraglacial river.

Figure 22 Transverse undulations of the glacier front. The glacier is adjusting to the topography of the push moraine.

Fabric elements

Vertical faults, thrust faults and folds are the elements determining the fabric of the push moraine. The fresh vertical faults form sharp-edged intersections with the topographical surface. Usually the width varies between 0.2 m and 0.8 m, the observed maximum being about 1.5 m. The horizontal component of the translation parallel to the fracture could rarely be determined. It amounted at the most to some decimetres, whereas the vertical component reaches in exceptional cases up to about 2 m.

The thrust surface, formed by the glacier overriding the push moraine, is seldom visible. It dips beneath the glacier by about 5° to 10°, the maximum measured being about 20°. Figure 22 shows the glacier front adjusting to the topography of the push moraine. The ice flows in the wave trough downward into a depression of the push moraine. This causes transverse tension crevasses on the glacier barely 50 m back from the front edge. In spring 1967, calving accidentally exposed a cut normal to the front edge of the glacier (Fig. 23). The bottom layer reveals a plicated fold intersected by partially healed ruptures. The axial surface bends slightly upwards and is roughly horizontal. The fold axis is probably parallel to the front of the glacier.

The thrust surface, separating the push moraine from undisturbed outwash, dips between 29° and 40°, the mean being 32°. The relative displacement along the overthrust in the dip direction usually measures about 4 m; the recorded maximum is about 8 m. The rupture area follows the boundaries of the rock particles which are sometimes scratched but otherwise intact. This means that the strength of the outwash is less than that of its non-ice

Figure 23 Plicated fold on a vertical cut normal to the glacier edge. The cliff exposes the bottom layer of the glacier and is about 10 m high. The photo was taken in May 1967 in the central part of the glacier front.

Figure 24 Outermost thrust of the push moraine. Aufeis, not older than 8 months, is incorporated into the push moraine. The photo shows the eastern part of the push moraine front in April 1968.

components. Figure 24 shows an active outermost thrust of the push moraine. The ice cover in the foreground consists of aufeis deposited in the preceding fall and originates from the Between River. Integrated in the thrust deformation this ice limits the utmost age of the thrust to about 8 months.

Figure 25 Folded sand probably of lacustrine origin. The overthrusting limb is practically undeformed and consists of outwash gravel about 8 m thick. A gravel layer of several decimetres cuts the folded sand horizontally.

The thrust surfaces inside the push moraine are mostly hidden underneath talus. The direct observations have to be interconnected by indicators such as breaks in the slope, sources or their remnants. Due to advanced decay the dip usually cannot be observed, normally it is less than 10°.

The anticlinal folds of the push moraine are fairly symmetrical with regard to the usually vertical axial plane. The anticlinal axis is horizontal or slightly dipping. Apart from the normal cylindrical development of the anticlines there are also dome and saddle structures (Fig. 14).

Small folds are observed in sand layers which are incorporated in the push moraine. Figure 25 shows some 4 m of folded, probably lacustrine sand covered with an outwash gravel block, about 10 m thick, which hardly shows any deformation. The axial plane of the sand folds dips at an angle of at least 28°, this is about 18° steeper than the stratification of the outwash block. A gravel layer some decimetres thick, cuts the folds horizontally. These folded sands occur in the north-western part of the push moraine (Fig. 11). The findings indicate that contrary to the brittle rupturing of outwash gravel to blocks, the frozen sand becomes plastically deformed if incorporated in the push moraine.

Kinematic and geometric data

Eleven cairns, some with a central pole, on the push moraine and 6 poles on the glacier were repeatedly surveyed over several years. All points are

254

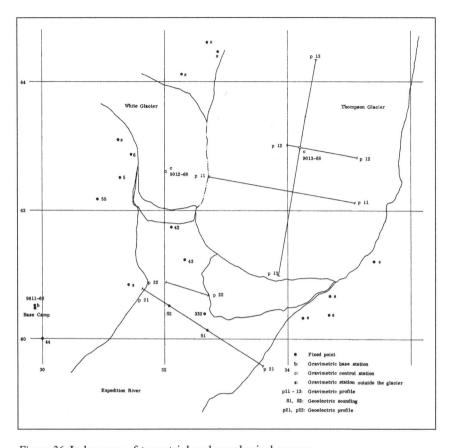

Figure 26 Index map of terrestrial and geophysical surveys.

located in the western part of the glacier snout and of the push moraine. They were surveyed by intersection from baselines which were linked to the triangulation network of the Expedition Area (Haumann, 1960) (Fig. 26). The local coordinates were computed and from these the displacement vectors and their rates calculated based on a method described by Bonyun (1966). The results are summarized in Figure 27 and listed in full in Appendix 4. The accuracy of the coordinates of the moving points was established from simultaneous intersection of fix point P332 in front of the push moraine. The standard deviation of the vertical coordinate amounts to 1.1 cm, the radius of a corresponding horizontal circle, measures 17 cm.

The 6 movement points on the glacier are 50 to 200 m from the top edge of the glacier cliff. The observation period of 258 days covers only winter conditions and does not include days of melt and run-off. The mean velocity amounts to 8.1 ± 0. 5 cm/day. Near the glacier edge the velocity decreases to

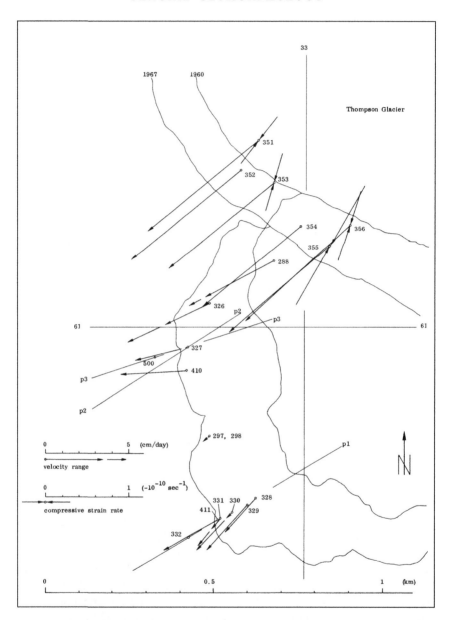

Figure 27 Surface velocities and strain rates at surveyed points of the Thompson Glacier and its push moraine.

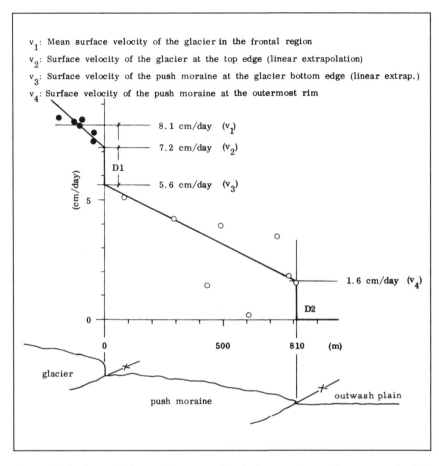

Figure 28 Surface velocity profile across the glacier – push moraine – outwash plain system.

about 7.2 cm/day (Fig. 28). There is no significant difference in velocity and direction between the 3 points near the unhindered glacier cliff (P351–P353) and those facing the push moraine (P354–P356). The velocity vectors are nearly parallel with an azimuth of 230 ± 1°. The divergence of the two outermost points up-glacier (P351 and P356), separated by a distance of some 360 m, is about 2.8°. The two vectors near the glacier edge and facing the push moraine (P354, P355), slant slightly up (−3° and −4°); the others are horizontal.

An attempt was made to determine the strain rate tensor at the 6 glacier points using rectangular strain gauge rosettes (Nye, 1959, p. 409). Significant readings of more than $0.2 \cdot 10^{-10}$ sec^{-1} could be taken at 4 points but only in one direction at each point. Parallel to the flow, the strain rates are compressive and vary between $-0.40 \cdot 10^{-10}$ and $-0.94 \cdot 10^{-10}$ sec^{-1}. As indicated by the slight divergence of the outermost velocities the strain rates normal to the flow seem to be slightly dilatative. Though the directions of the principal axis of the strain rate tensors are unknown, the compressive strain rates given above approximate minimum compressive values.

The mean velocity of the surveyed push moraine points amounts to 2.7 cm/day; the mean direction (azimuth 240 $\pm$ 18°) is almost parallel to the axis of the lower Expedition River valley (245°). This is 10° more than the mean azimuth of the velocity vectors of the glacier, confirming that the push moraine develops towards the centre of the valley. The dip of the vectors averages $-18°$ with a large spread of from $-73°$ to $+6°$.

Despite the fact that all measurements were made in the western, marginal part of the push moraine, an attempt was made to assemble a surface velocity profile across the entire system (Fig. 28): the mean velocities of the push moraine points measured between 1966 and 1968 are plotted against the distance from the 1967 glacier edge. The glacier velocities are those of the winter 1967–68. In a first approximation, the velocity distribution across the push moraine is assumed to be linear. There is a discontinuity in the horizontal velocity of the system at the glacier edge (D 1 in Fig. 28). The glacier velocity refers to the top edge of the cliff, the push moraine velocity, however, to the bottom edge. The difference between the two velocities amounts to 1.6 to 2.5 cm/day or 6 to 9 m/yr being of the same order of magnitude as the annual cliff ablation. The glacier velocity at the bottom edge of the cliff is less than that at the top edge and more than, or equal to, the push moraine velocity. Therefore, the glacier is unlikely to override the push moraine at the present stage. A second discontinuity in velocity occurs at the outer boundary of the push moraine (D 2 in Fig. 28). It amounts to about 1.6 cm/day and is the rate at which the push moraine is overthrusting the undeformed outwash. The velocity gradient across the push moraine yields a compressive strain rate of $-6 \cdot 10^{-10}$ sec^{-1} shortening it by about 4 cm/day.

The horizontal velocity of P288, the push moraine point situated closest to the glacier edge, varies seasonally (Fig. 29). Over a period of three years the mean winter velocity amounted to 4.6 $\pm$ 0.2 cm/day while the mean summer velocity reached 5.5 $\pm$ 0.2 cm/day. The ratio between the summer and winter velocities ranges from 1.15 to 1.25. Some 8.5 km up-glacier the corresponding ratio was found to be 1.14 in 1960–62 (Müller, 1963a, p. 70) and 1.24 in 1969–70 (A. Iken, personal communication). It is uncertain whether these velocity fluctuations occur in the snout of the glacier (the measurements answering this question have not been made). Since the increase of velocity during summer is due to enhanced bottom slip of the glacier (Müller and Iken, in press) the preservation of the velocity fluctuations down to the glacier tongue can only be expected if the glacier is not frozen to the bed at the tongue. The remarkable advance of the glacier and the sudden occurrence of a subglacial stream at the front favour bottom slip

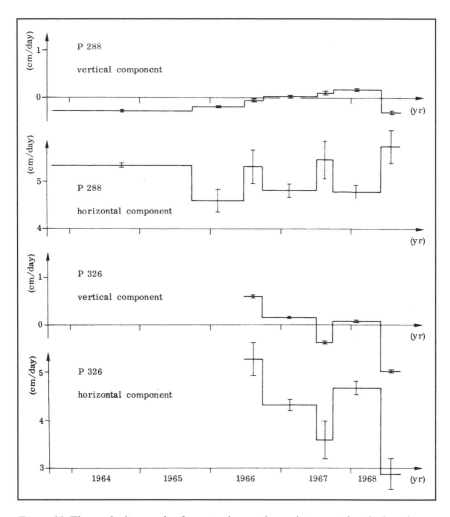

Figure 29 Time-velocity graph of two push moraine points near the glacier edge.

close to the tongue. Thus the velocity increase of P288 in summer could be due to either an increase in glacial velocity or a result of a decreased frictional resistance in the push moraine body, or both, caused by lubricating water and the effect of buoyancy. The seasonal velocity changes of P326 (Fig. 29) run contrary to those of P288 as the summer velocity is lower than that of winter. The push moraine at P326 may be considered to be slowing down with a short-lived velocity increase in 1967–68. There is also considerable difference in the seasonal behaviour of the vertical velocity component of the two points. While there is good agreement between horizontal and vertical velocity changes for P326, there is none for P288.

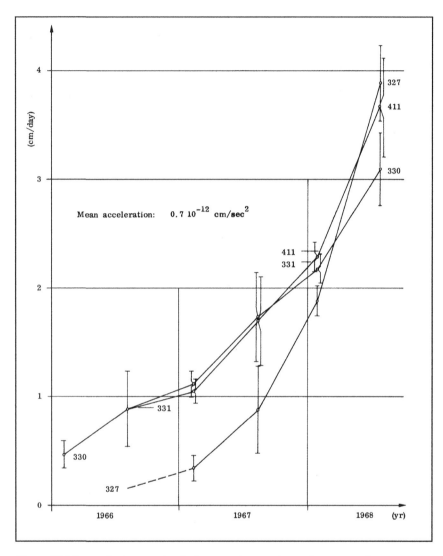

Figure 30 Time-velocity graph of three push moraine points near the outermost
edge.

The points P327, P330 and P331 became incorporated into the push
moraine sometime in 1965. From 1965 until 1968 they experienced an
acceleration of about $0.7 \cdot 10^{-12}$ cm/sec.^{2}, reaching a velocity of 3 to
4 cm/day (Fig. 30).

Three tachymeter profiles were surveyed covering the frontal portion of
the push moraine. The direction of the profiles (pl, p2 and p3, Fig. 27) runs
roughly parallel to or deviates only some 20° from that of the velocity

vectors in the area. Profile pl, repeatedly measured between 1966 and 1968, is shown in Figure 31. Profile p2, surveyed in 1966 and 1967, had to be replaced by profile p3 in 1968 because the fixed cairn which served as a point of origin became incorporated in the push moraine. Profile pl illustrates the frontal outermost portion of the push moraine, which ascends stepwise in blocks of outwash of 5 to 10 m in thickness and 40 to 80 m in width. Of the separating thrust surfaces I to V, I and II were directly observed, whereas III to V are based on an interpretation of the topographical shape. The cavity along thrust surface II measured some 25 by 25 m and up to 0.8 in height. The bottom exhibited striations indicating differential movement. The cavity collapsed in the summer of 1968. The dip of the thrust surfaces decreases towards the push moraine interior. The outermost rim shows a type of vaulting characteristic of the front portion of the push moraine (Fig. 33), a phenomenon which is occasionally still visible on heavily eroded blocks in the interior. Profiles p2 and p3 (Fig. 32) intersect an anticline running N-S along the western boundary of the push moraine (Fig. 14). In both years of observation the shape change was retrogressive during the summer.

The change of topographical shape is the sum of push moraine movement and morphological action on the surface. From the profiles it is evident that internal movement is the prevailing component.

Between 2 August 1960 and 14 August 1967 the basal area of the push moraine expanded from $(1.06 \pm 0.01)\ 10^6\ \mathrm{m}^2$ to $(1.24 \pm 0.01)\ 10^6\ \mathrm{m}^2$, i.e. it increased by a factor of 1.18. During the same time lapse it became shorter in the direction of flow over the entire width. The mean compressive strain amounts to -0.10 ± 0.07 which corresponds to a mean strain rate of about $-5 \cdot 10^{-10}\ \mathrm{sec}^{-1}$. The finding supports the strain rate obtained from the velocity gradient across the push moraine.

The volume of the push moraine cannot be estimated as the lower surface is unknown. However, an assessment of the mass raised above the level of the outwash plain can be carried out with good accuracy. On a topographical profile normal to the edge of the glacier, a straight horizontal line is drawn at the foot of the push moraine and one vertical to the glacier edge (see sketch, Fig. 34). The two lines, together with the upper surface of the push moraine, circumscribe an area (a) proportional to the volume of a slab with unity thickness. A series of parallel profiles, about normal to the edge of the glacier with an azimuth of 20° and at an interval of 200 m were measured using the 1960 and 1967 maps. The areas (a) were plotted for each profile. The resulting graph (Fig. 34) shows the transverse shift of the push moraine mass. This transverse mass change is asymmetrical, the maximum being at the western boundary. This reiterates the already noted fact that the push moraine is shifting towards the centre of the Expedition River valley. The volume of the mass raised above the outwash plain in 1960 thus calculated was found to be $(4.97 \pm 0.01)10^7\ \mathrm{m}^3$ and in 1967 $(5.30 \pm 0.01)10^7\ \mathrm{m}^3$,

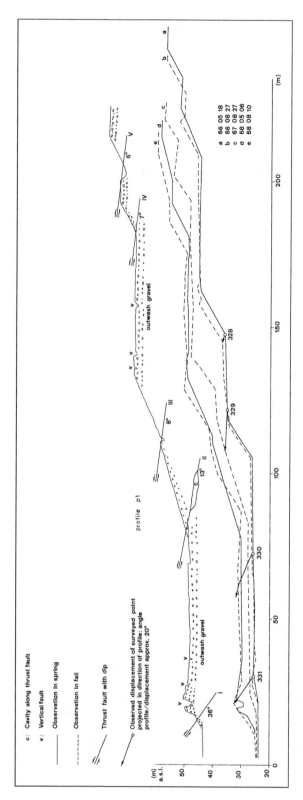

Figure 31 Tachymeter profile of the front edge of the push moraine.

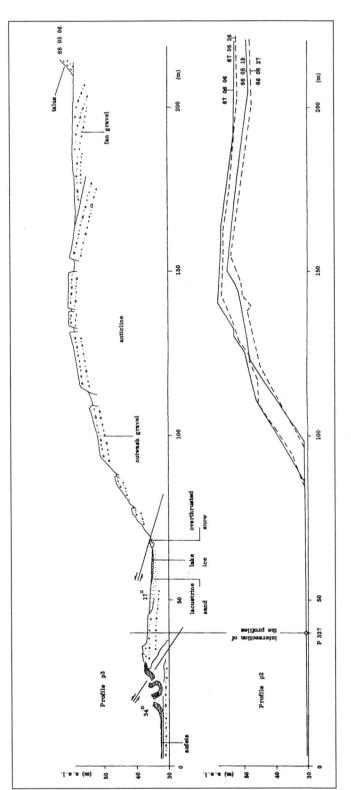

Figure 32 Tachymeter profile of the eastern front of the push moraine.

Figure 33 Outermost frontal rim of the push moraine (looking to the west) in early June 1968.

which means an increase of 7%. These values are the lower limits of the actual push moraine volume. In 1960 the mean height of the push moraine above the outwash plain measured about 47 m, in 1967 about 43 m.

In summary, between 1960 and 1967 the push moraine increased in volume and basal area indicating growth. Furthermore, it shrank in length, broadened, and slightly decreased in mean height above the outwash plain.

Gravity and geoelectric data

A study of the Thompson Glacier push moraine's asymmetric development may yield general information on the formation of push moraines. It seems unlikely that a western extension of the push moraine ever existed. Thus it is assumed that either the glacier acts asymmetrically or the substratum reacts asymmetrically or a combination of both is responsible for the irregularity. The dynamic behaviour of the glacier front area seems to be fairly uniform. Three hypotheses are illustrated in Figure 35 and were tested by geophysical methods.

Hypothesis a: The static load exerted by the glacier is unevenly distributed. Towards W the thickness of the ice is less than the critical value needed to cause the stress field necessary to overcome the strength of the ground.

Hypothesis b1: The strength of the ground on the western side is sufficient to resist the stresses caused by the glacier. The material is heterogeneous with respect to the properties controlling its strength. It is probable that the White Glacier extended towards the centre of the valley before it was forced

264

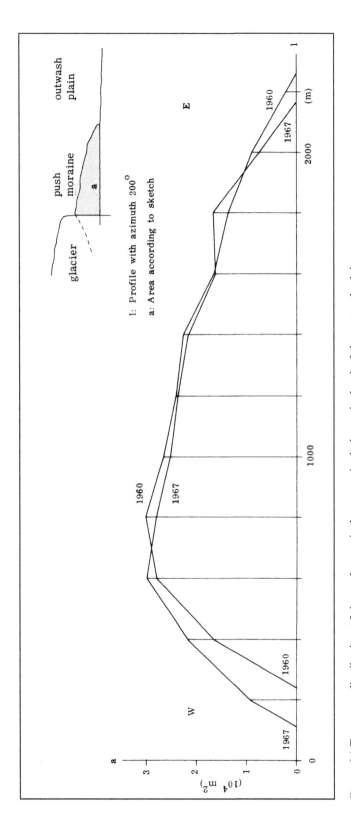

Figure 34 Transvere distribution of the push moraine's mass raised above the level of the outwash plain.

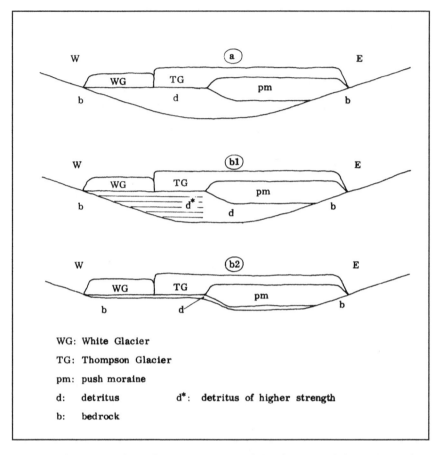

Figure 35 Some hypotheses for the asymmetrical development of the push moraine.

to change course by the advancing Thompson Glacier. The White Glacier would have preloaded and consolidated the sediments and thus increased their strength.

Hypothesis b2: The bedrock, being more resistant than the overburden, forms one or several terraces beneath the outwash. Marine terraces of fair to poor development exist along the Expedition River valley at levels which could render this hypothesis possible (Müller, 1963b, p. 171).

The hypotheses can also be combined. In addition, there are further possible speculations: a trigger needed to initiate the formation of a push moraine such as dead ice beneath the outwash perhaps only existed in the eastern side of the valley.

Hypothesis "a" can be tested through measurements of the glacier thickness. The gravity method which previously produced satisfactory results

(Weber, 1961; Beeker, 1963a) was chosen. The difference between glacier density and the mean density of the surrounding rocks gives rise to a gravity anomaly from which the shape of the glacier can be calculated. It was thought possible to distinguish between frozen detritus and bedrock in the underground (hypotheses "b1" and "b2") by geoelectrical sounding and profiling. The method is based on variations in the electrical conductivity of the material which mainly depends on mineral composition, fabric and temperature. Due to the ice matrix of the overburden the electrical resistivity is expected to be much higher than that of the bedrock. The findings of the experimental measurements are discussed below.

The gravity survey was carried out according to standard procedures (Gassmann and Weber, 1960, p. 54). The vertical gravitational component was determined relative to a gravity base (9,611–60, Fig. 26), with a Worden type gravity meter with a resolution of 0.01 mgl. The terrestrial survey was performed simultaneously; single points were resected and the tachymeter method applied along profiles. 137 stations were measured along the profile p11 to p13 on the Thompson Glacier snout and 9 stations in the surrounding area (Fig. 26). The data are presented in Appendix 5. The readings were corrected for time drift and height above ground, then reduced according to Bouguer (Gassmann and Weber, 1960, p. 58) yielding the total gravity anomaly. The terrain correction was obtained using the procedure outlined by Nagy (1966). From the 9 stations surrounding the glacier, the regional anomaly was approximated by a linear regression. The thus obtained regional gradient of 2.1 mgl/km N and −0.1 mgl/km E is supported by Becker's observations on the White Glacier (Becker, 1963a, p. 70). Subtraction of the regional anomaly from the total anomaly, yields the local anomaly assumed to result from the presence of the glacier. The error in the local anomaly varies between 0.9 and 1.6 mgl, and is mainly caused by errors in elevation, terrain correction and regional anomaly. The mean density of the rocks in the Expedition Area is 2.67 g/cm^3 (Becker, 1963a, p. 64); the density of ice 0.91 g/cm^3 resulting in a density contrast of 1.76 g/cm^3. The lower limit of the ice thickness is obtained assuming the glacier to be an infinite slab. By encircling the glacier snout with a vertical cylinder having its gravity station in the centre of the upper circular surface the upper limit is determined (Becker, 1963a, p. 55). The two thicknesses delimit the idealization error. In the longitudinal profile p13 (Fig. 38), the idealization error is illustrated by the curves l_s and l'_s; up-glacier it approaches about 10% of the glacier thickness. The Expedition Diapir (Fig. 8), mainly consisting of anhydrite gypsified at the surface, introduces an unknown error due to the assumption of a homogeneous regional density.

The lower transverse profile p11 (Fig. 36) is a test of hypothesis "a". The lower surface of the glacier seems to have a fairly circular shape, but towards the White Glacier it ascends in steps. These steps are more clearly recognizable in the profile of the ice thickness and coincide with the western

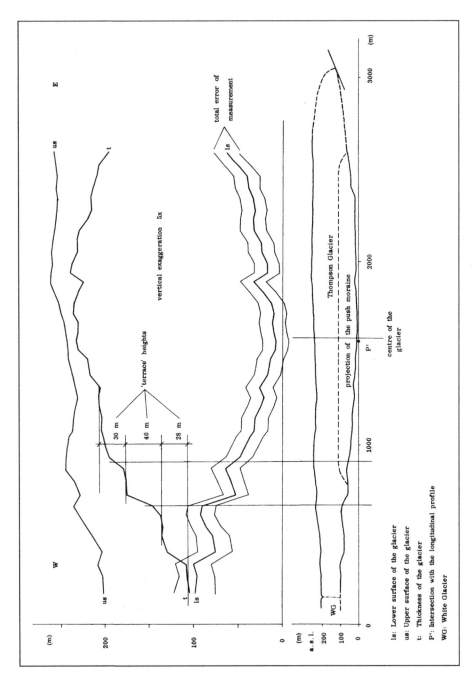

Figure 36 Lower transverse pro

boundary of the push moraine. The deepest point is about in the middle of the glacier while the ice is at its thickest slightly east of the centre.

The upper transverse profile p12 (Fig. 37) is incomplete because crevasses rendered measurements impossible. The deepest point of the lower surface of the glacier is shifted to the east of the centre by some 0.5 km. The lower surface ascends asymmetrically, towards the west the slope is 0.03, towards the east 0.15. The ice thickness increases from the eastern boundary, reaches a maximum at the deepest point and remains about constant as far as it could be observed.

Figure 38 shows the longitudinal profile p13. The lower surface intersects the extrapolated outwash plain at least 0.5 km up-glacier from the glacier front, and the sea level at least 2.7 km from the front. At the Eureka profile (8.5 km up-glacier) the glacier reaches about 50 m below sea level (Becker, 1963b, p. 98). The increase of the ice thickness up-glacier is approximately linear having a slope of 0.06. The lower surface undulates irregularly, the down-valley slope of the ridges being somewhat steeper. It is not known whether these undulations are transverse structures or not.

The Schlumberger configuration was used to determine apparent resistivities (Gassmann and Weber, 1960, p. 273). Current was measured with an ammeter (up to 1500 V DC were applied) with a fractional error of 3%, the potential by an electrometer with a fractional error of 2%. Steel nails, 30 cm long, served as electrodes and were driven into the frozen ground. To decrease the contact resistivity, the number of electrodes were increased or the nail holes were filled with brine. After the configuration was installed, the potential of the telluric field was measured. Then the artificial field was superimposed and readings noted after stabilization. At this stage the polarity was changed, readings retaken and the telluric potential measured again. The two resistivities, obtained thus, differed less than 3%. A twofold time dependence occurred. After the artificial field had been superimposed, current, potential and resistivity decreased. This polarization appeared within seconds to minutes and decreased the apparent resistivity in extreme cases by 30%. Trial soundings proved that at a given configuration length the apparent resistivity changes up to 50% within a few days. Daily temperature fluctuations cannot account for this but it may be caused by melt water. Though planned, it was impossible to complete the measuring programme before melting started. Lichtenecker's relation for heterogeneous bodies (Fritsch and Tauber, 1967, p. 8) shows that melting of 10% of ice in an ice-saturated gravel would lower its resistivity by about 60 to 80%. The presence of melt water complicates the interpretation, especially because the amount, extent and distribution are unknown.

As an additional complication mineralized groundwater may have occurred within the outwash. Sulphur springs emerge on the north shore of the Expedition River, about 2 km down-valley from the profiles (Beschel, 1963, p. 183). They are probably related to the Expedition Diapir (Fig. 8), which

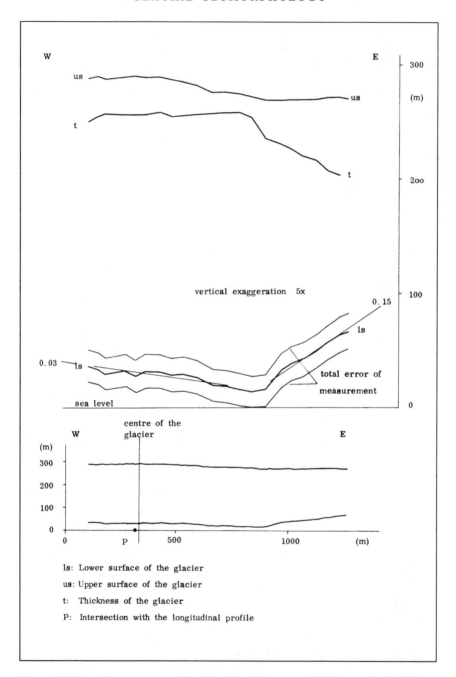

Figure 37 Upper transverse profile p12 of the Thompson Glacier as obtained from the interpretation of gravity measurements.

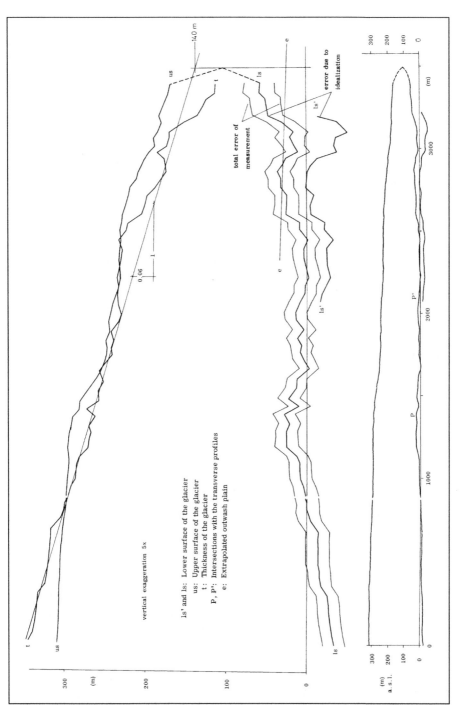

Figure 38 Longitudinal profile p13 of the Thompson Glacier snout as obtained from the interpretation of gravity measurements.

Table 3 Interpretation of the geoelectrical soundings.

| | S1 | | S2 | |
	thickness	*resistivity*	*thickness*	*resistivity*
layer 1	1 m	10^4 Ωm	2 m	$3 \cdot 10^4$ Ωm
layer 2	3 to 4 m	$4 \cdot 10^5$ Ωm	8 m	$6 \cdot 10^5$ Ωm
layer 3	∞ m	10^3 Ωm	∞ m	10^4 Ωm

extends underneath the region. Mineralized groundwater has a resistivity several orders of magnitude lower than that of frozen outwash and would strongly violate the assumption of a homogeneously layered ground.

The locations of the two geoelectrical soundings S1 and S2 are indicated on Figure 26. An interpretation assuming three homogeneous layers, is given based on the theoretical curves by Schlumberger (Table 3). This assumption seems to work for S1 but for S2 an approximation by five layers seems more appropriate (Fig. 39). The field evaluation suggested a configuration a half-width of 30 m or more to track variations in the resistivity by horizontal profiling at a probable depth of interest of about 10 m. Readings were taken at configurations of 30 m and 16 m half-width yielding the apparent resistivities r30 and r16 (Fig. 40). Generally r16 tends to be higher than r30. This tendency reverses itself in the middle of profile p22 and is in the corresponding western part of p21 somewhat similar to p22. Profile p21 rests on both sides on bedrock (gypsified anhydrite), the apparent resistivity being $0.5 \cdot 10^4$ to 10^4 Ωm. Resistivity increases symmetrically from both ends to $1.5 \cdot 10^5$ to $2 \cdot 10^5$ Ωm. In the middle of the outwash plain the resistivity varies abruptly, differing from all other variations which appear in both profiles; being more pronounced in r30 than in r16 indicates that the cause is at greater depth than that of the other variations. The connection of the two peaks is roughly parallel to the western boundary of the push moraine. The relatively high resistivity of pure ice (10^6 to 10^7 Ωm) suggests the interpretation of variations in resistivity as variations in ice content.

Radiocarbon data

Müller (1963b, p. 171) published the C-14 age of driftwood and marine shells found in the Expedition River valley. It can be assumed that these organic findings were deposited approximately at the time of death. It is necessary to distinguish information on samples which remained in situ after deposition from information on samples with additional displacement because of incorporation in the Thompson Glacier or the push moraine. For the Expedition River valley the data which remained at the original site of deposition indicate uplift over the last 8,000 years (Fig. 41). A splintered

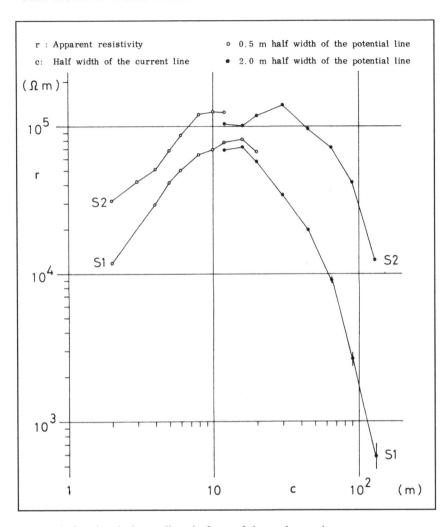

Figure 39 Geoelectrical soundings in front of the push moraine.

driftwood log, embedded in push moraine till near the glacier edge, was discovered in 1968. The location is indicated on Figure 11. The sample was dated by the Geological Survey of Canada and found to have a C-14 age of 5,690 ± 140 years B. P., (sample A in Fig. 41). The driftwood probably became buried slightly under the glacier surface and could thus be re-exposed by the frontal ablation and sedimented on to the push moraine. Figure 41 shows that the samples from the push moraine and the Thompson Glacier are supposed to have been deposited between about 30 and 35 m a.s.l. The two samples from the push moraine found close to the glacier edge

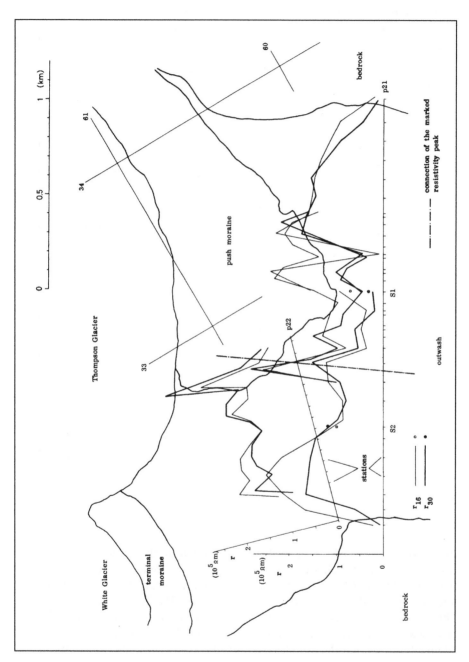

Figure 40 Geoelectrical horizontal profiles in front of the push moraine.

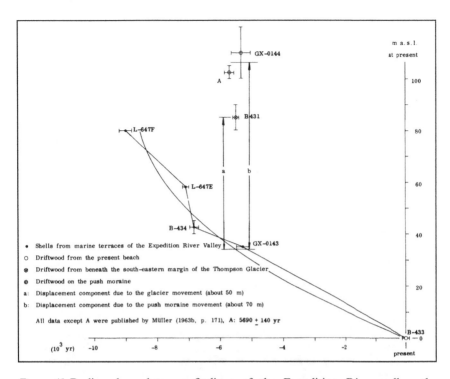

Figure 41 Radiocarbon data on findings of the Expedition River valley, the Thompson Glacier and the push moraine.

experienced an additional upward movement of some 70 m. The sample found underneath the Thompson Glacier about 1.5 km up-valley from the present glacier terminus seems to have been raised by some 50 m. Because the topography of the region now covered by the Thompson Glacier snout is not known the horizontal displacements of the samples cannot be estimated.

Comparison with the push moraine of the Finsterwalder Glacier

From air photographs numerous additional push moraines were known to exist on Axel Heiberg Island. Of these, the push moraine of the Finsterwalder Glacier was briefly visited in the fall of 1968. It is situated about 26 km west northwest of the Thompson Glacier push moraine.

The Finsterwalder Glacier is an outlet of the McGill Ice Cap with compound basins similar to that of the Thompson Glacier. In the ablation area it flows west southwest while in the snout region it partly coalesces with the southern branch of an outlet glacier which flows north of the Finsterwalder

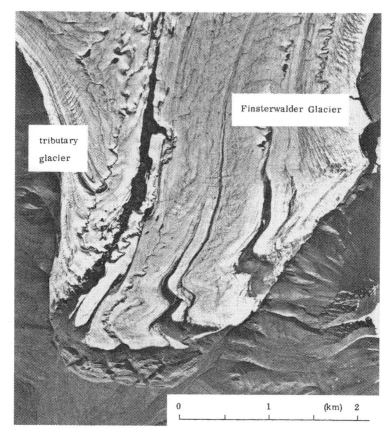

Finsterwalder Glacier

tributary
glacier

0   1   (km)   2

Figure 42 Vertical view of the Finsterwalder Glacier push moraine. (Expedition
        Photo, 14 August 1967).

Glacier (Fig. 42). The snout gradually narrows and is about 2 km wide
near the terminal. The front edge forms an upward curving slope typical of
stagnating or slightly retreating glaciers.

In front of the glacier, the valley narrows enclosing a triangle shaped
outwash plain which converges about 1.8 km ahead of the push moraine.

In vertical view the push moraine is an evenly developed band, 2.7 km
long and about 0.4 km wide along the glacier edge. It covers a base of about
0.6 km² which equals half the size of the Thompson Glacier push moraine.
The maximum height above the outwash plain is about 55 m. A belt of till
next to the glacier is about 80 to 120 m wide and partly extends on to the
glacier as a debris cover. The rest of the push moraine consists of stratified
outwash gravel similar to that of the Expedition River valley, only being
darker as a result of a higher content of shale components. The push
moraine ascends from the outwash plain in two block levels, one about 15 m

and the other about 25 m thick. The blocks are horizontal or dip slightly towards the glacier and they clearly underlie the till. A transverse structural trend is recognized by ruptures and terrace margins. The push moraine is in an advanced state of erosion: fractures appear as shallow grooves and no fresh ones were found while talus slopes nearly reach the top of the hills. Five rivers, originating as supra-glacial streams and perhaps following structural trends, cross the push moraine in V-shaped valleys. The slope of a straight line, drawn from the earlier position of the glacier front to the foot of the push moraine, is about 0.18. The Thompson Glacier push moraine, having a value of 0.12 for the same parameter, appears therefore compressively less strained.

In summary the Finsterwalder push moraine is smaller and steeper than that of the Thompson Glacier. It is nevertheless clear that it has a simpler construction consisting of two levels of outwash blocks underlain by thrust surfaces. The stagnation, or even retreat, of the glacier and the decaying forms indicate that active formation no longer takes place.

Chapter 3
The push moraine as a mechanical problem

The mechanical explanation of the push moraine is the quantitative description of the fields of force, stress, strain, strain rate and velocity. By direct measurements some data on velocity, strain and strain rate could be obtained. The mechanical problems concerning the push moraine are associated with the problems of the strength of materials. In soil mechanics quantitative methods have been developed (Terzaghi and Peck, 1967, p. 232). Though violating a long list of conditions they can be applied satisfactorily for certain predictions in practice. Treating the push moraine by analogy to the analysis of the slope stability we start with the following assumptions.

1) The geometry is two-dimensional.
2) The material is homogeneous and isotropic.
3) The total energy release takes place along a sliding surface which is a circular cylinder with a horizontal axis.
4) The Coulomb-Navier criterion of failure is valid.

   $s = c + n \tan \Phi$

   s: shearing strength
   c: cohesion
   n: normal stress (compressive) across the sliding surface
   Φ: angle of internal friction

These assumptions will be discussed at the end of the chapter guided by the result.

The forces introduced are outlined in Figure 43. Considering a slice of unit thickness and a small width b_i the weight W_i of the element i is

$W_i = gb_i(u_i q_g + v i q_o)$

g: gravity constant
q_g: density of glacier ice (approx. 0.9 g/cm^3)
q_o: density of outwash (approx. 2.0 g/cm^3).

At the lower end of the element i, along the surface of sliding, the vector $\vec{W}_i$ resolves into a normal component $\vec{N}_i$ and a parallel component $\vec{T}_i$. $\vec{N}_i$ contributes the frictional part of the shear strength $\vec{S}$. The length of $\vec{S}_i$ is

$$S_i = \frac{b_i}{\cos \alpha_i} c + W_i \cos \alpha_i \tan \Phi.$$

Adding up S_i along the sliding surface, the length of the force $R = \Sigma S_i$ which resists the sliding movement due to the shear strength is obtained. In the direction x

$$R_x \approx c\Sigma b_i + \tan \Phi \Sigma W_i \cos^2 \alpha_i.$$

The x-component of the sum of the gravitational (G) and glacial forces (H) is

$$D_x = G_x + H \approx \Sigma T_i \cos \alpha_i + H = \tfrac{1}{2}\Sigma W_i \sin 2\alpha_i + H.$$

To estimate the maximum horizontal compressive forces H_{max} exerted by the glacier on the sliding mass the directions of the principal stresses in the glacier are assumed to be horizontal and vertical. At a given depth d the vertical principal stress is assumed to be $q_g gd$ yielding the horizontal principal stress $q_g gd + 2t$, t being the shear stress. According to Nye (1961, p. 559), t varies between $0.5 \cdot 10^6$ and $1.5 \cdot 10^6$ dynes/cm^2 in glaciers, averaging 10^6 dynes/cm^2. Assuming an average and constant shear stress $\bar{t}$ H_{max} is obtained.

$$H_{max} = \tfrac{1}{2} q_g ga^2 + 2\bar{t}a.$$

Neglecting the term due to the shear stress the minimum horizontal compressive force H_{min} is found to be about 75% of H_{max} (Table 4).

The condition $\dfrac{R_x}{D_x}$ = minimal yields by trial and error one circle along which the sliding is most likely to occur. Since the parameters c and Φ are unknown the limiting cases are considered. By declaring $c = 0$ and giving Φ some arbitrary value the critical "Φ-circle", occurring with material of

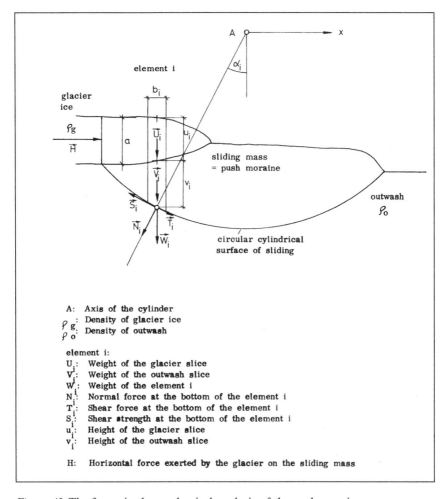

Figure 43 The forces in the mechanical analysis of the push moraine.

no cohesion, is obtained (Fig. 44). In the opposite case a "c-circle", which extends to greater depth than the former, is found. Common for all circles is the frontal point of the push moraine. The two circles emerge at this point with a dip of 32° and 35°; in the field this dip was found to vary between 29° and 40°. At the moment of failure R_x equals D_x. Thus

$$c\sum b_i + \tan \Phi \sum W_i \cos^2 \alpha_i - (\tfrac{1}{2}q_g a^2 + 2\bar{t}a + \tfrac{1}{2}\sum W_i \sin 2\alpha_i) = 0.$$

This linear relationship between c and $\tan \Phi$ is plotted on the top right-hand side of Figure 44. It describes the variation range possible for these two unknown parameters. The two straight lines, corresponding to the two

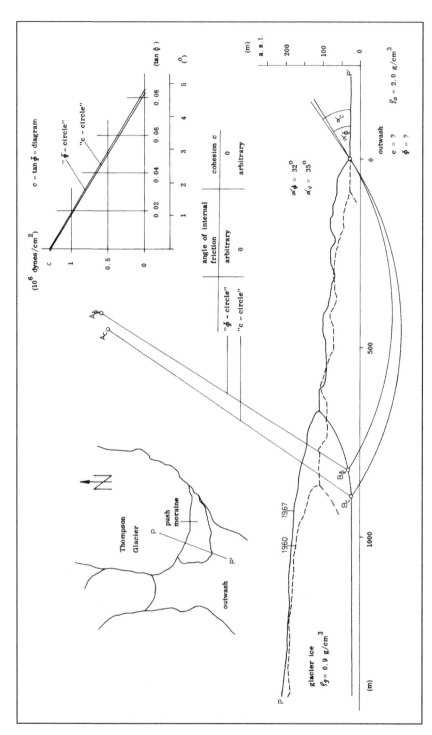

Figure 44 The critical sliding circles of the push moraine.

Table 4 The critical sliding circles of the push moraine.

		"*c-circle*"	"*Φ-circle*"
G_X	(10^{10} dynes)	6.6	6.7
H_{max}	(10^{10} dynes)	12.9	10.1
H_{min}	(10^{10} dynes)	9.9	7.5
tan Φ_{max}		0.082	0.084
Φ_{max}	(0)	4.7	4.8
c_{max}	(10^6 dynes)	1.28	1.29

G_X: Horizontal component of the gravitational force tending to drive a slice 1 cm thick of the sliding cylinder.
H_{max}:   Maximum glacial force driving a slice 1 cm thick of the sliding cylinder.
H_{min}:   Minimum glacial force driving a slice 1 cm thick of the sliding cylinder.
Φ_{max}:   Angle of internal friction if the material is assumed non-cohesive.
c_{max}:   Cohesion if the material is assumed cohesive.

circles, differ slightly. The maximum angle of internal friction, for material of no cohesion, amounts to about 5°; the maximum cohesion for material without friction is about $1.3 \cdot 10^6$ dynes/cm^2 (Table 4). The sediments exposed at the surface of the push moraine (mainly gravel and sands) can be assumed to have, in the unfrozen state, a cohesion of zero and an angle of internal friction of 30° to 40° (Appendix 2). The frictional part of the strength is accepted as being due to the particle to particle contact of the grains. The ice reduces the area of contact between the grains and thus the friction; it may introduce a certain cohesion. This agrees with the observation that the grains are often coated with ice. Non-cohesive sediments fitting the failure condition stated are not known. If the push moraine mainly consisted of well-graded fluvioglacial sediments which are non-cohesive materials in the unfrozen state these must have been reduced in strength. Cohesive materials such as clay might fit the failure condition but are unlikely to occur in the push moraine. Because the push moraine can be assumed to be frozen for the most part, the freezing process must have transformed the fluvioglacial sediments into cohesive sediments of reduced strength. These explanations suggest the push moraines of this type to be restricted to the zone of continuous permafrost.

It is difficult to say to what extent this idealisation distorts the conclusions. Assumption 1) is valid for a plane of symmetry. The sliding surface is bowl-shaped rather than cylindrical, therefore less shear strength is needed to keep the sliding mass stable, i.e. the straight lines on the *c*-tan Φ-diagram of Figure 44 move to the left. Assumption 2) is fairly justified for the surface, but no verification exists for depths of more than a few tens of metres. The assumption 3) of a sliding surface, neglecting the internal deformation, is probably the most contestable. The measured longitudinal strain-rate of about

$-5 \cdot 10^{-10}$ sec^{-1} clearly indicates internal deformation. Internal deformation consumes energy, a fact which should be taken into account when the shear strength is assessed. 4) The Coulomb-Navier criterion is generally accepted to be valid for failure by fracture.

In addition to the glacial stress field there is the stress field induced by thermal fluctuations. The seasonal temperature fluctuations penetrate some 10 metres into the ground (Fig. 9, Appendix 1); the daily fluctuations some decimetres. Thus, the "weakening" effect of the thermal stresses affects a thin crust of the push moraine and the outwash.

The crucial question remains whether or not the push moraine can be considered as a mass sliding on a well-developed surface and having negligible internal deformation or if it is rather a creeping mass where fracturing and overthrusting are restricted to a top layer of limited thickness. In the latter case, the concept of cohesion and internal friction might have no physical meaning and some yielding criterion would have to be applied. The push moraine would thus rather have to be thought of as a frontal continuation of the glacier consisting of dynamometamorphous rock, which essentially would be ice with a high detritus content. The observations of the overthrusting, however, favour the case discussed in detail.

The considerations which assume the push moraine to slide on a cylinder surface allow an educated guess of the total volume. According to Figure 44 it is about 2.4 to 2.9 times the minimum volume of mass raised above the level of the outwash plain. This minimum volume is about $5 \cdot 10^7$ m^3 and yields the total volume of about $12 \cdot 10^7$ to $15 \cdot 10^7$ m^3.

Chapter 4
The recent push moraines of Axel Heiberg and Ellesmere Islands

The push moraines which could be clearly identified on aerial photographs are listed in Figure 45 and Appendix 6. With the exception of one, the 19 push moraines of Axel Heiberg Island are located on the west side of the two inland ice caps. On Ellesmere Island there are 16 push moraines, most of them in the northwest of the island.

The summarizing statistics of Table 5 show some of the main characteristics of the push moraines of this part of the Arctic. The majority of the glaciers are outlet type of medium to large size. About half of the glaciers have an expanded foot. Most push moraines consist of frozen outwash. In vertical view they are half-moon shaped, the length along the front edge of the glacier being about 5 times the width. The structural trends are radial and transverse (circular-concentric). Most of the push moraines are in a state of decay. In most cases active formation no longer takes place because the glaciers are retreating. Thus the push moraine of the Thompson Glacier represents fairly well the "average" push moraine of this region.

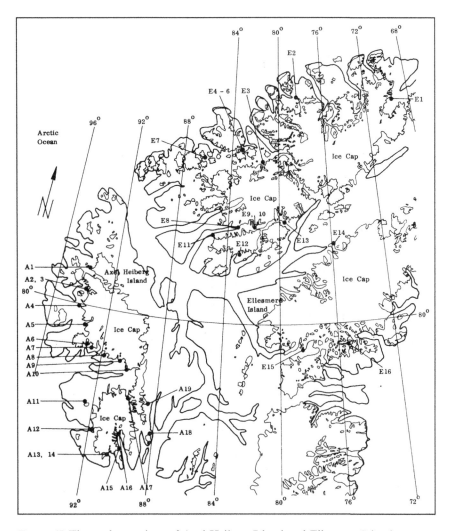

Figure 45 The push moraines of Axel Heiberg Island and Ellesmere Island.

Four main types of push moraines are illustrated in Figure 46. The push moraine of the Good Friday Glacier type (A 12) is most common. A well-developed push moraine is formed where the glacier overthrusts outwash. Where the glacier pushes on to bedrock only a rudimentary rim is produced. The radial and the circular-concentric structural trend are clearly noticeable. The glacier is advancing, the average being about 130 m/yr in the time interval 1959–64. During a possible surge the frontal advance increased to about 200 m/yr in 1964–67 and decreased to about 100 m/yr in 1967 (Müller, 1969). In 1967 the front edge reached the sea shore. The push moraine was

283

Table 5 Statistics of the push moraines of Axel Heiberg Island and Ellesmere Island.

					out of 35	%
glacier	type	1		outlet glacier	31	88
		2		valley glacier	2	6
		3		ice field	2	6
	frontal characteristics	4		semicircle	15	43
		5		expanded foot	16	46
		6		coalescing	4	11
	tongue activity	7		advancing	7	20
		8		stationary	7	20
		9		retreating	21	60
push moraine	material	10		outwash	33	94
		11		marine sediments	3	9
	shape			r (mean)	4.6	
				r (range)	(0.8 – 15.6)	
		12		± symmetrical	5	14
	structural trends	13		radial	11	31
		14		transversal-concentrical	22	63
	preservation state	15		good	13	37
		16		poor	22	63
situation glacier / valley	⊥	17		—[g]—o	9	26
		18		—{g}—o	5	14
	//	19		g)———o	8	23
		20		g) o	9	26
situation glacier / push moraine		21		)⊃ p	15	43
		22		)⊃ p	17	49

$$r = \frac{\text{length of the push moraine along the front edge of the glacier}}{\text{width of the push moraine in the centre}}$$

g: Glacier   o: Outwash plain   p: Push moraine

overridden until only a few semi-submerged rims remained. The push moraine of Ayles Fiord (E2) extends over the full width of the outwash plain. The transverse structural trend is dominant, the radial trends are absent or not visible on aerial photographs. The rudimentary push moraine of Strand Fiord (A10) is a structureless bulge in the centre of the

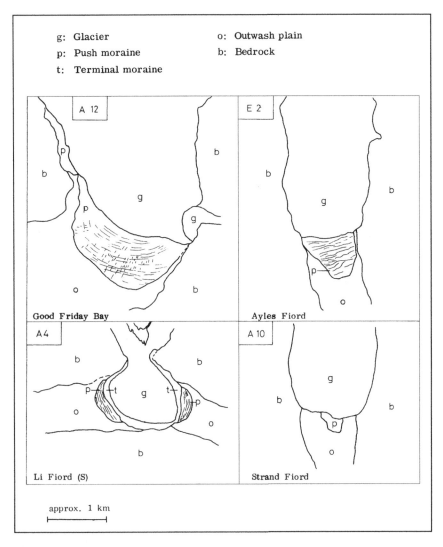

Figure 46 Variation of the push moraine types of Axel Heiberg Island and Ellesmere Island.

outwash plain. On both sides of the push moraine the glacier overthrusts outwash without the slightest effect. The push moraine of Li Fiord (S) (A4) has two lobes. The retreating glacier has an expanded foot which releases a terminal moraine.

Two of the 35 push moraines listed in Appendix 6 are uncertain cases: A 11 (Sand Bay) is possibly a landslide and A 17 (Wolf Fiord) a terminal moraine exhibiting a concentric structural trend.

Summary

A push moraine is a ridge pushed up in advance of a glacier. In shape it resembles a terminal moraine. Recent push moraines are known from the glaciated regions of the moderate latitudes of the northern hemisphere and the Arctic. Fossil push moraines are reported from the marginal zones of the Pleistocene glaciations.

The active push moraine of the Thompson Glacier on Axel Heiberg Island is situated in the zone of continuous permafrost. The Thompson Glacier, an outlet of the McGill Ice Cap, advances about 6.3 cm/day and bulldozes permanently frozen detritus to a push moraine. The ice velocities at the front, which partly forms a cliff, are about 7 to 8 cm/day and the surface strain rates in flow direction are about -10^{-10} to -10^{-11} sec^{-1}. An outwash plain extends in front of the glacier.

The push moraine is a half-moon shaped bulge about 2.1 km long, 0.7 km wide, and of a mean height above the outwash plain of about 45 m. It is asymmetrical in shape: the eastern end gradually narrows, the western convex outer edge turns abruptly towards the glacier. Judged from the surface the push moraine consists of about 95% by volume of outwash and fan gravel which are grey, coarse and well-graded gravels. Along the glacier edge till accumulates in a belt some 50 m wide. There are three structural trends: two are straight and orthogonal to each other and one is circular-concentric. At the surface the frozen sediments fracture to blocks about 5 to 10 m thick and tens of metres in lateral extent. The vertical faults and thrust faults are often filled with congelation ice. The fresh fractures are restricted to the frontal part of the push moraine. There are about 70 lakes, up to about 3,000 m^2 in area, covering the present push moraine. The predominant water supply of the push moraine is provided by glacial run-off being about $2 \cdot 10^6$ to $6 \cdot 10^6$ m^3/yr. The drainage system is closely related to the structures. There is no evidence that the water penetrates more than a few metres deep into the push moraine. The mean surface velocity of the push moraine is about 3 cm/day. At the front edge the surface velocity is about 1.6 cm/day and at the glacier edge about 5.6 cm/day. Close to the glacier edge it fluctuates seasonally, the summer velocity being about 1.2 times the winter velocity. The velocity vectors dip between $-73°$ and $+6°$, averaging $-18°$. In flow direction the strain rate is $-5 \cdot 10^{-10}$ to $-6 \cdot 10^{-10}$ sec^{-1} reducing the width of the push moraine about 4 cm/day. The acceleration of a portion newly integrated into the push moraine was about $0.7 \cdot 10^{-12}$ cm/sec^2. In 1960 the base area measured (1.06 ± 0.01) km^2 and the partial volume raised above the level of the outwash plain was $(4.97 \pm 0.01) \cdot 10^7$ m^3. Until 1967 the base area increased by a factor of 1.18 and the volume by one of 1.07. The total volume is estimated to be between $12 \cdot 10^7$ and $15 \cdot 10^7$ m^3. The gravity and geoelectric data indicate that the push moraine developed asymmetrically because of the heterogeneity of the valley filling and the glacier thickness.

286

As a supplement to the study of the Thompson Glacier push moraine in 1967 and 1968, a brief visit to the Finsterwalder push moraine was made in the fall of 1968.

The Finsterwalder Glacier, a retreating or stagnating outlet glacier of the McGill Ice Cap, situated about 26 km WNW of the Thompson Glacier, forms a push moraine consisting mainly of frozen outwash gravel. It appears to be simpler in construction and more compressively strained than that of the Thompson Glacier. The decaying forms indicate that there is no or little activity in this push moraine at present.

Theoretical considerations show that in the case of the Thompson Glacier push moraine the shear strength parameters' cohesion (c) and angle of internal friction (Φ) obey approximately the equation $c = (-1.57 \cdot 10^7 \tan \Phi + 1.3 \cdot 10^6)$ dynes/cm^2, $0 \leq c \leq 1.3 \cdot 10^6$ dynes/cm^2, $0 \leq \tan \Phi \leq 0.082$.

A study of the air photos of 19 push moraines on Axel Heiberg Island and 16 push moraines on Ellesmere Island showed that the Thompson Glacier push moraine may be considered as a representative example of the push moraines of the Canadian High Arctic.

Acknowledgements

Particular thanks are due to Professor F. Müller, Scientific Leader of McGill University's expeditions to Axel Heiberg Island, who made available ideas and the results of the field work carried out on the Thompson Glacier push moraine between 1959 and 1968 when the author's participation in the expeditions began. Professor Müller's supervision in the field and in the office, and his critical advice in the preparation of the thesis are appreciated. The assistance of Professor A. Gansser, Head of the Geological Department of the Swiss Federal Institute of Technology in Zürich, is gratefully acknowledged. Dr. J. Weber and Dr. P. Andrieux, Dominion Observatory, Ottawa, rendered aid in the preparation of the geophysical surveys. The author also wishes to express his appreciation for the help and co-operation of his expedition comrades: J. Blachut, G. and R. Desrocher, A. Iken, O. Langenegger, A. Ohmura, H. Snyder, M. Tallmann, D. Terroux and G. Young.

References

ANDREWS, R.H., 1964: Meteorology and heat balance of the ablation area, White Glacier, Canadian Arctic Archipelago – summer 1960. Axel Heiberg Island Research Reports, Meteorology No. 1, McGill University, Montreal, 107 p.

ANDRIEUX, M.P., 1966: Resultats préliminaires d'une campagne des mesures électriques sur des glaciers et calottes de glace du Canada. Société Hydrotechnique de France, Section de Glaciologie, Reunion des 24 et 25 février 1966.

BALTZER, A., 1896: Der diluviale Aaregletscher und seine Ablagerungen in der Gegend von Bern mit Berücksichtigung des Rhonegletschers. Beiträge zur Geologischen Karte der Schweiz, Lieferung 30, 162 p.

BAYROCK, L.A., 1967: Catastrophic advance of the Steele Glacier, Yukon, Canada. Boreal Institute, University of Alberta, Edmonton, Occasional Publication No. 3, 35 p.

BECKER, A., 1963a: On the determination of glacial depth. Ph.D. thesis, Department of Physics, McGill University, Montreal, 84 p.

—— 1963b: Gravity investigations. In: Müller *et al.*: Preliminary Report 1961–1962, Axel Heiberg Island Research Reports, McGill University, Montreal, p. 97–101.

BERGER, F., 1937: Die Anlage der schlesischen Stauchmoränen, Zentralblatt für Mineralogie, Geologie und Paläontologie, Abt. B, Nr. 1, p. 417–34 and 481–97.

BERSIER, A., 1954: Les collines de Noville-Chessel, crêtes de poussée glaciaire. Bulletin des Laboratoires de Géologie, Mineralogie, Géophysique et de Musée Géologique de l'Université de Lausanne, No. 109, 6 p.

BESCHEL, R.E., 1963: Sulphur springs at Gypsum Hill. In: Müller *et al.*: Preliminary Report 1961–1962, Axel Heiberg Island Research Reports, McGill University, Montreal, p. 183–7.

BLOCH, H., 1959: Ueber zwei bemerkenswerte Glaziär-Strukturen im Grundund Endmoränenbereich des Jungpleistozäns Mittel-Mecklenburgs. Neues Jahrbuch für Geologie und Paläontologie, Jahrgang 1959, p. 451–61.

BÖHM, A., 1901: Geschichte der Moränenkunde. R. Lechner, Wien, 334 p.

BONYUN, D.A., 1966: Computer evaluation of glacier surveying data. Unpublished report. Axel Heiberg Expedition, McGill University, Montreal.

BRINKMANN, R., 1953: Ueber die diluvialen Störungen auf Rügen. Geologische Rundschau, Bd. 41, p. 231–41.

BUBNOFF, S., 1956: Ueber glazigene Gesteinsdeformationen. Geologie, Jahrgang 5, p. 557–62.

CARLÉ, W., 1938: Das innere Gefüge der Stauch-Endmoränen und seine Bedeutung für die Gliederung des Altmoränengebietes. Geologische Rundschau, Bd. 29, p. 27–51.

CARSLAW, H.S. and JAEGER, J.C., 1959: Conduction of heat in solids. Clarendon Press, 2nd edition, Oxford, 510 p.

CHAMBERLIN, T.C., 1890: Boulder belts distinguished from boulder trains – their origin and significance. Bulletin of the Geological Society of America, Vol. 1, p. 27–31.

—— 1893: The nature of the englacial drift of the Mississippi Basin. Journal of Geology, Vol. 1, No. 1, p. 47–60.

CHAMBERLIN, T.C., 1894: Proposed genetic classification of Pleistocene glacial formations. Journal of Geology, Vol. 2, p. 517–38.

CHARLESWORTH, J.D., 1957: The Quarternary era. Vol. 1, Edward Arnold Ltd, London, 591 p.

CHARPENTIER, De I., 1841: Essai sur les glaciers. Imprimerie et Librairie de Marc Duclos, Lausanne, 363 p.

CLARK, S.P., 1966: Handbook of physical constants. Geological Society of America, 587 p.

CREDNER, H., 1880: Ueber Schichtenstörungen im Untergrund des Geschiebelehms, an Beispielen aus dem nordwestlichen Sachsen und angrenzenden Landstrichen. Zeitschrift der Deutschen Geologischen Gesellschaft, Bd. XXXII, p. 75–110.

DHAWAN, C.L., 1953: (Written discussion in Session 2: Laboratory investigation, including compaction tests, improvement of soil properties.) Proceedings of the

Third International Conference on Soil Mechanics and Foundation Engineering, Zürich, p. 127.

FELLENIUS, W., 1948: Erdstatische Berechnungen. Wilhelm Ernst und Sohn, Berlin, 48 p.

FLINT, R.F., 1940: End moraines of ice sheets. Zeitschrift für Gletscherkunde, Bd. XXVII, H. 1/2, p. 88–97.

—— 1957: Glacial and Pleistocene geology. John Wiley and Sons, New York, 553 p.

FRICKER, P.E., 1963: Geology of the Expedition Area, central western Axel Heiberg Island, Canadian Arctic Archipelago. Axel Heiberg Island Research Reports, Geology No. 1, McGill University, Montreal, 156 p.

FRIES, W., 1933: Tertiär und Diluvium im Grünberger Höhenrücken. Sonderdruck aus: Jahrbuch des Halleschen Verbandes für die Erforschung der mitteldeutschen Bodenschätze und ihre Verwertung, Bd. 12, neue Folge, 34 p.

FRITZSCH, V. and TAUBER, A.F., 1967: Die Veränderung der geoelektrischen Struktur des Untergrundes durch Mineralwässer. Acta Hydrophysica, Bd. XII, H. 1, p. 5–26.

FULLER, M.L., 1914: The geology of Long Island, New York. U.S. Geological Survey, Professional Paper, No. 82, 231 p.

GASSMANN, F. and WEBER, M., 1960: Einführung in die angewandte Geophysik. Verlag Hellweg, Bern, 284 p.

GRIPP, K., 1927: Glaziologische und geologische Ergebnisse der Hamburgischen Spitzbergen-Expedition 1927. Abhandlungen aus dem Gebiete der Naturwissenschaften, Naturwissenschaftlicher Verein Hamburg, Bd. XXII, H. 3/4, p. 145–250.

—— 1955: Eisbedingte Lagerungsstörungen. Geologische Rundschau, Bd. 43, p. 39–45.

GRIPP, K. and TODTMANN, E.M., 1926: Die Endmoränen des Green Bay Gletschers auf Spitzbergen; eine Studie zum Verständnis norddeutscher Diluvialgebilde. Geographische Gesellschaft Hamburg, Mitteilungen, Bd. 37, p. 43–75.

HAUMANN, D., 1960: Coordinates of ground control points. Photogrammetric Research Section, National Research Council of Canada, Ottawa, Mimeo Ms, 8 p.

—— 1963: Surveying glaciers on Axel Heiberg Island. The Canadian Surveyor, Vol. XVII, No. 2, p. 81–93.

HEIM, A., 1885: Handbuch der Gletscherkunde. Verlag J. Engelhorn, Stuttgart, 560 p.

HOCHSTEIN, M., 1965: Elektrische Widerstandsmessungen auf dem grönländischen Inlandeis. Meddelelser om Grønland, Bd. 177, Nr. 3, 39 p.

HOEN, E.W., 1964: The anhydrite diapirs of central western Axel Heiberg Island. Axel Heiberg Island Research Reports, Geology No. 2, McGill University, Montreal, 102 p.

HOPKINS, O.B., 1923: Some structural features of the plains area of Alberta caused by Pleistocene glaciation. Bulletin of the Geological Society of America, Vol. 34, p. 419–30.

JESSEN, A., 1928: Lønstrup Klint. Danmarks Geol. Undersøgelse (II), Nr. 49, 142 p.

JOHNSTRUP, F., 1874: Om Haevningsfaenomenerne i Møens Klint. J.H. Schultz, Københaven, 45 p.

KEILHACK, K., 1926: Das Quartär. In: Salomon, W.: Grundzüge der Geologie, Bd. II, E. Schweizerbart'sche Verlagsbuchhandlung, Stuttgart, p. 455–84.

KLEBELSBERG, R., 1948: Handbuch der Gletscherkunde und Glacialgeologie. Bd. I, Springer Verlag, Wien, 403 p.

KUNETZ, G., 1966: Principles of direct current resistivity prospecting. Gebr. Borntraeger, Berlin, 103 p.

KUPSCH, W.O., 1962: Ice-thrust ridges in western Canada. Journal of Geology, Vol. 70, p. 582–94.

LAMERSON, P.R. and DELLWIG, L.F., 1957: Deformation by ice push of lithified sediments in south-central Iowa. Journal of Geology, Vol. 65, p. 546–50.

LANG, H.D., 1964: Ueber glaziäre Stauchungen in den Mellendorfer und Brelinger Bergen nördlich von Hannover. Eiszeitalter und Gegenwart, Bd. 15, p. 207–20.

LEFFINGWELL, E. dek., 1919: The Canning River region, northern Alaska. U.S. Geological Survey, Professional Paper, No. 109, 251 p.

LLIBOUTRY, L., 1964: Traité de glaciologie. Tome 1 et 2, Masson & Cie, Paris, 1040 p.

LÜTSCHG, O., 1933: Beobachtungen über des Verhalten des vorstossenden oberen Grindelwaldgletschers im Berner Oberland. Verhandlungen der Schweizerischen Naturforschenden Gesellschaft, Thun und Jungfraujoch, II. Teil, p. 320–2.

MAAG, H., 1969: Ice dammed lakes and marginal glacial drainage on Axel Heiberg Island, Canadian Arctic Archipelago. Axel Heiberg Island Research Reports, McGill University, Montreal, 147 p.

MACKAY, J.R., 1956: Deformation by glacier-ice at Nicholson Peninsula, N.W.T., Canada. Arctic, Vol. 9, No. 4, p. 218–28.

—— 1959: Glacier ice-thrust features of the Yukon coast. Geographical Bulletin, No. 13, p. 5–21.

MÜLLER, F., 1963a: Surveying of glacier movement and mass changes. In: Müller et al.: Preliminary Report 1961–1962. Axel Heiberg Island Research Reports, McGill University, Montreal, p. 65–80.

—— 1963b: Radiocarbon dates and notes on the climatic and morphological history. In: Müller et al.: Preliminary Report 1961–1962. Axel Heiberg Island Research Reports, McGill University, Montreal, p. 169–72.

—— 1963c: Ablation measurements in 1962. In: Müller et al.: Preliminary Report 1961–1962, Axel Heiberg Island Research Reports, McGill University, Montreal, p. 37–46.

MÜLLER, F., 1963d: Englacial temperature measurements on Axel Heiberg Island, Canadian Arctic Archipelago. I.U.G.G. General Assembly of Berkeley, I.A.S.H. Publication No. 61, p. 168–80.

—— 1969: Was the Good Friday Glacier on Axel Heiberg Island surging? Canadian Journal of Earth Sciences, Vol. 6, No. 4, p. 891–4.

MÜLLER, F. and IKEN, A., in press: Velocity fluctuations and water regime of arctic valley glaciers. Proceedings of the Symposium on the Hydrology of Glaciers, September 7–13, 1969, Cambridge, England. I.A.S.H. Publication, approx. 10 p.

MÜLLER, F. and ROSKIN-SHARLIN, N., 1967: A high arctic climate study on Axel Heiberg Island, Canadian Arctic Archipelago – summer 1961. Part I General Meteorology. Axel Heiberg Island Research Reports, Meteorology No. 3, McGill University, Montreal, 82 p.

NAGY, D., 1966: The prism method for terrain corrections using digital computers. Pure and Applied Geophysics, Vol. 63, 1966/1, p. 31–9.

NYE, J.F., 1959: A method of determining the strain-rate tensor at the surface of a glacier. Journal of Glaciology, Vol. 3, No. 25, p. 409–419.

—— 1961: The flow of glaciers and ice-sheets as a problem in plasticity. Proceedings of the Royal Society, Ser. A, Vol. 207, No. 1091, p. 554–72.

PHILIPP, H., 1929: Die Wirkungen des Eises auf den Untergrund (Glazialerosion). In: Blank, E.,: Handbuch der Bodenlehre, Bd. 1, J. Springer, Berlin, p. 257–88.

PUGGAARD, C., 1851: Uebersicht der Geologie der Insel Möen. C.C. Jenni, Vater, Bern, 24 p.

QUERVAIN, de A., 1919: Ueber die Wirkungen eines vorstossenden Gletscher. Betrachtungen am oberen Grindelwaldgletscher. Heim-Festschrift, Vierteljahrschrift der Naturforschenden Gesellschaft in Zürich, Bd. LXIV, p. 336–49.

REDPATH, B.B., 1965: Seismic investigations of glaciers on Axel Heiberg Island, Canadian Arctic Archipelago. Axel Heiberg Island Research Reports, Geophysics No. 1, McGill University, Montreal, 26 p.

REINER, M., 1960: Lectures on theoretical rheology. North-Holland Publishing Company, Amsterdam, 158 p.

ROBITAILLE, B. and GREFFARD, C., 1962: Notes sur les materiaux terminaux du Glacier Thompson, Canada Arctique. Geographical Bulletin, No. 17, p. 85–94.

ROETHE, O., 1930: Zur Deutung ostdeutscher Braunkohlenfalten. Zeitschrift der Deutschen Geologischen Gesellschaft, Bd. 82, p. 498–506.

SCHOTT, C., 1933: Die Formengestaltung der Eisrandlagen Norddeutschlands. Zeitschrift für Gletscherkunde, Bd. 21, p. 54–98.

SCHULZ, W., 1965: Die Stauchendmoräne der Rosenthaler Staffel zwischen Jatzmick und Brohm in Mecklenburg und ihre Bedeutung zum Helpter Berg. Geologie, Jahrgang 14, p. 564–88.

—— 1966: Helpter Berg, Schmooksberg. Hohe Burg. Ein Vergleich dreier Stauchendmoränen Mecklenburgs. Geologie, Jahrgang 15, p. 174–87.

SHUMSKII, P.A., 1964: Principles of structural glaciology. Dover Publication Inc., New York, (translated from Russian by D. Kraus), 497 p.

TERZAGHI, K. and PECK, R.B., 1967: Soil mechanics in engineering practice. John Wiley and Sons, Inc., New York, 729 p.

WAHNSCHAFFE, F., 1882: Ueber einige glaziale Druckerscheinungen im norddeutschen Diluvium. Zeitschrift der Deutschen Geologischen Gesellschaft, Bd. 34, p. 562–601.

WEBER, J.R., SANDSTROM, N. and ARNOLD, K.C., 1960: Geophysical surveys on Gilman Glacier, Northern Ellesmere Island. I.A.S.H. Publication 54 (Extract), p. 500–11.

WEBER, J.R., 1961: Comparison of gravitational and seismic depth determination on the Gilman Glacier and adjoining ice-cap in Northern Ellesmere Island. Geology of the Arctic, University of Toronto Press, p. 781–90.

WEISSE, R., 1965: Entwurf einer Systematik der wichtigsten Endmoränen Norddeutschlands. Geologie, Jahrgang 14, p. 610–24.

WOLDSTEDT, P., 1955: Norddeutschland und angrenzende Gebiete im Eiszeitalter. K.F. Koehler Verlag, Stuttgart, 467 p.

WOLFF, W., 1927: Einige glazialgeologische Probleme aus dem norddeutschen Tiefland. Zeitschrift der Deutschen Geologischen Gesellschaft, Bd. 79, p. 342–60.

Maps

Canadian Department of Mines and Technical Surveys,
  1961: Eureka Sound, 2007, World Aeronautical Chart, scale 1:1,000,000

Canadian Department of Energy, Mines and Resources,
  1967: Sawer Bay, 39G,   scale 1:100,000
  1967: Dobbin Bay, 39H–29G,   ″
  1966: Glacier Fiord, 59E,   ″
  1966: Haig-Thomas Island, 59F,   ″
  1967: Middle Fiord, 59G,   ″
  1966: Greely Fiord West, 340B,   ″
  1967: Otto Fiord, 340C,   ″
  1967: M'Clintock Inlet, 340E-H,   ″
  1967: Bukken Fiord, 560A,   ″

McGill University, Axel Heiberg Expedition, and National Research
Council of Canada, Photogrammetric Research Section,
  1962: Thompson Glacier Region, Axel Heiberg Island, N.W.T., Canada,
                              scale 1:50,000
  1962: White Glacier, Thompson Glacier Region, Axel Heiberg
        Island, Canadian Arctic Archipelago,   scale 1:10,000
  1962: Thompson Glacier Snout, Axel Heiberg Island,
        Canadian Arctic Archipelago,   scale 1:5,000
  1964: White Glacier, Axel Heiberg Island, Canadian
        Arctic Archipelago, (in two parts)   scale 1:5,000

McGill University,
  1963: Expedition Area, Axel Heiberg Island, Canadian
        Arctic Archipelago,   scale 1:100,000
  Fricker, P.E., 1962: Expedition Area, Western Central
        Axel Heiberg Island, Preliminary Geological Map,
        approx.   scale 1:63,000
  Hoen, E.W., 1962: Geology with special reference to anhydrite
        diapirs, Central Western Axel Heiberg Island,
                         approx. scale 1:167,000

Appendix 1 Geothermal data of the Expedition River valley (borehole).

Thermistor	Depth [m]	Temperature [°C]											
		a	b	c	d	e	f	g	h	i	k	l	m
D 116	0.2	6.67		4.86		−17.95	−3.52	−2.03	4.77	9.40	8.02	5.03	7.90
E 117	3.3	−3.22	−3.22	−3.36		−19.20	−17.93	−15.92	−13.77	−10.20	−7.75	−6.09	−5.59
F 118	6.3	−0.23	−0.62	−10.48	−12.4	−15.59	−15.08	−15.04	−14.79	−14.34	−13.64	−12.61	−12.00
G 119	9.4	−0.21	−0.40	−13.07	−13.5	−13.06	−14.38	−14.65	−14.78	−14.89	−14.95	−14.87	−14.76
H 120	12.4	−7.74	−9.45	−13.55	−14.0	−13.11	−13.04	−13.13	−13.25	−13.35	−13.45	−13.56	−13.60
I 121	15.4	−3.00	−6.84		−12.0								
K 122	18.5	−8.80	−9.71		−12.6								
L 123	21.6	−10.62	−11.25		−13.3								
M 124	24.6	−11.81			−13.2								
N 132	27.7	−12.51	−12.74		−13.5								
P 126	30.7	−12.78	−12.98		−13.6								
Q 127	33.7	−13.19	−13.33		−13.8								

a 68 07 24 0600 Last complete set of readings, 61 hours after completion of the borehole

b 68 07 25 0000 Last set of readings before failure of the thermocable (at 12.4–15.4 m depth, 79–83 hours after completion of the borehole)

c 68 08 26 1830 Last readings in fall 1968, 865 hours after completion of the borehole

d 68 07 Extrapolated predrilling temperature, fall 1968 of the borehole

e 69 04 27
f 69 05 18
g 69 06 01
h 69 06 15
i 69 06 29   } Field readings and data compiled by F. Müller, A. Ohmura and J. Weiss
k 69 07 13
l 69 08 01
m 69 08 14

Accuracy: a – c, e – m   ±0.01°C
            d   ±0.3 °C

d) In a first approximation the predrilling temperature of the borehole is extrapolated. Drilling means insertion of a more or less linear heat source into the ground. Assuming the source to be of infinite linear extent and instantaneous heat release, the temperature in the axis is inversely proportional to time approaching equilibrium (Carslaw, H.S. and Jaeger, J.C., 1959, p. 258). Observing the cooling process over more than days (drilling time 8 days) and plotting the temperature data accordingly, the relationship proved to be reasonably satisfactory for thermistors deeper than 6 m.

Appendix 2 Sedimentological data of outwash gravel and lacustrine sand.

Sample s1: Outwash gravel, location see on Fig. 11
Sample s2: Lacustrine sand, location see on Fig. 11

d_1 (mm)	Cumulative frequency (% by weight)		d_2 (mm)	Analysis
	s1 ($w = 94.0$ kg)	s2 ($w = 1.710$ kg)		
200	100			
100	69.7			
60	60.3			hand
40	51.4			measurements
27	42.5			
12.7	28.9			
10.0		100	10.0	
4.0	17.4	99.8	4.0	
2.0	13.8	98.2	2.0	sieves
1.0	11.6	95.1	1.0	
0.25	4.8	47.8	0.25	
0.080	0.73	8.4	0.079	
0.057	0.51	6.6	0.056	
0.040	0.44	6.0	0.040	
0.026	0.36	5.4	0.025	araeometre
0.015	0.14	4.8	0.015	
0.0086	0.14	3.6	0.0084	
		2.4	0.0052	

	Outwash gravel	Lacustrine sand
i	10.2 % by weight ($w = 5.53$ kg)	22.4% by weight ($w = 1.17$ kg)
c	2.74 g/cm^3	2.74 g/cm^3
d_{max}	about 1.5 m	10 mm
l	sedimentary (mainly sandstone) 55% igneous (mainly basalt) 39% metamorphic (mainly quarzite) 6%	
r	rounded 9% subrounded 26% subangular 62% angular 3%	
Φ	40°	37°

d: grain diameter, the subscript of d refers to the sample number
w: sample weight
i: ice content
c: mean density of the fraction with a diameter less than 1 mm
d_{max}:   maximum diameter
l: lithology of the fraction with a diameter greater than 12.7 mm
r: roundness of the fraction with a diameter greater than 12.7 mm
Φ: estimated angle of internal friction (Dhawan, 1953, p. 127)

The field analysis of the samples has been performed by H. Snyder.

Appendix 3 Logarithmic probability of the cumulative frequency of outwash and lacustrine sand.

Appendix 4 Movement rates of surveyed points, Axel Heiberg Island, N.W.T.

STAKE	DATE		DIFFERENCES (CM)				TIME	RATES (CM/DAY)					DIP	ERRORS	
	FROM	TO	X	Y	Z	DIST	(DAYS)	X	Y	Z	DIST	DXY	(DEG)	r_h	r_v
288	630822	650823	-3508	-1781	-207	3939	731.0	-4.80	-2.44	-0.28	5.39	5.38	3.	0.01	0.01
288	650823	660520	-1095	-581	-52	1240	270.0	-4.06	-2.15	-0.19	4.60	4.59	2.	0.03	0.04
288	660520	660826	-451	-242	-7	520	98.0	-4.70	-2.47	-0.07	5.31	5.31	1.	0.07	0.31
288	660826	670603	-1188	-651	5	1354	281.0	-4.23	-2.32	0.02	4.82	4.82	-0.	0.03	0.44
288	670603	670825	-400	-212	8	452	83.0	-4.82	-2.55	0.10	5.46	5.45	-1.	0.08	0.27
288	670825	680501	-1012	-626	40	1190	249.0	-4.06	-2.51	0.16	4.78	4.78	-2.	0.03	0.05
288	680501	680812	-509	-298	-32	590	103.0	-4.94	-2.89	-0.31	5.73	5.73	3.	0.06	0.07
297	630822	660520	-107	2	48	117	1001.0	-0.11	0.00	0.05	0.12	0.11	-24.	0.29	0.05
298	660602	660826	-10	-8	-1	12	85.0	-0.12	-0.09	-0.01	0.15	0.15	4.	>1.0	>1.0
298	660826	670603	-4.3	-37	35	66	281.0	-0.15	-0.13	0.12	0.24	0.20	-32.	0.52	0.06
298	670603	670825	-21	-18	12	30	83.0	-0.25	-0.22	0.14	0.36	0.33	-23.	>1.0	0.18
326	660520	660826	-480	-192	58	520	98.0	-4.90	-1.96	0.59	5.31	5.28	-6.	0.07	0.04
326	660826	670603	-1127	-457	42	1216	281.0	-4.01	-1.63	0.15	4.33	4.33	-2.	0.03	0.05
326	670603	670825	-275	-116	-30	299	83.0	-3.31	-1.40	-0.36	3.61	3.60	6.	0.11	0.07
326	670825	680501	-999	-601	20	1166	249.0	-4.01	-2.41	0.08	4.68	4.68	-1.	0.03	0.11
326	680501	680812	-175	-240	-100	313	103.0	-1.70	-2.33	-0.97	3.04	2.88	19.	0.11	0.02
327	660520	660826	2	-4	15	15	98.0	0.02	-0.04	0.15	0.16	0.05	-73.	>1.0	0.15
327	660826	670603	-62	-4	72	95	281.0	-0.22	-0.01	0.26	0.34	0.22	-49.	0.36	0.03
327	670603	670825	-43	-3	58	72	83.0	-0.52	-0.04	0.70	0.87	0.52	-53.	0.47	0.04
327	670825	680501	-352	-97	290	466	249.0	-1.41	-0.39	1.16	1.87	1.47	-38.	0.07	0.01
327	680501	680812	-323	-67	226	399	103.0	-3.14	-0.65	2.19	3.88	3.20	-34.	0.09	0.01

328	650823	660520	−501	−537	315	799	270.0	−1.86	−1.99	1.17	2.96	2.72	−23.	0.04	0.01
329	660520	670826	−1062	−1245	78	1638	463.0	−2.29	−2.69	0.17	3.54	3.53	−3.	0.02	0.03
330	650823	660520	−52	−86	75	125	270.0	−0.19	−0.32	0.28	0.46	0.37	−37.	0.27	0.03
330	660520	660826	−26	−72	40	86	98.0	−0.27	−0.73	0.41	0.88	0.78	−28.	0.40	0.05
330	660826	670603	−194	−201	136	310	281.0	−0.69	−0.72	0.48	1.11	0.99	−26.	0.11	0.02
330	670603	670825	−90	−98	54	143	83.0	−1.08	−1.18	0.65	1.73	1.60	−22.	0.24	0.04
330	670825	680501	−328	−366	224	540	249.0	−1.32	−1.47	0.90	2.17	1.97	−25.	0.06	0.01
330	680501	680812	−258	−162	92	318	103.0	−2.50	−1.57	0.89	3.09	2.96	−17.	0.11	0.02
331	660520	660826	−32	−55	58	86	98.0	−0.33	−0.56	0.59	0.88	0.65	−42.	0.39	0.04
331	660826	670603	−158	−180	171	294	281.0	−0.56	−0.64	0.61	1.05	0.85	−36.	0.12	0.01
331	670603	670825	−77	−92	75	141	83.0	−0.93	−1.11	0.90	1.70	1.45	−32.	0.24	0.03
331	670825	680501	−306	−371	297	565	249.0	−1.23	−1.49	1.19	2.27	1.93	−32.	0.06	0.01
410	680501	680603	−129	−8	89	156	33.0	−3.91	−0.24	2.70	4.76	3.92	−35.	0.22	0.02
411	680603	680812	−223	−125	12	255	70.0	−3.19	−1.79	0.17	3.66	3.65	−3.	0.13	0.18
351	670825	680603	−1811	−1552	−4	2385	282.0	−6.42	−5.50	−0.01	8.46	8.46	0.	0.02	0.55
352	670825	680603	−1789	−1518	−33	2346	282.0	−6.34	−5.38	−0.12	8.32	8.32	1.	0.02	0.07
353	670825	680603	−1729	−1499	24	2288	282.0	−6.13	−5.32	0.09	8.12	8.11	−1.	0.02	0.09
354	670825	680603	−1619	−1324	132	2095	282.0	−5.74	−4.70	0.47	7.43	7.42	−4.	0.02	0.02
355	670825	680603	−1644	−1478	101	2213	282.0	−5.83	−5.24	0.36	7.85	7.84	−3.	0.02	0.02
356	670825	680603	−1666	−1635	5	2334	282.0	−5.91	−5.80	0.02	8.28	8.28	−0.	0.02	0.44

END READ – NO MORE DATA CARDS

r_h: Relative error of the horizontal components of the displacements and the rates
r_v: Relative error of the vertical components of the displacements and the rates

Appendix 5 Gravity data.

P: Gravity station
x, y, z: Local coordinates of P in m
r: Horizontal radial error
e: Error
g: Difference between the vertical gravitational component at the station P and the control station 9611-60 corrected for time drift and height above ground
g_l: Local anomaly in mgl

$$g_l = g + c_H + c_L + c_T - r$$

c_H: Correction for height above sea level (free air and plate correction combined, $c_H = 0.19498z$ (mgl)
c_L: Correction for latitude, $c_L = -0.0002929(x - 60,000)$ (mgl)
c_T: Correction for terrain, relative error 0.08
r: Reginal anomaly. It is approximated by a linear regression from the stations of the glacier's surrounding region (609, 610, 611, 613, 615, 617, 619, 622, 623), $r = 36.746 + 0.00212(x - 60,000) - 0.00009(y - 30,000)$ (mgl). The standard deviation of r is 0.41 (mgl).

t: Thickness of the glacier, assumed to be a stab of infinite extent, $\qquad t = \dfrac{1}{2\pi k d}\, g_l = q g_l$
k: Gravity constant
d: Density contrast, $d = 1.76g$ cm^{-3}
$\quad q = 13.558$ m mgl^{-1}
h: Height above sea level of the glacier's lower surface, $h = z - t$

P	x (m)	y (m)	r_{xy} (m)	z (m)	e_z (m)	g (mgl)	e_g (mgl)	c_T (mgl)	g_l (mgl)	e_{gl} (mgl)	t (m)	h (m)	e_t, e_n (m)
9611-60	60'501	29'861	0.1	196.5	0.1	0.0	0.0	3.57	0.0				
9012-68	62'728	32'047	3.0	204.0	0.6	-8.28	0.07	5.49	-19.30	1.0	262	28	14
9013-68	62'963	34'202	0.6	289.8	0.4	-37.11	0.11	4.82					
609	61'216	35'405	0.2	180.1	0.2	-4.04	0.10	8.50	0.38				
610	60'656	34'702	0.2	101.2	0.2	10.74	0.10	6.90	-0.53				
611	60'374	34'692	0.2	128.8	0.2	5.28	0.10	6.75	-0.09				

ID													
613	60'330	34'259	0.2	91.0	0.2	11.71	0.10	7.86	0.16				
615	60'824	31'428	0.2	64.0	0.2	21.93	0.02	4.65	0.46				
617	63'112	31'253	0.2	403.8	0.2	-43.85	0.02	9.02	-0.24				
619	64'120	32'271	0.2	202.7	0.2	-0.90	0.09	7.44	-0.42				
622	64'433	32'863	0.2	216.6	0.2	-2.76	0.09	8.23	0.40				
623	64'623	32'686	0.2	245.5	0.2	-8.53	0.09	8.56	0.23				
510	62'867	34'186	0.8	280.5	0.5	-34.84	0.15	4.48	-18.96	1.0	257	23	14
511	62'805	34'176	0.9	271.0	0.5	-32.53	0.15	3.81	-19.02	1.0	258	18	14
512	62'750	34'166	1.0	262.7	0.5	-30.48	0.15	3.77	-18.49	1.0	251	12	14
513	62'690	34'156	1.3	254.5	0.6	-28.57	0.15	2.96	-18.85	0.9	256	-1	12
514	62'635	34'147	1.	250.4	0.6	-27.74	0.15	3.44	-18.21	0.9	247	4	12
515	62'594	34'139	1.	247.7	0.6	-27.19	0.15	3.84	-17.68	1.0	240	8	14
516	62'549	34'132	2.	246.8	0.6	-27.12	0.15	3.78	-17.75	1.0	241	6	14
517	62'505	34'124	2.	248.2	0.7	-27.43	0.15	3.93	-17.53	1.0	288	11	14
518	62'453	34'115	2.	243.1	0.7	-26.29	0.16	3.45	-17.74	1.0	241	3	14
519	62'393	34'105	2.	244.5	0.7	-26.81	0.15	4.24	-17.05	1.0	231	13	14
520	62'351	34'098	2.	242.3	0.8	-26.29	0.15	4.27	-16.82	1.0	228	14	14
521	62'292	34'088	2.	239.3	0.8	-25.83	0.15	4.00	-17.08	1.0	232	8	14
522	62'239	34'079	2.	235.1	0.8	-24.85	0.15	3.58	-17.22	1.0	233	2	14
523	62'191	34'070	2.	231.4	0.9	-24.07	0.15	3.60	-17.01	1.0	231	1	14
524	62'142	34'062	2.	229.7	0.9	-23.90	0.15	3.50	-17.16	1.0	233	-3	14
525	62'078	34'051	3.	229.9	1.0	-23.94	0.15	3.52	-16.99	1.0	230	0	14
526	62'022	34'041	3.	230.7	1.0	-24.18	0.15	3.65	-16.81	1.0	228	3	14
527	61'967	34'032	3.	229.9	1.0	-24.08	0.15	3.68	-16.70	1.0	226	3	14
528	61'916	34'023	4.	228.3	1.0	-23.66	0.15	3.54	-16.98	1.0	290	-2	14
529	61'853	34'012	4.	229.8	1.0	-24.15	0.15	3.87	-16.33	1.1	221	8	15
530	61'813	34'006	4.	227.6	1.0	-23.67	0.15	3.88	-16.17	1.1	219	8	15
531	61'762	33'997	4.	226.7	1.0	-23.45	0.15	5.01	-14.87	1.2	202	25	16
532	61'709	33'988	4.	222.5	1.	-22.35	0.15	4.36	-14.62	1.1	198	24	15
533	61'648	33'977	4.	218.7	1.	-21.41	0.15	4.75	-14.38	1.1	195	24	15
534	61'602	33'970	4.	216.8	1.	-20.91	0.15	5.21	-13.68	1.2	185	31	16
535	61'552	33'961	4.	208.9	1.	-19.06	0.15	5.36	-13.10	1.2	178	31	16

Appendix 5 (cont'd)

P	x (m)	y (m)	r_{xy} (m)	z (m)	e_z (m)	g (mgl)	e_g (mgl)	c_T (mgl)	g_l (mgl)	e_{gl} (mgl)	t (m)	h (m)	e_t, e_n (m)
536	61'507	33'954	5.	206.5	1.	−18.17	0.15	5.26	−12.66	1.2	172	35	16
537	61'467	33'947	5.	202.3	1.	−17.25	0.15	4.96	−12.78	1.1	173	29	15
538	61'428	33'940	5.	198.7	1.	−16.42	0.15	4.71	−12.80	1.1	174	25	15
539	61'390	33'933	5.	193.4	1.	−16.25	0.15	4.80	−13.49	1.1	183	10	15
540	61'347	33'926	5.	186.0	1.	−13.49	0.15	4.23	−12.64	1.0	171	15	14
541	61'300	33'918	5.	187.3	1.	−13.81	0.15	4.46	−12.36	1.1	168	20	15
542	61'268	33'912	5.	188.8	2.	−12.63	0.15	4.07	−12.37	1.3	168	15	18
543	61'208	33'901	5.	177.7	2.	−11.12	0.15	4.74	−11.05	1.3	150	30	18
544	61'169	33'895	6.	180.6	9.	−12.02	0.15	6.20	−9.82	1.4	133	47	19
545	61'109	33'884	6.	175.8	2.	−10.62	0.15	6.16	−9.26	1.4	126	50	19
546	61'072	33'878	6.	172.3	2.	−9.67	0.15	6.09	−8.97	1.4	122	51	19
547	61'038	33'872	6.	169.8	2.	−8.85	0.15	6.41	−8.24	1.5	112	58	20
548	60'982	33'862	6.	167.3	2.	−8.13	0.15	6.12	−8.16	1.4	111	57	19
549	62'284	34'135	2.	240.7	0.8	−21.18	0.15	4.14	−17.00	1.0	230	10	14
550	62'275	34'190	2.	242.2	0.9	−26.52	0.15	4.15	−17.00	1.1	230	12	15
551	62'266	34'240	2.	245.6	0.9	−27.35	0.15	3.88	−17.41	1.0	236	10	14
552	62'257	34'295	2.	247.6	0.9	−27.85	0.15	4.04	−17.33	1.1	235	13	15
553	62'246	34'358	3.	252.2	1.	−28.84	0.15	4.56	−16.94	1.1	230	23	15
554	62'237	34'411	3.	256.5	1.	−18.83	0.15	4.06	−16.50	1.1	224	33	15
555	62'228	34'461	3.	256.3	1.	−29.66	0.15	3.80	−17.61	1.0	239	18	14
556	62'220	34'511	3.	254.7	1.	−29.07	0.15	3.63	−17.48	1.0	237	18	14
557	62'209	34'573	3.	255.2	1.	−28.92	0.15	3.81	−17.02	1.1	231	24	15
558	62'200	34'625	3.	253.8	1.	−28.21	0.15	3.92	−16.86	1.1	229	25	15
559	62'188	34'692	3.	251.6	1.	−27.60	0.15	3.98	−16.16	1.1	219	33	15
560	62'178	34'757	3.	249.4	1.	−26.78	0.15	3.85	−15.88	1.1	215	34	15
561	62'170	34'803	3.	248.0	1.	−26.27	0.15	3.53	−15.93	1.0	216	32	14
562	62'158	34'868	4.	247.0	1.	−25.75	0.15	3.36	−15.74	1.0	213	34	14
563	62'149	34'924	4.	241.6	1.	−25.66	0.15	3.48	−15.38	1.0	209	39	14

564	62'139	34'983	4.	247.6	1.	−25.64	0.15	3.55	−15.27	1.0	207	41	14
565	62'129	35'038	4.	248.8	1.	−25.39	0.15	3.75	−14.53	1.0	197	51	14
566	62'118	35'106	4.	250.7	1.	−25.41	0.15	4.06	−13.86	1.1	188	68	15
567	62'300	34'039	2.	238.2	0.8	−25.39	0.15	3.94	−16.94	1.0	230	9	14
568	62'308	33'992	2.	238.2	0.9	−25.30	0.15	4.62	−16.19	1.1	220	14	15
569	62'323	33'909	2.	233.8	0.9	−24.10	0.15	4.54	−15.99	1.1	217	17	15
570	62'336	33'828	3.	231.0	0.9	−22.41	0.15	4.26	−15.13	1.1	205	26	15
571	62'346	33'781	3.	228.6	1.	−21.88	0.15	3.71	−15.59	1.0	207	21	14
572	62'353	33'729	3.	231.5	1.	−22.35	0.15	4.07	−15.23	1.1	206	25	15
573	62'364	33'667	3.	233.8	1.	−22.68	0.15	4.27	−14.95	1.1	203	31	15
574	62'376	33'597	3.	235.3	1.	−22.64	0.15	3.91	−15.01	1.1	204	32	15
575	62'382	33'565	3.	239.4	1.	−23.49	0.15	4.06	−14.92	1.1	202	37	15
576	62'392	33'506	3.	241.8	1.	−23.81	0.15	4.21	−14.65	1.1	199	43	15
577	62'400	33'458	3.	243.3	1.	−23.84	0.15	4.23	−14.39	1.1	195	48	15
578	62'111	33'391	4.	243.4	1.	−23.51	0.15	5.09	−13.22	1.2	179	64	16
579	62'423	33'324	4.	234.8	1.	−20.90	0.15	4.40	−13.01	1.1	176	58	15
580	62'436	33'248	4.	227.8	1.	−18.67	0.15	3.67	−18.92	1.0	175	53	14
581	62'446	33'189	4.	235.3	1.	−20.26	0.15	5.99	−10.75	1.2	146	90	16
682	62'455	33'137	4.	229.2	1.	−18.40	0.15	5.98	−10.12	1.2	137	92	16
583	62'465	33'075	4.	221.1	1.	−15.96	0.15	5.28	−9.99	1.2	135	86	16
584	62'475	33'021	4.	214.3	1.	−14.06	0.15	4.70	−10.02	1.4	136	78	19
585	62'489	32'937	5.	204.3	2.	−11.30	0.15	4.40	−9.56	1.4	130	75	19
586	62'501	32'868	5.	202.4	2.	−12.41	0.15	6.06	−8.04	1.5	109	100	20
587	62'512	32'803	5.	204.3	2.	−10.84	0.15	5.50	−7.98	1.5	108	96	20
588	62'527	32'713	5.	203.1	2.	−11.40	0.15	6.66	−7.74	1.6	105	98	22
590	63'022	34'212	0.8	292.1	0.4	−37.69	0.15	4.50	−19.89	1.0	270	22	14
591	63'063	34'212	0.9	294.5	0.5	−38.22	0.15	4.77	−19.79	1.0	268	26	14
592	63'102	34'226	1.	295.7	0.5	−38.55	0.15	4.86	−19.88	1.0	270	26	14
593	63'157	34'235	1.	296.8	0.6	−38.80	0.15	4.92	−19.99	1.1	271	26	15
594	63'217	34'245	1.	295.7	0.6	−38.46	0.15	4.11	−20.82	1.0	282	13	14
595	63'254	34'252	2.	295.4	0.6	−38.36	0.15	3.92	−21.06	1.0	286	10	14
596	63'303	34'260	2.	296.5	0.6	−38.62	0.15	3.99	−21.15	1.0	287	10	14

Appendix 5 (cont'd)

P	x (m)	y (m)	r_{xy} (m)	z (m)	e_z (m)	g (mgl)	e_g (mgl)	c_T (mgl)	g_l (mgl)	e_{gl} (mgl)	t (m)	h (m)	e_t, e_n (m)
597	63'353	34'269	2.	297.5	0.6	−38.86	0.15	4.05	−21.26	1.0	288	9	14
598	63'399	34'277	2.	298.2	0.7	−39.06	0.15	3.87	−21.61	1.0	293	5	14
599	63'459	34'287	2.	299.3	0.7	−39.31	0.15	3.72	−21.93	1.0	297	2	14
624	63'502	34'295	2.	299.4	0.7	−39.38	0.15	3.70	−22.11	1.0	300	0	14
625	63'546	34'304	2.	300.7	0.7	−39.76	0.15	3.84	−22.21	1.0	301	0	14
626	63'594	34'313	2.	302.0	0.8	−40.03	0.15	3.70	−22.48	1.0	305	−3	14
627	63'663	34'327	2.	303.6	0.8	−40.44	0.15	3.08	−23.36	1.0	317	−13	14
628	63'712	34'336	3.	304.4	0.8	−40.55	0.15	3.16	−23.36	1.0	317	−12	14
629	63'765	34'347	3.	305.3	0.9	−40.81	0.15	3.24	−23.49	1.0	318	−13	14
630	63'814	34'356	3.	305.8	0.9	−40.88	0.15	3.27	−23.54	1.0	319	−13	14
631	63'861	34'365	3.	306.4	0.9	−40.98	0.15	3.33	−23.58	1.0	320	−13	14
632	63'903	34'374	3.	306.7	1.	−41.09	0.15	3.36	−23.70	1.0	321	−15	14
633	63'940	34'381	3.	306.6	1.	−41.02	0.15	3.35	−23.74	1.0	323	−15	14
634	63'989	34'390	3.	307.1	1.	−41.17	0.15	3.36	−23.91	1.0	324	−17	14
635	64'075	34'401	3.	307.2	1.	−41.20	0.15	2.89	−24.53	1.0	333	−25	14
636	64'105	34'413	3.	307.3	1.	−41.16	0.15	3.88	−24.63	1.0	334	−27	14
637	64'151	34'422	4.	307.4	1.	−41.19	0.15	2.92	−24.70	1.0	335	−27	14
638	64'219	34'435	4.	308.0	1.	−41.27	0.15	2.90	−24.84	1.0	337	−29	14
639	64'287	34'449	4.	308.7	1.	−41.37	0.15	2.80	−25.07	1.0	348	−31	14
640	64'345	34'460	4.	309.4	1.	−41.56	0.15	2.72	−25.34	1.0	344	−34	14

641	62'956	34'245	0.7	288.4	0.4	−36.81	0.15	5.19	−18.88	1.1	258	32	15
642	62'945	34'308	0.9	288.6	0.5	−37.00	0.15	5.29	−18.90	1.1	256	32	15
643	62'936	34'362	1.	286.2	0.5	−36.46	0.15	5.10	−19.00	1.1	256	29	15
644	62'928	34'410	1.	284.3	0.5	−35.96	0.15	5.18	−18.76	1.1	254	30	15
645	62'917	34'470	1.	281.5	0.6	−35.34	0.15	5.06	−18.78	1.1	255	29	15
646	62'905	34'539	2.	275.5	0.6	−34.73	0.15	4.48	−18.88	1.0	256	20	14
647	62'895	34'598	2.	275.8	0.7	−33.73	0.15	4.34	−18.93	1.0	257	14	14
648	62'885	34'660	2.	274.2	0.7	−33.16	0.15	3.95	−19.03	1.0	258	16	14
649	62'886	34'712	2.	271.4	0.7	−32.30	0.15	3.64	−19.00	1.0	258	14	14
650	62'866	34'772	2.	268.6	0.8	−31.24	0.15	3.44	−18.66	1.0	263	16	14
651	62'854	34'840	3.	268.3	0.8	−30.82	0.15	4.38	17.32	1.1	235	33	15
652	62'847	34'882	3.	269.3	0.8	−30.89	0.15	4.56	−17.00	1.1	230	39	15
853	62'838	34'933	3.	269.0	0.9	−30.54	0.15	4.60	−16.64	1.1	226	43	15
654	62'828	34'991	3.	269.6	0.9	−30.39	0.15	4.72	−16.21	1.1	220	50	15
555	62'820	35'040	3.	271.1	0.9	−30.50	0.15	4.93	−15.81	1.1	214	67	15
656	62'810	35'094	3.	271.2	1.	−30.19	0.15	5.08	−15.29	1.2	207	64	16
657	62'804	35'130	3.	290.1	1.	−29.67	0.15	5.05	−15.00	1.2	203	67	16
658	62'971	34'157	0.7	288.8	0.4	−36.67	0.15	5.05	−18.85	1.0	256	33	14
659	62'986	34'070	0.9	286.9	0.5	−35.95	0.15	4.67	−18.93	1.0	257	30	14
660	62'991	34'038	0.9	289.3	0.5	−36.46	0.15	4.81	−18.84	1.0	255	34	14
661	62'999	33'991	1.	287.6	0.5	−35.81	0.15	4.90	−18.45	1.0	250	37	14

Appendix 6 The push moraines of Axel Heiberg Island and Ellesmere Island.

Number	Local geographical name	Geographical coordinates latitude (±2')	longitude (±2')	Reference Aerial photograph trimetrogon	vertical	Map
						Axel Heiberg Island
A 1	Bunde Fiord	80° 25'	93° 45'	T 412 L – 32	A 16754 – 21/22 A 16861 – 49/50	9
A 2	Li Fiord (NW)	80° 10'	93° 50'	T 412 L – 134	A 16754 – 26/27 A 16755 – 43/44	9
A 3	Li Fiord (NE)	80° 10'	93° 42'	T 412 L – 134	A 16754 – 26/27 A 16755 – 43/44	9
A 4	Li Fiord (S)	79° 59'	94° 14'	T 412 L – 46	A 16186 – 94/95	5
A 5	Middle Fiord	79° 47'	93° 50'	T 489 C – 167	A 16755 – 52/53	5
A 6	East Fiord	79° 34'	93° 18'		A 16754 – 40/41 A 16755 – 57/58	5
A 7	Agate Fiord	79° 33'	92° 53'		A 16754 – 41/42	5
A 8	Finsterwalder Glacier	79° 27'	91° 35'		A 16836 – 126/127	16
A 9	Thompson Glacier	79° 24'	90° 34'	T 489 L – 34	A 16755 – 116/117 A 16864 – 36/37	12
A 10	Strand Fiord	79° 16'	90° 07'	T 440 R – 62	A 16755 – 113/114	16
A 11	Sand Bay	78° 48'	92° 10'	T 489 R – 142	A 16186 – 154/155	4
A 12	Good Friday Bay	78° 32'	91° 37'		A 16754 – 82/83	3
A 13	Surprise Fiord (N)	78° 24'	90° 55'		A 16836 – 6/7	3
A 14	Surprise Fiord (S)	78° 21'	90° 51'		A 16836 – 7/8	3

A 15	Glacier Fiord (S)	78° 27'	89° 59'		A 16'864 – 57/58	3
A 16	Glacier Fiord (N)	78° 39'	90° 05'		A 16'864 – 53/54	3
A 17	Wolf Fiord	78° 55'	89° 41'		A 16'859 – 10/11	3
A 18	Wolf/Skaare Fiord	78° 36'	88° 41'		A 16'860 – 124/125	3
A 19	Skaare Fiord	78° 57'	88° 17'		A 16'863 – 85/86	3
E 1	Clements Markham Inlet	82° 37'	69° 21'		A 16'607 – 123/124	10
E 2	Ayles Fiord	82° 41'	78° 17'	T 409 C – 13/14	A 16'693 – 210/211	8
E 3	Yelverton Bay (SE)	81° 54'	81° 20'	T 407 C – 33/34	A 16'785 – 96/97	10
E 4	Yelverton Bay (NW)	82° 05'	83° 24'	T 405 C – 214/215	A 16'785 – 191/192	10
E 5	Yelverton Bay (NE)	82° 05'	83° 15'	T 405 C – 214/215	A 16'785 – 191/192	10
E 6	Yelverton Bay (SW)	82° 03'	83° 09'		A 16'785 – 190/191	10
E 7	Phillips Bay	81° 55'	86° 19'	T 405 C – 16/17	A 16'760 – 99/100	7
E 8	Hare Fiord (NW)	81° 08'	83° 20'		A 16'724 – 29/30	7
E 9	Hare Fiord (NE)	81° 11'	81° 59'	T 407 R – 57	A 16'728 – 42/43	7
E 10	Hare Fiord (SE)	81° 10'	82° 09'	T 407 R – 57	A 16'785 – 114/115	7
E 11	Hare Fiord (SW)	81° 01'	85° 12'	T 405 C – 79	A 16'606 – 34/35	7
E 12	Oobloyah Bay	80° 51'	83° 06'		A 16'724 – 22/23	6
E 13	Chapman Glacier	81° 13'	79° 35'	T 408 C – 66	A 16'693 – 173/174	10
E 14	Antoinette Bay	80° 56'	76° 12'	T 490 C – 139	A 16'694 – 41/42	10
E 15	Canon Fiord	79° 42'	78° 43'	T 492 C – 97	A 16'691 – 10/11	1
E 16	Dobbin Bay	79° 48'	75° 24'		A 16'613 – 82/83	2

The list represents push moraines identified on aerial photographs.
The numeration begins with each island's northernmost push moraine and continues anticlockwise
The geographical coordinates roughly refer to the centre of the push moraine.

The classification of the push moraines of Axel Heiberg Island and Ellesmere Island due to the parameters given in Table 4.

											number of parameter												
	1	2	3	4	5	6	7	8	9	10	11	12	13	14	15	16	17	18	19	20	21	22	r
A 1	x				x			x		x		x				x	x				x		7.9
A 2	x	x			x				x	x				x		x	x					x	12.6
A 3	x				x				x	x				x		x		x			x		2.0
A 4	x			x	x				x	x					x		x		x		x		5.0 3.2
A 5	x			x	x		x			x						x			x		x		4.0
A 6	x	x			x				x	x						x		x				x	4.0
A 7	x			x			x			x					x				x		x	x	3.7
A 8	x				x	x			x	x		x		x	x					x		x	6.2
A 9	x			x			x			x				x	x					x	x		2.6
A10	x			x			x			x					x				x		x		1.8
A11	x		x						x					x		x							3.0
A12	x			x			x			x			x	x	x					x		x	3.5
A13	x			x					x	x			x	x	x		x					x	4.9
A14	x			x					x	x						x				x			
A15	x				x				x		x			x		x	x				x	x	15.8
A16	x					x			x	x		x	x	x	x							x	2.4

	1	2	3	4	5	6	7	8	9	10	11	12	13	14	15	16	
A17	×		×			×	×	×		×		×	×	×		×	
A18	×	×	×	×		×		×		×	?x	×		×	×		
A19	×	×	×		×	×	×	×		×		×	×	×	×	×	2.9
E 1	×			?x	×	×		×		×	?x	×		×		×	0.8
E 2	×	×	×			×	×	×		×	×	×		×		×	1.6
E 3	×	×	×	×		×	×	×		×	×	×	×	×		×	2.9
E 4	×	×	×	?x		×	×	×		×	?x	×		×	×		
E 5	×	×	×	×		×	×	×		×	×	×	×	×	×	×	1.8
E 6	×		×	?x		×	?x	×		×		×	×	×			
E 7	×	×	×	×		×	×	×	×	×	×	×		×		×	2.4
E 8	×	×	×			×	×	×		×	×	×	×	×		×	
E 9	×	×	×		×	×	×	×		×	×	×		×	×	×	9.5
E10	×	×	×		×	×	×	×		×	×	×		×	×	×	7.0
E11	×	×	×			×	×	×	×	×	×	×		×		×	
E12	×		×			×		×		×	×	×	×	×		×	
E13	×	×	×		×	×	×	×		×	×	×	×	×		×	4.2
E14	×	×	×			×	×			×		×		×	×	×	
E15	×	×	×			×	×	×		×	×	×		×	×	×	
E16	×	×	×		×	×	×	×		×	×	×		×		×	

12

MODERN ARCTIC GLACIERS AS DEPOSITIONAL MODELS FOR FORMER ICE SHEETS

G.S. Boulton

Source: *Quarterly Journal of the Geological Society of London* 128 (1972): 361–88.

Summary

Sedimentary sequences currently forming at the margins of Spitsbergen glaciers are identical in thickness and detail to many Pleistocene and pre-Pleistocene glacigenic sequences. The transport of considerable volumes of englacial debris leads directly to the predominance of supraglacial till deposition, giving hummocky till surfaces and till plains. The association of supraglacial outwash with flow till produces tripartite till/outwash/till sequences, and multi-till sequences, which are the result of a single glacier retreat phase.

Complex tectonic structures, often with systematic regional trends are described, which are not the result of ice pushing but of down-slope flow and collapse of supraglacial sequences above melting ice.

New classifications are suggested for ice-contact stratified deposits and till, both of which depend upon position of deposition, supraglacial, englacial or subglacial.

It is suggested that existing models for the interpretation of ancient tills and the sequences in which they lie are often too simple and lead to erroneous stratigraphic and palaeogeographic conclusions. Till is too often interpreted solely as lodgement till, and it is suggested that many Pleistocene and earlier sequences, currently thought of as products of repeated glacier advance and readvance, may be perfectly normal products of a single retreat phase by a glacier with a thick englacial debris load.

Ways of reconstructing the structural character of ancient ice margins are presented and it is also suggested that the thermal regimes of past ice sheets can be reconstructed from the nature of their deposits. The last Pleistocene ice sheet in Britain is thought to have been composed, at its maximum extent, of cold ice in the marginal zone and temperate ice in the internal zone.

1. Introduction

The conceptual model of glacial sedimentation most commonly used by stratigraphers is one in which till originates by subglacial deposition, although the existence of 'ablation till' on valley or piedmont glaciers (Sharp 1949) has led to the concept of a 'doublet' of subglacial lodgement till with a thin capping of 'washed' ablation till (Flint 1957). Interbedded sequences of till with stratified sediment are thus taken to represent a history of glacier advance and retreat.

This concept is probably accurate enough for most temperate glaciers (i.e. ice below the melting point). In such glaciers (e.g. Alps, Norway, Iceland), subglacially-derived debris tends to be transported in a thin basal regelation layer no more than a few centuries thick (exceptionally 0.5 m). Because of this low level of transport and the existence of basal melting beneath temperate glaciers, this debris is almost all deposited as lodgement till. This till is generally massive, with an absence of stratified horizons, except perhaps at the very top where it may be re-sorted by sub-aerial processes. Lodgement till shows a variety of surface forms. It may be fluted (Hoppe & Schytt 1953), it may have drumlins on its surface; during retreat of the ice margin it may be extruded from beneath the margin to form ridges which represent retreat stages (Price 1969); alternatively such ridges may be extruded from beneath the flanks of masses of dead ice (Stalker 1960).

However in Arctic glaciers in Spitsbergen—and this appears to be true in many other Arctic areas (Baffin Island, East Greenland)—the above simple picture is quite inadequate. Subglacially-derived debris may be transported at some considerable distance above the bed, probably a result of incorporation by subglacial freezing which is dominant beneath the marginal zones of these glaciers (Weertman 1961; Boulton 1970). As a result of this high level of transport and the absence of subglacial melting, relatively small amounts of lodgement till form, although where it is deposited it is similar to lodgement tills of temperate glaciers. A very large percentage of debris, perhaps as much as 70 per cent, is transported to the glacier terminus where during retreat it is released onto the glaciers surface. Because of this, a whole range of supra-glacial processes develop, producing many features commonly seen in areas occupied by former ice sheets. The descriptions which follow are intended to inform about this area of glacial sedimentation which has been somewhat neglected, and to show its relevance to the interpretation of ancient deposits.

Large glaciers are found in Spitsbergen on the islands of Vestspitsbergen, Nordaustlandet, Edgeøya and Barentsøya. These glaciers range in size from ice caps such as that of Austfonna on Nordaustlandet, 140 km in diameter, and areas of highland ice of similar size, to piedmont and valley glaciers with broad frontal zones, and traverse a great variety of hard and soft rocks. Temperature measurements such as those of Zinger *et al.* (1966), and Palosuo & Schytt (1960) indicate that the ice of these glaciers is cold in the outer terminal zones, whilst there is some evidence that the ice of the internal zones of some glaciers is temperate (Schytt 1969; Lliboutry 1965; Boulton 1970a).

2. Supraglacial deposition in Spitsbergen

(a) Hummocky till moraines and till plains

Many Spitsbergen glaciers transport considerable volumes of englacial debris which tends to occur in series of bands parallel to the glacier foliation (Plate 1). In these bands the debris content commonly varies from 10 per cent to 60 per cent by volume, although greater concentrations do occur. As the glacier surface is lowered by ablation, prominent debris bands become exposed on the glacier surface and give rise to thin sheet flows which provide an initial debris cover which, when more than 2–3 cm has accumulated, considerably slows down the ablation rate of the underlying ice. Continued ablation leads to further addition of till below this cover and if till accumulation reaches the depth of summer thawing (1–2 m) ablation of the underlying ice ceases, so that no more till accumulates. However this melt-water-soaked till is very unstable and readily flows down the glacier surface (Plate 2) and accumulates on low slopes, in hollows on the glacier surface, or upon other proglacial or supraglacial sediments. This material was described by Gripp (1929), a similar origin was inferred for certain Pleistocene tills in Massachusetts by Hartshorn (1958), and details of the origin and occurrence of these tills in Spitsbergen were described in a previous paper by this author (Boulton 1968). Flow tills often lose the finer-grained components from their surfaces as a result of winnowing by wind and water, but the mechanical composition of material below surface is generally identical to that of the debris in the underlying ice.

Variations in supraglacial till thickness lead to differential ablation of the underlying ice producing a very irregular topography of considerable relief (up to 30 m) in a relatively short time (Fig. 1). If englacial debris is distributed in distinct zones separated by relatively debris-free zones striking parallel to the glacier margin, the resultant variations in supraglacial till thickness lead to the development of a series of ice-cored ridges and intervening valleys, trending parallel to the active glacier margin (Boulton 1968). Ice-cored ridges such as these are termed 'controlled', in that they are geometrically related to structures in the underlying ice (Gravenor & Kupsch 1959). If, however, the greatest concentrations of debris in the glacier ice are more

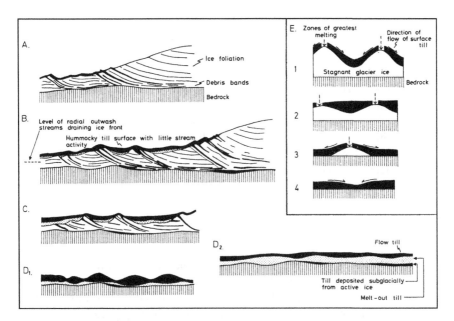

Figure 1 A-B-C-D₁: Development of a hummocky moraine topography as a result of predominantly supraglacial deposition. A-B-C-D₂ Development of a till plain, in this case the surface till is more fluid. E. Melting of buried ice beneath a fluid till to give a planar till sheet.

unevenly distributed, the derived supraglacial till shows more complex areal variations in thickness, and the hummocky topography which develops has an apparently random distribution of mounds, kettle holes and ridges. Such topographic features, are termed 'uncontrolled' forms.

The melting of buried ice in such features leads to one of two possible results, a hummocky till surface, which may be controlled or uncontrolled, or a till plain. Especially where buried ice relief is great, tills flowing down the flanks of ice-cores may fill up hollows and trenches between them to considerable thicknesses. Slow melting of these ice cores results in inversion of the topography and gives rise to a pattern of ridges and hummocks which reflect the original pattern of till-filled hollows (Figs. 1A, B, C, D₁ and Plate 3). This pattern may be controlled or uncontrolled, and forms essentially the inverted mould of the original ice-cored topography, and although some subsequent flow takes place on the flanks of hummocks, their relief and the absence of melting ice within them lead to some drying out and thus stabilisation of the till of which they are constructed.

Many tills however, are often waterlogged and therefore extremely fluid, especially where underlying ice is actively melting, and tend to flow down the gentlest slopes. Such tills induce the formation of complexes of low relief hummocks and mounds and of till plains (Figs. 1A, B, C, D₂ and Plate 4).

Differential ablation of the underlying ice produces continual readjustment of the ice surface in response to rapid changes of till thickness resulting from flow (Fig. 1E). This mechanism leads to lowering of the morainic slopes and gradual disappearance of ice-cores, the resultant being a till sheet with few surface irregularities, these latter being caused by englacial or subglacial till or fluvial accumulations, or bedrock irregularities.

Local drainage conditions are of fundamental importance in determining whether a supraglacial till sheet will retain a hummocky or low relief surface on ice melting. Those supraglacial moraines which lie in topographic depressions from which drainage is poor, tend to be of low relief, as water-soaked till is unstable even on very low slopes. Such tills often acquire characteristic flow structures.

Normally, a till sheet deposited at the surface during glacier retreat is sooner or later likely to be removed by fluvial erosion or covered by outwash deposits. However, there are many flat till plains and hummocky surface tills in Spitsbergen which have developed in the way shown in Figure 1 and are yet uncovered by bedded deposits. The major outwash streams flow radially out from the active glacier front and are often located along gorges deeply incised through ice-cored moraines and with steep gradients in which little net deposition takes place. The base levels for these streams lie in areas beyond the ice-cored moraine and well below the level of the moraine surface. Thus, because the supraglacial till sheets overlie a thick substratum of ice which maintains them above the level of outwash streams, they escape erosion by outwash streams or burial beneath their sediments. By the time the tills are lowered to the level of the outwash streams by melting of the underlying ice, the active glacier margin and its associated zone of outwash deposition is likely to have retreated sufficiently for them to remain uncovered by outwash sediments.

If supraglacial till has a relatively high water content, rapid flow may take place, and apparently stable slopes often fail, producing lobate flows of slumped till derived from arcuate slump scars. One such till flow on the moraine of Aavats markbreen in Oscar II Land, impounded sufficient drainage to form a large lake, and left a slump scar 0.5 km in diameter. In areas of very poor drainage many such mutually interfering lobate flows develop, individual flows being brought to a halt by neighbouring flows.

Sections cut into thick supraglacial till accumulations reveal two basic types of flow till. Where a buried glacier ice surface is melting down, the debris released from the glacier ice accumulates as till beneath the pre-existing overburden, be it till or outwash sediment. The newly released till tends to be unsorted and unstratified and it will retain this nature even during subsequent flow, as long as that till is not exposed to the atmosphere. However, downslope flow may translate parts of the till to the surface, where sorting by subaerial processes may occur. Thus, one often finds an upper, washed, and perhaps crudely stratified element, and a lower more massive element which has never been exposed subaerially (Plate 5). Thus the characters of the 'ablation till/basal till doublet' postulated by Flint (1957) may develop entirely supraglacially (see Boulton 1971). Where a flow till has acquired stratification, the subsequent folding of stratification during flow reveals the mode of flow (Plate 6a, Plate 10).

(b) Till/outwash sequences

Ice-cored moraines often form an effective barrier to the outward movement of meltwater from the glacier. Consequently, small lakes often develop within the morainic zone, and outwash streams are forced to follow sinuous outward courses between morainic ridges, with a resultant irregular pattern of deposition. Where gradients are low, widespread deposition of clays, silts, sands and gravels take place. Plate 7 and Figure 2 show the courses of streams draining the active margin of Elisebreen in Oscar II Land, and the sites of lakes. Sediments have accumulated along stream courses and in lake basins, and as these features lie between controlled ice-cored moraines, the resultant pattern of fluvial and lacustrine deposition is also controlled. The fluvial deposits sometimes lie in narrow and constricted channels, sometimes in broad open basins, and vary greatly in thickness. Such patterns of deposition are common and reflect the patterns of controlled or uncontrolled ice-cored moraines between which it takes place.

Movement of flow tills down the flanks of ice-cored moraines into adjacent troughs is often continuous. If outwash streams are active in these troughs, the till flowing into them tends to be disaggregated. However, if stream activity ceases, flow till may accumulate on top of the outwash deposits (cf. Plate 6b).

Where stream flow between ice-cored moraines has ceased, much of the surface is covered by till derived by flow from the adjacent ice-cored moraine ridges. Although a single flow of till rarely extends more than 0.5 km from the parent ridge, the coalescence of many flows may produce a single apparently continuous surface till often resting with a very sharp contact upon underlying outwash deposits. (Fig. 3) They form a permanent till capping on fluvial sediments only when fluvial activity has ceased in the channel. Stratified sediments immediately beneath flow tills often show an upward fining sequence representing final silting-up of the inter-moraine trough.

Where tills flow into lakes however, they may accumulate on the lake bed as recognisable flow tills, often interbedded with laminated lacustrine silts and clays (see Boulton 1968, fig. 14). If outwash activity at a particular locality is sporadic then inter-stratification of till and outwash may occur, producing multi-till sequence (see Plate 6b, and Boulton 1967, Figure 1d).

Sediments deposited by small streamlets on a till surface may be subsequently covered by till flows to produce stratified lenses within an apparently massive till (Plate 6c).

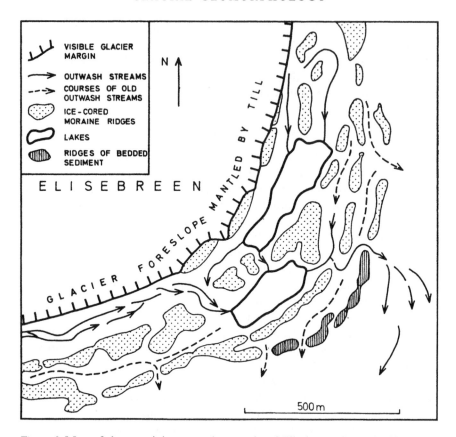

Figure 2 Map of the morainic arc at the margin of Elisebreen shown in Pl. 6. The outermost ridge is a ridge of sediments which themselves probably formed between ice-cored ridges which have now disappeared. If buried ice were to melt, topography would be inverted and ridges would form reflecting the pattern of outwash channels.

In the outer, inactive parts of areas of inter-moraine outwash deposition, many ice-cores have partially or completely melted out, the topography has been inverted and bedded outwash accumulations have become high points, and the original loci of ice-cored ridges have become elongate troughs and kettle holes. Many such ridges lying parallel to the glacier front form what on first acquaintance appear to be large terminal moraines. The stratified sediments are often overlain by thick and laterally continuous flow tills, and underlain by dead ice and the till derived from it. This tripartite sequence could easily be misinterpreted as the product of glacier advance, retreat, re-advance and second retreat.

Plate 8 shows a section in Oscar II Land, where a typical inter-moraine outwash accumulation has been affected by partial melting of underlying

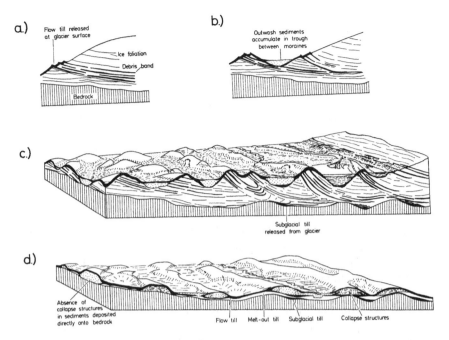

Figure 3 A series of hypothetical sections, elements of which are to be seen in many morainic zones in Spitsbergen, showing the development of a largely supraglacial sequence of tills and outwash, producing a controlled hummocky topography and a systematic stratigraphic sequence of till/outwash/till. Plates 1 and 7 show actual sections, and Figure 2 and Pl. 5 surface views in such morainic zones.

ice. These exposures of the internal structure of the old moraine belt are revealed in the flanks of a gorge occupied by an outwash stream. This stream probably partially follows the line of an old englacial tunnel, and cuts through both a series of controlled ice-cored moraine ridges lying parallel to the glacier margin, and the sediments lying in the troughs between these ridges. The sediments consist of an upper till, bedded outwash, (clay, silt and gravel) and lower till, all of which lie on dead glacier ice, and which are frozen to within an average of 2–4 m below surface. The glacier ice contains considerable volumes of debris entrained as a series of bands dipping up glacier at angles between 20° and 60°. In the troughs, the lower till is discontinuous with a maximum thickness of 3 m, the topmost 2 m of which locally contain lenses and layers of stratified silt and sand. This lower till is succeeded by a series of bedded sediments; in trough A (Plate 8b) these are coarse sands and gravels up to 15 m thick, whilst in trough B and further downstream in the section, they consist of sands and silts with gravel bands. The lower part of the bedded sequence includes rounded fragments

315

of till up to boulder size. The bedding of these sediments in trough A is warped into an anticline, the axis of which is parallel to the trend of the trough. High angle normal and reverse faults of similar trend, with down-throws away from the anticlinal axis of up to 2 m, are common. More complex tectonic structures often occur directly against the steep flanks of ice-cores. The upper till overlying the stratified sediments has a sharp planar base, there being no disturbance of the underlying sediments. This sequence can be explained as follows. The upper till, which is still actively forming, is a flow till which is derived from the debris-rich ice-cored moraine ridges which rise above the level of the stratified sediments in the intervening troughs. The stratified sediments were laid down at a much earlier stage when glacial drainage was constrained to flow between the ice-cored moraine ridges in much the same way as streams shown in Plate 7. Their internal tectonics are a result of anticlinal flexure by the differential ablation of the underlying ice-cores. The upper 2 m of the lower till, which includes bedded lenses and layers, is probably a flow till which formed the initial cover of the ice-cored moraine before fluvial deposits were laid down upon it, whereas the lower part of this till is probably a melt-out till (Boulton 1970b) which has been released from the underlying debris-rich ice after deposition of much of the overlying sequence. During further surface melting of the debris-rich ice-cores it is likely that yet more melt-out till will be released.

This interpretation is included in Figure 3 which illustrates how such a sequence develops, and the topography which results from final melting of buried ice.

(c) Structures in stratified sediments resulting from melting of buried ice

The structures which generally result from settlement over an ablating ice surface are folds, and collapse fractures. These tend to be high angle, sinuous (both in section and plan), normal or reversed faults with throws rarely exceeding 1–3 m although low angle rotational slump planes also occur. Both trend at right angles to the dip of the underlying ice surface, and in thus conforming to the pattern of ice-cored ridges and intervening hollows may be controlled or uncontrolled features. The amplitude of folds which develop in sediments overlying a buried ice surface is controlled by the net differential ablation since deposition of the sediments. If the amount of differential vertical movement is evenly distributed over a wide area, flexing with little faulting results. But considerable differential movement over a small area results in fracturing.

Melting of buried ice produces two main results. Firstly, general melting at the sediment/ice interface, or preferential melting of the ice-cored ridges at the flanks of channels, produce anticlinal flexuring of the sediments in the channel (e.g. Fig. 4a–b). If melting occurs during deposition, then dips will

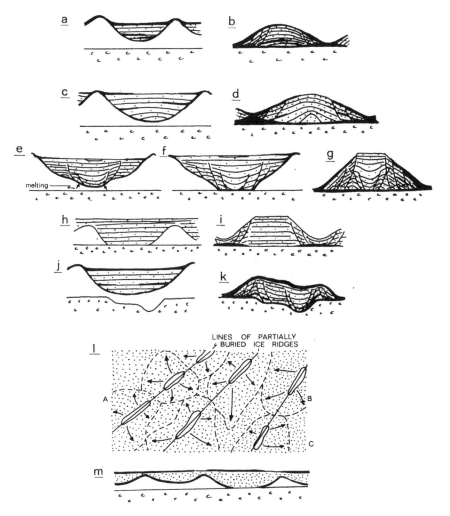

Figure 4 Hypothetical sections showing the development of typical structures in ice-contact sediments. Extent of flow till cover reflects prominence of ice cores, if they are entirely buried no overlying flow till forms.

a–b. Ice melting is post-depositional, bedding is thus parallel.

c–d. Contemporaneous melting produces greater dips at depth.

e–g. Local melting during deposition causes synclinal flexuring, final phase of flexuring is anticlinal. Flat tops result from subsidence to subglacial floor during deposition.

h–i. Pre-depositional erosion to subglacial floor produces a flat top.

j–k. Structures resulting from lowering onto an irregular substratum.

l. A hypothetical map showing a series of controlled ice-cored ridges almost completely submerged by outwash, the dashed lines show the extent of flow tills overlying outwash and derived from the ice-cores.

m. A section along the line A–B. Plate 8 is similar to the section which might occur along line AC.

increase with depth (Fig. 4c–d). Secondly, melting may be locally induced beneath a surface water body. This causes subsidence of supraglacial sediments as the base of the trough is lowered by local melting, and synclinal flexing and trough faulting result. This type of subsidence during deposition again results in downward steepening of dips (Fig. 4e–f). If both subsidence and deposition continue until the sediments are lowered onto the subglacial floor, a flat-topped deposit often of high relief results (Fig. 4f–g).

The thicknesses of sediments which accumulate in these supraglacial troughs are limited in the first case by the depth of the trough, and in the second by the total ice thickness.

It is clear from the explanation of the anticlinal flexuring shown in Plate 8, that in modern Spitsbergen, where the active layer of the permafrost rarely exceeds 3 m, the high thermal capacities of surface streams have the ability to induce melting beneath at least 20 m of gravels. Melting beneath lakes and streams often produces collapse (Fig. 4e–f). Another form of collapse is shown in Plate 9 which illustrates the surface of a series of silts and sands deposited in a supraglacial lake which drained between 1961 and 1965. Deflation has revealed the cores of folds, which have an amplitude of up to 5 m. Sections in the flanks of the stream channel which now cuts through these supraglacial sediments, shows that they accumulated upon an irregular ice surface, and that melting at the highest points of the ice surface has produced warping in the sediments. Synclines now coincide with high points of the buried ice surface and anticlines with low points. Dips increase downwards, and this, together with the fact that much of the silt is now frozen to within 2–3 m of surface, suggests that melting of the ice took place during deposition and for a short time immediately after the drainage of the lake.

A third type of collapse structure occurs when supraglacial sediments are lowered by melting of underlying ice onto an uneven subglacial floor. Fractures and flexures occur in the sediments immediately above the most irregular parts of this floor (Fig. 4j–k).

Clearly, the three types of structures described above may be super-imposed upon one another. In Fig. 4g, a set of anticlinal structures have been superimposed upon synclinal subsidence structures, and in Fig. 4k, anti-clinal structures have been superimposed upon structures related to the shape of the subglacial floor. Thus sporadic sedimentation and sporadic melting and collapse can produce a series of different sets of folds and faults and intraformational angular unconformities. Detailed study of these sediments can reconstruct the history of buried ice melting and supraglacial sedimentation. In many cases supraglacial streams cut down to the subglacial bed, removing both ice and till. Outwash sediments may then accumulate in such a trough, and will clearly show no signs of collapse, except at the margin where this lies upon flanking ice cores (Fig. 4h–i). In this case the accumulation has a permanent flat top. If this is overlain by flow till, the peculiar situation may arise in which the stratigraphic horizon along which the parent glacier lay is not marked by any sediment, although outwash sediments and flow till lie above this horizon.

The accumulations of supraglacial origin described above can be subdivided into 'ice-floored' deposits (Fig. 4a–b), and 'ice-walled' deposits. Amongst the latter there are some beneath which erosion cuts down to the bed before sediment accumulation (Fig. 4h–i), and some beneath which the contemporaneous melting of underlying ice lowers the sediments onto the bed (Fig. 4c–d).

In areas vacated by glaciers, the pattern of ice contact ridges directly reveals the pattern of outwash flow and accumulation, whilst intervening hollows reflect the positions of ice-cored ridges which are themselves an index of glacier structure.

(d) Structures in stratified sediments resulting from downslope movement under gravity

In addition to the deformation resulting from differential settlement over a melting ice surface, deformation also takes place by sliding or flow of sediments down an inclined surface. Downslope movement is not only common amongst tills, but it also takes place in thick supraglacial sequences.

A section exposed in a trough between two controlled ice-cored moraine ridges lying parallel to the active margin of Stubendorfbreen in Ny Friesland, revealed large scale folding and faulting in supraglacial sediments which seem to have slipped down a sloping ice surface (Fig. 5). The deposits involved in folding comprise (in downward sequence): two till beds separated

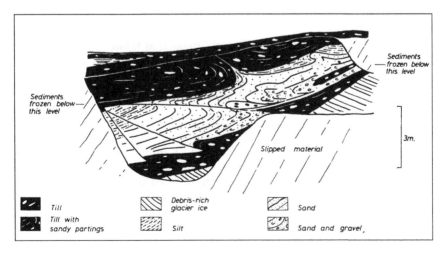

Figure 5 A supraglacial sequence deformed by flow down the flank of an ice-core.

by silts, probably representing two periods of flow till accumulation between which a small pool occupied the site; stratified silts, sands and gravels deposited by outwash streams flowing between ice-cored moraines; and a lower till which is probably flow till in its upper part and melt-out till (see p. 373) in its lower part. The sequence is now frozen to within an average of 2.5 m from surface. The folding probably occurred at some time when these sediments were unfrozen, perhaps shortly after deposition of the stratified silts, sands and gravels. The upper till, folded in its lower part, accumulated after folding of the lower sediments. The fold structures occur within a belt of 'controlled' moraine topography trending parallel to the active glacier front, and thus if such features are widespread within the supraglacial sediment mantle they are likely to maintain a consistent tectonic strike.

Plate 11 shows a series of sands with silt bands which have flowed downslope, probably in a water-logged condition. The flows have accumulated one above the other, and when sectioned parallel to flow, the bedding is shown to be isoclinally folded. The lower limb of each fold is attenuated by a thrust plane. Plate 6a shows a similar fold in a flow till stratified as a result of winnowing.

It should be stressed that although many such structures might suggest pushing by glacier ice, they merely represent superficial slumping. Though large numbers of glaciers have been investigated, this author shares with members of the 1959–1960 Polish Expeditions a general failure to find evidence for ice pushing. Ice pushed moraines (Stauchmoränen) have long been considered to be typical of Spitsbergen glacier margins after the classic work of Gripp (1929). However I would suggest that the majority of these features are supraglacial deposits of the type described in section 2B and that internal tectonics are a result of collapse and superficial flow rather than pushing.

3. Sediments directly deposited from an englacial position in Spitsbergen

(a) Melt-out tills

(i) Genesis

Slow melting of thick masses of debris-rich ice buried beneath supraglacial sediments often give rise to tills which retain some elements of their original englacial structure, and for which the term 'melt-out' tills has been suggested (Boulton 1970b). In Spitsbergen these ice masses melt from the top downwards and thus melt-out tills tend to be produced above a melting ice surface and below a confining overburden of other sediments. Without this sediment overburden, the newly released till would tend to creep down slope as a flow till. The overburden beneath which melt-out tills are released may be glacial outwash or a stabilised supraglacial flow till, and many flow tills grade down into melt-out tills. Thus, melt-out tills occur beneath the results of supraglacial deposition: e.g. hummocky outwash, hummocky till or a flat-lying till plain, but not beneath flat-lying outwash deposits. Melt-out tills could also be produced beneath stationary ice which is melting basally, in contrast to subglacial lodgement tills deposited from active ice. The changes in fabric and other textures produced in melt-out tills during the process of deposition has been described previously (Boulton 1970b).

No melt-out tills have been observed which have a foliation comparable to the stratification of the 'shear clays' of the north of England which Carruthers (1953) suggested originated as englacial 'banded dirts'.

The amounts of englacial debris within buried stagnant ice masses vary considerably from glacier to glacier, and thus the amount of potential melt-out till contained within these masses varies. Some glaciers which have advanced over soft wet sediments, such as Sefströmbreen in James I Land which advanced over an arm of the sea (Lamplugh 1911) and Nathorstbreen in Torell Land and Aavats markbreen in Oscar II Land, which move into fjords, contain pods and layers of till of great thickness. These tills contain high proportions of marine mud, and might potentially transport as much as 6–8 m of melt-out till.

(ii) Significance of thickness variations

It is possible that variations in thickness of recognisable melt-out tills could be used to reconstruct directly the pattern of stagnant ice ridges from which they have developed, and from this, inferences about local glacier structure can be made.

Buried stagnant ice masses vary considerably in thickness, and thus one might expect melt-out tills to show correlative thickness variations. Figure 1 shows a section in which stagnant ice lies on a bed of low relief and is overlain by supraglacial deposits. The stagnant ice has a surface which is cor-rugated on a large scale, these corrugations being originally produced by differential ablation beneath an irregular flow till cover. If, as often happens in uncontrolled topography, the debris content of the ice is uniformly dis-tributed, the bed of till which results from complete ice melting would also be corrugated. The ratio of thickness of till in ridges to thickness of till in troughs would be similar to the ratio of the original thickness of ice in ridges to thickness of ice in troughs. The slope of the ridge flanks would however be reduced, a 45° slope in ice containing 30 per cent debris would give a slope of 16° for the flanks of till ridges produced on ice melting.

As shown above (p. 363), controlled ridges tend to develop where debris is *not* uniformly distributed within the ice, but where it occurs in well-defined zones which are clearly separated by zones of lower debris content. This situation is illustrated in Figure 1 where the ice below controlled stagnant ice ridges contains a much greater debris concentration than the ice beneath troughs. The melt-out till produced by slow melting of such a buried mass would have a controlled corrugated surface whose pattern of ridges would be similar to that of the ridges in the parent stagnant ice. The slope of the ridge flanks would however be much steeper than those of uncontrolled melt-out ridges, as the proportion of till contained within controlled ice-cored ridges is greater than in the ice beneath intervening troughs. If the melt-out till lies on a flat substratum the shape of the surface will directly reflect thickness variations, but if the substratum is irregular the shape of the till surface will depend on this as well as on its own thickness variations. It is important in this context however to distinguish between flow tills and melt-out tills. The greatest thicknesses of flow tills tend to develop in positions which represent troughs between stagnant ice ridges, whereas the greatest thickness of melt-out tills will tend to develop in positions which represent stagnant ice ridges.

(b) Stratified deposits

Many of the stagnant ice masses beyond glacier margins are penetrated by complex tunnel systems, some actively in use, some abandoned. These tunnel systems are very much akin to those in limestone terrains, having irregular long profiles, siphon systems and sometimes corkscrew tunnels with vertical axes representing old moulins. Where dead ice lies at a low level at the glacier margin, tunnel systems are often full of water, stagnant or moving, and are controlled by local water pressure heads. Sediment filled channels commonly occur in active ice but open channels are of necessity ephemeral unless occupied by active streams. The presence or absence of sedimentary deposits along the floors of tunnels is probably largely controlled by the existence of a sediment supply. Many fast-flowing englacial streams transport no sediment, others tap surface moraine or englacial debris, or derive material from those points at which the tunnel intersects the glacier bed.

A large number of englacial stream sediments have now been observed in Spitsbergen. These tend to occur in linear channels which trend normal to the ice margin but which are sinuous with a constant wavelength and amplitude, presumably a result of the homogeneity of the ice in which they are cut. The materials of which they are composed are generally coarse-grained gravels and coarse sands, with few fine materials. Bedding is generally planar, high angle cross beds being rare. The channels in which sediments lie tend to have vertical walls. Down-melting of glacier ice slowly reveals the fluvial sediments contained within it, and the retaining ice walls disappear. The sediments can thus no longer retain a high angle lateral margin and so slumping down the flanks occurs (Fig. 6a–c). Further ice ablation leads

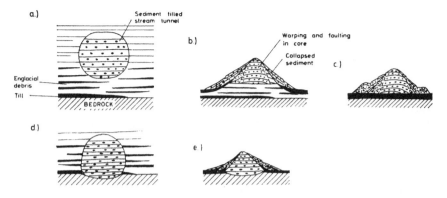

Figure 6 a–c. Deposition of englacial fluvial deposits.
d–e. Release from the glacier of subglacial fluvial deposits.

to the situation shown in Fig. 6b, where the stream deposits rest upon a wall of ice which they protect from ablation. Final melting of this irregular ice substratum leads to differential settlement of the overlying sediments and the consequent formation of fault planes and flexured bedding. Not only do fault planes and flexures strike parallel to long axis of the deposits as a result of transverse irregularities in the ice substratum, but they also strike normal to the long axis as a result of longitudinal irregularities. These stratified deposits of englacial origin often rest upon a till derived from the underlying ice. An overlying till of significant thickness derived from the melting of ice above the level of the stratified deposit is relatively rare because most of the debris which goes to form till lies at a low level in the ice. A till capping does however form where the deposit lies beneath medial moraine material, although this only occurs in valley glaciers or where nunataks pierce an ice cap (see Table II).

4. Sediments deposited subglacially in Spitsbergen

(a) Lodgement till

Although great thicknesses of till are released during retreat by glaciers in Spitsbergen, a relatively small percentage is subglacial lodgement till, a feature which contrasts strikingly with glaciers in some other geographic areas (e.g. Iceland) where most of the till is lodgement till (see p. 361). However this till is fluted and drumlinized as in other areas.

The paucity of till beneath active glacier soles in Spitsbergen (see also Gripp 1929; Klimaczewski 1960) contrasts strongly with the considerable volumes of potential till in dead ice masses and the ubiquity of thick flow tills at the surface.

323

(b) Stratified deposits

Although subglacial streams are visible along ice margins at many points in Spitsbergen, there are few instances of subglacially deposited ridges of fluvial sediments. Szupryczynski (1965) in a review of Polish work, cites only one instance, from the forefield of the Gås glacier, and only one has been encountered by this author, at the margin of Dronbreen in Nordenskiold Land. Here a subglacial tunnel based on bedrock and approximately 3 m high was almost completely filled by sands and gravels. Horizontal bedding in the sands and gravels intersected the vertical tunnel walls at ninety degrees. From this, and the other available observations, it is not yet possible to generalise about the nature of subglacial fluvial accumulations. If, however, one assumes that subglacial tunnels can fill up with fluvial sediments in the same way as englacial tunnels, then a plausible prediction of the main characteristics of such sediment accumulations can be generated. The restricted environment is similar to that of englacial tunnels, with high water velocities, but based on a stable substratum rather than on ice. Melting of the ice walls of the tunnel would presumably produce lateral slumping of the steep margin of the sediments, as in the case of englacial tunnels, but there need be no subsequent disturbance of the sediments in the core of the deposit, as this is not underlain by ice. Because of its position, at a lower level within the glacier, there is a greater chance of englacially-derived till being deposited on the fluvial materials, where it would be involved in slumping down the flanks. Some of the likely characteristics of subglacial fluvial deposits are reviewed in Table II, and a hypothetical diagram showing the acquisition of internal structure is shown in Fig. 6d–e.

5. Spatial variation in glacigenic sequences

Spatial variation in glacigenic sequences results on a regional, glacier-wide scale from the modes of erosion, glacier transport and flow, and on a local scale from the mode of deposition determined by the style of glacier retreat. On the regional scale, lateral and vertical variations in englacial debris content have been observed, the most fundamental systematic distinction being that between material derived supraglacially from nunataks and valley sides in the accumulation area and material derived from the glacier bed. Much of the supraglacially-derived material is highly angular, and some extremely large boulders occur within it (up to 30 m diameter) but it is often poor in fines, although when exposed at the glacier surface many rock types are rapidly comminuted. Such material is commonly continuous in the direction of glacier flow but it tends to lense out rapidly in the transverse direction. In a vertical sequence of supraglacially derived debris, the farthest travelled material is most common at the lowest point, the most locally derived at the top.

In contrast, subglacially-derived debris is generally subjected to a series of crushing processes which progressively increase the fine fraction. Abundant particles show signs of this crushing but there are also boulders and smaller grains of fluvial origin, and near the glacier bed stones bear heavily striated facets. A lateral and vertical variation in the composition of subglacially-derived debris is also often seen; this may be discontinuous variation with strikingly different debris types at different levels, or continuous gradation. A study of the englacial debris composition in the vertical terminal cliff of Aavatsmarkbreen in Oscar II Land showed that the downward variation in erratic type reflected the sequence of rock types over which the glacier had passed (Boulton 1970a), and raised the possibility that other basally derived englacial sequences may show similar systematic variation. The melt-out till derived from this englacial sequence retains a vertical variation in erratic content similar to that in the terminal ice cliff of the glacier, and also shows a longitudinal variation from furthest derived close to the glacier margin and most locally derived further from the glacier margin. However, the overlying flow till does not show a systematic vertical variation, although it retains the longitudinal variation, suggesting that the mode of deposition of the flow till has destroyed the vertical component of original variation.

The sequence of glacigenic sediments deposited at any one locality may contain any combination of the supraglacial, melt-out, and subglacial deposits described in section 2, 3 and 4, and the degree of lateral variation in these sequences depends on local factors which have governed the mode of decay of the glacier at that point. Uniform rates of deglaciation over a wide area of uniform topography may produce a widespread single style of deposition, whereas considerable variation in depositional style could be the result in an area of topographic diversity, or during jerky retreat. There are however in Spitsbergen three basic styles of deposition which produce patterns of sedimentary distribution which occur repeatedly and which I consider to be recognisable in many Pleistocene glacigenic sequences.

Firstly, there are the hummocky or planar till sheets (Fig. 1), in which the advance and retreat of the glacier is represented largely by till (mostly flow till and melt-out till) with minor stratified beds and lenses within it. It should be stressed that such till sheets have never been observed overlying a pre-existing constructional glacigenic topography: they tend to lie on smoothed and striated bedrock, and rarely on unlithified sediments.

Secondly, there are the till/outwash complexes in which lenses of outwash, often of considerable lateral extent, occur within till; beneath flow till; and above flow till, melt-out till, and subglacial till (Fig. 3). These sequences tend to occur in areas of hummocky topography. The layers and lenses of stratified sediment are sometimes very thin, and multiple till sequences, in which till beds are separated by stratified beds, are not uncommon. Lenses of outwash sediment within till may be several hundred metres across, or very small (Fig. 6c). Several patterns of sediment distribution resulting from these processes are shown in Fig. 3 and 4. A not uncommon feature of some flow-till covered outwash hummocks is that there is an upward-fining of the stratified sediment representing the slow dying and cessation of outwash activity, before flow tills accumulated above them. If the upper till were a readvance till, the very opposite sequence would be expected, a gradual coarsening of the deposit as more rapid discharges associated with proximity of the glacier front became dominant. The extent to which flow tills form a more or less complete covering over supraglacial outwash accumulations between ice-cored ridges depends upon the nature of these ridges. If the ridges form major topographic features rising above the outwash deposits, a thick, continuous covering of flow till upon outwash generally follows (Fig. 4a–b). On the other hand, if ice-cored ridges are more subdued, the cover of flow till is more sporadic (Fig. 4e–g). However, if ice-cored ridges fail to pierce supraglacial outwash but are completely covered by it, a hummocky topography results but without a flow till cover (Fig. 4h–i). The junction of upper till and lower till in positions which represent the original locations of ice cores is a fundamental necessity in these supraglacial sequences, as the upper till is essentially derived by flow from the ice-cores which are now represented by the lower till. However, in some drift terrains, only a certain number of cross-sections will reveal this contact as it only occurs at those points where the ice-core is undraped by outwash sediments, many cross-sections may show complete separation between upper and lower till (Fig. 4l–m). A second way in which complete separation may occur is seen in some morainic areas where topography has been inverted by melting of ice-cores. Fluvial erosion and deposition along the troughs may replace upper till by fluvial sediments, which are later covered by soliflucted till which had survived on the flanking ridges. The occurrence of englacial fluvial sediments is yet a third, though more fortuitous way in which complete separation between the upper and lower till may result.

It is important to note that the fact that flow tills tend to be derived from the upper part of the englacial debris load, and lodgement till from the lower part of this load, might lead to tills of systematically different composition produced by the same glacier retreat phase.

A third type of sequence is that in which there is little stagnant ice beyond the active glacier margin which retreats, as a relatively steep front with proglacial outwash beyond. Till may be preserved beneath this outwash or it may be entirely absent.

Lateral transitions between these three types of sequence are common.

6. Classification

In general, the most useful and fundamental classifications are those based on mode of genesis, providing that there are sufficient useable criteria to maintain these genetic distinctions. In the sediments described in preceding sections, the most distinctive and easily recognised characters are those acquired during the process of deposition, and thus I would follow Flint (1957) in suggesting that the most satisfactory classification is one based on mode of deposition. The two sets of deposits, tills and stratified outwash can be conveniently classified along parallel lines which depend on position of deposition. Thus each can be ascribed to a supraglacial, englacial or subglacial type.

(a) Till

Experience in Spitsbergen suggests that a very narrow definition is often difficult to apply. I would therefore suggest a definition of till as an aggregate whose components are brought together and deposited by the direct agency of glacier ice, which, though it may suffer post-depositional deformation by flow, does not undergo subsequent dis-aggregation and redeposition.

Under this definition and using the supraglacial, englacial, subglacial criterion, there are three basic types of till: (a) flow till, released supraglacially and undergoing subsequent deformation as a result of subaerial flow: (b) melt-out till, released either supraglacially or subglacially from stagnant ice beneath a confining overburden, and in which some of the original englacial features are preserved: (c) lodgement till, deposited beneath actively moving ice and undergoing deformation as a result of the shear imposed by this ice (Table 1).

The terms flow till and lodgement till have been widely used previously and I believe that the usage proposed here is consistent with previous usage, but the term melt-out till is newly proposed (Boulton 1970b) as a result of recent observations.

This scheme is a refinement of that used by Flint (1957) who divided tills into ablation tills and lodgement tills; but as has been demonstrated above, several rather different types of till fall into the ablation till category, and the vast bulk of these ablation tills have not had fines winnowed out as Flint suggests for his ablation tills. In view of this difference, the term ablation till

Table 1 Till classification.

Proposed classification		Existing classification
Supraglacial tills	{ Flow till { Melt-out till	Ablation till
Subglacial tills	{ Melt-out till { Lodgement till	Lodgement till

is not used. It will be noted that although melt-out tills might be deposited in either the supraglacial or subglacial position, I have kept them together as showing the effects of englacial transport rather than dividing them between the two categories. This separates them from flow tills, an important distinction from a stratigraphic viewpoint, for in a sequence of deposits left by a former glacier, melt-out tills together with lodgement tills represent the horizon occupied by that glacier. However, flow tills may occur at a higher level than the horizon of the parent glacier.

Within this scheme, individual units are capable of being further subdivided on the basis of good criteria. The contrast between the upper and lower parts of flow tills resulting from different modes of flow has already been noted (p. 365) and lodgement till can be deposited in several different ways and undergo a variety of post-depositional deformation.

(b) Ice-contact stratified deposits

An attempt to apply the existing nomenclature for ice-contact stratified deposits to these forms in Spitsbergen does not produce useful results. The two main types, eskers and kames, are normally defined principally in morphological terms. The former are taken to be sharp crested, often sinuous, ridges trending normal to the glacier front, whilst the latter are taken to be mound-shaped and often with flat tops. Unfortunately, this morphological distinction has been taken to imply a genetic distinction, and Flint (1957) does devote some space to consideration of the problem, how do eskers form, and how do kames form, as if their modes of formation are necessarily different. In fact, both morphological types occur in Spitsbergen in a supraglacial position, whilst most englacial deposits of any magnitude tend to have an 'esker-like' form. The problem of classification is also complicated by the fact that many Pleistocene 'moraines' are composed of stratified drift, and the idea that 'kames' can originate subglacially has of recent years become widespread (Gjessing 1960).

I would therefore suggest that because 'esker' and 'kame' forms can have a similar origin, which in turn can be similar to that of certain moraines, a different classification should be adopted based on position of deposition. In this, ice-contact stratified deposits are divided into subglacial, englacial and supraglacial types, on the basis not only of morphology but also internal structures, supraglacial deposits may be ice-floored or ice-walled. The criteria which in conjunction make these distinctions possible are reviewed in Table II.

Distinctions based upon this classification give valuable evidence of the nature of de-glaciation. Subglacial and englacial forms merely indicate the presence of ice, but give no clue as to its thickness or the position of the glacier front. Supraglacial forms indicate a zone of stagnant ice at the glacier margin.

Table 2 The characteristics of supraglacial, englacial and subglacial ice-contact stratified deposits occurring at the margins of modern glaciers.

	Supraglacial		Englacial	Subglacial	
	Ice-floored	Ice-walled (c.f. Clayton & Freers 1967)			
	Continuous or discontinuous ridges, single, or in groups which may be parallel or apparently randomly distributed. Often beaded. Also mounds, single or in complexes. The cross profiles of the ridges or mounds are smooth with rounded summits. Occurrence is not generally controlled by pre-existing surface topography. Often considerable relief and great volumes of material. Kettle holes may be very common.	As ice-floored deposits, except that a flat top is generally retained on ice melting. If however the bounding ice walls were steep and the trench in which it accumulated was narrow, collapse or ice-melting could destroy the flat-topped form. The flat-topped forms are particularly common on higher ground and near the margins of valleys, where they will be equivalent to "kame-terraces" under the existing nomenclature.	Generally single ridges. When underlying ice melts out, relief tends to be low. May have small kettles and surfaces pits. Relief and volume of material tend to be smaller than in supraglacial forms. Tend to be sinuous in plan with constant wavelength and amplitude.	Essentially sharp crested ridges which may also be single or associated with others in dendritic or anastomosing patterns. Marked break in slope at foot of ridge flank, ridges tend to be separated by ground of different character. In plan, no obvious regularity.	Topographic expression
	Coarsest to finest grain sizes. Products of high and low discharges may be intimately associated, fluvial gravels with lacustrine silts and clays. Some ridges may be composed dominantly of coarse materials, others of fine materials.	As ice-floored deposits.	Coarsest materials dominate. Sands and gravels deposited in narrow channels.		Materials of which composed.
	The full range of fluvial and lacustrine bedding structures.	As ice-floored deposits.	Planar bedding common. Internal core flanked by slipped material in which bedded is inclined at the angle of rest.		Bedding structures

Table 2 (cont'd)

Supraglacial		*Englacial*	*Subglacial*	
Till commonly occurs above and below these accumulations. It may occur within them, especially if the outwash sediments are lacustrine.	If the accumulation has built up after a period of erosion down to the glacier bed, an underlying till will be absent. If the outwash underlying ice was destroyed by melting beneath a cover of outwash sediment, an underlying till will remain.	An underlying till to be expected. An overlying till tends not to occur unless morainic debris lies at a very high level within the glacier.	Any underlying till would be likely to be destroyed. It may be overlain by a till which is likely to be involved in superficial slumping.	Association with till.
Tendency for original bedding to be anticlinally flexed reflecting surface, although local synclines may occur. Folding often less acute at higher levels. Normal faulting parallel to lateral flanks. Solifluxion structures common.	If the accumulation is built up after erosion to the glacier bed, sediments beneath the summit of the deposit will be undeformed and in the attitude of deposition. If the accumulation was lowered onto the bed after sedimentation had commenced, sediments beneath the summit will be synclinally flexed. In both cases, collapse at the flanks will be similar to that in ice-floored deposits.	Anticlinal warping of inner core, together with normal faulting with downthrows away from ridge crest. Some faults trend across the ridge.	No deformation of inner core.	Deformation structures.
Linear ridge most commonly parallel to the glacier margin, although they occasionally develop normal to it.	As ice-floored deposits.	As ice-floored deposits.	Most commonly occur trending normal to the glacier margin.	Orientation.

7. Applications to the Pleistocene

(a) Origin of till

One of the difficulties confronting those geologists dealing with Pleistocene and earlier glacigenic sequences has been the relative lack of data about the processes of deposition in modern glaciers and the types of stratigraphic sequence to which these processes give rise. Partly because of this lack of modern data, many authors are clearly not aware of a wide variety of glacial processes. Thus, a very simple model of glacial deposition has been generally used in which most fine grained tills are taken to be subglacial 'lodgement' tills, upon which a thin, washed ablation till may lie. As a result of this model, a basic premise in many works has been that unsorted till is lodgement till, and therefore that such features as till fabrics can be used to infer mode of lodgement and direction of ice movement (e.g. Andrews & Smith 1970). Furthermore successive till layers are therefore assumed to represent successive phases of glacier advance followed by retreat (e.g. West 1968, p. 97). In both the above instances it should first be determined that the till is lodgement till.

There are features of many Pleistocene sequences which have in the past convinced some workers that the simple model may often be in error. One such difficulty stems from the assertion by Hull (1864), oft-repeated since then, that the British drifts have a tripartite sequence of lower till, middle sands and upper till. Although there are many areas in which this sequence has been shown not to occur, it is well documented in others (Lleyn Peninsular, Cheshire, Durham, Lough Neagh area, etc.).

A typical sequence is summarised thus by Carruthers (1939):

> bottom till—resting on rock-head, was the only one which showed contacts providing such forward movement as would be expected from an advancing ice-sheet. Striation, shattering and incorporation of local rock-heads were common enough, and rafting was by no means infrequent. But with the other tills, those higher in the sequence, there was no disturbance whatsoever, the contacts were brilliantly sharp and there was no incorporation.

Several recurrent features of this type of sequence are not explicable by the hypothesis that both upper and lower tills are subglacial in origin.

1. The frequent contrast between upper and lower tills.
2. The absence or rarity of stratified outwash above the upper till when compared with the volume of 'middle sands.'
3. The nature of the lower boundary of the upper till. This is often a very sharp planar or interbedded junction with no sign of disturbance, unlike

331

the junction generally found when ice moves over unlithified bedded sediments.

4. The fact that the upper till often coats hummocky meltwater deposits as a thin skin. There are rarely constructional forms above this till.

5. The fact that the tripartite sequence is best developed in areas of hummocky kame and moraine topography.

6. The stratified deposits below many upper tills show an upward fining sequence, quite opposite to the effect which would be expected if the upper till were a readvance till.

Many workers who recognised these problems such as Goodchild (1875), Carvill Lewis (1894) and Carruthers (1939, 1948) proposed ingenious solutions. For instance Carruthers suggested that the clays which often occur between upper and lower till in many parts of northeastern England were laminated as a result of englacial shearing and that the upper till was an ablation till. Although I consider Carruthers' ideas, both of undermelting and shear-clay formation, to be almost entirely erroneous (see also *J. Glaciol. I*, 430–36), he at least recognised a problem and focused attention upon it, whereas subsequently most workers have opted for the simple model which is often quite untenable in the face of detailed observation.

Many of the above features, described from localities in Britain, can be paralleled by sequences from Spitsbergen described in sections 2, 3, 4 and 5. The best bases of comparison between current sequences of known genesis are detailed bed structures (cf. Boulton, 1968), and the patterns of spatial distribution of different lithologies. For instance, such patterns as those illustrated in Figures 3 and 4 and the bed structures typical of these till types, (Plate 6a–d) are to be found in the Lleyn Peninsular (e.g. Glanllynnau; Saunders 1968), in the Cheshire Basin, in Norfolk (Cox 1970), in the Vale of Eden (Goodchild 1875), and many other localities both in Britain and abroad. I would suggest in these cases that the lower till generally represents a combination of lodgement, melt-out, and flow elements, whereas the upper till or tills represent flow tills associated with supraglacial outwash sediments.

In addition there are many multi-till sequences similar to that shown in plate 6b, for which an origin as a result of simple fluctuation between outwash and flow till activity, might be considered, rather than complicated explanations of ice advance and retreat such as have been commonly used.

The concepts of supraglacial till deposition described in section 2 and 3 can also be applied to till plains and hummocky till deposits. Whilst in many cases till plains may be composed of subglacial lodgement till which has been preserved from outwash activity because outwash streams have cut deep restricted channels below the level of such plains, it is also possible that some have arisen in the way indicated in Fig. 2. The release of englacial debris can give rise to a thick but variable cover of flow till, thus preserving large masses of buried ice. Outwash then takes the lines of deep channels cut through the buried ice which is not therefore overlain by outwash sediments. Slow differential melting of buried ice together with flow of surface till produces a low flat-lying surface, and the loci of the original channels are masked by flow of till from the slowly disappearing ice cores, thus producing a flat surface.

There has also been a recent tendency to ascribe many constructional, ridged and hummocky till forms to subglacial formation. Although this is undoubtedly correct for many forms such as drumlins and fluted moraine, it is a very questionable procedure in the case of all mounds and ridges of till which do not have such regular distributions. For instance, some of the controlled and uncontrolled disintegration till ridges described by Gravenor & Kupsch (1959) from Western Canada can and do originate supraglacially; and many of the types of hummocky moraine described by Hoppe (1952) from Norrbotten in Sweden and attributed to hypothetical subglacial processes, for which there are as yet no modern analogues, are identical to moraines which form supraglacially at the margins of actual glaciers in Spitsbergen.

(b) Ice-contact stratified deposits

In most regions glaciated during the late Pleistocene, there occur stratified sediments for which an ice-contact origin can be confidently postulated. In many regions, these deposits form broad areas of sub-parallel ridges and mounds, often with intervening kettle holes, and are sometimes considered to lie parallel to ancient ice-front positions, although the mutual parallelism of ridges often breaks down. The sediments within these ridges, stratified gravels, sands, silts and clays, are frequently folded into series of anticlines and synclines and highangle normal faulting is common. Many mounds and ridges are draped by a thin superficial till sheet. Numerous examples of these ridges are described in the literature.

Of recent years, there has been a growing tendency to view many such areas of ice-contact stratified drift as subglacial in origin, and such names as kamemoraine, and esker complex, have been used to describe them. I would suggest, however, that the subglacial processes which have been postulated are other than those which would be invoked if the origin of modern forms were used as a guide. Many supposedly subglacial accumulations of Pleistocene age do not have the characteristics of known subglacial sediments, but are often identical to modern deposits of supraglacial origin which are described in this paper.

Several examples follow, which serve to illustrate this.

Hoppe (1963) describes a series of stratified deposits overlain by a thin till, with a kettled and channelled but otherwise flat-topped surface. He inferred that the surface till could not have been deposited by ice readvance as there is local evidence of rapid and continuous glacier retreat, and therefore concluded that the stratified sediments were deposited subglacially and that the surface till formed from englacial debris within the overlying ice. In fact the deposit is quite unlike any known subglacial deposit, and such widespread subglacial deposition seems rather implausible. Such deposits do however commonly form by deposition in hollows cut down to bedrock between ice-cored ridges, and the latter may be subsequently overlain by flow tills. Flat-topped ice contact deposits for which a similar origin could be suggested have been described from many areas (e.g. Vale of Eden, Trotter 1929; North Dakota, Clayton & Freers 1967).

It is interesting to note that many of the largest so-called end moraines in areas glaciated during the Pleistocene, are not composed of till but of outwash sediments. One such example is a deposit at Carstairs in Scotland, which consists of a series of parallel and sub-parallel undulating ridges and mounds with intervening kettle holes. These are constructed of sands and gravels, the bedding in which is warped into a series of broad anticlines and synclines roughly reflecting the surface topography. The bedded deposits are displaced by systems of high angle faults, the strikes of which tend to parallel the long axes of hummocks. Charlesworth (1926) originally termed this feature a moraine, whilst Goodlet (1964) referred to it as a kame, and Sissons (1961) and McLellan (1969) as a 'subglacial esker system'. This feature is in fact quite unlike any known subglacial ice-contact feature, because of its complex morphology, its size and its internal faulting. It is however identical in form and internal structure to many supraglacial fluvial deposits which have accumulated between and above ice-cored moraines. The similarity of the Carstairs deposit and many others which have been called end moraines to linear belts of ice-front ridges in Spitsbergen described on p. 367 is immediately striking. However, these latter are not moraines, although they are intimately associated with ice-cored moraines between which they have accumulated; and the overlying till is not a readvance till but a flow till derived from the ice-cores. Thus I would suggest that many of the Pleistocene features termed moraines are not moraines, but the inverted fluvio-lacustrine moulds of ice-cored moraines which have left no trace as their ice-cores have melted from beneath them. These original ice-cored moraines blocked outward drainage from the glaciers and thus determined important lines of outwash deposition. Indeed Worsley (1968, unpublished Ph.D. thesis) has noted the absence of outwash sediments immediately to the south of the Bar Hill—Wrexham moraine (Yates & Mosley 1967) compared with the concentration of bedded sediments within it. These Pleistocene features may simply be reflections of the fact, well illustrated at the margins of modern glaciers, that the presence of ice-cored moraines, blocking drainage, causes the immediate zone of the ice front to be the major zone of outwash deposition, leading to hummocky outwash accumulations on ice melting. Thus, such features as the relatively narrow Bar Hill–Wrexham moraine, the Escrick 'moraine' (Gaunt in press) and the Carstairs deposit, were quite clearly important lines of glacial drainage, and could be regarded as supraglacial 'eskers' which formed in a morainic zone. It may also be that the most impressive of all stratified ice-contact features which have been termed 'moraines', the Salpausselka system of southern Finland, may be best explained by a similar formula; as a series of fluvial and lacustrine features built up between ice-cored ridges which lay along relatively stable ice margins in successive retreat stages. The structure of these ridges is almost identical to that of supraglacial stratified ridges in Spitsbergen, and indeed Hyypa (1950) and Virkkala (1963) reach a very similar conclusion, that 'the ridges originated (sub-aerially) in a crevasse zone running roughly parallel to the glacier margin', although they did interpret the till which sometimes occurs on the ridge surface as a readvance till. A similar interpretation could also be applied to the many of the parallel fluvioglacial ice-contact features of Denmark and North Germany widely regarded as push-moraines (Stauchmoränen). The folding and faulting in some of these features is given as the evidence for ice-pushing and whilst this is convincing in many localities, there are many other exposures where collapse over buried ice and downslope flow is a more plausible explanation, especially in those cases where the deposits are capped by recognisable flow till. The presence in these areas of flat-topped ice-contact deposits (hat-shaped hills) I would suggest to be conclusive evidence for supraglacial deposition. In Germany and Denmark the distinction between 'push-moraines' and 'dead-ice topography' is largely made on morphological grounds, between parallel, controlled ridges and non-parallel, uncontrolled ridges, whereas on morphological grounds alone, both can represent dead ice topography.

(c) The nature of Pleistocene ice sheets

It was suggested in the introduction that because of the different levels of transport of subglacially-derived debris in temperate and sub-polar glaciers, the nature of the deposits left by these two different glacier types are very different. The relatively high level at which basally-derived debris can be transported in arctic glaciers makes widespread flow till deposition possible, and the associated suite of supraglacial deposits. However, temperate glaciers, with their low level load, tend only to produce subglacial lodgement till.

If the above relationship between the position of transport and thermal regime is accepted, then the interpretation of Pleistocene tills in section 7A and 7B as flow tills suggests the inference that the parent ice caps at this phase were composed of cold ice, at least in their outer zone. This conclusion is in fact hardly surprising in view of the evidence of contemporary permafrost beyond the margins of Pleistocene ice caps. I would further suggest that it should be possible to build up a picture of changing regime from the nature of the deposits of different stages of glaciation, and it is hoped to present further criteria for the inference of glacier regimes in a subsequent publication.

The above discussion is restricted to subglacially derived debris, for englacial debris derived supraglacially from nunataks and valley sides in the accumulation area might also give flow tills in temperate glaciers. However, large Pleistocene ice sheets, to which this discussion is directed, would have few nunataks and therefore little englacial debris derived from this source.

Acknowledgements

Financial support for this work from the Royal Society and the University of Keele is gratefully acknowledged. I am particularly indebted to W.B. Harland, R.H. Wallis, P. Maton and many others for support in the field. Some helpful comments on the manuscript were given by E.A. Francis.

8. References

ANDREWS, J.T. & SMITH, D.I., 1970. Statistical analysis of till fabric: methodology, local and regional variability. *Q. Jl geol. Soc. Lond.* **125**: 503–42.

BOULTON, G.S., 1968. Flow tills and related deposits on some Vestspitsbergen glaciers. *J. Glaciol.* **7**: 391–412.

—— 1970a. On the origin and transport of englacial debris in Svalbard glaciers. *J. Glaciol.* **9**: 213–29.

—— 1970b. The deposition of subglacial and melt-out tills at the margins of certain Svalbard glaciers. *J. Glaciol.* **9**: 231–45.

—— 1971. (in Goldthwait *ed.*) *Till. A Symposium.* Till genesis and fabric in Svalbard.

CARRUTHERS, R.G., 1939. On the northern glacial drifts. *Q. Jl geol. Soc. Lond.* **95**: 299–333.

—— 1947–1948. The secret of the glacial drifts. *Proc. Yorks. geol. Soc.* **27**: 43–57 and 129–72.

—— 1953. *Glacial drifts and the undermelt theory.* Newcastle.

CARVILL LEWIS, H. 1894. *Glacial geology of Great Britain and Ireland.* London.

CHARLESWORTH, J.K. 1926. The readvance, marginal kame moraine of the south of Scotland, and some later stages of retreat. *Trans. R. Soc. Edinb.* **55**: 25–50.

CLAYTON, L. & FREERS, T.F., 1967. Glacial geology of the Missouri Coteau. *North Dakota Geological Surveys, Miscellaneous Series,* **30**.

COX, F.C., 1970. (In Quaternary Research Association handbook)—unpublished Norwich.

FLINT, R.F., 1957. *Glacial and Pleistocene geology.* London (Wiley).

GAUNT, G.D. (in press). A temporary section across the Escrick moraine at Wheldrake, East Yorkshire.

GJESSING, J., 1960. Isavsmeltningstidens drenerings, dens forlop og Formdannende virkning i Nordre Atnedalen. *Ad Novas.* **3**: 1–492.

GOODCHILD, J.G., 1875. The glacial phenomena of the Eden valley and the western part of the Yorkshire Dale district. *Q. Jl geol. Soc. Lond.* **28**: 55–99.

GOODLET, G.A., 1964. The Kamiform deposits near Carstairs, Lanarkshire. *Bull. geol. Surv. Gt Br.* **21**: 175–196.

GRAVENOR, C.P. & KUPSCH, W.O., 1959. Ice disintegration features in Western Canada. *J. Geol.* **12**: 48–64.

GRIPP, K., 1929. Glaciologische und geologische Ergebnisse der Hamburgischen Spitsbergen Expedition—1927. *Abhandlungen aus dem Gebiet der Naturwissenschaften.* **22**: 146–249.

HARTSHORN, J.H., 1958. Flow till in south-eastern Massachusetts. *Bull. geol. Soc. Am.* **69**: 477–82.

HOPPE, G. 1952. Hummocky moraine regions with special reference to the interior of Norrbotten. *Geogr. Annlr.* **34**: 1–71.

—— 1963. Subglacial sedimentation with examples from north Sweden, *Geogr. Annlr.* **45**: 41–49.

—— & SCHYTT, V., 1953. Some observations on fluted moraine surfaces. *Geogr. Annlr.* **35**: 105–15.

HULL, E., 1864. The geology of the country around Oldham. *Mem. geol. Surv. U.K.*

HYYPA, E., 1951. Kuvia Salpausselan rakenteesta. On the structure of the First Salpausselka. *Geologi* 2–3.

LLIBOUTRY, L., 1965. *Traité de Glaciologie.* **2**: Paris (Nasson).

McLELLAN, A.G., 1969. The last glaciation and deglaciation of central Lanarkshire. *Scott. J. Geol.* **5**: 248–68.

PRICE, R.J., 1969. Moraines, sandar, kames and eskers near Breidamerkurjokull, Iceland. *Trans. Inst. Br. Geogr.* **46**: 17–43.

PALOSUO, E. & SCHYTT, V., 1960. Til Nordostlandet med den Svenska Glaciologiska Expeditionen. *Terresta.* **1**: 1–19.

SAUNDERS, G.E., 1968. A reappraisal of glacial drainage phenomena in the Lleyn Peninsula. *Proc. Geol. Ass.* **79**: 305–24.

SCHYTT, V., 1969. Some comments on glacier surges in eastern Svalbard. *Can. J. Earth Sci.* **6**: 867–73.

SHARP, R.P., 1949. Studies of super-glacial debris in valley glaciers. *Am. J. Sci.* **247**: 289–315.

SISSONS, J.B., 1961. The central and eastern parts of the Lammermuir Stranraer Moraine. *Geol. Mag.* **98**: 380–92.

STALKER, A., 1960. Ice-pressed drift forms and associated deposits in Alberta. *Bull. geol. Surv. Can.* **57**: 1–38.

SZUPRYCZYNSKI, J., 1965. Eskers and kames in the Spitsbergen area. *Geographia Polonica.* **6**: 127–40.

TROTTER, F.M., 1929. Glaciation of the Eastern Edenside, the Alston Block and the Carlisle Plain. *Q. Jl geol. Soc. Lond.* **88**: 549–807.

VIRKKALA, K., 1963. On ice-marginal features in Southwestern Finland. *Bull. Commn. géol. Finl.* **210**: 1–76.

WEERTMAN, J., 1961. Mechanism for the formation of inner moraines found near the edge of cold ice caps and ice sheets. *J. Glaciol.* **3**: 965–78.

WEST, R.G., 1968. *Pleistocene Geology and Biology.* London (Longmans).

WORSLEY, P. The geomorphic and glacial history of the Cheshire Plain and adjacent areas. Unpubl. Ph.D. thesis. Univ. of Manchester (1968).

YATES, B.M. & MOSELEY, F.M., 1967. A contribution to the glacial geomorphology of the Cheshire Plain. *Trans. Inst. Br. Geogr.* **42**: 107–25.

ZINGER, E.M., KORYAKIN, V.S., LAVRUSHIN, Y.A., MARKIN, V.A., MIKHAILOV, V.I., & TROITSKY, L.S., 1966. Issledovaniy lednikov Spitsbergens Sovietskoy ekspeditsiey letom 1965 goda. *Materialy Glyatsiologicheskikh Issledovaniy Khronika. Obsuzhdeniva.* **12**: 59–72.

Plate 1 Internal structure of an ice-cored ridge. Debris rich glacier ice (below line x–x¹) overlain by flow till derived from the englacial debris. At the left hand side bands of debris can be seen separated by clean ice layers, dipping up glacier at 60°. The a/b planes of blade and plate-shaped stones lie parallel to foliation. The section is 8 m high.

Plate 2 Debris bands cropping out on the glacier surface, striking parallel to the glacier margin, and giving rise to a series of supraglacial flow tills up to 2 m thick.

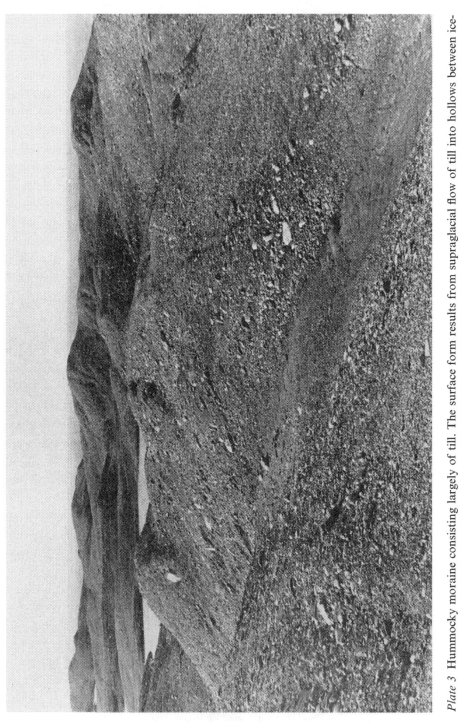

Plate 3 Hummocky moraine consisting largely of till. The surface form results from supraglacial flow of till into hollows between ice-cored ridges which have since melted. The ridges are controlled, lying parallel to the glacier margin.

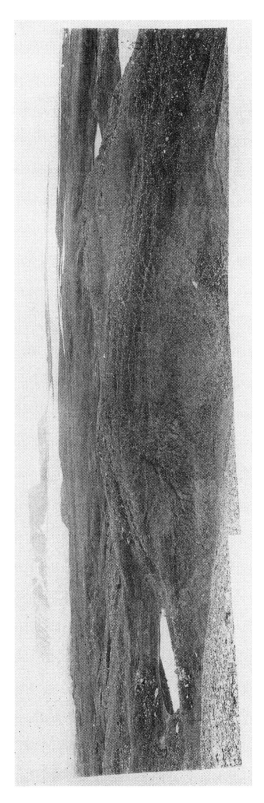

Plate 4 Till surface underlain by melting ice at the margin of Aavatsmarkbreen. Small kettle holes, developing as a result of melting of buried ice, are common, and cracks and furrows form as a result of ice melting and surface flow. The relief of this surface is low and is likely to result in a till plain on final melting of buried ice (see Fig. 1 D₂). Water body on left-hand side is 5 m across.

Plate 5 Section cut in undulating till surface revealing underlying debris rich ice. The till is 3.5 m thick and the upper part (1.5 m thick) shows stratified horizons produced by sub-aerial washing. The lower 2.0 m has never been exposed at surface and consists of melt-out till at base (partly frozen) and parautochtonous flow till. Base of till x–x^1.

Plate 6
a) A flow till with stratification produced by winnowing which has then been involved in rotational downslope flow. The section is slightly oblique to the fold nose.

b) A multi-till sequence of three flow tills separated by two gravel bands. Fluvial deposition and flow till accumulation have alternated at this locality to produce the sequence. Note the planar nature of the till-gravel junctions.

c) A contorted sand lens within flow till. The sand accumulated in a channel on a supraglacial flow till surface, it was originally a synclinal channel fill. Further flow till accumulated above the sand, and subsequent settlement and flow of the whole sequence resulted in the contortion of the lens. The surrounding flow till is massive with little apparent structure (M. Paul).

Plate 7 A series of ice-cored moraines lying parallel to the frontal margin of Elisebreen in Oscar II Land. Outwash streams are constrained to flow parallel to the ridges, but occasionally break through them. Two large lakes can be seen, held up between ice-cored ridges. Melt-water escapes through the morainic arc to the right (see Fig. 2).

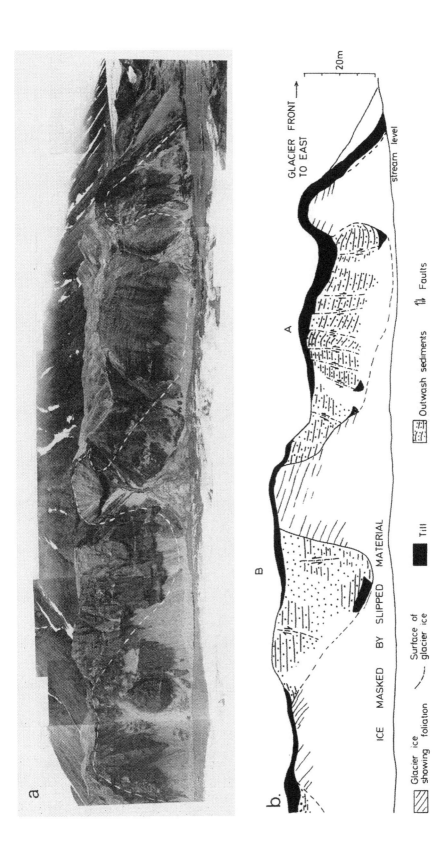

Plate 8 a–b Accumulations of outwash gravels between controlled ice-cored moraines. Much of the outwash is covered by flow till derived from the ice-cores, and underlain by till or glacier ice containing potential till.

Plate 9 Stratified silts, deposited in a supraglacial lake, and showing domes and basins reflecting melting of the underlying ice.

Plate 10 Vertical section cut parallel to the flow direction though a series of old lobate flows of fine sand. Folding and faulting are clearly shown by deformation of silt horizons. The flow direction was right to left. The lower limb of the uppermost isoclinal fold is truncated by a major shear plane. Height of section 2 m. Similar deformation occurs in flow till.

13

LANDSCAPES OF GLACIAL EROSION IN GREENLAND AND THEIR RELATIONSHIP TO ICE, TOPOGRAPHIC AND BEDROCK CONDITIONS

D.E. Sugden

Source: *Institute of British Geographers, Special Publication* 7 (1974): 177–95.

Abstract

A classification of landscapes of glacial erosion is given for Greenland (Table 1), together with a distribution map (Fig. 9). Well-developed *mountain valley glacier* landscapes are thought to have escaped inundation by the full Pleistocene Greenland ice sheets. Of the landscapes modified by ice sheets, *areal scouring* is thought to reflect the former widespread existence of basal ice at the pressure melting point, *selective linear erosion* the former existence of basal ice with only restricted zones at the pressure melting point, and landscapes with *little or no sign of erosion* the former existence of basal ice below the pressure melting point. The distribution of the various types of landscape is analysed at two scales. At a Greenland scale, areal scouring occurs mainly in the south-west and selective linear erosion mainly in the north and east (Fig. 10). This is thought to reflect the behaviour of the former Pleistocene ice sheet with greater activity and turnover of ice in the maritime south-west than in the continental north and east. The broad variation in bedrock altitude between west and east Greenland may also play a role at this scale through its influence on ice thickness. At more local scales, topographic form, bedrock altitude and rock permeability are variables which have caused local variations from the general trend.

An explanation of landscapes of glacial erosion created by ice sheets is an elusive prospect. Much of the field evidence lies in remote latitudes or worse still beneath the opaque ice sheets of Antarctica and Greenland. Partly as a result, not only is a simple morphometric description of forms rare, but the nature of the processes operating at the base of a glacier is recognized to be one of the most important unsolved problems in glaciology at present. Not surprisingly, understanding is at a less sophisticated level than in some other branches of geomorphology. D.L. Linton (1963a) has imaginatively and effectively applied a Davisian model of 'structure, process and stage' to the landscapes in west Antarctica and has stressed in particular the importance of evolutionary development. Elsewhere in the literature it is common to find landforms of glacial erosion grouped under such headings as 'upland', 'low relief', 'coastal', etc. Such terms lean heavily on only one of the many variables affecting glacial erosion, usually topography, and ignore others such as ice conditions and rock type. The use of such terms illustrates the lack of firm theory on which to hang a more meaningful classification. This paper approaches the problem from an initial classification of the landforms of Greenland based on their form. The forms are then viewed as elements of glacial systems and are related at the outset to two main types of glacier system: ice sheets and mountain valley glaciers.

Assuming adequate snow with conditions suitable for its accumulation, ice will build up over a land area in the form of a broad dome-shaped ice cap or ice sheet whose consistent shape reflects the basic flow properties of glacier ice. The crucial characteristics of such ice masses so far as erosion is concerned are that they submerge the underlying topography and that any erosional effects are superimposed onto the underlying land surface. Clearly for an understanding of the erosional effects of such an ice-sheet system it is important to examine the behaviour of the system as a whole. Thus in Greenland it is logical to begin a study of landforms eroded by ice sheets at a sub-continental scale. Mountain valley glaciation occurs wherever for some reason an ice sheet or ice cap cannot build up. Normally the reasons will reflect inadequate snow, insufficient time or more commonly topography whose overall gradient is so steep that it cannot accommodate the gentle slopes of an ice sheet or ice cap. Viewed at a Greenland scale such relief is relevant only in so far as it may represent those parts not submerged by an ice sheet.

The landforms considered in this paper lie in the ring of land between the edge of the present ice sheet and the coast (Fig. 1). Virtually the whole of the area was submerged by ice during the Pleistocene glacial maxima (B. Fristrup, 1966). A particular advantage of a study in this area is the nearness of the relatively well-known, contemporary ice sheet which may be used as a model for comparison with the ice sheets of the glacial maxima. A disadvantage of course is that little can be said about erosional landforms beneath the ice cover.

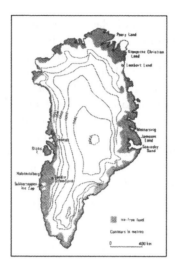

Figure 1 Location map of Greenland showing the ring of ice-free land and approximate ice-sheet surface contours.

Table 1 Classification of landform types.

System	Type of erosional landscape
Ice sheet	areal scouring
	selective linear erosion
	little or no sign of erosion
Mountain valley glaciers	integrated trough networks
	individual cirque (corrie) basins

The landforms

Before any sort of analysis is possible it is necessary to classify the landform types. The following classification is based on examination of the fine collection of Greenland air photographs held by the Geodetic Institute in København and on field observations in Greenland and Europe. Table 1 shows that initial subdivision is based on the distinction between landform associations developed beneath an ice sheet and those associated with mountain valley glaciers.

Ice sheets

Areal scouring describes topography everywhere affected by glacial erosion by an ice sheet (Fig. 2). This type of scenery has been well described by

Figure 2 Landscape of areal scouring in west Greenland. Reproduced with the permission (A. 481/73) of the Geodetic Institute, Denmark.

Linton (1963b) who called it 'knock-and-lochan' topography. It is characterized by the way in which structural lines in the bedrock have been exploited to form linear depressions with often complex and angular junctions. Often these depressions form rock basins and are studded with irregular lakes. The eminences between the depressions are moulded by ice action and often consist of bare, striated, rock surfaces. Sometimes the eminences are smooth on all sides but commonly they are shaped like roches moutonnées with a steep, plucked lee side; the detail of the plucked face is strongly related to the joint patterns in the rock (K.G. Pike, personal communication, 1973). The relief amplitude of such scenery is generally limited. For example, if one excludes large-scale undulations of several kilometres across, the area east of the Sukkertoppen ice cap in west Greenland has an amplitude of the order of 100 m. It is difficult to know how much scouring has been accomplished by ice in such areas. There are indications that frequently the ice has only etched a pre-existing surface and may have removed less than *c.* 100 m of rock debris. Over thousands of square kilometres in the area east of the Sukkertoppen ice cap, the main rivers maintain integrated drainage patterns reminiscent of non-glacial humid environments. These patterns occur regardless of the abundant signs of scouring and indeed the only exceptions are in the vicinity of distinct troughs. Since integrated patterns on such a scale are unlikely to have formed on a newly-created glacial surface without greater modification of the glacial forms, one is led to conclude that the ice was unable to change the land surface sufficiently to derange a pre-existing pattern. This view is confirmed by evidence from the summits between the rivers. Adjacent summits tend to be conformable in altitude and a generalized contour map of their elevations produces a surface which is conformable

Figure 3 Landscape of selective linear erosion, Nathorsts Land, east Greenland. Reproduced with the permission (A. 481/73) of the Geodetic Institute, Denmark.

with the direction of river flow. Both lines of evidence imply that a pre-existing surface has been modified by ice without major transformation. Similar conclusions have been reached elsewhere, for example by J.B. Bird in Arctic Canada (1967).

Landscapes with evidence of selective linear erosion are characterized by upper rolling slopes or plateau surfaces dissected by deep troughs. There are magnificent examples in east Greenland (Fig. 3) and a map of the trough pattern has been published elsewhere (D.E. Sugden, 1968). There are two types of trough in east Greenland. Those aligned approximately east–west (in the direction of ice flow) are long, slightly sinuous, narrow and deep. S. Funder (1972) notes that the depth of these troughs is of the order of 2,500–3,500 m while their width is about 5 km. Those aligned north–south (apparently at right angles to the overall direction of ice flow) follow lines of structural weakness and are straighter, shallower and wider. The upland plateau remnants between the troughs generally show little or no sign of glacial erosion and it is common to find the surface covered with regolith. In similar scenery in the Cairngorm Mountains in Scotland fragile pre-glacial remnants such as rotted rock and tors survive on the upland surfaces (Sugden, 1968). Characteristically the break of slope between the plateau and the trough is remarkably fresh and abrupt. This is particularly well brought out in Figure 3. It is relevant to note that such troughs are thought to have been cut beneath ice sheets. The case has been argued on morphological grounds in east Greenland (J.H. Bretz, 1935; H. W:son Ahlmann, 1941), in Scotland (Sugden, 1968) and in part of North America (K.M. Clayton, 1965). Often recent echo-sounding of modern ice sheets is revealing troughs beneath the ice (D.J. Drewry, 1972).

Figure 4 Landscape of mountain valley glaciation, Borggraven, east Greenland. Reproduced with the permission (A. 481/73) of the Geodetic Institute, Denmark.

Landscapes characterized by little or no sign of glacial erosion are devoid of areal scouring forms or troughs. Such areas commonly comprise coherent fluvial landscapes with smooth slopes covered in regolith. The main problem involved with discussion of this type of landscape is the difficulty of demonstrating whether or not such areas were covered by ice. In Greenland it is generally accepted that the whole continent was covered by ice during the Pleistocene maxima, with the possible exception of northern Peary Land (Fristrup, 1966).

Mountain valley glaciers

Perhaps the most spectacular erosional landscapes of all are those associated with erosion by mountain valley glaciers. The main features are brought out in Figure 4, which shows a network of glaciers overlooked by frost-etched mountain peaks. Characteristically the valley glaciers begin in an arcuate collecting ground and eventually several combine to form a major trunk glacier. Such glacier systems are restricted to local massifs and individually are rarely more than about a few tens of kilometres in length. Most valley glacier landscapes in Greenland support active glaciers today. Some, however, are free of ice and display many of the features well known in the Alps.

The final landscape category includes glacial landforms cut by isolated cirque glaciers. In such cases well-developed troughs are rare and the main feature is the cliffed cirque basin etched into a massif. As would be expected in a case where the cirque glaciers are isolated, there are commonly remnants of the initial surface of the massif between the cirques; this surface may form a plateau or gently undulating upland.

The categories mentioned above are intended to be no more than idealized models. In reality there is much overlap between the various groups and at times assignment into one category is difficult. For example, troughs may occur in areas of extensive areal scouring and in this paper such landscapes have been included in that category.

Theoretical considerations

Since landforms of glacial erosion are the result of processes operating at the interface between the sole of a glacier and the bedrock, it is logical to assume that the nature of the processes and the results of their operation will vary as glacier conditions and bedrock conditions vary.

Basal ice conditions

It is likely that basal slip is a necessary prerequisite for effective glacial erosion beneath an ice sheet. Further, whether or not basal slip occurs depends on a critical threshold related to the effective temperature of the basal ice. If the ice is at the pressure melting point basal slip may occur; if it is below the pressure melting point basal slip is unlikely (J. Weertman, 1957, 1964; G.S. Boulton, 1972). When basal ice is below its pressure melting point movement within the ice takes place within the lower layers of the ice mass rather than between ice and rock. When basal ice is at the pressure melting point movement may also involve basal slip between ice and rock. A film or layer of water is formed at the base and its relative abundance is thought to play a crucial role in reducing friction and influencing the rate of basal slip. The greater the quantity of basal water the greater the role of basal slip. This conclusion has been reached in a number of theoretical studies by L. Lliboutry (1965, 1968a) and Weertman (1957, 1964, 1966).

With regard to the landforming processes occurring beneath a glacier it is reasonable to suppose that the more favourable conditions are for basal melting the greater will be the opportunity for the transport of basal debris. There are two reasons for this. First, if one imagines a point on the ground beneath an ice sheet, the more readily ice slips over its base the greater the total amount of ice that will pass that point during a glaciation. Secondly, from an areal point of view favourable basal ice conditions are likely to mean that widespread areas of bedrock are affected by basal slipping. It is important to stress that this does not mean that the full potential of the ice for transport of material is necessarily used. Thus there are no grounds for expecting conditions most favourable for basal slipping to be associated with the greatest amount of glacial erosion, measured as the volume of debris removed. As recent work suggests, the process by which debris is entrained into the ice may be linked to rather sensitive thermal conditions (Boulton, 1972). However, assuming that there is debris in the sole of the ice it is fair to expect that areas subjected to widespread basal slip will bear extensive evidence of the ordeal.

Before attempting a geomorphological assessment of ice sheet erosion, it is clearly vitally important to review the conditions affecting temperatures at the base of ice sheets. Such basal temperatures are a response to heat obtained from three sources: (1) the base, (2) the surface, and (3) internal deformation.

(1) Geothermal heat is derived from the bedrock base and although it varies from place to place on the earth's surface is generally regarded as a constant providing $c.$ 38 cal/cm^2/year. In ice below the pressure melting point this gives a vertical gradient of 1°C per 44 m at the base of an ice sheet (G. de Q. Robin, 1955). If the ice is at its pressure melting point this heat cannot be absorbed into the glacier and will melt a layer of ice with an average thickness of 6 mm/year at the bottom of the glacier (W.S.B. Paterson, 1969).

(2) Sources of heat from the surface are affected by the depth of the ice, the initial temperature of the firn and the rate of accumulation (Robin, 1955). Since temperatures in ice sheets below the pressure melting point rise with increasing depth, the thicker the ice sheet the higher the basal temperatures will be (assuming similar surface temperatures). The temperature of the firn beneath the thin surface zone of seasonal variations is influenced by the seasonal climatic conditions at the surface. In areas of no summer melting the firn is close to the mean annual temperature. As the climate becomes more maritime with more snow and more summer melting the firn temperature rises rapidly until it reaches 0°C (Paterson, 1969). The rate of accumulation is important because it controls the vertical velocity of ice and thus the rate at which cold surface firn or ice is carried down into the glacier mass. High rates of accumulation tend to reduce the vertical temperature gradient within the ice.

(3) The third source of heat is derived from internal movement within the ice (W.F. Budd, 1969). In ice below its pressure melting point most heat is generated close to the bed in the zone of maximum deformation. For most purposes it can be regarded as an addition to the geothermal heat flux, with an ice velocity of 20 m/year approximately equivalent to the geo-thermal heat flux. In general the heat produced is likely to increase from the centre to a maximum near the equilibrium line of the ice sheet where ice velocities will tend to be greatest. Once slip over the bedrock has commenced, heat produced by that slip is equal to $S\tau/k$, where S = speed of slipping, τ = average shear stress at the glacier bed, and k = a conversion factor, energy to heat units (Boulton, 1972). This latter source of heat is of considerable importance for conditions at the ice/rock interface. Slipping will not occur till the ice has been raised to its pressure melting point. But then the very process of slipping releases more heat which ensures its continuation.

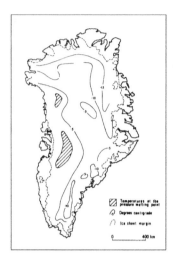

Figure 5 Calculated basal ice temperatures beneath the Greenland ice sheet assuming steady-state conditions. Reproduced by courtesy of D. Jenssen.

The basal ice temperatures of the contemporary Greenland ice sheet have been calculated from mathematical simulation models and give some guide to the main trends. Figure 5 shows the result of one such dynamic model where it is assumed that the ice sheet is in steady-state (D. Jenssen, personal communication, 1972). Details of the models and other alternatives are given by Budd, Jenssen and U. Radok (1971), where they are applied to the Antarctic ice sheet. In Greenland there are detailed variations in the pattern of basal ice temperatures depending on the model used and the exact values fed into the input data. However, the main pattern is consistent from model to model. Figure 5 shows that temperatures are at, or close to, the pressure melting point in a zone parallel to the west coast and under the central northern part of the ice sheet. Basal temperatures decrease towards the north-east. Other models show a similar pattern but with the zone of pressure melting more extensive and closer to the western edge of the ice sheet.

The pattern is consistent with what is known about the climate. Figure 6a is an estimate of precipitation totals over Greenland except the Thule area (S.J. Mock, 1967). With the exception of southern Greenland south of lat. 67°, precipitation is high near the west coast and falls off towards the north-east. Figure 6b shows the pattern of 10 m snow depth temperatures and is a reasonable approximation to mean annual temperatures. The pattern is largely a reflection of surface altitude, but latitude is also important with a tendency for temperatures to fall off towards the north (Mock and W. Weeks, 1966). Taken together these maps reflect a maritime climate in the south and west and a progressively more continental climate towards

356

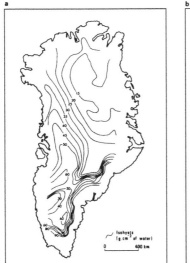

Figure 6 Some climatic data for the Greenland ice sheet. (a) Calculated isohyets in g. cm^{-2} of water (Mock, 1967) (b) Calculated isotherms (Mock and Weeks, 1966).

the north and east. The weather dynamics responsible for this trend are considered in detail by P. Putnins (1970). The maritime climate favours relatively high surface temperatures and a high turnover of ice with consequently a great deal of heat generated by internal deformation. Also, in view of the high velocities of many west Greenland glaciers, heat is also generated by basal slip. High basal ice temperatures are likely under such conditions. Towards the north-east where ice turnover is less and surface temperatures lower, basal ice temperatures tend to be lower.

The relatively high basal temperatures under the central northern ice dome where precipitation is low need further explanation. Here the highest parts of the ice dome lie over areas of low-elevation bedrock and ice thicknesses are high (Fig. 7). In this case ice thickness is the critical factor, as it is in areas of low basal temperatures in south-east Greenland which reflect thin ice.

Although these models apply to the present Greenland ice sheet and not to those of the Pleistocene maxima, there is no reason to suppose that the atmospheric circulation was very different in this area during the Pleistocene. So long as most depressions approached from the south and west and there was pack ice off the north and east coasts there would be a major contrast between the maritime west and south and the continental north and northeast. The main difference is likely to have been in the south-west where the

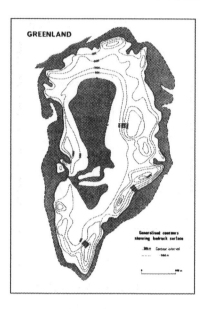

GREENLAND

Figure 7 Approximate bedrock topography of Greenland, showing low altitude of ice sheet base in central and north Greenland.

modern ice sheet is partially protected from a full maritime climate by coastal mountains. The contrast is well illustrated by comparing for example the high precipitation total at the coastal town of Holsteinsborg (355 mm) and the low total at Søndre Strømfjord at the edge of the contemporary ice sheet some 130 km inland (152 mm). During the Pleistocene maxima the ice sheet extended beyond the present coast and would have experienced a full maritime climate. Under these circumstances basal ice temperatures beneath the south-western periphery of the ice sheet are likely to have been higher than at present.

Bedrock variables

The shape and nature of the rock bed is an important variable affecting processes at the ice-rock interface. As Lliboutry (1968b) has pointed out, undulations in the bed affect ice thicknesses and thus lead to local variations in basal ice temperatures. It is possible to envisage a situation where an ice sheet covers an irregular land surface and the basal ice reaches the pressure melting point only over the sites of low-lying depressions and valleys. Clearly such a situation can be expected to favour irregular flow within the ice mass with a tendency towards highest velocities over the sites of valleys and depressions.

Rock permeability can play an important role in influencing the relative abundance of water existing at the ice/rock interface, assuming the ice is at the pressure melting point. If the rock is sufficiently permeable for water formed at the base to be evacuated through the rock then basal slipping may be greatly reduced (Boulton, 1972). The hydraulic conductivity of such beds is a function of rock permeability, thickness of the permeable beds and the hydraulic gradient beneath a glacier (Weertman, 1966). On a larger scale it will also be influenced by the attitude of the beds and in particular the ease with which water drainage may be evacuated in a horizontal direction.

Relationship of the landform associations to theory

If the theoretical considerations are accepted it is possible to suggest conditions under which the different landform associations occur. Areal scouring reflects slipping over the whole land surface. This is most likely to occur when conditions are such that the basal ice is at the pressure melting point almost everywhere, a situation that is to be expected in maritime climatic environments and where the ice is thick. Elsewhere, on a smaller scale, it is likely to occur where local factors favour local zones of ice at the pressure melting point, such as in local deeps in ice over depressions or in the vicinity of outlet glaciers where frictional heating occurs. Probably slipping is also favoured by abundant water at the base and thus is more likely to occur where rocks are impermeable rather than permeable (Boulton, 1972).

Areas with no sign of erosion are more likely to occur where conditions are the opposite. At a macro-scale these areas will tend to occur where climatic conditions are continental while more locally they will reflect patches of thin or largely stagnant ice and areas of highly permeable rocks.

Areas of selective linear erosion may be envisaged as an intermediate category where conditions are so marginal that basal slipping can only occur in places. Following R. Haefeli (1968) Figure 8 shows a hypothetical ice sheet with the ice just attaining pressure melting point over the sites of depressions. Assuming a depression is favourably orientated for evacuation of ice, basal slipping will occur in the position marked. This would generate sufficient heat to maintain the basal ice at the pressure melting point. Further slipping and selective erosion would increase the thickness of the ice at these points and accentuate the contrast further. It is quite possible to envisage such ice streams providing sufficient drainage for the ice sheet in these areas and thus there is no reason for the ice over the intervening plateaux to thicken and its basal layers to reach the pressure melting point. Under such circumstances the intervening plateaux can he regarded as local examples of landscapes with no sign of glacial erosion. Such selective linear landscapes would be favoured by climatic conditions midway between maritime and continental and more locally by relatively thin ice and/or irregular

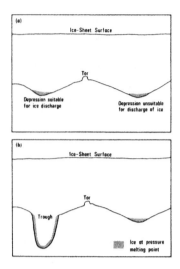

Figure 8 Diagram to illustrate the development of a landscape of selective linear erosion. (a) Ice sheet builds up and basal ice reaches the pressure melting point over the sites of depressions; (b) The depression on the left is suitable for ice flow and basal slipping occurs and erodes a trough. Ice will not slip over the adjacent upland areas unless the ice sheet thickens further (following Haefeli, 1968).

topography. In terms of rock type, local variations in permeability might be expected to lead towards a similar result.

Analysis of the landform patterns in Greenland

If the gist of the theoretical discussion is correct then it should be possible to test the conclusion against the evidence in Greenland, for one would expect certain types of landform association to occur in certain areas. Figure 9 is a map of the various landform types in Greenland. The map relies on the interpretation of several thousand vertical and oblique aerial photographs, together with field examination of part of east (Scoresbysund-Mestersvig) and west Greenland (Sukkertoppen and Umanak). Clearly the mapping of such a large area will contain mistakes especially in the less accessible and less known northern Greenland. Therefore for these areas the map relies also on interpretation and photographs in papers by workers who have visited them (e.g. W.E. Davies and D.G. Krinsley, 1961; R.L. Nichols, 1954, 1969; R.B. Cotton and C.D. Holmes, 1954; P.J. Adams and J.W. Cowie, 1953). Although there may be local areas where the classification can be disputed, it is hoped that the main pattern is substantially correct.

To facilitate analysis it is helpful to look at the problem at two scales. The macro-scale pattern is concerned with Greenland as a whole while the

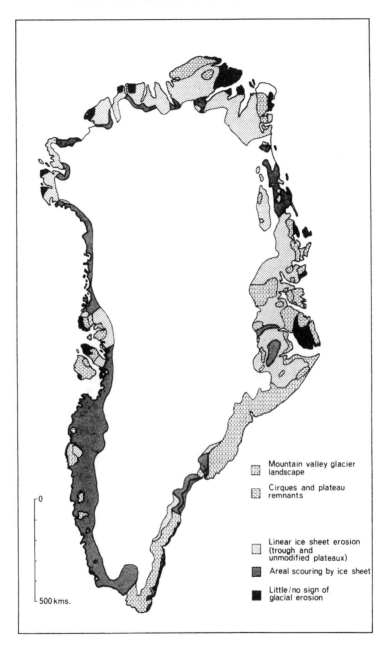

Mountain valley glacier landscape

Cirques and plateau remnants

Linear ice sheet erosion (trough and unmodified plateaux)

Areal scouring by ice sheet

Little/no sign of glacial erosion

0

500 kms.

Figure 9 A map of landscapes of glacial erosion in Greenland based mainly on examination of oblique and vertical air photographs. See text for description of landscape categories.

meso-scale pattern concerns local deviations from the Greenland pattern and is dealing with areas whose size is of the order of ten to a few hundred kilometres. Smaller areas are not considered in this paper.

The macro-scale pattern

Viewing Greenland as a whole there are several clear trends. Excluding for the moment areas of mountain valley glaciation and concentrating on landforms eroded by an ice sheet, the percentage of the land area classified into one or other ice-sheet categories can be calculated for unit areas. Such calculations were made for 105 squares representing land areas of 90 × 90 km and the results are plotted as linear trend surfaces for the two main categories of landform type. Higher order trend surfaces were avoided largely because of excessive distortion introduced by edge effects and the peripheral distribution of the data (J.C. Davis, 1973). Such problems are minimized by limiting interpretation to linear trend surfaces (D.J. Unwin, personal communication, 1973).

Figure 10a shows the linear trend surface for the percentage of the land area affected by areal scouring. The calculations involve only those landscapes affected by ice sheets. The fit of the surface is good: 55 per cent *RSS, Significance* 99.9 <F (Unwin, 1970). The surface brings out the broad trend and clearly shows how most of the available land area in the west and south has been subjected to areal scouringwhile the relative importance of this type of erosion falls off dramatically to the north-east, where values of less than 10 per cent are indicated. Figure 10b shows the linear trend surface for the percentage of available land area classified as landscapes of selective linear erosion. Again areas of mountain valley glaciation are excluded from the calculations. Although the fit of the surface is poorer it is still good:

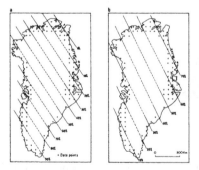

Figure 10 Linear trend surfaces showing for landscapes modified by ice sheets: (a) the percentage land area classified as areal scouring; (b) the percentage land area classified as selective linear. Areas of mountain glaciation have been excluded from these calculations.

31 per cent *RSS, Significance* 99.9 <F. It is worth pointing out that the main trend is complementary to that of areal scouring and that it varies in the opposite direction. Here values are highest in the north-east and lowest in the south-west.

Interpretation of such trends can only be tentative but the pattern is just what one would expect if the landscape categories are related to the former presence or absence of pressure melting at the base of the Pleistocene ice sheets. On a Greenland scale there are two variables which vary broadly in sympathy with the main trend of landform variation and which could account for this variation. One is the ice-sheet climate and the other is bedrock altitude. The variation of basal ice characteristics from south and west to north and east is shown on Figure 5 and is broadly similar to the variation in the amount of areal scouring (Fig. 10a). As suggested on page 184, during the Pleistocene maxima the similarity is likely to have been still closer, for the ice sheet in the south-west is likely to have had basal temperatures higher than those of today. The similarity of two maps is no proof of a causal link between the two. However at the macro-scale the pattern in Greenland is consistent with the hypothesis that areal scouring tends to occur beneath maritime areas of ice sheets while selective linear erosion or no apparent erosion tends to occur beneath ice which is nourished under a more continental regime. In addition, bedrock altitude happens to vary broadly from west to east in Greenland (Fig. 7). Although this trend does not mirror the landform trends exactly, it is likely to contribute to the overall landform pattern. A Kolmogorov-Smirnoff test was carried out to compare the average summit altitudes in thirty sample areas of both selective linear erosion and areal scouring. The test suggested that there was a significant difference between the two sets of altitudes (99 per cent level) and that there was a tendency for altitudes associated with selective linear erosion to be higher than those associated with areal scouring. This association is to be expected for, given a constant ice-sheet surface altitude, basal ice is more likely to be at the pressure melting point over low bedrock areas than over high altitude bedrock areas (Boulton, 1972). In the case of Greenland both the behaviour of the ice sheet and bedrock altitude contribute to the macro-scale pattern. The relative importance of the two could best be ascertained by examination of other areas where the two trends are known not to be approximately complementary.

Other variables can be regarded as relatively unimportant contributors to the macro-scale pattern unless they too can he shown to vary in a consistent and similar manner over Greenland as a whole. Thus structure and rock type are likely to be unimportant at this scale for they do not change consistently from west and south to north and east (K. Ellitsgaard-Rasmussen, 1970).

The distribution of areas of mountain valley glaciation forms a reasonably coherent pattern at the macro-scale. Well-developed landscapes in this

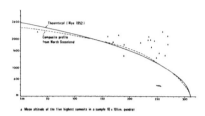

Figure 11 Surface profiles constructed for the maximum Pleistocene Greenland ice sheet(s) and their relationship to the altitude of areas of well-developed mountain valley glaciation. The profiles are drawn assuming the ice sheet extended over the continental shelf to a depth now marked by the −200 m submarine contour. The mean altitude of the five highest summits in each sample area is represented by one triangular symbol. Most of the summit areas are clearly above the profiles and are thus likely to have escaped inundation by the full ice sheet(s).

category occupy areas which are unlikely to have been covered by any Greenland ice sheet. Figure 11 attempts to show the relationship between the altitude of such landscapes and the surface of the reconstructed Pleistocene Greenland ice sheet(s). The ice-sheet profiles, one theoretical, assuming a horizontal base (J.F. Nye, 1952), and one a composite profile based on actual Greenland profiles, are drawn assuming that the seaward edge of the ice sheet was delimited by what is now the −200 m submarine contour. In most cases this contour is close to the top of the edge of the continental slope and is thus likely to be a reasonable approximation. In the north the limited knowledge of submarine conditions makes any reconstruction doubtful but in these regions it is probably as good an estimate as any (A. Weidick, personal communication, 1971). The altitudes of the mountain valley glacier landscapes were reached by averaging the altitudes of the five highest summits in 10 × 10 km squares chosen from a stratified random sample of areas of this landscape type. The altitudes are entered at the appropriate distance inland from the −200 m submarine contour.

It is notable that the summit altitudes of most areas of mountain valley glaciation rise decisively above both ice-sheet profiles. The main exception is in Kronprins Christian Land, which is a little-known area with few accurate altitudes or submarine contours. It is reasonable to suggest therefore that these areas represent land areas of Greenland which have not been effectively submerged by any Greenland ice sheet. As a result the upstanding massifs have been shaped by local valley glaciers throughout the glacial age.

It is important to note that this is a generalization applying only to areas of well-developed mountain valley glaciation with integrated valley glacier systems. There are many other areas where local valley glaciation has operated during 'inter-glacial' periods such as the present. Generally these

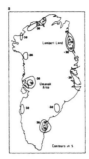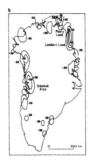

Figure 12 Positive and negative residuals from the linear trend surfaces of landscapes of (a) areal scouring and (b) selective linear erosion.

areas are characterized by cirque forms and rather simple valley patterns. The latter have been excluded from this discussion. In addition, the generalization ignores many factors which should be considered in a detailed examination of mountain glaciation. For example it is likely that the surface profile of an ice sheet is shallower and less convex in the vicinity of outlet glaciers (J.T. Buckley, 1969). Also glacio-isostasy will influence absolute summit heights.

The meso-scale pattern

The main residuals from the two trend surfaces highlight some of the areas where local factors override the general trend (Fig. 12). The residuals fall into three main groups:

(1) The Umanak area of the west coast which consists of a landscape of selective linear erosion (and of little sign of erosion) surrounded by a zone of areal scouring.

(2) The east coast of Greenland in approximately lat. 66° where there is more selective linear erosion and less areal scouring than expected.

(3) Areas in northern Greenland where there is more areal scouring and less selective linear erosion than would be expected. The most important anomaly here is the Lambert Land area.

In addition to these residual areas which tend to measure some hundreds of kilometres across there are much smaller-scale anomalies which can be picked out by visual comparison of the trend surfaces and the map of the original data (Fig. 9). These include examples of landscapes with no sign of glacial erosion, small areas of areal scouring surrounded by selective linear landscapes and *vice versa*.

365

Without far more detailed knowledge of Greenland it is impossible to explain individual anomalies at this stage. However, it is possible to highlight the various situations and to discuss the factors which could explain them.

Local contrasts in areal scouring and selective linear erosion

There are many instances where a contrast between a landscape of areal scouring and one of linear erosion is associated with a change in bedrock altitude. In northern and eastern Greenland the association commonly takes the form of a low-lying area of areal scouring surrounded by higher ground characterized by linear erosion, while in the west an upland marked by linear erosion may be surrounded by a zone of low-lying areal scouring. An example of the latter occurs in the Sukkertoppen ice-cap area. Here the surface of a plateau falls north-eastwards from an altitude of *c.* 1,700 m in the west to an altitude of 500–600 m in the east. The whole area was enveloped by a more extensive ice sheet flowing towards the west (Weidick, 1968; Sugden, 1971). The distinct contrast in landscape types is shown in Figure 13. Most of the lower plateau area is extensively scoured regardless of differences in rock type and structure. The western part of the plateau was sufficiently high and sufficiently close to the ice-sheet edge to escape inundation by the main ice sheet. As a result it now forms spectacular mountain valley glacier scenery. Between the mountain glacier scenery and the areal scouring is an

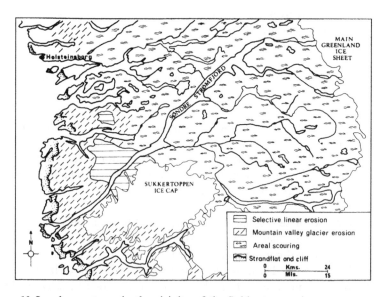

Figure 13 Landscape types in the vicinity of the Sukkertoppen ice cap.

366

area of high ground (now partly occupied by the Sukkertoppen ice cap) characteristic of selective linear erosion. The altitude and position of this bit of upland and the fact that it was relatively close to the ice-sheet edge suggest it was only thinly covered by ice. It is reasonable to postulate that basal ice over the upland was cooler than basal ice over lower ground further inland and may not have reached its pressure melting point over the uneroded plateau areas. Such an interpretation is strengthened by the fact that the changes in landscape types indicated occur regardless of the slight changes in bedrock type shown on the Tectonic Geological map of Greenland (Ellitsgaard-Rasmussen, 1970). It is tempting to suggest that the other main occurrence of selective linear erosion on the west coast, the Umanak area, is also related to the fact that this area is much higher than average, for the area contains some of the highest summits in north-west Greenland. However, it is also underlain by distinctive rocks.

In eastern and northern Greenland where selective linear erosion is dominant, areal scouring is confined to depressions, usually in the vicinity of troughs and outlet glaciers (Fig. 9). When the whole area was covered by the full Pleistocene ice sheets the basal ice in the vicinity of the depressions would have been warmer than basal ice over the adjacent upland. This would be due partly to the greater ice thickness and partly to the internal heat generated by ice flowage concentrated in the valleys at the expense of the upland areas.

Local areas of little or no glacial erosion

In Greenland all occurrences of landscapes with little or no sign of erosion are local in distribution and thus are likely to reflect critical local factors. If such areas simply reflect the lack of basal slipping then local factors will need to be sufficiently important to prevent such slipping.

Rock type is likely to be an important local factor for the reasons stated on p. 358. Although it is impossible to examine its role in any depth without detailed field and laboratory work, there are one or two suggestive relationships which are worth noting. In particular there is evidence which conforms to the expectation that outcrops of porous rocks will locally reduce glacial erosion. In east Greenland for example, the bulk of the areas with no sign of glacial erosion lie along a zone of relatively young marine sediments and volcanics, extending through time from Upper Permian to Tertiary (J. Haller, 1971). It is reasonable to assume that these relatively undisturbed rocks are more permeable than the older folded metamorphics immediately to the west. This is certainly true of parts of Jameson Land where the younger rocks include limestones and sandstones (E. Kempter, 1961). The association is even more striking when one realizes that, with the exception of a few fjord outlets, virtually all the remaining parts of the zone are devoid of icesheet forms. Indeed if it were not for minor etching by local cirque

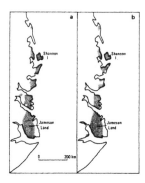

Figure 14 A visual correlation of (a) a zone of young (permeable?) rocks and (b) areas with little or no sign of ice-sheet erosion in East Greenland. (b) is derived from Figure 9, but areas of cirque erosion have been ignored.

glaciers virtually the whole area would have been mapped in the category of little or no erosion (Fig. 14). It may also be significant that the areas with no sign of glacial erosion in the vicinity of Disko Island lie on Tertiary volcanics while in Peary Land they occupy zones of young rocks (undifferentiated) and limestones. It is important to state, however, that some of these parts of Peary Land may never have been glaciated (Fristrup, 1966). Topography is an important local variable which is likely to have some effect on the distribution of landscapes of little or no glacial erosion. The position and altitude of a massif may he such that it is covered only thinly by ice, and basal ice temperatures may never rise to the pressure melting point. Special cases of this are the plateau areas between the troughs in landscapes of linear erosion. Also, topography may affect basal ice conditions through its role in causing ice to diverge or converge (Linton, 1963b). Such a role has been hinted at by reference to the channelling effect of depressions and valleys which drain more ice than surrounding areas and thus allow more internal heat to be generated. A narrow promontory or peninsula will fulfil the opposite role and thus tend, other variables being equal, to be less subject to erosion. In this context it may be significant that the sixteen main instances of little or no glacial erosion occur mainly on peninsulas, i.e. on sites where ice might be expected to diverge.

At this stage it is impossible to assess the relative role of different local factors in preventing basal slip. None the less it is interesting to consider their role in relation to the macro-scale pattern of basal slipping in Greenland. It is likely that variations in local conditions need to be less dramatic in the north and east to prevent basal sliding, than for example in the west and south. Thus it is reasonable to suppose that the existence of areas with no signs of glacial erosion in the vicinity of Umanak and Disko Island must represent a powerful combination of local factors. In this context it is interesting to note that the relevant areas lie on young basaltic (porous?) rocks, on peninsulas, and in general are high. It may be necessary for all these local factors to work together in order to override successfully the macro-scale tendency for basal slipping in the west.

Conclusion

This paper began with an initial classification of landscapes of glacial erosion based on morphological characteristics. On theoretical grounds it was then suggested that these variations could be attributed to different conditions in the basal layers of the ice. It was proposed that

(1) areal scouring is related to basal ice at its pressure melting point capable of slipping over a film of water at the ice/rock interface;
(2) landscapes of selective, linear erosion are related to situations where only zones of the basal ice are at the pressure melting point, and
(3) landscapes with no sign of glacial erosion are related to basal ice which is below the pressure melting point and thus does not slide at the ice/rock interface.

The distribution of the various landscape types was then analysed at two scales to test this hypothesis, apparently successfully. As a cautionary note it is important to stress that such a correlation on its own cannot prove a causal link between basal ice conditions and the resultant landforms. However, it can be stated that the hypothesis has survived a test in a significant part of the world's glaciated landscapes.

Acknowledgements

I am grateful to the Geodetic Institute in København for permission and space to work in their air photographs library and indebted to Dr A. Weidick for clarifying specific problems and providing some little-known literature on northern Greenland. Field-work in east and west Greenland was accomplished during the course of the Oxford University East Greenland Expedition, 1962, and the Aberdeen University West Greenland Expedition, 1968. It is a pleasure to acknowledge the help of all members of the expeditions and of all bodies whose support made them possible. I would like to thank all those who have kindly given valuable criticisms of early drafts of the paper and advice on technique, in particular A. Dawson, A. Gemmell, J. Gordon, P. Hamilton, D. Jenssen, J. von Weymarn and D. Unwin. Lastly, I thank the Carnegie Trust for the Universities of Scotland who partially financed a visit to København and contributed towards the cost of illustrations.

References

ADAMS, P.J. and J.W. COWIE (1953) 'A geological reconnaissance of the region around the inner part of Danmarks Fjord, Northeast Greenland', *Meddr Grønland* 111 (7)

AHLMANN, H.W:son (1941) 'Studies in North-East Greenland, 1939–40', *Geogr. Annlr* 23, 145–209

BIRD, J.B. (1967) *The physiography of Arctic Canada*

BOULTON, G.S. (1972) 'The role of thermal régime in glacial sedimentation', *Spec. Publ. Inst. Br. Geogr.* 4, 1–19

BRETZ, J.H. (1935) 'Physiographic studies in East Greenland' in L.A. BOYD (ed.) *The fiord region of East Greenland, Spec. Publ. Am. geogr. Soc.* 18, 161–266

BUCKLEY, J.T. (1969) *Gradients of past and present outlet glaciers*, Geol. Survey of Canada, Paper 69-29

BUDD, W.F. (1969) 'The dynamics of ice masses', *ANARE* Scientific Reports Ser. A. (4) Glaciology Publ. 108

BUDD, W.F., D. JENSSEN and U. RADOK (1971) 'Derived physical characteristics of the Antarctic ice sheet (Mark 1)', University of Melbourne Met. Dept. 18

CLAYTON, K.M. (1965) 'Glacial erosion in the Finger Lakes region (New York State, U.S.A.)', *Z. Geomorph.* 9, 50–62

COTTON, R.B. and C.D. HOLMES (1954) 'Geomorphology of the Nunatarssuak area' in *Final Report Operation Ice Cap 1953*, Program B. Dept. Army Project 9-98-07-002. Stanford Research Inst. 27–52

DAVIES, W.E. and D.G. KRINSLEY (1961) 'Evaluation of arctic ice-free land sites, Kronprins Christian Land and Peary Land, North Greenland, 1960', *Air Force Surv. Geophys.* 135

DAVIS, J.C. (1973) *Statistics and data analysis in geology*

DREWRY, D.J. (1972) 'The contribution of radio echo sounding to the investigation of Cenozoic tectonics and glaciation in Antarctica', *Spec. Publ. Inst. Br. Geogr.* 4, 43–57

ELLITSGAARD-RASMUSSEN, K. (1970) *Tectonic/geological map of Greenland*, Copenhagen, Geological Survey of Greenland

FRISTRUP, B. (1966) *The Greenland ice cap*, Copenhagen, Rhodos

FUNDER, S. (1972) 'Deglaciation of the Scoresby Sund fjord region, north-east Greenland', *Spec. Publ. Inst. Br. Geogr.* 4, 33–42

HAEFELI, R. (1968) 'Gedanken zur Problem der glazialen Erosion', *Felsmechanik u. Ingenieurgeol.* 4, 31–51

HALLER, J. (1971) *Geology of the East Greenland Caledonides*

KEMPTER, E. (1961) 'Die Jungpalaozoischen Sedimente von Süd Scoresby Land', *Meddr Grønland* 164 (1)

LINTON, D.L. (1963a) 'Some contrasts in landscapes in British Antarctic Territory', *Geogrl J.* 129, 274–82

LINTON, D.L. (1963b) 'The forms of glacial erosion', *Trans. Inst. Br. Geogr.* 33, 1–28

LLIBOUTRY, L. (1965) *Traité de Glaciologie*, 2, Paris, Masson

LLIBOUTRY, L. (1968a) 'General theory of subglacial cavitation and sliding of temperate glaciers', *J. Glaciol.* 7, 21–58

LLIBOUTRY, L. (1968b) 'Steady-state temperatures at the bottom of ice sheets and computation of the bottom ice flow law from the surface profile', *J. Glaciol.* 7, 363–76

MOCK, S.J. (1967) 'Calculated patterns of accumulation on the Greenland ice sheet', *J. Glaciol.* 6 (48), 795–803

MOCK, S.J. and W. WEEKS (1966) 'The distribution of 10 meter snow temperatures on the Greenland ice sheet', *J. Glaciol.* 623–41

NICHOLS, R.L. (1954) 'Geomorphology of south-west Inglefield Land' in *Final Report Operation Ice Cap 1953*, Program B. Dept. Army Project no. 9-98-07-002. Stanford Research Inst. 151–208

NICHOLS, R.L. (1969) 'Geomorphology of Inglefield Land, North Greenland', *Meddr Grønland* 188 (1)

NYE, J.F. (1952) 'A method of calculating the thickness of the ice-sheets', *Nature, Lond.* 169, 529–30

PATERSON, W.S.B. (1969) *The physics of glaciers*

PUTNINS, P. (1970) 'The climate of Greenland' in S. ORVIG (ed.), *World survey of climatology* 14, 3–128

ROBIN, G. DE Q. (1955) 'Ice movement and temperature distribution in glaciers and ice sheets', *J. Glaciol.* 2, 523–32

SUGDEN, D.E. (1968) 'The selectivity of glacial erosion in the Cairngorm Mountains, Scotland', *Trans. Inst. Br. Geogr.* 45, 79–92

SUGDEN, D.E. (1971) 'Deglaciation and isostasy in the Sukkertoppen ice cap area, West Greenland', *Arctic Alpine Res.* 4, 97–117

UNWIN, D.J. (1970) 'Percentage RSS in trend surface analysis', *Area* 2, 25–8

WEERTMAN, J. (1957) 'On the sliding of glaciers', *J. Glaciol.* 3, 33–8

WEERTMAN, J. (1964) 'The theory of glacier sliding', *J. Glaciol.* 5, 287–303

WEERTMAN, J. (1966) 'Effect of a basal water layer on the dimensions of ice sheets', *J. Glaciol.* 6, 191–207

WEIDICK, A. (1968) 'Observations on some Holocene glacial fluctuations in West Greenland', *Meddr Grønland* 165 (6)

14

NATURE OF ESKER
SEDIMENTATION

I. Banerjee and B.C. McDonald

Source: A.V. Jopling and B.C. McDonald (eds) *Glaciofluvial and Glaciolacustrine Sedimentation*, Tulsa, Okla.: Society of Economic Paleontologists and Mineralogists, Special Publication 23, 1975, pp. 132–54.

Abstract

Broad questions of esker sedimentation are reviewed in this paper. Two main environmental factors, nature of the conduit through which the esker stream flowed, and site of deposition, control esker sedimentation and commonly can be determined from the sedimentary succession. Interaction of these two factors permits definition of three different models of esker sedimentation: *open-channel, tunnel* and *deltaic.* Morphology of the esker ridge, sedimentary structures, facies relationships and paleocurrent variability are important parameters of proposed sedimentation models. The models are discussed on the basis of field data from eskers at Peterborough, Ontario and at Windsor, Quebec.

Sediments of the Peterborough esker were deposited largely in an *open channel* bordered laterally by ice walls. Backset beds related to antidunes are preserved at places. A common environment was *deltaic*, where dunes and ripples delivered sediment to avalanche faces; progressively downstream from the large foresets were regressive, sinusoidal, and progressive ripples, respectively. These in turn pass into graded beds and then into lacustrine rhythmites.

Tunnel sedimentation is illustrated by sediments in single steep-sided ridges in the Windsor esker. Sheetlike cross-bedded and parallel-bedded gravel and sand units persist downstream without facies change and are arranged in vertically stacked cycles that may be annual. Flow depth in the tunnel was 1 to 4 m and accumulation of sediment was accommodated by a melting upward of the ice roof.

Deltaic sedimentation is illustrated by beads in the Windsor esker that were deposited annually as subaqueous fans in the water body at the mouth of the subglacial tunnel. Cobble and pebble gravel at the proximal end of the bead intertongues over a few meters in a downstream direction with ripple-laminated fine sand, units of "structureless" fine and medium sand, and graded beds.

Introduction

An esker is a linear accumulation of gravelly and/or sandy stratified sediment that was deposited by a stream confined on both sides by glacier ice. In some cases, though not necessarily, the stream was also confined on the top and/or bottom by glacier ice.

Studies of eskers are numerous and have had a number of objectives. Most have been concerned with interpretation of late-glacial environments (*see* Boulton, 1972, and Flint, 1971, for many references). Others have studied the use of eskers as a drift prospecting medium (Cachau-Herreillat and LaSalle, 1971; Hellaakoski, 1931; Gillberg, 1968; Lee, 1965, 1968; Shilts, 1973), as a source of groundwater (De Geer, 1968; Parsons, 1970), and as a source of more fundamental data on fluvial sedimentary processes (Aario, 1971, 1972a and b; Allen, 1971); Shaw (1972) contributed a useful paper on the distributions of sedimentary facies in three glaciofluvial deposits in England. Eskers have only rarely been diagnosed in the bedrock record (Frakes and others, 1968).

The present study reviews briefly some characteristics of eskers that must be explained by any comprehensive theory of esker formation. These are based partly on literature review and partly on field study by McDonald of eskers in the District of Keewatin, northern Canada. The principal theories of esker formation are reviewed in terms of their sedimentary implications. Sedimentary structures and facies relationships in two eskers formed during late Wisconsinan deglaciation at Peterborough, Ontario, and at Windsor, Quebec have been studied in detail by the present authors, and these data are used to construct models of esker sedimentation.

Characteristics of eskers

Distribution

Eskers occur only in areas that have been glaciated. In North America they occur mostly south of about 72°N, suggesting that wet-based, or temperate, glaciers provide more favorable hydrological conditions for esker formation than do glaciers frozen to their bases.

Although isolated eskers are common, eskers tend to occur in swarms such as those in Scandinavia, Maine, Labrador, and the District of Keewatin. These are areas where large remnant Pleistocene ice sheets wasted away, and it seems probable that the esker swarms simply reflect widespread and rapid ablation having produced an abundant supply of meltwater within the glacier. Eskers in such swarms are commonly linked into networks in which the overall network pattern is parallel, a result of steep hydraulic gradients near the glacier termini having been oriented parallel to the surface slope of the glacier. Seldom, if ever, are more than two orders of tributaries (in the sense of Horton, 1945) linked to a trunk esker. In the District of Keewatin, eskers in swarms trend roughly parallel to each other and are regularly spaced about 13 to 14 km apart. This spacing, evidence of a well integrated meltwater escape system in the glacier, would be a function of the hydraulic conductivity within the glacier, the discharge required to maintain the channel against any tendency of the ice to close it, and the distance back from the glacier terminus over which the water collection system was integrated.

Eskers emerging from present-day glaciers have been reported by many workers (for example, Hartshorn, 1952; Howarth, 1971; Ives, 1967; Jewtuchowicz, 1965; Lewis, 1949; Meier, 1951; and Price, 1966, 1969). The glaciers are commonly valley glaciers, the eskers tend to be small, and ice cores make up a large proportion of the ridge volumes. The large Pleistocene continental ice sheets that formed the major esker systems of the world were hydrologically of such a larger scale than small glaciers existing today that it is debatable to what degree modern subaerially exposed eskers should constrain the interpretation of sedimentary sequences in the large "fossil" eskers. In addition, it is clear that many of the large Pleistocene eskers formed beneath ice sheets whose termini rested in deep bodies of standing water, a situation where it is difficult to study modern analogues.

Esker morphology

Morphologic characteristics, insofar as they can be related to a particular sedimentation history, must be considered in a study of the environments of esker deposition. Eskers may occur in association with other sedimentary and morphologic features that are closely related, both in location and origin, to the processes of esker formation. The entire "esker complex," then, includes the esker itself along with such features as marginal troughs, kettles, kames, and esker deltas.

Basic morphologic types of eskers are summarized in Table 1. The common morphological expression of an esker is that of a single, long, relatively narrow, sinuous ridge. The ridges may be sharp crested or flat topped. The single-ridge types, however, may be straight for long distances, while in other cases they seem to be truly meandering. The term "beaded esker" is applied to those reaches of eskers that consist of a series of separate, regularly

374

Table 1 Morphologic variation in eskers.

Basic morphology	Characteristics of esker complex	Deglaciation environment
1. Continuous single ridge; flanking outwash	Central esker ridge is highest element of complex; flanked progressively outward on each side by elongate marginal kettles, outwash terraces, and scalloped till plain; abrupt topographic discontinuities.	No standing water at glacier terminus.
2. Continuous single ridge; no flanking outwash	Ridge subdued by subsequent wave action commonly resulting in beach development on esker ridge; flanked by till.	Deposited within glacier but below level of standing water body at glacier terminus.
3. Broad ridge with multiple crests	May or may not be flanked by outwash; commonly multiple ridges form reticulate pattern.	Where broader than 200 to 300 m, probably in part subaerial and some may be part of interlobate moraine complex; narrower varieties flanked by outwash indicate no standing water at glacier terminus.
4. Beaded	Pronounced isolated beads are regularly spaced and flanked by till.	Deposited where esker stream entered standing water at glacier terminus.

spaced, roughly conical hills or "beads" of esker sediment. The beads are commonly 100 to 200 m diameter and 5 to 15 m high. Alignment of the beads forms an easily recognizable esker trend. Although occasionally successive beads are in contact with each other, more commonly they are separated by lower terrain underlain by till or by lacustrine or marine sand.

Morphologic variations are common along a single esker, especially between types 2 and 4 (table 1). Most long esker ridges have gaps in them that could result either from nondeposition or from postdepositional erosion. Postdepositional processes such as beach formation accompanying shallowing of the standing water body, solifluction, slumping, frost-cracking, and eolian activity commonly modify and subdue the original morphology and rework the surface sediments.

Dimensions of eskers can be summarized as follows: (a) *Height*—rarely more than 50 m but have been recorded to 80 m (Donner, 1965); (b) *Thickness of gravel*—225 m of esker gravel have been drilled in the Munro esker of northern Ontario although the esker stands only 30 m above surrounding terrain (Hobson and Lee, 1967); (c) *Width*—can be as large as

7 km for the broad flat-topped Munro esker complex, although generally widths of esker ridges are less than 150 m; (d) *Length*—variable from a few hundred meters to over 800 km, including gaps, for the Thelon esker extending westward from Dubawnt Lake in the District of Keewatin (Craig, 1964).

Theories of origin and implications for sedimentation

Beyond the general acceptance of eskers as ice-contact glaciofluvial features, little consensus exists with respect to theories of origin. In view of observations of modern eskers and variation in morphologic and sedimentologic types in Pleistocene eskers, it seems certain that environments of esker deposition varied considerably in detail.

Most theories can be grouped according to their principal concern: (a) *Formation and maintenance of the conduit*—the process by which meltwater was localized in specific channels; (b) *Position of esker sediments with respect to the glacier*—whether the esker stream occupied a subglacial, englacial, or supraglacial position; (c) *Nature of the conduit*—whether the esker stream flowed in an open channel or in a tunnel flowing full; and (d) *Site of deposition* —whether deposition took place inside the conduit before reaching the glacier terminus, or at the glacier terminus where flow conditions probably changed radically.

Formation and maintenance of the conduit

A few proposals have been made regarding the process by which conduits for meltwater originated and were maintained. Carey and Ahmad (1961) suggested that the water seepage pressure gradient in the saturated till at the base of a glacier may have been sufficient near the terminus to open a conduit in the till by piping. Once formed, this conduit would be maintained and enlarged by erosion and could form a subglacial tunnel localizing esker formation.

In Finland, Hyyppä (1954) and Härme (1961) have related esker trends to fault lines. They proposed that reactivation of old faults, perhaps accompanying isostatic rebound, resulted in lines of weakness in the glacier that then became loci for meltwater flow. It is possible, however, that fault lines are coincident with valleys and that the basic control on esker location is topographic.

Two important papers have recently discussed the flow of water within glaciers (Röthlisberger, 1972, and Shreve, 1972). Röthlisberger showed that: (a) subglacial water migrates down pressure gradients toward main flow arteries where velocity is greater. Addition of this water results in further concentration of water in main arteries; (b) a conduit will rapidly (days) increase its diameter in response to increased seasonal discharge, but closure of the conduit due to creep of ice during seasonally low discharge has a much longer time constant (months or years); and (c) a conduit may be straight, meandering, or even 'braided' depending upon the hydraulic gradient. Shreve (1972) argued that: (a) the stable mode of subglacial water movement is in tunnels; and (b) subglacial formation of eskers is favored by large discharge (i.e. widespread ablation) and low glacier-surface gradient.

Position of esker sediments with respect to the glacier

Subglacial, englacial, and supraglacial locations of esker streams have all been reported from modern environments. It is also probable that some eskers started to accumulate in tunnels and, after thinning and collapse of the ice roof, final sedimentation took place subaerially. Criteria that could be applied to determination of the position of the esker sediments with respect to the glacier include:

(a) Bedrock valleys

In some cases, drilling and seismic data indicate that eskers are located along the axes of buried bedrock valleys (Hobson and Lee, 1967; Hobson and Maxwell, 1968). Elsewhere bedrock surfaces, on which the eskers directly lie, have been scoured free of till for several tens of meters on either side of the esker ridge. In these situations it seems clear that the esker stream flowed at the base of the glacier and that erosion by water contributed sediment to the esker stream.

(b) Basal till overlying esker sediment

Where a basal till occurs at the highest stratigraphic level in an esker and where it can be shown that the till belongs to the same glacial episode as the esker, it is evident that the esker formed in an englacial or subglacial position;

(c) Deformed sequences

If deformation of an entire sequence can be related to an underlying ice mass that formed an integral part of the glacier (see McDonald and Shilts, this vol.), then that portion of the esker initially would have occupied either an englacial or supraglacial position. Price (1966, p. 123) has reported good evidence to show that as ice melts out from beneath esker gravel ridges, a ridge form can be preserved. The present authors, however, concur with Price that it is difficult to conceive of a sedimentary sequence being let down an appreciable distance, as underlying ice melted, without intense disruption of primary sedimentary structures; and

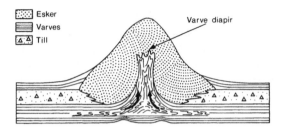

Esker
Varves
Till
Varve diapir

Figure 1 Diapiric intrusion of varved silt and clay upward into core of esker.

(d) Diapiric intrusion

In two eskers from southeastern Quebec diapirs of glacial-lake silt and clay have been extruded vertically upward several meters into the esker cores (fig. 1). The diapirs are tabular and can be traced for several tens of meters parallel to the esker axis. The primary laminations are vertical. The diapirs have been interpreted as the result of differential downward pressure exerted by the weight of the ice along the flanks of the eskers, causing lateral flow of the plastic clays and their extrusion upward into the conduit. Such esker streams were evidently flowing at the base of a glacier that had advanced over varved sediment.

Nature of the conduit

Stream sediments have been observed in channels with and without ice roofs in modern glaciers. Hydraulically, a tunnel not flowing full is simply an open channel. We will consider a "tunnel," therefore, to refer to a closed conduit flowing full, and an open channel to be distinguished by a free upper surface having air in contact with the water.

The observation that many eskers increase in altitude downstream, at least locally, led early workers to the hypothesis that flow in a tunnel under a hydrostatic head permitted the esker stream to flow up an adverse slope. Two assumptions underlie this hypothesis, both of which are subject to testing in the sedimentary sequence: (a) that the sediments were not deposited over ice and subsequently let down as the ice melted; such a lowering of the sediment pile would produce intense deformation in the sequence; and (b) that the esker is not significantly time transgressive, that is, that the sediments were not deposited at progressively later times in an upstream direction (fig. 2); this could be verified by detailed facies studies. Figure 2 depicts a series of subaqueous fans formed at the mouth of the channel (whether open, or a tunnel). When a number of such fans are built at the terminus of a retreating ice front they may be shingled on top of one another forming a time-transgressive linear ridge that rises, but is progressively older, in a downstream direction.

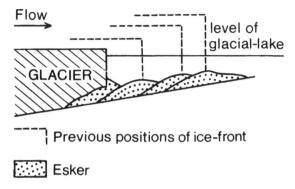

Figure 2 Deposition of esker sediment as a time-transgressive series of subaqueous fans not requiring subglacial flow up an adverse slope.

Lateral facies relationships that indicate overbank flooding from a central channel (Saunderson and Jopling, 1970; Shaw, 1972, p. 34) would support an open-channel flow model, as would the occurrence of such subaerial features as mud cracks. A greater abundance of silty clay lenses could be the product of frequent avulsion of subaerial channels resulting in backwater areas in which fines could settle.

Shaw (1972) concluded that three eskers in England had been deposited in open channels. He based his conclusions principally on the fact that a lateral fining of the fluvial sediment occurred about a central coarse zone. He observed that in contrast to nonglacial fluvial models, the vertical sequence indicated little lateral migration of the main channel, a fact that he attributed to low sinuosity of the esker streams. However, the occurrence of faults adjacent to the central coarse zone raises the possibility that the eskers have a complex origin, perhaps originating in subglacial tunnels that, after collapse or melting of the roof, had a final open-channel phase.

Sedimentary structures can complement other lines of evidence to distinguish tunnel from open-channel environments. Subglacial tunnels flowing full are gigantic natural pipes in which bed forms that depend on surface water waves cannot be stable. Such free-surface waves are necessary for the formation of antidunes (Kennedy, 1963), so the formation of antidune bedding would require an open channel. An exception could be made in the case of a density underflow that could sustain waves at the interface between the two fluids; it would be necessary to draw the distinction on the basis of sedimentary structures and facies relationships. In two localities in the Peterborough esker, for example, cross beds 1 to 2 m thick occur in gravelly coarse sand and dip consistently in a direction opposite to the general current flow (fig. 3). These have been interpreted as backset beds associated with antidunes and, by inference, with open-channel flow.

Figure 3 Backset bedding 1.5 m thick in Peterborough esker (locality 14, fig. 20). Flow direction was from right to left as indicated by underlying ripple-laminated sand, several paleocurrent measurements in other facies at this site, and its orientation with respect to the esker trend. Shovel is 50 cm long. (GSC 202287-A).

Experimental study of bed forms and sedimentary structures in pipes flowing full and up an adverse slope indicate that, as bed shear stress is increased over a sand bed, ripples give way to dunes, then to a plane bed, and then erosion of the entire bed results in sediment being transported in a heterogeneous suspension (McDonald and Vincent, 1972). An antidune bed mode was not observed. Dune height was observed to be $^1/_3$ to 1 times the hydraulic mean depth (cross-sectional flow area/wetted perimeter). Thus it seems possible to use dune height as a guide to water depth in pipe-flow systems, just as has been suggested for open channels (Allen, 1963). McDonald and Vincent (1972, p. 21) show that without making too serious assumptions regarding how much of an idealized cross-section is filled with sediment, or regarding the position of an observation with respect to the highest point in a circular-shaped roof, actual water depth would be 2 to 3 times the hydraulic mean depth. Thus, as a guide, actual water depth would be about 2 to 10 times dune height. Deposition in a tunnel could be inferred by calculation of a stream-flow depth that would indicate a stream surface significantly below that of an ambient water level imposed by an associated glacial lake.

Site of deposition

Sediment-laden water flowing through a conduit in a glacier can deposit sediment: (a) far back within the conduit; (b) subaqueously, that is, into a standing water body at the ice front; or (c) subaerially at the glacier terminus. Without respect to whether flow occurred in an open channel or in a tunnel, some conclusions regarding the site of deposition can be made from examination of the sediments.

Deposits within the conduit are supported laterally by the glacier. As the glacier melts, these supports are removed and the flanks of the esker collapse to a stable angle of repose (McDonald and Shilts, this vol.). This abrupt side topography with internal evidence of failure, characteristic of most eskers, is evidence that deposition took place in the conduit. The occurrence of flowtill or exceptional large blocks in the esker sequence also may result from the presence of glacier ice along the flanks of the esker stream. Sediment facies should be relatively persistent in longitudinal section.

Eskers can also be deposited right at the ice front where they owe their linear-ridge form to progressive retreat of the ice front and to the esker stream having been laterally constrained by glacier walls just upstream from the terminus. Discharge of the esker stream into a standing body of water would result in sudden decrease of competence with deposition of most of the sediment load just beyond the ice front. The sides of this deposit initially would be at a relatively stable angle and post-depositional collapse features would not be expected there. In transverse section, bedding surfaces should be roughly parallel to the topographic surface. From proximal to distal location in longitudinal section, the sediments rapidly should become finer grained, reflecting the rapid decrease of competence.

Subaerial discharge at the ice front would involve only removal of the constraining walls. Flow would expand into a conventional outwash stream system, the "esker" having been reduced simply to the highest point in the cross-profile of the outwash train. This point might also be marked by deposition of coarser components of the sediment load due to reduction of bed shear stress as flow expanded and shallowed at the terminus.

Nature of the conduit and sites of deposition can be combined into various environmental types of eskers as illustrated in Figure 4. It is apparent from the figure that there are three basic situations in esker sedimentation. When deposition takes place within the conduit the environment can either be (1) an open channel, or (2) a tunnel. When deposition takes place

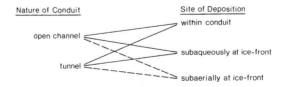

Figure 4 Models of esker sedimentation, based on nature of conduit and site of deposition. Dashed lines indicate poorly developed feature and improbable recognition of it as an esker.

subaqueously at ice front the environment is (3) a delta, irrespective of the nature of the conduit. The fourth situation "subaerial deposition at ice front" is probably difficult to recognize as an esker.

These three basic models: *open channel* (fluvial), *tunnel* and *deltaic* provide the fundamental framework of reference for discussion of esker sedimentation. Each of these models is characterized by typical facies arrays, sedimentary structure assemblages, and paleocurrent patterns. The basic aim of this paper is to analyze data provided by field studies of two eskers in Canada in terms of these proposed models.

Field examples

Facies types

Facies in the Peterborough and Windsor eskers have been differentiated on the basis of textures and sedimentary structures. Because most of the facies are common to both eskers, it is convenient to describe them before more specific discussion of facies relationships in the particular eskers. Facies types, and their symbols used in diagrams in the text, are summarized in Figure 5.

Gravel

The most abundant facies are sandy pebble and cobble gravel and pebbly sand. Commonly the gravel has a very coarse sand matrix and a stable framework, that is, coarser clasts touch and support each other. In some localities, however, open-work pebble gravel has been observed to grade laterally over 2 m within one bed to pebble gravel with a silty sand matrix and then to a pebble gravel with a disrupted framework.

Cross-bedding is common in gravel facies. Set thicknesses range up to 7 m (fig. 6). Commonly pebble gravel beds that at first glance appear massive or "structureless," exhibit a vague but definite cross-bedding with set thicknesses of 50 to 60 cm. These probably record downstream progradation of bar fronts. Backset bedding, interpreted as being related to antidunes, was observed in sandy pebble gravel at two sites in the Peterborough esker (fig. 3). Massive gravel, and gravel with vague bedding chaotically deformed and slumped (fig. 7), is present at the proximal ends of beads. Gravel that is located stratigraphically near the base of the esker sequence is commonly parallel bedded and poorly sorted. Thick, parallel-bedded gravel units have been observed that are graded and become more distinctly stratified upward in sequence (fig. 8). The presence of a poorly sorted fine matrix suggests that these units may have been deposited rapidly from a slurry, perhaps generated by a subaqueous slump a short distance upslope.

382

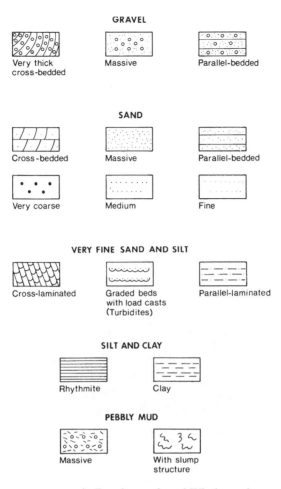

GRAVEL

Very thick cross-bedded

Massive

Parallel-bedded

SAND

Cross-bedded

Massive

Parallel-bedded

Very coarse

Medium

Fine

VERY FINE SAND AND SILT

Cross-laminated

Graded beds with load casts (Turbidites)

Parallel-laminated

SILT AND CLAY

Rhythmite

Clay

PEBBLY MUD

Massive

With slump structure

Figure 5 Facies types present in Peterborough and Windsor eskers.

Figure 6 Very thickly cross-bedded gravel, Peterborough esker. Regressive ripples were observed at the toes of these foresets. (GSC 202287-B).

383

Figure 7 Gravel with vague bedding intensely deformed near proximal end of bead in Windsor esker, locality 5. Flow right to left. (GSC 202287-C).

Figure 8 Gravel showing faint parallel bedding and overall grading, Peterborough esker. Shovel is 50 cm long. (GSC 146968).

Sand

Sand occurs in a variety of grain size and sedimentary structure combinations (fig. 5). Cross-bedded sand is an abundant facies and is produced either by dune trains on the bed or by individual prograding sandbar fronts. These commonly consist of downstream dipping foresets of medium to coarse sand with the coarsest material concentrated near the toe (fig. 9). Set thicknesses average 15 to 30 cm; usually the tops of the foresets have been truncated before deposition of the overlying set. The distinction between a dune and a bar front can be made at some localities by the use of related facies. Migration of dunes with superimposed ripples accompanied by appreciable aggradation of the bed produces sets of dune cross-bedded sand

Figure 9 Sequence of cross-bedded and parallel-bedded medium to coarse sand and ripple-laminated fine to very fine sand, Windsor esker. (GSC 202287-D).

Figure 10 Sets of cross-bedded sand separated by thin units of ripple-laminated sand indicating forward migration of dunes with superimposed ripples, Windsor esker. (GSC 202287-E).

separated by thin units of ripple-laminated sand (fig. 10). Bar fronts, on the other hand, may show a much more complex array of associated bed forms. Aario (1972b, his fig. 1) has exhumed the actual bedding surface for several meters in the vicinity of such a bar front and has produced admirable evidence to explain facies arrays that the present authors have observed in both the Peterborough and Windsor eskers. In a downstream direction the succession of bed forms and structures was: (a) dunes and/or ripples build

385

Figure 11 "Massive" or "structureless" sand unit deposited almost entirely directly from suspension, Windsor esker. (GSC 202287-F).

forward on the bed and deliver sand to the bar front leaving sets of cross-stratified sand on the bar top; (b) foresets of the bar front are straight or slightly concave upwards and occur, in the Windsor esker, in sets as thick as 2.2 m; (c) regressive ripples result from a backflow in the zone of flow separation and usually produce type B ripple lamination (Jopling and Walker, 1968) in the toesets; (d) the zone of zero velocity where the flow re-attaches to the bed produces a zone of polygonal nondirectional ripples. In section, these ripples have sharp crests and rounded troughs; as new sediment is deposited over these, sinusoidal ripples can be produced (the present authors have also verified this by exhumation of the bedding surface); and finally, (e) progressive ripples that migrate away from the bar front and produce type A or B ripple lamination.

"Massive" or "structureless" well sorted coarse, medium, or fine sand units (fig. 11) are also common in beads of the Windsor esker. Although not strictly "structureless," the absence of well defined lamination indicates the absence of a significant traction carpet. In most localities the units show a very vague parallel bedding. In some localities ripple lamination showing no stoss-side preservation passed upward into climbing ripples then faded out upwards into such a "structureless" sand unit. The angles of climb increased upwards to as much as 47 degrees before the distinct ripple lamination disappeared. This sequence is interpreted as recording an increase in the proportion of sediment falling directly from suspension onto the bed relative to the amount contributed by the traction load.

Parallel-bedded and -laminated coarse and medium sand (fig. 9) occurs associated with dunes and ripples. In most cases it appears from the associated facies to record a flat-bed condition in the lower part of the lower flow regime where the sediment was too coarse to permit development of ripples on the bed (Williams, 1967).

Figure 12 Ripple-laminated fine sand, Windsor esker. A climbing-ripple sequence (ripples have been exhumed) has been interrupted by the delivery of coarse sand that is parallel-laminated. (GSC 202287-G).

Figure 13 Ripple-laminated very fine sand showing gradual increase in stoss-side preservation upward in the sequence, Peterborough esker. (GSC 202287-H).

Very fine sand and silt

Fine and very fine sand admixed with silt is commonly ripple-laminated and is widespread in both the Peterborough and Windsor eskers (fig. 12). The degree of stoss-side preservation changes with the proportion of sediment falling from suspension. Commonly the transition is from type A upwards to type B (fig. 13) and results from the approach of a larger bedform. Regressive ripples (fig. 14) occur in the zone of flow separation downstream from the crests of larger bedforms and record a backflow in this zone. They are common downstream from bar fronts.

Many "hydroplastic" deformational structures are associated with very fine sand and silt units. Graded sand beds that overlie these fine-grained

Figure 14 Regressive ripples formed in the zone of backflow at the toe of a large tongue of disrupted-framework gravel in a bead of the Windsor esker, locality 6. General flow direction in the bead was right to left. (GSC 202287-I).

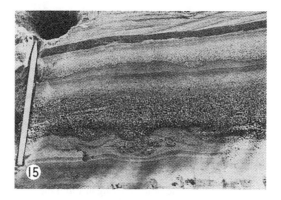

Figure 15 Graded sand bed with load structures at base, bounded by parallel-laminated very fine sand and silt, Windsor esker. (GSC 202287-J).

units commonly exhibit load structures at their bases (fig. 15). Where upward injection of the fluidized fines has kept pace with aggradation of the bed the tips of the diapirs are turned over in the direction of flow and the resulting flame structures are reliable paleocurrent indicators. Increase in bed velocity can create an oversteepened condition in fine material and produce downslope failure involving several laminations (figs. 16, 17).

Silt and clay

Rhythmically laminated sandy silt and clay (fig. 18) represent the most distal facies related to those eskers that are deposited in association with glacial

Figure 16 Slump structure in sandy silt, Peterborough esker. Convolute laminations are present in sand immediately upstream (to right). Ruler is 15 cm long. (GSC 202287-K).

Figure 17 Slump structure in fine sand, locality 10, Peterborough esker. This unit was traced 50 m downstream (leftward) into a diamictic varve. (GSC 202287-L).

lakes. In both the Peterborough and Windsor eskers, coarsegrained esker facies have been traced directly into rhythmite sequences.

Pebbly mud

Pebbles floating in a muddy matrix (fig. 19) form a distinctive, though not common, facies. All examples of this facies type observed in the Peterborough and Windsor eskers contain moderately well rounded pebbles that are not striated. Also they occur in tabular strata-bound units. This suggests that they are the product of a mudflow process that incorporated stream-worked pebbles, and that they are not till units.

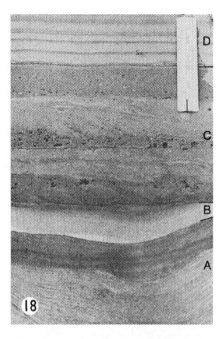

Figure 18 Sequence of fine-grained sediments, Windsor esker. (A) Very fine sand and silt with sinusoidal ripple lamination. (B) Graded bed of fine sand, probably a turbidite. (C) Three mud layers with floating clasts. (D) Graded, rhythmically laminated clay/silt couplets. Ruler is 15 cm long. (GSC 202287-M).

Figure 19 Pebbly mud in Peterborough esker. Shovel is 50 cm long. (GSC 202287-N).

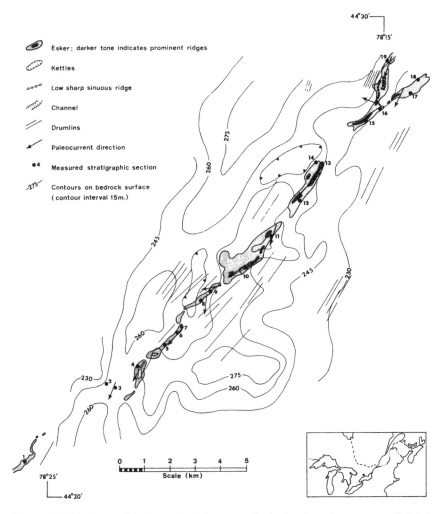

Figure 20 Peterborough esker and contours on the bedrock surface; unconsolidated sediments overlying the bedrock adjacent to the esker average about 10 m thick.

Peterborough esker

The Peterborough esker (fig. 20) lies north of Lake Ontario and just north of the town of Peterborough. It is approximately 25 km long, including numerous gaps, and locally is as much as 1 km wide. The esker is parallel to numerous well developed drumlins. Superimposed on the main broad ridge is a discontinuous prominent ridge and, in places, lower sharp sinuous ridges. The crest of the esker rises in a downstream direction from about

391

230 m a.s.l. to 275 m. Maximum relief of the ridge is 20 m. Numerous kettles and the occurrence of faults in the ridge sediments attest to its ice-contact origin.

Bedrock topography and thickness of the unconsolidated overlying sediments have been studied by Gagné and Hobson (1970). Ordovician limestone lies within. 25 m of the ground surface in the vicinity of the esker and is exposed within 1 km of the esker on either side. The position of the esker appears to be unrelated to the subjacent bedrock topography, rather it trends indiscriminately across bedrock valleys and ridges (fig. 20).

A large number of gravel pits provide well exposed sections through esker sediment. From these measured cross-sections a detailed picture of the stratigraphy has emerged. In the larger pits exposures continuous for 1 km permitted the tracing of individual beds. Sedimentary structures and facies relationships are summarized in Table 2.

Cross stratification is common at all scales up to 7 m (figs. 3, 6, 13). In most cases it represents the building forward of a deltaic front, bar front, or dunes. At localities 13 and 14, however, cross-beds of relatively poorly sorted gravel (fig. 3) are inclined in a direction opposite to the prevailing paleocurrent and have been interpreted as backset beds related to antidunes on the bed. This, in turn, provides evidence for open-channel flow since free-surface waves are necessary for their formation.

The most common facies array in the Peterborough esker involves deltaic, or gravel-bar, fronts and the facies associated with them. Figure 21 depicts a sequence of four vertically stacked successions produced by this environment as the bed aggraded. Very thick crossbeds mark the bar fronts and cross-bedded sands on top record dune trains in the shallower water there. Downstream the facies change first to regressively then progressively ripple-laminated fine sand, then graded beds with load casts possibly indicating that subaqueous slumps on the bar front generated turbidity flows, and finally to rhythmically laminated silt and clay of a lacustrine environment.

The downstream relationship of relatively coarse-grained esker sediments with rhythmically laminated very fine sand, silt and clay illustrates the important genetic relationship between glaciofluvial and glaciolacustrine sediments. The development of a turbidity current from a subaqueous slump at the delta front is vividly depicted by a single layer which changed character within a short distance (figs. 17, 22A and B). At locality 10, across 50 m in a downcurrent direction, a slumped silty sand layer (slump) passed in a downcurrent direction into a pebbly mud with clasts of silt and very fine sand (mudflow) and then into a graded very fine sand layer with silt clasts (turbidity current) which, in turn, formed part of a diamictic rhythmite sequence.

A general trend to finer grained sediments higher in the section was pronounced and is attributed to retreat of the glacier which resulted in progressively more distal and finer grained sediments at a particular site.

Table 2 Sedimentary structures in the Peterborough esker.

Structure	Morphology	Grain size	Facies relationships
1. Parallel bed	Thickness: 20 to 40 cm Form: Planar Tabular	Gravel to coarse sand	Interbedded with 2 or 3
2. Very thick crossbedding	Average thickness: 234 cm Form: Planar tabular	Gravel to coarse sand	Grades downstream to 4 to 5 Overlain by 3
3. Thick cross-bedding	Average thickness: 28 cm Form: Simple or planar tabular	Coarse to medium sand	Grades downstream to 5 Overlain by 3
4. Cross lamination	Average thickness: 3.5 cm Form: Ripple drift	Fine to very fine sand	Grades upstream to 2
5. Graded beds	Average thickness: 11 cm Form: Wedge-shaped, thins downstream, load structure at bottom	Very fine sand to silt	Grades upstream to 2 or 3 Grades downstream to 6
6. Rhythmic laminae	Average thickness: 3.0 cm Form: Tabular	Silt and clay	Grades upstream to 5

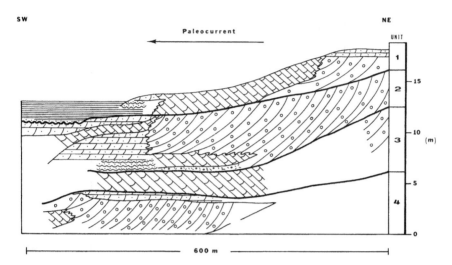

Figure 21 Four vertically stacked successions of bar-front sediments and associated downstream facies in the Peterborough esker, locality 10. Downcurrent facies change to finer sediments in conspicuous. For explanation of symbols see Figure 5.

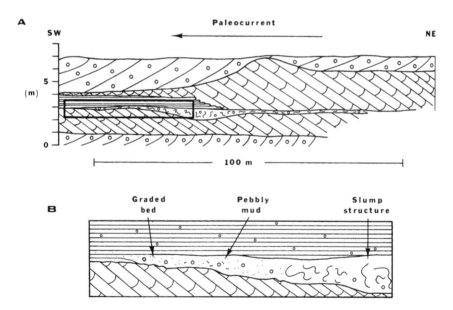

Figure 22 Change of facies in a single layer showing the transformation of a slump into a mudflow deposit and then to a graded rhythmite where final transport was by turbidity current; locality 10, Peterborough esker. For explanation of symbols see Figure 5. A, Stratigraphic framework. B, Enlargement of facies from area outlined in A.

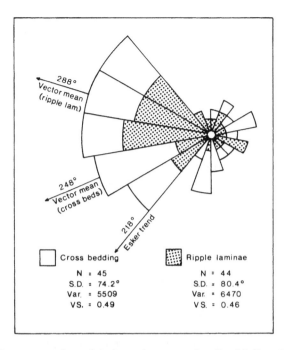

Figure 23 Paleocurrents from deltaic environment, locality 10, Peterborough esker. (N = number of observations; S.D. = standard deviation; Var. = variance; V.S. = vector strength.)

Paleocurrents in Peterborough esker

Paleocurrent directions shown on Figure 20 are vector means for each locality. They represent a total of more than 500 measurements of cross-strata dips. The overall vector mean is S 42°W, i.e. parallel with the esker trend; overall variance (based on the vector mean) is 4,723 and overall vector strength is 0.51. Variability of current direction within individual segments is commonly still higher. For example, the deltaic units depicted in Figure 21 have a variance and vector strength respectively of 5,509 and 0.49 for cross beds and 6,470 and 0.46 for ripple laminations (fig. 23). This variability is rather high for a unidirectional flow in a relatively short segment of a straight conduit but can be explained as being due to fanning out of flow at the delta head.

Environments of deposition in the Peterborough esker

From the above observations it is apparent that the Peterborough esker was deposited in an open channel which formed a delta at the ice front where it discharged into a lake. The facies arrangement in the esker is a strong evidence in support of the partly deltaic origin. Higher variance in the paleocurrent data may also be a result of deltaic mode of deposition caused by fanning out of flow at the channel mouth. Subaqueous slumps, mudflows, and turbidity currents are more likely to occur at the delta front. Therefore, the above features of the Peterborough esker might be said to characterize a deltaic model of esker sedimentation.

Windsor esker

The Windsor esker, 14 km in length, is situated near the bottom of the St. Francis River valley at the town of Windsor, Quebec (fig. 24). The esker was formed during final deglaciation of the Quebec Appalachians as the late Wisconsin ice front back-wasted northwestward down the gradient of the St. Francis River (McDonald, 1968). A glacial lake was impounded against the ice front during this retreat. In the vicinity of Windsor the depth of water at the ice front was 105 m.

The Windsor esker displays three different morphologic types: (a) Between localities 1 and 10 there are 16 beads, and localities 1 to 15 are mostly situated in beads. These beads are isolated from each other, each is 10 to 20 m high, and they have a regular spacing of 285 m; (b) From localities 16 to 22 and from 27 to 28 the esker consists of a fairly continuous single, steep-sided ridge, 10 to 20 m high, with sharp crest and uniform crest altitude; and (c) localities 23 to 26 are situated in a complex double ridge. No stream-deposited outwash terraces are present in the valley, rather, the esker is flanked by till and by very fine sand and silt deposited in the glacial lake.

Numerous bedrock exposures lie in the valley bottom and in the valley sides so it is clear that the Windsor esker lies in a bedrock valley. At locality 16, esker sediments lie directly upon slate; elsewhere bedrock is judged to lie within about 10 m below the base of examined esker sections.

Stratigraphy and facies relationships in the esker have been examined at 28 gravel-pit localities. All facies, summarized in Figure 5, are present in the esker. Graded beds and clay/silt rhythmites are much less common than in the Peterborough esker, but pebbly mud occurs much more frequently. "Structureless" sand units, not observed in the Peterborough esker, are common in beads of the Windsor esker. Because morphology of the esker is controlled largely by the shape of the primary depositional units, facies relationships will be discussed in terms of the three morphologic types.

Sediments within beads

The sediments in beads show a very rapid downstream facies change. Massive, cross-bedded, or parallel-bedded proximal gravel passes downstream by way of interfingering into a section of fine- to medium-grained sand that is either "structureless," or that shows climbing progressive ripples. At the

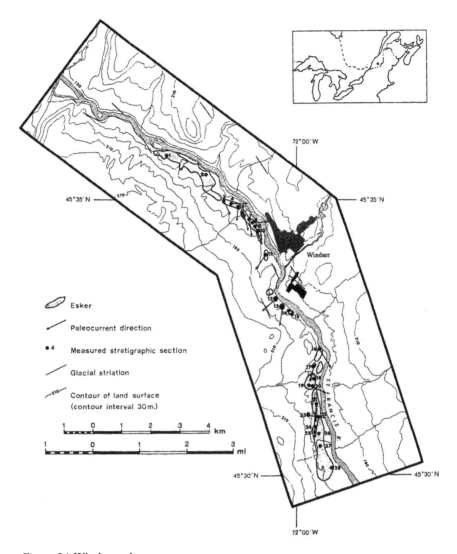

Figure 24 Windsor esker.

most distal portion of the bead very fine sand and silt rhythmites intertongue with and overlie these units. Thus the whole spectrum of the deposit illustrates rapid facies change at the mouth of the conduit with glaciofluvial sediments at one end and glaciolacustrine at the other, typifying a *deltaic model* of sedimentation.

Both longitudinal and transverse sections through a typical bead were exposed at locality 6 (fig. 25). Eighteen meters of vertical section at the upstream end of the bead are composed largely of "massive" to

397

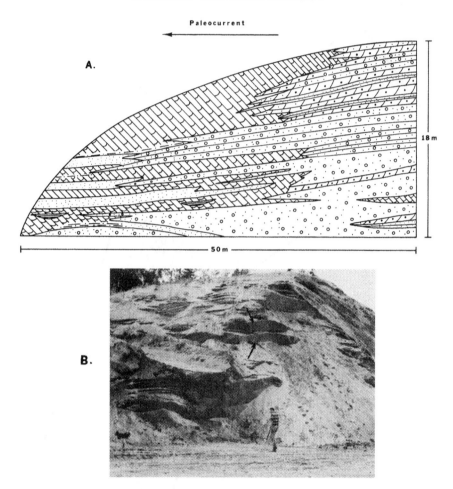

Figure 25 A, Longitudinal section through a typical bead in Windsor esker, locality 6. For explanation of symbols see Figure 5. B, Transverse section through bead at locality 6, Windsor esker. Prominent beds (see arrows) are "structureless" sand units. (GSC 202287-O).

parallel-bedded and cross-bedded pebble and cobble gravel. Proximal gravel facies are characteristically intensely deformed and faulted (fig. 7). Proximal gravel units are characterized by upstream dips of *ab* planes of pebbles. Within 25 m in a downcurrent direction this coarse-grained section passes, by way of large-scale intertonguing subunits, into a section of climbing ripples and "structureless" units of fine and medium sand. Scour-and-fill structures, regressive ripples (fig. 14), slump structures, pebbly muds, graded units, and rare frigites[1] occur near the downstream ends of large gravel tongues. Numerous "structureless" sand units, as thick as 75 cm, occur near

398

the distal end (fig. 25B). Load casts at their bases attest to their rapid deposition, probably by turbidity current. Very fine sand and silt rhythmites intertongue with and overlie the most distal portion of the bead. A transverse section through the centre of the bead at locality 9 is shown by McDonald and Shilts (this vol., their fig. 5). Collapse features are not common on the sides of beads. Rather major normal faults are commonly situated near the crest and strike parallel to the esker trend.

Sediments within single steep-sided ridges

In contrast to beads, single steep-sided ridges consist essentially of alternate tabular units of cross-bedded and parallel-bedded gravel and sand. Individual units persist for several tens of meters in longitudinal section with little change in thickness. In transverse section the subhorizontal units show only minor interfingering and are commonly truncated by high-angle reverse faults on the flank of the esker (McDonald and Shilts, this vol., their fig. 4 is from locality 21, Windsor esker). Scour-and-fill structures are not prominent. At most localities there is a tendency for the section to become finer grained upwards.

At locality 21 is a particularly good longitudinal exposure 60 m long (fig. 26A) through the core of a single steep-sided ridge that has a total continuous length of 640 m. The ridge consists of cyclical sedimentation units (four cycles exposed) that are persistent with little change in texture, structure, or thickness for at least 60 m in a downcurrent direction. Each cycle is characterized by cobble or pebble gravel at the base passing upward into sand or pebbly sand. Thickness of the cycles varies from 3.6 to 5.9 m. Average downcurrent dip of general bedding contacts is 8 degrees (N = 12). Cross-beds (fig. 26B) occur in tabular sets; straight foresets become tangential at the base and commonly are truncated at the top, although stoss-side preservation has been observed. Set thicknesses in sand subunits range from 17 to 53 cm and average 28 cm (N = 14); set thicknesses in gravel subunits range from 5 to 120 cm and average 40 cm (N = 25). The intermediate axes of ten rounded cobbles near the bases of units Ia, IIa, IIIa, and IVa averaged 11.2, 10.6, 14.3, and 7.6 cm, respectively. Pebbles in gravel foresets have long axes parallel to the flow direction and have their *ab* planes in the plane of the foreset. A thin (<50 cm) veneer of lacustrine sand mantles the ridge.

Sediments within complex double ridge

Double-ridge portions of the esker appear to be considerably more complex than do either single ridges or beads, and exposures were not extensive enough to unravel the complexity. Neither cycles nor broad textural changes were discernible. Lateral and vertical facies changes occur abruptly through both interfingering and gradation. Intensely deformed clay/silt rhythmites

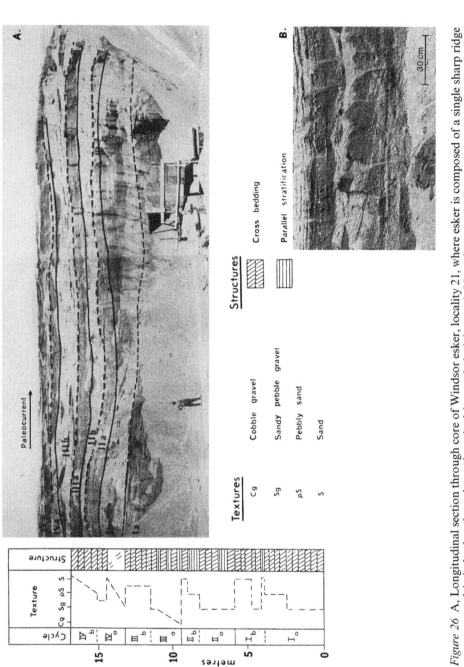

Figure 26 A, Longitudinal section through core of Windsor esker, locality 21, where esker is composed of a single sharp ridge and is judged to have been deposited in a subglacial tunnel. Note four vertically stacked sedimentation cycles. B, Parallel-bedded and cross-bedded pebbly sand from unit IIa, locality 21. (GSC 202287-P).

have locally been introduced from beneath up into the esker sequence. Paleocurrent measurements (locality 23, fig. 24) locally indicate flow at a high angle to the esker trend, suggesting that flow may have occurred between two adjacent meltwater arteries.

At locality 26, more than 10 m of undeformed parallel-bedded and cross-bedded cobble gravel is abruptly overlain by 5.3 m of silty very fine to medium sand that has been so intensely folded and faulted that only rarely can primary sedimentary structures be recognized, although gross subunits can be traced laterally for several meters. This is the only site observed in the Windsor esker where collapse of an ice roof with consequent lowering of superimposed lacustrine sediments seems to have affected the esker sediments.

Paleocurrents in Windsor esker

Paleocurrents were determined from measurements of azimuths of cross-beds and ripple laminations and lee directions of ripple trains. Because of their contrasting characters, beads and ridges were treated separately. Approximately 50 measurements were taken from each bead and each ridge. An attempt was made to collect data equally from all units exposed in the sequence. The data from a typical ridge and a typical bead are shown in Figure 27. In all cases vector means of paleocurrents lie close to the trend of the esker (fig. 24). Significantly, the beads and ridges differ markedly in their measures of paleocurrent variability. The variability values do not overlap. This difference between beads and ridges could facilitate environmental discrimination between a tunnel and a deltaic model as described later.

Paleocurrents from single ridges have variances (based on vector mean) of between 1,000 and 2,000 and vector strengths of about 0.80. The variability measures shown in Figure 27A were calculated separately for the cross-bed data there; the values were identical with those for the entire sample.

Paleocurrents from beads have variances generally between 3,000 and 4,000 but may go above 6,000. Vector strengths are less than 0.65. The breakdown of variability measures for particular sedimentary structures at locality 6 (fig. 27B) were, for general bedding, ripple laminations, and cross-bedding, respectively: variance = 4,062, 1,257 and 5,845; vector strength = 0.59, 0.83, and 0.43. Ripple laminations were measured mostly in the upper 15 percent of the section which may account for their low variability.

Paleocurrent measurements from one bead, locality 9, were plotted according to their stratigraphic position in the bead. This revealed that flow initially was deflected to the right of the mouth of the conduit and, after a significant sediment pile had accumulated there, flow was deflected toward the left. In addition to providing details of local sedimentation history, this points out that changing current systems must be recognized in an adequate sampling design.

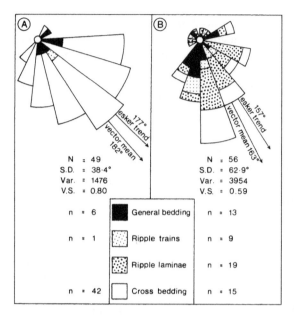

Figure 27 A, Paleocurrents from single ridge (tunnel) environment, locality 21, Windsor esker. B, Paleocurrents from bead (deltaic) environment, locality 6, Windsor esker. (N, n = number of observations; S.D. = standard deviation; Var. = variance; V.S. = vector strength.)

Environments of deposition in the Windsor esker

Sediments of the Windsor esker appear to have accumulated in two basically different environments of deposition: (a) at the points where flow in subglacial tunnels debouched into the glacial lake at the ice front; and (b) within subglacial tunnels.

Beads are clearly deposited in an environment characterized by rapid deceleration of flow. The proximal end of a bead is deposited in contact with ice and the distal end interfingers with lake-bottom sediments. Cross-bedded facies occur only at the proximal ends of the beads where they indicate the former presence of bars, and also possibly dunes, in a zone of rapidly varied flow with diverging flow lines. Preservation of even the most delicate sedimentary structures indicates that beads accumulated on the lake bottom and at the base of the glacier. A representation of this environment is shown in Figure 28. The relatively large variability of paleocurrent measurements results from expansion of flow as the lateral constraint of the tunnel walls is removed. Each bead represents a complete sedimentologic system, thus preventing correlation of sedimentation units from bead to bead.

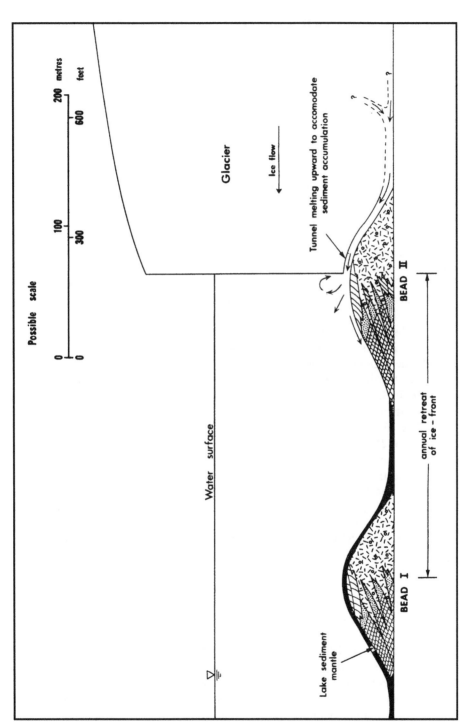

Figure 28 Generalized model of esker-bead formation.

These beads are subaqueous fans, or deltas, similar to the "ose centra" or "submarginal deltas" of De Geer (1940) who considered them to represent annual depositional events at the ice front. An annual periodicity for the beads near Windsor is also probable. The ice front retreated from the International border to the St. Lawrence Lowland, a distance of 100 km, in 400 C^{14}-years (Gadd and others, 1973). This gives an average retreat rate of 250 meters per year. The average spacing of thirteen successive beads in the Windsor esker is 285 m.

A subglacial-tunnel model alone fits the evidence from the single steep-sided ridges and, to some extent, from the complex double ridge. The glacial lake stood 105 m deep against the ice front, yet only at locality 26 is there evidence, albeit restricted to the upper 5.3 m of section, for pervasive deformation in the sediments. Elsewhere, even the most delicate primary sedimentary structures are preserved. This, and the presence of varve diapirs in sediments of the double ridge indicates that the esker stream flowed at the base of the glacier. The downstream persistence of high stream competence is evident from sediments at locality 21 (fig. 26A). In addition, low variability of paleocurrent indicators would reflect a unidirectional flow laterally constrained by ice walls of the tunnel. Faults in the flanks of these ridges record subsequent melting of these ice walls.

Several interesting questions arise, such as the depth of flow in the tunnel, the significance of the sedimentary cycles exposed at locality 21 (fig. 26A), the distance back from the ice front that escape of meltwater was localized in basal tunnels, and the manner in which these basal tunnels extended themselves headward as the ice front wasted back.

Evidence from pipe-flow experiments has been cited from McDonald and Vincent (1972) to show that actual water depth in the tunnel would be 2 to 10 times dune height. At locality 21, bed-form heights of 40 cm throughout a vertical section 16 m high indicate flow depths of 1 to 4 m. Water depth in the lake at the terminus was 105 m. This is strong corroborative evidence for flow having occurred in a tunnel. The absence of scour-and-fill structures indicates a relatively constant aggradation. Alternation of cross-bedding and parallel-bedding indicate (McDonald and Vincent, 1972) that hydraulic mean depth was maximized during this aggradation. It is suggested that accumulation of sediment occurred by low transverse bars migrating down in the tunnel, and that this deposition was accommodated by a melting upward of the ice roof. Thus a sedimentary sequence 15 to 20 m thick could accumulate even though flow depth at any one time was less than 4 m.

Each of the four distinct cycles at locality 21 (fig. 26A), records an initial high stream competence that persisted downstream but that gradually decreased through time, leaving a unit that became finer grained upward. It is tempting to relate this to a cyclical fluctuation of discharge that, in glaciers, could either be diurnal or annual. Thickness of the units, and the observation that at least four cycles underlie a ridge 640 m long where annual retreat of the ice front was about 250 m, favor an annual periodicity.

The question of how far back from the terminus of an active glacier did a tunnel extend at any one time is difficult to answer from sedimentary structure data without an unusually long and good exposure. The ridge of which locality 21 is part is 640 m long and quite uniform in external morphology. This might provide a minimum figure. A study of the occurrence of certain rock types in the bedrock compared with the occurrence of pebble lithologies in the Windsor esker indicates that this distance may be more in the order of 3 to 4 km (Shilts and McDonald, 1975). The length and time-transgressive character of most eskers indicates that subglacial tunnels extend themselves headward as the ice front recedes. The details of this extension are unknown.

Discussion

Sedimentary features of the two eskers studied have permitted interpretation of the basic factors of esker sedimentation: the nature of the conduit and the site of deposition. Facies arrangements in the two eskers were interpreted in terms of three basic models of esker sedimentation: open channel (fluvial), tunnel, and deltaic. An attempt will be made here to summarize the characteristics of each model:

(a) Open-channel model

Esker streams that flow in open channels would almost certainly be braided due to steep slopes and high bedload discharge. It is believed that the Peterborough esker formed partly in an open channel. The principal evidence for this is the presence of backset beds which have been related to antidunes and to waves on a free water surface. Lateral facies relationships that indicate lateral fining (Shaw, 1972) or overbank flooding from a central channel (Saunderson and Jopling, 1970) may also be present in an open-channel deposit although, as Allen (1965, p. 163) has pointed out, overbank deposits are only rarely preserved in a braided environment. Depositional units would be lenticular in transverse cross-section and more tabular in longitudinal section.

(b) Tunnel model

Single steep-sided ridges in the Windsor esker are believed to be deposited in subglacial tunnels. The characteristics of this model are: (i) tabular and longitudinally persistent units of parallel- or cross-bedded sand and gravel; (ii) absence of finer sediments such as very fine sand, silt, and clay; and (iii) low variability of paleocurrent directions (variance based on vector mean lies between 1,000 and 2,000; vector strength is about 0.80).

(c) Deltaic model

Many stratigraphic units in the Peterborough esker and the sediments in the beads of the Windsor esker are believed to be of delta origin. The characteristic features are: (i) rapid downstream facies change through gradation and interfingering from proximal gravel to lake-bottom rhythmites composed of silt and clay; (ii) occurrences of deposits caused by slump, mudflow, and turbidity current; (iii) large variability of paleocurrent directions (variance greater than 3,000; vector strength less than 0.65); and (iv) typical morphological expression as separate beads, as in the Windsor esker. However, when the deltaic units are juxtaposed they form a continuous ridge as in Peterborough.

Despite the widely held view that the associated glacier be stagnant in order for prominent eskers to form, there is nothing in the mechanics of formation that requires this. The view arose from consideration of the amount of meltwater required and the thought that flowing ice would destroy esker accumulations. Beaded eskers indicate, however, that the associated ice fronts were well defined and subject to regular backwasting. Accumulation in beads took place right at the ice front and was not subject to destruction by flowing ice. Even within the glacier, the ability of flowing water to maintain a conduit far exceeds that of the ice to close it by creep (Röthlisberger, 1972).

Similarly the presence of a standing water body at the terminus is not necessary, but it increases the chances for preservation of an esker by immediately placing the newly emerged esker in a low energy environment, removed from the erosive potential of proglacial braided streams.

Eskers are time-transgressive features with the downstream portion being the oldest. This is clear from the process of bead formation and also would offer an easy explanation for the formation of very long eskers such as the Thelon esker.

Additional detailed sedimentary studies of eskers are required, especially from a wide range of deglacial environments, in order to outline more fully the environments of formation and how these can be recognized. These will add to the usefulness of eskers in studies related to drift prospecting, groundwater, and regional geologic history.

Conclusions

1. Eskers are the products of a highly organized meltwater-flow system within the glacier and are common where an abundant supply of meltwater was available during deglaciation. They do not require that the glacier was stagnant, nor that there was a body of water ponded at the ice front.

2. External morphology of the esker complex is controlled to a large degree by the shape of the primary depositional units; both are closely

related to the nature of the conduit, the site of deposition, and to whether or not a standing body of water abutted the ice front.

3. Different combinations in the nature of the conduit (open channel or tunnel) and the site of deposition (within conduit or at ice front (define three basic models of esker sedimentation: *open channel, tunnel*, and *deltaic.* Each model is characterized by typical sediments, facies arrays and paleo-current patterns.

4. Eskers are time-transgressive, the downstream portion being the oldest; they appear to have formed either at the ice front or within 3 to 4 km of it during deglaciation.

5. The eskers examined in detail were deposited in contact with the glacier base; diapiric intrusion of underlying rhythmites upward into esker sediments locally supports this conclusion. Only rare components of the eskers showed intense deformation attributable to wholesale collapse as underlying ice melted.

6. The Peterborough esker was deposited, at least partly, in an open channel constrained laterally by ice walls. Backset beds are related to antidunes in the channel. Dunes and ripples delivered sediment to bar and delta fronts. The main foresets pass downstream into regressive ripples, sinusoidal ripples, progressive ripples, graded beds with load casts, and finally to rhythmically laminated silt and clay. This facies arrangement typifies the deltaic model.

7. A further type of deltaic model is represented by beads in the Windsor esker that were deposited annually where the subglacial tunnel delivered sediment to the lake abutting the ice front. Cobble and pebble gravel at the proximal end of the bead intertongues over a few meters in a downstream direction with ripple-laminated fine sand, units of "structureless" fine and medium sand, and graded beds, providing clear evidence of rapid deceleration of flow.

8. Examples of the tunnel model are provided by sediments in single, steep-sided ridges of the Windsor esker. Sheet-like cross-bedded and parallel-bedded sand and gravel units are persistent downstream. They occur in vertically stacked cycles that may have an annual periodicity. Flow depth in the tunnel was 1 to 4 m, and accumulation of sediment was accommodated by a melting upward of the ice roof.

9. Paleocurrent variability has proven useful in discriminating between depositional environments. Paleocurrent patterns in tunnels have a vector strength about 0.80 and a variance (based on vector mean) of 1,000 to 2,000. Paleocurrent patterns in esker deltas have vector strengths less than 0.65 and variances of greater than 3,000.

Acknowledgments

Thanks are extended to R.A. Edwards, John Finnie, R.A. McGinn, and G.V. Minning who assisted in the collection of field data between 1968 and 1970. The authors are grateful for useful comments on this manuscript that have been offered by G.M. Ashley and H.C. Saunderson.

Note

1 *Frigites* is a term used by Barbour (1913) to mean a well defined mass (of boulder- or smaller size) of unconsolidated sediment, usually well stratified, the attitude of stratification and/or component grain size of which differs noticeably from the enclosing unconsolidated sediments; called a frigite where interpreted as having been emplaced as a clast while frozen.

References

AARIO, RISTO, 1971, Syndepositional deformation in the Kurkiselkä esker, Kiiminki, Finland: Geol. Soc. Finland Bull., v. 43, p. 163–172.

——, 1972a, Associations of bed forms and paleocurrent patterns in an esker delta, Haapajärvi, Finland: Ann. Acad. Scient. Fennicae, ser. A, pt. III, Paper 111, 55 p.

——, 1972b, Exposed bed forms and inferred three-dimensional flow geometry in an esker delta, Finland: Proc. 24th Internat. Geol. Cong., sect. 12, p. 149–158.

ALLEN, J.R.L., 1963, Asymmetrical ripple marks and the origin of water laid cosets of cross-strata: Liverpool and Manchester Jour. Geology, v. 3, p. 187–236.

——, 1965, A review of the origin and characteristics of recent alluvial sediments: Sedimentology, v. 5, p. 89–191.

——, 1971, A theoretical and experimental study of climbing-ripple cross-lamination, with a field application to the Uppsala esker: Geografiska Annaler, v. 53, ser. A, p. 157–187.

BARBOUR, E.H., 1913, A minor phenomenon of the glacial drift in Nebraska: Nebraska Geol. Survey, v. 4, p. 161–164.

BOULTON, G.S., 1972, Modern Arctic glaciers as depositional models for former ice sheets: Jour. Geol. Soc. (London), v. 128, p. 361–393.

CACHAU-HERREILLAT, F., AND LASALLE, PIERRE, 1971, The utilization of eskers as ancient hydrographic networks for geochemical prospecting in glaciated areas: Geochemical Exploration, Spec. V. 11, p. 121.

CAREY, S.W., AND AHMAD, NASEERUDDIN, 1961, Glacial marine sedimentation: First Internat. Symposium on Geol. of Arctic, Proc., v. 2, p. 865–894.

CRAIG, B.G., 1964, Surficial geology of east-central District of Mackenzie: Geol. Survey Canada Bull. 99, 41 p.

DE GEER, GERARD, 1940, Geochronologia Suecica Principles: Kungl. Svenska Vetenskapsakademiens Handlingar, ser. 3, v. 18, 367 p.

DE GEER, JAN, 1968, Some hydrogeological aspects on aquifers, especially eskers: p. 73–87 *in* Ground Water Problems, (*ed.*) E. Eriksson, Y. Gustafsson, and K. Nilsson, Pergamon Press, New York, 223 p.

DONNER, J.J., 1965, The Quaternary of Finland: p. 199–272 *in* The Quaternary, (*ed.*) K. Rankama, v. 1, Interscience, New York.

FLINT, R.F., 1971, Glacial and Quaternary geology, John Wiley and Sons, New York, 892 p.

FRAKES, L.A., FIGUEIREDO, P.M. DE, F., AND FULFARO, VINCENTE, 1968, Possible fossil eskers and associated features from the Paranã basin, Brazil: Jour. Sed. Petrology, v. 38, p. 5–12.

GADD, N.R., McDONALD, B.C., AND SHILTS, W.W., 1973, Glacial recession in southern Quebec: Geol. Survey Canada Paper 71–47, 19 p.

GAGNÉ, R.M., AND HOBSON, G.D., 1970, A hammer seismic survey of an esker north of Peterborough, Ontario: *ibid.*, Paper 70-1B, p. 91–99.

GILLBERG, GUNNAR, 1968, Lithological distribution and homogeneity of glaciofluvial material: Geol. Fören. Förh., v. 50, p. 189–204.

HÄRME, MAUNU, 1961, On the fault lines in Finland: Bull. Comm. Geol. Finlande, no. 196, p. 437–444.

HARTSHORN, J.H., 1952, Superglacial and proglacial geology of the Malaspina Glacier, Alaska, and its bearing on glacial features of New England: Geol. Soc. America Bull., v. 63, p. 1259–1260.

HELLAAKOSKI, AARO, 1931, On the transportation of materials in the esker Laitila: Fennia, v. 52, no. 7, 41 p.

HOBSON, G.D., AND LEE, H.A., 1967, Thickness of drift, Lebel, Gauthier, Boston, and McElroy townships: Geol. Survey Canada Map 11–1967.

HOBSON, G.D., AND MAXWELL, F.K., 1968, Hammer seismograph overburden and bedrock investigations, Cochrane, Ontario: *ibid.*, Paper 68-1A, p. 78.

HORTON, R.E., 1945, Erosional development of streams and their drainage basins: hydrophysical approach to quantitative morphology: Geol. Soc. America Bull., v. 56, p. 275–370.

HOWARTH, P.J., 1971, Investigations of two eskers at eastern Breidamerkurjökull, Iceland: Arctic and Alpine Research, v. 3, p. 305–318.

HYYPPÄ, ESA, 1954, Åsarnas uppkomst: Geologi, v. 6, p. 45.

IVES, J.D., 1967, Glacier terminal and lateral features in northeast Baffin Island: illustrations with descriptive notes: Geographical Bull., v. 9, p. 106–114.

JEWTUCHOWICZ, STEFAN, 1965, Description of eskers and kames in Gåshamnoyra and on Bungebreen, south of Hornsund, Vestspitsbergen: Jour. Glaciology, v. 5, p. 719–725.

JOPLING, A.V., AND WALKER, R.G., 1968, Morphology and origin of ripple-drift cross-lamination, with examples from the Pleistocene of Massachusetts: Jour. Sed. Petrology, v. 38, p. 971–984.

KENNEDY, J.F., 1963, The mechanics of dunes and antidunes in erodible-bed channels: Jour. Fluid Mechanics, v. 16, p. 521–544.

LEE, H.A., 1965, Investigation of eskers for mineral exploration: Geol. Survey Canada Paper 65–14, 17 p.

——, 1968, An Ontario kimberlite occurrence discovered by application of the glaciofocus method to a study of the Munro esker: *ibid.*, Paper 68-7, 3 p.

LEWIS, W.V., 1949, An esker in process of formation: Böverbreen, Jotunheimen, 1947: Jour. Glaciology, v. 1, p. 314–319.

McDONALD, B.C., 1968, Deglaciation and differential postglacial rebound in the Appalachian region of southeastern Quebec: Jour. Geology, v. 76, p. 664–677.

McDONALD, B.C., AND VINCENT, J.-S., 1972, Fluvial sedimentary structures formed experimentally in a pipe, and their implications for interpretation of subglacial sedimentary environments: Geol. Survey Canada Paper 72–27, 30 p.

MEIER, M.F., 1951, Recent eskers in the Wind River Mountains of Wyoming: Iowa Acad. Science, v. 58, p. 291–294.

PARSONS, M.L., 1970, Groundwater movement in a glacial complex, Cochrane District, Ontario: Canadian Jour. Earth Sci., v. 7, p. 869–883.

PRICE, R.J., 1966, Eskers near the Casement Glacier, Alaska: Geografiska Annaler, v. 48A, p. 111–125.

——, 1969, Moraines, sandar, kames and eskers near Breidamerkurjökull, Iceland: Institute British Geographers Trans., public. 46, p. 17–43.

RÖTHLISBERGER, HANS, 1972, Water pressure in intra- and subglacial channels: Jour. Glaciology, v. 11, p. 177–203.

SAUNDERSON, H.C., AND JOPLING, A.V., 1970, Glaciofluvial sedimentation of the Brampton esker, Ontario: Geol. Survey Canada Paper 70-1A, p. 200–201.

SHAW, JOHN, 1972, Sedimentation in the ice-contact environment, with examples from Shropshire (England): Sedimentology, v. 18, p. 23–62.

SHILTS, W.W., 1973, Drift prospecting: geochemistry of eskers and till in permanently frozen terrain: District of Keewatin; Northwest Territories: Geol. Survey Canada Paper 72–45, 34 p.

SHILTS, W.W., AND McDONALD, B.C., 1975, Dispersal of clasts and trace elements in the Windsor esker, southern Quebec: *ibid.*, Paper 75-1A, p. 495–499.

SHREVE, R.L., 1972, Movement of water in glaciers: Jour. Glaciology, v. 11, p. 205–214.

WILLIAMS, G.P., 1967, Flume experiments on the transport of a coarse sand: U.S. Geol. Survey Prof. Paper 562B, 31 p.

15

PROGLACIAL FLUVIAL AND LACUSTRINE ENVIRONMENTS

M. Church and R. Gilbert

Source: A.V. Jopling and B.C. McDonald (eds) *Glaciofluvial and Glaciolacustrine Sedimentation*, Tulsa, Okla.: Society of Economic Paleontologists and Mineralogists, Special Publication 23, 1975, pp. 22–100.

Abstract

This paper reviews the hydrology and hydraulics of high energy, particularly proglacial, riverine and deltaic environments, and discusses some of the consequences for resultant patterns of sediment movement and deposition.

The hydrology of proglacial rivers is under strong thermal influence and exhibits a singular pattern of flow, both seasonally and diurnally. Moderate flood flows are common. Sediment is frequently entrained and deposited, so that rapid evolution of fluvial sedimentary features occurs on outwash plains. The possibility exists for extraordinary jökullhlaup floods to occur in front of many glaciers. The hydraulic behavior of proglacial rivers features frequent upper regime flow and rapid adjustment of channel resistance to accommodate the wide variations in discharge and sediment transport.

Sediment entrainment is reviewed in some detail, and the concepts of "overloose" and "underloose" boundary are introduced. Sediment transport theory is reviewed and recommendations made for assessing total sediment yield. The sediment transport in proglacial rivers is anomalously high by comparison with that in nonglacial environments, because of the large volumes of drift delivered to the glacier margin. Flood deposits on outwash plains and surficial patterns of sediment texture are described. The character of the surface is conditioned by selective deposition of sediment in a simple, aggradational context.

Depositional bedforms are classified as *small* forms (scale controlled by flow depth or lesser flow dimensions) and *large* forms (scale controlled by channel width). The former reflect purely local flow conditions, whereas the latter are influenced by the total flow pattern of the river. The persistence of bedforms as sedimentary structures in the stratigraphical record is considered. Gravels commonly exhibit only rudimentary plane bedding and imbrication, whereas fine materials commonly feature a wide variety of sedimentary structures. This is a consequence of the vertical distance available for deposition as compared to particle size, and the energy status of the depositional environment.

River channels in coarse, noncohesive materials are wide and shallow, so that boundary resistance to flow is high. When large volumes of sediment are being transported in flood, total resistance may become too high to permit passage of the water plus sediment load; deposition and selective scour then produce narrower, deeper channels that are hydraulically more efficient. The morphological result is the occurrence of channel braiding, which is a frequent characteristic on outwash. The long profiles of proglacial rivers and outwash surfaces are concave upward as a consequence of persistent aggradation.

Where proglacial rivers enter standing water bodies, classical, high-angle deltas develop. Sediment transported as bed load is deposited on the delta surface, or is deposited on the foreset wedge by avalanching over the delta lip. Turbidity flows (underflows) and slumps move coarse material farther into the water body. Fine-grained material is carried in suspension into the standing water and settles to the bottom to form varves.

Examples illustrating the application of principles and characteristic conditions are drawn from the literature on glaciofluvial and glaciolacustrine environments.

Introduction

The purpose of this paper is to describe characteristic features of riverine and lacustrine behavior in the proglacial environment, in order to provide a context for studies of glaciofluvial and glaciolacustrine sedimentology. A brief description is given of hydrological conditions, and then a review is made of pertinent aspects of river hydraulics, sediment transport, and of riverine and deltaic sedimentation.

The conditions that produce the usual coarse, clastic sedimentary deposits of the proglacial zone depend on an abundant source of detrital material and on a hydrological regimen characterized by sufficiently frequent high flows so that significant quantities of material can be moved. These conditions are apt to be duplicated in several other environments as well. In particular, periglacial, high mountain, and semiarid regions often present very similar fluviatile effects. Hence, the range of relevant literature is by no means restricted to that describing proglacial conditions; indeed, a large proportion of all work on fluvial hydraulics, sedimentology, and geomorphology may be instructive in one way or another. Consequently, this paper cannot possibly represent an exhaustive review. The attempt has been, rather, to treat the main principles underlying sediment transport and deposition in high energy, fluviatile environments. Much of the knowledge in this area remains empirical or, at best, quasi-theoretical; questions that have not been satisfactorily resolved have been indicated.

The paper is mainly illustrated from the writers' own work. However, important results and instructive examples are drawn from a wide range of literature. Standard notations are not normally defined in the text: the reader should refer to the list of notations following the text.

Proglacial hydrological regimen

Pattern of runoff

Glaciers represent natural storage reservoirs that retain a greater or lesser proportion of a year's total precipitation as snow, releasing water for run-off only during the warm summer period. Hence, almost an entire year's precipitation runs off in a few weeks or months of summer. In the longer term, glaciers may redistribute precipitation inputs from year to year, or over periods of many years according to whether the mass budget of the glacier is positive or negative in any year. During major glacial periods very large volumes of water are stored on the land for thousands of years, and then released for runoff as the ice caps wane. The distribution of runoff from glacierized basins, then, bears little or no relation to the variation of precipitation (Meier and Tangborn, 1961; Meier, 1964).

Streamflow in proglacial rivers, conditioned as it is by snow and ice melt, is highly variable and highly seasonal, Figure 1 presents some typical hydrographs. It is apparent that runoff is extremely sensitive to weather events, and that a remarkable difference can be introduced between seasons by different weather patterns.

Near the glacier, there are five periods of runoff during the year: (a) breakup; (b) nival (snow-melt) flood; (c) late summer; (d) freezeback, and (e) winter. The breakup period begins when snowmelt commences in spring. However, most meltwater is reabsorbed into the snowpack, with much of it refreezing in the early part of the period. The normal course of breakup is for the river channels to turn to slush, and then for flow to begin over channel ice. Only after flow is well established does the winter ice on the river bed break up. During this period, the length of "connected" channel is short and restricted to low elevations. An extended period of snowmelt or, more efficiently, a heavy rainstorm is necessary to flush out an appreciable length of channel and initiate significant runoff. Temperate glaciers with deep snowpacks may develop extensive drainage under the snow, which is not apparent until well into the nival flood.

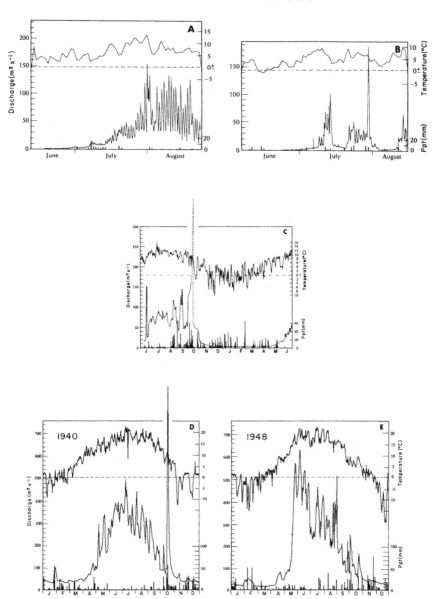

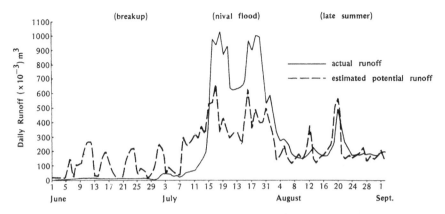

Figure 2 The effect of runoff delay due to water storage on the glacier during the early part of the melt season. The estimated potential runoff (total non-refreezing water generated) was computed using a recession-regression model derived by analysis of temperature and runoff patterns of late summer. Data are from Mikkaglaciären, Sarek, northern Sweden, for 1957 (after Stenborg, 1970, p. 27, reproduced by courtesy of *Geografiska Annaler*).

The nival flood follows the establishment of a large area of connected drainage. Then, much of the meltwater stored in the glacier is drained off, so that anomalously high flows, by comparison with daily melt, occur for some time. Stenborg (1970) has illustrated the net effect of this additional redistribution of runoff on the annual hydrological regimen (fig. 2). During

Figure 1 (*opposite*) Daily runoff in proglacial rivers, with mean daily temperature and daily precipitation data (precipitation is plotted as a bar graph on the base of the diagram): A, Lewis River, Baffin Island, 1963. Seasonal runoff from Lewis Glacier, an outlet glacier of Barnes Ice Cap, in arctic Canada. Drainage area is 205 km², of which 89 percent is glacierized. The season was warm and the hydrograph is dominated by runoff from glacier ice ablation with a strong diurnal pattern (temperature and precipitation records for Lewis River, 1 km from the glacier). B, Lewis River, 1964. The season was cool and stormy, and the hydrograph is dominated by runoff from snowmelt and precipitation. C, Austurfljót, Hornafjördur, Iceland, 1951–1952. Runoff for a one-year period from Hoffellsjökull, an outlet glacier on the south side of Vatnajökull. Drainage area is 200 km², of which about 67 percent is glacierized. Meteorological data for Holar, 17 km south of the glacier (data from Arnborg, 1955, p. 190, reproduced by courtesy of *Geografiska Annaler*). D, Lillooet River, Coast Mountains, British Columbia, 1940. Runoff from a major valley outlet glacier of alpine snowfields in midlatitude mountains. Drainage area is 2,200 km², of which about 9 percent is glacierized. Meteorological data for Pemberton Meadows, 64 km south of the glacier. Note the prominent effect of autumn rains in the hydrograph. E, Lillooet River, British Columbia, 1948. This was a year of particularly high summer runoff, the result of exceptional ablation in the upland snowfields.

the late summer, runoff continues more or less in accordance with melt generation. This reflects several developments on the glacier: (a) the early season "heat deficit" in the near surface zone of the glacier has been overcome (that is, the temperature of the snowpack and near surface ice has risen to 0°C), (b) the surface drainage network has extended far up the glacier and drained slush ponds, and (c) internal and subglacial drainage ways have opened up.

During the nival flood and late summer periods, storm precipitation may occur as rain over a considerable area of the glacier. This produces an abrupt runoff peak, since most of the water runs directly off the impermeable ice surface in channels with low resistance. This is particularly true for arctic glaciers whose drainage is entirely surficial; temperate glaciers possessing internal drainage via crevasses, moulins, and tunnels may impose some delay on runoff (Stenborg, 1969).

The freezeback period occurs in autumn after melt has ceased on the glacier. Recession flow is maintained for some time as channelways (particularly englacial passages) drain, and as groundwater levels around the glacier and on the outwash plain decline. In the Arctic, the freezeback may be very rapid, because all the flow is restricted to the ground surface by the presence of permanently frozen ground below, but rivers draining temperate glaciers may maintain a low, steady rate of flow, fed mainly by groundwater drainage, throughout the winter (Stenborg, 1965).

Farther from the glacier, nonglacial hydrological processes may severely alter the streamflow pattern. The graphs of Figure 1 illustrate regimen for rivers draining areas with between 9 percent and 89 percent ice cover. At 9 percent glacierization, the records for Lillooet River are very similar to those of a normal, cool temperate climate with considerable winter snow, except for the continuation of regular high flows throughout late summer.

Coarse alluvium normally admits flow of water below the surface, so that, on outwash plains there may be considerable exchange of water between channel flow and subsurface water. Where consistent abstraction of water from channel flow occurs, there may be a significant effect on sedimentation. The extent of subsurface seepage can be estimated using Darcy's law for flow in a porous medium (see *Notations*, following *References*, for explanation of Symbols):

$$Q = \frac{k\gamma}{\mu} AS_f \qquad (1)$$

Hjulström (1955) made a direct calculation of subsurface discharge on the Hoffellssandur, using field measurements of S_f and hydraulic conductivity:

$$K = k\gamma/\mu \qquad (2)$$

Table 1 Permeability of sands.

Material	k *(darcys)*	K *(ms⁻¹@5°C)*
Very fine sand (well sorted)	9.9	6.2×10^{-5}
Medium sand (very well sorted)	2.6×10^2	1.6×10^{-3}
Coarse sand (very well sorted)	3.1×10^3	2.0×10^{-2}

Data from de Wiest (1965).

He found values of K varying between 9.0×10^{-12} meters per second and 1.2×10^{-3} meters per second, and subsurface discharges of order 0.3 cubic meters per second. This amounted to about 0.4 percent of the summer period mean surface flow of 80 cubic meters per second. If maintained through the winter, however, it would constitute a significant proportion of the mean surface flow of about 2 cubic meters per second (data on surface flow from Arnborg, 1955). A calculation was carried out for the sandur in Sarvalik arm of Ekalugad Fiord, Baffin Island (Church, 1972). No direct measurements of permeability were made here; however, it was assumed that the fine materials in the sandur deposits would control its value. These consist of moderately sorted coarse to medium sands, with no notable changes downsandur. Table 1 gives a range of probable values for k and K. Computations on the data for medium sand indicated a range of subsurface flows between 0.2 cubic meters per second and 0.3 cubic meters per second. This would amount to less than 1 percent of the seasonal mean flow of 40 cubic meters per second. Because permafrost restricted the subsurface flow depth to less than 2 m, a good estimate of the available flow section is included in this computation. It appears, in sum, that the total subsurface flow is hydrologically negligible.

Where consistent abstraction of water from the channel occurs, declining competence of the stream may result in general sediment deposition; the effect is usually associated with streams in semiarid areas. Abstraction of water upstream and discharge to channels downstream, as reported by Hjulström (1955), provide a filter mechanism to trap fine sediment on the sandur that would normally be washed directly away.

Flow events

Descriptions of relationships between runoff from glaciers and weather have been given by, amongst others, Mathews (1964a), Østrem (1964, 1966), Anonymous (1967), Lang (1968) and Gudmundsson (1970).

During normal weather periods, snow and ice melt is influenced by the diurnal fluctuation of available heat, so that runoff is also markedly periodic. As an example, Figure 3 illustrates the diurnal stage variation at Lewis River. Gudmundsson and Sigbjarnarson (1972) showed by cross-spectral analysis

417

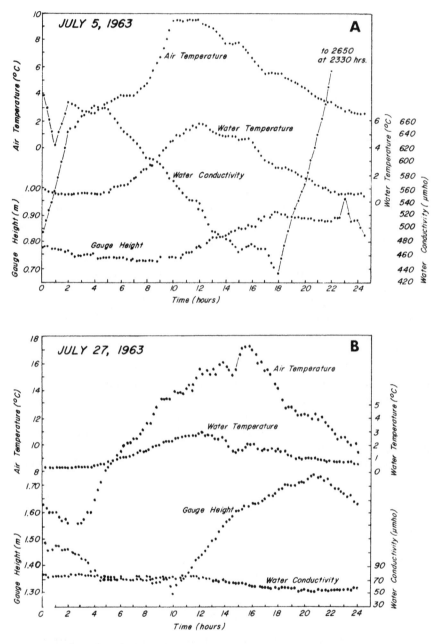

Figure 3 Diurnal variation of air temperature and water characteristics during 24-hour periods at Lewis River, Baffin Island. Note the lag between highest air temperature and highest water stage, representing the time of concentration and travel for the daily "flood wave" through the drainage system. A, during spring snow-melt. B, during a period of rapid glacier ice melt (after Church, 1972, fig. 21, reptoduced by courtesy of *Geological Survey of Canada*).

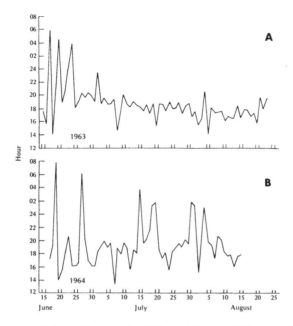

Figure 4 Time of peak flow (phase angle of diurnal flow cycle) in a proglacial stream: A, Lewis River, 1963; a season dominated by glacier ice ablation. B, Lewis River, 1964; a season dominated by storm precipitation. Corresponding flow hydrographs are shown in Figures 1A and 1B.

that daily temperature fluctuations accounted for most of the fluctuation in runoff at Tungnaá, on the western margin of Vatnajökull, Iceland. Precipitation accounted for some very low frequency effects. Hence, the runoff may be treated as a periodic function:

$$Q = \bar{Q} + A \sin \theta(t) \tag{3}$$

This function accounts for virtually all the daily variance, indicating that shorter periods are unimportant. The periodic phase angle, θ, represents an approximation of peak flow time: Figure 4 shows graphs of θ for melt-dominated and storm-dominated runoff periods, illustrating the striking difference in regularity of occurrence.

Small-scale variations in flow do occur within the diurnal period, an example of which is given in Figure 5. The irregular perturbations are apparently the result of frequent ice falls and slush avalanching on the glacier that block the channels, followed by release surges of water after a short period. Sometimes large amounts of ice or slush are carried down the river (fig. 6); the large volume of ice effectively damps the turbulence of flow and thereby exerts a negative effect on sediment transport.

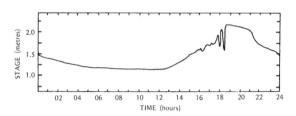

Figure 5 Automatic gauge trace at Lewis River, July 26, 1964. The gauge is 1 km from the glacier.

Figure 6 Heavy load of slush and ice being carried by the proglacial Lewis River following slush avalanching on the glacier. View upstream at the gauge section, $Q \sim 20$ cubic meters per second.

Proglacial rivers are subject to extraordinary floods, which may occur when bodies of water stored in or around the glacier drain away. Ice-marginal ponds or lakes, crevasse ponds, water held in cavities within or under the glacier, large slush ponds on the glacier surface, or water held in headwater stream channels by ice or snow dams, may drain more or less suddenly when a barrier gives way. Such jökullhlaups (Icelandic: "glacier bursts") often exhibit a characteristic hydrograph, rising steadily to a peak and then dropping abruptly (fig. 7). Various workers have speculated that this is due to the gradual enlargement of ice-walled drainage ways (Arnborg, 1955) using heat energy generated by the flow (Liestøl, 1956; Gilbert, 1971), so

420

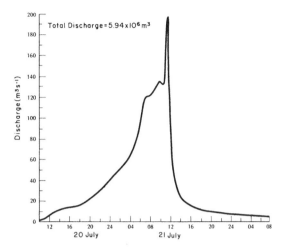

Figure 7 Jökullhlaup in South River, Ekalugad Fiord, Baffin Island, on July 20–22, 1967, when an ice dam burst, causing catastrophic lake drainage. Of the total discharge during 30 hours, 4.80×10^6 cubic meters can definitely be ascribed to lake drainage (data from Church, 1972, p. 38, reproduced by courtesy of the *Geological Survey of Canada*).

that the capacity of the channel is greatest when the water at the source becomes exhausted. Recent papers by Röthlisberger (1972) and Shreve (1972) have developed the theory of water movement in glaciers, including description of the development of flow conduits and drainage networks.

Where an ice-dammed lake persists for many years, jökullhlaups may become seasonal events (Kerr, 1934; Marcus, 1960; Stone 1963; Meier, 1964). Certain of the Icelandic sandur plains are subject to occasional, truly catastrophic floods when subglacial volcanic eruptions melt vast quantities of ice (Thorarinsson, 1953). Tryggyason (1960) suggested that earthquakes associated with the volcanism may aid the draining. Klebelsberg (1948) noted drainings or increased ablation associated with volcanism in Spitzbergen, Alaska, Ecuador, and Kamchatka.

During jökullhlaup floods, prodigious volumes of sediment can be moved and material of anomalously large size may be displaced (Hjulström, 1952; Thorarinsson, 1953; and Birkeland, 1968). Even relatively small slushflows may move considerable sediment (Washburn and Goldthwait, 1958), and ice rafting of material may be locally important. In large floods, entire river channels may be displaced, with consequent rearrangement of the surface morphology of the outwash plain. Arnborg (1955) has illustrated the rapid attenuation of jökullhlaup peak flow as the flood wave moves down sandur: this would produce a rapid decline in the competence of the flood and hence promote graded deposition of the sediment load downstream.

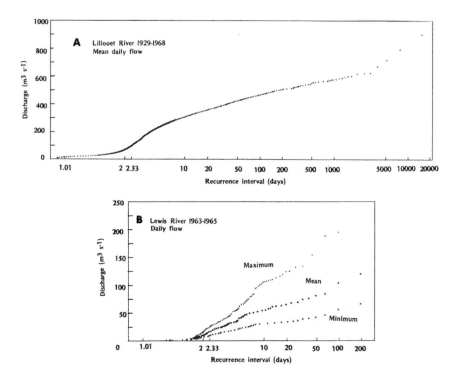

Figure 8 Total daily duration series for two proglacial rivers: A, Lillooet River, British Columbia. The gauging station is 77 km downstream from the main glacier. B, Lewis River, for all days with significant flow (except for about 15 days in late August, 1964) during three years. Note that this plot does not represent an unconditional flow frequency diagram, since the degree of serial correlation amongst individual flows in the short series precludes the assignment of unconditional probabilities.

Clague (this vol.) has systematically assessed the nature and effects of recorded jökullhlaups, and developed simple empirical relationships to indicate the magnitude of concomitant sediment transport.

Significance of hydrological regimen

The significance of hydrological regimen for the development of proglacial fluvial deposits lies in the magnitude and frequency characteristics of flows that are generated, because this affects sediment transport. Figure 8 illustrates daily duration series for two contrasting proglacial regimes. It is apparent, first of all, that the flows do not represent homogeneous populations of events. Flows of modest magnitude mainly occur early or late in the season, when freezing temperatures restrict runoff to the lowest reaches of the watershed. Similarly, the condition of the snowpack early in the season will affect runoff generation. On the other hand, the high flows of the nival flood and late summer periods occur when a high proportion of the watershed is contributing water to runoff. Because optimum melt rates may be approached on many days in this period, rather similar runoff often occurs, so that the upper portion of the duration plot is flattened. Superimposed upon these considerations is the contribution of summer or autumn rainstorms to runoff generation. Indications are that, on arctic glaciers, the largest flows generated by normal weather events are those produced by warm summer rains, because the specific rate of runoff generation may then considerably exceed optimum melt rates. In temperate regions the situation is probably similar, although peak melt rates probably approach more closely the usual precipitation intensities. Runoff from rainstorms is of disproportionate importance for sediment transport, because flow is generated directly on moraine and other ice-edge deposits from which most of the sediment is derived. Finally, at the upper end of the duration series, there may appear anomalously large flows representing jökullhlaup runoff.

The marked diurnal flow variation has the effect of producing a "complete flood event" in each 24 hours, which becomes steadily attenuated as it moves farther from the glacier. Sediments are entrained on rising stage, moved downstream with the flood wave, and then are deposited again on falling stage when the water wave outruns the sediments. Large volumes of material are moved sporadically downchannel and dumped into temporary storage in channel bars along the sides of the channel, or otherwise spilled over onto the outwash surface where the rivers overtop their banks.

In sum, glacier runoff can be expressed as some function of thermal conditions (as they influence melt), precipitation characteristics, and recent flow history. The exact functional relationship is complex and is not likely to be stable, because the effect of changing snow-cover conditions through the season will be important. There is a wide range of flows in each year, but with a relatively high proportion of moderate flood flows that should be capable of moving a considerable amount of sediment. Hence, the processes of sediment entrainment, motion, and deposition are frequently repeated through the summer season, imparting the possibility for rapid development and destruction of sedimentary deposits, frequent change of the river channel pattern, and rapid evolution of the sandur surface. There is always the possibility for extraordinary floods to occur.

River channel flow

Description of open channel flow

Flow in open channels has been described in many textbooks (Rouse, 1938, 1950; Chow, 1959; Albertson and Simons, 1964; Henderson, 1966; Sellin, 1969). Briggs and Middleton (1965) have discussed fluid mechanics with particular reference to the formation of primary sedimentary structures. This section reviews those aspects of the subject which are of importance for understanding fluvial processes in high-energy environments.

Flow in open channels is characterized by two dimensionless flow ratios: Reynolds number, $R_e = vL/v$, and Froude number, $F_r = v/\sqrt{Lg}$. The symbol L is some characteristic length, usually flow depth or hydraulic radius for open channels. R_e expresses the ratio of inertial to viscous forces in the flow. For $R_e < 500$, viscous forces significantly condition the flow, which is *laminar*. For $R_e > 750$, inertial forces dominate and flow is *turbulent*. Transitional regimes occur between these limits. Turbulent flow occurs in all natural open channels. The Froude number represents the ratio of inertial to gravitational forces. For $F_r < 1.0$, gravitational forces dominate and the flow is "tranquil" or "subcritical"; this is the normal circumstance in most rivers. For $F_r > 1.0$, inertial forces dominate, and flow becomes "rough" or "supercritical." Upstream wave propagation (hence, information transfer) cannot occur in supercritical flow. This distinction is also important in the wave mechanics of bedforms developed on the noncohesive bottom of the stream. For low values of F_r (referred to in this context as *lower regime*), gravitational waves at the surface of flow are reflected in wave formation on the bed in opposite phase; for high values of F_r (referred to as *upper regime*), wave trains are in phase on the surface and bed.

Consider, first, the total flow properties of the river (*see* Figure 9 for definitions). The forces acting on a unit volume of fluid are given by the Euler equation for incompressible, frictionless flow:

$$\frac{\partial}{\partial s}(p + \gamma z) + \rho a_s = 0; \quad a_s = \frac{dv}{dt} + v\frac{dv}{ds} \tag{4}$$

Steady flow ($dv/dt = 0$) may be described by the Bernoulli equation, taking the integral, with respect to distance along a streamline, of equation (4):

$$p/\gamma + v^2/2g + z = H \tag{5}$$

where H is the total energy and z is channel elevation above the given base level (for the energy computation). The reciprocal relationship between dynamical pressure, p, and velocity, v, will determine whether flow is converging (v increasing; p decreasing), or diverging; this is an important consideration for sediment entrainment and deposition. The value of H decreases continually downstream as energy is consumed to overcome flow resistance.

If longitudinal slope is small, vertical acceleration of flow will be 0 and $p/\gamma = d$ will represent water depth. Then $p/\gamma + z$ will define the water surface elevation above datum, that is, the hydraulic grade line (however, for steep channels or near drawdowns to overfalls, this is no longer the case).

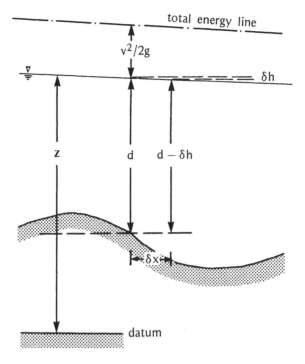

Figure 9 Definition sketch for parameters of flow.

Specific energy (energy per unit width of the stream) is defined as:

$$E = d + v^2/2g \qquad (6)$$

where E is referenced to the channel bed as datum. Generally, the division of energy depends on the channel configuration. The relationship between E and d is given by a specific-head diagram: it shows that for any value of E there are two possible values of d. When flow passes through some change in channel configuration (transition), the adjustment is that determined by the most accessible or nearest value of d (fig. 10). The relationship yields minimum $E = E_c$, corresponding to d_c for a given flow. At this juncture $d_c = {}^2/_3 E_c$ and $F_r = 1.0$, that is, flow is *critical*. Flow is a maximum for given E at critical flow, in which case the channel is operating at maximum efficiency.

Flow may pass smoothly from subcritical to supercritical, but the reverse does not generally occur (fig. 11); experiments show that abrupt discontinuities of the water surface normally characterize supercritical to subcritical flow transitions, with considerable energy loss occurring in the process. Such a transition is called a *hydraulic jump*. The position of the jump is not necessarily stable, but in real channels the resistance effects localize it.

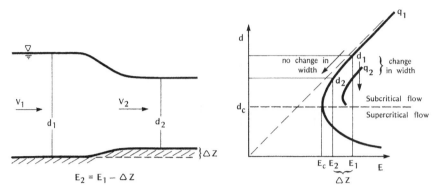

Figure 10 Flow transition past an obstruction in the channel. Of the two possible solutions on the *E-d* curve, the one nearest the initial state will always be adopted.

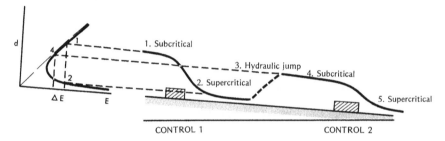

Figure 11 Occurrence of a hydraulic jump between two channel controls. The upstream control effects a transition to supercritical flow, which meets the subcritical backwater from the downstream control at the jump. In a natural channel, riffles and pools may produce this sequence.

Flow transitions in real channels include changes in channel configuration, as over channel bars or around bends and changes in boundary material properties that determine resistance to flow. Because flow conditions change here (*see* fig. 10), sediment transport conditions which depend on flow parameters are also affected.

The foregoing discussion has been couched in terms of two-dimensional flow. It generalizes to real channels, with channel area becoming the appropriate geometrical term. The detailed flow pattern is dominated by the turbulent nature of flow. Turbulent flow is characterized by continuous, irregular interchange of fluid masses within the flow, leading to diffusion of fluid and transfers of momentum across the flow. In open channels, flow disturbances range from secondary circulation induced by eddies whose mean size approaches that of the channel width, down to microscale velocity fluctuations. Because large eddies arise from interaction with the channel boundaries, and therefore tend to be persistent, flow variations arising from them are termed "eddy structure." Eddy structure is very important in determining the nature of some primary sedimentary structures.

Flow over a boundary is conventionally divided into two regions: the *free-flow* region, far removed from the boundary, and the *boundary layer*, where flow is affected by the boundary. There is a *laminar sublayer* very close to the boundary as well. However, for river channels large roughness elements protrude into the turbulent flow region and the laminar sublayer is then of no practical importance. Flow in river channels is treated as boundary layer flow, that is, the flow is everywhere affected by the boundary.

Flow over a boundary is subject to boundary resistance, which induces shear in the fluid. For turbulent flow the mass exchange induced by eddy and turbulence effects promotes rapid momentum transfer so that the simple shear stress equation for laminar flow is modified as follows:

$$\tau = (\mu + \eta)dv/dy \tag{7}$$

where η is the "eddy viscosity," a surrogate term to account for the turbulent effects. As $\eta \gg \mu$, it is usual to describe the shear by:

$$\tau = \eta dv/dy \tag{7a}$$

Because of turbulent transfers, mean velocity is more uniform in a turbulent shear flow than in laminar flow. Theoretical and experimental investigations by Keulegan (1938) led to application of the logarithmic flow law to describe the velocity profile in real open channels:

$$v/v_* = 8.5 + (2.3/\kappa) \log (d/D) \tag{8}$$

where $v_* = \sqrt{\tau_o/\rho}$ and τ_o is the shear stress at the bed; D is some measure of the height of roughness elements at the boundary, hence (d/D) is a measure of relative roughness.

The effect of turbulent velocity fluctuations near the boundary is to induce changes in the shear and pressure stresses at the boundary, thereby affecting sediment entrainment and deposition. Where flow velocities are relatively high and the boundary is sharply curved, flow separation may occur, producing reverse flows and wakes. Flow separation plays an important role in determining the form of some primary sedimentary structures, such as dunes and cross-stratification (Simons, and others, 1965a; Jopling, 1965a).

Flow resistance

A resistance equation is derived by balancing the resisting shear force at the channel boundary against the force propelling the flow. Although superficially a simple matter, this is always complicated by the irregularity of river channels, which means that the distribution of shear stress around the boundary is non-symmetrical. The existence of secondary flows provides further complications. The logarithmic flow law (eq. 8) is a theoretically correct expression of flow resistance effects for two-dimensional flow; the law however, is difficult to generalize to mean flow parameters in real channels.

Though shear stress distribution may not be known, an equation for resistance based on mean shear stress may be written as follows:

$$\gamma A dz - \tau_o P dx + \rho A v \frac{dv}{dx} dx = 0$$

The first term represents the propulsive force for the fluid element, where $A = \bar{d}w$ is the flow area, provided that the slope is sufficiently small that pressure distribution is hydrostatic; the second represents the flow resistance due to shear stress along the boundary of wetted perimeter, P; the third term represents flow acceleration, due to imbalance of the first two forces in nonsteady flow, = 0 for steady flow. This expression reduces to:

$$\tau_o = \gamma R \frac{d}{dx} [z + v^2/2g] \tag{9}$$

$$= \gamma R S_f \tag{9a}$$

where R is hydraulic radius, A/P, and S_f is the energy grade; for steady flow $S_f = S_o$, the hydraulic gradient. For most river channels $R \sim \bar{d}$.

Dimensional analysis of the situation leads to the similar result that $\tau_o = \alpha \rho v^2$, where α is a dimensionless number depending on boundary roughness, channel shape, and the Reynolds number. These factors are the ones affecting shear distribution. Hence:

$$v = \sqrt{gRS/\alpha} \approx \sqrt{gaS/\alpha} \tag{10}$$

where the approximate formula holds in a wide channel. Evidence for the behavior of α can be deduced by comparison with the Darcy-Weisbach resistance coefficient, ff, in the similar pipeflow resistance equation (Henderson, 1966, p. 92–94), but the results are strictly applicable only to channels comparable in size to pipes. Experimental results of Nikuradse (1933) showed that, for fully turbulent flow in circular pipes

$$ff = 0.113(k_s/R)^{1/3} \tag{11}$$

where k_s is equivalent grain roughness of the boundary (ASCE Hydromechanics Committee, 1963). This yields, from equation 10, $v \propto \sqrt{gRS/ff}$ or $v \propto R^{2/3}S^{1/2}k_s^{1/6}$.

Independent field evidence in natural channels is summarized by the formula:

$$v = R^{2/3}S^{1/2}/n \approx \bar{d}^{2/3}S^{1/2}/n \qquad (12)$$

where n is a total flow resistance coefficient. This is the Manning formula, the generally accepted resistance equation for open channels. It should be emphasized that this result is entirely empirical. However, its form is confirmed by the previous result: replacement of k_s there by D, the grain size of sediment particles on the stream bed, completes the formal comparison, provided $n \propto D^{1/6}$. Empirical evidence for this was provided by Strickler (1923), who found that $n = 0.04D^{1/6}$ ($D > 8$ mm). Figure 12 illustrates the variation of n in a proglacial channel: the resistance coefficient decreases at higher flows as the relative roughness of the boundary, D/d, decreases.

The foregoing provides an alternative relationship to that of equation (8) for resistance due to a particulate boundary. In natural channels, however, there are normally several additional sources of flow resistance (Inglis, 1952): (a) that due to ripples, dunes, bars and other transitory bedforms (Einstein

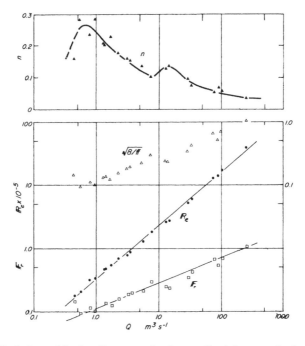

Figure 12 Variation of hydraulic properties in an alluvial outwash channel, Lewis River, Baffin Island. The section is located where the river passes through a moraine, and is stabilized by large boulders on the bed. Note that the values $\sqrt{8/ff}$ represent an inverse measure of resistance to flow.

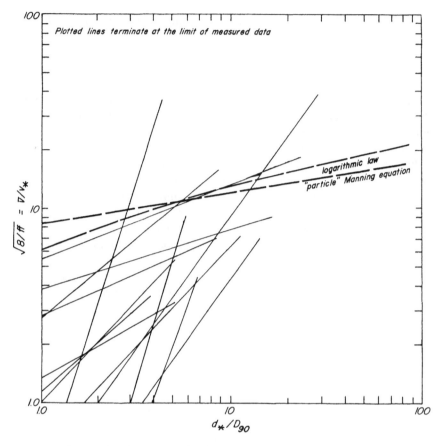

Figure 13 Observed resistance relationships for Baffin Island outwash channels with high sediment transport capability, computed under the assumption of a constant value for D_{90} in the channel bed (after Church, 1972, p. 73, reproduced by courtesy of the *Geological Survey of Canada*).

and Barbarossa, 1952; ASCE Hydromechanics Committee, 1963), (b) that due to channel curvature (Bagnold, 1960) projections into the flow, structures, *etc.*, and (c) that due to "spill" associated with excessive deformation of the streamlines when flow discontinuities occur (Leopold and others, 1960). The last source is only important when $F_r > 0.5$, whereas the second changes very slowly. It is principally the change in the first factor, combined with changes in boundary particle characteristics when sediment transport occurs, that produces rapid changes in total flow resistance as flow increases in real channels. In general, resistance decreases much more rapidly (fig. 13) than indicated by the Manning equation or the logarithmic law of Keulegan, suggesting that a more general power law relationship of form $\bar{v} = C_1 R^{C_2} S^{1/2}$ is probably appropriate. This can be represented as:

$$\frac{\bar{v}}{v_*} = C_3\left[\frac{d}{D}\right]^{C_4} = \sqrt{8/ff} \qquad (13)$$

For the "particle Manning equation" (that is, Manning equation when all resistance to flow is provided by particles on the channel boundary) $C_3 = 8.4$ and $C_4 = 1/6$. The relationship to the Darcy-Weisbach coefficient is derived from the equation $ff = 8gdS/v^2$; $v = \sqrt{8gdS/ff}$, assumed to hold for open channels. Values of $\sqrt{8/ff}$ are plotted in Figure 12 for comparison with n. Figure 13 plots resistance relationships in this form for a variety of alluvial channels.

Once the bed of the stream becomes "live," standard formulae can no longer be applied in order to partition resistance. The "live bed" conditions which pertain at high flow levels in high energy streams allow reduction in resistance to values lower than those indicated by stable bed roughness elements at lower flows; this obtains because the moving bed of sediment, largely sand, will provide an apparently smooth water-sediment zone near the bed. Blench (1963) even suggested that regime channels moving high sediment loads might exhibit boundary characteristics sufficiently smooth to agree with an extended Blasius equation (equation for hydraulically smooth boundaries; Chow, 1959; Henderson, 1966, p. 93). Figure 14 indicates a supposed form of channel behavior under live bed conditions.

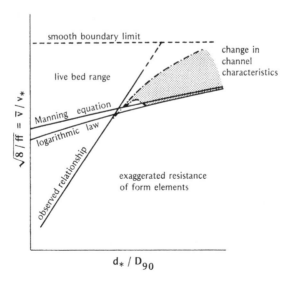

Figure 14 Hypothesized form of resistance relationship for channels with movable beds.

Pipe flow: subglacial channels

Flow in subglacial channels may not have a free surface, in which case it may be analogized to flow in pipes (*see* Chow, 1959, amongst others, for discussion of pipe flow). Several problems exist in applying pipe flow laws to englacial or subglacial tunnels. First, the tunnel is likely to have an irregular section. However, the pipe flow formulae continue to hold approximately, provided that the effective tunnel diameter is chosen to be that of a hydraulically equivalent, round pipe. Second, sources of flow resistance in the tunnel are likely to include bends, contractions, and form-roughness on the walls, but it is not feasible to determine these elements in detail. Hence, it is not likely that a suitable value for the resistance coefficient can be determined. Third, drainage may be occurring through more than one separate or interconnected tunnel; for example, Marcus (1960) noted five outlets for Tulsequah Lake and two tunnels were seen at the toe of Salmon Glacier after the 1967 draining of Summit Lake (Gilbert, 1971).

Hence, it is common to fall back on empirical formulae, such as a modified Manning formula (Mathews, 1973), with

$$n = \frac{R^{2/3}S^{1/2}}{v} = \frac{(d/4)^{2/3}S^{1/2}}{v} \tag{14}$$

where d is here the tunnel diameter. For smooth ice walls, n may have a value of <0.01. However, form resistance elements will increase total resistance by some factor. Table 2 reports some observed values of n for *river ice covers*, and Figure 15 shows a graph relating total flow resistance to ice and bed resistance. A total resistance value of $n\sim0.02$ appears to represent a reasonable criterion when there is no further information. By introducing the continuity equation, $Q = \pi d^2 v/4$, the pipe diameter, d, for a subglacial channel can be determined (Mathews, 1973) as follows:

$$d = 2.4\left[\frac{Qn}{S^{1/2}}\right]^{3/8} \tag{15}$$

Here d is expressed in terms of the parameters most likely to be known or estimated. The equation is relatively insensitive to errors in the estimates.

Nezhikhovskiy (1964), Carey (1967), and Larsen (1969) also presented methods for computing flow in an ice-covered river, using roughness coefficients separately assigned for ice and streambed surfaces which might be applicable in certain circumstances for subglacial tunnels on bedrock (that is, fixed boundary; *see* also, fig. 15).

Table 2 Observations of Manning roughness coefeicien is for ice cover on rivers.

Site	Value	Remarks	Source
St. Croix River, Wisc.	0.0135	nearly smooth ice	Carey, 1966
St. Croix River, Wisc.	0.028	ripple amplitude on underside of ice, 2–3 cm, wave length, 20–30 cm	Carey, 1966
Kilforsen, Sweden	0.019	ripple amplitude ≤4 cm, wave length ≤36 cm	Larsen, 1969
Gallejaur Headrace, Sweden	0.025 −0.027		Larsen, 1973

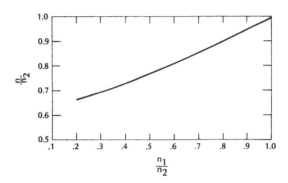

Figure 15 Total (composite) Manning coefficient, *n*, as a function of partial coefficients for ice cover and stream bottom. $n_1 < n_2 < 0.04$; n_2 applies to the rougher of the two boundaries (after Larsen, 1969, p. 57, reproduced by courtesy of Boston Society of Civil Engineers).

Sediment movement: hydraulic considerations

The theory of sediment transport is the least satisfactorily developed area of open-channel hydraulics. Reviews of the subject are given in Rouse (1950), and by Vanoni and others (1960), Henderson (1966), Raudkivi (1967), Graf (1971), and Yalin (1972). Reviews emphasizing sedimentological consequences of sediment movement are given by Briggs and Middleton (1965) and Brush (1965b). The present discussion will emphasize certain empirical results of importance for interpreting high-energy fluviatile deposits (*see* also ASCE Task Committee, 1971b).

Sediment load transported by a stream is made up of three parts: solution load (not discussed here), wash load, and bed-material load. *Wash load* consists of fine clastic materials which, once entrained, can normally be carried through the channel system in suspension. *Bed-material load* consists of coarse clastic materials, normally occurring in the channel bed, which are sporadically moved, usually along the bed, during high flows. The actual volume of wash load moved may be limited by supply, but there is always a supply of bed material available for transport according to the capability of the flow.

Two major mechanisms exist for clastic sediment movement: (a) movement of material *in suspension*, that is, in such a way that the (immersed) weight of the sediment is supported by water currents, and (b) movement of material *as bed load*, that is, by rolling, sliding, or skipping along the stream bed, so that the weight is primarily carried by the solid bed of the river. This division is recognized in measurements of sediment in motion, and in most theories about sediment transport. It should be realized, however, that the distinction is purely arbitrary: what material comprises suspended load and bed load will depend on the stream's competence; hence, ultimately, on flow level. In particular, the behavior of sand size particles (63 to 2,000 microns) is subject to changing flow conditions. Finer materials almost always constitute either wash or suspended load, whereas coarser materials are usually bed-material or bed load.

The stream must continually do work in overcoming frictional contacts between moving sediment and the bed in order to maintain bed load in motion. Hence, there should be, theoretically, a well defined functional relationship between the force exerted by the stream at the bed and the quantity of material moved as bed load, provided only that the supply of sediment is not limited. Since the stream is almost always competent to move wash load, the discharge of wash load will be limited by the supply of material. Therefore, a functionally close relationship between stream power and sediment transport can be expected to apply only to the movement of bed material. The relationship may be obscured in situations where the movement of wash load at the bed is sporadically important. This has conditioned theoretical approaches to sediment transport.

Sediment entrainment

This subject has been comprehensively reviewed by Leliavsky (1955) and by an ASCE Task Committee (1966). The entrainment of a single, noncohesive particle at the bed of a stream can be considered as a simple mechanical problem (Jeffreys, 1929; White, 1940). Bed velocities necessary to initiate motion can be determined by balancing moments of the fluid forces of drag and lift on the particle against the particle's submerged weight. It is evident from Bernoulli's equation (eq. 5) that zones of flow convergence, where velocity increases and pressure decreases, will be areas of enhanced entrainment capability by comparison with areas of flow divergence, where velocity decreases and pressure increases.

White considered that a particle would be entrained when the mean drag and weight moments were equal. Assuming a spherical particle, he expressed this condition as

$$v_{*c}^2 = \frac{2\tan\theta}{3C'}\left[\frac{\rho_s'}{\rho}\right]gD \tag{16}$$

where C' is a proportional drag coefficient that depends on the exposure of the grain at the bed and $\rho_s' = \rho_s - \rho$. Alternatively, this may be expressed as

$$\tau_c = C\rho_s'gD\tan\theta \tag{16a}$$

where

$$C = 2/3C'.$$

The equation is easily modified to take account of non-zero bed slope. Realizing that $D^3 \propto$ particle weight and also $v_*^2 \propto D$, we see that $v_* \propto$ (particle weight)$^{1/6}$, which is the classical sixthpower law.

Helley (1969) analyzed the situation to take account of lift force, which for a spherical particle is given as $0.18\rho v_*^2(\pi D^2/4)$, 0.18 being the lift coefficient of Einstein and El-Samni (1949). Helley also used a more general, elliptical particle and a refined analysis of particle dynamical centroid, taking into account the appropriate turning radii. He actually measured v_* as $v_{0.6c}$ (that is, velocity at a distance $0.6 \times c$-axis dimension above the bed), and considered varying specific gravity and varying shape factors (fig, 16) of particles. Experimental results for particles ranging from 15 to 60 cm in

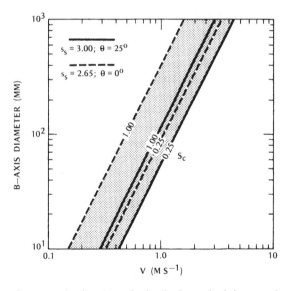

Figure 16 Entrainment criterion (near-bed velocity at incipient motion) for a range of particle specific gravity (s_s), angle of repose (θ), and shape factor $S_c = C/\sqrt{ab}$ (diagram modified after Helley, 1969, p. 11, reproduced by courtesy of the *United States Geological Survey*).

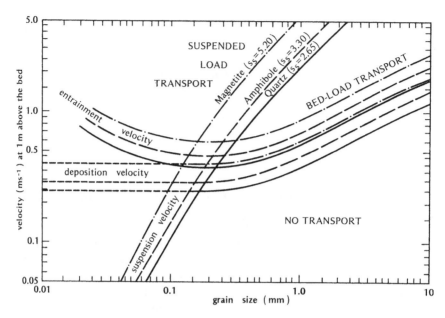

Figure 17A Relationship amongst water velocity, grain size and density, and sediment
movement (diagram redrawn after Sundborg, 1967, p. 337, and Ljunggren
and Sundborg, 1968, p. 134, reproduced by courtesy *Geografiska Annaler*).

diameter agreed reasonably well with the theory. The particles were placed
on the stream bed and were not constrained in any way.

Benedict and Christensen (1972) and Cheng and Clyde (1972) have
reported more detailed experimental work on the problem. Hjulström
(1939) presented a diagram showing critical erosion velocity for fine materials which was considerably elaborated by Sundborg (1956, 1967), who took
grain density into account (fig. 17A).

Shields (1936) approached the problem by dimensional methods, and
Raudkivi (1967; p. 20–23) and Gessler (1971) reviewed his analysis. He
re-expressed equation (16) as

$$\frac{1}{\psi_c} \equiv \frac{\tau_o}{\rho_s' g D} = C \tan \theta$$

It was found that C is not a constant, but depends on the "particle Reynolds
number," $R_{e_*} = D v_* / v$. This is reasonable when it is recalled that C represents a drag coefficient. The plot of $1/\psi$ gainst $\Phi \propto D v_* / v$ constitutes the
classical Shields diagram.

For small materials, the value of the entrainment function varies, since the
particles exist within the laminar sublayer near the boundary. Because
the sublayer thickness $\delta \approx 11.6 v / v_*$, we have

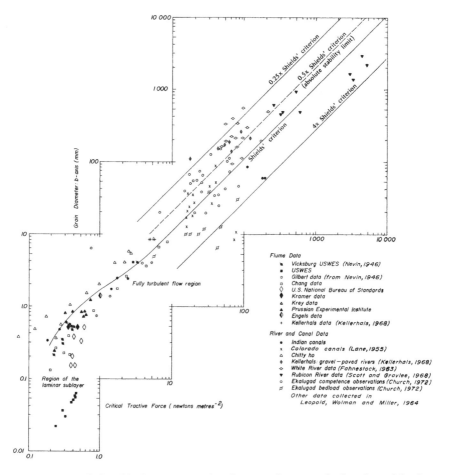

Figure 17B Relationship between tractive force at the streambed and particle size for incipient motion ($\tau_o = \tau_c$).

$$R_{e_*} = 11.6D/\delta \qquad (17)$$

Hence at $R_{e_*} \approx 10$, $D = \delta$, and $1/\psi$ has a minimum value. Below this the particles subsist entirely within the laminar layer and are progressively more difficult to move. For fully turbulent flow about a particle ($R_{e_*} > 600$), the entrainment function becomes constant; $1/\psi \sim 0.056$. Since $v_* = \sqrt{\tau_o/\rho}$, and since the same result was found for all reasonable values of ρ_s' the only free parameters remaining are tractive force, τ_o, and grain size D. The direct relationship between them is given in Figure 17B.

There is wide scatter of empirical data about Shields' results, and a variety of reasons why this may be expected. From the theoretical point of view, there are two major problems. Barr and Herbertson (1968) have discussed sediment transport in the light of similitude criteria. For the case where ρ'_s is substantially constant and the channel is wide (that is, two-dimensional flow), a generalized "Shields relationship" of the form

$$\Pi[R_{e_*}, 1/\psi, \Phi, d/D] = 0; \quad \Phi = \frac{q_s}{v_* D}$$

was found. By initially combining energy grade and flow depth in the shear stress, Shields had ignored at least one degree of freedom in the problem, which reappears in the apparently important scale factor d/D. The effect of this factor is demonstrated in the equation of velocity profile (eq. 8), which shows $v/v_* \propto \log d/D$, so that total flow depth and mean flow velocity enter into the relationship between v_* and D. In any event, this probably accounts for a portion of the scatter observed in the figure, which then is a two-dimensional representation of a three-dimensional surface.

Some scatter may be expected to occur in the simple nature of Shields' convention for the threshold condition. Movement of material does not begin abruptly, but rather as the sporadic movement of isolated grains. As the flow increases, such movements become more and more common until ultimately there is general motion on the bed. Shields chose as the "threshold condition" that stage at which "continuous movement occurs on the bed"—this implies that there is continually at least some material in motion. The stage at which this occurs is considerably higher than the stage of incipient motion, which has been estimated by various workers to occur at approximately $1/\psi = 0.03$. Gill (1972) has indicated that even lower figures may be contemplated. Some workers have pointed out that large materials, which are exposed to the flow, often move first at values of $1/\psi$ considerably lower than the supposed critical values, whereas small particles, which are sheltered by the large ones, require higher values of $1/\psi$ to be set into motion (Pantelopulos, 1955 and 1957; Egiazarov, 1964; Neill, 1968). Actual entrainment of particular particles will no doubt be strongly influenced by the previous history of deposition and scour.

Entrainment at the bed depends not on mean velocity and/or pressure conditions, but on instantaneous values. Many workers have reported that velocity fluctuations as much as twice the mean are common near the bed (White, 1940; Kalinske, 1947; Einstein and El-Samni, 1949), leading to 4× fluctuations of lift and drag. This permits entrainment of material that is as much as 4× as large as might be expected from Shields' criterion. For large materials, this is the result of direct impingement of turbulent eddies onto the particles. For small materials which lie in the laminar sublayer, Sutherland (1967) proposed a mechanism of periodic disruption of the sublayer by such eddies. Thompson (1965) indicated that the actual magnitude of velocity and pressure fluctuations may be as much influenced by a flow obstruction shedding wake eddies as by purely hydrodynamic factors. Lyles and Woodruff (1972) have presented evidence to show that turbulence intensity (hence velocity and pressure fluctuations) is influenced by the size, shape, and arrangement of bed elements. Hence, in a natural stream bed containing a mixture of particle sizes, the size and frequency of the larger protuberances on the bed control eddy intensity and entrainment characteristics, as was first postulated by White (1940) in his "dominant particle" theory. Grass (1970) has experimentally investigated the variation in shear stress distribution at the bed that arises from such fluctuations, and the variation in critical tractive force for actual entrainment of fine sand grains.

An alternative approach to this problem, suggested by Neill (1968), is to recognize that entrainment can only be dealt with in terms of some form of time-space average of displacements, so that $f(\Phi, \psi) = 0$ becomes a three-parameter function $f(\Phi, \psi, N) = 0$. Practical application of this idea is hindered by the difficulty of keeping track of the numbers of very small grains in mixtures. Grigg (1970) has reported experimental results for the behavior of individual sand grains.

A further cause of scatter about the theoretical criterion is the condition of the boundary. A non-cohesive bed may exist in three states:

(a) *Normal boundary*: materials resting in a non-disperse state, without imbrication, and with the usual packing arrangement (by *usual packing* is implied a random arrangement of grains, such as one might expect to find in a pile of sand dumped from a hopper—neither entirely "close" nor "open");

(b) *Overloose boundary*: materials resting in a disperse or dilated state, normally due to the presence of a large volume of water within the sediment; also materials with completely "open" packing. Such "quick" sediments are common in river channels which have been recently active, and have been observed to occur in sizes up to medium gravels, with a matrix of coarse sand;

(c) *Underloose boundary*: materials resting in a state of close packing or of imbrication. This is the common condition in gravel and cobble-floored streams.

Most experimental work has been done with normally loose boundaries, and it is probably fair to state that Shields' criterion refers to this state. On the contrary, observations in nature apply mainly to one of the other two states. Clearly, the dislodgement of materials at an overloose boundary will require less than the indicated tractive force by virtue of the reduced solid friction, and it will require greater than the indicated force at the underloose boundary. This is seen in Figure 17B, where observations of moving cobbles in the White River, and of marked cobbles set on top of the channel bed at Ekalugad Fiord, plot well above Shields' relation. In both cases the cobbles would behave as overloose materials. Conversely, observations of tractive

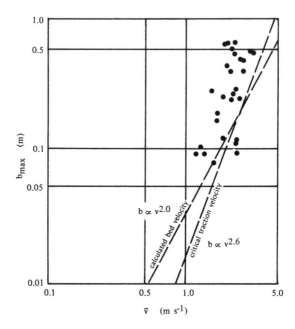

Figure 18 Relation between particle size and mean flow velocity, as observed by Fahnestock (1963), White River, Mt. Rainier. The functional relationships were derived by Nevin (1946). The variation in particle size with velocity appears to correspond to a *"2.6-power law"* more closely than to the traditional *"2-power law"* (or *6-power law* for particle weight). Diagram modified from Fahnestock, 1963, p. 29 and reproduced by courresy of the *United States Geological Survey*.

force necessary for particle movement on the bed of some Colorado irrigation canals, in Indian canals, and at Ekalugad Fiord, plot below the theoretical relationship, indicating an underloose boundary condition.

The competence of a stream to maintain material in motion, once entrained, may considerably exceed that which would be indicated from tractive force criteria. Little direct study has been made of the competence of glacial streams; however, Fahnestock (1963) has reported some interesting extreme data (fig. 18). It is well known that, for live-bed conditions in high energy streams, large boulders may move on a sand carpet, possibly supported by dispersive stresses in the sand. Fahnestock and Haushild (1962) carried out flume experiments which showed that large boulders (D/d up to 1.0) could be moved over a sand bed in upper regime flow.

Very fine materials (<63 microns) may possess cohesive properties; particles of clay size (<2 microns) certainly will. Such materials are not characteristic in high-energy environments, but may occur in backwater areas, or in certain overbank zones after high floods. For such materials, electrochemical surface forces between particles are more important than

440

gravity forces in controlling erodibility. There is no need for entrainment to occur on a particle-by-particle basis. Erodibility of cohesive sediments has been related to various gross properties of the material, including shear strength (Sundborg, 1956; Dunn, 1959), dispersion ratio, plasticity index and moisture content (Smerdon and Beasley, 1959; Moore and Masch, 1962). The range of results has been synthesized by an ASCE Task Committee (1968), and reviewed by Partheniades and Paaswell (1970). Various materials appear to behave radically differently depending on their physical and electro-chemical state. The cohesive forces are imperfectly understood. Since they may also control the behavior of fine materials in suspension, they also affect the transport and deposition of the material.

Naleds

A singular form of channel-bed disturbance, which may affect the stability of sandur channels and sediment entrainment in permafrost regions, is the occurrence of *naleds*. These "ice mounds" (fig. 19) develop beneath the bed of the channel after freezeback in the autumn. Water left in the channels and the channel bed freezes downward and ultimately meets the frost table below. Some zones ultimately become closed, isolated cells of unfrozen water and sediment below the surface. As freezing continues, static pressure grows in the remaining zone of unfrozen material. If pressure is great enough the frozen "roof" is forced upward, or may be ruptured completely so that water flows out and freezes above, forming *aufeis*.

Individual structures vary in dimension from 2 or 3 m in diameter and a few cm in elevation to about 50 m in length and several metres in height. Where rupture occurs, extensive *aufeis* may develop if the water supply is persistent (fig. 19C). Some naleds are composed of completely clean ice, probably indicating that they developed from a reservoir of water trapped between river ice *above* and the frozen bed *below*. Other forms appear to have lifted individual large cobbles or isolated masses of material from the bed (fig. 19A), suggesting that the freezing plane was near the bed itself. In still other cases the entire river bed has been lifted up (fig. 19B), indicating that the freezing plane was *below the bed*.

These structures may severely derange a section of channel bed which was quite stable before, and their persistence after commencement of flow in the spring may lead directly to a rear-rangement of the channel pattern. They also pick up a great deal of material—some of it very large—that must be moved at least a small distance if the naled ultimately melts into flowing water.

Suspended sediment

The theory of transport of suspended sediment has recently been reviewed by Ippen (1971) (*see* also Vanoni, 1946; ASCE Task Committee, 1963). It is

Figure 19 Development of naleds and aufeis: A, Individual naled in South River, Ekalugad Fiord, Baffin Island, which has picked up large cobbles from the streambed, B, Sandy section of South River bed which has been heaved up in its entirety about 1.5 m. C, Extensive aufeis on the Canning River, Alaska. These develop where persistent springs from below permafrost continue to flow throughout the winter.

assumed that suspended sediment diffusion is equivalent to momentum diffusion in turbulent flow. In order to maintain the suspended load, upward diffusion of material must be equivalent to settling:

$$c_{ss}w = -\frac{\eta dc_{ss}}{\rho dy} = -\varepsilon\frac{dc_{ss}}{dy} \tag{18}$$

where c_{ss} is the concentration of sediment. Integration of this equation after substitution for the kinematic eddy viscosity, ε, of the expression $\kappa v_* y(1 - y/d)$ (see standard references for this development) yields:

$$\frac{c_{ssy}}{c_{sso}} = \left[\frac{y_o}{y}\frac{(d-y)}{(d-y_o)}\right]^{w/\kappa v^*} \tag{18a}$$

which gives sediment concentration at any depth, y, relative to concentration at some reference depth y_o. The necessity to know concentration at this depth is predicated on the fact that the suspended-sediment load depends on availability of sediment as well as on flow. Figure 20 shows typical suspended-sediment curves; c_o is usually taken as a maximum measured value of c very close to the bed.

Two problems arise in the assessment of suspended-sediment transport via this equation:

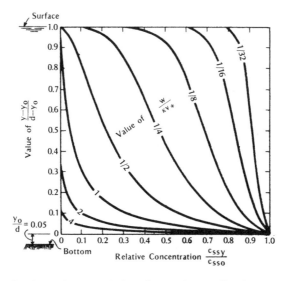

Figure 20 Distribution of suspended sediment in profile for several values of $w/\kappa v_*$ (after ASCE Task Committee, 1963, reproduced by courtesy of the *American Society of Civil Engineers*).

(1) w is a function of temperature (Straub, 1955; Straub *et al.*, 1958; Colby and Scott, 1965), because it decreases with increasing water viscosity. This may have a major effect on the profile of suspended-sediment concentration, allowing sediment transport to change by a factor of as much as 2× over a range of about 20°C. Glacial meltwater streams are generally cold and have a high transport efficiency. The effect is most pronounced for fine sands (80 to 90 microns), but appears to be unimportant for materials <62 microns and >250 microns.

(2) κ depends on sediment concentration (Ippen, 1971).

Particle size of suspended materials in glacial streams appears to vary over a range larger than that between 1 mm and 0.001 mm. The size of smallest material will probably depend on local geology, and upon weathering and glacial comminution. Amongst coarser materials, the distinction between bed and suspended load is arbitrary, since material may move on the bed for some distance, be swept into suspension where turbulence is very intense, and be deposited again on the bed downstream. In very exceptional floods, materials larger than 8 mm may move in long saltation jumps. Maximum mean concentration of suspended sediment measured in glacial streams also appears to be largely a function of geology (table 3) and of climate (values in the Arctic are generally low).

Diurnal variation in suspended-sediment concentration is illustrated in Figure 21. Peak sediment concentration normally leads peak discharge by a few hours (Fahnestock, 1963; Østrem and others, 1967); that is, there is

Table 3 Some maximum observed mean concentrations of suspended sediment in proglacial streams.

River	Datum	Local Geology	Source
Small melt-streams	39,000 ppm	volcanic	Klimek, 1972
Skeidarársandur, Iceland White River, Washington	17,200	volcanic	Fahnestock, 1963
Athabasca Glacier	8,000–10,000 (estimated)	sedimentary	Mathews, 1964b
Isortoq. Fd., West Greenland	9,750	—	Jensen, 1881
Ekalugad Fd., Baffin Island	4,800	granite-gneiss	Church, 1972
Chamberlin Glacier, Alaska	2,800	sedimentary	Rainwater and Guy, 1961
Norwegian glaciers	2,600	granite-gneiss	Pytte, 1969
Lewis River, Baffin Island	1,740	granite-gneiss	Church, 1972
Decade Glacier, Baffin Island	1,060	granite-gneiss	Østrem, Bridge and Rannie, 1967

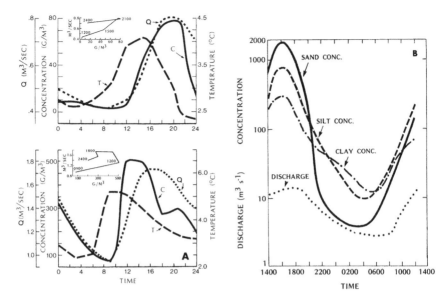

Figure 21 A, Diurnal patterns of suspended sediment concentration, stream discharge, and water temperature for Decade River, Baffin Island, July 28 (upper) and August 4 (lower), 1966. *Inset*, the relationship between stream discharge and suspended-sediment concentration, showing hysteresis over the 24-hour period (data by courtesy of W.F. Rannie). B, Diurnal pattern of suspended-sediment concentration and stream discharge downstream from Chamberlin Glacier, Alaska. Note that the three size fractions maintain phase, but amplitude varies—finer material exhibits more conservative behavior (after Rainwater and Guy, 1961, p. 9, reproduced by courtesy of the *United States Geological Survey*).

hysteresis in the sediment concentration *versus* stream discharge relationship. Reasons adduced for this are greater availability of wash load during the rising limb of a flood (Leopold and Maddock, 1953) from overland flow, or (more appropriate in the present context) from freshly slumped channel banks; or from different hydraulic conditions between rising and falling limbs of the hydrograph (Skibinski, 1965), with higher flow velocities and greater water-surface slope on rising stage leading to greater shear stress at the bed and greater sediment entrainment. The effect of changing water temperature in modulating this relationship appears to be relatively slight.

Practical measurement of suspended-sediment transport may be tedious, unless there is a more or less unlimited supply of wash-load material available to the river. Such a condition is not infrequently approximated in proglacial rivers, where material freshly melted from ice, moraine erosion, and renewed entrainment of outwash deposits provide abundant sources. In such circumstances, suspended-sediment discharge relationships of the form $c_{ss} = pQ^i$ are

445

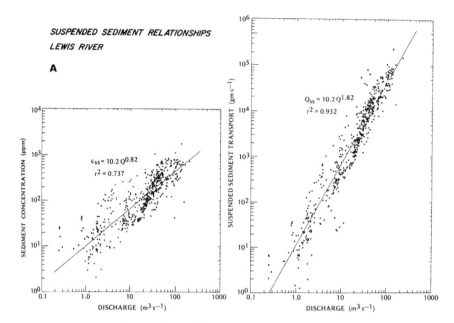

Figure 22A Suspended sediment *versus* discharge relationships for Lewis River, Baffin Island. Note that the rating relationship between sediment discharge and water discharge, although useful for computing total sediment transport, is statistically spurious, since it involves a correlation between $c_{ss}Q$ and Q. Standard error of estimate is the same as that derived from the left hand plot, despite the apparently improved relationship (after Church, 1972, fig. 34 (part), reproduced by courtesy of the *Geological Survey of Canada*).

found to hold (fig. 22), though there remains considerable scatter in the data. Straub (1935), and Campbell and Bauder (1940) had previously found such relationships, whereas Leopold and Maddock (1953) included them as part of their "hydraulic geometry." Rannie (written comm.) has shown that some of the remaining variance can be accounted for by considering the hysteresis effect that accompanies change in discharges (fig. 22B).

A study of various possible effects introducing additional variance into the data of Figure 22A (and data gathered from several other proglacial rivers in Baffin Island) included the rate of change of discharge, conductivity of water (influencing sediment flocculation), water temperature, and length of time (number of days) since the last rainstorm (P). Only the last factor was consistently significant. This reflects the importance, even for these rivers, of the fluctuating supply of wash load derived from overland flow or channel-bank seepage following rains. The general relationship gave

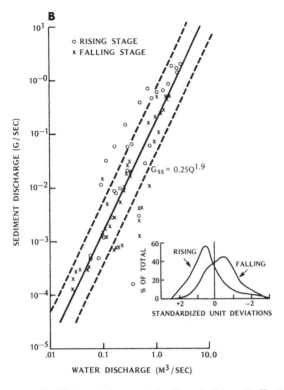

Figure 22B Suspended sediment rating curve for Decade River, Baffin Island, showing the discrimination between rising stage and falling stage measurements. Two separate ratings could be used to give more precise results (data by courtesy W.F. Rannie).

$$c_{ss} = pQ^{j}P^{r} \qquad (19)$$

where $0.75 < j < 1.5$; $-0.20 > r > -0.60$ approximately.

Bed load

In most rivers, the bulk of material in transport is suspended load. In some, however, the general prevalence of mechanically weathered materials means that most particles are of sand size or larger; here bed-load transport dominates by virtue of sediment character. Proportionally high bed-load transport rates are also characteristic of steep gravel-bed channels. Proglacial rivers generally are of this type, at least in their proximal reaches.

The theory and measurement of bed-load sediment transport are less well developed than almost any other aspect of open channel hydraulics. There is, as yet, no completely valid theory and no satisfactory method for routine

447

field measurement. This is especially true for gravel-bed rivers: by far the most effort has been expended on the study of sediment in sand-bed rivers. Recent reviews include those of the ASCE Task Committee (1971a, 1971b) and Bogardi (1972).

The problem of measurement has made little progress since Colby's (1963) review. Measurements using trap-type samplers (Hollingshead, 1971) are carried out, and some attempts have been made to carry out routine measurements (Stichling and Smith, 1968). Stichling (1969) reviewed available instruments. However, problems of sampler interference with flow, variable sampler efficiency in different flows, and the difficulty of designing a sampler that will trap, with equal efficiency, a wide range of material sizes, have proven difficult. Ultrasonic, pressure, and impulse methods have all been investigated. Special structures can sometimes be constructed to abstract sediment from flow (Klingeman and Milhous, 1970) or put it all into suspension for easier measurement (Hubbell and Matejka, 1959). Radioactive tracer techniques may be useful in high energy environments (*see* Crickmore, 1967, for discussion of techniques). None of these methods is routine, and all are likely to be difficult to apply in the unstable fluviatile environment of a proglacial stream. Fahnestock (1963) and Church (1972) have presented some measurements to show competence of proglacial streams, but no complete measurement of bed-material transport has been achieved at significant flow levels.

Due to the persistent difficulty of measuring bed-load transport, a great deal of attention has been paid to the development of computational formulae. Most bed-load formulae are based more or less explicitly on tractive force criteria, and have been arrived at by empirical or semi-rational means. The earliest was the Du Boys equation, put forward *ca.* 1880:

$$q_{bs} = C_s \tau_o (\tau_o - \tau_c) \tag{20}$$

where C_s is a material parameter (Straub, 1935). Of the many variations on the Du Boys formula, the most important was that of Shields:

$$\frac{g_{bs}}{\rho q S} = 10(1/\psi - 1/\psi_c) \tag{21}$$

where S refers to the river bed slope in this instance. The formula is dimensionally homogeneous and fits a wide range of data reasonably well.

Schoklitsch (1934) presented a variation expressed in terms of discharge (see Shulits, 1935):

$$q_{bs} = \Sigma p_i \frac{25.3}{D_i^{1/2}} S^{3/2}(q - q_{ci}) \tag{22}$$

where $q_{ci} = 0.638 \, D_i/S^{4/3}$ is the flow at which material of diameter D_i will begin to move, and p_i is the proportion of material of size D_i on the bed. The formula has been popular because it directly relates sediment transport to discharge.

Beginning in 1942, H.A. Einstein published a series of studies which led to his formulation of the "bedload function" (Einstein, 1950), $\Phi = q_{bs}/wD$, where w is the fall velocity of the grains in transport. For high transport rates, his bed-load transport formula was shown by Brown (1950) to be equivalent to

$$\Phi = 40(1/\psi)^3 \tag{23}$$

Though derived for uniform grain sizes, it is claimed that the formula can be extended to mixtures if D_{35} is chosen as the representative grain size. Equation 23 ignores the Shields threshold condition $1/\psi_c = 0.056$, since at high transport rates the correction is minor and the approximation using only $1/\psi$ suffices. Kalinske (1947) and Graf and Acaroglu (1968) have developed functionally similar formulae.

Einstein's analysis was elaborated considerably in 1950 and a complex series of computations was outlined for estimating sediment transport. Two salient features are important. First, Einstein developed his procedure so that estimates of total sediment load are made (*see* Colby and Hubbell, 1961, for modified computational procedures). This draws attention to the fact that, merely in considering an "entrainment function," no real distinction is made between material that may remain in bed load and material that may move into suspension. Further, so long as zero net flux is maintained between material entering into and settling out of suspension, this difference need not be detectable. Second, Einstein distinguished clearly amongst several mechanisms that contribute to flow resistance and hence to dissipation of the power of the stream. These include: (a) direct frictional resistance, or "skin" resistance, of bed and banks; (b) form resistance of channel bed structures, such as ripples or bars; and (c) structural resistance of channel patterns, such as bends. Only the first factor, since it alone impinges directly on the grains at the bed, should be considered in sediment entrainment. Einstein accordingly divided total flow resistance into "grain resistance" and "bar resistance," and considered only the former.

Another approach to bed-load transport considerations was that of Meyer-Peter and Müller (1948). Extensive laboratory data led them to propose an equation which can be put in the form

$$g_{bs} = 8.57 \frac{Sd}{(nk_r)^{3/2}} - 0.67 D_m^{3/2} \tag{24}$$

where

449

D_m = the effective diameter of the live bed material mixture, usually the D_{50} value;

k_r = coefficient of particle friction with a smooth bed,
= $26/D_{90}^{1/6}$

n = Manning roughness coefficient.

This formula also recognized the distinction between particle friction, k_r, and total resistance n. Chien (1954) has shown that equation (24) can be reduced to:

$$\Phi = 8(1/\psi - 0.047)^{3/2} \qquad (24a)$$

The constant, 0.047, represents the critical entrainment value $1/\psi_c$. This is slightly less than the value indicated by Shields for fully turbulent flow.

The question of assessing sediment transport in the zone of sporadic movements (below the conventional threshold value) and in the zone of "discrete" movements above threshold but below live bed conditions, has previously been handled in several ways. Einstein introduced a correction function for bed-load transport rates at low entrainment intensities, based on the observation that movements are apparently random—presumably in reflection of the apparent space-time randomness of critical stress fluctuations. The theory is not well established, and an alternate approach, exemplified by Meyer-Peter and Müller, has been to adjust the transport equation by an empirical factor representing the necessary entrainment value for sporadic movement to begin.

Bagnold (1966, 1968) has examined the physics of sediment motion from a basic mechanical standpoint. He pointed out that transport of a fluid-sediment mixture downchannel requires the expenditure of energy to overcome friction. The energy is supplied from the loss of potential energy as the stream flows downslope, and is equal to the power supply per unit length of stream:

$$\omega = \bar{d}\bar{v}gS(\rho + \bar{c}_{ss}\rho_s) \qquad (25)$$

per unit width. Ignoring the second term (which is for the solids carried in the flow, and is usually a small proportion of the total flow), this becoms $\omega = \tau_o\bar{v}$. It is not surprising then, that tractive force formulae alone do not produce universal results. Application of this concept led Bagnold (1966) to the result:

$$q_s = \frac{\omega}{\gamma(s_s - 1)}\left[\frac{e_b}{\tan\theta} + 0.01\frac{\bar{v}}{w}\right] \qquad (26)$$

where

e_b = transport efficiency (generally $0.10 < e_b < 0.15$)

$\tan\theta$ = solid friction coefficient = $\tau/(s_s - 1)gDc_o$

and

c_o = static volume concentration of material ~ 0.65.

The last term ($\bar{v}/w$) refers to suspended load. If we ignore it, we have for bed-load transport:

$$q_{bs} = \frac{\omega}{\gamma(s_s - 1)} \cdot \frac{e_b}{\tan\theta}$$ (26a)

The formula is not generally operational. It can be expressed as follows:

$$\frac{g_{bs}}{\bar{v}D} = \frac{\tau_o}{\gamma(s_s - 1)D}\left[\frac{e_b}{\tan\theta}\right]$$ (26b)

that is, $\Phi = f(1/\psi)$, where the functional relation is seen to depend on transport efficiency and the solid friction coefficient. These factors are related to bed condition and load availability.

Coarse materials

It is worthwhile to consider the transport of gravels separately from that of fines. Gravels will almost always be moved on the bed as bed-material load, whereas fines may readily move into suspension, given sufficiently powerful flow. In such circumstances a total sediment-transport computation is probably more appropriate, and some factors that influence the balance between suspension of sediment and movement on the bed must be considered.

The formulae discussed above were all developed with the aid of flume experiments utilizing sands and gravels (usually well sorted), generally in the range 0.25 to 30 mm (*see* ASCE, 1971b for further details). The Shields, Schocklitsch, and Meyer-Peter and Müller formulae were developed after experiments with very little or no suspended load, whereas the Einstein formula has been tested for a wide range of flow conditions. These formulae are thought, then, to represent conditions of dominant bed-load transport of coarse materials.

Hollingshead (1971) has critically tested the two most sophisticated formulae (Meyer-Peter and Müller, and Einstein) in a gravel-bed river. The problem with the formulae appears to lie not in their basic functional character, but in the matter of applying them to a real channel, by the use of mean flow properties, in order to obtain shear stress or shear velocity. These characteristics vary in detail in the channel and may not be applicable to the section of channel in which bed movement actually occurs. As an alternative procedure, Hollingshead proposed the following:

(a) determine effective width, w_e, over which bed sediment movement occurs as a function of discharge (Hollingshead used hydrophones to determine the zone of active transport);

(b) for the "effective" portion of the channel only, determine $\bar{v}_e$ and $\bar{d}_e$ as a function of discharge;

(c) determine the following parameters for various flow stages:

$$\text{densimetric Froude number } F\tau' = \frac{\rho_f \bar{v}_e^2}{\rho_s' g \bar{d}_e}$$

roughness ratio, d_e/D_{50}

"effective" discharge $Q_e = w_e \bar{d}_e \bar{v}_e$;

(d) using $F\tau'$ and the roughness ratio, bed-load concentration is determined from Figure 23; and

(e) the concentration is used with Q_e to compute total bed-load discharge for each stage.

This procedure utilizes specific information about sediment-transport conditions, hence is more likely to yield reasonable results than uncritical use of formulae. The chief limitation lies in the question of whether the values of c_{bs} are applicable, derived as they are from flume data. Even if this should become a problem, the procedure for determining "effective" flow parameters, when applied to the usual formulae, should yield improved results. An alternative procedure is to realize that the critical tractive force in formulae of the type given in equation (21) may vary from river to river according to the entrainment conditions. Hence, a study of riverbed morphology, or observation of conditions when bed motion is just beginning, may allow a more reasonable estimate of τ_c.

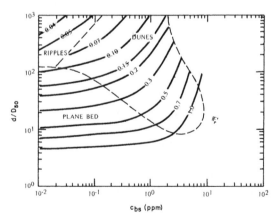

Figure 23 Relationship amongst $F\tau'$, c_{bs}, d/D_{50} (derived from flume data; after Hollingshead, 1971, p. 1830, reproduced by courtesy of the *American Society of Civil Engineers*).

Fine materials

Formulae explicitly developed for use with sand-bed rivers include that of Engelund and Hansen (1967; cited in ASCE, 1971a), which can be put in the form:

$$\Phi = 0.05 \left(\frac{v}{v_*} \right)^2 (1/\psi)^2 \tag{27}$$

This formula is basically a tractive force formula, modified by a velocity term. Toffaletti (1969) has presented a formula for total discharge of bed sands in large rivers, and the results appear to accord well with observations for sand-bed rivers (ASCE, 1971a).

Colby and Hubbell (1961) have presented a modified Einstein procedure for total sediment discharge computation that has been extensively tested in sand-bed streams with good results. The procedures are complex and the reader should consult the references for an explanation.

Colby (1957, 1964a), using empirical data, has developed a method for estimating sediment discharge in sand-bed streams. Required data are: (a) mean width (w), depth (d), velocity (v) of the stream; (b) measured mean suspended-sediment discharge concentration $\bar{c}_{ss}$; and (c) measured discharge concentration $\bar{c}_m$ of *bed* sediment. The latter figure is derived from the suspended-sediment sample by application of a bed-sediment distribution curve. If this is unknown, the limiting size between wash and bed sediment load is arbitrarily accepted as 62 microns. The computational steps proceed via a series of nomograms. Colby (1964b) has elaborated this procedure greatly, in particular paying attention to effects of varying sediment concentration and water temperature. The procedure is based on velocity as the basic measure of stream power rather than tractive force, and no threshold value is considered.

Conventional bed-load transport equations may be expected not to apply when the stream-bed is rippled or duned. In such circumstances, the assumptions about uniformly representative shear or velocity parameters for the bed will not apply. Simons and others (1965b) investigated bed-load movement over ripples and dunes in a flume. Their results seem to be substantially limited in applicability to two-dimensional flows with relatively coarse sands. No field tests are known.

Load distribution

Discussion of the time distribution of bed-load sediment transport is made difficult by the relative paucity of direct measurements (Østrem, this vol.). Computational formulae relate the sediment discharge to some hydraulic parameters which are ultimately functions of water discharge. The presence of a finite threshold in most formulae effects an adjustment in the distribution of bed-load transport by comparison with that of water, so that most of the sediment movement takes place in a more restricted time than does the flow.

Most of the equations presented above can be reduced, using the appropriate factors of hydraulic geometry, to equations of the type:

$$q_{bs} = f\left[\frac{d}{D}, S\right]$$

$$\approx f(Q) \tag{28}$$

because relative roughness varies with discharge. As is pointed out by the ASCE Task Committee (1971a), most available data seem to fit relations of the form:

$$q_{bs} = C_1 Q^{C_2} \tag{28a}$$

The fit is good for sand-bed rivers, with low or negligible threshold values, and then $1 < C_2 < 2$ as a rule. For gravel-bed streams, at discharges somewhat above the threshold, the relation holds approximately with $0 < C_2 < 1$. The implication is that the rate of transport increases less rapidly in gravel-bed streams than in sand-bed ones, probably because "live bed" conditions do not develop so readily in the former.

Recommendations for computing bed-load sediment transport

It is evident that no computational procedure presently available represents all the dimensions of significance that may affect transport. If formulae must be resorted to, care should be taken to select ones that were developed under circumstances which conform as reasonably as possible to the field situation (since most formulae are based on flume data to begin with, this may be a difficult criterion to fulfil). In all events, computations should be based on actual measurements of sectional hydraulic properties, rather than on mean values, and field corroboration of results should be sought.

ASCE (1971a) and Tywoniuk (1972) have recently reviewed available procedures and made some recommendations. For conditions likely to occur in proglacial streams, the following may be considered. For gravel-bed streams:

(a) If measurements are available, use modified Einstein procedure;
(b) in absence of measurements, use Einstein, Meyer-Peter and Müller or Schocklitsch formulae, with adjusted threshold parameter; and

454

(c) for simple calculations, Schocklitsch equation is recommended for its convenience.

For sand-bed streams:

(a) if measurements are available, use Colby's procedure;
(b) in absence of measurements use Engelund-Hansen formula.

Hollingshead's (1971) means of partitioning the section into effective and non-effective zones should be applied if practicable.

If large bedforms are present, none of the convential formulae may apply. If reasonable measurements or computations can be made, simple, power-law relations may be sought as the most precise means warranted by the results for extending the data to compute total sediment discharge.

Total sediment yield

Little comparative information is available on the relative importance of suspended and bed load in proglacial rivers, but Table 4 summarizes some results using computed bed-load transport values. The streams are arctic, gravel-bed rivers. Table 5 shows examples of the proportion of total seasonal runoff and sediment transport achieved cumulatively on the highest 1, 2, 3, 4 and 5 days of flow or transport. Interesting features of the table include the very high concentration of sediment movement events achieved by storm-dominated hydrographs as opposed to melt-dominated ones (*cf.* Lewis River 1964 as opposed to Lewis River 1963), and the extremely high concentration of transport in single remarkable events (*cf.* The South River, 1967, records, within which the *jökullhlaup* of 20–22 July stands out). These are far from the most extreme results on record, but they do illustrate that a major proportion of total sediment transport takes place in a very restricted period of time.

No data are available to assess the magnitude of long-term fluctuations in sediment supply from glacierized watersheds on an annual basis. Spatial variation in fluvial sediment yield from glacierized watersheds depends, first,

Table 4 Comparative contributions of suspended and bed-load sediment transport in certain proglacial rivers.

River	Sediment Type	Suspended Load	Bed load
Lewis River 1963	gravel	18%	82%
Lewis River 1964	gravel	22%	77%
South River, Ekalugad Fd. 1967	gravel	10%	90%

Data from Church, 1972.

455

Table 5 Proportional distribution of discharge/sediment transport for high flow events.

River	Days of Record	Q	Cumulative Proportional Discharge/ Transport During Highest Flow Days				
			1	2	3	4	5
Lewis, 1963	70	Q	.053	.092	.130	.166	.200
		Q_s	.097	.157	.213	.265	.312
Lewis, 1964	64	Q	.116	.199	.269	.335	.396
		Q_s	.296	.451	.553	.648	.727
Lewis, three seasons (inc. 1965)	206	Q	.032	.059	.081	.102	.122
		Q_s	.078	.135	.179	.220	.255
South, 1967	66	Q	.103	.155	.200	.245	.288
		Q_s	.466[a]	.562	.635[a]	.702	.767

[a] Jökullhlaup of July 20–22.

on the position at which measurements are taken. Sediment yield from glacial deposits is usually considerably higher than normal yield for an area (Church and Ryder, 1972), and the farther one moves from the glacier, the smaller the specific yield becomes as the influence of the glacier becomes less important. Borland (1961) noted this effect in south-central Alaska. His data have been reworked and presented in Figure 24 in a form that clearly shows the effect of increased distance from the glacial source of sediment. A simple effect of increasing watershed size is confounded in these results; Schumm (1963) indicated that specific sediment yield $\propto$ (drainage area)$^{-0.15}$ approximately.

Other factors of importance in affecting total sediment yield include the following: (a) geology, which will determine the resistance of rocks to glacial erosion, hence the volume of detritus produced; (b) climate, which will determine the length of runoff season, character of spring melt period, effect of summer rainstorms, etc.; and (c) glacier regimen, which will determine the relative volume of runoff to precipitation at any one time; also the flow rate and (to some extent) erosive capacity of the glacier.

The broad relationship between glaciation and fluvial sediment movement has been considered by many writers. Broecker and others (1968) have shown that gross sedimentation rates are generally much higher during glacial periods than during interglacial times. Tricart (1952), Frye (1961), and Embleton and King (1968; p. 406–419) have discussed various relationships between glacier activity and outwash sedimentation.

Because of the great variability of local conditions, it is unlikely that systematic relationships between glacial activity and fluvial activity will be apparent in the short term at any one glacier. In particular, because meltwater

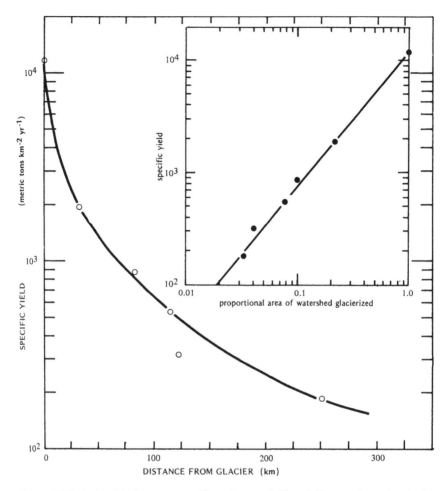

Figure 24 Relationship between specific sediment yield and distance from the glacier for stations in south-central Alaska (data from Borland, 1961).

flows through moraine and over previous glaciofluvial deposits, there is not likely to be a close relationship between glacial erosive activity, and concurrent downstream fluvial sediment movement or sedimentation at time scales of less than several decades. For design and comparative purposes, Table 6 presents some reported values of proglacial sediment yield rates.

Sediment transport in subglacial streams

Subglacial streams may flow either in tunnels with a free-air surface, or in closed conduits. In the latter case, pipeflow conditions prevail. For low sediment concentration, the sediment transport mechanics remains similar

Table 6 Indicated proglacial sediment yield rates.

Area	Datum (as metric tons/km²/yr)	Remarks
A. Arctic		
Lewis River, Baffin Island	735	Three years of record of sediment transport. 204 km²: 89% ice-covered. Granite-gneiss terrain. (Church, 1972).
Ekalugad Fiord, Baffin Island	400–790	High estimate: one year's record of sediment transport; low estimate: gross sedimentation over 2,000 years. 384 km²; 21% ice-covered; Granite-geniss terrain (Church, 1972).
Colville River, Brooks Range, Alaska	140	50,000 km². One season's record. (Arnborg and others, 1967) Minor perennial ice. Limestone, sandstone, shale.
B. Subarctic Mountains		
Erdalsbreen, Norway	1,300	16 km²: 70% ice-covered. 1967–69 mean. No bed-load data. Granite-gneiss terrain. (Østrem and others 1970).*
Vesledalsbreen, Norway	140	As above. 7.2 km²: 56% ice-covered. (Østrem and others, 1970).
Austre Memurubre, Norway	330	As above. 15.9 km²: 56% ice-covered. (Østrem and others, 1970).
Nigardsbreen, Norway	330	57 km²: 75% ice-covered. 1968–69 mean. Total sediment transport. (Østrem and others, 1970) Granite-gneiss terrain.
Rapaälven, Sweden	265	684 km²: 12% ice-covered. Igneous and metamorphic terrain. Subarctic. (Axelsson, 1967).
Hoffellsjökull, Iceland	4,290–8,220	313 km²: recent volcanic terrain. Data recomputed from Thorarinsson (1939).
C. Glacial Mountains		
Austrian watershed	1,500	165 km²: 47% ice-covered. (Lanser, 1958).
Venter-Ache, Vent, Austria	1,150–1,520	165 km²: ice-free, but high rate of moraine erosion. (Moossbrugger, 1958).
Upper Indus, Karakoram Ra.	2,900	190,000 km². (Hewitt, 1968). Sedimentary rocks.
Susitna River at Denali	1,909	2,200 km²: 21% ice-covered: 33.4 km from nearest glacier. (Borland, 1961).
Susitna River at Gold Creek	185.5	16,100 km²: 3.4% ice-covered: 250 km from nearest glacier. (Borland, 1961).
Southern-central Alaska mountains	11,500	At glacier snouts. (Borland, 1961). *Estimate only.*

* Data of Østrem and others, 1970, have been adjusted to distribute eroded material over the entire contributing watershed. Their report assigns it all to the ice-covered area.

to that for open channels. However, for high sediment concentration the fluid may be modified so that it behaves in non-Newtonian fashion. No comprehensive theory for such flows exists. Raudkivi (1967), ASCE (1970), Bain and Bonnington (1970, Chap. 1), and Graf (1971) present summary reviews of relevant concepts applicable to sediment transport in pipes.

For a pipe of given size, transport conditions will depend upon flow velocity, sediment concentration, and particle size distribution. Figure 25 indicates the four principal transport regimes as a function of particle size characteristics and flow velocity. Situations of practical interest in subglacial sediment transport will be restricted to the first three regimes. The major problem of interest is to determine the conditions under which deposition or erosion of bed material occurs within the pipe.

The discrimination between depositing and eroding conditions has been determined to be of the form

$$v_c = K\sqrt{2gd(s_s - 1)} \tag{29}$$

where d is the pipe diameter and K depends on concentration of sediment (by volume) and particle size (Durand-Condolios relation). Minimum values of head loss occur at v_c somewhat in excess of the minimum velocity at which no deposit occurs in the pipe (fig. 26; in the studies upon which these results were based, smoothwalled pipes of regular geometry were employed).

For total sediment transport in a closed pipe, Graf and Acaroglu (1968) developed a tractive force type relationship from a variety of empirical data:

$$\Phi = 10.4(1/\psi)^{2.5} \tag{30}$$

which corresponds with their similar relationship for open channel flow.

Application of these results to flow in glacial tunnels presents several problems. The fore-going analysis properly applies to horizontal, regular pipes of small diameter. In inclined pipes, head losses become larger, and in irregular pipes there is no obvious way to make calculations of losses to determine what the transport regime may be locally for a given set of flow conditions. Qualitatively different features may develop in very large pipes for certain flows.

Sediment deposition

Sediment deposition occurs under the reverse set of conditions than sediment entrainment. Deposition in most fluvial environments proceeds in more or less orderly fashion, as the flowing water sorts the material and deposits it in certain geometrical forms. In a perturbed flow field, locally selective erosion and deposition creates distinct bedforms. A bedform is "any

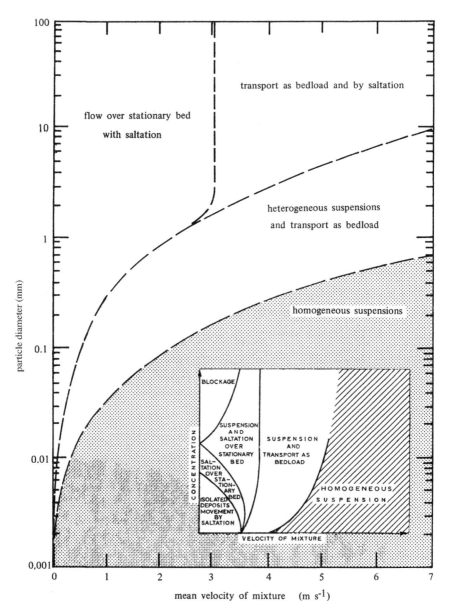

Figure 25 Principal transport regimes for sediment (s.g. = 2.65) in a closed conduit; *inset*, schematic effect of varying sediment concentration with constant particle size and specific gravity (from Raudkivi, 1967; after Newitt and others, 1955). The shaded area is of no practical importance for flows in nature (after Raudkivi, 1967, p. 316–317, reproduced by courtesy of *Pergamon Press*; based on results of Newitt and others, 1955, and reproduced by courtesy of the *Institution of Chemical Engineers*).

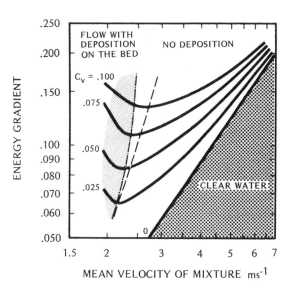

Figure 26 Discrimination between eroding and depositing conditions in a horizontal closed conduit. Critical velocity is little affected by concentration and particle size for $D > 0.5$ mm. The results are affected by grain size; here the diagram applies to conditions for coarse sands (after Raudkivi, 1967, p. 318, reproduced by courtesy of *Pergamon Press*; based on results of Durand, 1953, reproduced by courtesy of the *International Association of Hydraulic Research*).

deviation, from a flat bed, that is readily detectable by eye or higher than the largest sediment size present in the parent bed material" (ASCE Task Force, 1966). The geometry of bedforms is related to the character of the flow, and not to the character of the initial disturbance that produced the perturbed field. The migration of bedforms and accumulation of bed sediment generates sedimentary structures in the stratigraphical record. There is a wide range in the scale of sedimentary structures. For discussion purposes, the scales will be divided arbitrarily into small scale forms (characteristic dimension less than the characteristic flow dimension), and large scale forms (all other forms).

The characteristic dimensions of sedimentary structures are usually height or length; flow characteristic dimensions are normally channel depth or width. The purpose of the distinction is simply to discriminate between those structures whose formation can be considered in light of the local flow only (that is, without reference to total channel configuration), and those whose size is of the same order of magnitude as that of the channel itself, and whose morphology is consequently determined by the total flow in the channel.

Because this paper is intended to discuss principles of hydrology and hydraulics applicable to proglacial sedimentary environments. The following discussion will emphasize hydraulic conditions accompanying sediment deposition: detailed sedimentological descriptions will not be undertaken. For that reason, as well, only primary sedimentary structures will be considered; post-depositional changes will not be discussed.

There is no comprehensive review of fluvial sediment deposition. Middleton (1965) presents a recent collection of research results, and Allen (1970a) has provided a general introductory study to processes of sedimentation. ASCE Task Force (1966) has reviewed nomenclature, and Pettijohn and Potter (1964) have provided a geologically-oriented reference description of forms. By far the most attention has been paid to the development of small primary forms in sand.

Small forms

Fine materials

Considerable attention has been paid to the development of sedimentary forms in silt and sand, as these change readily and have major influence on the resistance to flow in the channel (ASCE, 1971c). A summary classification of bed forms that develop in sand has resulted from the work of Simons and his collaborators (1963, 1965a, 1966). They considered only transverse bedforms, however. Allen (1970a) also described longitudinal forms or "parting lineations" (*see* also Allen, 1964; Jopling, 1964) which form at very small transport rates, as well as sand ribbons that occur when the supply of movable sediment is limited. The Simons classification is given in Figure 27 and, with extension to include parting lineations and sand ribbons, in Table 7. Figure 28 summarizes the findings of Simons' group on the relationship between bedform occurrence and mean hydraulic properties, including an indication of the effects on flow resistance of bedforms. In rivers with cohesive sediment boundaries (silt or clay), erosional scallops and flutes may play a similar role (Allen, 1969a, 1971).

Yalin (1972) has shown by dimensional arguments that bedforms are related to flow characteristics as follows:

$$\frac{\lambda}{d} = f(R_{e^*}, d/D, F_r) \tag{31}$$

Very small features (ripples, parting lineations) appear to be independent of mean flow properties (Raudkivi, 1963), and experimental evidence reveals no strong, systematic relationship with R_{e^*} for ripples, whence

$$\lambda_R = C \cdot D \tag{31a}$$

462

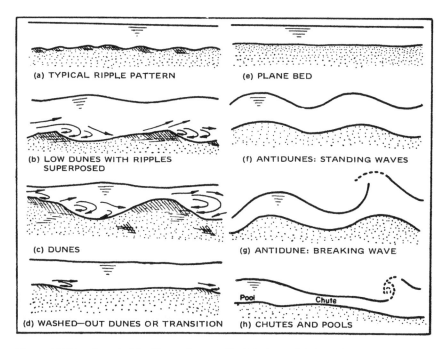

Figure 27 Depth-scaled bedforms in an alluvial channel (after Simons and others, 1965a).

Table 7 Classification of bedforms in fine materials.

Flow Regime	Bedform	Bed-Material Transport Concentration	Mode of Sediment Transport
		(ppm)	
Lower regime	Parting lineations	very low	Occasional grain movements
	Sand ribbons	low	in discrete steps
	Ripples	10–200	
	Dunes	200–2000	Continuous grain movements in discrete steps
Transition	Washed out dunes	1000–3000	Continuous movements
Upper regime	Plane bed	2000–6000	"Live bed"
	Antidunes	>2000	
	Chutes and pools	>2000	

Mainly from Simons and others, 1965.

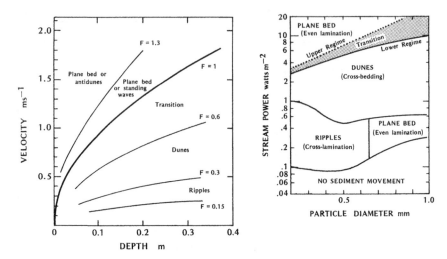

Figure 28 Relationships between depth-scaled bedforms and hydraulic variables in the stream (after Simons and others, 1963, reproduced by courtesy of *American Society of Civil Engineers*, and further data of Simons and others, plotted in Allen, 1968a, p. 167, reproduced by courtesy of *Sedimentology* and by permission of the authors).

Experimental results plotted by Yalin (1964) for ripples gave

$$\lambda_R \sim 1000D \tag{31b}$$

Allen (1968a) has empirically investigated the relationship between bedform geometry and the total flow geometry within which they occur. His results may be summarized as follows:

—for parting lineations: $\lambda_L \leq 0.025$ m (32a)
(wavelength transverse to flow)
—for ripples: $\lambda_R \leq 0.60$ m (32b)

From equations (31b) and (32b) it appears that such features are restricted to fine sands ($D < 0.5$ mm). This suggests that the features develop in essentially smooth flow. Nevertheless, they must be related to characteristics of the bed-flow interaction. Sundborg (1956) and Williams and Kemp (1971) observed that ripples were initiated by the occurrence of elements of order $h = 2$ to $3D$ in height on the bed. Presumably these features are higher than the laminar sublayer, that is, $\delta < h$. From equation (17), separation eddies will form and the perturbation will grow if

$$h > \delta = 11.6D/R_{e*}$$

In such cases the separation distance is of order 100 h. Yalin (1972) pointed out that it would be reasonable for λ_R to be of order 3× that distance; that is, $\lambda_R \sim 600$ to $900D$. This compares with the constant in equation (31b).

For parting lineations, Allen observed empirically that

$$\frac{\lambda_L}{D} = 100/R_{e*} \tag{32c}$$

Since transverse circulation is necessary for parting lineations to form, it is not surprising that R_{e*} appears in the relationship.

Allen has made a detailed qualitative study of turbulent flow effects near the bed and has related the phenomena to the morphology of current ripples. The basic flow over a rough bed is characterized by unstable motion due to spatial variability of fluid-boundary shear stresses, resulting from the irregular geometry of the boundary, and to sediment motion. The result is perturbation in flow properties (velocity, turbulence intensity, eddy structure and shear) in the longitudinal (two-dimensional flow) and transverse (three-dimensional flow) cases. The kinematic results include quasi-harmonic longitudinal motion that may lead to areas of flow separation over a sufficiently rough bed, and transverse, rotational flows characterized by alternate separation and reattachment of streamlines at the boundary. The three-dimensional result is pairs of spiral vortices. By plotting streamlines of water-sand movement at rippled boundaries, Allen (1969b) showed directly that ripples are the response of a movable boundary to such flows.

Dunes occur in both rivers and pipes. Hence, free surface effects cannot be important in their formation and equation (31) must have the form

$$\frac{\lambda}{d} = f(R_{e*}, d/D) \tag{31c}$$

For sufficiently large values of R_{e*} and d/D, such as are prevalent in real rivers, the relationship is insensitive, and we have

$$\lambda = C \cdot d.$$

Disturbances in the structure of turbulent flow appear to be responsible for the conditions near the bed that localize dunes. The largest eddy disturbances of relevance will be those for which the diameter $\sim d$ (fig. 29). These will lead (Yalin, 1972, p. 217–221) to

$$\lambda_D = 2\pi d \tag{31d}$$

Yalin (1964), summarizing experimental results, found

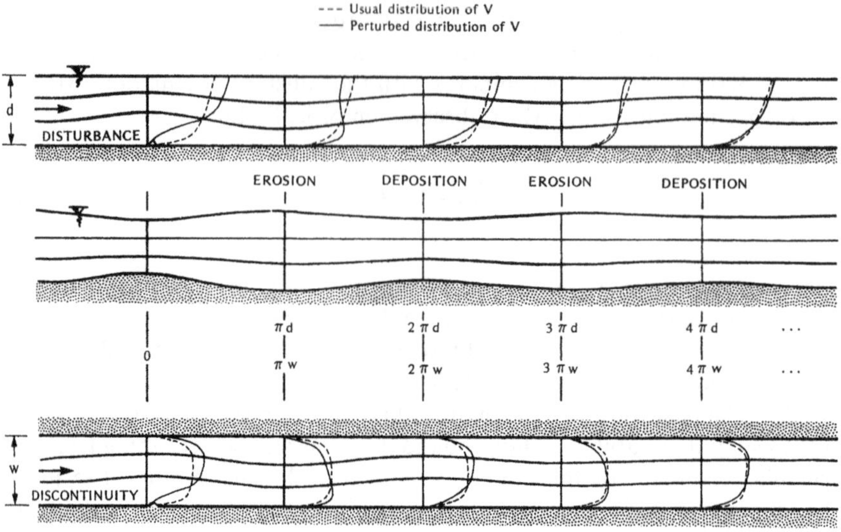

Figure 29 Propagation of perturbations in channel flow and their effect in producing dunes (upper) or bends (lower) in the river (after Yalin, 1971,
p. C13-6, reproduced by courtesy of *International Association of Hydraulic
Research*).

$$\lambda_D \approx 5d \qquad\qquad (31e)$$

Sundborg (1956) gave excellent field descriptions of dunes ("transverse bars")
in the River Klarälven. Allen (1968a), summarizing field evidence, found

—for dunes: $\qquad\qquad \lambda_D = 1.16d^{1.5+}$ $\qquad$ (32d)

—for sand ribbons: $\qquad \lambda_S = 1.35d^{1.3}$ $\qquad$ (32e)
$\qquad\qquad\qquad\qquad\qquad$ (wavelength transverse to flow)

The discrepancy between field and laboratory data is as yet unresolved.

Allen asserts that his description of bed-flow interaction can be extended
to apply to duned beds, mainly in view of superficial morphological similarity
of the forms. Certainly, they are geometrically similar. The observed slower
rate of growth of dunes relative to wavelength (Allen, 1970a) is probably
due to significant influence of total flow depth on the larger forms. Yalin
(1972) gave a physically based discussion of the geometry of ripples and
dunes. However, there are few critical data.

Antidunes occur only in free surface flows where gravity waves occur,
that is, they occur for $F_r \geq 1.0$, and occur more or less in phase with flow
perturbations. Hence

$$\frac{\lambda}{D} = f(F_r) \tag{31f}$$

Kennedy (1961) showed theoretically that the wavelength of antidunes is given by

$$\frac{\lambda_A}{d} = \frac{2\pi v^2}{gd} \tag{33}$$

whence, for $F_r \sim 1$, we have

$$\lambda_A = 2\pi d \tag{33a}$$

Allen's empirical equation is

$$\lambda_A = 6.3d \quad (F_r \sim 1.0). \tag{32f}$$

Antidunes are the bed response to two-dimensional, gravitational instability of the free surface near $F_r = 1$. They act kinematically, and can move in either direction or remain static. (Williams (1967) detected antidune development in flume studies at $F_r \geq 0.85$, and B.C. McDonald (written comm.) reports the features for $F_r \geq 0.75$).

Several studies have attempted to explain river bedforms by hydrodynamic analysis, as for example, the work of Exner (1920). Essentially, he imposed a solution of salutory form on the continuity equations of water and sediment transport. Kennedy (1963) developed the equations of two-dimensional potential flow over a moving, sinusoidal bed. The resulting theory relates bed waves to surface waves in the flow, which is regarded as the fundamental mechanism for generating bedforms. These results were discussed in detail by Raudkivi (1967). Reynolds (1965) further generalized Kennedy's results by providing solutions for non-ideal fluids (that is, fluids with friction) in two dimensions, and an extension to three dimensions for potential flows. Results were quoted in terms of the limiting Froude numbers for the development of dunes and antidunes. Raudkivi (1963) studied flow over ripples and in their wake, and gave a physical description of their formation. Mercer and Haque (1973) have recently developed an analytical model for ripple formation based on irrotational flow with a lee-side eddy, and have confirmed the usefulness of their results experimentally.

The large geometry range coupled with the certain involvement of flow fluctuations in the generation of bedforms make it clear that a range of scales of flow fluctuation is important, leading to a hierarchy of bedforms. Descriptive work by Nordin and Algert (1966) and by Hino (1968) has shown that there is indeed a range of frequencies present in apparently

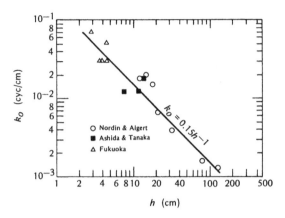

Figure 30 Relationship between lower limit ("critical") wave number of the "–3 power law," and depth of flow for ripple spectra (various data collected by Hino, 1968, p. 573, reproduced by courtesy of Cambridge University Press).

uniform bedforms (these studies investigated dunes). The spectrum of wave numbers exhibited the proportionality

$$S(k) \propto k^{-3} \tag{34}$$

where k is wave number (cycles/unit length) (Hino, 1968). From a variety of data (fig. 30), it has been shown that the lower limit of this –3 power range is given by the expression

$$k_o \simeq 0.15/d \tag{35a}$$

that is,

$$\lambda \simeq 2\pi d \tag{35b}$$

as would be expected for depth-scaled features. Squarer (1970) has studied conventional hydraulic properties—such as flow resistance and bedform celerity—in terms of the spectral statistics of the bed.

From the sedimentological point of view, sediment sorting during transport is of importance. Particles of differing size, shape, and specific gravity exhibit different thresholds for erosion and deposition (fig. 17) and move at different transport velocities once in motion (Sundborg, 1967; Meland and Norrman, 1969). Sorting occurs in three phases: (a) varying hydrodynamic response to the mean flow, in suspension and in traction; (b) preferred entrainment and deposition of certain fractions during transport across the stoss side of ripples or dunes; and (c) sorting during avalanche or slump transfer of material on the lee side of bed forms. Brush (1965b) provided a review of this subject, and Jopling (1960, 1963, 1964, 1965a) and McKee (1957, 1965) have contributed detailed experimental studies. Brush (1965a) has provided an extensive bibliography of further sedimentological work on primary fluvial structures, and recent work is quoted elsewhere in this volume. Hence this topic will not be reviewed further here.

Formation of bedforms in closed channels has received relatively little attention. Experimental studies by Craven (1952), Acaroglu and Graf (1968) and McDonald and Vincent (1972) defined a sequence of forms similar to those for open channels, except that, instead of an antidune phase, heterogeneous or homogeneous suspension of the sediment load occurred. McDonald and Vincent found that:

(a) $v_w \propto v^3$, where v_w is bed wave celerity
(b) $1/3R < h < R$; $0.1d < h < 0.5d$
(c) the system tends to maximize the hydraulic radius by adjustment of bed height and bedform.

The resulting structures are superficially very similar to those deposited in open channels.

Coarse materials

Relatively little information is known about the primary structure of coarse materials. Individual elements constitute "finite roughness elements" on the bed, capable of playing the same role, hydraulically, as ripples or small dunes in sand-bed rivers. C.M. White, in his classical study of the threshold of particle motion, noted that "dominant" grains appear to be present on a streambed consisting of large, noncohesive elements. These exposed grains provide most of the fluid drag. The number of such grains per unit area of the bed will be r/D^2, where r is a "packing factor." Henderson (1966) gave $r \sim 0.15$. This leads to the result that the area of streambed dominated by each grain is related to the (projected) grain area as follows:

$$\frac{A_B}{A_G} = 8.5 \tag{36}$$

(calculation based on the assumption of circular areas).

Gravel features corresponding to dunes have been reported in the deposits left behind by catastrophic lake drainings by Theil (1932) and Pardee (1942), though in these cases the features were huge. They have also been reported on a more representative scale in gravel-bed rivers by Galay (1967). Some miscellaneous evidence for their occurrence in flumes has been noted (Galay, 1967).

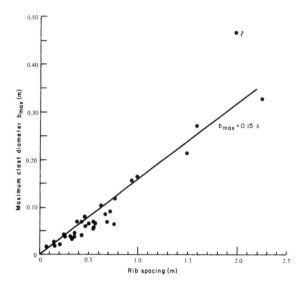

Figure 31 Transverse rib spacing versus mean maximum *b*-axis size of constituent stones: observations from Peyto outwash and nearby areas, Alberta (after McDonald and Banerjee, 1971), and from outwash fans in Alaska (Boothroyd and Ashley, this vol.).

McDonald and Banerjee (1971) identified a feature in shallow gravel channels, which they called transverse ribs—"a series of regularly spaced pebble, cobble or boulder ridges oriented transversely to the flow. Commonly, though not in every case, a marked grain size sorting has occurred with coarser particles making up the ridges" (p. 1289–1290). On pebbly outwash these take the form of "pebble ripples," whereas in steep, cobble channels, they form prominent steps that turn the channel into a sequence of pools and cascades. Data in Figure 31 indicate that the ratio of stone size (*b*-axis diameter) to rib spacing is approximately 0.15. Boothroyd (1970) has suggested that the features may represent relict antidunes. They are certainly restricted to sites in upper flow regime, and it is apparent from the fact that the ribs often sit on sand or silt, that the constituent stones have been transported into place. Using the ratio $D/\lambda{\sim}0.15$ in conjunction with equation (33a), gives $D/2\pi d{\sim}0.15$, or $D/d{\sim}1.0$. The very sparse data that exist on the maximum size of boulders transported in a flow of given depth (Fahnestock and Haushild, 1962; Fahnestock, 1963, his table 8) suggest that, in fact, the largest boulders are of the same size as flow depth. In both cases the boulders moved over a sand bed in upper regime flow in the presence of plane bed or standing waves only. This supports Boothroyd's hypothesis.

470

There is no well documented information on the development of full-scale antidunes in gravel. Fahnestock (1963) reported that, in White River, at Mount Rainier, antidunes occurred in the channel when large volumes of sand were being washed through the reach, and that boulders moved through the antidunes on the surface of the sand (possibly buoyed up by dispersive pressure). Boothroyd and Ashley (this vol.) provide evidence of antidune formation in gravel channels.

Large forms: channel bars

These forms generally take channel width as the scaling parameter. They occur in all sediment types, from clay to boulders, and a wide range of observations has been made of their character. A simple, descriptive classification recognizes three categories of river bars:

I.   Bars produced by riverbed deformation:
   1.   Symmetric shoals
      a.   longitudinal bars
      b.   linguoid bars
   2.   Asymmetric shoals
      a.   left diagonal bars (flow diverted to left)
      b.   right diagonal bars (flow diverted to right)
II. Bars produced by channel adjustment:
   1.   Point bars
   2.   Lateral bars (also "side bars")
   3.   Triangular bars, at channel junctions
III. Bars produced by nonfluvial elements.

Bars represent important roughness elements in the channel, and they evolve in response to changed flow and sediment-transport conditions in such a way as to provide adjustments of resistance to flow. Their localization is determined by the major secondary currents in the river. Whilst the material comprising some bars moves during floods, the bars themselves may be relatively stable, so long as the entire channel does not undergo a major shift. They constitute the primary bedform which leads to evolution of the stream channel (Quraishy, 1944; Keller, 1972). In unstable channels, though, individual bars may shift radically and rapidly (Smith, 1971).

Symmetric shoals, or "longitudinal bars," are simple, longitudinally-extended shoals in the centre of the channel (fig. 32A). They develop from differential scour and accumulation of sediment across a wide channel: the deep channels near the sides are aided in formation by the development of secondary currents as a result of wall drag. If they develop to the point of inducing lateral erosion by the main channels, then they develop into linguoid bars.

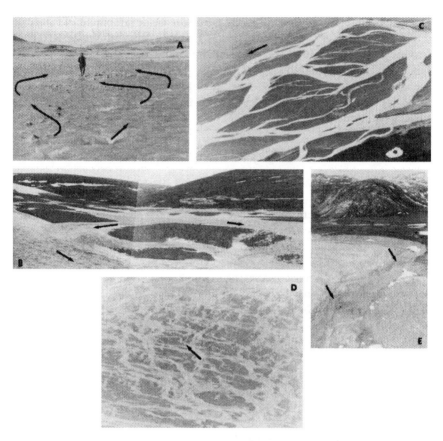

Figure 32 Types of river channel bars: A, Longitudinal bar on gravel outwash (Lewis sandur, Baffin Island). Flow lines indicated. B, Linguoid bars in a gravel-bed river (Lewis sandur, Baffin Island). C, Gravel-veneered linguoid bars, essentially composed of sand (Maktak sandur, Baffin Island). View from 1200 m above surface. D, Linguoid bars in sand in area of aggradation (Maktak sandur; Baffin Island). View from 1200 m. E, Left and right diagonal bars in a gravel-bed river (Ekalugad sandur, Baffin Island). Flow is toward the bottom of the picture.

Linguoid bars (Allen, 1968b; Collinson, 1970) (figs. 32B, D), also called "diamond bars" or "spool bars," are particularly associated with braiding, since downcutting around the bar results in the development of two channels and leaves a central island (Brice, 1964). Leopold and Wolman (1967), in providing excellent descriptions of the development of such bars both in the field and in a flume, designated them as the basic unit of channel anastomosis. Krigström (1962) showed how complex braiding patterns can be built up by superposed spool bars on successively larger scales, and Williams and Rust (1969) proposed a hierarchical division of bar components. Some workers have attempted to differentiate between linguoid sand bars and gravel spool bars. The former are frequently submerged, shift relatively rapidly, and reveal varied sedimentary structures. The latter tend to be more permanent and their structures are more rudimentary. The differences appear to stem from the nature of the sediment, the strength of its deposits, and the manner of sediment movement. There is no hydraulic reason to distinguish between them.

Asymmetric shoals (fig. 32E) contribute significantly to lateral erosive activity in the stream by deflecting the flow from a straight course downstream. Diagonal bars (also called "transverse bars") may be associated with the development of meanders when they are found, as is often the case, in successive alternating form downstream. In gravel channels, however, they often provide much more abrupt flow deflections than would be associated with true meandering. Nevertheless, they appear to develop from a pattern of alternating zones of scour and deposition on the channel bed in the presence of abundant sediment load. In sand-bed rivers, they may take the form of tabular sand "sheets," with alternating lee faces. Wolman and Brush (1961) and Stebbings (1964) have described large "sheet" waves of similar (scaled) dimensions in flume experiments with coarse sands. Znamenskaya (1962) has repeated the description for natural channels. Wertz (1963) reported finding alternating zones of boulder accumulation and sand fill in the mid-course sections of some Arizona mountain-front arroyos. The diagonal bars, which are the more common riverine form, may develop from a linguoid bar when one of the branches comes to completely dominate the other (Leopold and Wolman, 1957, their figs. 34 and 35), though development in the opposite direction can occur as well (fig. 34, *ibid.*).

Bars produced by channel adjustment are more strictly depositional in form. They occur in slack-water locations where sediment load is deposited on falling stage. Lateral bars, in particular, are sites of temporary storage of fine material that moves in suspension. These bar forms may have local significance for flow adjustments (large lateral or point bars may serve to reduce flow width) but they are largely scoured out during high flows.

Point bars are important in channel evolution of meandering rivers, and have been extensively studied (Fisk, 1944; Sundborg, 1956; Davies, 1966). Krigström (1962) has also described point bars on channel bends in braided, gravel rivers, but the form appears to be considerably different from the classical one. The bar usually develops some distance from the convex river bank, and exhibits a steep shoreward face, so that a deep channel may remain between bar and bank. Smith (in press) gives a detailed description of the form.

Occasionally, a riffle is situated squarely across a stream where a pool spills into a major set of rapids. Such locations are usually controlled by the presence of anomalously large boulders and are probably not determined by normal fluvial processes, as are the other bar types.

A wide range of observations has been made on the spacing of macroform elements and their relative sedimentary properties. Neill (1969) has described large "waves" or bars in the bed of Red Deer River, with wavelength approximately $2w_s$ (he took all major forms on echo sounder traces into account). Various workers, including Leopold, Wolman, and Miller (1964), have indicated that meander lengths tend to vary between 10 and 16 channel widths, providing a spacing of 5 to 8 widths between crossover points. Similarly, the distance between successive riffles in straight or irregular channels has been reported as 5 to 7 widths, as has the spacing of gravel concentrations in sand bed arroyos of the U.S. Southwest (Leopold and Miller, 1956). Maddock (1969) has demonstrated remarkably clear sets of alternating sand banks in rectified sections of the Colorado and Rio Grande Rivers, with a spacing of $2\pi w_s$. Speight (1965, 1967) carried out spectral analyses of meander wavelengths on Australian rivers and found that a great variety of spectral peaks occurred. Spectral analysis of bed and surface elevations through channels in the braided outwash streams of Baffin Island (Church, 1972) yielded a similar range of spectral peaks, with the dominant ones falling in the range 5 to 8 w_s.

On the basis of this evidence, it is safe to conclude that a range of macroform wavelengths exists in natural channels. A simple approach to the spacing of elements may be made by considering the nature of wave propagation from some disturbance. The boundary provides any number of such disturbances, particularly at channel contractions and expansions. In supercritical flow, flow velocity, v, will be greater than wave celerity, c, hence a series of stationary shock fronts will develop (fig. 33). Henderson (1966) gave an analytical discussion of shockwave propagation. Shock-related wave forms are probably restricted to relatively small features in upper regime flow. The irregularity of channel configuration would produce interference effects that would restrict the propagation of such effects over large areas.

Where secondary currents develop in the flow, harmonic motion may impose recognizable wavelengths in the channel. Simple repetition of the argument for dune spacing (*see* p. 55 above), where flow width is now the scaling parameter (fig. 29), gives

$$\lambda = 2\pi w_s \qquad (37)$$

Dominant wavelengths (as multiples of channel width) appear to run in multiples of π: $2\pi w_s$, $4\pi w_s$, $6\pi w_s$, ... Possibly, odd-integer wavelengths occur also, there being two riffles per meander length in comparison with an undifferentiated pool-riffle sequence. It appears, also, that the scale of flow affects the result. Very small streams and boulder torrents appear to have dominant wavelengths approximately 2 to $3.5w_s$, whereas the Colorado River has its dominant forms close to $8\pi w_s$.

Figure 33 Development of standing wave in supercritical flow ($c < v$) and diamond pattern of standing waves in a gravel-bed channel.

Channel bars store sediment in transit. The material found on bar surfaces is usually smaller than the bed pavement found in the channel locally, testifying to its transported origin (*see* Nordin and Beverage, 1964, for comments on conditions in a sand-bed river; Church, 1972, for gravel-bed). Although large volumes of material are set in motion by high flows, it is significant that individual cobbles in the bed-material load do not appear to move very far at any one time (Tricart and Vogt, 1967). Emmett and Leopold (1965) pointed out that, although scour seemed to be a general rule on a channel bed during a flood, material moved only short distances, so that sediment discharge was far less than the generality of movement would suggest. Hence, much material is stored for long periods in the channel.

The nature of individual cobble movements during flood flows might lead to some understanding of the rate of material movement through the bars (and, incidently, to an estimate of bed-load sediment discharge). For gravel rivers in Baffin Island, however, the results of observing the movements of marked cobbles were very irregular (Church, 1972). It appears that, once dislodged, individual cobbles maintain their progress over the bed until caught by a protruding cobble. The result is a highly imbricated bed structure. Such interception appears to be random and the pattern of individual movements is similarly random. Nir (1964) also reported inconclusive results from a study of cobble movements in ouadis. Apparently it requires a very large experiment to determine if any meaningful pattern does exist in cobble movements, and only one such experiment has ever been done (Leopold, Emmett and Myrick, 1966; Langbein and Leopold, 1968), in a desert arroyo. Two interesting results appeared: (a) there was an apparent tendency for cobbles to be less mobile when they were densely distributed on the bed, as they are in riffles, and (b) movements tended to be from riffle to riffle, possibly a consequence of the fact that the cobbles moved over a sand bed in upper regime flow, so that there was little chance for them to be intercepted and forcibly lodged against other cobbles en route.

Langbein and Leopold (1968) advocated the occurrence of kinematic waves to account for this behavior, based on their observations of cobble mobility. The Baffin gravel data showed individual movement lengths much smaller than the distance between the dominant bars, so that several displacements may occur in each stage of a stone's journey. However, this is not important: the second conclusion of Langbein and Leopold is more or less implied (over the long term) from the first.

Channels and channel bars in sand-bed streams are usually much more active than ones in gravel-bed rivers, that is, sand moves to a greater or lesser extent in all but the lowest flows. Consequently, bars evolve rapidly. Smith (1971) has studied the evolution of individual bars in a sand-bed river. The development of channel bars in gravel is usually much more sporadic, as the necessary high flows for sediment movement are less frequently reached. However, development can be very rapid when it occurs (Smith, in press).

Sedimentary structures

Williams and Rust (1969) and Rust (1972) have provided facies models for braided river deposits on the basis of studies of Donjek River sediments (Yukon Territory, Canada), as has Smith (1970) from Platte River sediments (nonglacial environment). The most important feature of the classifications is the distinction between proximal environments, characterized by nearly featureless or very rudimentary parallel bedding in gravels, and distal environments, characterized by the variety of sedimentary structures found in fluvial sand bars. The latter frequently prograde by frontal avalanching of sediment, leaving fine cross-sets, and preserving a variety of dune and ripple structures occurring on the bar surfaces.

In proximal environments a wide range of grain sizes is present and bedding remains rudimentary (fig. 34) except in certain backwater locations where relatively good sorting and well developed bedding may be found: such zones are often destroyed by renewed erosion later on as the main channels swing across the sandur. Channel side bars, which may be preserved, are often the sites of the best developed bedding. Most sandur bedding is horizontal, as would be expected from the frequency of upper regime flows (Harms and Fahnestock, 1965), the low silt-clay content of the sediment (Schumm, 1960b), and the shallow flow depth relative to sediment size. However, Costello and Walker (1972) have reported foresets in gravel in Pleistocene outwash in what was interpreted as a local "delta" filling into an abandoned channel with 0.5 to 3 m of available relief.

The possibility for sets of cross-strata to develop presumably depends on the availability of sufficient space (depth, in this case) for the processes to occur that produce the form; that is, if total flow depth is of the same order of magnitude as *particle size*, one would not expect cross-bedding to occur since there would simply not be room for sediment to pile up to produce lee-side slumping and settling. No critical data seem to be extant on limiting relative roughness for cross-stratification: the writers are inclined to suspect that cross-beds will be rare if $D/d > 0.1$. This value of particle relative roughness is frequently exceeded in high-energy streams carrying coarse material. Bar structures that develop in gravel deposits appear, in detail, to be usually erosional.

Poor sorting and rudimentary bedding result from rapid deposition of material from transport (Fahnestock, 1963; Stewart and LaMarche, 1967). The initial deposition of very coarse materials leads to discontinuous, horizontal bedding. Imbrication is common and preferred fabric may develop in the deposits (Church, 1972; Rust, this vol.). Entrapment of fines later on amongst the coarse materials tends to obscure whatever structure the original deposits may have had and to render stratification less visible (*see* also Ore, 1964). In large, continuously aggraded deposits, considerable variation in sediment characteristics may often be found in an extensive section, no doubt reflecting considerable variation in sedimentary environment over a long period of time.

In fines, foreset bedding commonly develops at the distal, prograding end of linguoid and diagonal bars (Collinson, 1970). Jewtuchowicz (1953) has described extensive lamination, cross-bedding and channel dunes preserved in sands in Polish Pleistocene sandurs. Similarly, Doeglas (1962) noted festoon structures, and both parallel and oblique foreset laminations in sand and fine gravel deposits of the torrential Ardeche River in France. Ore (1964) and Smith (1970) have described foreset bedding in both contemporary and

Figure 34 Rudimentary bedding in outwash deposits: A, Early post-Pleistocene gravel outwash at Ekalugad Fiord, Baffin Island. B, Channel bar structure on active outwash plain, Lewis River (direction of flow indicated). C, Section in proximal gravel outwash, Chilliwack Valley, British Columbia (Pleistocene age, direction of flow indicated): detail, cross-stratified sands preserved in the gravel. The strata dip upsandur.

ancient braided channel deposits. Cross-stratification was more typically associated with sand deposits in distal sites than with proximal gravels and was quite rare at sites where rapid aggradation was occurring. Hence, fore-set bedding appears to be characteristic of tabular deposits of reworked channel material, and sorting is relatively good in such an environment. Similarly, trough-fill structures (Harms and Fahnestock, 1965) are typical of such conditions rather than of rapidly aggrading sites.

Sediments appear to fine upwards and downstream on linguoid bars, so long as their depositional history remains simple. However, changes often are complicated by several episodes of erosion and deposition. Channel-adjustment bars commonly show better sediment sorting than do the other channel bedforms for the reason that they occur in a more restricted envir-onment, (that is, in backwater areas). Here, lateral gradation of sediment size may occur, within reasonably well defined beds, which may be preserved for a long time as the channel shifts away from the site.

Flood deposits

Although the great majority of river flows are contained within the chan-nels, with sediment movement essentially continuous through the channel system, periodic flooding distributes material over a wide area. Major shifts in channel position, which are of great importance in determining the locus of sedimentation, are themselves usually associated with floods. The incid-ence of flooding is variable; on a very active outwash plain, a large area of the surface may be flooded for several days in each summer season. On the other hand, a relatively inactive surface may experience only one or two such events in a season—or perhaps none—and then only a small portion is actually flooded. Over much of the surface the water may be too shallow and move too slowly to effect notable bed-load movement, although considerable movement and redeposition of fines may occur. However, in the region close to major active channels, large volumes of gravel and/or sand may be moved.

Since channels may shift quite rapidly, any position on the recently active sedimentary surface can expect to be reworked and to receive fresh sediment within a period of a few tens of years. The sedimentary deposits asso-ciated with floods take three major forms: channel fill, levees, and surface veneer.

Channel fill deposits (fig. 35A) occur normally as the last stage of scour-and-fill processes that accompany flooding. On the falling stage, large volumes of material carried near the bed are dumped in the channel as the river loses its capacity to transport sediment. Deposits develop in areas of flow divergence which, because of changes in the flow pattern that may have occurred during the flood, may not be the same as they were before the flood commenced. A channel fill area will continue to act as a zone of flow

Figure 35 Flood deposits of proglacial rivers: A, Channel fill deposit: Ekalugad outwash plain, Baffin Island. B, Gravel levee, Ekalugad outwash plain. C, Sand levee, Maktak sandur, Baffin Island.

divergence and sediment deposition until the river successfully cuts a channel through or around it. Such zones may develop into linguoid bars.

Levees are not well-defined on gravel-bed rivers. However, where overbank flow occurs greatly elongated areas of river bank deposit occur (fig. 35B). They are often laterally graded, with very coarse gravel and cobbles being dumped at the channel edge and finer materials occurring farther away. The elongate gravel bars along the river might then be termed levees. These overbank deposits are often associated with channel fill and may result in superelevation of the river above the general surface. The channel fill and levee deposits may cause a shift in the channel location by avulsion. Low levees are common along sandbed streams (fig. 35C).

480

Surface veneer appears across wide areas after inundation. Material deposited commonly varies from fine gravel to sand. Often the sand is rippled. Subsequently, wind may deflate the deposit and leave a coarse lag gravel.

Sediment sorting occurs locally on such deposits, the downstream sides of the bars characteristically having finer, better sorted material (fig. 36). It is probable that this feature, and in fact much of the "structure" possessed by the deposits, is a result of washing of materials and adjustment of the

Figure 36 Sorting of sediments on the surface of outwash deposits: A, fining of sediments away from channel axis in flood deposited channel fill, Ekalugad sandur. Divergent flow directions superimposed. B, View downstream on linguoid bar on Maktak sandur, showing changing surficial sediments.

surface to flow on declining stage, and that it does not represent conditions during the height of the flood when sediment movement is general.

Surface sediment textural patterns

Sandurs comprise extensive areas of sand and gravel—perhaps even boulder fields—completely devoid of vegetation in their recently active parts. The surface appearance of a sandur is that of a gravel or sand plain, with frequent ribbons or "islands" of finer material where former channels or "quiet-water" bars existed. Much fine material is deflated by wind, but immediately under the surface there is always an abundant sand matrix, and well-sorted sediments of any size are very rare. Traces of former channels are prominent on the surface, and may lend up to one meter of local relief. The pattern of the former channels indicates the general direction of sediment movement.

Krumbein (1941b) stated that the fundamental attributes of clastic sediment particles were size, shape, roundness, surface texture, petrology, and orientation in situ. In considering the mean character of deposits, the variance of these characteristics also becomes important. Sneed and Folk (1958) attributed these characteristics to: the initial shape of the particles as liberated from the parent rock; internal structural characteristics of the rock; rock size (that is, several of the form parameters appear to depend on the size of the clast); distance of transport; the transporting agent; and finally to chance events in the course of transport. It seems reasonable to expect sediment characteristics to change downsandur as the result of sorting by flowing water. It is apparent, however, that other factors, such as lateral proximity to the rivers, relative topographic elevation (locally), extent of water sorting or wind deflation that has occurred since the main depositional events, additional sediment sources, and so on, will also affect the textural variations of surficial deposits.

Pertinent studies of the modification of coarse clastic sediments under fluvial transport include those of Krumbein (1940, 1942a) in aridland canyons of the American southwest, Blissenbach (1954), Bluck (1964) and Lustig (1965) on semiarid alluvial fans in the same region, Plumley (1948) on the Black Hills terrace gravels, Sneed and Folk (1958), and Pittman and Ovenshine (1958) on the character of river cobbles. Only two detailed studies have been made on active outwash surfaces, those of Bradley and others (1972) on the Knik River (Alaska) outwash, and Church (1972) on Baffin Island sandurs. By reviewing some of their results, major features of outwash will be illustrated.

Coarse materials

The size of coarse materials (>8 mm) has been found to decrease down-sandur in a number of studies (fig. 37). Church mapped the first four

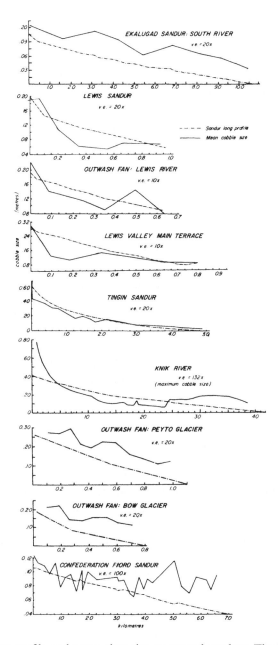

Figure 37 Long profile and mean clast size on several sandurs. The lower diagram presents a rectilinear profile with no significant change in clast size; this surface is apparently graded and not undergoing active aggradation.

483

moment measures (mean, standard deviation, skewness, kurtosis) and found significant patterns for the first two moments only (fig. 38). Total size reduction is about 60 percent in 10 km. Bradley and others (1972) found about 87 percent reduction in the size of coarse clasts in 26 km on the Knik River outwash. Experimental abrasion of sample Knik clasts in a Kuenen tank produced only 8 percent size reduction in an equivalent travel distance, whence the authors concluded that sorting processes accompanying selective deposition accounted for most of the change.[1] These observations are accounted for by simple selective deposition of the coarsest materials during floods as diverging flood waters lose competence downsandur (Russell, 1939).

A contrast appears between the character of the materials on the sandur surface and those in the low-water channel. At Ekalugad Fiord, it was found that grain size in South River did not decrease downsandur nearly as much, nor as regularly, as on the surface (fig. 39). This is because the distribution of grain sizes on the river bed is derived from a different sort of process than that for the surface, that is, the channel bed consists largely of lag material from which finer materials have been removed. The efficacy of elutriation will be determined by local channel slope, and in particular by the pool and riffle sequence, rather than by the general slope of the river. Hence, the end result is quite different from the end result on the adjacent sandur surface.

Pashinskiy (1964) and Meland and Norrman (1969) have presented evidence that particle shape affects transportability in fluvial environments. The study of variations in grain shape is somewhat confused by the variety of shape indices that have been used. Clast roundness may be measured by

$$R = 2r/a: \text{ Cailleux roundness } (0 \leq R \leq 1)$$

Increase in roundness is an expected consequence of the abrasion sustained by the stones in moving downsandur. Krumbein (1941a) showed that roundness initially increases very rapidly, so that it is not surprising that a change should be detected even over a few kilometers, as was reported by Church (1972). The most commonly presented particle shape measures are

$$S_w = (bc/a^2)^{1/3}: \text{ Krumbein's intercept sphericity (Wadell sphericity)}$$

Bradley and others (1972) investigated S_w and

$$S_p = (c^2/ab)^{1/3}: \text{ maximum projection sphericity.}$$

Helley (1969) used the equivalent Corey Shape Factor, $S_c = c/\sqrt{ab}$. All are distributed over the range 0–1. The superiority of S_p as a measure of hydraulic behavior has been demonstrated by Sneed and Folk (1958). They also showed that grain size was significantly correlated with their sphericity

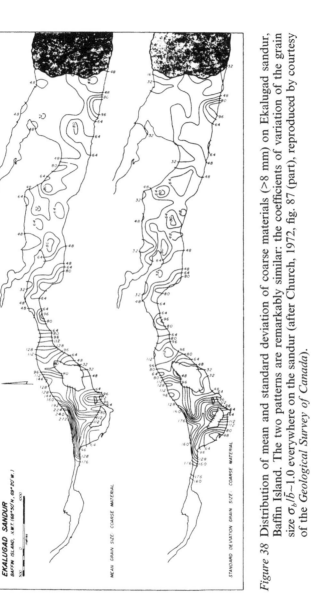

Figure 38 Distribution of mean and standard deviation of coarse materials (>8 mm) on Ekalugad sandur, Baffin Island. The two patterns are remarkably similar: the coefficients of variation of the grain size $\sigma_b/\bar{b} \sim 1.0$ everywhere on the sandur (after Church, 1972, fig. 87 (part), reproduced by courtesy of the *Geological Survey of Canada*).

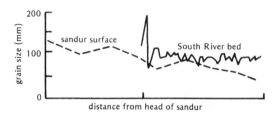

Figure 39 Comparison of grain size characteristics on the sandur surface and in South River bed, Ekalugad Fiord, Baffin Island.

measures—implying that clast behavior should be investigated only for restricted size ranges.

Church found a systematic change in sphericity at Ekalugad Fiord, whereas the results of Bradley and others are difficult to interpret. Thus at Knik River the two measures of sphericity act in opposite fashion. On balance there appeared to be a slight increase in S_p downsandur.

Axis ratios may also be used as indices of particle shape. Church used the Zingg (1935) ratios (Krumbein and Sloss, 1963). Bradley and others used Folk and Sneed's (1958) form ratios. Again, the results of Church were remarkably uniform. Bradley's results indicated that the order of *decreasing* mobility was: discs, rollers, compact (spheroid). Abrasion tank studies again were used to show that selective transport was indeed the major reason for this distinction.

The literature (reviewed by Bradley and others, 1972) is somewhat confused on the subject of the significance of particle form in hydraulic behavior. Most experimental studies (for example, Krumbein, 1942b; Stringham and others, 1969) have considered settling behavior; traction was investigated by Krumbein (1942b) and by Helley (1969). Helley's results indicate that equant clasts are more easily entrained than others, whereas Krumbein noted that elongated clasts outrun equant ones, once in motion. Meland and Norrman (1969) reported that, although there appears to be a well defined relationship between particle shape and settling velocity, there is only a poor correlation between shape and transport velocity in contact with the bed. Clearly, this is an area in which far more experimental work is required, commencing, probably, with consistent definition. The present writers advocate the use of R, S_p (or S_c) and the Sneed-Folk axial classification.

Sediment sorting by lithology clearly occurs. Church found a modest shift in proportions amongst four gneissic rock types in 10 km of transport. The shift reflected the greater resistance to weathering processes on the outwash of quartz gneisses. Bradley and others (1972), with a more definitely contrasted suite of rocks and a 26-km reach, found similar results, with greywackes and strongly foliated rocks declining in importance at the expense of dark volcanic and quartz rocks. Their study of distinctive clasts from several lateral sources further showed that lateral mixing of sediments (across the sandur) is very restricted. Hence, distinctive lithologies may safely be used as indicator trains for some distance. Ehrlich and Davies (1968) observed remarkable gradients of abundances of several lithological types across a short transect (1.5 km) at the terminus of Mendenhall Glacier, Alaska, which included moraine as well as outwash. They concluded that both transport distance and intensity of attritional processes contributed toward the enrichment of mechanically strong detrital types.

Cobble fabric patterns on flood deposits and bar surfaces have been studied by several workers. Lane and Carlson (1954) noted that the bed cobbles of irrigation canals in Colorado had a preferred long-axis orientation normal to the flow direction, as did Doeglas (1962) for torrential river gravels. On the other hand, Krumbein (1940) noted that flood gravels in the San Gabriel canyon showed a preferred orientation parallel to the flow direction. Church (1972) found no clear pattern: some orientation patterns showed weak maxima across the direction of flow, whereas some showed maxima parallel to the direction of water flow. More recently, Rust (1972), Boothroyd and Ashley, and Rust (this vol.) have shown that selection of appropriate clasts as indicators can result in strong and consistent transverse-orientation data.

It would appear that material that moves by being rotated or rolled along will take up a transverse orientation with respect to the flow, so that it rotates about the two lesser axes, whereas material that is shoved or pushed over the bed, or within a disperse layer, will take up a position parallel to the flow, so as to present a minimal drag profile to the moving medium. The former type of movement is probably typical of low rates of transport, and the latter typical of more general movement. In the rapidly fluctuating flows of sandur rivers both forms of cobble motion are probably common, and so strongly bimodal orientation patterns—or sometimes random patterns— may reasonably be expected.

Fine materials

Church (1972) investigated the characteristics of fine materials on Ekalugad sandur (8 mm $> D > 0.063$ mm). Several characteristic grain-size distributions were identified, indicating distinct local environments of deposition. These were as follows (fig. 40): (a) logarithmic-normal distributions of sizes, (b) distribution truncated in fines, (c) distribution truncated in coarses, and (d) bimodal distribution. The logarithmic distribution of sizes is that which would be expected to occur purely by chance in a deposit abstracted randomly from a population of materials with a finite size limit (Rogers and others, 1963). Processes of erosion and rock disintegration produce such a result. Basal moraine or earth-flow deposits probably preserve such a "primary distribution" of materials since no sorting processes occur (Elson, 1961; Tricart and Vogt, 1967). It is also possible that on the sandur certain

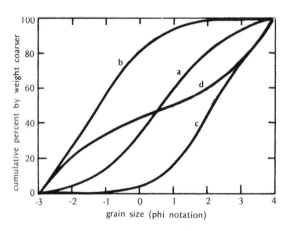

Figure 40 Type cumulative distributions of grain sizes for fine materials on Ekalugad sandur, Baffin Island. See text for classification.

flood deposits reflect a total lack of selective processes, being the result of rapid deposition on declining stage when competence is not critical. Truncated distributions result from deposition in special environments. A paucity of fines indicates selective sorting in an "active" environment, the result being a lag deposit from which fines have been removed. Such a process is common in channels and on channel bars, and in fine deposits may also be a consequence of eolian winnowing. Paucity of coarse material indicates deposition of fine materials in a "passive" environment—in quiet water. Very little coarse material finds its way to such a spot. Such areas are by no means restricted to the distal end of the sandur, but can occur in channel backwaters or sandur surface areas away from active currents. Bimodal curves are mainly the result of mixed depositional environments on the sandur. In considering the entire range of grain sizes, it was clear at Ekalugad Fiord that a relative paucity of materials occurs in the 0.5- to 8-mm range. Slatt and Hoskin (1968) found a similar situation on the Norris glacier outwash, and this is a common feature of fluvial sediments.

No significant gradients were detected in any textural property over the sandur as a whole. This is to be expected inasmuch as high floods are presumably competent to carry all sizes in the fines range at any place on the sandur. However, it was found that the proportion of heavy minerals declined systematically downsandur, so that some sorting was imposed on this fraction of fines.

No detailed study is available from a sand outwash. It is to be expected that, in such an environment, size sorting of sands occurs in conformity with the observation that materials whose size range spans the competence of the stream downsandur will become sorted.

Channel form

Introduction

Channel form is the outcome of the adjustments amongst the magnitude and properties of the water and sediment load, and the properties of the channel boundaries. Factors involved are summarized in Table 8. Available relationships amongst the variables include the following:

(1) continuity: $Q = w_s d_* \bar{v}$
(2) resistance relationship: not universally formulated for open channels

Table 8 Factors involved in river channel formation in alluvial material.

	Dependence[1]	
Variable	*Short Term*	*Long Term*
Physical constants		
g acceleration of gravity	I	I
Material properties		
T fluid temperature	I	I
ρ fluid density	D	D
v fluid kinematic viscosity	D	D
ρ_s sediment particle density	I	I
w wash particle settling velocity	N	D
D size distribution characteristics	I	D
σ_D of bed materials[2]		
— structure of bed and bank	I	D
material deposits		
Mass characteristics of flow system		
Q water discharge	I	I
G_{bs} bed material transport	D	I
c_w wash material concentration	I	I
Dynamical characteristics of flow system		
resistance factor[3]	D	D
$\bar{v}$ flow mean velocity	D	D
Geometrical characteristics of channel		
S energy gradient (slope)	I	D or I
P or w_s channel width	I/D	D
R or d* channel mean depth	D/I	D
X_i planimetric geometry parameters	I	D

[1] I = independent; N = depends only on other material properties; D = dependent. Short term indicates a time scale comparable with time of travel through the reach; long term indicates a time scale consistent with the establishment of regime conditions in the channel.
[2] Or bed strength characteristics for cohesive materials. For the case where bed and bank materials are different, separate descriptive parameters will be required to characterize bank conditions.
[3] Includes the effects of bedforms.
  Based partly on ASCE (1971c).

(3) definition of a friction coefficient ff: $8/ff = v^2gRS$
(4) sediment transport relation: not universally formulated
(5) channel shape factor: not universally formulated
(6) relationship between flow and sediment properties, and channel planimetric form: not formulated
(7) relationships amongst fluid properties: $\rho = v(T)$; $v = v(T)$.

For the short term, there are a sufficient number of relations available to specify flow conditions, provided the independent parameters are known. For the long term this is not the case. Langbein (1964) has suggested that this leaves the river with some degrees of freedom in its adjustment to imposed conditions (*see also* ASCE, 1971c, for a more complete discussion).

Some of the necessary relationships are not universally formulated. In any case, it is not clear that aggrading (or degrading) rivers need conform to relationships that are usually well defined for graded (Mackin, 1948) or "regime" (Blench, 1969) channels, though many of these non-equilibrium channels change sufficiently slowly or sporadically that they do conform to regime criteria for considerable periods of time. Proglacial channels are frequently characterized by aggradation. For these reasons, some simple descriptive, partial relationships will be investigated in this section in an attempt to develop a qualitative idea of channel behavior.

Coarse, clastic materials are noncohesive. Hence, their resistance to erosion by flowing water is invested almost entirely in the form and weight of individual particles. In materials with a wide range of grain sizes, erosion of the channel bed and bank is selective, so that gradually a lag deposit of the coarsest materials comes to armor the channel bed and sides. So long as sediment transport into the reach from upstream is sufficiently small that continual aggradation does not occur, the channel will become stable for flows below the competence level of the lag material. Imbrication and packing may further raise the threshold of competence for flows capable of altering the channel shape. Most of the channel adjustment is by bank erosion, because an added gravity component raises the total stress on sidewalls at any flow level, and because imbricated deposits that may be underloose with respect to shear from above may nevertheless be overloose when exposed to shear from the side. Hence, additional sediment is often derived from the banks first, and channels in noncohesive material tend to be very wide with respect to depth (Schumm, 1960a). The tendency for such channels to widen excessively is an important aspect of the development of braiding.

Channels can be dichotomized into two groups: single and multiple channelways. Single channels have a variety of forms ranging from the tumbling flow and pool and riffle sequences of small, steep streams, through varying degrees of sinuosity to fully developed classical meanders. Various writers have pointed out that pools and riffles, or talweg meanders, can be

490

detected in almost all "straight" reaches. Leopold and others (1964) and Henderson (1966) have concluded that the quasi-regular spacing of pools and riffles and flow deflections represents the morphological consequences of essentially similar flow mechanics through the entire range of channels, and that they can be treated similarly.

The braiding that characterizes multiple channelways clearly indicates a regimen different from that found in single channels. Multiple channels are found in two distinct situations. Firstly, there are high-energy streams flowing in noncohesive sediments that carry heavy sediment loads and commonly exhibit braiding; this is a common situation in proglacial streams. Secondly, and by contrast, there are channel divisions that commonly occur in deltaic situations where the energy available to the stream is rapidly diminishing.

Meandering, pool-and-riffle structure, and braiding may be found super-imposed upon each other in a channel. Taken collectively they represent channel adjustments to the imposed water and sediment load, and to the strength of the materials in which the channel is excavated, so that a "graded" profile results (in the sense of Mackin, 1948). A series of model studies carried out by Schumm and Khan (1972) clearly distinguished the conditions of braiding from those of meandering (fig. 41).

In the following discussion, conditions in braiding channels will be emphasized, but it will not be possible to isolate this subject from general considerations of channel behavior.

Channel geometry

Outwash channels are generally composed of coarse, noncohesive materials and are characteristically wide and shallow. Their shape varies from trape-zoidal form to that of an extreme scalene triangle. The latter form occurs where the talweg is near one bank. Midchannel longitudinal bars are common, even in single channels (that is, the bar remains submerged at all normal flow levels).

Distributions of width-depth ratios, w_s/d_*, are illustrated (fig. 42) for several samples of sandur channels. Summary statistics of the Baffin channels are given in Table 9. Stebbings (1964) found a limiting geometry beyond which channels that transport high amounts of bed load begin to braid. Of course, when a channel splits into two new ones, the mean geometry for the two new channels becomes much more conservative than it was for the single original channel—other things being equal, the value of w_s/d_* should, initially, be approximately halved. Amongst the data presented, however, the mean width/depth ratio for nonbraided channels is not appreciably greater than for braided channels. It seems that anabranches rapidly adopt a sectional mean geometry that is very similar to that of the undivided parent channel.

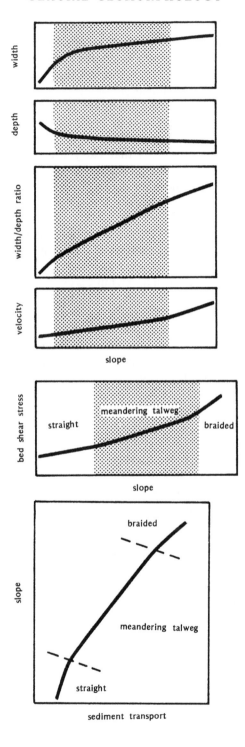

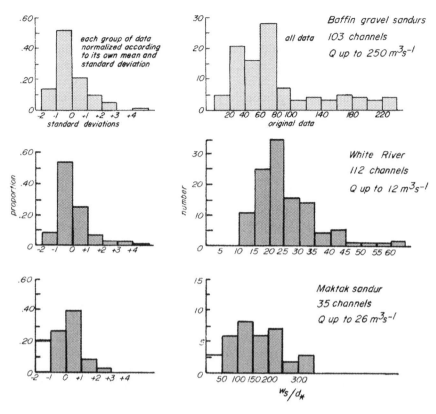

Figure 42 Distribution of the ratio w_s/d_* for outwash channels. For braided streams, each channel unit is treated as a separate channel (White River data, after Fahnestock, 1963, Table 7; Baffin Is. data after Church, 1972, p. 72, reproduced by courtesy of *Geological Survey of Canada*).

Table 9 Mean values of w_s/d_* in braiding and nonbraiding reaches.

River	Braiding	N	Nonbraiding	N
Lewis	86.6	12	71.2	8
Ekalugad Fiord, North River	122.6	6	85.1	5
Middle	—	—	119.3	4
South	69.8	15	84.5	27
Others (small channels)	40.4	4	44.9	6
All	81.3	37	82.1	50

Figure 41 (*opposite*) Discrimination amongst river channel patterns on the basis of the imposed variables, sediment load and slope. Scales have been eliminated, because direct quantitative correspondence to conditions in real channels has not been confirmed. The results are from model studies by Schumm and Kahn, 1972, p. 1763–1764 and 1769; reproduced by courtesy of the *Geological Society of America*, and by permission of the authors).

Channel geometry is closely associated with the nature of bed and banks. For an increasingly wide channel, the shear on the bed is increased at the expense of that on the bank. Wolman and Brush (1961) pointed out that to pass a given fluid discharge along a given slope requires a specified cross-sectional area for a given particle roughness (determined by the nature of the bed material). A given specific discharge requires a certain velocity and depth of flow in order to maintain bed stability (or to pass the imposed bed load). If the resulting stresses exceed bank strength, then channel widening will occur until stability is reached. In natural channels, bed and/or bank armoring confuses this picture.

The sectional hydraulic geometry, in the context of Leopold and Maddock (1953), is presented for several typical gravel-bed outwash streams in Table 10, along with comparative information from other environments. Unfortunately, no information is available for sand channels. In outwash channels, width and depth are relatively conservative, and a proportionately greater amount of the adjustment to changing discharge is taken up by changing velocity than is the case in many other channels. These results confirm the findings of Schumm and Khan (fig. 41).

The high rate of change of velocity is associated with the relatively large and rapidly increasing sediment load that the streams carry at high flows: high competence to move the large fractions of the bed load is maintained in channels in which width and depth changes are relatively minimized with respect to velocity. This indicates that resistance decreases more rapidly as discharge increases than is the case in most channels. Figure 13 showed power-law resistance equations computed for several proglacial outwash channels under the assumption that D_{90} is constant. The results show much more rapid decline in resistance than would occur under either Manning or logarithmic resistance laws. Three main reasons may contribute to this effect:

(a) rapidly increasing sediment discharge at higher flows may contribute significantly to damping of turbulence and increased mass transfer in the section (Vanoni and Nomicos, 1960);
(b) at progressively higher flows an increasing proportion of water flows "straight through" the pattern of pools and riffles that exists at low flow, greatly reducing the form resistance associated with these elements; and
(c) at high flows, a "live" bed presents lower boundary resistance than occurs with a stable bed at low flows.

The first reason is presumably much less important than the other two.

Change in form resistance cannot explain entirely the behavior of the sandur channels. The effect of live-bed conditions is almost always to decrease the effective particle diameter at the bed (the bulk of the sediment

494

Table 10 Comparison of exponents in sectional hydraulic geometry.[1]

River	$b(w_s)$	$f(d)$[2]	$m(\bar{v})$	$b + f(A)$	Source
Baffin sandur means	0.22	0.31	0.48	0.52	Church (1972)
$\alpha = 0.05$ confidence range on Baffin sandur means	0.195– 0.245	0.276– 0.344	0.435– 0.525	0.475– 0.565	
Gauge 1: Hoffellssandur	0.17	0.39	0.44	0.56	Arnborg (1955), data (computed by present writers)
Bowser River: Berendon Glacier, British Columbia	0.23	0.30	0.48	0.52	Water Survey of Canada data (computed by present writers)
Ephemeral streams in semi-arid United States	0.29	0.36	0.34	0.66	Leopold and Miller (1956)
U.S. Midwest (mean conditions)	0.26	0.40	0.34	0.66	Leopold and Maddock (1953)
All U.S. (158 stations)	0.12	0.45	0.43	0.57	Leopold, Wolman and Miller (1964)
Flume channels ($D_{50} = 0.67$ mm)	0.50	0.39	0.16	0.84	Wolman and Brush (1961)
Non-cohesive river: Langbein theory	0.50	0.23	0.27	0.73	Langbein (1964)
Regime channels	0.50	0.36	0.14	0.86	Simons and Albertson (1963) (referred to P, R instead of w_s, d_*: exponent for v deduced by writers)
White River, Mt. Rainier (112 channels)	0.38	0.33	0.27	0.73	Fahnestock (1963)
Maktak sandur, Baffin Island (35 channels)	0.64	0.17	0.19	0.81	Church (not previously published)

[1] $w_s = aQ^b$; $d_* = cQ^f$; $\bar{v} = kQ^m$; $A = acQ^{b+f}$.

[2] d is variously defined, but is normally A/w_s, as in this study.

transported is sand). Hence, values of d_*/D_{90} will increase more rapidly than indicated under the assumptions by which Figure 13 was constructed. The relationships for sections with live bed conditions will have much lower slopes than those indicated, and hence might conform more closely to the Manning or logarithmic equations. When a combination of decreasing form resistance and decreasing particle resistance operates over a range of flow conditions, the results may be very complex. Figure 14 indicated a probable net result of the changes that may occur.

Proglacial rivers have not been studied systematically for downstream changes in hydraulic geometry. Indeed, most do not have a system of regularly confluent, major tributaries associated with them on their proximal reaches of active outwash sedimentation, so that no downstream hydraulic geometry in the usual sense could be contemplated.

Maktak sandur data and Fahnestock's (1963) results from the White River are of some interest. Both sets arose from haphazard measurements in many channels (the rapid rate of change of the channels precluded a more systematic program), with the results plotted all together (fig. 43). Because the sections were of varying size, the outcome is in some respect equivalent to that of a sequence of observations made downstream in a "normal" river. The implication of the White River exponents is that the three dependent factors adjust themselves coequally to changes in discharge. As the channels occur in noncohesive alluvium, Langbein (1964) subsequently made the interpretation that this outcome represented an expected coequal distribution of variance due to discharge variation in absence of any physical constraints on the channel.

The Maktak sandur is a sand outwash, and these results represent the only ones available from sand rivers. They are anomalous by comparison with all other results; unfortunately the manner of their collection and the great amount of scatter in the graphs make it difficult to interpret the outcome.

It is possible to study the mutual variation of various geometric and hydraulic parameters along a channelway at some constant flow level. This has been done for six sections on South River, Ekalugad Fiord. The flow level chosen was 100 cubic meters per second, close to the maximum flow that could pass through the normal channelways. The sections are spread over about 5 km, but with irregular spacing. The following empirical relationships were found:

$$\bar{v} = 56 W_s^{-0.8} \quad r^2 = 0.96 \tag{38a}$$

$$\bar{v} = 3.4 d_*^{2.1} \quad r^2 = 0.41 \tag{38b}$$

Wide, shallow sections have lower mean velocities—these are, presumably, zones of relatively high flow resistance.

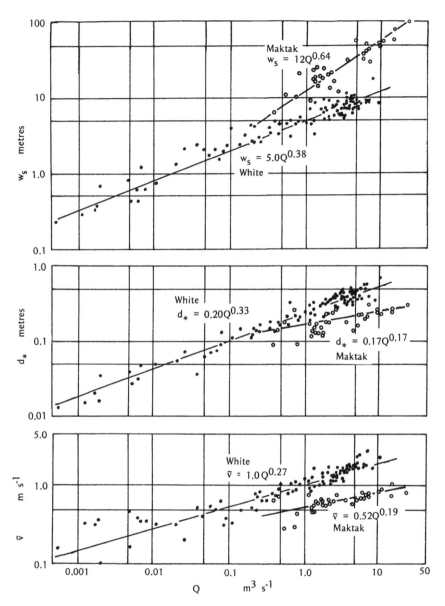

Figure 43 Relation of w_s, d_*, $\bar{v}$ to Q for White River gravel channels (closed points) and Maktak sandur sand channels (open circles). Data for White River from Fahnestock, 1963, p. 13, reproduced by courtesy *U.S. Geological Survey*.

$$S = 45D_{90}^{5.0} \qquad r^2 = 0.73 \qquad\qquad (38c)$$

$$\frac{\bar{v}}{v_*} = 0.2 \left[\frac{d_*}{D_{90}}\right]^{2.0} \qquad r^2 = 0.68 \qquad\qquad (38d)$$

Slope is closely related to roughness element size, and the resistance equation remains very sensitive to changing relative roughness throughout the system.

The pool and riffle sequence

Channel bars have been discussed above as large bedforms. Taken together, they form a pool-riffle sequence down a channel that serves to regulate energy expenditure of the flow (Yang, 1971). They influence channel form and hydraulics by acting as flow controls. Small deposits have little effect since the contraction in flow area is small; however, as they grow in size the contraction in flow area increases until the available energy ($E = A/w + Q^2/2gA^2$ from equation (6), where d is the depth of flow referred to the lowest point in the section) is just sufficient to pass the flow through the contracted section. Flow will then be critical. If the deposit develops beyond this, specific energy must increase further, and this is done by increasing $d = d_{crit}$, producing an upstream backwater. Flow across the riffle at the contraction is supercritical: energy is released in flow resistance across the riffle, and in a "plunge pool" at the lower side of the riffle, or in a hydraulic jump downstream after a period of supercritical flow (fig. 11). The riffle acts as a broad-crested weir.

The important implication of flow over a riffle or bar is the effect on sediment transport. As flow approaches a bar from upstream, the channel becomes shallow. Accordingly, relative roughness increases—flow velocity may also decrease in the area behind the riffle (eqs. 38a and b). Sediment being carried as bed load will then be deposited. At the same time the channel usually becomes considerably wider in order to maintain flow area. This is a zone of flow divergence and decreasing rate of specific energy expenditure. In this situation boundary resistance becomes very great. From the Manning equation, $v = d^{2/3}S^{1/2}/n$, it is seen that, for a given value of n, the most efficient channel is a deep one. It appears, then, that the deeper channel below a bar is more efficient than the wide, shallow one above, so that once the bar is established, the channel will "draw down" water across it, into a zone of flow convergence where the specific rate of energy expenditure increases greatly. On the bar face, where flow becomes supercritical, erosion may occur, and a chute or trench will begin to work upstream along the bar. In this way characteristic "bar and chute" structures form. At higher flows the critical point will move upstream across the front of the bar, and general erosion of the bar lip will occur.

The growth of a chute below or beside the bar (Wertz, 1963) is the key feature of the development of a distinctive morphological form. At a diagonal bar the chute develops toward one side of the channel, whereas at a linguoid bar a chute develops on either side. Minor chutes then develop through the edge of the bar, expedited by the drawdown phenomenon, some of which may later become major channelways. It is often not clear whether the chutes are erosional, or merely non-aggraded channel areas downstream of a depositional bar front (Stebbings, 1964).

Braiding and the channel pattern

A variety of reasons has been cited for the development of channel braiding. Lane (1957) indicated two major categories of braiding: that due to steep slope, and that due to aggradation. Both Lane (1957) and Leopold and Wolman (1957) indicated discriminant functions which divided meandering from braided channels according to slope. Henderson (1963) refined this criterion by taking grain size of bed material into account, and Kirkby (1972) has explored a hydraulic basis for the discrimination. Chien and others (1961) have suggested a more complex functional criterion which takes into account flow variability as well. Schumm and others (1971) made a systematic study of variations in river channel behavior as slope changed.

Common to almost all discussions of the cause of braiding is a heavy sediment load. In particular, appreciable bed-load transport seems to be a necessary part of the mechanics for anastomosis in high energy channels (Hjulström, 1952; Fahnestock, 1963). In these circumstances, local overloading (Stebbings, 1964) or undercompetence (Leopold and Wolman, 1957) determined by variation in slope, channel depth and width, discharge fluctuation, or irregularity in the channel, lead to bar development and possibly eventually to braiding. Inability of the river to move its sediment load also appears to be the basis for braiding in deltaic environments. This circumstance often occurs also on alluvial fans where the specific energy available to the stream is often less than in the single channel upstream from the apex of the fan. Price (1947) and Rubey (1952) explicitly designated channel anastomosis as "braiding," in the classical sense of the term, for those streams in which there is an appreciable reduction of energy gradient.

Braiding has most often been associated with environments of aggradation. Leopold and Wolman (1957) asserted, however, that braided channels can exist in equilibrium in response to particular combinations of hydraulic variables. Lane (1957) and Brice (1964) indicated that braided channels may also be maintained under degrading conditions.

Bank erodibility has been cited as a cause of braiding (Fisk, 1944; Mackin, 1956; Brice, 1964): where the banks are very weak, excessive lateral erosion may lead to very wide channels in which shoaling occurs on central bars, and so to the development of multiple channels. Shen and Vedula (1969) reported an experimental study in which excessive channel widening reduced the shear force on the bed of the channel to the point where braiding occurred and the stream developed narrower and deeper channels through which to move the sediment load. Engelund and Skovgaard (1973), in an extension of Callander's (1969) work, demonstrated from a stability analysis for three-dimensional flows that braiding should occur in a sufficiently wide channel. Doeglas (1951) ascribed braiding in some Pleistocene rivers to rapid, large variations in runoff. Whilst this phenomenon is undoubtedly a part of the regimen of many braided rivers, it seems insufficient as a cause by itself.

Common to most of these considerations is the possibility that actual stream power may at some point fall below that which is necessary to move the water/sediment load through the channel. Approached in another way: at some stage or another, channel resistance becomes too great to pass the imposed load.

It is clear that, in some sandur channels, extension of the indicated hydraulic geometry to high, but still feasible, flow levels results in flow resistance requirements *smaller* than those that could be attained in the surveyed channel (fig. 13). A "live bed" condition may allow the limit apparently set by bed conditions at low flow to be exceeded, but friction cannot decrease below a smooth boundary limit. Channels in which the water/sediment load may become too large undergo bed changes at very high flow in order that the channel can continue to convey the imposed load. In such a situation it is necessary to increase depth (eq. 12), or to decrease the effect of roughness elements at the boundary. This may be achieved in several ways: (a) general scour in the channel bed, leading to greater depths; (b) backwater development leading to overbank flooding; and (c) selective scouring at certain points in the channel, so that a deeper and narrower channel is created with the ability to sustain higher flow velocities; this normally occurs when the sediment load of the stream is too great to permit general scour and is accompanied by deposition of material. The latter adjustment is the common one in rivers carrying very high sediment loads in noncohesive material. It results in the development of flow chutes or channel braiding, where a wide, shallow channel is divided into two (or more) steeper ones able to convey the water and sediment load in the face of the imposed boundary resistance. These morphological developments often emerge and become sharpened only on falling stage—at very high stage the channel continues to operate as a single channelway.

Brice (1964) in his study of braided sections of the Loup Rivers in Nebraska arrived at the same conclusion, as did Maddock (1969), from a general consideration of hydraulic geometries. This process, which leads to the division of a channel as the result of hydraulic and/or sedimentary processes within the channel, may be termed *primary anastomosis*.

Table 11 Mean exponents of hydraulic geometry in braiding and nonbraiding reaches.

Channel State[1]	N	w_s	d_*	$\bar{v}$	A	C^2 *(Resistance Equation)*
Braiding	5	.244	.232	.524	.476	1.685
Propensity to braid	7	.181	.276	.543	.457	1.160
Nonbraiding	2	.130	.440	.430	.570	0.545[2]

[1] Divided channels were treated as a single unit in calculations. "Propensity to braid" indicates the presence of a prominent mid-channel bar, emergent at low flow, but no well defined channel division.

[2] The individual values of 0.04 and 1.05 are highly disparate. The high value of 1.05 is the result of the very large boulders present in the reach which act, in effect, as form resistance elements.

In order to describe hydraulic conditions associated with primary anastomosis, mean exponents are presented from Baffin Island gravel outwash sections, classified according to the channel state (table 11). Width increases most quickly, and depth least quickly, in the braided. reaches. In both cases, reaches with a "propensity" toward braiding are in an intermediate position, but in the latter case they are significantly closer to braiding than to straight reaches in their behavior. As a result, increase in velocity is greater in the braiding and "near-braiding" reaches than in the stable, single channels. Resistance declines much more rapidly in these sections as well (from much higher initial values due to the large relative roughness and relatively great wetted perimeter).

A second form of braiding on an established outwash surface is *secondary anastomosis*. There are many old channel scars and dry channels left behind when the river has shifted radically. Such channels are often reoccupied at high flows when the present main channels overflow (Krigström, 1962; Williams and Rust, 1969; Church, 1972).

The evolution and ultimate stabilization of a channel division will depend on how it handles the water and sediment load carried into it. In natural situations, conditions are very variable (Hjulström, 1952; Axelsson, 1967). Often aggradation occurs at the head of the less efficient branch, and degradation at the head of the more efficient one, perhaps abetted by unbalanced division of the sediment load, or by contrasting flow resistance conditions between two channels (Knighton, 1972). If the process becomes sufficiently unbalanced, one channel may well ultimately capture the entire flow. Flow-division angles do not seem to be important (Axelsson, 1967).

The channel pattern of braiding rivers appears to be highly unstable (Fahnestock, 1963; Chien and others, 1961), but in fact the main channel may remain relatively stable over a number of years, and shift much more slowly than individual anabranches (Krigström, 1962; Church, 1972).

By contrast to the major channels, minor channels may exist only very ephemerally. Their pattern of existence appears to follow a sequence: (a) initial scour when overflow occurs onto the sandur surface from a major channel, (b) establishment of a more or less stable channel with a lag-gravel bed and sides, and (c) infilling with sand (fig. 44). Because these channels are usually established at a higher level than the bed of the main channel, little bed load may be diverted (this is particularly true of channels that persist due to secondary anastomosis) and so the channel fill that appears during waning floods may consist of relatively well sorted sands that were carried in suspension down the main channel. The channels may go through an entire cycle of existence in one flood, but are usually reoccupied during many subsequent high flows. The result is a ribbon deposit of sand across the sandur.

Long profile

Proglacial rivers are so named because they receive most of their water and sediment input from the glacier. As the river flows in its own alluvium, the channel conditions should be essentially uniform from the glacier to some downstream point where it is joined by a major tributary, encounters some geological control, or debouches into standing water. An outwash plain that encounters such control points can be analyzed in terms of two (or more) successive "uniform reaches."

If the river were at grade (Mackin, 1948), then it would develop a rectilinear profile through a homogeneous reach. However, rivers on active outwash plains are characterized by excessive sediment loads at high flow so that aggradation occurs by way of material being deposited along the reach.

Models to describe the aggradation can be derived by considering transient solutions of the equation

$$\frac{\partial z}{\partial t} = \frac{\partial [\text{Sediment transport}]}{\partial x} \tag{39}$$

Culling (1960), Scheidegger (1965, 1970), and Devdariani (1967) have done this using a variety of more or less plausible assumptions about sediment transport, which generally lead to diffusion equations dominated by

Figure 44 (*opposite*) Evolution of minor channels on a sandur surface (flow toward camera): A, Cobble paved, scoured channel. Flow largely maintained from groundwater discharge during low flow. B, Channel choked with sand deposited from suspended load of major channel, which overflowed into minor channels during flood. C, Sand fill in a former minor channel.

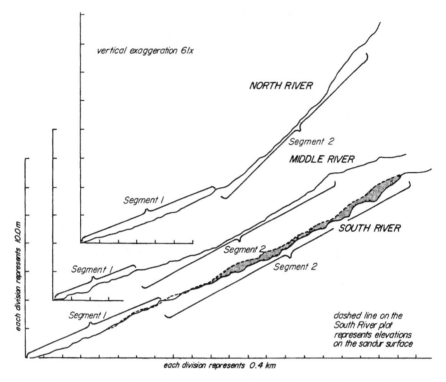

Figure 45 Thalweg profiles of three outwash rivers at Ekalugad Fiord, Baffin Island. Second-order parabolas fit the long profile very well in two segments, which are clearly distinguishable on the ground. The distinction derives from regrading in single channels occurring upstream, and extensive braiding downstream (after Church, 1972, p. 78, reproduced by courtesy of the *Geological Survey of Canada*).

exponential terms. Deposition is accompanied by size sorting, because stream competence is related to gradient by the relationship $D \propto \tau_o \propto d_* S$. As S decreases, either d_* must increase or τ and D decrease. In fact, d_* usually decreases in areas of deposition and D decreases sharply (fig. 37).

Figure 45 shows the profiles of three Baffin Island sandur rivers, with two clearly distinguishable reaches in each one corresponding to a prominent change in channel habit on the ground where sedimentation is especially active (fig. 46). It is evident that simple models of continual aggradation cannot adequately describe such profile development in detail, since, unless available relief can be readily increased, deposition must reduce the overall gradient available to the river, at least in some portion of the course (fig. 47). Whereas the enhanced stream power may permit this during very high flows, development of a "stepped profile" appears to be the usual

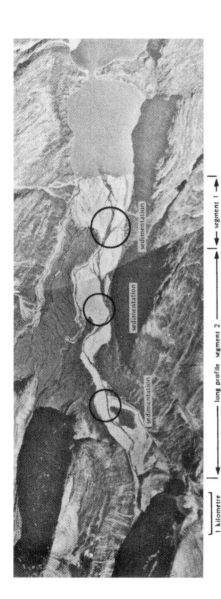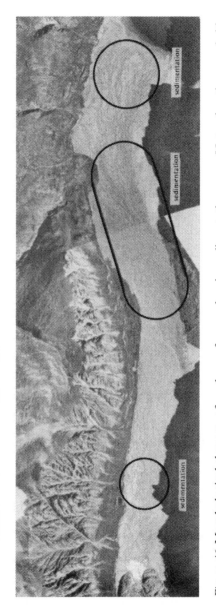

Figure 46 Morphological elements of sandur surfaces and major sedimentation zones. Note, also, the remarkable contrast of the channel form: *Top*, Ekalugad sandur, Baffin Island, a gravel sandur (A17019-137/157). *Bottom*, Maktak sandur, Baffin Island, a fine gravel/sand surface (A17012-91/124). (Photographs by courtesy Canada Dept. Energy, Mines and Resources, National Air Photo Library.)

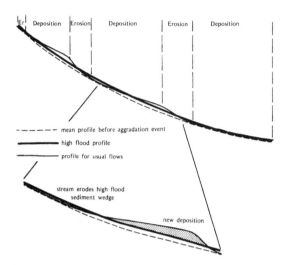

Figure 47 Schematic diagram of detailed pattern of aggradation.

hydraulic response to the situation, so that sediment is deposited in specific zones where the competence of the river is exceeded, and stream power is concentrated on the steepened, downstream face of such a zone where renewed entrainment moves some material further through the system. Such a stepped profile may be represented in the major channel bar sequence of the river, but morphological evidence suggests that much larger areas are involved (fig. 46). Extensive depositional areas constitute major zones of flow divergence, characterized by channels which become very wide and shallow.

The localization of such areas may be subject to changing patterns of runoff and sediment supply, which in turn may be affected by climatic and glacial fluctuations. For example, a reduction in glacier size and activity following an amelioration in climate may reduce the volume of sediment delivered to an outwash surface, and perhaps alter the characteristics of runoff. Aggradation may be reduced or ended in the proximal zone when the river is capable of handling the reduced sediment load; sediment may even be entrained here and moved farther downsandur as the stream regrades its proximal course to that consistent with the changed regimen. Fahnestock (1969) advanced this explanation for the appearance of Slims River in front of Kaskawulsh Glacier, as did McDonald and Banerjee (1971) for the Peyto outwash plain, and Lustig (1965) has indicated a similar sedimentary response to changed climatic conditions on alluvial fans in semiarid environments. Nevertheless, moderate trenching on proximal portions of outwash surfaces (reported by Arnborg, 1955; Krigström, 1962; and Church, 1972)

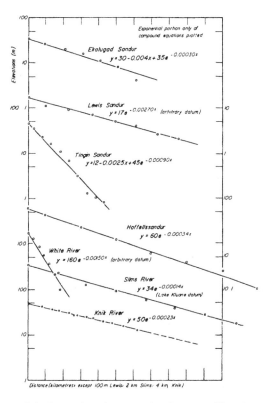

Figure 48 Exponential plots of various sandur long profiles. Irregularities can be attributed to control points that affect the detailed form of the surfaces.

may be quite normal, because the concentrated effects of high floods may determine the general surface level.

The long profile of the sandur is sharply concave, similar to that of the rivers. Figure 48 presents the profiles of several sandurs. At Lewis Valley, the proximal end of the sandur (outside the active river channel) passes into little altered ablation moraine which continues right up onto the glacier. At Skeidarárjökull (Thorarinsson, 1939) and Breidamerkurjökull (Price, 1969) in Iceland, the active sandur surface runs directly onto the ice. Ekalugad sandur provides an example of a compound profile with two regular profile segments occurring. Four of the surfaces are reasonably well approximated by simple exponential equations (*see* Krumbein, 1937; Blissenbach, 1954; Bluck, 1964, for similar results from alluvial fans). The two east Baffin sandurs, Ekalugad and Tingin, are represented by the sum of linear plus exponential functions (the fit is not ideal, the consequence perhaps of considering segmented profiles as simple forms).

507

Deltaic and lacustrine environments

Flow conditions

Backwater

Channel form and process are altered in backwater conditions at the head of a stationary water body. Aggradation on the bed due principally to the deposition of bed load extends upstream some distance from the lake, along with reduced flow velocities. Consideration of the Bernoulli equation (eq. 5) in the differential form appropriate for nonuniform flow:

$$\frac{dH}{dx} = \frac{d}{dx}\left(z + d + \frac{v^2}{2g}\right) = -S_f$$

leads directly to an expression for the longitudinal stream profile and energy relation (*see* for example, Henderson, 1966, Chap. 4):

$$\frac{dy}{dx} = \frac{S_o - S_f}{1 - F_r^2} \tag{40}$$

For the condition of backwater in subcritical flow above a lake, the water depth, d, is greater than that of uniform flow, d_o, as determined from the Manning equation. As the water depth increases toward the stationary water body, ignoring the effect of sediment deposition, F_r and $S_f \to 0$ and $dy/dx \to S_o$; that is, the water surface is asymptotic to the horizontal surface of the undisturbed water body (fig. 49). Upstream, the influence of backwater decreases and uniform flow conditions are approached; that is, as $S_o \to S_f$, $dy/dx \to 0$ and $d \to d_c$. Rao and Sridharan (1966), extending theoretical studies of this backwater curve by Vallentine (1964), showed that the distance upstream that backwater is influential bears a complex relation to channel cross-section shape, Froude number, and channel roughness.

Whereas most of the fine-grained suspended material passes through the backwater zone as wash load, much of the bed load is deposited, especially in the initial stages of delta development or during a period of high controlling (lake) water level. This has the result of lowering the bed slope and raising the water surface and stream-bed elevations (fig. 49). Harrison (1952) proposed that the rate of sedimentation could be calculated with an equation of the form

$$\frac{dz}{dt} = \frac{1}{\bar{w}\gamma_s}\frac{dq_s}{dx} \tag{39a}$$

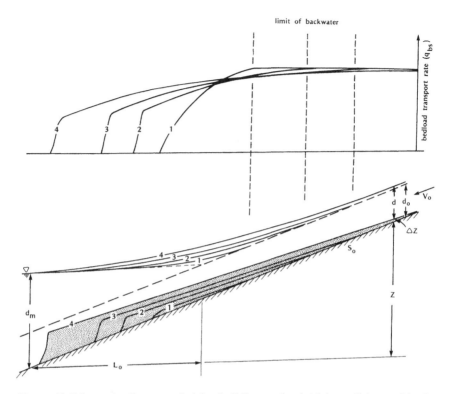

Figure 49 Schematic diagram of delta building under initial conditions of back-water (after Harrison, 1952, p. 209, reproduced by courtesy of Institute of Hydraulic Research, University of Iowa).

At the limit of backwater the rate of bed-load transport begins to decrease from that characteristic of uniform flow. In the initial stages of delta building dq_s/dx would change uniformly, leading to maximum deposition at the intersection of the uniform flow surface and the level surface of the water body. As the delta builds, the front forms a sharp break of slope over which the remaining bed load is dumped. Vincent's (1966) flume experiments verified this.

In an established delta with relatively steady control, and as sedimentation progresses, the channel or channels apparently change to accommodate the changed profile so that most of the sediment continues to be carried to the delta lip before being deposited. Examples from the Colorado River at Lake Mead (Lara and Sanders, 1970) and Lillooet Lake, British Columbia, illustrate the resulting morphology (fig. 50). In both cases, although there was substantial delta advance between the two periods of measurement (Colorado, 10 km; Lillooet, 0.75 km), there was no measurable aggradation of the river bed. At Lillooet delta a lowering of the average controlling lake

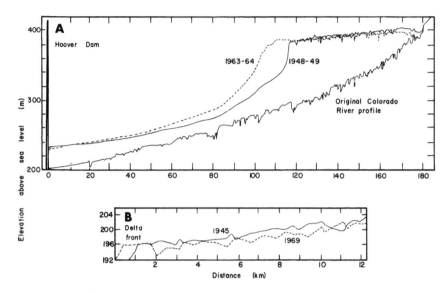

Figure 50 Profiles of (A) Colorado River at Lake Mead (after Lara and Sanders, 1970, p. 155) and (B) Lillooet River at Lillooet Lake, showing that the greatest accumulation on an established delta occurs at the delta front and on the foreset beds. Lake Mead data reproduced by courtesy of the *United States Bureau of Reclamation.*

elevation by two meters in 1952 is reflected in a comparable lowering of river-bed level upstream, except in the vicinity of the delta. Here aggradation as the delta builds forward is assisted by a three- to four-meter annual stage change in the lake. Clearly, as the delta front advances, some aggradation must occur to maintain sufficient slope to transport water and sediment to the river mouth. But since slopes in backwater are usually very low, particularly in larger rivers, considerably more deposition occurs on the delta front than on topset beds and flood plain.

Jet analogy

Suspended sediment that passes through the backwater zone is carried to the lake beyond. A number of workers have found the analogy of the free jet useful in interpreting flow conditions beyond the delta front. Albertson and others (1950) investigated the diffusion of a jet discharging into an infinite volume of the same fluid from a slot of finite thickness and infinite width (a two-dimensional jet) and from a circular orifice (a three-dimensional jet). Assuming the velocity at the orifice to be constant across the entire opening as water moves out, shear along the jet boundary gives rise to turbulent eddies with resulting lateral mixing of the jet. As the surrounding fluid is entrained and accelerated, the jet fluid is slowed and the volume of the jet

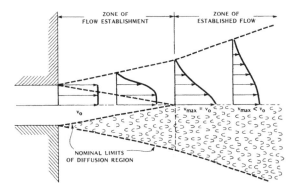

Figure 51 Schematic representation of axial jet diffusion. Dashed lines delimit the edges of the jet (after Albertson and others, 1950, p. 640, reproduced by courtesy of the *American Society of Civil Engineers*).

is increased. This inward and outward diffusion persists to the limit of the "zone of flow establishment" (fig. 51) at which point only axis velocity remains equal to the original jet velocity. Beyond, in the "zone of established flow" continued entrainment slows the velocity.

A number of assumptions are made in the development of the theory of jet flow which may not be met in the case of a river discharging into a body of standing water (Crickmay and Bates, 1955). It is assumed that the jet is submerged and free to expand without boundary interference; that the reservoir is sufficiently large that return flow is not necessary to replace fluid entrained, so that fluid in the reservoir is stationary; that the fluid in the jet and in the reservoir are of the same density; and that the velocity at the jet outlet is uniform across the orifice.

Using the analogy of the jet, Bates (1953) proposed that deltas formed under three conditions of entrance flow:

(a) inflow more dense (hyperpycnal flow), entering the water body as a plane jet on the bottom (underflow);
(b) inflow equally dense (homopycnal flow), entering the water body as an axial jet (interflow); and
(c) inflow less dense (hypopycnal flow), entering the water body as a plane jet on the surface (surface flow).

Bates suggested that each flow type would result in a characteristic delta morphology. Although some success has been achieved in modeling delta conditions using the jet analogy (Bonham-Carter and Sutherland, 1967, 1968; Farmer, 1971), the assumptions are so frequently not met in actual situations that severe modification is necessary before the jet analogy can be of further use in interpreting actual deltaic situations.

511

Effects of the lake

Bottom flow

As early as 1885 Forel recognized that a sediment-charged river may plunge beneath clearer lake water and flow down the lake bottom as a clearly defined flow. Since then the universal significance of turbidity currents[2] has been recognized in both marine and lacustrine environments, and the theory of stratified flow in fluids has been developed in fields as diverse as geology and meteorology and applied to turbidity currents.

Harleman (1961, 1963) provided a summary of studies of the theory of stratified flow. The more dense underflow is subjected to gravitational acceleration according to the difference in density between the flow and the surrounding medium:

$$g' = \frac{\gamma'}{\rho_d} = \frac{g(\rho_d - \rho_1)}{\rho_d} \tag{41}$$

(commonly $g' \lll g$).

Considering only the simple two-layer system, the energy relation and densimetric Froude number ($F_r' = v_d/\sqrt{gd_d}$) are analogous to free-surface flow conditions. Where the depth of the overlying fluid is great compared to that of the underflow, the velocity of return flow caused by the entrainment of water along the shear boundary is low. In this case the mean velocity of uniform underflow may be calculated from the equilibrium equation:

$$\tau_o + \tau_i = (\rho_d - \rho_1)gd_dS \tag{42}$$

as:

$$\bar{v}_d = \sqrt{8g\frac{d_dS}{ff(1 + \alpha)}} \tag{43}$$

where $\alpha = \tau_o/\tau_i$.

By analogy with free-surface flow, the friction factor ff may be obtained from the Moody diagram using a hydraulic radius of $4d_d^*$. Limited experimental evidence suggests that $\alpha \approx 0.43$ (Bata and Bogich, 1953). In the case of frictionless basal boundary conditions (Hinze, 1960) or when a flowing grain layer shears out from under a turbidity current (Sanders, 1965), $\alpha \lll 1$. Middleton (1966), in his laboratory experiments, found a weak relationship between ff and Reynolds number for conditions of varying slope, and between $ff_i/ff_o = \alpha$ and densimetric Froude number. However, the work of Dick and Marsalek (1973) indicates that the calculation of ff_i alone does not depend on the densimetric Froude number.

For nonuniform flow the same procedure, with modifications of the type discussed above, is applied to the differentiation of the energy equation to express the friction slope, S_f, in the same way as for open channel flow. Similarly, continuity equations may be developed for unsteady flow (Harleman, 1961; Elwin and Slotta, 1969). Analytical studies of the behavior of turbidity currents with respect to resistance relations, velocity distributions, and the characteristics of the currents have been made by, amongst others, Levi (1959, 1965), Hinze (1960), Levi and Kylyesh (1960), and Chawla (1964).

The analogy between stratified and open channel flow breaks down in the area of interfacial mixing, because of the great differences in density and different phases of the fluids. Keulegan (1949) offered an analysis for the stability of the internal waves between the density current and the surrounding water. If one assumes that: (a) the depth of water and thickness of the density current are large compared to the internal wavelengths (λ), (b) the wave height is small compared to the wavelength, and (c) the flow is irrotational, then the basic equation for the celerity of interfacial waves (Harleman, 1961, his equation 26.24) can be reduced to:

$$c = \frac{v_d}{2} + \frac{1}{2}\left[\frac{g'\lambda}{\pi} - v_d^2\right]^{1/2} \tag{44}$$

It is generally assumed that internal waves begin to break concomitant with mixing when the right hand portion of equation (44) becomes undefined; that is, when $V_d^2 > g'\lambda/\pi$. Thus the critical velocity is expressed as:

$$c_{crit} = v_{d(crit)}^2/2 \tag{44a}$$

Keulegan's (1949) experiments showed a wide variation, but $c = 0.8v_d$ approximated the actual relation.

However, when the gravitational driving force is small $(\rho_d - \rho_1)/\rho_d \to 0$; that is $g' \to 0$) viscous and surface-tension forces become more significant in determining the nature of mixing. Using the theory of pressure distribution on waves, Keulegan (1949) derived a stability criterion for the breaking of waves:

$$\theta = (v_d g')^{1/3}/v_d \tag{45}$$

(see also Monish, 1938, for a similar development based on Prandtl's equation for continuous density change between two currents). Keulegan's flume experiments produced values of θ in the range 0.14 to 0.24 with a mean of 0.18. Ippen and Harleman's (1952) work confirmed this value (note that $\theta < 0.18$ implies no mixing). They also found that when the turbidity current is not thick, wave breaking begins when $\lambda \approx \pi d_a$.

However, most authors have noted that *turbidity* currents both in the laboratory and the field generally maintain themselves for long distances without appreciable mixing. Density currents are often weaker (g' small), so that mixing may be more significant (Bell, 1947).

River water flowing along the lake bottom may carry even coarse-grained fluvial sediment far into the lake. Studies on Rhone delta (fig. 52A) showed underflow from a well-defined river mouth carrying sand in a single channel with obvious levees to the deepest part of the lake some 15 km from the river mouth. At Lillooet Lake, British Columbia, distributaries from a large glacial stream wander over the delta surface (fig. 53), but although underflow frequently transports silt and fine sand 5 to 6 km into the lake, its effect on surficial sediment patterns is not nearly as pronounced (fig. 52B).

Interflow

Very commonly deep glacial lakes show strong thermal and sediment-density stratification (Mathews, 1956; Gilbert, 1972). Inflowing river water at a density between those of surface and bottom lake water may flow beneath the lake water down the delta front until it reaches water of almost the same density. Here it will spread out through the lake as an interflow. Sustained periods of interflow are probably not common in proglacial lakes, because large diurnal fluctuations in the volume, suspended sediment load, and temperature of inflowing water will cause large daily fluctuations in its density. At Lillooet Lake in August 1971, during a period of clear warm weather which caused large diurnal fluctuations in inflow, interflow along the thermocline was observed nearly every day, interspersed with periods of strong underflow (Gilbert, 1973). Both underflow and interflow are known to occur in Lac Léman as well (Houbolt and Jonker, 1968).

Interflow should be intermediate between underflow and surface flow in its ability to transport coarse sediment fractions beyond the delta front, but little is known of this and verification is difficult, particularly from stratigraphic evidence.

Surface flow

If the density of the inflowing river water is less than that of the body into which it flows, the river water will spread on the surface. Studies of glacial streams flowing to the heads of a number of British Columbia fiords (*see*, for example, Pickard, 1953, 1961; Tabata, 1954; Tabata and Pickard, 1957) showed a layer of fresh water divided from the salt water below by a strong

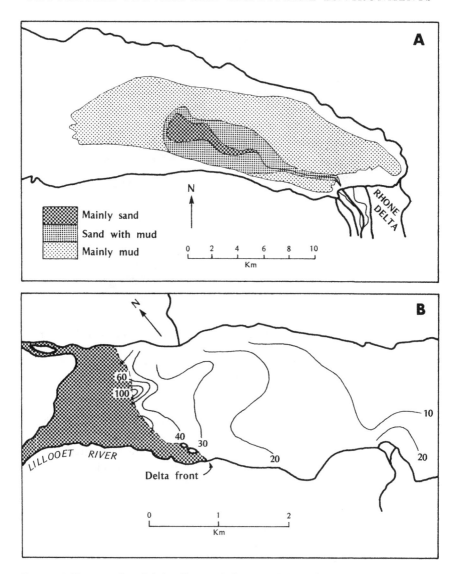

Figure 52 Texture of surficial sediment: A, in Lac Léman (after Houbolt and Jonker, 1968, p. 135, reproduced by courtesy of the *Royal Geological and Mining Society of the Netherlands*). B, Lillooet Lake; the isolines are of geometric mean grain size in microns, based on fifty arbitrary bottom samples.

halocline at 5- to 50-m depth (depending on flow, tidal and wind conditions). Surface flow has also been observed on fresh water lakes when glacial sediment inflow is low (Houbolt and Jonker, 1968) and occasionally during periods of high sediment inflow (Mathews, 1956). Sustained periods of surface flow have been noted on Lillooet Lake, British Columbia, in winter

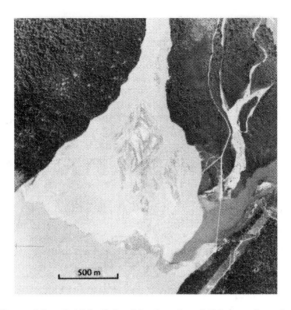

Figure 53 Lillooet delta, British Columbia showing shifting surface channels, debris line indicating the meeting of river and lake currents and the plunging of river water beneath the clearer lake water. Lockwood Survey Corp air photograph 42128, July 3, 1960. Lillooet River discharge is 309 cubic meters per second (see recurrence series, Figure 8A).

and early spring when glaciofluvial suspended-sediment concentration is lower than that residual in the lake from the previous melt season. At this time of year the lake is near isothermal at 2 to 3°C, while inflowing river water is close to 0°C, further increasing the density difference. Unlike underflow, once surface flow passes beyond the effect of backwater, it is driven forward by inertia alone. Thus velocity and sediment transporting capacities are low, and the flow is affected by diffusion as well as by current patterns in the water body and wind on the surface.

Coarse-grained sediment is deposited as a bar at the river mouth (Bates, 1953), especially where a wedge of salt water protrudes under the river water (Farmer, 1971). Fine-grained sediment may settle slowly to the bottom from surface flow, although silt particles are probably deposited from suspension relatively quickly (Ashley, this vol.). Bradley (1965) proposed vertically cascading density currents as a means of depositing the finest rock flour and clay whose settling time in deep water bodies would otherwise be measured in years. The fine-grained sediment is spread more uniformly as bottomset beds in the water body, and a sharper textural boundary is seen between these beds and the foreset beds composed of bed load and coarse suspended-sediment discharge than is the case when bottom flow predominates.

Lake currents

Entrainment of lake water, particularly by underflows, necessitates its replacement at the delta front. Water may be brought toward the delta on the lake surface to meet the inflowing river water (Grover and Howard, 1938). Where the two currents plunge beneath the lake surface at the delta front a line of debris frequently develops (fig. 53). Studies by Bell (1947) showed a pattern of two cells associated with underflow and interflow. In British Columbia fiords, Pickard and Rogers (1959) noted surface flow down the fiord induced by river inflow, as well as one or more circulation cells beneath. This relatively simple pattern was severely disrupted by wind and tides. At Lillooet Lake, surface-return velocities of as much as 0.25 meters per second have been observed during calm conditions, but moderate winds down the lake (5 to 10 meters per second), are sufficient to reverse the flow and break up the debris line even during periods of strong underflow. These complex current patterns may be responsible for transportation of slowly settling sediment particles far beyond the delta front before deposition takes place.

In restricted water bodies, the river inflow itself may establish currents that affect sediment distribution. Tasiujaq Cove, at the end of Ekalugad sandur, is a nearly enclosed water body isolated from the main fiord by a moraine. Here, inflowing South River water, which enters the southwest corner of the cove, frequently establishes a counterclockwise circulation of surface water (flowing over salt water beneath), which is reflected in the offshore extension of coarse sediment deposits opposite the river mouth.

Sediment deposition

Gilbert's (1885) classical description of the tripartite structure of a delta remains the standard from which variations are described and explained. Topset beds of channel and overbank deposits build forward over steeply dipping foreset beds which, in turn, rest on thin bottomset beds composed of fine-grained material that was carried in suspension far beyond the delta front (fig. 54).

Initial deposition: foreset beds

Allen (1968c) summarized a series of experiments (Allen, 1965; Jopling, 1960 through 1967) on the deposition in the lee of dunes and small deltas:

(1) the rate of deposition and the mean grain size of particles decreases outward from the delta;

(2) the gradients of these parameters with respect to distance from the delta crest vary inversely as the ratio of fluid flow velocity to grain fall velocity;

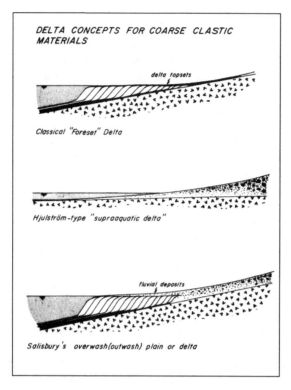

Figure 54 Delta morphometry resulting from different conditions of sedimentation. In the *classical* delta (Gilbert, 1885) tabular foreset beds represent most of the accumulation of sediment. The sediment is dropped by the diminishing stream current as it enters the standing water body. Hjulström (1952) defined a *supraaquatic* delta, after studies of glaciofluvial valley fills in Scandinavia and Iceland, as a distinctive form resulting from rapid sedimentation and high competence in certain valley-fill environments (particularly proglacial ones). Deposition in shallow water would appear to be requisite. In high energy environments, as are common in proglacial streams, fluvial sedimentation may occur on top of the subaqueous deposits as the delta builds forward (Salisbury, 1892). In the extreme case a fan-delta may be formed that combines features of the preceding types.

(3) discontinuous avalanching of oversteepened foreset deposits leads to alternating coarse- and fine-grained layers near the toe of the delta;

(4) as sediment discharge is increased, avalanching of grains becomes more frequent, eventually being continuous, and the bottomset deposits increase in size relative to the foreset.

Jopling (1965a) calculated deposition on the delta front from a consideration of the concentration of grains suspended in the inflowing stream, and the grain-size distribution and fall velocities, assuming quasi-laminar flow.

This "path-line method" was used by Bonham-Carter and Sutherland (1967) in their computer simulation of deltaic sedimentation. Allen (1968c) pointed out that this approach does not account for the distribution of bed load on the delta front, which may be particularly significant, especially with reference to sliding material on oversteepened foresets. He showed that the assumption of quasi-laminar flow is an oversimplification, particularly in the mixing region between inflowing and lake water. The treatment of the "dispersion of grains on the lee of a sand body [as] essentially a problem of diffusion in a free-turbulence shear flow and not one of ballistics" (Allen, 1968c, p. 622) makes theoretical solution impossible, but Allen's experimental results showed that

$$\frac{W_x}{W_L} = \left(\frac{x}{L}\right)^{-m} \tag{46}$$

where the exponent, $m \propto w/v$. In a later paper, Allen (1970b) developed a theory for the frequency of avalanche events on a delta front. If dW/dx is assumed to be constant, the angle of repose of the foreset deposits increases to some critical value when failure occurs and a wedge of material slides down, leaving the slope near the initial (smaller) angle. The frequency of the slide events is a function of the rate of slope buildup, the difference between the slope necessary to cause failure and the slope after failure occurs, the time taken for a slide to occur, and the rate of sediment supply during the period of sliding. As sediment supply rate increases, slides become more frequent until sliding is continuous. On larger deltas and where sediment supply to the foreset zone is small, avalanching will be infrequent so that the period may be approximated by only the rate of slope change and the angle through which the change must occur before failure occurs, that is:

$$P \approx P_\infty = \frac{\Delta\phi_s}{dS/dt} \tag{47}$$

Further, as Allen pointed out, large infrequent slumps are interspaced with more numerous, small slides which "freeze" on the foreset face to build up as a thick, oversteepened deposit which must eventually fail.

On steeply dipping foreset beds reported at the mouths of distributaries on several high-energy glacial deltas (Fulton and Pullen, 1969; Gilbert, 1972) these small-scale phenomena may mainly account for the distribution of sediment, but this zone is of small vertical extent (less than 15 m) in comparison with the zone below, which rests at a lower angle (3 to 5 degrees) and on which large slumps occur very infrequently. Those events are ultimately set off by sediment that has moved in small slides from the steep foresets above.

Initial deposition: bottomset beds

On very small deltas, Allen (1968c) observed that appreciable quantities of bed material reach the bottomset beds by slipping down the foreset slope. However, in most large glaciomarine or lacustrine deltas the true bottomset beds result from transport of sediment in suspension well beyond the direct influence of the river and the zone of slumping on the delta front. Sediments are characteristically fine grained, although lenses of sand may occur.

Since early studies of Pleistocene varved sediments in the nineteenth century, the methods of sediment transport and the origin of the alternate light and dark layers have been the subject of considerable study. Following De Geer's (1912) proposal that each couplet represented the deposition during one year, varves have been used extensively to date late and post-Pleistocene events in Europe (Antevs, 1928, 1946), and in North America (Antevs, 1925, 1928; Hughes, 1965), and to interrelate the two (De Geer, 1926; Antevs, 1928). It has been shown that "varves" are indeed annual deposits on the basis of agreement with radiodating techniques (Antevs, 1957), seasonal distribution of pollen grains (Terasmae, 1963), and by virtue of the correlation between varve thickness and climatic or discharge parameters (Shostakovich 1931, 1936; Granar, 1956).

De Geer (1912) proposed that varves resulted from deposition by turbidity currents in the melt season followed by settling from suspension of the fine-grained sediments during winter. The role of turbidity currents was doubted (Antevs 1925; 1951), partly because they had not been observed in the field (Johnston, 1922; Kindle, 1930). However, Kuenen (1951) showed that density differences based more on sediment concentration than on temperature were indeed great enough in fresh water lakes for underflows to be significant. Subsequent studies support this view (*see* for example, Lajtai, 1967). Agterberg and Banerjee (1969, p. 647) consider varves to be composed of "three genetically dissimilar parts:

> 1. the silt ("summer") part was deposited in a relatively short period by a turbidity current. Variations between successive layers may be strong and a single layer will show a strong decrease in thickness away from the source. When multiple graded units are present this may mean successive turbidity currents generated in the same year or pulsations within a single turbidity current.
> 2. the clay ("winter") layer consists of two parts: Part 2a deposited by the turbidity current after stagnation, and Part 2b deposited by slow, continuous settling from suspension.

As shown by Gilbert (1972), turbidity underflow in a near-glacier lake where distinct varves are forming is not a continuous event through the melt season. Preliminary results suggest that in some years significant underflow

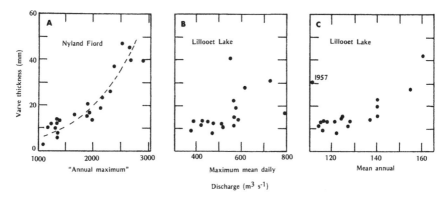

Figure 55 Varve thickness plotted against: A, "annual discharge maximum," Nyland Fjord, Sweden (Granar, 1956, p. 656, reproduced by courtesy of *Geologiska Foreninger*.) B, maximum mean daily flow Lillooet Lake, British Columbia, and C, mean annual flow, Lillooet Lake. Lillooet thickness data are the averages of varves in four cores located four to seven kilometers from the delta front.

may occur not at all or only once or twice, lasting a few hours during the four-month melt season, whereas in other years significant underflow may occur during as much as half of the melt season.

Granar (1956) in a study of sedimentation in Nyland Fiord, Sweden, found a good correlation between the "annual discharge maximum" and varve thickness (fig. 55A). His maximum is not defined, but is assumed to be maximum mean daily inflow. At Lillooet Lake, British Columbia, only 30 percent of the variance is varve thickness could be accounted for by regression against maximum mean daily discharge (fig. 55B). Other hydrologic parameters produced better regression results but none significantly better than mean annual discharge (fig. 55C). The value for the year 1957 is clearly anomalous; the daily discharge throughout the melt season was, for the most part, well below average, but a single very large flood in September may have washed much of the fine sediment accumulated on the delta surface and prodelta slope to the depths of the lake in a single, unusually powerful turbidity flow.

This event suggests the difference between the processes of sedimentation in salt and fresh water. Kuenen (1951) proposed that, in salt water, varve deposition occurred not by underflow, but by flocculation and settling of sediment spread in the near-surface water. Thus it might be expected that varve thickness would be related more closely to sediment supply and thus to discharge. In fresh water lakes, on the other hand, the occurrence of turbidity current flow, and therefore more rapid deposition, is dependent not only on discharge, but on the temperature of river and lake water, on the thermal structure of the lake, on the quantity of sediment suspended in the lake from antecedent events, and on the dissolved sediment concentrations in lake and inflowing water. If it is the case that turbidity currents are generally intermittent and difficult to predict from minimal historical records, then varve-like laminations in fresh-water sediments may be more difficult to interpret correctly, because periods of slow deposition of fine-grained sediments may alternate with turbidity-current deposits within one year (Ashley, this vol.; Banerjee, 1973).

Secondary sediment deposition

Slumps

A number of studies of glacial deltas have shown that slumping may be of significance in redistributing the coarse-grained sediment piled up from bed-load transport at river mouths. Following Terzaghi's analysis (1950, 1955, 1956), the resistance to shear (τ) associated with slumping of a sediment is approximately:

$$\tau = c + (\gamma_s z - p_e) \tan \theta \tag{48}$$

The cohesion, c, is the result of molecular interaction between solid particles and varies from near zero in sand to at least 500 to 800 kilograms per square meter in fine-grained sediments. At the sediment surface $z = 0$ and $\tau = c$. The angle of internal friction, θ, varies from 20 to 45 degrees, generally being highest in densely packed, coarse sand. The symbol p_e refers to the pore water pressure in excess of that due to hydrostatic pressure (γz). It has limits 0 and γz_s. This excess pore water pressure may be a particularly significant factor in fiord-head sedimentation where sudden changes in hydrostatic pressure are brought about by large tidal ranges. Terzaghi (1956) attributed a slump on a post-Pleistocene delta in Howe Sound, British Columbia, to a sudden increase in excess pore water pressure as the tide ebbed (γz decreased). In fresh water lakes, although surface-elevation changes between low and high flow periods may be considerable, the change is slow, so that excess pore water pressure is not likely to become large enough to be a significant factor in slumping.

The stability of slopes in unconsolidated material is the subject of considerable engineering study (*see* for example, Moore, 1961; Noorany and Gizienski, 1970). In the case of a long slope of uniform composition, the vertical pressure, p, at any point is given by:

$$p = \gamma z_s \tag{49}$$

This can be resolved into a normal stress

$$\sigma = p \cos (s) \tag{49a}$$

and a shear stress

$$\tau = p \sin (s) \tag{49b}$$

When these values are plotted (fig. 56) with the empirically determined shear envelope of the sediment, the point at which failure will occur can be determined; that is, at some critical pressure (AC in fig. 56) failure must occur. Clearly, if the slope of the sediment, s is low and/or the angle of internal friction, θ, is high, failure may never occur regardless of the depth or amount of sediment accumulated.

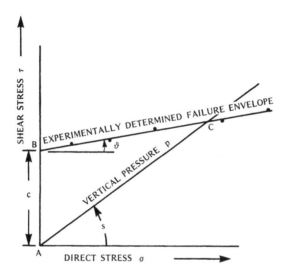

Figure 56 Shear *versus* normal (direct) stress diagrams showing the conditions for failure when slope, S, is greater than the internal friction angle, θ. BC is determined experimentally in shear tests, AC represents the critical vertical pressure (thus the critical depth of accumulation for failure $z = \rho/\gamma \cos S$) (after Moore, 1961, p. 346, *Journal of Sedimentary Petrology*).

Slumping has been most extensively studied on large marine deltas, for example the Mississippi River delta (Shepard, 1955; Walker and Massingill, 1970) and the Fraser River delta (Mathews and Shepard, 1962; Tiffin and others, 1971), as well as on submarine slopes not associated with rapid delta building (Uchupi, 1967; Moore and others, 1970). Mathews (1956) referred to small slumps on the steep sides of Garibaldi Lake, British Columbia, associated with the discharge of glacial streams. Slow slumping in sediment deposited at the mouth of Columbia River in Upper Arrow Lake is thought to be responsible for redistribution of large volumes of material partially derived from glaciers (Fulton and Pullen, 1969). In Lillooet Lake, bed material is deposited in steeply dipping foreset beds extending only to depths of 8 to 10 m. Below a sharp break at this depth the slope decreases to less than 5 degrees. Mounds of slumped material cover this surface, increasing in height to 7 m at 120-m depth. Here the slump mounds abruptly give way to nearly flat lying bottomset beds beyond. Slumps, in this case, are probably irregular and infrequent events.

Slumping is not known on some large, rapidly growing deltas, as for example the Colorado Delta in Lake Mead (Lara and Sanders, 1970) and the Rhone Delta in Lac Léman (Houbolt and Jonker, 1968). In the latter case, underflow is sufficiently frequent and powerful to carry coarse-grained bed material to the deepest parts of the lake, thus preventing the build-up of over-steepened foreset beds (fig. 52A).

Acknowledgments

Some of the field results not previously reported were obtained during work under a National Science Foundation (U.S.A.) grant to J.T. Andrews, University of Colorado, Boulder (Church), and a National Research Council (Canada) grant to H.O. Slaymaker, University of British Columbia, Vancouver (Gilbert). Karen Ewing drafted many of the figures, and Heather Hamel typed the manuscript. W.F. Rannie contributed valuable discussion to the section on sediment transport, and B.C. McDonald made many valuable suggestions in the sections on sedimentation.

References

ACAROGLU, E.R., AND GRAF, W.H., 1968, Sediment transport in conveyance systems—Part 2: The modes of sediment transport and their related bed forms in conveyance systems: Internat. Assoc. Sci. Hydrology Bull., v. 13, p. 123–125.

AGTERBERG, F.P., AND BANERJEE, INDRANIL, 1969, Stochastic model for the deposition of varves in glacial Lake Barlow-Ojibway, Ontario, Canada: Canadian Jour. Earth Sci., v. 6, p. 625–652.

ALBERTSON, M.L., DAI, Y.B., JENSEN, R.A., AND ROUSE, Hunter, 1950, Diffusion of submerged jets: Am. Soc. Civil Engineers Trans., v. 115, p. 629–697.

ALBERTSON, M.L., AND SIMONS, D.B., 1964, Fluid mechanics: Section 7 in Chow, V.T. (ed.), Handbook of applied hydrology, McGraw-Hill, New York.

ALLEN, J.R.L., 1964, Primary current lineation in the Lower Old Red Sandstone (Devonian), Anglo-Welsh Basin: Sedimentology, v. 3, p. 89–108.

——, 1965, Sedimentation in the lee of small underwater sand waves—an experimental study: Jour. Geology, v. 73, p. 95–116.

——, 1968a, The nature and origin of bed-form hierarchies: Sedimentology, v. 8, p. 161–182.

——, 1968b, Current ripples; their relation to patterns of water and sediment motion: North-Holland Publishing Company, Amsterdam, 433 p.

——, 1968c, The diffusion of grains in the lee of ripples, dunes and sand deltas: Jour. Sed. Petrology, v. 38, p. 621–632.

——, 1969a, Erosional current markings of weakly cohesive mud beds: *ibid.*, v. 39, p. 607–623.

——, 1969b, On the geometry of current ripples in relation to stability of fluid flow: Geografiska Annaler, v. 57A, p. 61–96.

——, 1970a, Physical processes of sedimentation: George Allen and Unwin, London 248 p.

——, 1970b, The avalanching of granular solids on dune and similar slopes: Jour. Geology, v. 78, p. 326–351.

——, 1971, Bed forms due to mass transfer in turbulent flows: a kaleidoscope of phenomena: Jour. Fluid Mechanics, v. 49, p. 49–63.

AMERICAN SOCIETY OF CIVIL ENGINEERS, COMMITTEE ON HYDRO-MECHANICS, 1963, Friction factors in open channels: Am. Soc. Civil Engineers Proc., v. 89, no. HY2, p. 97–143.

——, COMMITTEE ON SEDIMENTATION, TASK FORCE ON BED FORMS IN ALLUVIAL CHANNELS, 1966, Nomenclature for bed forms in alluvial channels: Am. Soc. Civil Engineers Proc., v. 92, no. HY3, p. 51–64.

——, TASK COMMITTEE ON EROSION OF COHESIVE MATERIALS, 1968, Erosion of cohesive sediments: Am. Soc. Civil Engineers Proc., v. 94, no. HY4, p. 1017–1049.

——, TASK COMMITTEE ON PREPARATION OF SEDIMENTATION MANUAL, 1963, Suspension of sediment: Am. Soc. Civil Engineers Proc., v. 89, no. HY5, p. 45–76.

——, 1966, Sediment transportation mechanics: initiation of motion: Am. Soc. Civil Engineers Proc., v. 92, no. HY2, p. 291–314.

——, 1970, Sediment transportation mechanics: J. Transportation of sediments in pipes: Am. Soc. Civil Engineers Proc., v. 96, no. HY7, p. 1503–1538.

——, 1971a, Sediment transportation mechanics: H. Sediment discharge formulas: Am. Soc. Civil Engineers Proc., v. 97, no. HY4, p. 523–567; with discussion, v. 97, no. HY9, p.1573–1576; v. 98 (1972), no. HY1, p. 284–290; no. HY2, p. 388–397; no. HY10, p. 1869–1872.

——, 1971b, Sediment transportation mechanics: Fundamentals of sediment transportation: Am. Soc. Civil Engineers Proc., v. 97, no. HY12, p. 1979–2022.

——, 1971c, Sediment transportation mechanics: F. Hydraulic relations for alluvial streams: Am. Soc. Civil Engineers Proc., v. 97, no. HY1, p. 101–141.

ANONYMOUS, 1967, Hydrology of the Lewis Glacier, North-Central Baffin Island, N.W.T., and discussion of reliability of the measurements: Geog. Bull., v. 9, p. 232–261.

ANTEVS, ERNST, 1925, Retreat of the last ice sheet in Eastern Canada: Geological Survey Canada Memoir 146, 142 p.

——, 1928, The last glaciation: Amer. Geog. Soc. Research Ser. 17, 292 p.

——, 1946, Review of Sigurd Hansen: Varvity in Danish and Scanian late-glacial deposits: Jour. Geology, v. 54, p. 336–340.

——, 1951, Glacial clays in Steep Rock Lake, Ontario, Canada: Geol. Soc. America Bull., v. 62, p. 1223–1262.

——, 1957, Geological tests of the varve and radiocarbon chronologies: Jour. Geology, v. 65, p. 129–148.

ARNBORG, LENNART, 1955, Hydrology of the glacial River Austurfljót: The Hoffellssandur—A glacial outwash plain: Geografiska Annaler, v. 37, p. 185–201.

——, WALKER, H.J., AND PEIPPO, J., 1967, Suspended load in the Colville River, Alaska, 1962: *ibid.*, v. 49A, 131–144.

AXELSSON, V., 1967, The Laitaure Delta: *ibid.*, v. 49A, p. 1–127.

BAGNOLD, R.A., 1960, Some aspects of river meanders: U.S. Geol. Survey Prof. Paper 282-E, p. 135–144.

——, 1966, An approach to the sediment transport problem from general physics: *ibid.*, Prof. Paper 422-I, 37 p.

——, 1968, Deposition in the process of hydraulic transport: Sedimentology, v. 10, p. 45–56.

BAIN, A.G., AND BONNINGTON, S.T., 1970, The hydraulic transport of solids by pipeline: Pergamon, Oxford, 257 p.

BANERJEE, INDRANIL, 1973, Sedimentology of Pleistocene glacial varves in Ontario, Canada: Geol. Survey Canada Bull. 226, pt. A, p. 1–44.

BARR, D.I.H., AND HERBERTSON, J.G., 1968, Similitude theory applied to correlation of flume sediment transport data: Water Resources Research, v. 4, p. 307–315.

BATA, G.L., AND BOGICH, K., 1953, Some observations on density currents in the laboratory and in the field: Internat. Assoc. Hydraulic Research, 5th Cong. Proc., Minneapolis, p. 387–400.

BATES, C.C., 1953, Rational theory of delta formation: Am. Assoc. Petroleum Geologists Bull., v. 37, p. 2119–2162.

BELL, H.S., 1947, The effect of entrance mixing on the size of density currents in Shaver Lake: Am. Geophys. Union Trans., v. 28, p. 780–791.

BENEDICT, B.A., AND CHRISTENSEN, B.C., 1972, Hydrodynamic lift on a streambed: Chap. 5 *in* Shen, H.W., (*ed.*), Sedimentation, Water Resources Pubs., Fort Collins, Colorado.

BIRKELAND, P.W., 1968, Mean velocities and boulder transport during Tahoe-age Floods of the Truckee River, California-Nevada: Geol. Soc. America Bull., v. 79, p. 137–142.

BLENCH, T., 1963, *in* discussion of Henderson, F.M., Stability of alluvial channels: *Loc. Cit.*

——, 1969, Mobile-bed fluviology: Univ. Alberta Press., Edmonton 221 p.

BLISSENBACH, E., 1954, Geology of alluvial fans in semi-arid regions: Geol. Soc. America Bull., v. 65, p. 175–189.

BLUCK, B.J., 1964, Sedimentation of an alluvial fan in Southern Nevada: Jour. Sed. Petrology, v. 34, p. 395–400.

BOGARDI, J.L., 1972, Fluvial sediment transport: p. 183–259 *in* Chow, V.T. (*ed.*), Advances in hydroscience, v. 8, Academic Press, New York.

BONHAM-CARTER, C.G., AND SUTHERLAND, A.J., 1967, Diffusion and settling of sediments at river mouths—a computer simulation model: Symposium on the Geological History of the Gulf of Mexico, Antillean Caribbean Region, Gulf Coast Assoc. Geol. Soc. Trans., v. 17, p. 326–338.

——, 1968, Mathematical model and FORTRAN IV program for computer simulation of deltaic sedimentation: Kansas State Computer Contr. 25, 56 p.

BOOTHROYD, J.C., 1970, Recent braided-stream sedimentation, South-Central Alaska (Abst.): Am, Assoc. Petroleum Geologists Bull., v. 54, p. 836.

BORLAND, W.M., 1961, Sediment transport of glacier-fed streams in Alaska: Jour. Geophys. Research, v. 66, p. 3347–3350.

BRADLEY, W.C., FAHNESTOCK, R.K., AND ROWEKAMP, E.T., 1972, Coarse sediment transport by flood flows on Knik River, Alaska: Geol. Soc. America Bull., v. 83, p. 1261–1284.

BRADLEY, W.H., 1965, Vertical density currents: Science, v. 150, p. 1423–1428.

BRICE, J.C., 1964, Channel patterns and terraces of the Loup Rivers in Nebraska: U.S. Geol. Survey Prof. Paper 422-D, 41 p.

BRIGGS, L.I., AND MIDDLETON, G.V., 1965, Hydromechanical principles of sediment structure formation: p. 5–16, in Middleton, G.V. (ed.), Primary sedimentary structures and their hydrodynamic interpretation, Soc. Econ. Paleontologists and Mineralogists Spec. Pub. 12, 265 p.

BROECKER, W.S., TUREKIAN, K.K., AND HEEZEN, B.C., 1958, The relation of deep sea sedimentation rates to variations of climate: Am. Jour. Sci., v. 256, p. 503–517.

BROWN, C.B., 1950, Sediment transportation: p. 769–857 in Rouse, H. (ed.), Engineering Hydraulics, Wiley, New York.

BRUSH, L.M., JR., 1965a, Experimental work on primary sedimentary structures: p. 17–24 in Middleton, G.V., (ed.), Primary sedimentary structures and their hydrodynamic interpretation, Soc. Econ. Paleontologists and Mineralogists Spec. Pub. 12, 265 p.

——, 1965b, Sediment sorting in alluvial channels: p. 25–33 in Middleton, G.V. (ed.), Primary sedimentary structures and their hydrodynamic interpretation, Soc. Econ. Paleontalogists and Mineralogists Spec. Pub. 12, 265 p.

CALLANDER, R.A., 1969, Instability and river channels: Jour. Fluid Mechanics, v. 36, p. 465–480.

CAMPBELL, F.B., AND BAUDER, H.A., 1940, A rating curve method for determining silt discharge of streams: Am. Geophys. Union Trans., v. 21, p. 603–607.

CAREY, K.L., 1966, Observed configuration and computed roughness of the under-side of river ice, St. Croix River, Wisconsin: U.S. Geol. Survey Prof. Paper 550-B, p. 192–198.

——, 1967, Analytical approaches to computation of discharge of an ice-covered stream: ibid., Prof. Paper 575-C, p. 200–207.

CHAWLA, M.D., 1964, A study of the thickness of density currents in reservoirs: M.Sc. thesis, Univ. of Alberta, 220 p.

CHENG, E.D.H., AND CLYDE, C.G., 1972, Instantaneous hydrodynamic lift and drag forces on large roughness elements in turbulent open channel flow: Chap. 3 in Shen, H.W., (ed.), Sedimentation, Water Resources Pubs., Fort Collins, Colorado.

CHIEN, N., 1954, Meyer-Peter Formula for bed-load transport and Einstein bed-load function: U.S. Army Corps of Engineers, Missouri River Div., Sediment Ser. no. 7, 23 p.

CHIEN, N., ZHOU, W., AND HONG, R., 1961, The characteristics and genesis analysis of the braided stream of the Lower Yellow River: Acta Geographica Sinica, v. 27, p. 1–27 (in Chinese, English summary, p. 26–27).

CHOW, V.T., 1959, Open-channel hydraulics: McGraw-Hill, New York, 680 p.

CHOW, V.T., (ed.), 1964, Handbook of applied hydrology: McGraw-Hill, New York.

CHURCH, M., 1972, Baffin Island sandurs: a study of arctic fluvial processes: Geol. Surv. Canada Bull. 216, 208 p.

CHURCH, M., AND RYDER, J.M., 1972, Paraglacial sedimentation: a consideration of fluvial processes conditioned by glaciation: Geol. Soc. America Bull., v. 83, p. 3059–3072.

COLBY, B.R., 1957, Relationship of unmeasured sediment discharge to mean velocity: Am. Geophy. Union Trans., v. 38, p. 707–717.

——, 1963, Fluvial sediments—a summary of source, transportation, deposition and measurement of sediment discharge: U.S. Geol. Survey Bull. 1181-A, 47 p.

——, 1964a, Practical computation of bed-material discharge: Am. Soc. Civil Engineers Proc., v. 90, no. HY2, p. 217–246.

——, 1964b, Discharge of sands and mean velocity relationships in sand-bed streams: U.S. Geol. Survey Prof. Paper 462-A, 47 p.

COLBY, B.R., AND HUBBELL, D.W., 1961, Simplified methods for computing total sediment discharge with the modified Einstein procedure: U.S. Geol. Survey Water Supply Paper 1593, 17 p.

COLBY, B.R., AND SCOTT, C.H., 1965, Effects of water temperature on the discharge of bed material: U.S. Geol. Survey Prof. Paper 462-G, 25 p.

COLLINSON, J.D., 1970, Bedforms of the Tana River, Norway: Geografiska Annaler, v. 52A, p. 31–56.

COSTELLO, W.R., AND WALKER, R.G., 1972, Pleistocene sedimentology, Credit River, Southern Ontario: A new component of the braided river model: Jour. Sed. Petrology, v. 42, p. 389–400.

CRAVEN, J.P., 1952, The transportation of sand in pipes: 1. full-pipe flow: Univ. Iowa 5th Hydraulics Conf., Eng. Bull. 34, p. 67–76.

CRICKMAY, C.H., AND BATES, C.C., 1955: in discussion of delta formation: Am. Assoc. Petroleum Geologists Bull., v. 39, p. 107–114.

CRICKMORE, M.J., 1967, Measurement of sand transport in rivers with special reference to tracer methods: Sedimentology, v. 8, p. 175–228.

CULLING, W.E.H., 1960, Analytic theory of erosion: Jour. Geology, vol. 68, p. 316–344.

DAVIES, D.K., 1966, Sedimentary structures and subfacies of the Mississippi River point bar: Jour. Geology, v. 74, p. 234–239.

DE GEER, GERARD, 1912, A geochronology of the last 12,000 years: Internat. Geol. Cong. 1910, Compte Rendu 11, p. 241–253.

——, 1926, On the solar curve as dating the ice age, the New York moraine, and Niagara Falls through the Swedish timescale: Geografiska Annaler, v. 8, p. 253–283.

DEVDARIANI, A.S., 1967, The profile of equilibrium and a regular regime: Soviet Geography, Rev. and Trans., v. 8, p. 168–183.

DE WIEST, R.J.M., 1965, Geohydrology: Wiley, New York, 366 p.

DICK, T.M., AND MARSALEK, J., 1973, Interfacial shear stress in density wedges: 1st Canadian Hydraulics Conf., Edmonton, Proc., p. 176–191.

DOEGLAS, D.J., 1951, Meanderende en verwilderde rivieren: Geol. en Mijnbouw, v. 13, p. 297–299.

——, 1962, The structure of sedimentary deposits of braided streams: Sedimentology, v. 1, p. 167–190.

DUNN, I.S., 1959, Tractive resistance of cohesive channels: Am. Soc. Civil Engineers Proc., v. 85, no. SM3, p. 1–24.

DURAND, R., 1953, Basic relationships of the transportation of solids in pipes—experimental research: Internat. Assoc. Hydraulic Research, 5th Cong. Proc., Minneapolis, p. 89–103.

EGIAZAROV, I.V., 1964, Effects of nongraded sediment and of accumulation of large bed material in the channel on the movement and discharge of sediment: Soviet Hydrology, 1964, no. 4, p. 379–409.

EHRLICH, R., AND DAVIES, D.K., 1968, Sedimentological indices of transport direction, distance, and process intensity in glacio-fluvial sediments: Jour. Sed. Petrology, v. 38, p. 1166–1170.

EINSTEIN, H.A., 1942, Formulas for the transportation of bedload: Am. Soc. Civil Engineers Trans., v. 107, p. 561–573.

——, 1950, The bedload function for sediment transportation in open channel flows: U.S. Dept. Agriculture Tech. Bull. no. 1026, 70 p.

——, AND EL-SAMNI, E.A., 1949, Hydrodynamic forces on a rough wall: Reviews of Modern Physics, v. 21, p. 520–524.

——, AND BARBAROSSA, N.L., 1952, River channel roughness: Am. Soc. Civil Engineers Trans., v. 117, p. 1121–1146 (with discussion).

ELSON, J.A., 1961, The geology of tills: 14th Canadian Soil Mechanics Conf. Proc., Natl. Research Council Canada, Assoc. Comm. Soil and Snow Mechanics Tech. Mem. 69, p. 5–17.

ELWIN, E.H., AND SLOTTA, L.S., 1969, Stratified reservoir currents Part I: Entering streamflow effects on currents of a density stratified model reservoir: Oregon State Univ. Eng. Experiment Station Bull. 44, Corvallis, Oregon, 97 p.

EMBLETON, C., AND KING, C.A.M., 1968, Glacial and periglacial geomorphology: Arnold, London, 608 p.

EMMETT, W.W., AND LEOPOLD, L.B., 1965, Downstream pattern of riverbed scour and fill: p. 399–409 in Federal Interagency Sedimentation Conf. Symposium, Pt. 2, Sediment in Streams, U.S. Dept. Agriculture Misc. Pub. 970.

ENGELUND, F., AND HANSEN, E., 1967, A monograph on sediment transport in alluvial streams: Teknisk Verlag, Copenhagen, 62 p.

ENGELUND, F., AND SKOVGAARD, O., 1973, On the origin of meandering and braiding in alluvial channels: Jour. Fluid Mechanics, v. 57, p. 289–302.

EXNER, F.M., 1920, Zur Physik der Dünen: Vienna Acad. Sci. Proc., v. 129, p. 929.

FAHNESTOCK, R.K., 1963, Morphology and hydrology of a glacial stream—White River, Mount Rainier, Washington: U.S. Geol. Survey Prof. Paper 422-A 70 p.

——, 1969, Morphology of the Slims River: p. 161–172 in Bushnell, V.C., and Ragle, R.H. (eds.), Icefield Ranges Research Project, Scientific Results, v. 1, Am. Geog. Soc. and Arctic Inst. North America.

——, AND HAUSHILD, W.L., 1962, Flume studies of the transport of pebbles and cobbles on a sand bed: Geol. Soc. America Bull., v. 73, p. 1431–1436.

FARMER, D.G., 1971, A computer simulation model of sedimentation in a salt wedge estuary: Marine Geology, v. 10, p. 133–143.

FISK, H.N., 1944, Geological investigations of the alluvial valley of the Lower Mississippi River: Mississippi River Comm., Vicksburg, Miss., 78 p.

FOREL, F.A., 1885, Les ravins sous-lacustres des fleuves glaciaires: Acad. Sci. Comptes Rendus, Paris, v. 101, p. 725–728.

FULTON, R.J., AND PULLEN, M.J.L.T., 1969, Sedimentation in Upper Arrow Lake, British Columbia: Canadian Jour. Earth Sci., v. 6, p. 785–790.

FRYE, J.C., 1961, Fluvial deposition of the glacial cycle: Jour. Geology, v. 69, p. 600–603.

GALAY, V.J., 1967, Observations of bed-form roughness in an unstable gravel river: Internat. Assoc. Hydraulic Research, 12th Cong. Proc., v. 1, No. A11, p. 85–94.

GESSLER, J., 1971, Beginning and ceasing of sediment motion: Chap. 7 in Shen, H.W. (ed.), River mechanics, Water Resources Pubs., Fort Collins, Colorado.

GILBERT, G.K., 1885, The topographic features of lake shores. U.S. Geol. Survey Annual Report, v. 5, p. 104–108.

GILBERT, R., 1971, Observations on ice-dammed Summit Lake, British Columbia, Canada: Jour. Glaciology, v. 10, p. 351–356.

——, 1972, Observations on sedimentation at Lillooet Delta, British Columbia: p. 187–194 in Slaymaker, H.O., and McPherson. H.J. (eds.), Mountain geomorphology, Tantalus Press, Vancouver.

——, 1973, Processes of underflow and sediment transport in a British Columbia mountain lake: Natl. Research Council of Canada, Assoc. Comm. Geodesy and Geophysics, Subcomm. Hydrology, Proc. Hydrology Symposium, no. 9, p. 493–507.

GILL, M.A., 1972, in discussion of ASCE Task Committee, 1971b, Sediment Transportation Mechanics: Fundamentals of sediment transportation: Loc. Cit.

GRAF, W.H., 1971, Hydraulics of sediment transport: McGraw-Hill, New York, 513 p.

——, AND ACAROGLU, E.R., 1968, Sediment transport in conveyance systems (Part I): Internat. Assoc. Sci. Hydrology Bull., v. 13, p. 20–39.

GRANAR, L., 1956, Dating of recent fluvial sediments from the estuary of the Angerman River (the period 1859–1950): Geol. Fören. Forh., Stockholm, v. 78, p. 654–658.

GRASS, A.J., 1970, Initial instability of fine bed sand: Am. Soc. Civil Engineers Proc., v. 96, no. HY3, p. 619–632.

GRIGG, N.S., 1970, Motion of single particles in alluvial channels: Am. Soc. Civil Engineers Proc., v. 96, no. HY12, p. 2501–2518.

GROVER, N.C., AND HOWARD, C.S., 1938, The passage of turbid water through Lake Mead: Amer. Soc. Civil Engineers Trans., v. 103, p. 720–790.

GUDMUNDSSON, G., 1970, Short-term variations of a glacier-fed river: Tellus, v. 22, p. 341–353.

——, AND SIGBJARNARSON, G., 1972, Analysis of glacier run-off and meteorological observations, Jour. Glaciology, v. 11, p. 303–318.

HARLEMAN, D.R.F., 1961, Stratified Flow: p. 26-1–26-21 in Streeter, V.L. (ed.), Handbook of fluid dynamics, McGraw-Hill Book Co., New York.

——, 1963, Sediment transportation mechanics: Density currents: Am. Soc. Civil Engineers Proc., v. 89, no. HY5, p. 77–81.

HARMS, J.C., AND FAHNESTOCK, R.K, 1965, Stratification, bed forms and flow phenomena (with an example from the Rio Grande): p. 84–115 in Middleton, G.V. (ed.), Primary sedimentary structures and their hydrodynamic interpretation, Soc. Econ. Paleontologists and Mineralogists Spec. Pub. 12, 265 p.

HARRISON, A.S., 1952, Deposition at the head of reservoirs: 5th Hydraulic Conf. Proc., State Univ. of Iowa Studies in Engineering, Bull. 34, p. 199–225.

HELLEY, E.J., 1969, Field measurement of the initiation of large bed particle motion in Blue Creek near Klamath, California: U.S. Geol. Survey Prof. Paper 562-G, 19 p.

HENDERSON, F.M., 1963, Stability of alluvial channels: Am. Soc. Civil Engineers Trans., v. 128, p. 657–720 (with discussion).

——, 1966, Open Channel Flow: Macmillan, New York, 522 p.

HEWITT, K., 1968, The freeze-thaw environment of the "Karakoram Himalaya": Canadian Geographer, v. 12, p. 85–98.

HINO, M., 1968, Equilibrium-range spectra of sand waves formed by flowing water: Jour. Fluid Mechanics, v. 34, p. 565–573.

HINZE, J.O., 1960, On the hydrodynamics of turbidity currents: Geol. en Mijnbouw, v. 39, p. 18–25.

HJULSTRÖM, F., 1939, Transportation of detritus by moving water: p. 5–31 in Trask, P.D. (ed.), Recent marine sediments: Am. Assoc. Petroleum Geologists, Tulsa.

——, 1952, The geomorphology of the alluvial outwash plains (sandurs) of Iceland, and the mechanics of braided rivers: Internat. Geog. Union, 17th Cong., Proc., Washington, p. 337–342.

——, 1955, The groundwater: The Hoffellssandur—a glacial outwash plain: Geografiska Annaler, v. 37, p. 234–245.

HOLLINGSHEAD, A.B., 1971, Sediment transport measurements in gravel river: Am. Soc. Civil Engineers Proc., v. 97, no. HY11, p. 1817–1834.

HOUBOLT, J.J.H.C., AND JONKER, J.B.M., 1968, Recent sediments in the eastern part of the Lake of Geneva (Lac Léman): Geol. en Mijnbouw, v. 47, p. 131–148.

HOWARD, C.S., 1953, Density currents in Lake Mead. Internat. Assoc. Hydraulic Research, 5th Cong. Proc., Minneapolis, p. 355–368.

HUBBELL, D.W., AND MATEJKA, D.Q., 1959 Investigations of sediment transportation, Middle Loup River at Dunning, Nebraska: U.S. Geol. Survey Water-Supply Paper 1476, 123 p.

HUGHES, O.L., 1965, Surficial geology of part of the Cochrane District, Ontario, Canada: p. 535–565 in Wright, H.E., and Frey, D.G. (eds.), International Studies of the Quaternary, Geol. Soc. America Spec. Paper 84.

INGLIS, C.C., 1952, in discussion of Einstein, H.A., and Barbarossa, N.L., River channel roughness: Loc. Cit.

IPPEN, A.T., 1971, A new look at sedimentation in turbulent streams: Jour. Boston Soc. Civil Engineers, v. 58, p. 131–163.

——, AND HARLEMAN, D.R.F., 1952, Steady-state characteristics of subsurface flow: U.S. Bur. Standards Circular 521, 79 p.

JEFFREYS, H., 1929, On the transportation of sediment by streams: Cambridge Phil. Soc. Proc., v. 25, p. 272–276.

JENSEN, J.A.D., 1881, Beretning om reisen og de geografiske forhold: in Part V of Expedition Til Holsteinsborgs og Egesmundes Distrikter, 1879, Meddelelser om Grønland, no. 1, p. 113–139.

JEWTUCHOWICZ, S., 1953, La structure du sandre: Soc. Lett. Sci. Lodz Bull., Classe 3, Sci. Math, et Nat., v. 14, no. 4, 29 p.

JOHNSTON, W.A., 1922, Sedimentation in Lake Louise, Alberta, Canada: Am. Jour. Sci., 5th Ser., v. 4, p. 376–386.

JOPLING, A.V., 1960, An experimental study on the mechanics of bedding: Unpubl. Ph.D. Thesis, Harvard University, 358 p.

——, 1962, Mechanics of small scale delta formation, A laboratory study: Natl. Shallow Water Research Conf. Proc., Natl. Sci. Foundation, Washington, D.C., p. 291–295.

——, 1963, Hydraulic studies on the origin of bedding: Sedimentology, v. 2, p. 115–121.

——, 1964, Laboratory study of sorting processes related to flow separation: Jour. Geophys. Research, v. 69, p. 3404–3418.

——, 1965a, Laboratory study of the distribution of grain sizes in cross-bedded deposits: p. 53–65 *in* Middleton, G.V. (*ed.*), Primary sedimentary structures and their hydrodynamic interpretation, Soc. Econ. Paleontologists and Mineralogists Spec. Pub. 12.

——, 1965b, Hydraulic factors controlling the shape of laminae in laboratory deltas: Jour. Sed. Petrology, v. 35, p. 777–791.

——, 1966, Some applications of theory and experiment to the study of bedding genesis: Sedimentology, v. 7, p. 71–102.

KALINSKE, A.A., 1947, Movement of sediment as bedload in rivers: Am. Geophys. Union Trans., v. 28, p. 615–620.

KELLER, E.A., 1972, Development of alluvial stream channels: A five-stage model: Geol. Soc. America Bull., v. 83, p. 1531–1536.

KELLERHALS, R., 1967, Stable channels with gravel-paved beds: Am. Soc. Civil Engineers Proc., v. 93, no. WW 1, p. 63–84.

KENNEDY, J.F., 1961, Stationary waves and antidunes in alluvial channels: California Inst. Technology, W.M. Keck Lab. of Hydraulics and Water Resources, Rept. KH-R-2, 146 p.

——, 1963, The mechanics of dunes and antidunes in erodible-bed channels: Jour. Fluid Mechanics, v. 16, p. 521–544.

KERR, F.A., 1934, The ice dam and floods of the Talsekwe, British Columbia: Geog. Rev., v. 24, p. 643–645.

KEULEGAN, G.H., 1938, Laws of turbulent flow in open channels: Jour. Research Natl. Bureau Standards [U.S.], v. 21, p. 707–741.

——, 1949, Interfacial instability and mixing in stratified flows: *ibid.*, v. 43, p. 487–500.

KINDLE, E.M., 1930, Sedimentation in a glacial lake: Jour. Geology, v. 38, p. 81–87.

KIRKBY, M.J., 1972, Alluvial and non-alluvial meanders: Area, v. 4, p. 284–288.

KLEBELSBERG, R.v., 1948, Handbuch der Gletscherkunde und Glazialgeologie: Vienna, 1028 p.

KLIMEK, K., 1971, Wspolczesnk procesy fluwialne I Rzezba Rowniny Skeidararsandur (Islandia): Prace Geograficzne, no. 94, 139 p. (Present-day fluvial processes and relief of the Skeidararsandur Plain (Iceland); English summary, p. 129–136).

KLINGEMAN, P.C., AND MILHOUS, R.T., 1970, Oak Creek vortex bed-load sampler: Am. Geophys. Union, 17th Pacific Northwest Regional Mtg., Univ. Puget Sound, Tacoma, 11 p.

KNIGHTON, A.D., 1972, Changes in a braided reach: Geol. Soc. America Bull., v. 83, p. 3813–3822.

KRIGSTRÖM, A., 1962, Geomorphological studies of sandur plains and their braided rivers in Iceland: Geografiska Annaler, v. 44, p. 328–346.

KRUMBEIN, W.C., 1937, Sediments and exponential curves: Jour. Geology, v. 45, p. 577–601.

——, 1940, Flood gravel of San Gabriel Canyon, California: Geol. Soc. America Bull., v. 51, p. 639–676.

——, 1941a, The effect of abrasion on the size, shape and roundess of rock fragments: Jour. Geology, v. 49, p. 482–520.

——, 1941b, Measurement and geological significance of shape and roundness of sedimentary particles: Jour. Sed. Petrology, v. 11, p. 64–72.

——, 1942a, Flood deposits of Arroyo Seco, Los Angeles County, California: Geol. Soc. America Bull., v. 53, p. 1355–1402.

——, 1942b, Settling velocity and flume behavior of non-spherical particles: Am. Geophys. Union Trans., v. 23, p. 621–632.

——, AND SLOSS, L.L., 1963, Stratigraphy and sedimentation: Freeman, 2nd Ed., San Francisco, 660 p.

KUENEN, PH.H., 1951, Mechanics of varve formation and the action of turbidity currents: Geol. Fören. Förh., Stockholm, v. 73, p. 69–84.

LAJTAI, E.Z., 1967, The origin of some varves in Toronto, Canada: Canadian Jour. Earth Sci., v. 4, p. 633–639.

LANE, E.W., 1955, Design of stable channels: Am. Soc. Civil Engineers Trans., v. 120, p. 1234–1260.

——, 1957, A study of the shape of channels formed by natural streams flowing in erodible material: U.S. Army Corps of Engineers, Missouri River Div., Sediment Ser. no. 9, 106 p.

——, AND CARLSON, E.J., 1954, Some observations on the effect of particle shape on the movement of coarse sediments: Am. Geophys. Union Trans., v. 35, p. 458–462.

LANG, H., 1968, About relations between glacier runoff and meteorological factors observed on the glacier and outside: Internat. Union Geodesy and Geophysics 14th Gen. Assembly, Bern, 1967, Internat. Assoc. Sci. Hydrology Pub. no. 79, p. 429–439.

LANGBEIN, W.B., 1964, Geometry of river channels: Am. Soc. Civil Engineers Proc., v. 90, no. HY2, p. 301–312; discussion, v. 90, no. HY5, p. 277–286, no. HY6, p. 332–342.

——, AND LEOPOLD, L.B., 1968, River channel bars and dunes—theory of kinematic waves: U.S. Geol. Survey Prof. Paper 422-L, 20 p.

LANSER, O., 1958, Réflexions sur les débits solides en suspension des cours d'eau glaciaires: Internat. Assoc. Sci. Hydrology Bull., v. 4, p. 37–43.

LARA, J.M., AND SANDERS, J.I., 1970, The 1963–64 Lake Mead Survey: U.S. Dept. Interior, Bureau of Reclamation Rept. REC-OCE-70-21, 172 p.

LARSEN, P., 1969, Head losses caused by an ice cover on open channels: Jour. Boston Soc. Civil Engineers, v. 56, p. 45–67.

——, 1973, Hydraulic roughness of ice covers: Am. Soc. Civil Engineers Proc., v. 99, no. HY1, p. 111–119.

LELIAVSKY, S., 1955, An introduction to fluvial hydraulics: Constable, London, 257 p.; also Dover Publ., 1966.

LEOPOLD, L.B., AND MADDOCK, T., JR., 1953, The hydraulic geometry of stream channels and some physiographic implications: U.S. Geol. Survey, Prof. Paper 252, 57 p.

LEOPOLD, L.B., AND MILLER, J.P., 1956, Ephemeral streams—hydraulic factors and their relation to the drainage net: *ibid.*, Prof. Paper 282-A, 37 p.

LEOPOLD, L.B., AND WOLMAN, M.G., 1957, River channel patterns, braided, meandering, and straight: *ibid.*, Prof. Paper 262-B, 85 p.

LEOPOLD, L.B., WOLMAN, M.G., AND MILLER, J.P., 1964, Fluvial processes in geomorphology: W.H. Freeman, San Francisco, 522 p.

LEOPOLD, L.B., EMMETT, W.W., AND MYRICK, R.M., 1966, Channel and hillslope processes in a semiarid area, New Mexico: U.S. Geol. Survey Prof. Paper 352-G, 61 p.

LEOPOLD, L.B., BAGNOLD, R.A., WOLMAN, M.G., AND BRUSH, L.M., JR., 1960, Flow resistance in sinuous or irregular channels: *ibid.*, Prof. Paper 282-D, p. 111–134.

LEVI, I.I., 1959, Theory of underflow in storage reservoirs; Internat. Assoc. Hydraulic Research, 8th Cong. Montreal, v. 2, p. 8-C-1 to 8-C-26.

——, 1965, New problems in the theory of bottom current in reservoirs: Izvestiia vsesoiuznogo nauchno issledovatel' skogo instituta gidrotekhniki, no. 78, p. 71–82 (translated 1969 from Russian by D.L. King, Hydraulics Branch, Office of Engineering Reference, Denver, Colorado, 18 p.)

——, AND KYLYESH, H.P., 1960, The laws governing the motion of highly turbid currents in reservoirs: Trans. M.E. Kaliniena Polytech. Institute, Leningrad, No. 208 (translated 1965 from Russian by J.B. Nuttall and L.E. Gads, Univ. Alberta, 21 p.)

LIESTØL, O., 1956, Glacier dammed lakes in Norway: Norsk Geografisk Tidjskrift, v. 15 (1955), p. 122–149.

LJUNGGREN, P., AND SUNDBORG, Å., 1968, Some aspects on fluvial sediments and fluvial morphology. II. A study of some heavy mineral deposits in the valley of the river Lule Älv: Geografiska Annaler, v. 50A, p. 121–135.

LUSTIG, L.K., 1965, Clastic sedimentation in Deep Springs Valley, California: U.S. Geol. Survey Prof. Paper 352F, p. 131–192.

LYLES, L., AND WOODRUFF, N.P., 1972, Boundary-layer flow structure: Effects on detachment of noncohesive particles: Chap. 2 *in* Shen, H.W., (*ed.*), Sedimentation, Water Resources Pubs., Fort Collins, Colorado.

MACKIN, J.H., 1948, Concept of the graded river: Geol. Soc. America Bull., v. 59, p. 463–512.

——, 1956, Cause of braiding by a graded river: *ibid.*, v. 67, p. 1717–1718 (abs.).

MADDOCK, T., JR., 1969, The behaviour of straight, open channels with movable beds: U.S. Geol. Survey Prof. Paper 622-A, 69 p.

MARCUS, M.G., 1960, Periodic drainage of glacier-dammed Tulsequah Lake, British Columbia: Geog. Rev., v. 50, p. 89–106.

MATHEWS, W.H., 1956, Physical limnology and sedimentation in a glacial lake: Geol. Soc. America Bull., v. 67, p. 537–552.

——, 1964a, Discharge of a glacial stream: Internat. Union Geodesy and Geophys. 13th Gen. Assembly, Berkeley, California, 1963, Internat. Assoc. Sci. Hydrology Pub. no. 63, p. 290–300.

——, 1964b, Sediment transport from Athabasca Glacier, Alberta: *ibid.*, Pub. no. 65, p. 155–165.

——, 1973, Record of two jokullhlaups: Union Géodésique et Géophysique Internationale, Assoc. Internationale d'Hydrologie Scientifique, Commission de Neiges et Glaces, Symposium on the hydrology of glaciers, Cambridge, 1969, A.I.S.H. Pub. 95, p. 99–110.

——, AND SHEPARD, F.P., 1962, Sedimentation of Fraser River Delta, British Columbia: Am. Assoc. Petroleum Geologists Bull., v. 46, p. 1416–1443.

MCDONALD, B.C., AND BANERJEE, INDRANIL, 1971, Sediments and bed forms on a braided outwash plain: Canadian Jour. Earth Sci., v. 8, p. 1282–1301.

——, AND VINCENT, J.S., 1972, Fluvial sedimentary structures formed experimentally in a pipe, and their implications for interpretation of subglacial sedimentary environments: Geol. Survey of Canada, Paper 72–27, 30 p.

MCKEE, E.D., 1957, Flume experiments on the production of stratification and cross-stratification: Jour. Sed. Petrology, v. 27, p. 129–134.

——, 1965, Experiments on ripple lamination: p. 66–83 in G. Middleton, (ed.), Primary sedimentary structures and their hydrodynamic interpretation, Soc. Econ. Paleontologists and Mineralogists Spec. Pub. 12.

MEIER, M.F., 1964, Ice and glaciers: Section 16 in Chow, V.T. (ed.), Handbook of Applied Hydrology, McGraw-Hill, New York.

——, AND TANGBORN, W.V., 1961, Distinctive characteristics of glacier runoff: U.S. Geol. Survey Prof. Paper 424-B, Art. 7, p. 14–16.

MELAND, N., AND NORRMAN, J.O., 1969, Transport velocities of individual size fractions in heterogeneous bed-load: Geografiska Annaler, v. 51A, p. 127–144.

MERCER, A.G., AND HAQUE, M.I., 1973, Ripple profiles modelled mathematically: Am. Soc. Civil Engineers Proc., v. 99, no. HY3, p. 441–459.

MEYER-PETER, E., AND MÜLLER, R., 1948, Formulas for bed-load transport: Internat. Assoc. Hydraulic Research, 2nd Mtg., Proc., App. 2, p. 1–26.

MIDDLETON, G.V., (ed.), 1965, Primary sedimentary structures and their hydrodynamic interpretation: Soc. Econ. Paleotologists and Mineralogists Spec. Pub. 12, 265 p.

——, 1966, Experiments on density and turbidity currents—II Uniform flow of density currents: Canadian Jour. Earth Sci., v. 3, p. 627–637.

MONISH, B.H., 1938, in discussion of Grover, N.C., and Howard, C.S., The passage of turbid water through Lake Mead: Am. Soc. Civil Engineers Trans., v. 103. p. 751–755.

MOORE, D.G., 1961, Submarine Slumps: Jour. Sed. Petrology, v. 31, p. 343–357.

MOORE, T.C., VAN ANDEL, T.H., BLOW, W.H., AND HEATH, G.R., 1970, Large submarine slide off northeastern continental margin of Brazil: Am. Assoc. Petroleum Geologists Bull., v. 54, p. 125–128.

MOORE, W.L., AND MASCH, F.D., 1962, Experiments on the scour resistance of cohesive sediments: Jour. Geophys. Research, v. 67, p. 1437–1449.

MOOSBRUGGER, H., 1958, Le charriage et le débit solide en suspension des cours d'eau de montagnes: Internat. Union Geodesy and Geophys., 11th Gen. Assembly, Toronto, 1957, Internat. Assoc. Sci. Hydrology Pub. no. 43, p. 203–231.

NEILL, C.R., 1968, Theoretical discussion of sediment transport: Alberta Research Council File Rept. Edmonton, 29 p.

——, 1969, Bed forms in the Lower Red Deer River, Alberta: Jour. Hydrology, v. 7, p. 58–85.

NEVIN, C.M., 1946, Competency of moving water to transport debris: Geol. Soc. America Bull., v. 57, p. 651–674.

NEWITT, D.M., RICHARDSON, J.F., ABBOTT, M., AND TURTLE, R.B., 1955, Hydraulic conveying of solids in horizontal pipes: Inst. Chemical Engineers Trans., v. 33, p. 93–110.

NEZHIKHOVSKIY, R.A., 1964, Coefficients of roughness of bottom surface of slush-ice cover: Soviet Hydrology, Selected Papers no. 1964/2, p. 127–149.

NIKURADSE, J., 1933, Strömungsgesetze in rauhen Rohren: Versuchsanstalt Deutsche Ingenieurwirtschaft, Forschungsheft 361 [*Translation*: Laws of flow in rough pipes, NACA. Tech. Mem. 1292, 1950, 61 p.]

NIR, D., 1964, Les processus érosifs dans le Nahal Zine pendant les saisons pluvieuses: Annales de Geographie, v. 73, p. 8–20.

NOORANY, I., AND GIZIENSKI, S.F., 1970. Engineering properties of submarine soils: state of the art review: Am. Soc. Civil Engineers Proc., Soil Mechanics and Foundations Division, v. 96, p. 1735–1762.

NORDIN, C.F., AND ALGERT, J.H., 1966, Spectral analysis of sand waves: Am. Soc. Civil Engineers Proc., v. 92, no. HY5, p. 95–114.

NORDIN, C.F., AND BEVERAGE, J.P., 1964, Temporary storage of fine sediment in islands and point bars of alluvial channels of the Rio Grande, New Mexico: U.S. Geol. Survey Prof. Paper 475-D, p. 138–140.

ORE, H.T., 1964, Some criteria for recognition of braided stream deposits: Univ. Wyoming Dept. Geology Contr., v. 3, p. 1–14.

ØSTREM, G., 1964, Glacio-hydrological investigations in Norway: Jour. Hydrology, v. 2, p. 101–115.

——, 1966, Mass balance studies on glaciers in western Canada, 1965: Geog. Bull. v. 8, p. 81–107 (with discussion by W.S.B. Paterson and reply, p. 383–389).

——, BRIDGE, C.W., AND RANNIE, W.F., 1967, Glacio-hydrology, discharge and sediment transport in the Decade Glacier area, Baffin Island, N.W.T.: Geografiska Annaler, Ser. A. v. 49, p. 268–282.

ØSTREM, G., ZIEGLER, T., AND EKMAN, S.R., 1970, Slamtransportundersökelser i Norske Breelver, 1969: Norges Vassdragsdirektoratet, Hydrologisk Aydeling, Oslo, Rapport nr. 6/70, 68 p.

PANTELOUPULOS, J., 1955, Note sur la granulométrie de charriage et la loi du débit solid par charriage de fond d'une mélange de matériaux: Internat. Assoc. Hydraulic Research, 6th Cong. Proc., Paper D-10, 11 p.

——, 1957, Etude expérimentale du mouvement par charriage de fond d'une mélange de matériaux: *ibid.*, 7th Gen. Mtg., Proc. Paper D-30, 24 p.

PARDEE, J.T., 1942, Unusual currents in Glacial Lake Missoula, Montana: Geol. Soc. America Bull., v. 53, p. 1569–1599.

PARTHENIADES, E., AND PAASWELL, R.E., 1970, Erodibility of channels with cohesive boundary: Am. Soc. Civil Engineers Proc., v. 96, no. HY3, p. 755–772.

PASHINSKIY, A.F., 1964, Experience of the study of alluvial deposits of the Psezuapse River: Soviet Hydrology, 1964, no. 3, p. 156–173.

PETTIJOHN, F.J., AND POTTER, P.E., 1964, Atlas and glossary of primary sedimentary structures: Springer-Verlag, Berlin, 370 p.

PICKARD, G.L., 1953, Oceanography of British Columbia mainland inlets: II Currents: Fisheries Research Board [Canada], Progress Report (Pacific), no. 97, p. 12–13.

——, 1961, Oceanographic Features of inlets in the British Columbia Mainland Coast: Jour. Fisheries Research Board [Canada], v. 28, p. 907–1016.

——, AND RODGERS, K., 1959, Current measurement in Knight Inlet, British Columbia: *ibid.*, v. 16, p. 635–678.

PITTMAN, E.D. AND OVENSHINE, A.T., 1968, Pebble morphology in the Merced River (California): Sedimentary Geology, v. 2, p. 125–140.

PLUMLEY, W.J., 1948, Black Hills terrace gravels: A study in sediment transport: Jour. Geology, v. 56, p. 526–577.

PRICE, R.J., 1969, Moraines, sandar, kames and eskers near Breidamerkurjökull, Iceland: Inst. British Geographers Trans., no. 46, p. 17–37.

PRICE, W.A., 1947, Geomorphology of depositional surfaces: Am. Assoc. Petroleum Geologists Bull., v. 31, p. 1784–1800.

PYTTE, R. (ed.), 1969, Glasiologiske undersökelser i Norge, 1968: Hydrologisk Avdeling, Norsk Hydrologisk—og Electrisitetsvesen, Oslo, Rapport no. 5/69, 149 p.

QURAISHY, M.S., 1944, The origin of curves in rivers: Current Sci., v. 13, p. 36–39.

RAINWATER, F.H., AND GUY, H.P., 1961, Some observations on the hydrochemistry and sedimentation of the Chamberlin Glacier Area, Alaska: U.S. Geol. Survey Prof. Paper 414-C, 14 p.

RAO, N.S.L., AND SRIDHARAN, K., 1966, Characteristics of M1 backwater curves: Am. Soc. Civil Engineers Proc., v. 92, no. HY6, p. 131–139.

RAUDKIVI, A.J., 1963, Study of sediment ripple formation: ibid., v. 89, no. HY6, p. 15–33.

——, 1967, Loose boundary hydraulics: Pergamon, Oxford, 331 p.

REYNOLDS, A.J., 1965, Waves on the erodible bed of an open channel: Jour. Fluid Mechanics, v. 22, p. 113–133.

ROGERS, J.J., KRUEGER, W.L., AND KROG, M., 1963, Size of naturally abraded materials: Jour. Sed. Petrology, v. 33, p. 628–632.

RÖTHLISBERGER, HANS, 1972, Water pressure in intra- and subglacial channels: Jour. Glaciology, v. 11, p. 177–203.

ROUSE, HUNTER, 1938, Fluid mechanics for hydraulic engineers: McGraw-Hill Engineering Societies Monograph, New York, 422 p.; (also Dover (1961).

——, 1950 (ed.), Engineering Hydraulics: Wiley, New York, 1033 p.

RUBEY, W.W., 1952, Geology and mineral resources of the Hardin and Brussels Quadrangles (in Illinois): U.S. Geol. Survey Prof. Paper 218, 179 p.

RUSSELL, R.D., 1939, Effects of transportation on sedimentary particles: p. 32–47 in Trask, P.D. (ed.), Recent marine sediments, Am. Assoc. Petroleum Geologists, Tulsa.

RUST, B.R., 1972, Pebble orientation in fluvial sediments: Jour. Sed. Petrology, v. 42, p. 384–388.

SALISBURY, R.D., 1892, Overwash plains and valley trains: Geol. Survey of New Jersey, Annual Rept. of The State Geologist, Section 7, p. 96–114.

SANDERS, J.E., 1965, Primary sedimentary structures formed by turbidity currents and related resedimentation mechanisms: p. 192–219 in Middleton, G.V. (ed.), Soc. Economic Paleontologists and Mineralogists Spec. Pub. 12, 265 p.

SCHEIDEGGER, A.E., 1965, On the dynamics of deposition: Internat. Assoc. Sci. Hydrology Bull., v. 10, p. 49–57.

——, 1970, Theoretical Geomorphology: 2nd Edit., Springer-Verlag, Berlin, 435 p.

SCHOKLITSCH, A., 1934, Der Geschiebetrieb und Geschiebefracht: Wasserkraft und Wasserwirtschaft, v. 39, p. 37.

SCHUMM, S.A., 1960a, The shape of alluvial channels in relation to sediment type: U.S. Geol. Survey Prof. Paper 352-B, p. 17–30.

——, 1960b, The effect of sediment type on the shape and stratification of some modern fluvial deposits: Am. Jour. Sci., v. 258, p. 177–184.

——, 1963, The disparity between present rates of denudation and orogeny: U.S. Geol. Survey, Prof. Paper 454-H, 13 p.

——, AND KHAN, H.R., 1972, Experimental study of channel patterns: Geol. Soc. America Bull., v. 83, p. 1755–1770.

SCHUMM, S.A., KHAN, H.R., WINKLEY, B.R., AND ROBBINS, L.G., 1971, Variability of river patterns: Nature (Physical Science), v. 237, p. 75–76.

SCOTT, K.M., AND GRAVLEE, G.C., JR., 1968, Flood surge on the Rubicon River, California—Hydrology, hydraulics and transport: U.S. Geol. Survey Prof. Paper 422-M, 40 p.

SELLIN, R.H.J., 1969, Flow in channels: Macmillan, London, 149 p.

SHEN, H.W., AND VEDULA, S., 1969, A basic cause of a braided channel: Internat. Assoc. for Hydraulic Research, 13th Cong. Proc., Kyoto, Japan, v. 5–1, p. 201–205.

SHEPARD, F.P., 1955, Delta-front valleys bordering the Mississippi Delta: Geol. Soc. America Bull., v. 66, p. 1489–1498.

SHIELDS, A., 1936, Anwendung der Ähnlichkeitsmechanik und der Turbulenzforschung auf die Geschiebebewegung: Preussische Versuchsanstalt für Wasserbau und Schiffbau [Berlin], Mitteil. no. 26, 26 p. (Translation by W.P. Ott and J.C. Van Uchelen, Application of similarity principles and turbulence research to bed-load movement: U.S. Dept. Agriculture, Soil Conservation Service, Cooperative Lab., California Inst. Technology, 70 p.)

SHOSTAKOVICH, V.B., 1931, Die Bedeutung der Untersuchung der Bodenablagerungen der Seen für einige Fragen der Geophysik: Internat. Ver. Theor. Angew., Limnol Verg. Stuttgart, v. 5, p. 307–311.

——, 1936, Geschichtete Bodenablagerungen der Seen als Klima-ann: Meteorologische Zeit., v. 53, p. 176–182.

SEREVE, R.L., 1972, Movement of water in glaciers: Jour. Glaciology, v. 11, p. 205–214.

SHULITS, S., 1935, The Schoklitsch bed-load formula: Engineering, v. 139, p. 644–646 and 687.

SIMONS, D.B., AND ALBFRTSON, M.L., 1963, Uniform water conveyance channels in alluvial material: Am. Soc. Civil Engineers Trans., v. 128, p. 65–107.

SIMONS, D.B., AND RICHARDSON, E.V., 1963, Forms of bed roughness in alluvial channels: Am. Soc. Civil Engineers Trans., v. 128, p. 284–302, (with discussion).

SIMONS, D.B., RICHARDSON, E.V., AND NORDIN, C.F., JR., 1965a, Sedimentary structures generated by flow in alluvial channels: p. 34–52 in Middleton, G.V. (ed.), Primary sedimentary structures and their hydrodynamic interpretation, Soc. Econ. Paleontologists and Mineralogists, Spec. Pub. 12.

——, 1965b, Bedload equation for ripples and dunes: U.S. Geol. Survey Prof. Paper 462-H, 9 p.

SIMONS, D.B., RICHARDSON, E.V., AND GUY, H.P., 1966, Summary of alluvial channel data from flume experiments, 1956–61: U.S. Geol. Survey Prof. Paper 462-I, 96 p.

SKIBINSKI, J., 1965, Bedload transport at flood time: Internat. Assoc. Sci. Hydrology, Symposium on River Morphology, Pub. no. 75, p. 41–47.

SLATT, R.M., AND HOSKIN, C.M., 1968, Water and sediment in the Norris Glacier outwash area, Upper Taku Inlet, Southeastern Alaska: Jour. Sed. Petrology, v. 38, p. 434–456.

SMERDON, E.T., AND BEASLEY, R.P., 1959, Tractive force theory applied to stability of open channels in cohesive soils: Univ. Missouri Agric. Exp. Stn., Research Bull. no. 715, Columbia, Mo.

SMITH, N.D., 1970, The braided stream depositional environment: Comparison of the Platte River with some Silurian clastic rocks, North-Central Appalachians: Geol. Soc. America Bull., v. 81, p. 2993–3014.

——, 1971, Transverse bars and braiding in the Lower Platte River, Nebraska: Geol. Soc. America Bull., v. 82, p. 3407–3420.

——, 1974, Sedimentology and bar formation in the upper Kicking Horse River, a braided meltwater stream: Jour. Geology, v. 82, p. 205–223.

SNEED, E.D., AND FOLK, R.L., 1958, Pebbles in the Lower Colorado River, Texas: A study in particle morphogenesis: Jour. Geology, v. 66, p. 114–149.

SPEIGHT, J.G., 1965, Meander spectra of the Angabunga River: Jour. Hydrology, v. 3, p. 1–15.

——, 1967, Spectral analysis of meanders of some Australian Rivers: p. 48–63 in Jennings, J.A., and Mabbutt, J.A., (eds.), Landform studies from Australia and New Guinea, Australian Natl. Univ. Press, Canberra.

SQUARER, D., 1970, Friction factors and bedforms in fluvial channels: Am. Soc. Civil Engineers Proc., v. 96, no. HY4, p. 995–1018.

STEBBINGS, J., 1964, The shapes of self-formed model alluvial channels: Inst. Civil Engineers Proc., v. 25, p. 485–510 (with discussion, v. 26, 1964, p. 225–232.)

STENBORG, T., 1965, Problems concerning winter runoff from glaciers: Geografiska Annaler, Ser. A, v. 47, p. 141–184.

——, 1969, Studies of the internal drainage of glaciers: ibid., Ser. A, v. 51, p. 13–41.

——, 1970, Delay of runoff from a glacier basin: ibid., Ser. A, v. 52, p. 1–30.

STEWART, J.H., AND LAMARCHE, V.C., JR., 1967, Erosion and deposition produced by the flood of December, 1964, on Coffee Creek, Trinity County, California: U.S. Geol. Survey Prof. Paper 422K, 22 p.

STONE, K.H., 1963, Alaskan ice-dammed lakes: Assoc. Am. Geographers Annals, v. 53, p. 332–349.

STRAUB, L.G., 1935, Missouri River: United States 73rd Congress, 2nd Session, House Doc. No. 238, App. 6, p. 1032–1235.

——, 1955, Effect of water temperature on suspended sediment transport in an alluvial river: Internat. Assoc. Hydraulic Research, 6th Gen. Mtg., Proc. Paper D-25, 5 p.

——, ANDERSON, A.G., AND FLAMMER, G.H., 1958, Experiments on the influence of temperature on the sediment load: U.S. Army Corps of Engineers, Missouri R. Div., Sediment Ser. No. 10, 36 p.

STICHLING, W., 1969, Instrumentation and techniques in sediment surveying: Natl. Research Council Canada, Assoc. Comm. Geodesy and Geophysics, Subcomm. Hydrology, Hydrology Symp. Proc., no. 7, p. 81–140.

——, AND SMITH, T.F., 1968, Sediment surveys in Canada: Canada Dept. Energy, Mines and Resources, Inland Waters Branch, Tech. Bull. 12, 17 p.

STRICKLER, A., 1923, Beiträge zur Frage der Geschwindigkeitsformel und der Rauhigkeitszamen für Strome, Kanäle, und geschlossene Leitungen: Mitteil. des Amtes für Wasserwirtschaft, Bern, no. 16, 77 p.

STRINGHAM, G.E., SIMONS, D.B., AND GUY, H.P., 1969, The behavior of large particles falling in quiescent liquids: U.S. Geol, Survey Prof. Paper 562-C, 36 p.

SUNDBORG, AKE, 1956, The River Klarälven: A study of fluvial processes: Geografiska Annaler, v. 38, p. 127–316.

——, 1967, Some aspects on fluvial sediments and fluvial morphology. I. General views and graphic methods: ibid., v. 49A, p. 333–343.

SUTHERLAND, A.J., 1967, Proposed mechanism for sediment entrainment by turbulent flows: Jour. Geophys. Research, v. 72, p. 6183–6194 (with discussion, v. 73, p. 4778–4779).

TABATA, S., 1954, Oceanography of British Columbia Mainland Inlets. V. Salinity and temperature of waters of Bute Inlet: Fisheries Research Board [Canada], Progress Report (Pacific), no. 100, p. 8–11.

——, AND PICKARD, G.L., 1957, The physical oceanography of Bute Inlet, British Columbia: Jour. Fisheries Research Board [Canada], v. 14, p. 487–520.

TERASMAE, J., 1963, Notes on palynological studies of varved sediments: Jour. Sed. Petrology, v. 33, p. 314–319.

TERZAGHI, K., 1950, Mechanism of landslides: p. 83–123 in Paige, S. (ed.), Application of geology to engineering practice (Berkey Vol.), Geol. Soc, America.

——, 1955, Influence of geological factors on the engineering properties of sediments: Econ. Geology, 50th Anniversary vol., p. 557–618.

——, 1956, Varieties of submarine slope failures: Harvard Soil Mechanics Series [Cambridge], no. 52, 41 p.

THEIL, G.A., 1932, Giant current ripples in coarse fluvial gravel: Jour. Geology, v. 40, p. 452–458.

THORARINSSON, S., 1939, Hoffellsjökull, its movements and drainage: Chap. 8 in Vatnajökull—Scientific Results of the Swedish-Icelandic Investigations, 1936–1938: Geografiska Annaler, v. 21, p. 189–215.

——, 1953, Some new aspects of the Grimsvötn problem: Jour. Glaciology, v. 2, p. 267–275.

THOMPSON, S.M., 1965, The transport of gravel by rivers: 2nd Australian Cong. Hydraulics and Fluid Mech. Proc., Auckland, N.Z., p. A259–A274.

TIFFIN, D.L., MURRAY, J.W., MAYERS, I.R., AND GARRISON, R.E., 1971, Structure and origin of foreslope hills Fraser Delta, British Columbia: Canadian Petroleum Geol. Bull., v. 19, p. 589–600.

TOFFALETTI, F.B., 1969, Definitive computations of sand discharge in rivers: Am. Soc. Civil Engineers Proc., v. 95, no. HY1, p. 225–248.

TRICART, J., 1952, Accumulation glaciaire, fluvioglaciaire et périglaciaire: L'exemple de la Durance: IVe Congres du Quaternaire Actes: v. 1, p. 48–56.

——, AND VOGT, H., 1967, Quelques aspects du transport des alluvions grossiers et du faconnement des lits fluviaux: Geografiska Annaler, v. 49A, p. 351–366.

TRYGGYASON, E., 1960, Earthquakes, jökullhlaups and subglacial eruptions: Jökull, Ar. 10, p. 18–22.

TYWONIUK, N., 1972, Sediment discharge computation procedures: Am. Soc. Civil Engineers Proc., v. 98, no. HY3, p. 521–540.

UCHUPI, E., 1967, Slumping on the continental margin southeast of Long Island, New York: Deep Sea Research, v. 14, p. 635–639.

VALLENTINE, H.R., 1964, Characteristics of the backwater curve: Amer. Soc. Civil Engineers Proc., v. 90, no. HY4, p. 39–49.

VANONI, V.A., 1946, Transportation of suspended sediment by water: Am. Soc. Civil Engineers Trans., v. 111, p. 67–102.

VANONI, V.A., BROOKS, N.H., AND KENNEDY, J.F., 1960, Lecture notes on sediment transportation and channel stability: California Inst. Technology, W.M. Keck Lab. of Hydraulics and Water Resources, Rept. KH-R1, Various pagination.

VANONI, V.A., AND NOMICOS, G.N., 1960, Resistant properties of sediment-laden streams: Am. Soc. Civil Engineers Trans., v. 125, p. 1140–1175 (with discussion).

VINCENT, I.J., 1966, Bed and water-surface deformations at the mouth of a sediment transporting river feeding a reservoir: Internat. Assoc. Scientific Hydrology Pub. 70, p. 518–523.

WALKER, J.R., AND MASSINGILL, J.V., 1970, Slump features on the Mississippi Fan, northeastern Gulf of Mexico: Geol. Soc. America Bull., v. 81, p. 3101–3108.

WASHBURN, A.L., AND GOLDTHWAIT, R.P., 1958, Slushflows: Geol. Soc. America Bull., v. 69, p. 1657–1658 (abs.).

WERTZ, J., 1963, Mechanism of erosion and deposition along channelways: Arizona Acad. Sci. Bull., v. 2, p. 146–163.

WHITE, C.M., 1940, Equilibrium of grains on the bed of a stream: Roy. Soc. London Proc., Ser. A., v. 174, p. 322–334.

WILLIAMS, G.P., 1967, Flume experiments on the transport of a coarse sand: U.S. Geol. Survey Prof. Paper 562-B, 31 p.

WILLIAMS, P.B., AND KEMP, P.H., 1971, Initiation of ripples on flat sediment beds: Am. Soc. Civil Engineers Proc., v. 97, HY4, p. 505–522.

WILLIAMS, P.F., AND RUST, B.R., 1969, The sedimentology of a braided river: Jour. Sed. Petrology, v. 39, p. 649–679.

WOLMAN, M.G., AND BRUSH, L.M., JR., 1961, Factors controlling the size and shape of stream channels in coarse, noncohesive sand: U.S. Geol. Survey Prof. Paper 282-G, 28 p.

YALIN, M.S., 1964, Geometrical properties of sand waves: Am. Soc. Civil Engineers Proc., v. 90, no. HY5, p. 105–119.

——, 1971, On the formation of dunes and meanders: Internat. Assoc. Hydraulic Research, 14th Cong., Paris, v. 3, p. C13-1 to C13-8.

——, 1972, Mechanics of sediment transport: Pergamon, Oxford, 290 p.

YANG, C.T., 1971, Formation of riffles and pools: Water Resources Research, v. 7, p. 1567–1574.

ZINGG, T., 1935, Beitrag zür Schotteranalyse: Schweizerische Mineralogische und Petrologische Mitteilungen, Bd. 15, p. 39–140.

ZNAMENSKAYA, H.S., 1962, Calculations of dimensions and speed of shifting channel formations: Soviet Hydrology, no. 2, p. 111–115.

Notations

Overbar indicates mean value.

Where a symbol has more than one meaning, the particular meaning will be clear in context or defined each time it is used.

SI units are used throughout, except on a few occasions when other authors' results are being quoted.

SYMBOL	DESCRIPTION	DIMENSIONS
a_s	acceleration of water flow (along a streamline)	$[LT^{-2}]$
$a, b, c,$	principal axis lengths of sediment particles	$[L]$
A	cross sectional area of flow	$[L^2]$
	amplitude of diurnal discharge wave (eq. 3)	$[L^3T^{-1}]$
A_B, A_G	area of streambed (eq. 36); area of grain	$[L^2]$
c	cohesion (eq. 48)	$[ML^{-1}T^{-2}]$
	celerity of internal waves in stratified flow	$[LT^{-1}]$
	sediment concentration (see q for further explanation of subscripts)	$[ML^{-3}]$
c_o	static volume concentration of material (eq. 26)	$[O]$
C	a constant	
	drag coefficient	$[O]$
d	depth of flow	$[L]$
	pipe diameter	$[L]$
d_d	depth of underflow (density current)	$[L]$
d_*	mean flow depth, computed as A/w_s	$[L]$
D	measure of roughness element height at flow boundary; grain size of particles (usually equivalent to b-axis diameter)	$[L]$
e_b	transport efficiency	$[O]$
E	specific energy (of stream)	$[ML^{-2}T^{-2}]$
f	function of	
ff	Darcy-Weisbach flow resistance coefficient	$[O]$
F_r	Froude number	$[O]$
F_r'	densimetric Froude number	$[O]$
g	gravitational acceleration	$[LT^{-2}]$
g'	reduced gravitational acceleration in a density flow (eq. 41)	$[LT^{-2}]$
g_{bs}/G	specific/total sediment transport rate (see q for further explanation of subscripts)	$[MT^{-1}L^{-1}]/[MT^{-1}]$
h	bed form amplitude	$[L]$
H	total energy (of stream)	$[ML^2T^{-2}]$
j	exponent in suspended sediment discharge equation $c_{ss} = pQ^i$	
k	intrinsic permeability	$[L^2]$
	wave number	[cycles/L]
k_r	coefficient of particle friction on a smooth bed in equation 24, $= 26/D_{90}^{1/6}$	
k_s	equivalent grain roughness of boundary (eq. 11)	$[L]$
K	hydraulic conductivity	$[LT^{-1}]$
	parameter in Durand-Condolios relation (eq. 29)	
L	characteristic length	$[L]$
n	Manning total flow resistance coefficient	$[L^{-1/3}T]$
N	an arbitrary probability density distribution for frequency of particle entrainments at a river bed sample size	
p	a proportion (equation 22)	
	coefficient in suspended sediment discharge relation $c_{ss} = pQ^i$ fluid pressure	$[ML^{-1}T^{-2}]$
p_e	pore water pressure (eq. 48)	$[ML^{-1}T^{-2}]$

SYMBOL	DESCRIPTION	DIMENSIONS
P	number of days since the last rain (eq. 19)	
	wetted perimeter	$[L]$
	period of delta front avalanching	$[T]$
$P\infty$	period of delta front avalanching assuming instantaneous failure	$[T]$
q	specific discharge of water (discharge per unit width)	$[L^3T^{-1}L^{-1}]$
q_s/Q_s	specific/total discharge of sediment	$[L^3T^{-1}L^{-1}]/[L^3T^{-1}]$
q_{bs}/Q_{bs}	specific/total discharge of bed load	$[L^3T^{-1}L^{-1}]/[L^3T^{-1}]$
q_{ss}/Q_{ss}	specific/total discharge of suspended sediment	$[L^3T^{-1}L^{-1}]/[L^3T^{-1}]$
Q	total discharge of water	$[L^3T^{-1}]$
r	minimum radius of curvature on the principal axis plane of a clast	$[L]$
	packing factor	$[O]$
	exponent of P in equation 19	
r^2	coefficient of determination in least squares regression relationship	
R	Cailleux roundness	$[O]$
	hydraulic radius, $= A/P$	$[L]$
R_e	Reynolds number	$[O]$
R_{e*}	"particle" Reynolds number, $Dv*/v$	$[O]$
s	generalized displacement	$[L]$
	slope in degrees; $S = \tan s$	$[O]$
s_s	specific gravity of sediment particle	$[O]$
S	river slope (water surface slope)	$[O]$
	spectral density (eq. 34)	
S_c	Corey particle shape factor, $= c/\sqrt{ab}$	$[O]$
S_f	friction slope; energy gradient	$[O]$
S_o	hydraulic gradient (water surface slope in open channel flow)	
S_p	maximum projection sphericity $= \sqrt[3]{c^2/ab}$	$[O]$
S_w	Krumbein's intercept sphericity, $= \sqrt[3]{bc/a^2}$	$[O]$
t	time	$[T]$
T	temperature	
v	flow velocity	$[LT^{-1}]$
v_d	velocity of underflow	$[LT^{-1}]$
v_*	friction velocity or shear velocity, $= \sqrt{\tau_o/\rho}$	$[LT^{-1}]$
v_w	bedwave celerity	
w	channel width, flow width	$[L]$
	particle settling velocity	$[LT^{-1}]$
w_s	surface width of channel	$[L]$
W	weight of material deposited on the delta front at distance X or reference distance L from the delta lip (eq. 46)	$[M]$
x, y, z	cartesian co-ordinates	$[L]$
y	depth below the surface of flow	$[L]$
z	depth of overburden (eq. 48)	$[L]$
	height above arbitrary datum	$[L]$
α	dimensionless number (eq. 10)	$[O]$
	ratio of shear stresses (eq. 43)	$[O]$

SYMBOL	DESCRIPTION	DIMENSIONS
γ	specific gravity of water	$[MLT^{-2}]$
γ_s	specific gravity of sediment	$[MLT^{-2}]$
γ'	$= \gamma_s - \gamma$, effective specific gravity of submerged sediment	$[MLT^{-2}]$
δ	depth of laminar sublayer	$[L]$
ε	kinematic eddy viscosity, $= \eta/\rho$	$[L^2T^{-1}]$
η	dynamic "eddy viscosity"	$[ML^{-1}T^{-1}]$
θ	periodic phase angle of daily runoff (eq. 3)	$[O]$
	angle of repose of sediment particles	$[O]$
	stability criterion for internal waves (eq. 45)	
	stability parameter for underflow mixing	
$\tan\theta$	solid friction coefficient	$[O]$
κ	von Karman's constant, $\simeq 0.4$	$[O]$
λ	bed form wavelength	$[L]$
	internal wavelength	$[L]$
v	kinematic viscosity of water, $= \mu/\rho$	$[L^2T^{-1}]$
v_d	kinematic viscosity of underflow	$[L^2T^{-1}]$
μ	dynamic viscosity of water	$[ML^{-1}T^{-1}]$
ρ	mass density of water	$[ML^{-3}]$
ρ_d	density of underflow (eq. 41)	$[ML^{-3}]$
ρ_l	density of lake water (eq. 41)	$[ML^{-3}]$
ρ_s	density of sediment	$[ML^{-3}]$
ρ_s'	effective density of submerged sediment	$[ML^{-3}]$
σ	normal stress	$[ML^{-1}T^{-2}]$
	standard deviation	
τ	shear stress	$[ML^{-1}T^{-2}]$
τ_c	critical tractive force; tractive force at the threshold of motion	$[ML^{-1}T^{-2}]$
τ_i	interfacial shear in underflow	$[ML^{-1}T^{-2}]$
τ_o	shear stress at the bed	$[ML^{-1}T^{-2}]$
$\Delta\phi_s$	dilatation angle (difference between angle just before and just after failure occurs) (eq. 47)	$[O]$
Φ	bed-load function	$[O]$
$\dfrac{1}{\psi}$	entrainment function $(= \tau/\rho_s'gD)$	$[O]$
$\dfrac{1}{\psi_e}$	critical entrainment function	$[O]$
ω	stream power	$[ML^3T^{-3}]$

Notes

1 Note that the results of Bradley and others are based on measurement of the largest available clasts, whereas the results of Church are based on mean size of a random sample. Church reported very high correlation between mean and maximum sizes, however; hence the results appear to be truly equivalent. Sampling procedures are detailed in the authors' reports.

2 In accordance with definitions usually accepted (Howard, 1953, p. 356–357; Middleton, 1965, p. 248; Middleton, 1966) underflows that owe their density difference with the surrounding water to their content of "suspended" sediment rather than to temperature or dissolved solid load differences are referred to as turbidity currents.

16

A FRAMEWORK FOR THE INVESTIGATION OF MEDIAL MORAINE FORMATION

Austerdalsbreen, Norway, and Berendon Glacier, British Columbia, Canada

N. Eyles and R.J. Rogerson

Source: *Journal of Glaciology* 20(82) (1978): 99–113

Abstract

Morphology of medial moraines on Austerdalsbreen, Norway, and Berendon Glacier, British Columbia, depends upon englacial debris supply. Major sub-types of this "ablation-dominant" model are related to the zone of debris entrainment relative to the firn line, and the manner of entrainment.

On Austerdalsbreen, debris derived from extraglacial bedrock slopes is entrained via crevasses at the confluence of two ice-cap outlet glaciers below the firn line. Revelation of crevasse-bound debris generates a distinct ice-cored morphology which is destroyed as crevasse bottoms are revealed down-glacier.

On Berendon Glacier ice streams coalesce above and below the firn line. Above the firn line, debris from extraglacial rock outcrops, subnival and subglacial zones, undergoes seasonal sedimentation with snowfall, and extends throughout the ice depth. Distinct moraine morphology in the terminal zone is related to continuing debris supply. Most debris is transported at depth near the glacier base.

An "ice-stream interaction" model where medial moraines formed below the firn line from the confluence of ice streams with large lateral moraine load are morphologically controlled by flow, explains morphology on the Berendon Glacier in the main confluence zone only. Down-glacier, this moraine becomes "ablation dominant". A minor "avalanche-type" model is also recognized.

Introduction and objectives

Medial moraines are marked and well-defined features of many glaciers, predominantly compound glaciers formed from the joining of two or more ice streams from basins in the accumulation zone. They are revealed in varying states of development over and along the time-transect of the glacier surface in the ablation zone.

This paper proposes several models for the origins and morphological development of medial moraines as described in the literature and as exemplified in two field investigations, one on Austerdalsbreen, Norway (Eyles, 1976), and one on the Berendon Glacier, British Columbia (Rogerson and Eyles, unpublished).

Previous work

A concise account of early ideas on the origins of medial moraines appears in Charlesworth (1957, Vol. 1, p. 406–07). He ascribes first recognition of their true origin to B.F. Kuhn in 1787, clarified over fifty years later by the works of Agassiz, Charpentier, and Godeffroy.

Heim (1885, p. 345–46) recognized that medial moraines need not be formed at the junction of ice streams through the union of lateral moraines, but may also be formed in compound firn basins and not revealed until further down-glacier. Salisbury, in a classic paper (1894) made a full exploration of glacier debris systems and included the further example of medial moraines formed from subglacial rock bosses as first suggested by Tyndall (1872). Thus, early ideas on medial moraine origins were quite comprehensive.

Detailed field examination of medial moraine morphology was less systematic although Salisbury described the widening of medial moraines down-glacier in response to continuing englacial sediment supply. Hess (1907[a], [b]) determined the supply of englacial and subglacial debris to medial moraines of the Hintereisferner, but concentrated more on the significance of the medial moraines in the general erosion of glacierized basins than to their morphological development.

A large literature discusses minor relief features on medial moraines without relating them to the type of origin of the moraine as a whole. Dirt-cones, rock tables and supraglacial melt-stream activity have been well researched (Agassiz, 1840; Forbes, 1859; Ray, 1935; Lewis, 1940; Sharp, 1949; Swithinbank, 1950; Wilson, 1953; Streiff-Becker, 1954; Krenek, 1958; Lister, 1958; Kozarski and Szupryczński, 1971; Drewry, 1972; Knighton, 1973): a literature on dirt cones has, for instance, existed since 1750 (Thorarinsson, 1960).

More recent work has drawn attention to the important relationships between medial-moraine morphology and the causal factors of origin, sediment supply, and sediment thickness. Rapp (1960) demonstrated that the medial moraine of Templefjorden, Spitsbergen is entirely supraglacial being essentially transported talus. Ives and King (1955) reached similar conclusions concerning the medial moraine of Morsárjökull, Iceland. Loomis (1970) working on the large medial moraine formed at the junction of the north and central arms of the Kaskawulsh Glacier, Yukon Territory, also considered debris to be predominantly supraglacial, as did Lister (1958) on Britannia Gletscher, north-east Greenland. On Kaskawulsh Glacier, moraine morphology was considered in terms of differential ablation of debris-laden and clean ice. Young (1953) had previously recognized a direct correlation between moraine height and thickness of debris cover on the medial moraines of Breiðamerkurjökull. Loomis considered that constant down-glacier moraine width on the Kaskawulsh Glacier did not reflect lateral compression between the confluence ice streams; low rates of englacial sediment supply provide a sufficient explanation. Small and Clark (1974) related the development of medial moraine morphology on the lower Glacier de Tsidjiore Nouve to a careful account of sediment incorporation and supply. Debris is entrained in crevasses at the confluence of two ice falls below the firn line. A lower limit of englacial debris supply can be recognized: cessation of englacial debris supply results in the degeneration of relief, hastened by extensive new crevassing. Publication by Small and Clark coincided with field work on Austerdalsbreen which exhibits a similar pattern of moraine development (Eyles, 1976).

Models

Schematic models of medial-moraine formation and morphological development were constructed during, and have been refined since, work on the Berendon Glacier. They serve as a useful introduction to a detailed study of the medial moraines of both Austerdalsbreen and Berendon Glacier. Medial moraines can be seen (Fig. 1) to be formed of debris from a variety of sources, deposited on or in the glacier by a number of routes, and revealed down-glacier with varying morphological consequences. Important variables in the recognition of discrete models and sub-models therefore become:

1. moraine origin with reference to the firn line;
2. source and character of debris supply to or into the glacier;
3. source and character of debris supply to the moraine, when distinct from 2;
4. morphological development of the moraine including changing shape, height and breadth.

These variables form a framework for detailed examination of the moraines of both glaciers.

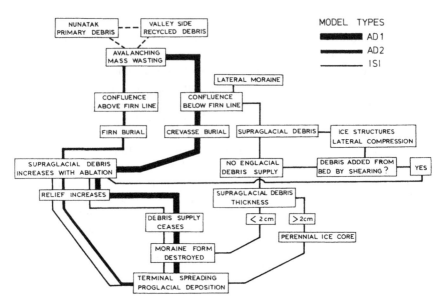

Figure 1 Schematic models of medial-moraine formation.
AD1: "Ablation-dominant" model; below firn-line sub-type.
AD2: "Ablation-dominant" model; above firn-line sub-type.
Is I: "Ice-stream interation" model.
The figure of 2 cm was derived from Berendon Glacier.

The total number of types and sub-types of medial moraines recognized could be considerable, but if the emphasis is directed to the relationship between sediment supply and morphological development, two important models emerge. Debris comprising medial moraines may be held englacially and revealed down-glacier by ablation, or be largely supraglacial and therefore not dependent on ablation for surface expression. Moraines formed from englacially transported debris are here termed "ablation dominant" (AD) since they depend on ablation for surface expression. Moraines formed with immediate supraglacial expression at the confluence of large ice streams, often by the joining of two lateral moraines, are here described by the "ice-stream interaction" model (ISI). A third, minor model, the "avalanche type" is also recognized and described (AT). The ablation dominant model has three sub-types dependent on whether englacial material is incorporated below the firn line, via crevasses (AD1: ablation dominant, below firn-line sub-type); above the firn line through the sedimentation of annual snow and debris layers (AD2: ablation dominant, above firn-line sub-type); or subglacially from an ice-covered rock knob (AD3: ablation dominant, subglacial rock-knob sub-type). This last sub-type was recognized on neither Austerdalsbreen nor Berendon Glacier, but is included from the reliable accounts of its existence elsewhere.

Field methods

Field methods on both glaciers were designed to elucidate the important relationship between englacial sediment supply and changing moraine morphology.

On Austerdalsbreen in 1974 a long profile of the moraine was surveyed by undergraduates of the University of Leicester and the ablation rates of moraine zones of varying debris cover were measured to examine the relationship between debris and moraine relief.

On the Berendon Glacier a more extensive programme was undertaken which included detailed sampling of sediments along the moraines and at nunatak rock walls in the firn basins and further experimental ice-melt studies reported elsewhere (Rogerson and Eyles, unpublished). Of particular interest to the investigation of models was the construction of debris clearance sites, where supraglacial debris was removed from moraines over areas of up to 500 m^2 to permit observation of the quantity of englacial sediment revealed by ablation at the surface. In addition the repeated tacheometric measurement of 50 stakes along the line of the medial moraines, allowed determination of ice velocity and strain-rate. In the absence of any marked correlation between the form of medial moraines and local ice velocity and strain-rate these measurements will only be briefly discussed. Detailed description of field methods and derived velocity and strain-rate data are to be found in Eyles (unpublished).

The "ablation dominant" model:
below firn-line sub-type (AD1)—Austerdalsbreen

Austerdalsbreen lies on the southern margin of Jostedalsbreen, Europe's largest ice cap. The glacier is fed by two ice falls, Odinsbreen and Thorsbreen both about 800 m high and 300 m wide (Fig. 2). A third ice fall, Lokebreen, to the west, has thinned and ablated back to the regional snow line which traverses the ice falls approximately 1,600 m a.s.l. Ice velocity along the medial line of the ice falls decreases from approximately 2,000 m year^{-1} at the apex (King, 1959), and inversely correlates with an increase in ice depth from less than 40 m in the ice falls to greater than 120 m below the junction (Ward, 1961).

Transverse surface waves immediately below the ice falls merge down-glacier into a double arcuate system of ogives with an alternation of diffuse surficial debris and clear ice bands. The lower albedo of the former results in alternating troughs and waves with a relief amplitude of 1 m. King and Lewis (1961) invoked wind-blown dust, longitudinal attenuation and high melt-season ablation rates to account for differences in ice types comprising the ogive suite. The medial moraine becomes a marked morphological feature in the same zone as systematic ogive banding emerges.

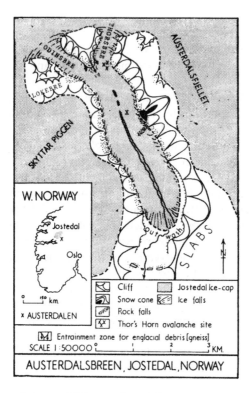

Figure 2 Location of Austerdalsbreen, Jostedal, Norway. × is that point where moraine height reaches a maximum (12 m) relative to bare ice.

Dynamics and debris throughput of the AD1 model

1. The regional snow line lies 1,600 m a.s.l., close to the apex of the ice falls. None of the debris incident upon the margins of the two ice streams becomes bedded in firn, except where small units of firn fill shaded crevasses.

2. Bedrock material is derived by rockfalls and slides from an extensive outcrop between Odinsbre and Thorsbre ice falls, and is englacially entrained at the margins of the ice streams via deep crevasses. In addition, some debris is buried by occasional ice avalanches from the "Thor's Horn" avalanche site (Fig. 2). For such crevasse-bound debris a lower depth of englacial penetration can be suggested which may be well-defined (Small and Clark, 1974). Penetration beyond this general limit may occur (Glen and Lewis, 1961) although it is unlikely that entrained material penetrates to the base of the glacier. There is little contact between subglacial load and crevasse-bound debris for the compressive flow of the main glacier trunk at the base of the ice falls and resultant increasing ice depth (Glen, 1956; Ward, 1961) effectively sever such contact.

550

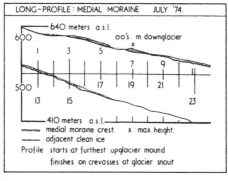

(a)

(b)

Figure 3

(a) Long profile of Austerdalsbre medial moraine surveyed in July 1974. Note the absence of a positive moraine relief in the terminal area.

(b) The moraine viewed from the snow cone (Fig. 2). Note the base of the ice falls and irregular debris mounds along the medial line. Moraine morphology collapses down-glacier of that point at which moraine height is greatest relative to bare ice. Note also the winter ogive waves and dirty summer troughs. Composite photograph taken in July 1974.

3. Englacial debris is progressively revealed down-glacier (Fig. 3). Discontinuous debris mounds along the medial line, with no observed systematic relationship to ogive bands, merge down-glacier to form a typical medial moraine in response to continuing englacial debris supply. The early appearance of isolated mounds approximately 500 m below the confluence may represent debris entrained in shallower crevasses on the ice falls. No evidence exists to suggest that shear planes have influenced sediment supply to the moraine. Shearing activity in the manner described elsewhere by Bishop (1957) and Boulton (1967) does not occur.

Approximately 1,200 m below the ice falls the englacial supply of debris ceases. Distinct beading of the lower part of the medial moraine, in harmony with ogive banding, does not represent seasonal differences in the quantity of englacial material but mass movement of moraine sediment into summer ogive troughs (Eyles, 1976). The insignificant quantity of englacial debris in ogives has been well known since the studies of Huxley on Ghiacciaio della Brenva (Huxley, 1857), and the contribution of ogive debris to the medial moraine on Austerdalsbreen can be ignored. Elsewhere however, as for instance on Mer de Glace in the French Alps, an irregular discontinuous medial moraine exhibits large ice-cored debris mounds clearly derived from summer ogive bands.

4. Differential ablation between clean and debris-covered ice results in the formation of an ice-cored moraine 800 m down-glacier from the confluence (Fig. 3). The moraine increases to a maximum height of 12 m almost 1,200 m below the confluence. This greatly exceeds the 9 m of ice melt experienced in an average melt season on adjacent clean ice (King, 1959). The ice-core is thus perennial and possesses differential relief at the commencement of each melt season (Hannell and Ashwell, 1959). Down-glacier, height decreases rapidly and lateral attenuation of surface debris dominates over vertical development of the moraine (Fig. 3). The zone in which relief collapses is crevasse-free, unlike the equivalent zone described by Small and Clark (1976) on the lower Glacier de Tsidjiore Nouve. Width increases from 40 m in the vicinity of maximum relief to over 200 m in the terminal zone. In this zone the growth of ice cores is terminated by the lateral dispersion of debris over the flanks of the core. Topographic inversion occurs as accelerated ice-melt rates characterize the exposed crests of the ridges where debris is lost by sliding over the ice-core flanks. With debris covers of <1 cm, ice-melt rates were accelerated by up to 30% resulting in a depressed moraine morphology in the terminal zone (Fig. 3).

The "ablation dominant" model:
above firn-line sub-type (AD2)—Berendon Glacier

Berendon Glacier (lat. 56° 15' N., long. 130° 5' W.) has a drainage basin area of 53 km^2 within the Boundary Ranges of the northern Coast Mountains of British Columbia (Fig. 4). Its location adjacent to copper-concentrating facilities of the Granduc Operating Co. has attracted much attention (Untersteiner and Nye, 1968; Fisher and Jones, 1971). The glacier, which is receding at present, consists of two major tributaries, north and south arms, which coalesce 2.2 km above the glacier terminus and give rise to a broad medial moraine which will be considered later in this paper. Either side of this moraine prominent ice-cored medial moraines ablate out of north and south arm ice in the terminal area. The development of these moraines is explained with reference to the "ablation dominant" model: above firn-line sub-type (AD2).

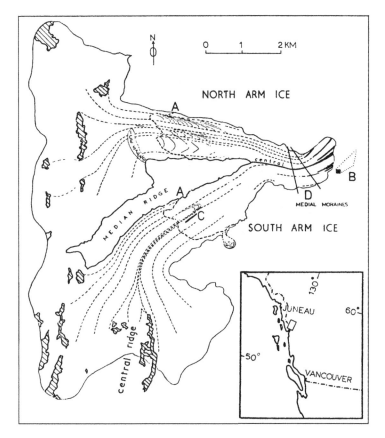

Figure 4 Berendon Glacier, its location and medial moraines. Medial moraines in the terminal area of north and south arms can be traced up-glacier above the firn line (A) to individual accumulation basins where bedrock nunataks emerge. The location of Granduc Mill is depicted (B), as is the site of the "avalanche-type" moraine (C). D marks the terminal ice-falls.

Dynamics and debris throughput of the AD2 model

1. Debris septa can be traced up-glacier from distinct medial moraines north and south of the central moraine (Fig. 4). The septa appear immediately below the firn line and clearly relate to junctions in flow units above the firn line. Distinctive lithologies present in the moraines of south arm permit accurate extrapolation from the firn line to known bedrock outcrops up-glacier.

2. Source areas for debris septa can be related to subaerial sites of rock falls in the firn basins. Usually an upper ice and snow carapace is separated from the main firn mass in the valley bottom by free faces in bedrock up to

Figure 5 An avalanche of ice and snow bringing down bedrock debris from a nunatak rock-wall in the firn basin of south arm. These dirty avalanches are the subaerial debris input for medial moraines.

500 m high. Fresh, primary debris is generated subglacially by the upper ice carapace, appearing as well-defined debris bands in the glacier sole. This debris is released by avalanching to the firn below (Fig. 5) and undergoes compaction and sedimentation with seasonal snow loads.

Source areas of sediment must also exist subnivally and subglacially below the level of the main firn mass, where they may border on or merge with subglacial debris proper. No observations of such sites were possible in 1975.

3. Following confluence of ice flow units in the firn basin, debris is revealed at the firn line in distinct debris septa. The supply of englacial debris remains very low until very near the terminus. Debris septa revealed at the firn line are in the order of 2–3 m wide, varying little down-glacier until the zone near the terminus is reached. Consequently no ice-cored morphology is present, indeed enhanced ablation may occur on a minor scale, although this is confused with the effects of surface melt water.

Below the ice falls in the terminal area a great increase in the quantity of surface debris demonstrates an increased englacial debris supply to the moraines. A marked release of englacial debris at the surface is apparent from observations of those areas cleared of all debris. Below the terminal ice falls, well-defined ice-cored moraine ridges demarcate component ice-flow units; three on south arm ice but only one on north arm ice where morainal septa are relegated to a lateral position up-glacier by the thinning of lesser ice-flow units (Fig. 4). Moraine width increases down-glacier to 10 m before sediments merge and are subsequently consumed down splaying crevasses and the steep terminal ice front. Debris thickness along the moraine ridges seldom exceeds a few centimetres, and ice cores are seasonal, being accentuated during the summer but degraded in the fall, when ice melt is assisted by the better radiation absorption of debris over bare ice. Maximum moraine relief was 2.5 m and occurred within 150 m of the ice margin.

Clearly, the late development of distinct ice-cored medial moraines may be attributed to deep englacial transportation of debris from the firn basin. The bulk of medial moraine debris is transported through the glacierized basin at depth and is only revealed close to the terminus. While englacial debris extends throughout the depth of the ice, as suggested by de Martonne (1925–26), depicted by Sharp (1948), and demonstrated by Battle (1951), the distribution with depth is not constant. This may in fact represent two populations of sediment issuing from the firn basin; one derived subnivally and subglacially, and one derived subaerially such as from avalanches. The former unit may be part of the subglacial load proper. High rates of ice melt (6–9 m year^{-1}) and severe compressive strains measured along the moraines in the terminal area (Eyles, unpublished) may also promote higher concentration of englacial debris, after the manner proposed by Small and Clark (1976). It is not thought however, that these processes alone can account for the great increase in englacial sediments there.

The avalanche-type model (AT)

The Berendon Glacier exhibits only one major ice-stream junction below the firn line. Medial moraines visible up-glacier therefore may be supposed to result from junctions above the firn line. This, however, may not always be the case. A moraine on south-arm ice is evidently unrelated to any junction and is an example of the avalanche type.

On south arm a multi-ridged longitudinal moraine almost 1,000 m long is unrelated to surrounding AD2 septa. By its position close to the firn line the debris has been only shallowly entrained and may be related to an exceptional rock fall from the ridge dividing the western and eastern firn basins. The moraine stops abruptly down-glacier (Fig. 6) and thus bears strong similarity to the "moraines de boulement" of Agassiz (1840). On Castner Glacier, Alaska, Nielson and Post (1953) attributed a similar unusual moraine to the sudden collapse of a rock spire, and another is described from Highway Glacier on Baffin Island (Ward, 1955). This moraine type is unimportant in the glacier debris system unless avalanching persists for many years resulting in englacial continuity of debris.

Figure 6 "Avalanche-type" medial moraine on south-arm ice, looking up into the firn basin.

The "ice-stream interaction" model (ISI)

Distinct ice structures are known to develop where large valley outlet glaciers converge (Sharp, 1960; Brecher, 1969; Anderton, 1970; Loomis, 1970). In these zones, medial moraines are often formed by the merging of supraglacial lateral moraines. It is suggested that in many instances the character of ice flow such as lateral compression between merging ice streams, longitudinal strain-rate and ice velocity determine the morphology of the medial moraine rather than the nature of the englacial debris supply. This is termed the "ice-stream interaction" model, and while it appears to be better represented in the literature quoted, the model can be tested on Berendon Glacier where north and south arms combine to form the central medial moraine (Figs 1, 4 and 7).

1. Most of the debris supplied to the central medial moraine comes from the northern margin of south arm. At least two sources of debris are observed:

(a) morainal septa relegated to a lateral position by the thinning of lesser ice-flow units from the firn basin;
(b) recycled glacial debris from slopes of the median ridge down to the point of confluence.

In effect debris originates both above and below the firn line.

2. and 3. Moraine debris is entrained in crevasses of the south-arm ice fall, and in the fewer chevron crevasses along the southern margin of north

Figure 7 The central medial moraine ("ice-stream interaction" type) with Granduc Mill in the terminal area. North-arm ice to the left. The longitudinal zone of shear along the glacier contacts (giving rise to the northern debris band) is visible as are transverse debris ridges on south-arm ice.

arm. The crevasses penetrate to the bed on north arm, and ingested debris is contributed to the confluence as sub-marginal load. On south arm, in contrast, the crevasses above the confluence do not penetrate to the bed. With ablation and revelation of crevasse-bound debris, transverse till ridges are found in the confluence zone where they compose much of the south arm debris contribution to the medial moraine (Fig. 7). The confluence medial moraine consists of three lithologically distinct debris bands. The southern and central bands (volcanic conglomerate and tuff) are part of south arm ice and constitute the bulk of the debris. The northern band (argillite and siltstone) is formed by the shearing up of ingested submarginal moraine debris under the combined influences of severe lateral compression and differential ice-arm velocities (Fig. 7). Towards the terminus the three bands cannot be distinguished as the result of passage through the terminal ice falls. Debris clearance sites on the central moraine indicate that englacial debris supply above the terminal ice falls is absent (Fig. 8). Even the northern debris band, where debris is sheared up from submarginal zones providing maximum observed debris density, rapidly exhausts its supply. Below the terminal ice falls, the moraine is characterized by an englacial debris supply

Figure 8 A clearance zone across the entire central medial moraine, 40 m long and 3.5 m wide, at the end of August 1975. Note the absence of freshly revealed englacial debris, and the ice-cored mounds developed where debris was dumped during clearance. Elsewhere supraglacial debris thickness is insufficient to allow ice cores to develop.

of increasing quantity. The emergence of a further debris band of distinct colour and lithology (volcanic conglomerate) well below the terminal ice falls is significant for it possesses no extraglacial outcrop in the firn basins and has been clearly derived subglacially or subnivally. In this zone subglacial debris is added to medial moraine sediments.

4. The morphology of the moraine is best described with reference to three zones, the first in the immediate confluence area, and the others down-glacier; the second above and the third below the terminal ice falls.

Zone 1

In the first zone a distinct moraine morphology is found in response to the contrasting character of confluent ice streams. North-arm material sheared to the surface as described and comprising the northern debris band is generally 2–3 m wide and ice-cored, increasing in height down-glacier to a maximum relief of 1.5 m.

Ice-cored transverse debris ridges are common in the immediate confluence zone on south-arm ice but are rendered arcuate and lose their distinct form only 150 m down-glacier (Fig. 7).

No differential relief prevails on south-arm ice between clear and debris-covered ice as a result of the low depth of supraglacial debris. Severe lateral compression between the two arms has an important effect on the morphology of the moraine. Surface wave forms whose origin is in part due to severe compression at the base of south-arm ice fall have a wave-length of ≈60 m which approximates annual flow velocities. The lowermost wave crest is found along the contact zone, and consequently the surface of south-arm ice is higher than adjacent north-arm ice (Fig. 7).

Zone 2

The distinct morphology of the moraine in the confluence zone is lost down-glacier. Whilst individual debris bands can still be identified, they are not ice-cored and relief is indistinct. Rock tables and scattered dirt-cone groups add diversity. Moraine width (40 m) remains constant over a distance approaching 1,000 m. The immature morphology can be clearly related to the low volume of englacial debris as observed at clearance sites. In addition, differential velocity between north- and south-arm ice has ceased 800 m down-glacier of the confluence and is replaced by unified flow. Indeed it is significant that clearance of debris from these sites resulted in the development of an ice core over those areas where debris had been dumped (Fig. 8).

Zone 3

Below the terminal ice falls the release of englacial debris generates a more mature ice-cored morphology. The late appearance of the distinctly coloured volcanic conglomerate debris band generating a further ice-cored moraine ridge, is identical to the situation described in the AD2 model. A splaying crevasse system and terminal calving disrupts moraine form in the terminal zone where debris from englacial and subglacial sources has increased the width of the moraine to over 100 m. In all three zones of the moraine no correlation can be established between ice velocity and longitudinal strain-rate (Eyles, unpublished).

In conclusion, the morphological development exhibited by the central medial moraine below Zone 1 can be accommodated by the "ablation-dominant" model: complexity introduced by lateral compression of north and south arms is not a persistent determinant of moraine morphology down-glacier of the confluence zone. The moraine contains debris bands developed both above and below the firn line in the manner described for both sub-types of the "ablation-dominant" model (Fig. 1).

Conclusions

Two major models of medial moraine formation, the "ablation-dominant" and the "ice-stream interaction" model explain the development of medial moraine morphology exhibited on two temperate valley outlet glaciers.

(A) The "ablation-dominant" model relates moraine morphology to the nature of englacial debris supply and on this basis two major sub-types are formulated: (1) below firn-line sub-type, and (2) above firn-line sub-type. A third, minor sub-type is formed from subglacial rock knobs, but was not observed or studied on Austerdalsbreen or Berendon Glacier.

(1) Where moraines are formed below the firn line at the margins of ice-cap outlet glaciers (Austerdalsbreen), moraine debris is derived subaerially from nunatak areas in the confluence zone. Such debris is precipitated into crevasses and comes to occupy only a shallow englacial position. The quantity of englacial debris below the base of the deepest crevasses is small and is attenuated further by the thickening of ice in response to compressive strains at the base of the ice fall. Crevasse-bound debris is revealed down-glacier and an ice-cored moraine ridge generated. With the cessation of englacial debris supply as crevasses ablate out, moraine morphology collapses. "Annual" and "perennial" ice cores can be recognized.

(2) Where moraines are formed above the firn line (Berendon Glacier: north and south arms) debris released from rock walls is precipitated onto the firn surface where it undergoes sedimentation with seasonal snowfall. As a result, subaerially derived debris extends throughout ice depth, from the glacier surface to the bed, where additional debris derivation takes place. With the merging of flow units, a distinct vertical column of debris, a medial moraine or debris septum, is generated. The distribution of englacial debris with depth is not constant and the bulk of moraine debris is transported near the bed and only revealed in the terminal zone where it is associated with a developing ice-cored morphology. Subglacial debris is added to moraine sediments in the terminal zone. An ice-core is not present up-glacier, supraglacial debris quantities remain low, and moraine width remains constant. This is related to the decline in the quantity of englacial debris above the bed. This upper debris is derived entirely from extraglacial rock slopes and upper ice carapaces; the clarity with which medial moraines and debris septa are revealed immediately below the firn line reflects the rate of erosion in these areas.

(B) Medial moraines are generated in many instances below the firn line by the confluence of large valley outlet glaciers transporting lateral moraine debris. Medial moraine width is seen to remain constant down-glacier until the terminal zone is reached. An "ice-stream interaction" model of moraine formation relates unchanging moraine width to lateral compression between outlet glaciers and the complex patterns of ice flow found in these zones.

On Berendon Glacier the "ice-stream interaction" model is only substantiated in the immediate confluence area of two large valley outlet arms (north and south arms). Unchanging moraine width can be related to the low rate of englacial debris supply. Moraine morphology down-glacier of the immediate confluence zone is more clearly explained by reference to the "ablation-dominant" model, since debris septa contributing to the lateral moraine load of the arms are formed both above the firn line and below the firn line via crevasses.

(C) The "avalanche-type" model which accounts for medial moraines not formed at ice or firn junctions and with little or no englacial sediment supply, is regarded as a minor model. Such moraines may be quite transient features on a glacier surface.

Acknowledgements

The research on Berendon Glacier was financed by an Environment Canada, Water Research Incentives Grant (5043-5-5-103 1975). Granduc Operating Company of Vancouver provided generous base-camp support. Work on Austerdalsbreen benefitted from field discussion with Dr T.D. Douglas. Jan Williams assisted in the preparation of the manuscript. Charles Auger of Stewart, B.C., worked as field assistant on the project. The authors gratefully acknowledge all the above for their unfailing support.

References

Agassiz, L. 1840. *Études sur les glaciers.* Neuchâtel, H. Nicolet. 2 vols. [Translated by A.V. Carozzi, London, Hafner, 1967.]

Anderton, P.W. 1970. Deformation of surface ice at a glacier confluence, Kaskawulsh Glacier. (*In* Bushnell, V.C., *and* Ragle, R.H., *ed. Icefield Ranges Research Project. Scientific results, Vol. 2.* New York, American Geographical Society; Montreal, Arctic Institute of North America, p. 59–76.)

Battle, W.R.B. 1951. Glacier movement in north-east Greenland, 1949. *Journal of Glaciology*, Vol. 1, No. 10, p. 559–63.

Bishop, B.C. 1957. Shear moraines in the Thule area, northwest Greenland. *U.S. Snow, Ice and Permafrost Research Establishment. Research Report* 17.

Boulton, G.S. 1967. The development of a complex supraglacial moraine at the margin of Sørbreen, Ny Friesland, Vestspitsbergen. *Journal of Glaciology*, Vol. 6, No. 47, p. 717–35.

Brecher, H.H. 1969. Surface velocity measurements on the Kaskawulsh Glacier. (*In* Bushnell, V.C., *and* Ragle, R.H., *ed. Icefield Ranges Research Project. Scientific results. Vol. 1.* New York, American Geographical Society; Montreal, Arctic Institute of North America, p. 127–43.)

Charlesworth, J.K. 1957. *The Quaternary era, with special reference to its glaciation.* London, Edward Arnold.

Drewry, D.J. 1972. A quantitative assessment of dirt-cone dynamics. *Journal of Glaciology*, Vol. 11, No. 63, p. 431–46.

Eyles, N. 1976. Morphology and development of medial moraines: comments on the paper by R.J. Small and M.J. Clark. *Journal of Glaciology*, Vol. 17, No. 75, p. 161–62, 164–65. [Letters.]

Eyles, N. Unpublished. Medial moraines as part of a glacier debris system, their formation and sedimentology. [M.Sc. thesis, Memorial University, Newfoundland, Canada, 1976.]

Fisher, D.A., *and* Jones, S.J. 1971. The possible future behaviour of Berendon Glacier, Canada—a further study. *Journal of Glaciology*, Vol. 10, No. 58, p. 85–92.

Forbes, J.D. 1859. *Occasional papers on the theory of glaciers.* Edinburgh, Adam and Charles Black.

Glen, J.W. 1956. Measurement of the deformation of ice in a tunnel at the foot of an ice fall. *Journal of Glaciology*, Vol. 2, No. 20, p. 735–45.

Glen, J.W., *and* Lewis, W.V. 1961. Measurement of side-slip at Austerdalsbreen, 1959. *Journal of Glaciology*, Vol. 3, No. 30, p. 1109–22.

Hannell, F.G., *and* Ashwell, I.Y. 1959. The recession of an Icelandic glacier. *Geographical Journal*, Vol. 125, Pt. 1, p. 84–88.

Heim, A. 1885. *Handbuch der Gletscherkunde.* Stuttgart, J. Engelhorn.

Hess, H. 1907[a]. Die grösse des jahrlichen Abtragres durch Erosion in Firnbecken des Hintereisferners. *Zeitschrift für Gletscherkunde*, Bd. 1, Ht. 3, p. 355–56.

Hess, H. 1907[b]. Über der Schuttinhalt der Innenmoränen einiger Oetztaler Gletscher. *Zeitschrift für Gletscherkunde*, Bd. 1, Ht. 2, p. 287–92.

Huxley, T.H. 1857. Observations on the structure of glacier ice. *London, Edinburgh and Dublin Philosophical Magazine and Journal of Science*, Fourth Ser., Vol. 14, No. 93, p. 241–60.

Ives, J.D., *and* King, C.A.M. 1955. Glaciological observations on Morsárjöktull, S.W. Vatnajökull. Part II: regime of the glacier, present and past. *Journal of Glaciology*, Vol. 2, No. 17, p. 477–82.

King, C.A.M. 1959. Geomorphology in Austerdalen, Norway. *Geographical journal*, Vol. 125, Pts. 3–4, p. 357–69.

King, C.A.M., *and* Lewis, W.V. 1961. A tentative theory of ogive formation. *Journal of Glaciology*, Vol. 3, No. 29, p. 913–39.

Knighton, A.D. 1973. Grain-size characteristics of superglacial dirt. *Journal of Glaciology*, Vol. 12, No. 66, p. 522–24. [Letter.]

Kozarski, S., *and* Szupryczński, J. 1971. Ablation cones on Sidujökull, Iceland. *Norsk Geografisk Tidsskrift*, Bd. 25, Ht. 2, p. 109–19.

Krenek, L.O. 1958. The formation of dirt cones on Mount Ruapehu, New Zealand. *Journal of Glaciology*, Vol. 3, No. 24, p. 310, 312–14.

Lewis, W.V. 1940. Dirt cones on the northern margins of Vatnajökull, Iceland. *Journal of Geomorphology*, Vol. 3, No. 1, p. 16–26.

Lister, H. 1958. Glaciology (3): glacial prehistory or the evidence of debris. *(In* R.A. Hamilton, *ed. Venture to the Arctic.* Harmondsworth, Penguin Books Ltd., p. 200–09.)

Loomis, S.R. 1970. Morphology and ablation processes on glacier ice. *(In* Bushnell, V.C., *and* Ragle, R.H., *ed. Icefield Ranges Research Project. Scientific results. Vol. 2.* New York, American Geographical Society; Montreal, Arctic Institute of North America, p. 27–31.)

Martonne, E. de. 1925–26. *Traité de géographie physique. Quatriéme ed.* Paris, Armand Colin. 2 vols. [Seventh edition, Paris, 1948.]

Nielsen, L.E., *and* Post, A.S. 1953. The Castner Glacier region, Alaska. *Journal* of *Glaciology*, Vol. 2, No. 14, p. 276–80.

Rapp, A. 1960. Talus slopes and mountain walls at Tempelfjorden, Spitsbergen: a geomorphological study of the denudation of slopes in an Arctic locality. *Norsk Polarinstitutt. Skrifter*, Nr. 119.

Ray, L.L. 1935. Some minor features of valley glaciers and valley glaciation. *Journal of Geology*, Vol. 43, No. 3, p. 297–322.

Rogerson, R.J., *and* Eyles, N. Unpublished. The form of medial moraines, their sediments, and terminal ice melt studies, Berendon Glacier, B.C. [Report to Glaciology Division, Environment Canada, 1976.]

Salisbury, R.D. 1894. Superglacial drift. *Journal of Geology*, Vol. 2, No. 6, p. 613–32.

Sharp, R.P. 1948. The constitution of valley glaciers. *Journal of Glaciology*, Vol. 1, No. 4, p. 182–89.

Sharp, R.P. 1949. Studies of the supraglacial debris on valley glaciers. *American Journal of Science*, Vol. 247, No. 5, p. 289–315.

Sharp, R.P. 1960. *Glaciers.* Eugene, Oregon, University of Oregon Press.

Small, R.J., *and* Clark, M.J. 1974. The medial moraines of the lower Glacier de Tsidjiore Nouve, Valais, Switzerland. *Journal of Glaciology*, Vol. 13, No. 68, p. 255–63.

Small, R.J., *and* Clark, M.J. 1976. Morphology and development of medial moraines: reply to comments by N. Eyles. *Journal of Glaciology*, Vol. 17, No. 75, p. 162–64. [Letter.]

Streiff-Becker, R. 1954. The initiation of dirt cones on snow; comments on J.W. Wilson's paper. *Journal of Glaciology*, Vol. 2, No. 15, p. 365–66, 367.

Swithinbank, C.W.M. 1950. The origin of dirt cones on glaciers. *Journal of Glaciology*, Vol. 1, No. 8, p. 461–65.

Tarr, R.S., *and* Martin, L. 1914. *Alaskan glacier studies.* Washington, D.C., National Geographic Society.

Thorarinsson, S. 1960. Glaciological knowledge in Iceland before 1800: a historical outline. *Jökull*, Ár 10, p. 1–18.

Tyndall, J. 1872. *Forms of water. Second edition.* London, C. Kegan Paul.

Untersteiner, N., *and* Nye, J.F. 1968. Computations of the possible future behaviour of Berendon Glacier, Canada. *Journal of Glaciology*, Vol. 7, No. 50, p. 205–13.

Ward, W.H. 1955. Studies in glacier physics on the Penny Ice Cap, Baffin Island, 1953. Part IV: the flow of Highway Glacier. *Journal of Glaciology*, Vol. 2, No. 18, p. 592–99.

Ward, W.H. 1961. Experiences with electro-thermal ice drills on Austerdalsbre, 1956–59. *Union Géodésique et Géophysique Internationale. Association Internationale d'Hydrologie Scientifique. Assemblée générale de Helsinki, 25–7—6–8 1960. Commission des Neiges et Glaces*, p. 532–42. (Publication No. 54 de l'Association Internationale d'Hydrologie Scientifique.)

Wilson, J.W. 1953. The initiation of dirt cones on snow. *Journal of Glaciology*, Vol. 2, No. 14, p. 281–87.

Young, R.A. 1953. Some notes on the formation of medial moraines. *Jökull*, Ár 3, p. 32–33.

17

A THEORETICAL MODEL OF GLACIAL ABRASION

B. Hallet

Source: *Journal of Glaciology* 23(89) (1979): 39–50.

Abstract

Preliminary results of a quantitative model of glacial abrasion are presented. The analysis, which is constructed within a framework of modern glaciological views of processes near to the bed, is aimed at modeling abrasion under a temperate glacier whose basal layers contain only occasional rock fragments. It does not simulate abrasion by debris-rich ice or by subglacial drift. Calculations of abrasion-rates reduce to evaluations of the forces pressing rock fragments against the glacier bed and of the rates at which they are moved along the bed. The estimated viscous drag induced by ice flow toward the bed due to basal melting is generally the dominant contribution to this contact force. Although the analysis shares several important elements with the pioneering study of Boulton ([°1974]), sufficient fundamental differences in the modeling lead to distinctly different conclusions. Several new results are noteworthy: (1) other parameters being equal, abrasion will tend to be fastest where basal melting is most rapid, (2) glacier thickness does not affect abrasion through its influence on basal pressures, and (3) lodgement of rock fragments is only possible if the sliding velocity is very low, equivalent to the rate of basal melting.

Introduction

Erosion by glaciers is conventionally attributed to at least three basic mechanisms: abrasion, quarrying, and subglacial fluvial activity. For glaciers in calcareous terrains, subglacial dissolution of the bed is also a significant erosive mechanism, which often leaves a clearer imprint on the glacier bed than abrasion (Hallet, 1976). There are numerous discussions of the relative importance of these erosional mechanisms in the literature (for summary see

Flint, [°1971]; Andrews, [°1975]; Vivian, 1975), but they all suffer from a lack of experimental data.

A fundamental understanding of the mechanics of glacial abrasion is essential for the study of erosion by glaciers. The intent of this paper is to present preliminary results of a quantitative model of glacial abrasion constructed in the context of modern glaciological views of processes occurring close to the bed. This model shares several important elements with the pioneering and comprehensive study made by Boulton ([°1974]). He discusses the principal facets of glacial erosion and presents observations and theoretical considerations of the abrasion process and its dependence on the size distribution of debris in basal ice. In the present paper, a more complete view is developed of abrasion by single particles lodged in the ice; I arrive at quite different conclusions because there are fundamental differences in the modeling, particularly in the evaluation of forces between rock fragments and the glacier bed. This analysis models the abrasion under temperate glaciers with basal layers which contain only occasional rock fragments, as is often observed in subglacial cavities and as is reflected by the common scarcity of debris on extensive rock surfaces exposed by glacial retreat. There are, however, numerous examples of glaciers with considerable debris in basal ice (Boulton, [°1974]) and even with a subglacial drift (Kamb and others, 1979); the model does not simulate abrasion by such glaciers because it neglects particle interactions and the effect of debris on sliding.

It has long been recognized that glacial abrasion is due to rock fragments forced against, and dragged along, the glacier bed by sliding ice. The rate of abrasion depends primarily on the effective force with which individual fragments are pressed against the bed, the flux of fragments over the bed, and the relative hardness of rocks in the ice and of the bed. In the model presented here, the rate of abrasion is taken to be proportional to the flux of rock particles that make contact with the bed, and to the particle–bed contact forces. Effects of relative rock hardnesses and the geometry of contact areas are incorporated in an effective constant of proportionality which is to be determined empirically. The analysis yields estimates of the rate of abrasion for a glacier bed of arbitrary, but low, roughness by rock fragments of a particular size. The more realistic case, in which rock fragments which vary widely in size abrade the bed, could be modeled by considering the effective abrasion-rate due to each narrow size range of particles and by summing the individual rates to obtain a total as in the study of Boulton ([°1974]). The interdependence of the concentration of fragments dragging on the bed and the rate of basal sliding and, hence, the rate of abrasion will be discussed in a subsequent paper.

In the abrasion process, striations are formed by the motion of rock asperities that locally deform and fracture the rock of the glacier bed and remove the debris. The deformation and fracture of rock requires large stress differences between the contact point and areas immediately adjacent

565

to it. Where abrasion occurs, rock fragments are likely to be completely enveloped by pressurized ice or water except where they contact the bed. This implies that the basal pressure does not affect the stress differences which are important in the striation process, because it contributes equal stresses to contact areas and to the adjacent zones. The effective force of contact is then independent of basal pressure and, hence, of the glacier thickness; it is only dependent on the buoyant weight of rock particles and on the viscous drag induced by ice flow towards the glacier bed due to basal melting and longitudinal straining of basal ice. For particular rock types and concentrations of rock particles in the ice, calculations of the abrasion-rates reduce to evaluations of the effective force pressing particles against the bed and of the particle velocity. These assumptions render this model markedly different from and considerably more realistic than the analysis of Boulton ([°1974]).

After a brief introduction to the quantitative characterization of glacial abrasion, the analysis presented here proceeds from an evaluation of the contact force between rock fragments and the glacier bed to an estimate of the rate at which rock fragments are transported along the bed, and then to an expression for the rate of abrasion. Finally, the abrasion of a simple sinusoidal bed is discussed to shed light on the relative rates at which bed irregularities of different sizes would disappear under sustained abrasion.

Quantitative characterization of glacial abrasion

The rate of abrasion of the bed depends on the flux of particles in contact with the bed, on the force with which rock particles press against the bed, and on the relative hardness of the bed and the abrasive particles. There-fore, the simplest general expression for the rate of abrasion $\dot{A}$ of any portion of the bed, is

$$\dot{A} = \alpha D F^n, \tag{1}$$

where α and n are empirical constants, D is the number of particles dragging along the bed per unit time and unit length perpendicular to the ice flow direction, and F is the effective force pressing particles against the bed. Simple experiments to simulate the abrasion of a rock surface by single rock fragments suggest that Equation (1), which is similar to that used by Boulton ([°1974]), is valid. These, and similar experiments, can provide numerical values for α and n that depend on the hardnesses, and hence on the lithologies of the striator and the rock which is being striated, and on the shape of the striator point. Reliable and realistic values of these constants are, however, difficult to determine empirically because the experimental results are likely to depend sensitively on the experimental apparatus. Rock-to-rock friction experiments designed to study faulting (e.g. Dieterich, 1972), show that accumulations and sudden releases of stress, which are responsible for the stick–slip behavior, are sensitively dependent on the effective stiffness of the apparatus as well as on the pressure, temperature, and rates of displacement. Because such behavior is probably related to the fracturing, chipping, and gouging of striated rocks, realistic experiments to simulate abrasion should be conducted under conditions representative of those at the glacier bed, using fragments embedded in temperate ice moving at reasonable rates. A few experiments of this type have been conducted by Boulton ([°1974]) in subglacial cavities and have yielded very interesting data on rates of abrasion. More such experiments ought to be conducted in a variety of subglacial situations, and special care ought to be given to determining the force with which particles press against the bed. The results of Mathews (1979) will be difficult to apply to actual abrasion by temperate glaciers because his experiments were conducted at temperatures well below zero, the sliding velocities were very high, and the confining pressures were low.

The particle flux D is the product of the particle velocity v_p and the number of rock asperities protruding from the basal ice per unit area of the glacier sole, a measure of the rock concentration C_r. Thus, the average thickness of rock removed by abrasion per unit time can be expressed as

$$\dot{A} = \alpha C_r v_p F^n. \tag{2}$$

All parameters in this expression are, in general, functions of position along the bed. In the analysis, expressions for v_p and F will be obtained for particles of specific sizes. The abrasion-rate due to each particle-size fraction can be obtained using the surface concentration of particles of that size range. The total abrasion-rate can then be obtained by summing individual rates.

Contact force between fragments and the glacier bed

The density difference between rocks and ice leads to a downward force on rock fragments equal to their buoyant weights. Basal melting, as well as longitudinal extension of basal ice, causes the ice to flow toward the bed, thereby moving fragments to the ice–rock interface and forcing them against the bed. An effective viscous drag is exerted on these particles as temperate ice moves around them by deformation and regelation. Where basal freezing occurs along lee surfaces, the ice flows away from the bed and decreases the force pressing particles against the bed; it could also lift particles from the ice–rock interface.

The drag on a particle, as temperate ice moves around it by regelation as well as by deformation, is analogous to the drag on a sphere moving slowly through temperate ice, a problem studied by Watts (unpublished). He shows that the drag force per unit cross-sectional area $\bar{\sigma}$ on a rigid isolated sphere forced through temperate ice, with Newtonian viscosity, is

$$\bar{\sigma} = \frac{4\eta R v_r}{(R_*^2 + R^2)},\tag{3}$$

where η is the effective viscosity of the ice, v_r is the velocity of the particle relative to the ice, R is the radius of the spherical particle, and R_* is a parameter, analogous to the transition wavelength in the bed-slip theory, which is referred to as the critical radius at which the drag per cross-sectional area is greatest. The motions of smaller and larger spheres through temperate ice are made easier by melting–regelation and viscous deformation, respectively. For rock particles, R_* is typically about 11 cm (Watts, unpublished). It is noteworthy that the drag expression reduces to $4\pi\eta R v_r$ for a large sphere when regelation is ineffective ($R_*^2 \ll R^2$). Stokes' law for the viscous drag on a slowly moving sphere yields a similar result, but $1\frac{1}{2}$ times as large because, unlike the temperate-ice case, a no-slip boundary condition is imposed on the sphere surface.

The evaluation of the force exerted on a particle against the glacier bed as ice flows toward the bed is a problem which is similar to, but much more complicated than, that of drag on an isolated particle forced through an infinite body of temperate ice. Morris (1979) recently showed that the analogous problem of the motion of a rigid cylinder near a rigid boundary through temperate ice cannot be solved with the boundary conditions conventionally used in glacier sliding and other regelation problems. Moreover, additional complications arise near to the bed because ice flow around particles is affected by steep velocity gradients associated with active sliding over a rough bed, the irregular shape of rock fragments in basal ice, and the likelihood of their being in close proximity to, and hence interfering with, other rocks. Nevertheless, it appears reasonable to assume, as a first approximation, that for basal ice with a fairly low debris concentration, the drag on a fragment at the glacier bed will be of the same order of magnitude as that on an isolated fragment forced through temperate ice with a velocity equal to the mean velocity around the fragment in basal ice. Making this approximation, the viscous drag on particles at the glacier bed can be expressed by Equation (3).

More refined calculations do not seem warranted, especially in view of the fact that even for the simplest problem involving regelation—the motion of wires forced through temperate ice—the drag predicted by the same theory that leads to Equation (3) can differ greatly from observed values (Drake and Shreve, 1973).

Using Equation (3), the force with which a rock fragment presses against the bed can be approximated as

$$F \approx \frac{4}{3}\pi R^3(\rho_r - \rho_i)g\cos\theta + \frac{4\pi\eta R^3}{R_*^2 + R^2}v_n.\tag{4}$$

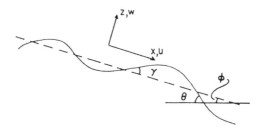

Figure 1 Coordinate system. In a vertical cross-section parallel to the ice flow, the gently-undulating glacier bed slopes down-valley at a mean angle ϕ from the horizontal.

The fragment is taken to be equivalent to a sphere of radius R, ρ_r and ρ_i, are the densities of rock and ice, g is the gravitational acceleration, v_n is the velocity component of the ice normal to the bed (positive toward the bed). The coordinate system used in these calculations is shown in Figure 1. The local down-glacier inclination of the bed θ depends on its overall slope ϕ averaged over tens of meters, as well as on the local slope γ, which is equal to $\tan^{-1} dz/dx$, of the rough surface. θ can be characterized as

$$\theta = \phi - \gamma = \phi - \tan^{-1} \frac{dz}{dx}, \tag{5}$$

where $z(x)$ is the two-dimensional glacier bed profile. The first term in Equation (4) represents the buoyant weight of the fragment and the second is the approximate force contribution due to ice flowing toward the bed with a velocity v_n. The ice velocity normal to the bed can be expressed as

$$v_n = w_{re} + w_0 + w_s, \tag{6}$$

where w_{re}, w_0, and w_s are the normal velocity components due respectively to regelation sliding, uniform melting (resulting from geothermal heating and sliding friction), and longitudinal straining. Assuming that, for ice with low debris content, glacier sliding is not appreciably affected by the rock fragments in the ice, w_{re} can be expressed in terms of the velocity components parallel to and perpendicular to the overall trend of bed, u and w, as obtained in glacier-sliding theories. In terms of the formulation due to Nye (1969),

$$w_{re} = -w \cos \gamma + u \sin \gamma, \tag{7}$$

where γ is the deviation of the local bed slope from the overall trend of the glacier bed. Geothermal heating and sliding friction are assumed to cause fairly uniform melting of the basal ice; reasonable estimates of the resulting melting-rate are on the order of 10 to 100 mm a^{-1} for normal glaciers (Weertman, 1969).

In addition to the plastic deformation associated directly with glacier sliding, basal ice may, in general, be straining on a large scale. The mean convergence rate of ice toward the bed due to large-scale uniform longitudinal extension of basal ice is

$$w_s = L \frac{dU}{dx} \cos \gamma \tag{8}$$

(assuming that the ice is incompressible and that the deformation is two dimensional), U is the mean sliding velocity and L is the distance from the bed. For this abrasion problem a suitable L is on the order of the radius of the abrading fragments.

Particle velocity

A particle embedded in basal ice will move when the viscous drag due to ice flowing past the particle at a velocity v_r together with the down-slope component of the buoyant weight exceeds the resistive frictional drag due to the fragments rubbing against the bed, that is, when

$$\frac{4\pi\eta R^3}{R_*^2 + R^2} v_r + \frac{4}{3}\pi R^3 (\rho_r - \rho_i)g \sin \theta \geq \mu F, \tag{9}$$

where μ is the effective coefficient of friction. Once in motion a rock fragment will tend to move at a rate v_p such that the equality in Equation (9) is satisfied; the viscous forces urging the particle forward will then be exactly compensated by the frictional drag of the particle on the bed. The velocity component of the ice tangential to the bed is

$$v_t = u + w \sin \gamma. \tag{10}$$

The following equation for the particle velocity is derived from Equation (9), noting that $v_r = v_t - v_p$,

$$v_p = v_t - [\mu F - \frac{4}{3}\pi R^3 (\rho_r - \rho_i) \sin \theta] \frac{(R_*^2 + R^2)}{4\pi\eta R^3}. \tag{11}$$

A general expression for the abrasion-rate can be obtained using Equations (2), (4), and (11) together with values of the ice velocity components locally tangential and normal to the bed, v_t and v_n respectively. These can be readily obtained using Nye's bed-slip theory and Equations (6), (7), (8), and

(10). Consider a temperate glacier sliding over an arbitrary bed whose profile parallel to the ice flow is defined as

$$z(x) = \varepsilon \sum_{n=1}^{\infty} a_n \sin nk_0 x, \qquad (12)$$

where ε is a dimensionless number much smaller than one which is used to ensure that the roughness of the bed is sufficiently low for the theory to be valid, a_n are the Fourier amplitudes of the bed harmonics, and k_0 is the wave number. Assuming a Newtonian viscosity for the ice, the velocity component in the x direction, parallel to the overall trend of the bed is to first order (Nye, 1969, P. 454)

$$u = U + \varepsilon U z \sum_{n=1}^{\infty} a_n \beta_n (nk_0)^2 \exp(-nk_0 z) \sin nk_0 x, \qquad (13)$$

where U is the mean sliding velocity. The velocity away from the bed, the z component, is

$$w = \varepsilon U \sum_{n=1}^{\infty} a_n \beta_n (nk_0) \exp(1 + nk_0 z) \exp(-nk_0 z) \cos nk_0 x, \qquad (14)$$

where

$$\beta_n = \frac{k_*^2}{k_*^2 + (nk_0)^2},$$

k_* being the transition wave number, a fundamental material constant whose value is characteristically around 10 m^{-1} using a value for the effective viscosity of the ice of 10^5 Pa a (1 bar a).

In as much as the ice velocities vary with distance from the bed, the rate of ice flow around particles will not be uniform. A formal treatment of the drag on a particle as straining temperate ice moves past it does not, however, seem warranted. Instead, as a first approximation, the viscous drag is considered as equivalent to that due to ice flow with uniform velocity past a spherical inclusion, using a representative value of the ice velocity averaged over an ice thickness equal to the particle diameter. It is more convenient, however, to evaluate simply u and w at $z = R$, rather than to use the spatial average i.e.

$$\bar{w} = \frac{1}{4R^2} \int_0^{2R} \int_0^{2R} w(x, z) \mathrm{d}x \, \mathrm{d}z,$$

but this could be done with little difficulty in the numerical model, discussed later.

The concentration of particles in the ice C_r as a function of position along the bed is not known. However, on intuitive grounds, it is expected to be highest where the flow of debris-laden ice toward the bed is fastest. One could, therefore, postulate that

$$C_r = C_0 + Cv_n, \tag{15}$$

where C_0 is the average debris concentration at the glacier sole and C is a proportionality constant dependent, in part, on the volumetric concentration of particles in the basal ice.

As can be seen from Equation (11), an increase in the contact force F reduces the velocity v_p of abrading particles. In fact, given a sufficiently large F, particles would be lodged against the bed, abrasion could cease, and rock debris would be deposited on the bed. Therefore, as was clearly recognized by Boulton ([°1974]), abrasion-rates cannot be assumed to increase monotonically with increasing particle–bed contact forces.

The flow of ice toward the bed pressing rock fragments against the bed is generally the most important factor controlling the contact force. Figure 2 shows the relative magnitude of each of the components contributing to the

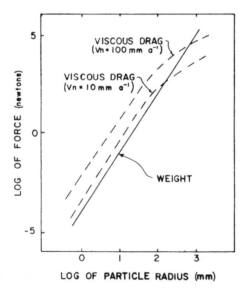

Figure 2 Relative importance of the buoyant weight and the viscous drag, which control the effective contact force between fragments and the bed. The viscous-drag contribution is shown for two relatively low values of the basal melting-rate. The effective viscosity of the ice was taken to be 10^5 Pa a (1 bar a).

contact force. Except for abrading rock fragments exceeding 0.2 m in diameter, the viscous-drag contribution is greater than the contribution due to the buoyant weight of fragments and basal pressures, even for very modest values of the normal velocity components. The curves in Figure 2 were obtained assuming that ice flows toward the bed at 10 and 100 mm a^{-1}. Normal geothermal heating of a temperate glacier will by itself melt sufficient basal ice to induce a convergence of the ice and the bed of 10 mm a^{-1}; local rates several orders of magnitude larger than this can be associated with sliding over small obstacles (<0.5 m) on the glacier bed.

No direct data are available on actual fragment–bed contact forces. They can, however, be estimated from considerations of the widths of striations which reflect the size of the contact areas between rock fragments and the bed. The size of contact areas is in turn indicative of the contact force because, despite the wear on the fragments, actual contact is likely to be established only over small areas just large enough to support the load and to resist crushing of the asperities. The contact force will, therefore, tend to be about equal to the product of the crushing strength of the rock fragment and the area of contact. The widths of over a hundred striations were measured on specimens of Paleozoic limestone from the vicinities of Green Bay, Wisconsin, U.S.A. and Athabasca Glacier, Alberta, Canada. The results, shown in Figure 3, suggest that the majority of fragments were in light contact with the bed. The

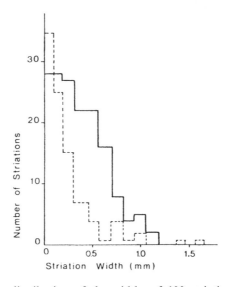

Figure 3 Frequency distribution of the widths of 100 striations on each of two limestone slabs from glaciated carbonate terrains. These reflect small areas of contact and, hence, relatively small contact forces between rock fragments and the glacier bed. The solid and dashed curves correspond respectively to samples from the vicinity of Green Bay, Wisconsin, U.S.A. and of Athabasca Glacier, Alberta, Canada.

wider striations are one millimetre wide; this width corresponds to a contact area of about $F/\sigma_0 = 1$ mm^2 and to a contact force of 10^2 N, using a crushing strength representative of limestones: $\sigma_0 = 10^2$ MPa (10^3 bar) (Handin, 1966, p. 273–74). This force is small but reasonable as it is equivalent to the buoyant weight of a rock particle, with a radius of about 10 cm, immersed in ice. Deep erosional gouges and particularly arcuate cracks on glaciated rock surfaces indicate much higher contact forces (Johnson, unpublished) presumably associated with the motion of very large rock fragments.

The model is suitable for calculating the abrasion-rate at every point of an arbitrary surface of low roughness and could, therefore, be used to simulate glacial abrasion and its spatial variability; starting with an initial surface, its evolution under abrasion could be simulated. One would start by defining or measuring the initial topography, Fourier analyzing it, computing the abrasion-rate at every point, and then displacing each point to represent an episode of abrasion, finally iterating through time.

An important motivation for modeling abrasion of rock surfaces is to understand better how the bed roughness changes when subjected to glacial erosion. This question is of particular importance in the context of glacier sliding and it has recently been addressed by Johnson and others (1976) on the basis of spectral power analyses of several profiles measured on former glacier beds. The rates at which bedforms of different sizes wear down through time, which control the evolution of the bed-roughness spectrum, can most readily be analyzed by simulating abrasion over simple sinusoids of different sizes.

Abrasion of sinusoidal beds

The velocity components tangent and normal to the bed, v_t and v_n respectively, are the main parameters needed in order to evaluate abrasion-rates. With a low-slope approximation (i.e. $\cos \gamma = 1$ and $\sin \gamma \approx \tan \gamma = dz/dx$) and the bed surface described by $z = a \sin k_0 x$, we use the u and w velocity components of Nye (1969) to get

$$v_t = U\left\{1 + \frac{ak_*^2 k_0^2}{k_*^2 + k_0^2} \exp\left(-k_0 R\right)[R \sin k_0 x + a(1 + k_0 R) \cos^2 k_0 x]\right\},$$

(16)

and

$$v_n = Uak_0 \cos k_0 x \left\{\frac{k_*^2}{k_*^2 + k_0^2} \exp\left(-k_0 R\right)[Rak_0^2 \sin k_0 x - (1 + k_0 R)] + 1\right\}$$

$$+ w_0 + R\frac{dU}{dx}.$$

(17)

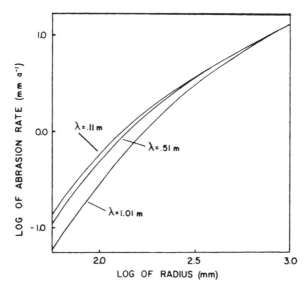

Figure 4 Dependence of the abrasion-rates on the effective radius of fragments and on the wavelength λ of bed undulations. The fragment concentration was taken to be one per square metre independent of size. No correction was made for the tendency for small particles to be more numerous than larger ones. The ice velocity and effective viscosity were 100 m a^{-1} and 10^5 Pa a (1 bar a).

To obtain a velocity which is representative of the average velocity around the particles of radius R, the velocities were calculated for a height above the bed of $z = R$. A numerical scheme was used to derive these velocities from which basis the contact force, the particle velocity, and finally the abrasion-rates were obtained. Figure 4 shows the computed abrasion-rates for different fragment sizes. The rate represents an average of rates calculated near the tops of the undulations, it is an intermediate value representative of the rate of abrasion of the entire surface. The abrasion coefficient α was taken to be on the order of 10^{-10} Pa^{-1} and $n \approx 1$; these values were based on the preliminary results of rough experiments in which conical limestone samples striated saw-cut slabs of the same limestone. Because the model is limited to abrasion by sparse debris in the basal ice, the concentration of fragments projecting from the glacier sole was chosen arbitrarily to be one per square meter. This is probably equivalent to about one-tenth of the effective debris concentration at Glacier d'Argentière, as deduced from an abrasion experiment that Boulton ([°1974], p. 467) conducted there. An aluminum plate fixed to the bed under the glacier, which is sliding at ≈ 250 m a^{-1}, was traversed and deeply striated by about 15 fragment asperities over a width transverse to the flow direction of ≈ 60 mm during a 30 d period. With the possibility that some fragments may contact the bed simultaneously at as many as three asperities, it appears reasonable that about ten fragments actually traversed and contacted the aluminum plate. Assuming that the mean particle velocity is approximately equal to the sliding velocity, the flux of fragments and their concentration are 3 mm^{-1} a^{-1} and 10 m^{-2} respectively.

The rate of abrasion at a point does not determine the rate at which bedforms change in shape through time as a result of abrasion. Rather, a comparison is needed between the abrasion-rates calculated at the crest and at the troughs of bed undulations. The damping coefficient D, defined as the crest rate minus the trough rate normalized with the amplitude of the bed undulation, is a useful measure of the rate at which bedforms would disappear under abrasion. When $D = 0$ the abrasion is uniform, and the bedform is preserved; when $D > 0$ sustained abrasion wears down bed irregularities. As shown in Figure 5, D decreases with decreasing particle size and with increasing wavelength, except for very small particles and wavelengths. Hence, large rock fragments are most effective in wearing down bedrock bumps because the troughs suffer little wear. In fact, the rock fragments can be so large that they cannot make contact with the troughs, in which case only the

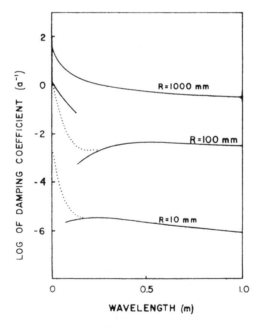

Figure 5 The rate at which irregularities in the glacier bed are damped by abrasion, as a function of the radius of the particles and the wavelength of irregularities parallel to the ice flow direction. The dotted lines suggest how the damping actually varies in the wavelength range where the numerical model yields an unrealistic discontinuity (see text). Parameters used in calculations are as indicated for Figure 4.

crests will be abraded. The discontinuities in the damping curves (Fig. 5) are unrealistically abrupt. However, they do represent the transition between abrasion by large particles, which tends to be restricted to the crests of bed undulations of relatively short wavelengths, and abrasion by small particles, which can take place over the entire surface of larger bed undulations. Dotted lines in Figure 5 were drawn to suggest how the damping actually varies with wavelength.

Conclusions

This model yields interesting new insights into the abrasion process, its controls, and its spatial variability. The viscous drag, induced by ice flow towards the bed due to basal melting, is generally the dominant process forcing particles against the bed. The glacier thickness appears to have no influence on abrasion through its control of basal pressures because they do not affect the differences in stress at or near the fragment–bed contact areas, yet it is these differences that control the rock deformation and fracture which are responsible for abrasion. For large particles ($R > 0.5$ m) the buoyant weight can contribute very significantly to the fragment–bed contact force.

The absolute rates of glacial abrasion can only be estimated roughly because few appropriate data are available to calculate abrasion coefficients and the concentration and size distribution of debris in basal ice. The spatial variability of abrasion can be calculated more reliably; it is such that irregularities of the glacial bed would invariably tend to disappear under sustained abrasion. Irregularities that are relatively short in the direction of ice flow tend to be particularly vulnerable to abrasion by large fragments (Fig. 5). The roughness of the glacier bed can only be maintained if other subglacial erosional processes effectively roughen the bed. The patterns of melting and freezing associated with regelation sliding are bound to affect greatly the patterns of abrasion because the flow of ice towards the bed appears to be the primary control of the contact force and, hence, of the abrasion-rate. For relatively low contact forces, abrasion ought to be strongest on surfaces facing up-glacier, particularly near their crest, and weakest along lee surfaces. This leads to the characteristic truncation of bed obstacles (Fig. 6) and to maximum abrasion near corners as observed by Boulton ([°1974]). Inasmuch as both the particle–bed contact force and the force urging particles along the bed result principally from viscous drag, the speed at which particles will be dragged along the bed depends simply on the relative magnitude of the ice velocity components parallel and perpendicular to the bed. Assuming, for simplicity, that the coefficient of rock-to-rock friction is about one, the force along the bed on a particle would about equal the particle–bed contact force. This implies that the relative velocity between the sliding ice and the rock ought to be approximately equal to the

Figure 6 A heavily abraded surface truncates a more steeply-inclined lee surface. Note contrasting abrasion character across the distinct break in slope. Abrasion is most pronounced where basal melting is fastest. Former ice flow direction is indicated by a 30 cm ruler.

melting-rate. Therefore, particles ought to remain in motion, entrained in the ice, except when the sliding velocity slows to a low value equal to the melting-rate (10 mm a^{-1}). This conclusion is not in accord with Boulton's ([°1974], p. 52) suggestion that particles could lodge against a bed at sliding velocities up to 100 m a^{-1}. His results apparently stem from a considerable overestimate of the particle–bed contact force which he views as depending dominantly upon the "effective normal pressure" around the fragments at the glacier base. If particles start to lodge, they would effectively increase the roughness of the glacier bed, thereby inducing continued lodgement. Deposition of lodgement till may, therefore, represent a fairly short-lived event when sliding velocity slows to rates equivalent to the mean basal melting-rate, perhaps as in the waning phase of a glaciation.

On the larger scale, in areas where the geothermal heat flow or frictional heating are high or where the ice is extending, both the contact forces and the particle concentrations would tend to be relatively high. For a given mean particle velocity, abrasion ought to be relatively rapid in these areas. On intuitive grounds, Lliboutry (1975) suggests that over-deepenings characteristic of glacier valleys are due to erosion by ice heavily laden with debris in these areas. However, if the debris concentration becomes too high in the basal ice, the sliding would be impeded and the flux of particles against the bed would be decreased, thereby slowing abrasion. Accordingly, relatively high sliding velocities as well as high debris content must be responsible for continued excavation of valley over-deepenings. It is stressed that the greater ice thickness in these areas would not favor lodgement of debris against the bed nor inhibit active abrasion there.

This analysis has been further developed to examine the interesting interdependence of the sliding velocity, the concentration of fragments projecting from the glacier sole, and the abrasion-rates. The resistance to glacier sliding induced by rock fragments per unit area of the bed is simply the product of the frictional resistance of one fragment moving against the bed μF and the surface concentration of particles in the basal ice C_r. Even for low debris concentrations, this rock-to-rock frictional resistance is significant compared to the normal range of shear stress at the base of temperate glaciers. These results will be presented in a subsequent paper.

Acknowledgements

I am pleased to acknowledge the competent assistance of Robert Anderson and Brian Aubry in several facets of this work. I gained a great deal from discussing glacial abrasion with them, as well as with Charles Raymond, Joseph Walder, Ray Fletcher, Chester Burrous, and Robert Rein. This manuscript was completed while I was visiting the Quaternary Research Center at the University of Washington in Seattle. I greatly appreciate the technical assistance I received from the Center, and particularly from Judy Hartman. Support for this work was provided by National Science Foundation Grant EAR77-13631.

References

Andrews, J.T. [^{c}1975.] *Glacial systems. An approach to glaciers and their environments.* North Scituate, Mass., Duxbury Press. (Environmental Systems Series.)

Boulton, G.S. [^{c}1974.] Processes and patterns of glacial erosion. (*In* Coates, D.R., ed. *Glacial geomorphology.* Binghamton, N.Y., State University of New York, p. 41–87. (Publications in Geomorphology.))

Dieterich, J.H. 1972. Time-dependent friction in rocks. *Journal of Geophysical Research*, Vol. 77, No. 20, p. 3690–97.

Drake, L.D., *and* Shreve, R.L. 1973. Pressure melting and regelation of ice by round wires. *Proceedings of the Royal Society of London*, Ser. A, Vol. 332, No. 1588, p. 51–83.

Flint, R.F. [^{c}1971.] *Glacial and Quaternary geology.* New York, etc., John Wiley and Sons, Inc.

Hallet, B. 1976. Deposits formed by subglacial precipitation of $CaCO_3$. *Geological Society of America. Bulletin*, Vol. 87, No. 7, p. 1003–15.

Handin, J. 1966. Strength and ductility. (*In* Clark, S.P., *jr.*, ed. *Handbook of physical constants.* New York, Geological Society of America, Inc., p. 223–90.)

Johnson, C.B. Unpublished. Characteristics and mechanics of formations of glacial arcuate abrasion cracks. [Ph.D. thesis, Pennsylvania State University, 1975.]

Johnson, C.B., *and others.* 1976. Glacier bed roughness: spectral analysis of two terrains, [by] C.B. Johnson, J. Melosh, and [W.] B. Kamb. *Eos. Transactions, American Geophysical Union*, Vol. 57, No. 12, p. 1000–01.

Kamb, W.B., *and others*. 1979. The ice–rock interface and basal sliding process as revealed by direct observation in bore holes and tunnels, by [W.] B. Kamb, H.F. Engelhardt, and W.D. Harrison. *Journal of Glaciology*, Vol. 23, No. 89, p. 416–19.

Lliboutry, L.A. 1975. Loi de glissement d'un glacier sans cavitation. *Annales de Géophysique*, Tom. 31, No. 2, p. 207–25.

Mathews, W.H. 1979. Simulated glacial abrasion. *Journal of Glaciology*, Vol. 23, No. 89, p. 51–56.

Morris, E.M. 1979. The flow of ice, treated as a Newtonian viscous liquid, around a cylindrical obstacle near the bed of a glacier. *Journal of Glaciology*, Vol. 23, No. 89, p. 117–29.

Nye, J.F. 1969. A calculation on the sliding of ice over a wavy surface using a Newtonian viscous approximation. *Proceedings of the Royal Society of London*, Ser. A, Vol. 311, No. 1506, p. 445–67.

Vivian, R.A. 1975. *Les glaciers des Alpes Occidentales*. Grenoble, Imprimerie Allier.

Watts, P.A. Unpublished. Inclusions in ice. (Ph.D. thesis, University of Bristol, 1974.]

Weertman, J. 1969. Water lubrication mechanism of glacier surges. *Canadian Journal of Earth Sciences*, Vol. 6, No. 4, Pt. 2, p. 929–42.

18

A MODEL FOR SEDIMENTATION BY TIDEWATER GLACIERS

R.D. Powell

Source: *Annals of Glaciology* 2 (1981): 129–34.

Abstract

Sampling of sediment from the fjord floor in front of tidewater glaciers in Glacier Bay, Alaska, has provided information about processes in this restricted glacimarine (Dreimanis 1979) setting. Sediment sampling, in conjunction with oceanographic and glacial dynamics data, has also enabled the discrimination of sediment types and their facies associations. Deposits are strongly controlled by: sea-water characteristics, position and sediment discharge of melt-water streams, iceberg calving, and rate of glacierfront retreat.

Five distinct facies associations have been found to reflect glacier-fjord regimes. The facies associations and ice-fjord conditions responsible for them form the basis for constructing a preliminary model for glacimarine sedimentation by tidewater glaciers. The model can be used to predict (i) rapid retreat of an actively calving ice front, (ii) slow retreat or stabilization of a calving ice front at a channel constriction, (iii) stabilization of a melting (very rarely calving) ice front when the glacier base is near tidewater elevation, and (iv) large outwash delta progradation into a fjord when the ice front retreats onto land. This model can be used to interpret facies associations found in a stratigraphic record.

Introduction

This study presents results from data collected by grab sampling and gravity coring in front of 11 glaciers in Glacier Bay National Monument, southeast Alaska. Five distinct facies associations reflecting different glacier-fjord regimes have been distinguished, and formulated into a sedimentary facies model based on processes operating in a temperate tidewater glacimarine (Dreimanis 1979) setting.

Glacimarine lithofacies

The major lithofacies are described below.

Morainal bank lithofacies is composed of a chaotic mixture of diamicton, gravel, rubble, and sand built into a large ice-contact bank during slow glacier-front retreat. Sediment is contributed from calving icebergs, en- and subglacial melt-out, and subglacial streams. Small morainal banks (*push moraines*) of gravel, rubble, and diamicton are formed by very minor winter advances of an ice front. Such deposits are preserved and are mainly identifiable in a regime where the ice front undergoes rapid summer retreat. *Isolated piles of gravel and rubble* are contributed to mud and reworked diamicton on the fjord floor from bergs calving from a rapidly retreating glacier front and, to a lesser degree, from bergs overturning in the more distant iceberg zone.

Diamicton lithofacies may be produced by several processes: subglacial and melt-out till deposition, sediment gravity flows from fjord floor "highs", and the mixing of silt and clay (from melt-water streams) and coarse-grained sediment (from numerous dirty icebergs). When icebergs are less numerous and cleaner, the silt and clay contain a lower proportion of coarse-grained sediments and an *iceberg-zone mud* lithofacies is produced. When a glacier front is terrestrial, and not contributing icebergs with a coarse-grained sediment fraction to the sea, a *marine-outwash mud* lithofacies is produced.

Underflows may be generated from subglacial melt-water discharges if sediment loads are sufficiently high; they carry away sand and form thin laminae within mud which is deposited during periods of low stream discharge when the underflows are suppressed. The underflow will gradually dissipate as momentum and density decrease, finally rising buoyantly to become an interflow or overflow. Laminae of fine- to medium-grained sand (mm thick), having sharp top and bottom contacts, are produced in a rhythmic layering with mud when the site is influenced by underflows. Thinner laminae, one or two grains thick, of very fine-grained sand and silt are deposited from interflows and underflows distant from the efflux. The resultant lithofacies is *laminated sand and mud*.

Sediment gravity flows and *turbidity current channels* are commonly produced from morainal banks and large outwash deltas. Deltas form as melt-water streams enter the sea in both lateral and central positions from the ice face. *Central deltas* may form under both fast and slow rates of ice retreat, and can produce *esker-type forms* when ice retreats rapidly. The largest *lateral deltas* build across the ice face and down-fjord when the base of the ice is near tidewater. When the glacier front has entirely retreated from the sea, *outwash deltaic and braided stream* lithofacies are produced.

Large tidal ranges and high mud concentrations within the sea are conducive to *tidal-flat* formation (some are ephemeral, being destroyed by winter erosion).

The sediment in different glaciers in Glacier Bay is derived predominantly from acidic intrusives and metasediments and is generally similar mineralogically. The degree of comminution by glaciers is therefore assumed constant so that particle-size distributions are primarily a function of depositional processes and comparable between different glaciers.

Some of the lithofacies, such as diamicton, iceberg-zone mud, and laminated sand and mud, are common to three or more facies associations. Such facies associations are defined by lithofacies combinations and their relationships. Other lithofacies such as morainal banks, marine outwash mud, and tidal-flat mud are unique, and therefore define a facies association.

Fjord glacimarine facies associations

Facies Association I: Facies of rapidly retreating tidewater glaciers with ice fronts actively calving in deep water (Fig. 1)

The following description of facies associations is based on findings from Muir Glacier and inferences made from other glaciers studied. Facies deposited close to such an ice front consist of reworked subglacial till, subglacial stream gravel and sand, scattered and dumped coarse-grained supraglacial debris, and low icepush moraines. Farther away from the ice

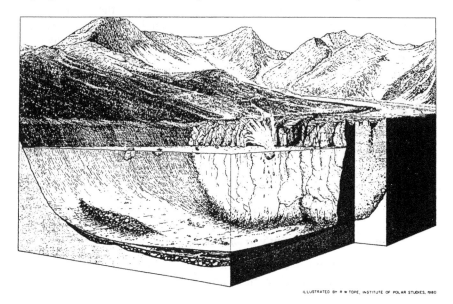

ILLUSTRATED BY R W TOPE, INSTITUTE OF POLAR STUDIES, 1980

Figure 1 Rapidly retreating tidewater glacier actively calving in deep water. Sediment facies shown are Facies Association I.

front, iceberg-zone mud accumulates and intertongues with the interbedded sand and mud lithofacies.

Ice calves actively during rapid summer ice retreat, but markedly less actively during winter. Water is deep (hundreds of meters), so wind-generated water waves will not influence sediment on the fjord floor, and icebergs calving from above sea-level and associated turbulence will not influence the fjord floor.

Subglacial till is exposed on the fjord floor and may be reworked by tidal currents, stream underflows, and sediment gravity flows. Therefore, a newly exposed fjord floor has a thin gravel and sand layer due to minor reworking of compacted subglacial till. A small amount of glacial basal debris may melt out at the fjord floor against the ice face.

If substantial supraglacial debris is dumped off the ice as it calves, it will remain as piles of rubble and gravel on the fjord floor. The area close to the ice face is a highly turbid brackish-water environment devoid of organisms except for many shrimp and rare polychaetes.

Subglacial melt-water streams introduce gravel and sand into the marine environment directly at the base of the water column. If highly sediment-charged, they produce laminated sand and mud lithofacies. Gravel and sand may remain as an ice-moulded ridge (an esker-type form), but large fans will not be produced because of the mobile ice front. Mud of the iceberg zone accumulates in the ice-proximal basin, but, if retreat is rapid, sufficient turbulence is produced within the water column to maintain mud in suspension accompanied by transport away from the ice front.

During winter months, when calving slows drastically, the ice front may remain in one position or even advance (meters or hundreds of meters). Stream activity decreases and fine-grained sediment settles quietly out. Unlike the lacustrine environment, the more open system in a fjord may remove such clays. The most characteristic sediments are small subaqueous end moraines deposited against the ice face, approximately normal to the fjord walls. Sediment incorporated in the ridges can originate in two ways: (i) melting of debris-rich basal ice by the water, or (ii) ice-push. The former may be important, but the rate of ice melting decreases from summer because of the decrease ($\approx 2°C$ for Muir Inlet) in bottom sea-water temperature. The result would be ridges similar in morphology to De Geer, washboard or crossvalley moraines.

Facies Association II: Facies of a slowly retreating tidewater glacier with ice front actively calving predominantly in shallow water (Fig. 2)

Retreat of an ice front decreases dramatically or may virtually cease at a channel constriction created by fjord-floor highs or wall protuberance, but calving can still continue at a relatively rapid rate. The resulting Facies Association II consists of ice-proximal coarse grained morainal banks with ice-contact lateral and central fan deltas. Subaqueous sediment-gravity flows

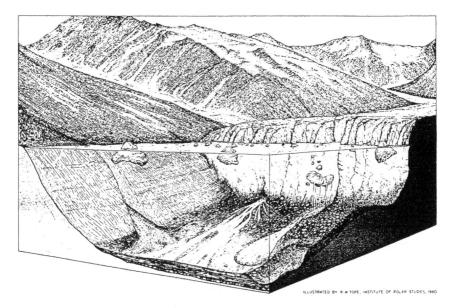

ILLUSTRATED BY R W TOPE, INSTITUTE OF POLAR STUDIES, 1980

Figure 2 Slowly retreating tidewater glacier actively calving in shallow water. Sediment facies shown are Facies Association II.

are common down the bank foreslope producing intertonguing sand layers within iceberg-zone mud deposited more distally (e.g. Riggs, Grand Pacific, Margerie, Lamplugh, and Johns Hopkins glaciers).

Under such conditions, subaqueous morainal banks in an ice-contact position are constructed of coarse-grained clastic sediment (Fig. 2). Sediment is contributed by melting out of basal and englacial debris and also by dumping of supraglacial debris as ice calves. Diamicton textures may be present from the melting-out process, especially when the bank is initially constructed in relatively deep water. As the bank builds into shallower water it comes under a higher energy regime created by calving events, tidal currents, and, perhaps, wind-generated water waves. This yields a lag of sand- to cobble-size sediment on the bank surface with rare pockets of diamicton texture in depressions, which may form from subaqueous sediment gravity flows such as debris flows and slumps. Rarely, a diamicton is formed by mud from overflows from melt-water streams, combined with ice-rafted coarser material dropped from numerous bergs floating in front of the ice face. Subglacial till may be plastered on to the bank during minor winter ice-front advances.

Subglacial subaqueous streams discharge laterally and centrally from the ice face and build deltas or fans on the bank surface. Lateral deltas build along the ice face and down fjord walls. Stream discharges and sediment concentrations are occasionally high enough to form underflows that transport sand into the pro-deltaic environment. Gravel is rolled down the fore-slope of deltas built by such high discharge streams.

Other lateral deltas are formed under lower energy conditions by the accumulation of mud on the fore-slope due to flocculation and sedimentation from the turbid overflow plumes produced as the stream enters the sea-water and remains buoyantly at the sea surface. Rarely, interflows may form if the sea-water has a marked pycnocline. More often, interflows are produced by the sinking of the turbid overflow of an ebb tide beneath the inflowing water of the following flood tide. When deltas of this type are adjacent to an ice front with supraglacial debris, a diamicton is produced by coarser-grained material contributed from icebergs.

Occasionally, central subglacial streams have sufficiently high sediment concentrations to produce underflows that transport sand away from the ice front while gravel is left near it. Structureless sand and inter-laminated sand and mud lithofacies are produced.

These central streams can build ice-contact subaqueous fans on the bank surface and the surface of the fan may eventually become intertidal if the stream stays in one position for several melt-water seasons and melts a channel vertically up the ice face. Often, however, the position of a subglacial-stream discharge changes from one melt-water season to the next and a series of small overlapping fans may form along the bank in front of the ice.

Slumping and sediment gravity flows are common down the bank fore-slope (angles ≤20°). Sloping fjord walls direct slumps into a central trough which is elongated and parallel to the walls, and extends farther away from the ice face as turbidity-current channels develop from the slumps. Slumping may originate from high sedimentation rates and consequent slope over-steepening, or from calving events when bergs reach or come very close to the bank surface. Sediment is finer-grained down the fore-slope.

In more distant areas from the ice front, iceberg-zone mud accumulates at a rapid rate ($\geq$4.4 m a^{-1}) in the ice-proximal fjord-floor basin. This mud laps onto the morainal bank, and pro-bank sand layers (turbidites or underflow deposits) intertongue with the iceberg-zone mud facies. The thickness of the mud on the floor of a fjord basin is determined by the length of time that the ice remains at the head of the basin and by the volume of the basin. Most mud is deposited in the ice-proximal basin whereas down-fjord basins receive much smaller amounts of mud. Therefore, each isolated basin will have its own level of mud infill, unless it is full in which case mud is transported across the sill into the next basin by sediment-gravity flows.

Facies Association III: Facies deposited by a slowly retreating or advancing tidewater glacier rarely calving into shallow water (Fig. 3)

When ice fronts retreat or advance into shallow water they may end in a protected bay environment. Resulting Facies Association III consists of iceberg-zone mud deposited close to the ice front away from stream

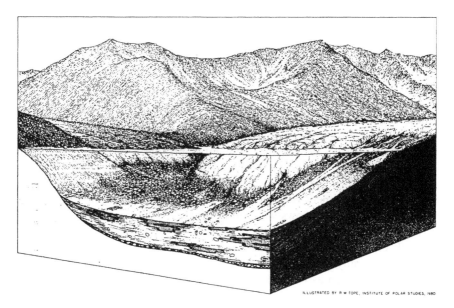

ILLUSTRATED BY R W TOPE, INSTITUTE OF POLAR STUDIES, 1980

Figure 3 Slowly retreating tidewater glacier rarely calving in shallow water. Sediment
facies shown are Facies Association III.

discharges and in more distant areas. Lateral streams dominate the environment close to the ice front, introducing gravel and sand over large fan
deltas, and the sand intertongues with mud in more distant locations (e.g.
McBride, Lituya, and North Crillon glaciers).

If this protected embayment occurs at the head of a fjord and the glacier
stabilizes or perhaps stagnates, then ice loss by surface melting may be of
similar or greater magnitude than calving. This is the opposite situation to
that described previously for other facies associations.

Very large lateral fan deltas build along the ice face and down-fjord. A
glacier can then override these deposits and advance. Discharges from the
deltas and from subaqueous subglacial streams introduce sand-size material
to the mud facies. Structureless fine- to coarse-grained sand beds are produced near subaqueous discharges. This sand changes laterally into muddy
sand and finally into interlaminated sand and mud. This latter interlaminated
sediment is also common in pro-deltaic areas of lateral discharges and gravel
rolls down the lateral delta fore-slopes.

Farther away from the ice front, the flows introduce occasional laminae
of fine-grained sand. Thicker sand layers in more distal positions exhibit
normal grading and resemble Bouma sequences of unit A turbidites.
X-radiographs of distal mud show interflow deposits of thin laminae of
very fine-grained sand and silt. Temporal variations in particle size, due to
fluctuations in discharge and sediment loads of melt-water streams from

587

where the plumes originate, produce these laminae. Iceberg-zone mud has varying amounts of ice-rafted debris (IRD) with generally less than 10% pebble-sized clasts, especially in more distal fjord floor basins. Any supraglacial debris is dumped in piles on the fjord floor mud close to the ice front. If ice loss is dominated by surface melting rather than calving and the ice is clean of debris (Reid Glacier), then very little IRD may be produced and better-sorted mud accumulates. If the ice has large volumes of supraglacial debris (Carroll Glacier), a diamicton texture with clasts comprising at least 10% of the sediment is produced in areas away from stream discharges. These conditions of dominance of melting rather than calving and large volumes of supraglacial debris would the most favorable of all environments described for production of debris flows from the ice, although no large flows were observed.

In protected basins with shallow entrance sills, black layers are produced, possibly on a seasonal basis, due to the occurrence of periodic anaerobic conditions. Generally, however, black layers and patches occur irregularly with depth but are often associated with turbidite layers.

Facies Association IV: Facies of a turbid outwash fjord (Fig. 4)

When the front of a glacier retreats from the sea and becomes terrestrial, large outwash delta deposits build up and prograde out into the fjord. The

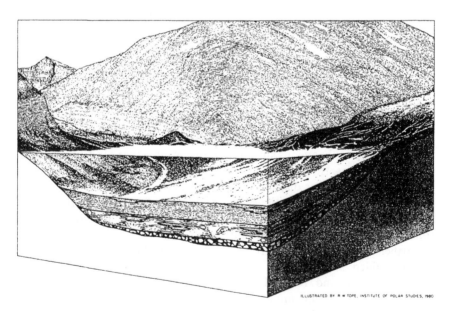

ILLUSTRATED BY R W TOPE, INSTITUTE OF POLAR STUDIES, 1980

Figure 4 Turbid outwash fjord where the ice front is terrestrial. Sediment facies shown are Facies Association IV.

resulting Facies Association IV consists of coarse-grained fluvial deposits on the delta surface and has sand on its foreslope. This sand intertongues with marine outwash mud (glacial flour silt and clay with very minor coarse-grained component) in areas more distant from the ice front.

The outwash delta progrades out into the fjord producing similar facies to the deltas formed in Facies Associations II and III. If high peak discharges occur and large quantities of fine-grained sediment are available on the delta plain for fluvial transport, continuous underflows may be produced from the melt-water stream discharges as they enter fjord water during high summer flow. As the delta progrades, coarse-grained glacifluvial sediment on the delta surface is deposited unconformably over the delta foreset sands and gravels and signifies the late stages of glaciation in an area.

Mud is the dominant sediment type in more distal areas. Very little IRD is present within the mud as only very few and small bergs are introduced via melt-water streams over the outwash delta. The resultant sediment is marine outwash mud which consists of glacial flour silt and clay from glacial melt-water streams. This marine outwash mud may be cut by channels (e.g. Queen and Rendu inlets) floored with rhythmically bedded silt and very fine-grained sand, interpreted as turbidites. The erosional-depositional turbidity current channels are cut by the underflows produced by melt-water streams discharging from the outwash delta, or from slumps on the delta fore-slope.

Facies Association V: Facies of shallow-water environments distant from ice fronts

During deposition of Facies Associations I to IV, tidal flats commonly develop in shallow water. Facies Association V consists of tidal-flat muds and braided stream and beach sands (e.g. Dundas Creek sections and Dundas Bay tidal flats).

Tidal flats develop on coarse-grained clastic morainal sediment or glacimarine diamicton as the land is uplifted isostatically during ice retreat. If a site is distant from the ice, the tidal environment has abundant life that leaves its remains in the mud. Icebergs may ground on the flats during ebb tide and contribute coarser-grained clastic sediment. Farther offshore, diamicton accumulates and may contain barnacles attached to clasts dropped from temporarily stranded icebergs (e.g. at Ptarmigan Creek). Winter sea-ice may also freeze-in beach pebbles which are rafted away during break-up.

With continued isostatic recovery, the flats may become subaerially exposed, and shrubs and, later, trees will become established. If the ice readvances, this land will again be drowned by the sea, the forest will die, and stumps will remain to be buried during renewed mud deposition (e.g. Dundas Bay).

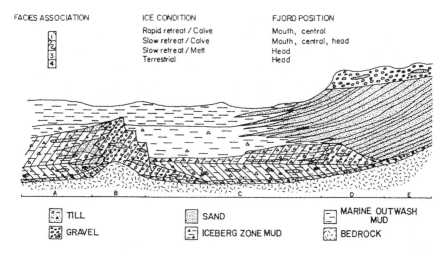

FACIES ASSOCIATION ICE CONDITION   FJORD POSITION

Rapid retreat / Calve Mouth, central
Slow retreat / Calve Mouth, central, head
Slow retreat / Melt   Head
Terrestrial   Head

TILL   SAND   MARINE OUTWASH MUD

GRAVEL   ICEBERG ZONE MUD BEDROCK

Figure 5 Hypothetical section of glacimarine sediment showing general sedimentary facies and facies associations.

Interpretation of a hypothetical section

A hypothetical sediment section has been drawn (Fig. 5) using the model above. At interval A, the ice front ended in deep water and retreated rapidly by calving. Subglacial till deposited beneath the ice was exposed to the sea and slightly reworked. Facies Association I sediment was deposited. The ice front retreated to B where it was slowed by a channel constriction but the front still calved actively. Facies Association II was deposited with build-up of a morainal bank. Eventually the ice front retreated from the constriction into deeper water (C). Ice retreat was again rapid and Facies Association I was once more deposited. Eventually, the ice front reached the head of the fjord (D) and ice retreat slowed. At first, the front still calved actively to produce Facies Association II sediment, but eventually the main method of ice loss changed from calving to surface melting and Facies Association III was deposited. Finally, at E, the ice front became completely terrestrial and a large outwash delta built out over all previous facies, and Facies Association IV sediment accumulated.

Conclusion

This paper presents a tentative sedimentation model using lithofacies produced under a tidewater glacimarine regime which can be used to interpret older glacimarine sequences by application of Walthers' Law. It should also be useful for predicting conditions and deposits in other glacimarine regimes. Not all lithofacies or facies associations described will be present, especially in ancient sequences, and some may be present by themselves. Companion papers presenting basic data and more detailed field descriptions are in preparation (also see Powell, unpublished).

Acknowledgements

This study was carried out at the Institute of Polar Studies (IPS), Ohio State University (OSU), under the guidance of Dr Ken Stanley. The author wishes to thank US Geological Survey scientists, Drs P. Carlson, B. Molnia, and A. Post, for information and the National Park Service for logistics. Funding was provided by IPS, the Arctic Institute of North America, the Geological Society of America, and Sigma Xi and the Orton Hall Fund of OSU.

References

Dreimanis A. 1979. Commission on genesis and lithology of Quaternary deposits (INQUA). *Boreas* 8(2): 254

Powell R.D. Unpublished. Holocene glacimarine deposition by tidewater glaciers in Glacier Bay, Alaska. (Ph D thesis, Ohio State University, 1980)

19

SEDIMENT DEFORMATION BENEATH GLACIERS

Rheology and geological consequences

G.S. Boulton and R.C.A. Hindmarsh

Source: *Journal of Geophysical Research* 92(B9) (1987): 9059–82.

Abstract

Experiments beneath Breidamerkurjökull in Iceland have led to development of flow laws for the subglacial till, relating strain rate to shear stress and effective pressure and assuming either Bingham fluid or nonlinearly viscous fluid behavior. Water pressures in the till are less than ice pressures and it is suggested that this may lead to infiltration of ice into the sediment, which inhibits sliding at the ice/sediment interface. Where water pressures are equal or near to ice pressures, infiltration does not occur and sliding may result. A one-dimensional theory of subglacial deformation is developed in which the empirical flow law is coupled with a model of subglacial hydrology and consolidation. This predicts stable states in which subglacial sediment either does not deform or a dilatant deforming horizon forms with positive effective pressures at the ice/bed interface or unstable states where zero or negative effective pressures are predicted. Time dependent analyses show that response times following perturbations of the system may be of the order of one thousand years and thus that unsteady behavior may be normal on glaciers flowing over unlithified sediment beds. It is suggested that the natural variability of material properties in subglacial sediment beds leads to the development of drumlins on the glacier bed. It is suggested that unstable deformation at zero or negative effective stress leads to "piping" in subglacial sediments at the glacier terminus and the growth of sediment-floored, subglacial tunnels. Their frequency is that which is sufficient to draw down subglacial water pressures so as to prevent unstable deformation. Where they discharge large water volumes, subglacial sediments flow laterally toward them producing "tunnel valleys." This sediment is then removed by water flowing along the axial tunnel. Tunnel valleys can be regarded as the equivalent in soft sediment areas of eskers in bedrock areas.

1. Introduction

It is easiest to describe the nature of glacier beds when the glaciers which formerly hid them have disappeared. The extensive areas uncovered by glacier retreat in the last hundred years demonstrate that at least the outer zones of these glaciers were predominantly underlain by unlithified sediment. The same is true of the mid-latitude areas covered during the last glacial period by the ice sheets of Europe and North America. Their central zones overlay a bedrock of ancient shields (20–30%) but much of the surrounding subglacial zones (70–80%) were composed of soft, unlithified or poorly lithified sediments of Quaternary or Tertiary age, resting on a variety of older, harder rocks. Though there are arguments for believing that a relatively larger proportion of hard bedrock may underlie most large modern glaciers [*Boulton and Jones*, 1979], these observations still lie in stark contrast to the assumption, underlying theoretical analyses of glacier/bed interactions used as a basal boundary condition for glacier flow over the last 30 years, of a hard, unyielding subglacial bed.

We believe that in areas where glaciers overlay soft sediment there is ubiquitous evidence that these sediments underwent strong shear deformation. In some areas, this is very easily demonstrated by obviously tectonic structures. However the more intensive the attenuation produced by shear strain, the less easy it is to identify. *Boulton* [1987] suggests that many laminated tills and some massive tills found in lowland regions invaded by the mid-latitude ice sheets are products of this process.

It has been argued by *Boulton and Jones* [1979] that if high water pressures are sustained in subglacial sediments by melting of the glacier sole, effective pressures in them may be sufficiently low to make the sediments much softer than ice, which can thereby play a fundamental role in glacier dynamics. Under these circumstances there will be a very strong interaction between the glacier and its sediment bed which will not only determine the behavior and response of the glacier to atmospheric conditions, but also the form, structure and dispersal of sediments.

The analysis of *Boulton and Jones* [1979] is a steady state analysis which does not incorporate a flow law for subglacial sediments. This reflects a general problem that whereas soil mechanics engineers are concerned with the mechanical behavior of sediments up to the point of failure, the sediment flow processes of most concern to geologists reflect behavior after failure. Quantitative sediment flow laws are difficult to derive from laboratory experiments because of the problems of sustaining steady conditions for large strains. However, shear deformation of sediment in a confined subglacial environment provides an ideal and relatively steady natural "shear box" test, provided it can be adequately monitored.

The aim of this article is to develop a sediment flow law from natural experiments on subglacial shearing; to establish how hydrological conditions develop which control the nature and rate of subglacial sediment deformation; and to examine some glaciological and geological implications of the inter-action between a glacier and its deforming sediment bed.

2. Experiments to determine the response of subglacial sediments to glacier movement

Breidamerkurjökull is a major southern outlet of Europe's largest icecap, Vatnajökull (Figure 1a). Extensive areas of subglacially deposited till have been revealed beyond the glacier margin by up to 4 km of retreat during this century. *Boulton et al.* [1974] observed a widespread two-layer structure in this till, of an upper horizon of high void ratio resting on a lower, denser horizon with a platy structure. It was suggested that this resulted from deformation of the upper layers of the till, an hypothesis subsequently con-firmed by experiment [*Boulton*, 1979]. A series of experiments undertaken in 1977, 1977/1978, 1980 and 1983 permit a relatively full description of the deformation process and its relation to glacier movement.

The 1980 experiment gave the fullest indication of subglacial conditions and is thus described in some detail. From tunnels excavated within the basal ice (Figure 1b) series of strain markers and oil-filled piezometers were inserted through small-diameter drill holes into the subglacial till [*Boulton*, 1979]. These access holes were then plugged and armoured piezometer capillaries fed into the tunnel through pressure seals in the plugs. The experiment lasted for 136 hours, during which time the piezometers and water pressures in nearby moulins were continuously monitored. The three-dimensional displacement of ice in the tunnels and on the glacier surface and the glacier surface ablation rate were measured at 6-hour intervals by surveying the movement of wooden pegs in the ice. At the end of the experi-ment, cores of the subglacial till were taken by drilling through the tunnel floor, water was pumped from the subglacial bed through a large access hole, a section was excavated in subglacial sediments and studied in detail, and the location of strain markers and piezometers noted (Figure 2).

The granulometry of the till, sampled to 1 m below the glacier sole, is shown in Figure 3. It is relatively constant throughout the section. There is, however, a sharp change in void ratio at between 0.4 m and 0.6 m below the glacier sole where values drop from 0.55–0.6 to 0.4–0.45 (Figure 4a). The upper, low-density horizon is relatively massive in appearance, and con-trasts strongly with the dense well-jointed till of the lower horizon. The joint planes are subhorizontal and frequently bear slickensides. The displacements

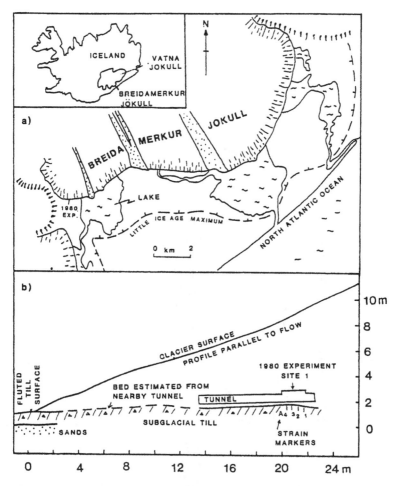

Figure 1 (a) The location of the experimental site at West Breidamerkurjökull. (b) The location of the access tunnel and site 1 in the 1980 experiment.

of strain markers show that relatively high strain rates occur within the upper, low-density horizon, and very small strain rates within the lower horizon. We refer to the upper horizon as the A horizon and the lower as the B horizon.

There appears to be a strong rheological contrast between the A and B horizons. In the former, deformation appears to occur by continuous deformation of the whole derived from the A horizon, had a volume of about 13 m³. From the extent of consequent tunnel floor collapse it was estimated that the extruded till was derived from an area of 30–40 m². We suggest that local increase of piezometric gradient during pumping produced an increase in seepage pressure sufficient to overcome the resistance of the till mass and

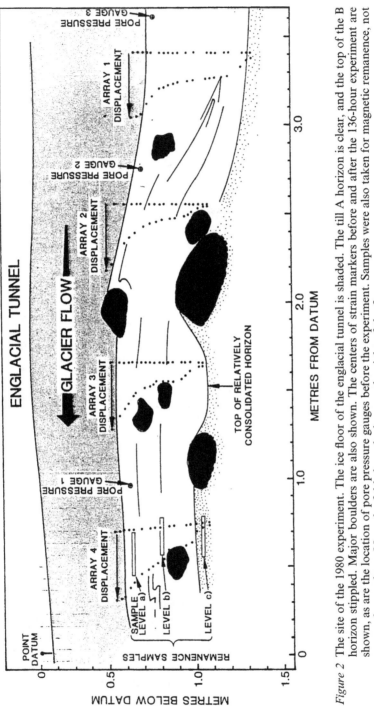

Figure 2 The site of the 1980 experiment. The ice floor of the englacial tunnel is shaded. The till A horizon is clear, and the top of the B horizon stippled. Major boulders are shown. The centers of strain markers before and after the 136-hour experiment are shown, as are the location of pore pressure gauges before the experiment. Samples were also taken for magnetic remanence, not reported upon here. Note the folded sand wisps on the right of the figure.

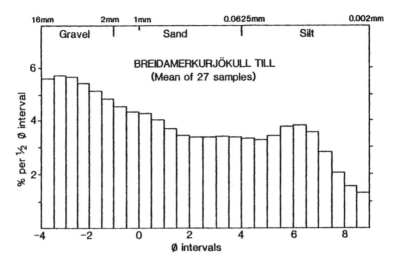

Figure 3 Granulometry of the till (A and B horizons) at the West Breidamerkurjökull site.

thus to produce "piping." After this experiment, remote determinations of strain were made using an auger with a detachable head (Figure 6). This was driven into the B horizon from an englacial tunnel. The wire holding the auger head into its housing was then unclamped and the auger rod withdrawn from the head. The wire from the head was then led through a wooden cylinder with internal and external pressure seals and connected to a reel and chart recorder. As the glacier sole moved in relation to the B horizon, the wire was pulled over the reel and its displacement recorded. The position of the B horizon was readily established by its resistance to auger penetration. Although the precise form of the velocity profile could not be determined, it was possible to measure the relative velocity between the glacier sole and the B horizon. Measurements of basal velocity u_b and of pore water pressure p_w in the A horizon and of pressures in nearby moulins during the 1980 experiment showed that u_b peaks lagged those of p_w, which in turn lagged behind the moulin pressure (see also Figure 5).

3. Flow laws for the Breidamerkurjökull till

The experiments conducted between 1977 and 1983 have yielded a series of estimates of pore water pressure, strain rates, and shear stresses in the deforming subglacial till. A first estimate of the rheological behavior of this till has been made. We have sediment mass, while the slickensided joints in the latter probably reflect discontinuous movement along well-defined planes. Several well-defined, acutely folded, sandy wisps occur in the A horizon. Because of their granulometric similarity to the sands which locally occur

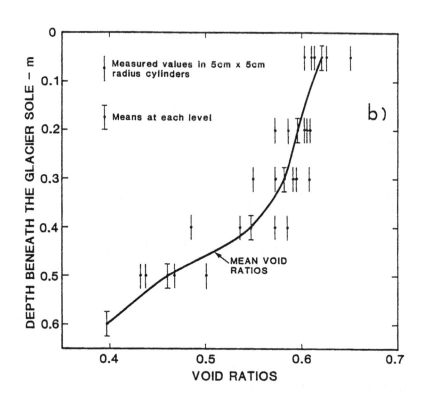

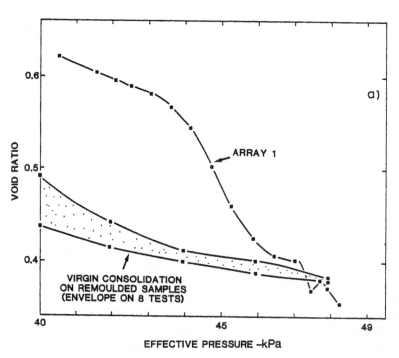

below the till, we suggest that these were folded into the till at some upglacier point where the till/sand interface deformed.

The displacement of the glacier surface immediately above the site during the period of the experiment was 0.42 ± 0.02 m. It is difficult to establish whether significant slip occurs at the glacier sole, though it seems theoretically unlikely because of the high roughness of the sediment surface on which numerous large spheroidal particles are found [*Boulton*, 1975]. Assuming this, the displacement due to sediment deformation is between 0.38 m and 0.35 m, between 95% and 80% of the forward movement of the glacier.

The geometry of deformation is complex. Between array 1 and array 2, the two dimensional discharge of deforming sediment parallel to glacier flow decreases, while between 2 and 3 and 3 and 4 it increases. While a small part of this is accounted for by a measured rise in the elevation of the glacier sole in the vicinity of array 2, much must be the product of net transverse extensional strain between arrays 1 and 2 and net transverse compressive strain between arrays 2 and 4. We suggest that this reflects enhanced flow around the large boulder at 2.0 m from datum. Measurements of pore pressure by gauge 2 appear to reflect this process. During the experiment it moved by about 0.30 m in a direction in which the base of the rapidly deforming horizon rose steeply against a large 0.5 m boulder (Figure 2). In the early part of the experiment pore water pressures measured at this piezometer show a rise followed by a later fall compared with the pressure fluctuations measured in others (Figure 5). We suggest that this reflects a normal stress increase caused by the movement of the glacier and the mobile A horizon against the rising surface of the B horizon. Initially this is borne by the pore water, producing an increased pore water pressure. The water then begins to drain towards the flanking low pressure areas and the till is able to consolidate and pore water pressures decrease (Figure 5).

In situ measurements of pore water pressure and measurements of the density (void ratio) of the till at the end of the experiment, but before pumping, permit us to assess the state of consolidation of the till. Figure 4b shows a plot of void ratio e versus effective pressure p_e for in situ samples, compared with the results of one-dimensional drained consolidation tests in the laboratory where the remoulded sample is permitted to consolidate to an equilibrium

Figure 4 (*opposite*) (a) Void ratios beneath the glacier sole near to site 1 (1980). The junction between A and B horizons lies at approximately 0.45 m below the glacier sole. (b) Virgin consolidation curve from remoulded samples of Breidamerkurjökull till (the test cell had a diameter of 20 cm and a height of 10 cm; all stones larger than 2 cm were removed) compared with the in situ state of consolidation beneath the glacier sole near array 1 (Figure 2) at the beginning of the experiment. The uppermost point on the latter curve is from a sample centered 5 cm beneath the glacier sole, and successive points are from samples spaced at 5 cm intervals. The difference between the two curves is an index of the degree of dilatation due to shear.

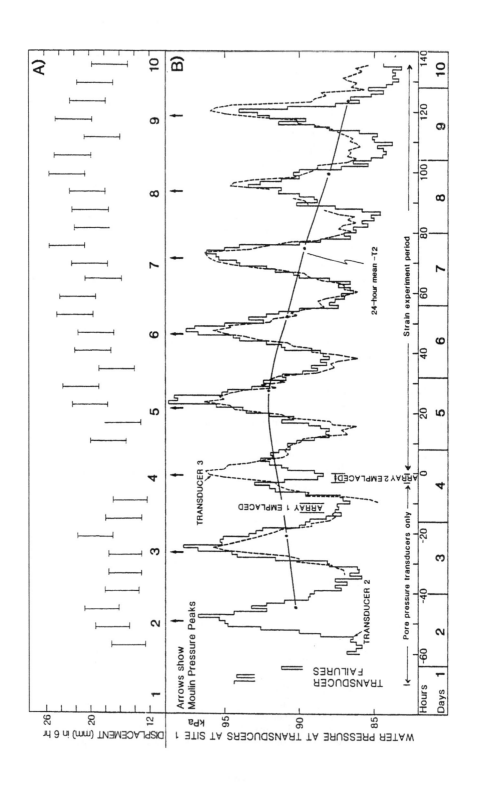

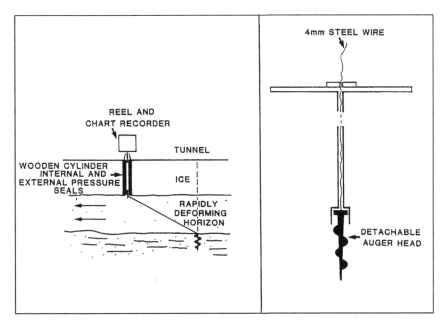

Figure 6 Use of the detachable head auger to determine mean strain rate in the A horizon.

state in relation to a series of effective pressure increments. The degree of underconsolidation, reflected in the difference in void ratio Δe between these two curves diminishes with depth in the A horizon. Samples from the B horizon appear to be slightly overconsolidated. This is reflected in the strain rates, which are greatest when Δe is large. Sampling problems did not permit a more precise relationship between Δe and strain rate to be established. The change in the state of consolidation at the junction between A and B horizons, from underconsolidated to overconsolidated, reflects a change in the response to stress, from deformation of the whole sediment grain skeleton to movement along well-defined planes in an otherwise resistant sediment mass.

Figure 5 (*opposite*) (a) Six-hour ice displacements (0000–0600, 0600–1200, 1200–1800, 1800–2400) measured in the tunnel floor immediately above array 2 (1980). (b) Pore water pressure record at transducers 2 and 3 (Figure 2). The whole strain array had been deployed at the initial time (4 P.M., day 4). Note the relatively steady values at transducer 3 and the rise and fall measured by transducer 2. This is interpreted as a rapid stress buildup at a rate greater than the drainage rate at transducer 2 as the A horizon moves over underlying boulders, leading to increased p_w. Subsequently p_w decreases as the drainage rate exceeds the rate of stress build up. Note the lag of 1–1.5 hours in water pressure peaks measured by the transducers behind those measured in moulins about 100 m from the site. The latter showed a 20–30 kPa pressure fluctuation.

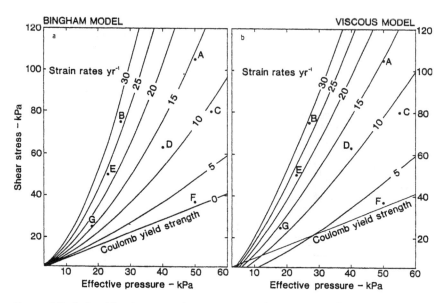

Figure 7 Relationships between shear stress, strain rate and effective pressure at Breidamerkurjökull, to which Bingham fluid (a) and non-linearly viscous fluid (b) models have been fitted. The Mohr-Coulomb failure criterion for the till has also been plotted. The letters indicate data points. Measured mean strain rates in yr^{-1} for these points are A, 14.17; B, 27.8; C, 9.35; D, 12.08; E, 24.58; F, 2.12; and G, 12.16.

The potential instability of the A horizon was revealed at the end of an experiment in 1980 in which an access hole to the subglacial bed had been constructed from an englacial tunnel. Half an hour's pumping had dried out the till near to the access hole, when a partially dried till mass was driven up through it and into the englacial tunnel, followed by a fluidized till mass. The expelled till, which we assume to have been plotted the average strain rate measured in the A horizon of the till; the average effective pressure ($p_e = p - p_w$) derived from measured pore water pressure p_w averaged through the period of the experiment and through the whole A horizon, and from calculated overburden pressure p, and the calculated average shear stress (Figure 7). We recognise that the most satisfactory way of obtaining a rheological relation is by direct mathematical modeling of the process and fitting to such a model rheological and other constitutive relations which describe the dilatant and hydrogeological properties of the material. This involves complex mathematical analysis which is best based on the results of simpler models such as that presented in this paper.

Two rheological models were fitted to this data, one of a nonlinear Bingham fluid the other a nonlinearly viscous fluid. In the former, we assume that the Coulomb failure criterion of $\sigma = c + p_e\eta$ (where σ is the yield strength, c cohesion, and η the internal friction coefficient) delimits the failure surface.

Observations of the till's behavior suggests that below the failure surface it behaves as an elastic solid and above, when $\sigma \geq c + \eta p_e$, as a viscous fluid. (For the Breidamerkurjökull till $c = 3.75$ kPa and $\eta = 0.625$); [Boulton at al., 1974].

Bingham fluid model (Figure 7a)

For a Bingham fluid, we assume that the strain rate will depend on the extent to which shear stress τ exceeds Coulomb strength σ, and on effective pressure. We define θ_B by $\theta_B = (\tau - \sigma)^n p_e^m$ and assume a rheological relation of the form $\dot{\varepsilon} = A_B \theta_B$, where $\dot{\varepsilon}$ is the strain rate and n, m and A_B are empirically determined constants. The best fit was obtained by the law

$$\dot{\varepsilon} = 7.62(p_e/10^5)^{-1.25}([\tau - \sigma]/10^5)^{0.625} \tag{1}$$

where $\sigma = p_e \eta + c$, in which $c = 3750$ Pa, $\eta = 0.625$, $\dot{\varepsilon}$ has units yr^{-1} and stresses are in Pa. The correlation coefficient $r = 0.986$, which must be partially due to the paucity of data points and the number of fitting parameters.

Nonlinearly viscous fluid model (Figure 7b)

In this model we assume that strain rate depends simply on τ and p_e. We define θ_v by $\theta_v = \tau^n p_e^m$ and assume a rheological relation of the form $\dot{\varepsilon} = A_v \theta_v$. The best fit was obtained by the law

$$\dot{\varepsilon} = 3.99(p_e/10^5)^{-1.80}(\tau/10^5)^{1.33} \tag{2}$$

where $\dot{\varepsilon}$ has units yr^{-1} and stresses are in Pa. The correlation coefficient $r = 0.987$.

These are important relationships, as flow laws for natural sediments have proved difficult to establish. They are applicable to materials of similar granulometry (Figure 3) not only beneath glaciers but in any geological situations where flow or creep can occur. The Bingham model is useful in that it incorporates a conventional Coulomb stability state which is an important descriptor of the mechanical behavior of many natural sediments. The nonlinearly viscous model is probably more useful in modeling in general and in particular for the long term behavior of many clay-rich sediments, where creep can occur at stresses below the Coulomb yield strength.

4. Interactions at the glacier/sediment interface: ice infiltration and sliding

The experiments at Breidamerkurjökull demonstrate unequivocally that the water pressure a few centimeters below the glacier sole is less than the ice pressure, and that there is little or no slip at the glacier/sediment interface. The force balance at the interface requires that

$$p_i = (1 - n)p_s + np_w$$

where p_i is the ice overburden, p_s the sediment pressure, p_w the water pressure, and n the porosity. If $p_i > p_w$, intergranular contacts will bear a greater proportion of the ice load so that $p_s > p_w$. The sediment pressure is the average pressure within the sediment grains, the averaging being carried out over a volume somewhat larger than the grain size. The pressure disequilibrium at the interface between ice and interstitial water will cause the ice to penetrate into interparticle voids thus achieving the necessary momentum transfer. A steady state can, however, be maintained by melting at the ice penetration front.

Use of Darcy's law with a typical ice viscosity predicts extremely small infiltration velocities; this of course does not include the effect of regelation, which must be very efficient around silt size particles. The driving stress will depend on the water pressure within the sediment; high permeabilities will let the ice penetrate faster and allow water to drain faster, leading to a higher driving stress. The infiltration velocity need not be equal to the basal melting rate in a steady state as internal melting within the basal layers may produce substantial meltwater [*Carol* 1947; *Robin* 1976; *Fowler* 1984].

Most tills are diamictons, and we expect those in which clasts dominate to transmit a large part of the applied stress through contacts between the large particles. Thus the ice load may be unequally partitioned between the large and small grains, thereby permitting ice to expel matrix from the voids between large clasts. We thus expect clast-dominated diamictons, such as that at Breidamerkurjökull, to behave like a gravel and to permit significant ice infiltration. Such infiltration has been observed at several subglacial sites and must inhibit sliding at the glacier/sediment interface, ensuring that movement occurs either in the subglacial sediment or in the ice or in both. However, even if there is no significant infiltration, we would expect the whole upper surface of the sediment to be in close contact with the ice when $p_w < p_i$, thereby offering shear resistance that will inhibit sliding unless the sediment grain size is very small [*Boulton*, 1975].

5. A one-dimensional model of subglacial sediment deformation

In analyzing the response of an unlithified material to glacier overriding we assume a stratum of given thickness and permeability overridden by a temperate glacier melting basally. If there is drainage into an underlying aquifer, there will be a potential difference across it. We arbitrarily set the value of the potential ψ_Q at the base of the stratum. We adopt a strict one-dimensional approach and thus do not permit water to be drained along the stratum horizontally. For cases where the stratum is underlain by an aquifer, it is easy to show that the bulk of the drainage will be vertical. When the stratum is underlain by an aquiclude, we are forced to assume no vertical potential gradient in order to be consistent with the assumption of one-dimensional flow.

If the shear forces imposed by the overriding glacier are strong enough to disturb the sediment structure, a dilatant A horizon will develop. In order to understand the way in which this stratum responds to changes in external conditions we have developed a vertical one-dimensional model of deformation within the system. In modelling its behavior we derive mass and momentum balances for both phases in association with constitutive relationships. The resultant nonlinear equations have four unknowns; water pressure, sediment pressure, porosity, and velocity of the mineral skeleton. Subscripts s and w refer to sediment and water phases, respectively. Quantities without subscription indicate total (mixture) values. We consider a glacier moving over a horizontal sediment surface which is fixed in space, and define a co-ordinate system (x,z) where x is horizontal in the direction of ice-sheet flow and z points upwards.

For continuity, the following mass balances must be satisfied:

For the water

$$\frac{\partial(n\rho_w)}{\partial t} + \frac{\partial(w_w n\rho_w)}{\partial z} = 0 \tag{3}$$

For the sediment

$$\frac{\partial[(1-n)\rho_s]}{\partial t} + \frac{\partial[w_s(1-n)\rho_s]}{\partial z} = 0 \tag{4}$$

where n is porosity, w the vertical velocity, ρ the density and t the time.

Remembering that w_w is the pore water velocity, the statement of Darcy's Law is

$$w_w = -\frac{k}{n\mu}\left\{\frac{\partial p_w}{\partial z} + \rho_w g\right\} + w_s \tag{5}$$

where K is the permeability in area units, μ the viscosity, p the pressure and g the acceleration due to gravity. In the derivation of these equations we apply mixture theory [Bowen, 1976]. Note that in our definition of Darcy's law we do not use the barycentric velocity but the physically more reasonable sediment velocity.

We introduce a hydraulic potential ψ defined by

$$\left.\begin{array}{c} \dfrac{\partial \psi}{\partial z} = \dfrac{\partial p_w}{\partial z} + \rho_w g \\[2mm] \psi(z = T) = p_i \end{array}\right\} \tag{6}$$

where T represents the top of the deforming layer (a function of time) and p_i is the ice overburden. We further suppose that $K = K(n)$ and that viscosity μ is constant.

The compressibilities γ are defined by

$$\gamma_w = \frac{1}{\rho_w} \frac{\partial \rho_w}{\partial p_w}, \quad \gamma_s = \frac{1}{\rho_s} \frac{\partial \rho_s}{\partial p_s} \tag{7}$$

By expanding (3) and substituting in (5) and (7) to remove ρ_w, ρ_s, and w_w, and ignoring the effect of vertical compression on Darcy Flow we obtain

$$n\gamma_w \frac{\partial p_w}{\partial t} + \frac{\partial n}{\partial t} - \frac{K}{\mu} \frac{\partial^2 p_w}{\partial z^2} - \frac{1}{\mu} \frac{\partial K}{\partial n} \frac{\partial n}{\partial z} \left\{ \frac{\partial p_w}{\partial z} + \rho_w g \right\}$$

$$+ n \frac{\partial w_s}{\partial z} + w_s \frac{\partial n}{\partial z} + w_s n \gamma_w \frac{\partial p_w}{\partial z} = 0 \tag{8}$$

For the case of no dilation ($\partial n/\partial t = 0$, $w_s = 0$) equation (8) reduces to the normal groundwater flow equation.

By expanding (4) we obtain

$$(1 - n)\gamma_s \frac{\partial p_s}{\partial t} - \frac{\partial n}{\partial t} + (1 - n) \frac{\partial w_s}{\partial z} - w_s \frac{\partial n}{\partial z} + (1 - n) w_s \gamma_s \frac{\partial p_s}{\partial z} = 0 \tag{9}$$

The momentum balances are obtained from consideration of each phase and a mixture total. We assume that the shear stress in the water is zero, which allows us the simplification that the sediment partial shear stress is symmetric [Bowen, 1976]. We also assume simple shear. In the z direction, ignoring acceleration,

For **water**

$$\frac{\partial (n p_w)}{\partial z} + n \rho_w g + \hat{p}_w = 0 \tag{10}$$

For **sediment**

$$\frac{\partial ([1 - n] p_s)}{\partial z} + (1 - n) \rho_s g + \hat{p}_s = 0 \tag{11}$$

For the **mixture**

$$\hat{p}_w + \hat{p}_s = 0 \tag{12}$$

where $\hat{p}$ is the interphasic rate of momentum transfer. Combining (10)–(12) to eliminate $\hat{p}$, we obtain

$$n\frac{\partial p_w}{\partial z} + p_w\frac{\partial n}{\partial z} + (1-n)\frac{\partial p_s}{\partial z} - p_s\frac{\partial n}{\partial z} + [n\rho_w + (1-n)\rho_s]g = 0 \tag{13}$$

The final relation is the porosity constitutive relation defined in terms of the effective pressure p_e, which is defined by

$$p_e = p - p_w = (1-n)(p_s - p_w) \tag{14}$$

and effective strain rate $\dot{\varepsilon}$ (note that in our assumption of simple shear we ignore the strain rate component $\dot{\varepsilon}_{zz}$, thus $\dot{\varepsilon} = \dot{\varepsilon}_{xz}$). As will be seen, the assumption that the vertical longitudinal strain rate is less than the shear strain rate appears to be good, though a careful analysis would be required to find the range of cases under which this is true. We use a porosity constitutive relation of the form

$$n = n(\dot{\varepsilon}, p_e) \tag{15}$$

By differentiating this relation with respect to time we obtain the derivative

$$\frac{dn}{dt} = \frac{\partial n}{\partial \dot{\varepsilon}}\frac{\partial \dot{\varepsilon}}{\partial t} + \frac{\partial n}{\partial p_e}\frac{\partial p_e}{\partial t} \tag{16}$$

and use the following relationship in (8) to account for the advection of porosity by the moving sediment skeleton:

$$\frac{dn}{dt} = \frac{\partial n}{\partial t} + w_s\frac{\partial n}{\partial z} \tag{17}$$

Equations (8, 9, 13, 15 or 16) together form a set of four coupled nonlinear partial differential equations. The bottom of the layer is at $z = 0$, while the top is at $z = T(t)$, $dT/dt = w_s(z = T)$; thus the problem is of moving boundary type.

Equation (8) is parabolic and requires two boundary conditions. At the base $\psi(z = 0) = \psi_Q(t)$ where ψ_Q is the aquifer potential, while at the surface

$$m = \frac{K}{\mu}\frac{\partial \psi}{\partial z}$$

607

where m is the basal influx of water. Equation (13) is first-order elliptic and requires one boundary condition for the sediment pressure p_s; this is implicitly defined by

$$p_i = np_w + (1 - n)p_s$$

where p_i is the ice overburden.

Equation (9) is first-order hyperbolic and requires one boundary condition:

$$w_s = 0 \quad \text{at } z = 0$$

While (17) is first order hyperbolic, it only requires a boundary condition if material is entering or leaving the system, as the moving boundaries are defined by w_s.

Finally, we define the strain rates $\dot{e}_{xz}$ by the rheological relations (1) or (2). In practice, we only use the nonlinearly viscous model as the calculations with the Bingham model were inordinately expensive.

The approach we use to describe the phenomenology of dilating, shearing sediments is by no means the only one (see *Clarke*, [this issue] for an alternative approach). *Iverson* [1985] has discussed the correct formulation of flow laws for fully dilatant sediments, but the development of a full constitutive theory requires experimental and mathematical investigations of the relations between dilating and volume-preserving sediment deformation styles.

Since these partial differential equations are nonlinear, numerical methods are needed. We use the finite difference advecting mesh method described by *Hindmarsh et al.* [1987]. The moving domain is mapped onto a fixed domain normalised by the thickness T. Thus the finite difference points convect in space. Second order spatial differences are used in conjunction with a backward difference marching scheme. Newton-Raphson iteration is used and the resulting Jacobian is inverted by a preconditioned conjugate gradient scheme. It was found that in order to avoid spurious modes in the solutions, equations (9) and (13) (both of first order) had to be solved using space-centered grids.

6. Steady state modes of subglacial deformation and their hydrological context

We now use the theory developed section 5 to establish the conditions under which steady state deformation of subglacial sediments will occur. We consider a fine-grained relatively impermeable stratum, which might be till, overlying an undeformable bed. This is either impermeable or appreciably more permeable than the upper stratum. In the former case there is no drainage through the base of upper stratum and, in consequence, no vertical potential gradient within it. In the latter case we prescribe an interstitial water potential in the aquifer which is related to the larger scale hydrological system beneath the glacier.

There are three possible modes of response of a deformable stratum to glacier overriding: (1) no deformation, B horizon only, (2) deformation of an upper layer of the stratum only, A and B horizons; and (3) deformation of the whole stratum, A horizon only.

We discuss the hydrological conditions under which each of these modes will develop and show that under certain limiting conditions stable deformation modes develop. Beyond those limits instabilities arise which must be resolved by recourse to other processes, discussed in section 10.

We assume that the porosity and permeability of each stratum (n_A, n_B, and K_A, K_B respectively) are constant with depth and that in the A horizon they are independent of strain rate.

In a subglacial stratum, water pressures will be determined by a gravitational component and a component due to the potential gradient ($d\psi/dz$) driving any water flow through the stratum.

In a steady state, melting that occurs at the glacier sole at a rate m must be discharged at the same rate across the stratum's lower boundary. From Darcy's law

$$\psi_0 = \frac{\mu m}{K} + \psi_Q \tag{18}$$

where $\psi_0 = \psi(z = T)$ and ψ_Q is the aquifer potential at the base of the stratum.

In the basal ice, $\psi_i = p_i$, where p_i is ice pressure; this defines a datum for the potential. The potential difference across the ice/sediment interface is $\psi_i - \psi_0$; this is the driving stress for infiltration.

The pressures in the stratum are as follows:

Overburden pressures p

At the top of the stratum $p(z = T) = p_i$. At the base of the stratum, in which bulk density $\rho = (1 - n)\rho_s + n\rho_w$, $p(z = 0) = p_i + \rho g T$

Water pressures p_w

At the top of the stratum (from 18) $p_w(z = T) = \psi_0 = \psi_Q + \mu m T/K$. At the base of the stratum $p_w(z = O) = \psi_Q + \rho_w g T$.

Effective pressure $p_e = p - p_w$

At the top of the stratum $p_e(z = T) = \Delta\psi - \mu m T/K$ where $\Delta\psi = \psi_i - \psi_Q$. At the base of the stratum $p_e(z = O) = \Delta\psi + GT$ where

$$G = (1 - n)(\rho_s - \rho_w)g \tag{19}$$

(The same result may be derived rigorously from relations 8–17)

We now consider the case of a dilatant A horizon and an undisturbed B horizon of permeability K_A, K_B respectively.

We define

$$\chi = \mu m / K_B \tag{20}$$

$$\beta = K_A / K_B \tag{21}$$

Thus, from Darcy's law,

$$\psi_0 - \psi_Q = \chi \left[T - T_A \frac{\beta - 1}{\beta} \right]$$

where T_A is the thickness of the A horizon, and the effective pressure at the glacier sole is given by

$$p_e(z = T) = \psi_i - \psi_0 = \Delta\psi - \chi \left[T - T_A \frac{\beta - 1}{\beta} \right] \tag{22}$$

In the A horizon

$$\frac{dp_w}{dz} = -\rho_w g + \frac{\chi}{\beta}$$

while

$$\frac{dp}{dz} = -\rho g$$

thus

$$\frac{dp_e}{dz} = \frac{dp}{dz} - \frac{dp_w}{dz} = -G - \frac{\chi}{\beta} \tag{23}$$

As deformation ceases at the base of the A horizon ($z = T - T_A$), We assume that the Mohr-Coulomb equation is satisfied so that

$$\tau = \eta \left[p_e(z = T) - \frac{dp_e}{dz} T_A \right] + c \tag{24}$$

We assume the glacially induced shear stress τ, applied on the sediment surface to be constant with depth. This will be true where ice surface slopes are small and the thickness of the deformable layer is much less than the thickness of the glacier; we expect these conditions to be satisfied away from the margin. Near to the margin, the variation of shear stress with depth must be derived by solution of the coupled twoor three-dimensional momentum balance equations.

Substituting (22) and (23) into (24), we obtain

$$T_A = \frac{\kappa - \Delta\psi + \chi T}{G + \chi} \tag{25}$$

where $\kappa = (\tau - c)/\eta$.

There are thus four possible behavioral states for a subglacial deformable stratum, defined by equation (25) and shown in Figure 8: Case I, if $T_A < 0$, no failure occurs. There is only a B horizon. Case II, if $T_A < T$ and $\psi_0 < p_i$, both A and B horizons develop under stable conditions. Case III, if $\psi_0 > p_i$, the effective pressure at the glacier sole is negative. Case IV, if $T_A > T$, the problem is ill-posed. This may mean that either case (iii) is realised, or that the whole of the stratum deforms.

Case III, if indeed achievable, must be unstable, and we expect. It to be resolved either by acceleration of the glacier flow as basal friction is lost, or by drainage of the water in piping events.

From (25) we see that for failure to occur,

$$\Delta\psi \leq \kappa + \chi T \tag{26}$$

By setting $T_A = T$ on the left hand side of (25) we find that to avoid failure of the whole layer

$$\Delta\psi \geq \kappa - TG \tag{27}$$

To avoid negative interfacial effective pressures, we set the left hand side of (22) to zero and substitute for T_A using (25). We find

$$\Delta\psi > \frac{\chi^2 T + [\kappa(1 - \beta) + \beta TG]\chi}{G\beta + \chi} \tag{28}$$

The permeability ratio $\beta = K_A/K_B$ does not appear in (25) as for a given potential drop $\Delta\psi$, an increase in β leads not only to a lower effective pressure at the interface but also to a change in effective pressure gradient $(G + \chi)$, which exactly counterbalances it.

In Figure 9 we have plotted the response of a subglacial sediment to changing values of critical effective stress for failure ($\kappa = (\tau - c)/\eta$), potential

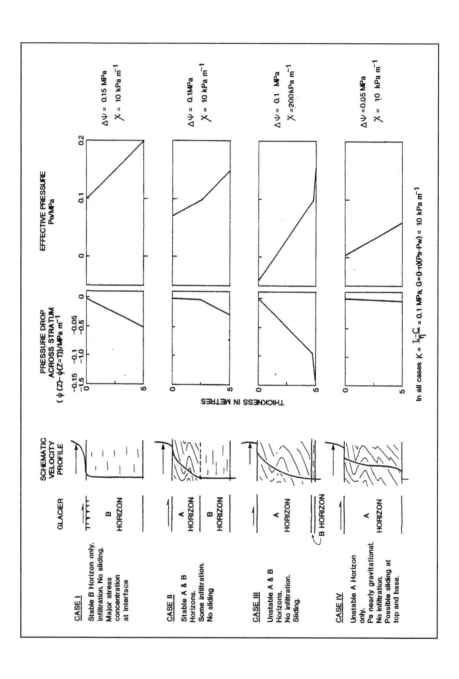

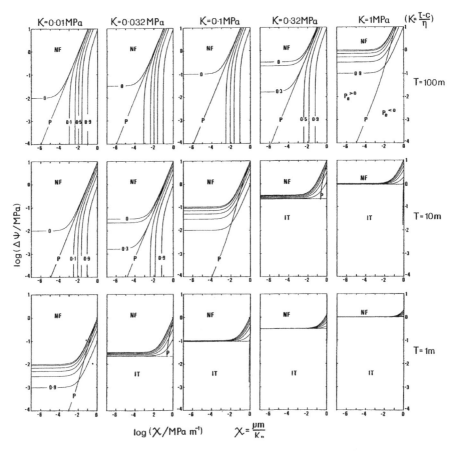

Figure 9 The deformational response of the subglacial bed to changing magnitudes
of critical stress for failure $\kappa = (\tau - c)/\eta$, potential drop across the stratum
$\Delta\psi$, potential gradient across the undilated stratum $\chi = \mu m/K_B$, and stratum
thickness T. NF indicates the no-failure field (Figure 8 case I): IT (insufficient
thickness) the field where the problem becomes ill-posed (Figure 8 case IV):
the contour lines 0–0.9 show the proportion of the stratum comprising the
A horizon, whilst the line P separates the case of stable A and B horizons
(Figure 8, case II) to the left from the case of unstable A and B horizons (Figure 8 case III) to the right. Strong deformation and instability is favored by
high values of χ and low values of κ, t and $\Delta\psi$.

Figure 8 (*opposite*) Four different modes of deformation of subglacial beds related
to changes in potential across the stratum $\Delta\psi$ and potential gradient in the
undeformed sediment $\chi = \mu m/K_B$ with constant shear stress. In cases I to
III an increase in melting rate is accommodated by formation and thickening
of a high permeability A horizon, but case III is unstable as the effective
pressure is negative. This will lead to liquifaction of the bed and zero shear
resistance. Case IV shows the effect on case II of decreasing sediment
permeability. Even with a fully dilatant bed, the water produced cannot be
discharged through it, leading to zero effective pressures at the glacier
sole. Complete deformation of the stratum may permit it to slide over its
substratum.

drop across the stratum ($\Delta\psi$), the potential gradient in the undilated stratum ($\chi = \mu m / K_B$), and stratum thickness (T).

We define four principal cases, the conditions in each of which are represented schematically in Figures 8a–d. The combination of κ, T, χ, and $\Delta\psi$ under which each of these conditions will occur is shown in Figure 9.

Case I B horizon only, $0 < (\tau - c)/\eta < p_e$ (Figure 8)

In this case the base potential ψ_Q is relatively small and/or the potential drop across the stratum $\Delta\psi$ is small, so that the difference between the water pressure immediately beneath the glacier sole and the ice pressure is greater than the critical normalised stress $\kappa = (\tau - c)/\eta$ required to cause failure. There is thus no failure in the subglacial sediments. This case is favored by low basal melting rates, low values of the base potential, low basal shear stress, and high sediment strength. If the sediments are of high permeability, we expect significant ice infiltration and no sliding.

Case II Stable A and B horizons $0 < p_e < (\tau - c)/\eta$ (Figure 8)

In this case, effective pressure at the interface is less than the critical normalized stress $(\tau - c)/\eta$, resulting in deformation, dilation, and an upper layer of higher permeability. Without this dilation and consequent permeability increase, p_e at the glacier sole could be zero or negative. This case is favored by higher m, ψ_Q, and τ and lower c, η, and K_B compared with case 1.

There may be significant ice infiltration at the sediment surface, thus inhibiting sliding.

Case III Unstable A and B horizons
$p_e < 0 < (\tau - c)/\eta$ (Figure 8)

In this case, even though the upper part of the sediment dilates, ψ_Q or $\Delta\psi$ is so large that the water pressure is equal to the ice pressure. In this region the system is particularly sensitive to small increases in basal melting rate or small decreases in permeability. It is clearly unstable. No ice infiltration is likely. Sliding is possible at the interface, and the sediment offers little resistance to glacier movement. While this situation may occur locally, it is obviously not sustainable over large areas. The weakness of the sediment will render it highly susceptible to a piping event which will reduce water pressures and restore stability (see Section 10).

Case IV Unstable A horizon only $T_A > T$ (Figure 8)

Here the problem is ill-posed, but we may consider qualitatively what happens as the system responds to changes which cause the thickness of the A horizon to attain that of the whole stratum. Since the completely dilated stratum may be unable to discharge meltwater while sustaining positive values of effective pressure, it may be in an unstable state corresponding to case III. If it is in a stable state, and if the frictional coefficient of the interface between the sediment and the material on which it rests is sufficiently low, sliding could occur at the lower interface.

If the sediment rests on an impermeable bed, there is no vertical drainage, the potential drop ($\Delta \psi$) must be zero, and water pressure gradients will be entirely gravitational. Water must be discharged horizontally. For a fine-grained stratum, discharge from a relatively small length of flow line will normally be great enough to produce zero effective pressures and cause instability [*Boulton and Jones*, 1979].

The steady-state velocity of the glacier sole (u_b), when its movement is due entirely to deformation in the underlying sediments, can be computed from:

$$u_b = 2 \int_0^T \dot{\varepsilon}_{xz}[p_e(z),\tau] \, dz$$

An example is the last time step in the velocity profile in Figure 10, which is close to a steady-state profile.

The potential drop across the stratum $\Delta \psi$ is largest when the stratum is undeformed and undilated, which gives the sediments their lowest value of permeability K_B. For given values of K_B, melting rate m, thickness T, and $\kappa = (\tau - c)/\eta$, Figure 9 indicates qualitatively how the deformation velocity increases as $\Delta \psi$ decreases. This decrease is produced by the increasing thickness of the dilating, more permeable A horizon in response to an increase in the potential pressure at the base of the stratum (ψ_Q). Eventually this process will lead to liquifaction of the whole stratum which will cease to offer resistance to glacier movement, resulting in some instability such as piping or a catastrophic flow event. For a constant value of $\Delta \psi$, increase of $\mu m / K_B$ leads to the same result.

7. Time-dependent response of subglacial deformation to changing external controls

The one-dimensional finite difference model described above (Section 5) was used to calculate the time-dependent response of a deforming till to changes in the controlling parameters. The parameters investigated were applied shear stress, aquifer potential, basal melting rate and till permeability. The fixed parameters were

$$\rho_w = 1{,}030 \text{ kg m}^{-3} \qquad \rho_s = 2{,}700 \text{ kg m}^{-3}$$

$$\mu = 0.0018 \text{ Pa s}$$

$$\gamma_w = 5.0 \times 10^{-10} \text{ Pa}^{-1} \qquad \gamma_s = 3.0 \times 10^{-11} \text{ Pa}^{-1}$$

The phenomenological relations $K = K(n)$ and $n = n(\dot{\varepsilon}, p_e)$ were not strictly based on empirical data but were defined so as to represent data obtained from Breidamerkurjökull [*Boulton et al.* 1974]. Let n_s be the porosity of the A horizon when dilation is fully developed (assumed independent of the effective pressure), n_u the porosity at zero effective pressure and zero strain rate, and n_c the porosity at zero strain rate and high effective pressure. We define the porosity by the following relation

$$n = (n_u - n_c) \exp\left(-\beta_1 \dot{\varepsilon}\right) \exp\left(-\beta_2 \rho_e\right) + n_s - (n_s - n_c) \exp\left(-\beta_3 \dot{\varepsilon}\right)$$

We set $n_s = 0.4$, $n_u = 0.33$, $n_c = 0.27$, with β_1, β_2, β_3 so chosen that when $\dot{\varepsilon} = 30 \text{ yr}^{-1}$, $\exp\left(-\beta_1 \dot{\varepsilon}\right) = \exp\left(-\beta_3 \dot{\varepsilon}\right) = 0.01$, and when $p_e = 0.2$ MPa, $\exp\left(-\beta_2 p_e\right) = 0.01$.

Permeability was defined by a function of the form:

$$K = \exp\left[(n - \hat{\gamma})/\delta\right]$$

and two sets of values of $\hat{\gamma}$ and δ were determined for a low-permeability case (Model 1) and a relatively high-permeability case (Model 2) (see Table 1). In all cases the overburden pressure was set to 4.498 MPa (= 500 m ice).

The results of this parameter study are summarised below:

Change in base potential ψ_Q

Figure 10a shows the response of the low permeability model. Increased water pressures at the base of the stratum diffuse upwards through the stratum, leading to a decrease in effective pressure and an increase in strain rate and horizontal velocity. Swelling of the stratum, indicated by vertical velocity and porosity changes occurs both directly because of the increase in ψ_Q and indirectly by the consequent dilation due to shear strain. Because of this latter effect, swelling is greatest in the upper part of the stratum. The instantaneous thickness of the stratum is indicated by the position of the top of each curve. Note that the response time for equilibrium is of the order of 6,000 years, which is a function of the low sediment permeability.

For the high permeability model (Figure 10b) the response time is very much shorter.

The increase in ψ_Q was spread over a 10-year period, and the residual lag time to equilibrium at the end of this period was of the order of 3–4 years.

Table 1 Porosity/Permeability Pairs Defining the Exponential Relation Between the Permeability K and the Porosity n.

	Model 1		Model 2	
	n	K, m^2	n	K, m^2
n_c	0.27	10^{-18}	0.27	10^{-15}
n_s	0.40	10^{-17}	0.40	10^{-14}

The relation has the form $K = \exp[(n - \hat{\gamma})/\delta]$.

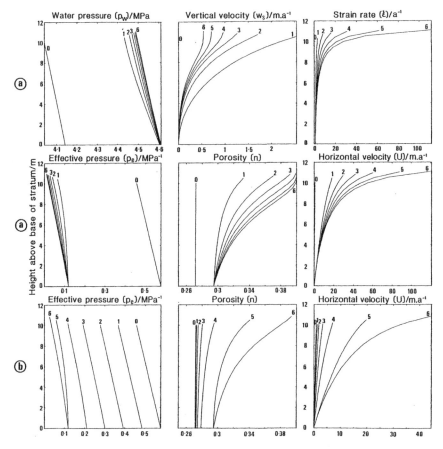

Figure 10 (a) Response of low permeability stratum ($K(n = 0.27) = 10^{-18}$ m^2, Model 1, see Table 1) to an instantaneous increase in base potential ψ_Q from 4.04 to 4.50 MPa. 1000-year time steps. ($\tau = 50$ kPa, $m = 0$ m yr^{-1}) (b) Response of high permeability stratum ($K(n = 0.27) = 10^{-15}$ m^2, Model 2, see Table 1) to increase in base potential ψ_Q from 4.04 to 4.50 MPa over a 10 yr period. 2-year time steps. ($\tau = 50$ kPa, $m = 0$ m yr^{-1}).

Figure 12c shows the effect of a decrease in the base pressure ψ_Q. This leads to an expulsion of water from the consolidating lower part of the stratum, leading to an increase in effective pressure at the base and a decrease at the top, thus producing increased strain rates in the upper part and decreased strain rates in the lower part of the stratum.

Change in shear stress

In Figure 11a we start from the final state shown in Figure 10a and increase the shear stress from 50 to 100 kPa. The upper part of the stratum is already

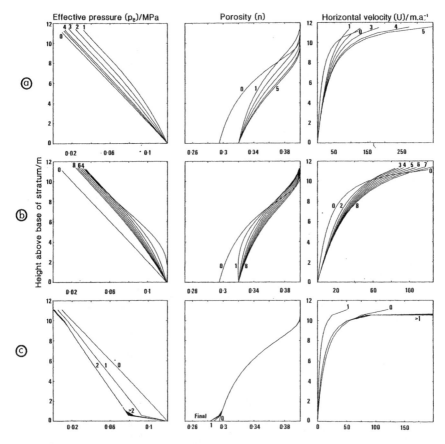

Figure 11 (a) Response of low permeability stratum to an increase in shear stress τ from 50 to 100 kPa. Starts from final state in Figure 10a. 400-year time steps. ($\psi_Q = 4.50$ MPa, $m = 0$ m yr^{-1}). (b) Response of high permeability stratum to increase in shear stress from 50 to 100 kPa. Same starting point as Figure 11a. 0.1-year time steps. ($\psi_Q = 4.50$ MPa, $m = 0$ m yr^{-1}). (c) Response of low permeability stratum to reduction in shear stress τ from 50 to 25 kPa. 0.2-year time steps. ($\psi_Q = 4.47$ MPa, $m = 0$ m yr^{-1}).

dilating to its maximum extent. As increased deformation and dilatation extend downward due to increased shear stress, the middle part of the stratum, which is most effected by this, shows the greatest increase in porosity and thus draws in water from the upper and lower horizons. The effective pressure in the upper horizon thus increases, with the result that the strain rate, more sensitive to p_e than τ, decreases, as does velocity. As porosity progressively equilibrates to the new stress, strain rates and velocities increase and exceed initial values, although only after a period of the order of 1,000 years.

Similar features are shown in the high-permeability model (Figure 11b) but with a much shorter response time. Note that the strong drainage from the lower part of the stratum into the newly dilatant middle section produces a non-monotonic change in effective pressure.

In this and in the following examples, the distribution of swelling and consolidation changes is indicated by changes in porosity.

In Figure 11c we investigate the response of the stratum to a reduction in shear stress from 50 to 25 kPa. This leads to a reduction in strain rate. The consequent decrease in dilatation produces lower porosity, leading to increased water pressure, decreased effective pressure and increased strain rate. This increase of strain rate beyond its original value has occurred in 0.4 years. The collapse of the dilatant lower layers expels water into the upper part of the stratum thus enhancing the strain rate.

Changes in basal melting rate

In Figure 12a we show the effect of increase in basal melting rate on the low permeability case from a starting point in which there is no deformation. Increased water pressure develops at the top of the stratum, thus permitting failure to occur. As this extends downward, a situation develops in which a stable A and B horizon state is achieved. Note that considerable changes in n, p_w, and p_e occur in a zone of small strain below an upper high strain rate horizon. The latter is our A horizon. The small strain zone at the top of the B horizon, which we term the B_1 horizon, is frequently represented in geological sequences by drag folds and faults [*Boulton*, 1987].

Figure 12b shows the collapse of the dilatant A horizon and the reduction in p_w and n which occurs when the melting rate is again decreased. The time scale of these changes is relatively small, of the order of 20 years. The reason for this rapid response compared with the low permeability examples above, where the base potential ψ_Q is increased, is that the latter depend upon drainage through the whole thickness of the undilated stratum, whereas in the former case changes occur almost entirely in the more permeable, dilatant A horizon, whose structure is opened and closed in response to changes in the melting rate.

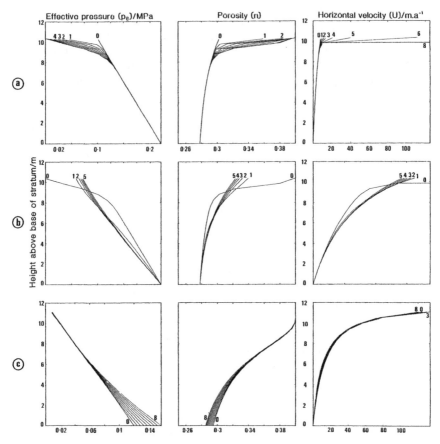

Figure 12 (a) Response of low permeability stratum to increase in melting rate m to 0.01 m yr⁻¹. 600-year time steps. ($\psi_Q = 4.40$ MPa, $\tau = 50$ kPa). (b) Starting from final state of Figure 12a, response of low permeability stratum to reduction in melting rate from 0.01 to 0 m yr⁻¹. 600-year time steps. ($\psi_Q = 4.40$ MPa, $\tau = 50$ kPa) (c) Starting from final state of Figure 10a, response of low permeability stratum to a fall in the base potential (ψ_Q) from 4.50 to 4.04 MPa. 1,000-year time steps. ($\tau = 50$ kPa, $m = 0$ m yr⁻¹).

Changes in the overburden pressure p_i

No results are shown here, as in the cases we computed, changes in the overburden led to immediate instability. In a case where the overburden pressure was increased, consolidation occurred, which led to water being expelled both upward and downward, resulting in negative effective pressures near the surface. In a case where the overburden was decreased, the aquifer potential was maintained at a constant value; the fact that overburden pressures decreased while water pressures remained constant led to negative effective pressures at the surface. We are not suggesting that a change in the overburden pressure will necessarily lead to an instability, as aquifer pressure and ice sheet geometry will be strongly coupled. However, the relation between the timescale of change of overburden pressure and aquifer potential will have an important influence on stability.

Computed diffusion times $\gamma \mu T^2/K$ for these models suggest relaxation times substantially less than those computed above. This is due to the fact that the water storage capacity of the system is governed by dilation and is therefore several orders of magnitude larger than the compressibility γ.

A one-dimensional, time-dependent analysis such as we present illustrates the general response of the sediment deformation system. However, such an analysis cannot fully couple the behavior of the glacier with that of the bed. For example, although we have demonstrated that an increase in shear stress can produce an initial fall followed by an increase in strain rate over periods of the order of 10^3 years in impermeable sediments, a fall in deformation velocity will cause shear stress buildup if the ice flux is to be maintained by the glacier, resulting in a shortening of the time scale of change. Initial two-dimensional analysis suggests, however, that very significant leads and lags, of the order of 10^3 years, can occur in the system if fine-grained impermeable sediments are involved. As mid-latitude ice sheets appear to have shown significant variation on this time scale [*Boulton et al.* 1985], we suggest that where these ice sheets overlay unlithified sediments, time-dependent bed effects may have strongly influenced glacier dynamics and the development of subglacial sediments and landforms.

8. Drumlins as products of small-scale variations in bed lithology

Subglacial sediment deformation has been suggested as a likely process of formation for drumlins, e.g. *Smalley and Unwin* [1968]. while this explanation has been popular, it has not gone unquestioned, and has been strongly challenged by those who believe that many drumlins with cores of fluviatile sediment are the result of subglacial fluvial deposition [e.g. *Shaw*, 1983]. We wish to suggest how drumlins, including those with a fluviatile core, may develop by subglacial deformation.

Sedimentary sequences formed in terrestrial environments tend to show strong small to medium-scale variability in properties such as permeability, internal friction, cohesion (functions of granulometry and mineralogy), and thickness, all of which play an important role in governing sediment response to glacier overriding. We suggest that detormation of such sedimentary sequences could lead to drumlin formation as a result of the heterogeneity of their strain response. As an extreme example, consider a mass of relatively coarse-grained permeable material within an otherwise relatively fine grained impermeable sediment. Figure 13 exemplifies such a situation, showing an ice marginal fan system surrounded by a fine-grained impermeable sediment mass. The coarse-grained permeable sediment would drain basal meltwater readily to produce relatively high effective pressures, thus permitting infiltration, inhibiting both deformation (B horizon only) and sliding. It would form an undeforming or slowly deforming mass within an otherwise rapidly deforming material and would contribute a large part of the frictional resistance of the bed to glacier movement. The fine-grained impermeable sediment would impede drainage to produce relatively low effective pressures, permit little infiltration, tend to deform strongly with a well-developed A horizon, might permit sliding, and would offer relatively little frictional resistance to glacier movement.

Apart from a narrow zone near the terminus, most of a glacier exhibits a regime of longitudinal tensile strain. The consequent attenuation will tend, as in metamorphic rocks, to isolate the more resistant masses as boudins. The high strain rates between the resistant boudins and the rapidly flowing flanking material would tend to incorporate fragments from the margins of the boudins in the more rapid flow, progressively developing streamlined forms (Figure 13).

We suggest that the subglacial boudins, whose development we have inferred, are represented by drumlins, the distinctive streamlined subglacial bedform. We suggest that variation in elongation from barchan shaped to egg-shaped to spindle-shaped to highly elongate flute forms depends upon the incremental strain within the original materials from which the drumlin developed. The drumlin core might be deformed or quite undeformed, depending on its mobility. This core could be equivalent to our B horizon, and might be sheathed by an A horizon, or this may have been completely removed. Although the coarse and fine sediment surfaces may originally have been on a similar level, the larger strain rates in the fine-grained sediment would produce relative attenuation and thinning compared to the coarse sediment boudin, which would thus tend to be exposed above the level of surrounding sediment. Precisely the same effect would be achieved if a lithologically uniform stratum were underlain by locally highly permeable beds or where the stratum thinned over a buried, undeforming till or step. In these cases the boudin would be fixed rather than mobile.

Figure 13 (*opposite*) Hypothesis of drumlin formation based on the contrasting rheological response of materials of different permeability to subglacial shear forces. (a) A proglacial distribution of coarse-grained outwash on an otherwise fine-grained surface. (b–c) Progressive development of drumlins from the coarse-grained masses. They are analogous to boudins in highly deformed rocks. Originally transverse lines in the coarse sediment illustrate the deformation of the coarse-grained masses.

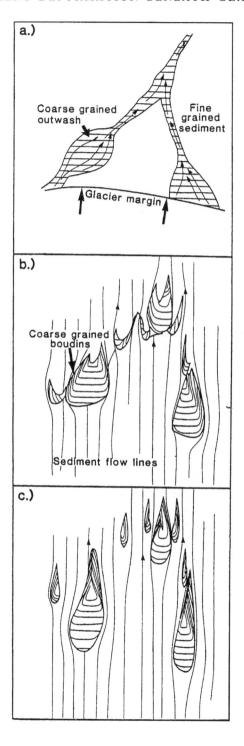

Longitudinal variations in the deformation velocity of subglacial sediments will inevitably produce transverse components of flow in bed sediments [*Boulton* 1979] and complex patterns of flow in basal ice. If the scale of the perturbation is of similar order to the ice thickness, we expect it to be reflected in glacier surface flow patterns. It should certainly be observed near glacier margins. Such a phenomenon might explain the inferred presence [*Whillans et al.* this issue] of elongate, mobile masses of relatively stiff ice on the surface of ice Stream B in Antarctica, as a reflection of a zone of mobile, but relatively stiff sediment on its bed.

9. Stability of the subglacial deforming system

It is clear from Figures 10–12 that the nature of subglacial deformation and the deformation velocity of the glacier can be very sensitive to small changes in glacier stress, in basal melting rates or in the relatively independent subglacial aquifer pressure. However, we suggest that there are important negative feedbacks which tend to restrain the system from departing far from a relatively steady mode of operation in which sediment deformation is maintained at the level required to sustain the basal velocity needed to discharge the flux of ice through the glacier system. We have shown that instabilities may be produced when effective pressures in subglacial sediments fall near to or below zero, thereby reducing frictional resistance to glacier movement, and leading to very high basal velocities (Figures 10–12). These will discharge ice from the glacier reservoir so as to reduce shear stresses and the water flux produced by frictional heating, which in turn will lead to a reduction in the rate of sediment deformation, causing the glacier to decelerate.

If, on the other hand, a perturbation causes a decrease in the deformation rate so that the necessary ice flux can no longer be sustained by the available shear stresses, glacier thickness and surface slope will increase, thus increasing shear stresses, water production, and the deformation rate in the bed. The system is thus driven back toward its former mode of operation.

Such perturbations and adjustments of a sensitive system may be absorbed within a large ice mass, such as an Antarctic ice stream, in such a way as to produce only small effects far from the perturbation, but in a relatively small valley glacier, the acceleration produced by local increase of subglacial water pressure, or increase in local mass balance, could generate a surge which affects the whole system.

We stress most strongly our view that the deformation of subglacial sediments is merely one amongst several glacier movement mechanisms. It is quite compatible with steady glacier movement, though, as other mechanisms, it can lead to instabilities. We reject the recent assumption by *Shoemaker* [1986] that bed deformation necessarily leads to instability.

10. Piping, sediment-floored subglacial channels and the origin of tunnel valleys

Hydrological conditions beneath glaciers which are melting basally and moving over deforming sediment beds are likely to lie between two extremes: (1) deforming sediment overlies an impermeable substratum, and (2) it overlies an aquifer with a high hydraulic transmissibility (i.e., the product of permeability and the thickness).

If the melting rate is m, the water discharge in subglacial beds Q at any point along a flow line at a distance x from the initiation of melting will be $Q = mx$. To sustain this discharge the horizontal pressure gradient in a layer of thickness D beneath the glacier must be:

$$\frac{dp_w}{dx} = \frac{Qu}{KD}$$

For stability, the value of dp_w/dx must not exceed the pressure gradient due to the ice mass. Taking the case of Breidamerkurjökull, this maximum stable value is approximately 500 Pa m^{-1} in the terminal 10 km. The distance from ice divide to terminus is 3.5×10^4 m and the basal melting rate is estimated as 0.015 m yr^{-1}. To maintain the resultant discharge in the terminal zone through subglacial sediments alone would require these to have a hydraulic transmissibility of 1.5×10^{-11} m^3 in order to avoid instability. This would require 1.5×10^6 m of clay-rich till ($K = 10^{-17}$ m^2), 150 m of coarse sandy till ($K = 10^{-13}$ m^2) or 0.15 m of coarse gravel ($K = 10^{-10}$ m^2). This demonstrates [cf. *Boulton and Jones*, 1979] that for any large glacier undergoing large scale basal melting, a normal thickness of relatively fine-grained deforming subglacial sediments cannot drain the meltwater flux in the absence of a substantial underlying aquifer. Without this, the instabilities of cases III and IV in Figure 9, associated with zero or negative effective pressures, will develop. We believe that under these circumstances, for a glacier to retain a relatively stable state, subglacial channels must exist to drain the excess water and permit stable deformation of the sediments. *Shoemaker* [1986] has recently suggested that channels will develop with a frequency which drains sediments sufficiently to prevent deformation. In our theory, they must only prevent *unstable* deformation and negative effective pressures. We first discuss how such channels might develop and then their stable operation.

Channel development – piping

Consider a temperate glacier overlying a sediment bed of low hydraulic transmissibility and expanding from a small nucleus in the early part of a glacial period. The subglacial water catchment is initially small, and water production may be small enough to be discharged by groundwater flow alone. As the glacier grows, water discharge from it increases, and the piezometric gradient required to drive the flow may begin to produce water pressure values which match ice overburden pressures. This is clearly most likely to happen first in the terminal zone where aggregate discharges and hydraulic gradients will be highest. If interparticle friction is small, due to low effective pressures, sediment may liquify, and because of its unconfined position near to the ice margin, flow of liquified sediment into the proglacial zone may occur, producing subglacial "pipes" (Figure 14c). These would form drainage conduits which draw down potentials ψ_Q so that zero or negative effective pressures do not occur at the glacier sole, and stable deformation can occur. As the glacier enlarges, the subglacial drainage basin becomes larger, and channels need to sustain ever larger discharges to maintain stability at the glacier sole.

If a large glacier whose sole was formerly below the melting point undergoes a change of regime which initiates subglacial melting, the location of the point of initiation of melting will determine the consequences. If melting is initiated at the terminus and progressively extends upglacier, the system can remain stable as piping can develop at the unconfined margin to produce channels. If it is initiated far from the terminus, high subglacial water pressures will build up which cannot be relieved by flow through the distal frozen-bed zone. In this case we would expect a major instability to develop as the ice in the inner zone, with little drag at its bed, presses ever more strongly against the restraining ice in the outer frozen zone. The surges of Bråsvelbreen in Spitsbergen [*Schytt*, 1969], and of Trapridge Glacier [*Clarke et al.*, 1984] may have been caused by such behavior.

The stability of soft sediment-floored tunnels

Rock-floored subglacial tunnels have been discussed at length. In these the closure of the tunnel due to ice creep is balanced by melting [*Röthlisberger*,

Figure 14 (*opposite*) The origin of sediment-floored subglacial tunnels and tunnel valleys. (a) The contrast between small low-discharge and large high-discharge tunnels. In the latter case, sediment creeps towards the tunnel and is removed by water flow as it intrudes into the tunnel, thus intitiating tunnel valleys. (b) A positive feedback effect (equation 33) ensures that once excavation of a tunnel-valley has begun, further deepening will be concentrated in the same place. (c) As the glacier grows, groundwater flow can no longer discharge all subglacial meltwater and piping causes tunnel formation at the glacier terminus (top and middle). As it grows further, the enlarging tunnel and its tunnel valley draw down the water table so as to prevent instability (bottom). (d) Meltwater is initially discharged by groundwater flow beneath the growing glacier (top). As channels develop to prevent unstable deformation (middle), groundwater catchments also develop, and the larger, distal parts of tunnels produce tunnel valleys.

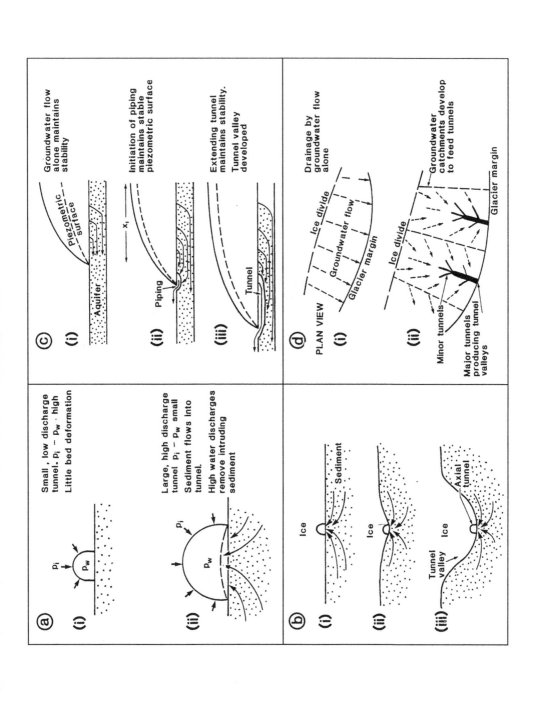

(a)

(i) Small, low discharge tunnel. $p_i - p_w$. high
Little bed deformation

p_i
p_w

(iii) Large, high discharge tunnel $p_i - p_w$ small
Sediment flows into tunnel.

High water discharges remove intruding sediment

p_i
p_w

(b)

(i) Ice
Sediment

(ii) Ice

(iii) Tunnel valley
Ice
Axial tunnel

(c)

(i) Groundwater flow alone maintains stability

Piezometric surface

Aquifer

(ii) Initiation of piping maintains stable piezometric surface

x_i

Piping

(iii) Extending tunnel maintains stability.
Tunnel valley developed

Tunnel

(d) PLAN VIEW

(i) Drainage by groundwater flow alone

Ice divide
Groundwater flow
Glacier margin

(ii) Groundwater catchments develop to feed tunnels

Ice divide
Ice divide

Minor tunnels
Major tunnels producing tunnel valleys
Glacier margin

1972]. Sediment-floored tunnels have not hitherto been considered. In these, closure due to sediment creep would be balanced by removal of sediment by water flow in the tunnel.

In the theories developed by *Röthlisberger* [1972] and by *Weertman* [1972], it is assumed that the weight of the ice is entirely borne by a thin water film. In the case of ice overlying a permeable stratum, in which water pressure can be substantially less than the ice pressure, the theoretical treatment is quite different. It is important to recognize that a subglacial conduit in hydraulic contact with a subglacial aquifer is no different to a sink in standard potential theory, and that there is no collection width for such channels analogous to that in *Weertman's* [1972] theory for subglacial conduits overlying an impermeable bed, where drainage may only occur in a layer between the ice and the bed.

Analysis of the stability of a sediment-floored tunnel requires simultaneous solution of the flow of ice, sediment, and water. Water flow in the channel will depend upon the size of the catchment required to maintain stability of the sediment, the transmissibility of subglacial sediments (in turn partly dependent on effective pressure and strain rate in the sediments) and the pressure in the tunnel. The stress and flow fields within the ice and sediment will be coupled to the water pressure through the dependence of sediment response on effective pressure. Water pressure in the tunnel will be a boundary condition for groundwater flow in the catchment. A bridging effect will produce a stress concentration beneath the tunnel walls.

Nye [1953] derived an expression for the closure rate of a rock-floored ice tunnel:

$$\zeta_i = A \frac{(p - p_w)^n}{n^n} \tag{29}$$

where $\zeta_i = \dot{r}/r$ (r = tunnel radius), p is the basal ice pressure, and the ice flow law is of the form

$$\dot{\varepsilon} = A\tau^n$$

If the water pressure in the sediments beneath the tunnel is the same as the water pressure in the tunnel, the pressure drop which causes closure ($p - p_w$) is the same for both ice and sediments. Using a nonlinearly viscous model (cf. equation 2) for the sediments e.g.:

$$\dot{\varepsilon} = A_V p_e^j \tau^k$$

$A_V p_e^j$ then becomes equivalent to A in equation (29) and the tunnel closure rate for sediments becomes:

$$\zeta_s = A_V \frac{(p - p_w)^{j+k}}{k^k} \tag{30}$$

If $j + k < 0$ (as is the case at Breidamerkurjökull), intrusion of sediment into the tunnel diminishes with decreasing water pressure, as sediment viscosity increases with driving stress $(p - p_w)$ at a greater than linear rate. Tunnel closure due to creep of ice is, however, increased under these conditions. Figure 15 shows the rates of tunnel closure due to sediment creep in relation to rates of closure due to ice creep for different values of driving stress/ effective pressure $(p - p_w)$.

The unknown in this theory is the driving stress $p - p_w$. This could be evaluated in the same way as in *Röthlisberger's* [1972] theory for subglacial conduits overlying an impermeable bed. In our theory, fluviatile erosion of the sediment is analogous to ablation of tunnel walls in Röthlisberger's theory.

If water pressure in a tunnel is high due to high stream discharges, ice creep will be inhibited, but sediment creep enhanced. Sediment will therefore be extruded into the tunnel and removed by the high stream discharges in the tunnel. Continued sediment creep toward the tunnel, with the highest

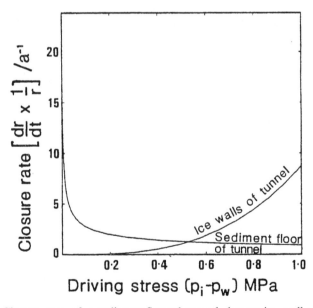

Figure 15 Closure rate of a sediment-floored tunnel due to ice wall closure and sediment floor intrusion. Provided that water discharges are sufficient to remove the intruding sediment, tunnel valley formation is favored by lower driving stresses. As higher water discharges are favored by larger driving stresses, intermediate values of driving stress are most likely to produce tunnel valleys.

creep rates near the tunnel, would lead to a general lowering of the sediment surface near to the tunnel by the complementary processes of creep and fluvial erosion (Figures 14a–b). We suggest that such a process may explain the so-called "tunnel valleys" [e.g., *Hunsch*, 1979; *Kuster and Meyer*, 1979] found in mid-latitude areas covered by ice sheets during the Pleistocene. These are linear depressions, parallel to the direction of glacier flow, typically 50–200 m in depth, up to several kilometers in width and several tens of kilometers in length. They are underlain by unlithified sediment sequences.

It is quite clear that they could not have formed as simple subglacial conduits of this width and depth other than as the products of catastrophes. In the uniformitarian theory which we advocate, they are produced by progressive lowering of the sediment surface as a result of sediment flow toward one or more conduits situated along the tunnel valley axis (Figures 14a–b), and indeed some Danish tunnel valleys have eskers, the remains of blocked tunnels, along their axes [*Sjörring*, 1979]. (*It is argued later that tunnel valleys will also be favored locations for enhanced erosion resulting from extensional strain in the sediments.*) Many of these valleys are filled by sedimentary sequences similar to those found in modern proglacial lakes and probably reflect proglacial sedimentation as the tunnel valleys were exposed by glacier retreat.

If water pressure in a tunnel is low, due to relatively minor water discharges, ice closure rates will tend to be large, although sediment creep will be minor, resulting in little bed erosion.

The occurrence of the two conditions referred to above will depend on the extent and nature of the groundwater catchment. In subglacial beds of uniform transmissibility, small ineffective tunnels will tend to develop in the upper part of the catchment, while highly active tunnels will tend to produce tunnel valleys in the outer parts of catchments (Figure 14a). The effectiveness of tunnel valley development could vary strongly along a drainage path according to the transmissibility of subglacial beds; local overdeepening of tunnel valleys occurring where transmissibility is low.

Location and spacing of sediment-floored tunnels and tunnel valleys

Let us consider a series of parallel tunnel valleys of simple cross-profile, described by $V(y)$ and separated by a distance 2ξ, where y is horizontal and orthogonal to the axes of the tunnel valleys. We assume the width of the valleys to be negligible compared with their spacing, their depth V_m, to be significantly less than the aquifer thickness D, and the one dimensional groundwater flow equations to apply in which the vertical potential gradient in the aquifer is zero. We set $V(y = 0) = 0$, so defining a co-ordinate frame moving down with the valley axis as it is eroded, and where, by definition, the maximum height of the plateau is $V(y = \xi) = V_m$. The equation for steady flow is:

$$-\frac{KD}{\mu}\frac{d^2\psi}{dy^2} + m = 0 \tag{31}$$

where the melting rate, m, is a constant. The boundary conditions are $\partial\psi/\partial y = 0$ at $y = \xi$ (by symmetry) and $p_i - \psi = \Delta\psi$ at $y = 0$, where $\Delta\psi$ is an arbitrary potential difference. Integrating (31) twice, we obtain

$$\psi(y) = -\frac{y(y - 2\xi)}{\Omega} + p_i - \Delta\psi \tag{32}$$

where $\Omega = 2KD/\mu m$ and may be regarded as an index of the ability of groundwater flow to discharge the meltwater production of the ice-sheet in question. Table 2 gives Ω as a function of K, D, and m.

The effective pressure at the ice/sediment interface is

$$p_e[y,z = V(y)] = \Delta\psi + \Gamma V(y) - \psi(y)$$

or, using (32),

$$P_e[y,z = V(y)] = \Delta\psi + \Gamma V(y) + \frac{y(y - 2\xi)}{\Omega} \tag{33}$$

where $\Gamma = (\rho_w - \rho_i)g$ and assuming that $dp_i/dz = \rho_i g$.

This analysis goes further than that of *Shoemaker* [1986] in that topography is considered. It will be shown below that retention of this term is important.

Table 2 The Ability of the Aquifer to Drain the Meltwater $\Omega = 2KD/\mu m$ as a Function of Aquifer Permeability K, Aquifer Depth D and the Melting Rate m.

	K/m^2			
D/m	10^{-15}	10^{-14}	10^{-13}	10^{-12}
		$m = 0.01$ yr^{-1}		
10	10^{-1}	1	10	10^2
100	1	10	10^2	10^3
1,000	10	10^2	10^3	10^4
		$m = 0.1$ yr^{-1}		
10	10^{-2}	10^{-1}	1	10
100	10^{-1}	1	10	10^2
1,000	1	10	10^2	10^3

Units in m^2 Pa^{-1}.

Equation (33) indicates that effective pressure at the ice/bed interface is increased on bed elevations and diminished in depressions, while it decreases farther from a tunnel as a result of increasing potential. On an irregular surface where the decreasing effective pressure near the margin of an expanding glacier approaches zero, we expect unstable conditions to occur at low points in the topography or in areas of locally low transmissibility, where water pressures would be high. Thus we expect conduit initiation, which may lead to tunnel valley development, to be controlled by topography and geology.

We expect channels to be sufficiently close to drain subglacial beds to the extent that stable effective pressures ($p_e > 0$) are maintained at the groundwater drainage divides between adjacent channel catchments but not so close that the discharge into them from groundwater is inadequate to keep the channel open. An upper bound to the spacing can be calculated by setting the left hand side of (33) to zero and substituting ξ for y and noting that $V(\xi) = V_m$:

$$\xi^2 = \Omega(\Delta\psi + \Gamma V_m) = \frac{2KD(\Delta\psi + \Gamma V_m)}{m\mu} \tag{34}$$

Table 3 shows the relationship between channel half-spacing ξ and Ω; a high Ω indicates that groundwater flow is better able to discharge the meltwater.

The spacing of tunnels is largely determined by the melting rate and the transmissibility of subglacial beds. Good aquifers require fewer valleys to drain them sufficiently to maintain stability. For a 1-km-thick stratum of permeability 10^{-14} m^2 underlain by an aquiclude, tunnel spacings (2ξ) of 20 km are required. The spacing is determined by the initiation of channelling as glaciers expand in size (Figure 14d). Subsequent deepening to produce tunnel valleys produces general lowering of water pressures and an increase in effective pressures, and thus the spacing predicted by equation (34) refers

Table 3 Tunnel Valley Half Spacing ξ as a Function of Tunnel Valley Depth V_m, Potential Difference Between Water and Ice at the Tunnel $\Delta\psi$ and Ω, Tabulated in Table 2.

	$\Omega/$		m^2 Pa^{-1}				
$\Delta\psi + \Gamma V_m$	10^{-2}	10^{-1}	1	10	10^2	10^3	10^4
0.01	0.01	0.032	0.1	0.32	1	3.2	10
0.1	0.032	0.1	0.32	1	3.2	10	32
1	0.1	0.32	1	3.2	10	32	100
10	0.32	1	3.2	10	32	100	320

Units of ξ in km. and $\Delta\psi$ in MPa.

to the initial topography. A possible effect of changing subglacial trans-missibility may be reflected in the contrast between spacings of tunnel valleys of the order of 20 km in the southern North Sea region, underlain by thick relatively permeable beds [*Cameron et al.*, 1985] and in Poland, where, in areas underlain by impermeable Tertiary clays, spacings are of the order of 1–5 km [*Pasierbski*, 1979].

Equation (33) shows that in passing from a tunnel valley to the drainage divide on the adjacent plateau, the effective pressure at the sediment surface will first rise to a maximum high on the flanks of the valley and then decrease toward the divide, where potentials are highest. The term $\Delta\psi$ in this rela-tion, equivalent to $p - p_w$ in relation (30), can only be determined once an integrated theory of tunnel closure and fluviatile erosion is developed.

It is unlikely, however, that divide effective pressures will approach this value, since we expect fluctuations in melting rate and other unsteady effects discussed in section VII to bring periodically effective pressures to zero, leading to conduit formation and valley initiation. Equation (34) thus provides an upper bound to channel spacing.

Apart from sporadic events at the divide, we generally expect to find the lowest effective pressures and the weakest sediments at the conduits, with sediment creeping toward them. Since the larger the channel spacing, the larger the discharge of water down the channel, we should expect widely spaced channels to be more effective at eroding tunnel valleys. The wider spaced tunnel valleys of the North Sea are much larger than the closer packed valleys of northern Poland [*Cameron et al.*, 1985; *Pasierbski*, 1979].

Kinematic lowering, that is, lowering due to the dominant glacially imposed regime of extensional strain [*Boulton*, 1987], will tend to be con-centrated where the sediment is weakest along the axes of the tunnel valleys. This will add a longitudinal stretching component of erosion to that effected by the complementary processes of lateral sediment flow toward the conduit and fluvial transport within it. However, as tunnel valleys are of significant size on an ice sheet scale, the fact that they have a weaker bed will lead to the ice surface in their vicinity being drawn down, reducing the applied stress and, in consequence, the velocity. Determination of the relat-ive importance of kinematic erosion along valley floors and plateaux requires three-dimensional modeling of the integrated system.

As we have argued above, consistently low effective pressures along the drainage divide will lead to the formation of new valleys. However, episodic low effective pressures of duration insufficient for valley inititiation could result in some interesting effects depending upon the behavior of sediments at very low effective pressures alluded to in section 6, cases III and IV. If high water pressures unrelieved by conduit drainage were to develop at the drainage divide, it is conceivable that water-filled, sediment-floored cavities might develop, which most probably would extend in the direction of lowest water pressure gradient, the direction of glacier flow. These would be quite different from known types of subglacial rigid-walled cavities which depend upon upstream bedrock knobs or water-filled bedrock hollows for their location and which are extremely unlikely to develop in unlithified sediments at low effective pressure. The absence of a conduit would prevent there being significant water flows sufficient to remove coarse-grained sediment, although fine-grained sediments could settle from suspension.

Perturbations of the system might also lead to periodic upstream migration of the heads of conduits and consequent episodes of fluviatile deposition on deformation tills succeeded by further glacial deposition.

Since the development of tunnel valleys leads to plateaux with stiffer beds, we conclude that tunnel valleys may be essential to the stability of large ice sheets flowing over deformable beds and that the mid-latitude Pleistocene ice sheets may have been unable to maintain steady state conditions until sufficiently deep tunnel valleys had been excavated. It is interesting to note that in the North Sea areas occupied by European ice sheets, the largest ice sheets, those of Elsterian and Saalian age, excavated the deepest tunnel valleys. An analysis of the evolution of these features in relation to sediment bed drainage is thus important in understanding the evolution of the ice sheet as a whole.

We suggest that the style of subglacial drainage completely changes in passing from ice sheet bed areas of rigid bedrock to deformable sediment. In the former case we expect rock-floored Röthlisberger or Nye channels, which, when blocked by sediment, give rise to eskers. In the latter case, we expect tunnel valleys to develop as a product of sediment-floored channels, although the channel itself, when blocked, may give rise to an esker. This contrast is certainly reflected in Europe, where the bedrock areas of Scandinavia show frequent and often very long esker systems aligned parallel to former glacier flow, whereas the peripheral soft sediment zones of northwest USSR, Poland, Germany, Denmark, Holland, and the North Sea show tunnel valleys parallel to former glacier flow.

Acnowledgments

Much of this work was done with the assistance of generous grants from the Natural Environment Research Council (NERC grant GR3/5253) and the Royal Society. The assistance of many students of the University of East Anglia is gratefully acknowledged, as is that of P. Robinson (University of Auckland), F. Eybergen (University of Amsterdam), and Flosi and Sigidur Björnsson of Kvisker, Iceland.

634

References

Boulton, G.S., Processes and patterns of subglacial sedimentation: A theoretical approach. *Ice Ages Ancient and Modern* edited by A.E. Wright and E. Mosely. p. 7–42, Seel House Press, Liverpool, 1975.

Boulton, G.S., Processes of glacier erosion on different substrata, *J. Glaciol.* 23, 15–38, 1979.

Boulton, G.S., A theory of drumlin formation by subglacial sediment deformation in *Drumlins: a Symposium, 1985* edited by J. Rose and J. Menzies, Balkema, Rotterdam, in press, 1987.

Boulton, G.S., D.L. Dent and E.M., Morris Subglacial shearing and crushing, and the role of water pressures in tills from south-east Iceland. *Geografiska Annaler* 56A, 3–4, 135–145. 1974.

Boulton, G.S. and A.S. Jones 1979, Stability of temperate ice sheets resting on beds of deformable sediment. *J. Glaciol.* 24, 29–43. 1979.

Boulton, G.S., G.D. Smith, A.S. Jones and J. Newsome, Glacial geology and glaciology of the last mid-latitude ice sheets. *J. Geol. Soc. London*, 142, 447–474, 1985.

Bowen, R.M., Theory of Mixtures, in *Continuum Physics*, Vol. III, *Mixtures and EM Field Theories*, edited by A.C. Eringen, p. 1–127, Academic, Orlando, Florida, 1976.

Carol, H., The formation of roches-moutonnées. *J. Glaciol.*, 1, 57–59, 1947.

Cameron, T.D.J., C. Laban and R.G.D. Schüttenhelm, Indefatigable, sheet 53°N–0°E, scale 1/250,000, series Quaternary geology, Br. Geol. Surv. and Rijks Geol. Dienst, Ordnance Survey, Southampton, 1985.

Clarke, G.K.C., S.G. Collins and D.E. Thompson, Flow, thermal structure, and subglacial conditions of a surge-type glacier. *Can. J. Earth Sci.* 21, 232–240, 1984.

Clarke, G.K.C., 1987, Subglacial till: A physical framework for its properties and processes, *J. Geophys. Res.*, this issue.

Fowler, A.C., On the transport of moisture in polythermal glaciers. *Geophys. Astrophys. Fluid Dyn.*, 28, 99–140, 1984.

Hindmarsh, R.C.A., L.W. Morland, G.S. Boulton and K. Hutter. The unsteady plane flow of ice sheets: a parabolic problem with two moving boundaries, *Geophys. Astrophys. Fluid Dyn.*, 39, 183–225, 1987.

Hunsch, W., Rinnen an der Basis des glaziären Pleistozäns in Schleswig-Holstein. *Eiszeitalter Ggw.* 29, 173–178, 1979.

Iverson, R.M., A constitutive relation for mass-movement behavior. *J. Geol.* 93, 143–160, 1985.

Kuster, H. and K.D. Meyer, Glaziäre Rinnen in mittleren und nordöstlichen Niedersachen. *Eiszeitalter Ggw.* 29, 135–156, 1979.

Nye, J.F., The flow law of ice from measurements in glacier tunnels, laboratory experiments and the Jungfraufirn borehole experiment. *Proc. R. Soc. London, Ser. A* 239, 113–133, 1953.

Pasierbski, M., Remarks on the genesis of subglacial channels in Northern Poland. *Eiszeitalter Ggw.* 29, 189–200, 1979.

Robin, G., Is the basal ice of a temperate glacier at the pressure melting point? *J. Glaciol.* 16, 183–196, 1976.

Röthlisberger, H., Water pressure in inter- and subglacial channels. *J. Glaciol.* 11, 177–203, 1972.

Schytt, V., Some comments on glacier surges in eastern Svalbard. *Can. J. Earth Sci.* 6, 867–873, 1969.

Shaw, J., Drumlin formation related to inverted erosion marks. *J. Glaciol.* 29, 461–479, 1983.

Shoemaker, E.M., Subglacial hydrology of an ice sheet resting on a deformable aquifer. *J. Glaciol.* 32, 20–30, 1986.

Sjörring, S., Tunneltäler in Dänemark. *Eiszeitalter Ggw.* 29, 179–188, 1979.

Smalley, I.J. and D.J. Unwin, The formation and shape of drumlins and their distribution and orientation in drumlin fields. *J. Glaciol.* 7, 377–400, 1968.

Weertman, J., General theory of water flow at the base of a glacier or ice sheet. *Rev. Geophys.*, 287–333, 1972.

Whillans, I.M., J. Bolzan and S. Shabtaie, Velocity of ice streams B and C, Antarctica. *J. Geophys. Res.* this issue.

20

TILL BENEATH ICE STREAM B

3. Till deformation: evidence and implications

R.B. Alley, D.D. Blankenship,
C.R. Bentley and S.T. Rooney

Source: *Journal of Geophysical Research* 92(B9) (1987): 8921–9.

Abstract

Most of the velocity of ice stream B near the Upstream B camp (UpB), West Antarctica, appears to arise from deformation of a seismically detected, subglacial till layer that averages 6 m thick. Available evidence indicates that the entire thickness of this till layer is deforming and is eroding subjacent bedrock into flutes parallel to ice flow and hundreds of meters across. The resulting till flux beneath UpB is equivalent to an average erosion rate of about 0.4 mm yr^{-1} in the catchment area and suggests that till deltas tens of kilometers long have been deposited at the grounding line during the Holocene. Such deltas should be characterized by partial ice-till decoupling across a water film and by a small ice-air surface slope; they may have been discovered by recent geophysical work.

1. Introduction

It is clearly recognized that till beneath glaciers can deform and can make a significant contribution to glacier dynamics, both for small glaciers and for continental ice sheets [e.g., MacClintock and Dreimanis, 1964; Engelhardt *et al.*, 1978; Boulton, 1979; Boulton and Jones, 1979; Clarke *et al.*, 1984; Clayton *et al.*, 1985]. Most studies of the modern ice sheets have not considered the possibility of a deforming bed, however. Recent seismic experiments conducted at the Upstream B camp (UpB) on the upper part of ice stream B on the Siple Coast of West Antarctica suggest that subglacial till is important there [Blankenship *et al.*, 1986, this issue; Rooney *et al.*, this issue]. These studies show that the ice stream near UpB is underlain by a layer of water-saturated, unconsolidated material of about 40% porosity in which the water pressure is within 50 kPa of the overburden pressure. The thickness of the layer varies from ≤2 m to about 13 m. The upper surface of the layer is smooth; the lower surface of the layer is smooth parallel to ice flow, but transverse profiles show that the lower surface of the layer is carved into flutes hundreds of meters across oriented parallel to ice flow. This lower surface is an angular unconformity on lithified sedimentary rocks.

Alley et al. [1986] argue that such a thin, unconsolidated layer beneath an active, wet-based glacier must be till and that deformation within this till accounts for most of the ice stream velocity at UpB. Here we present further development of the arguments from Alley et al. [1986] that the velocity of the ice stream arises from till deformation. We then use a till continuity argument to estimate the erosion rate beneath the catchment area of ice stream B and the deposition rate of deltas at the grounding line. We also discuss characteristics of such deltas, evidence supporting their existence, and possible implications for the sedimentary record of the Ross Embayment. In the companion paper [Alley et al., this issue] we develop a numerical model of ice stream flow on deforming till.

2. Evidence for till deformation

2.1. Porosity

Shear deformation of sediment, including till, causes dilation and increase of porosity. By studying recently deformed till exposed by the retreat of Breidamerkurjökull in Iceland, G.S. Boulton and coworkers [Boulton and Dent, 1974; Boulton et al., 1974; Boulton and Paul, 1976] found that subglacial shear deformation of till causes porosity of about 40%. Within a year following the end of deformation, the dilated structure collapses and porosity decreases to less than 30% while still water-saturated; undeformed lodgement till beneath the sheared layer also exhibits porosity less than 30%. The seismically determined porosity of about 40% averaged over a till thickness of about 8 m at a point near UpB [Blankenship et al., 1986; this issue] thus indicates that the till there is deforming.

2.2. Basal shear stress and till strength

As a second demonstration that the till beneath UpB is deforming, we now estimate the basal shear stress there and show that it exceeds the expected till strength. The basal shear stress can be calculated from force balance considerations constrained by glaciological observations. Whillans [1987] provides an elegant general treatment of force balance on the flow line from the ice divide through ice stream B to the front of the Ross Ice Shelf; here we conduct a simple stress balance calculation leading to a specific estimate of basal shear stress at UpB.

638

The driving stress for ice flow τ is balanced by the basal shear stress τ_b, by gradients in longitudinal deviatoric stresses $2G$, and by the stress from side drag S, so that

$$\tau = \tau_b + 2G + S \tag{1}$$

If ice thickness h is measured vertically, then τ and $2G$ are given by the familiar formulas [e.g., Paterson, 1981, p. 100]

$$\tau = \rho_i g h \tan \alpha \tag{2}$$

$$2G = \frac{2\partial(\bar{\sigma}'_x h)}{\partial x} \tag{3}$$

where ρ_i is the density of ice, g is the acceleration of gravity, $\tan \alpha$ is the ice-air surface slope, $\bar{\sigma}'_x$ is the depth-averaged longitudinal deviatoric stress, and x is the horizontal coordinate. If drag from the two sides is distributed evenly across the ice stream width w, then

$$S = \frac{2\tau_s h}{w} \tag{4}$$

where τ_s is the depth-averaged shear stress at the side margin. At UpB, $\tan \alpha \approx 2.7 \times 10^{-3}$ and $h \approx 1,050$ m [Shabtaie et al., this issue], so $\tau \approx 25.3$ kPa. Experiments are in progress to constrain all other terms in equation (1) [Whillans, 1987], but results are not available yet. However, we can place limits on likely values.

Looking first at S, ice often is approximated as a perfect plastic with yield strength $\tau_s = 100$ kPa [e.g., Paterson, 1981, p. 154]. However, preferred fabrics, fracture, and strain heating should occur at ice stream margins and reduce τ_s [Hughes, 1977], perhaps to 10 kPa or less. (Measurements of shear strain parallel to the ice stream margins at a point near UpB not immediately adjacent to a margin suggest that $\tau_s \approx 40$ kPa there if all other stresses are neutral [Whillans, 1987]; including the softening effect of longitudinal deviatoric stresses would reduce τ_s to perhaps 10 kPa there.) Taking $w = 33$ km [Shabtaie et al., this issue] as the total ice stream width between the slow moving ridges on either side, we then find that $0.6 \leq S \leq 6.4$ kPa.

A similar estimate can be made for $2G$. Longitudinal deviatoric stresses are tensional near the head of the ice stream [Whillans, 1987] but become almost neutral near the grounding line [Thomas and MacAyeal, 1982; Jezek, 1984]. If the stresses decay smoothly to zero along the length, $\ell \approx 300$ km, of the ice stream, then

$$2G \approx \frac{2}{\ell}(\bar{\sigma}'_x h)_o \qquad (5)$$

where the subscript zero indicates the head of the ice stream. Young [1981] estimates that $\bar{\sigma}'_{xo} \approx 50$ kPa, and I.M. Whillans (personal communication, 1986) used continuity considerations to estimate that $\bar{\sigma}'_{xo} \approx 100$ kPa, where $h_o \approx 1,300$ m. Taking $\bar{\sigma}'_{xo} = 150$ kPa should overestimate $2G$ and gives $2G = 1.3$ kPa. On the other hand, Vornberger and Whillans [1986] have identified open crevasses transverse to ice flow near UpB, and passive seismic studies suggest that these crevasses are opening further [Blankenship et al., 1987]. This indicates that $\bar{\sigma}'_x$ is strongly tensional at UpB; $\bar{\sigma}'_x$ could be as large or larger there than at the head of the ice stream. (Weertman [1977] has shown that the depth to which crevasses open depends on crevasse spacing and on the longitudinal stress. Short-pulse radar studies near UpB show that crevasse depths correspond to about 25 m of ice and that crevasses are about 150 m apart (S. Shabtaie, personal communication, 1985). This gives $\bar{\sigma}'_x \approx 120$ kPa from the Weertman theory, or slightly larger than the estimates at the head of the ice stream.) Thus $2G = 0$ is possible. ($2G$ also could be negative, which would serve to strengthen the conclusion we reach below.)

If we take $\tau = 25.3$ kPa, $0 \leq 2G \leq 1.3$ kPa, and $0.6 \leq S \leq 6.4$ kPa in equation (1), then the limiting cases give $17.6 \leq \tau_b \leq 24.7$ kPa and basal shear stress balances between 70% and 98% of the driving stress. Because of the high probability of shear softening at the ice stream margins and because of the evidence for strongly tensional stresses at UpB, we believe that τ_b is near the higher end of this range of values.

The basal shear stress must exceed till strength for till deformation to occur. Estimation of till strength is discussed by Alley et al. [1986]. Briefly stated, they show that any basal shear stress in excess of the strength of dilated till will cause sufficiently large stress concentrations [Boulton et al., 1974] to mobilize lodged till. Till can be approximated as a Coulomb material [e.g., Boulton et al., 1974], so that failure and deformation occur when the shear stress exceeds the shear strength of dilated till, τ_d, given by

$$\tau_d = \Delta P \tan \phi_d + C \qquad (6)$$

where ΔP is the excess of overburden pressure over water pressure in till, and $\tan \phi_d$ and C are the internal friction and cohesion of till, respectively.

Rheological properties of till probably vary with grain size and other factors [Clarke, this issue], but sufficient data are not available to constrain this variation accurately. We thus use the data from Breidamerkurjökull and from Blue Glacier to estimate $\tan \phi_d$ and C at UpB, but remember that our lack of direct measurements at UpB introduces some uncertainty.

In situ measurements by Boulton *et al.* [1974], using a vane shear apparatus in a dilated till exposed by the retreat of Breidamerkurjökull in Iceland, yielded total shear strengths for till of 3–8 kPa. Vane shear tests use faster strain rates than natural subglacial deformation and so overestimate the strength of till [Walker, 1983], and the strength measured in a vane shear test includes a contribution from both cohesion and internal friction. Cohesion in deformed till thus is less than 3–8 kPa and may be zero. The observations of Engelhardt *et al.* [1978] also indicate that cohesion is zero or small. We thus set $C = 0$ in equation (6). (Our conclusion that till at UpB is deforming is unchanged for any $C < 15$ kPa.)

At Blue Glacier, measurements show that $\Delta P \approx 1100$ kPa and that the basal shear stress $\tau_b \approx 230$ kPa [Engelhardt *et al.*, 1978]. Till is deforming there, so $\tau_b \geq \tau_d$. From equation (6) we then find that $\tan \phi_d \leq 0.21$. A similar calculation for Breidamerkurjökull yields a less restrictive upper bound on $\tan \phi_d$. However, laboratory measurements on till from there indicate that $\tan \phi_d$ is between 0.17 and 0.50 [Boulton and Jones, 1979], and $\tan \phi_d$ should be smaller than this at in situ strain rates beneath glaciers [Walker, 1983]. Based on the data from Blue Glacier and also from Breidamerkurjökull, we consider 0.2 ± 0.1 as the best estimate of the upper limit for $\tan \phi_d$.

The value of ΔP has been determined seismically for UpB and is about 50 ± 40 kPa [Blankenship *et al.*, 1986, this issue]. Taking $\tan \phi_d \leq 0.2$, this gives $\tau_d \leq 10$ kPa, although τ_d could be as large as 20 kPa. The best estimate of the basal shear stress, calculated above, is about 25 kPa, although it could fall as low as 18 kPa. Thus this analysis implies that the till at UpB is deforming, although the result is not unequivocal.

2.3. Water balance

The third argument favoring till deformation at UpB comes from a water balance calculation. Weertman [1969] calculated that a water film of about 5 mm thickness or greater is needed to allow sliding at ice stream velocities over a rigid substrate, and Weertman and Birchfield [1982] calculated that sufficient water generation can occur to produce such a water layer near the grounding line of a West Antarctic ice stream. If water-lubricated sliding over a rigid bed is to explain ice streams, however, then a water layer at least 5 mm thick must exist at the head of an ice stream where fast sliding first begins and where less water is available. For ice stream B this would require average production of 2 mm yr^{-1} of water over the entire catchment [Alley *et al.*, 1986; Weertman and Birchfield, 1982]. However, as summarized by Alley *et al.* [1986], the model results of Budd *et al.* [1971] show that this much melting is unlikely for the expected geothermal flux in West Antarctica [Gow *et al.*, 1968; Rose, 1979], results from the Byrd Station core indicate that some parts of the catchment area of ice stream B probably experience net basal freezing rather than melting [Gow *et al.*, 1979], and some water produced may be diverted from a Weertman film into subglacial sedimentary rock that has been shown by seismic measurements to exist [Bentley and Clough, 1972; Rooney et al., 1987, this issue] or into subglacial channels [Bindschadler, 1983]. It thus is unlikely that enough water is present near the head of the ice stream to cause rapid water-lubricated sliding, so that most ice stream velocity must arise from till deformation in the upper reaches of the ice stream. Near the grounding line the calculation of Weertman and Birchfield [1982] still is valid, and below we discuss the possibility that the ice stream is controlled by till deformation in its upper reaches and by a combination of till deformation and water-lubricated sliding in the lower reaches. Water balance calculations cannot exclude the possibility of significant sliding between ice and till at UpB, but based on the arguments in section 4.1, we consider such decoupling to be unlikely.

3. Till flux

3.1. Thickness of deforming layer

We next show that the entire thickness of subglacial till near UpB is deforming. As noted in section 2.1, till exhibits a dilated, deforming state and a collapsed, nondeforming state. Observations show that deforming till rests directly on lithified bedrock in some cases [MacClintock and Dreimanis, 1964] but that deforming till and bedrock are separated by collapsed till or other nondeforming but unlithified materials in other cases [MacClintock and Dreimanis, 1964; Boulton, 1979]. Two lines of evidence indicate that deforming till contacts bedrock directly at UpB.

First, the seismic results of Blankenship et al. [1986, this issue] apply to the average properties over a till thickness of about 8 m at one point near UpB and indicate that this entire thickness is dilated and thus deforming. Rooney et al. [this issue] show that the till averages about 6 m thick over a larger area near UpB. If the deforming thickness indicated by the work of Blankenship et al. [1986, this issue] is characteristic of a larger region, then deforming till reaches bedrock at least in most areas near UpB.

Also, glaciological considerations suggest that the entire thickness of till is deforming. Deformation of till causes stress concentrations [Boulton et al., 1974] that increase with strain rate. Alley et al. [1986] argue that such stress concentrations in deforming till will mobilize subjacent, lodged till until bedrock is intersected or until the simple strain rate $\dot{e}$ falls to the minimum value $\dot{e}_1$ needed to cause sufficient stress concentrations to mobilize more till. At Breidamerkurjökull a deforming layer of till about 0.5 m thick overlies lodged till [Boulton, 1979], so we expect that $\dot{e} = \dot{e}_1$ there. The velocity at the top of the till is about 15 m yr^{-1} [Boulton, 1979], giving $\dot{e}_1 \approx 30$ yr^{-1}. At Blue Glacier a till layer about 0.1 m thick with a velocity of about 4 m yr^{-1} at its upper surface rests on bedrock [Engelhardt et al., 1978], giving $\dot{e} \approx 40$ yr$^{-1} \geq \dot{e}_1$.

If most ice velocity at UpB arises from till deformation, as argued in sections 2.3 and 4.2, then if we assume that till at UpB is similar to till at Breidamerkurjökull and thus that $\dot\varepsilon_1 = 30$ yr^{-1}, the basal ice velocity of 450 m yr^{-1} at UpB would deform till to a depth of 15 m, slightly deeper than the deepest till observed there and much deeper than the average till. Glacial geological evidence from the St. Lawrence valley shows that the entire thickness of a till layer can deform to at least 10 m depth with localized deformation along zones of weakness at greater depth [MacClintock and Dreimanis, 1964], so this is not an excessive thickness for deforming till. (Notice that if $\dot\varepsilon_1 \approx 30$ yr^{-1}, then the 8 m of dilated till observed by Blankenship *et al.* [1986, this issue] requires that over half of the basal ice velocity must arise from till deformation.)

Glacial geological evidence shows that deforming till in contact with bedrock will abrade that bedrock [MacClintock and Dreimanis, 1964]. Together with the evidence above that deforming till contacts bedrock at UpB, this indicates that ice stream B is eroding its bed. Rooney *et al.* [this issue] found that the seismically observed base of the till beneath UpB is an angular unconformity on lithified sedimentary rocks, that this surface is fluted parallel to ice flow, and that the positions of flutes may be influenced by lithologic contrasts in the underlying rock. We now hypothsize that these flutes are being produced by erosion beneath deforming till and thus may be characteristic erosional features of deforming till (see also MacClintock and Dreimanis [1964]). This may have implications for interpretation of glacial flutes observed elsewhere [Wardlaw *et al.*, 1969; Goldthwait, 1979]. Although bedrock character may influence the formation and location of flutes, it is not clear why flutes are stable forms. Among other possibilities, the narrow rock ridges between flutes may represent regions of water piping and partial ice-till decoupling. Three-dimensional flow patterns in till [Boulton, 1979] also may be important.

3.2. Erosion rate

The results above can be used to calculate the till flux beneath UpB and thus the steady state erosion rate in the catchment area and the deposition rate at the grounding line. Observations at Breidamerkurjökull [Boulton, 1979] and Blue Glacier [Engelhardt *et al.*, 1978] show that there is little natural slip across the upper and lower boundaries of till, and as noted in sections 2.3 and 4.1, we consider that this also is true at UpB. The ice velocity at UpB is about 450 m yr^{-1} [Vornberger and Whillans, 1986] and is nearly independent of depth [Alley, 1984; Budd *et al.*, 1984]. The till thickness h_b is about 6 m [Rooney *et al.*, this issue], and the porosity n is about 0.4 [Blankenship *et al.*, 1986, this issue]. Shear stress varies little through the till because the till is so thin (6 m) compared to the ice (1,050 m). The horizontal velocity in till thus should vary almost linearly with depth, giving an average till velocity of $\bar{u}_b = 225$ m yr^{-1}. The rock flux Qr through ice stream B2 (the branch of ice stream B on which UpB is located) at UpB where the fast moving ice is $w \approx 30$ km wide [Shabtaie and Bentley, 1987] then is

$$Q_r = h_b \bar{u}_b (1 - n)w = 24 \times 10^6 \text{ m}^3 \text{ yr}^{-1} \tag{7}$$

The area over which the till is generated can be calculated from the information from Shabtaie and Bentley [1987]. The total area of the ice stream B system is 159×10^3 km^2, of which 23×10^3 km^2 lie closer to the grounding line than UpB. If we make the reasonable assumption that half of the catchment area feeds each branch of ice stream B, then the area feeding till past UpB is 68×10^3 km^2. A mean erosion rate $\lambda_r \approx 0.4$ mm yr^{-1} over this area is required to maintain steady state.

An erosion rate of this magnitude for a wetbased catchment area is reasonable based on available evidence. From review of the literature, Boulton [1979] reports that glacial erosion rates up to 1 mm yr^{-1} are common. To assess the rate of glacial abrasion, which along with plucking or quarrying causes most subglacial erosion, Boulton [1979] placed rock plates of different types flush with subglacial bedrock in direct contact with debris-laden basal ice at Breidamerkurjökull. Measured abrasion rates varied with subglacial conditions and with rock type (limestone was abraded more rapidly than basalt) and ranged from 0.1 to 2 mm yr^{-1}.

It is worth noting here that the water flux in the till at UpB, Q_w, occurs almost entirely by advection in the deforming till rather than by conduction through the till [Alley et al., 1986]. (Taking the hydraulic conductivity of deforming till to be 10^{-6} m s^{-1} [Boulton et al., 1974] and the pressure gradient arising from the slope of the ice-air surface to be about 25 Pa m^{-1}, the conductive water velocity is less than 0.1 m yr^{-1} [Alley et al., 1986]; by comparison, the advective water velocity is the till velocity $\bar{u}_b \approx 225$ m yr^{-1}.) The advective water flux is given by

$$Q_w = h_b \bar{u}_b nw \approx 16 \times 10^6 \text{ m}^3 \text{ yr}^{-1} \tag{8}$$

This requires generation of a thickness of water λ_w per unit time over the catchment and upstream part of the ice stream given by $\lambda_w \approx 0.2$ mm yr^{-1}. This is almost an order of magnitude less water than is needed to allow fast velocities by Weertman sliding.

3.3. Deposition rate

The till flux beneath UpB also requires the deposition of "till deltas," or morainal banks [Powell, 1984], at the grounding line. If we assume that the till generation rate λ_r calculated above applies to the entire length, $1 \approx 300$ km, of the ice stream and to the catchment, then the rock flux at the grounding line is $Q_{rg} \approx 60 \times 10^6$ m^3 yr^{-1}. If we assume that deltaic deposits have porosity $n_d \approx 0.25$, fill water of depth $h_d \approx 100$ m [Greischar and Bentley, 1980] and are distributed over a width $w_d \approx 100$ km [Shabtaie and Bentley, 1987], then the rate of grounding line advance $\dot{g}$ is given by

$$\dot{g} = \frac{Q_{rg}}{(1 - n_d)w_d h_d} \approx 8 \text{ m yr}^{-1} \qquad (9)$$

Thomas and Bentley [1978] estimate that post-Wisconsinan sea level rise caused grounding lines of Siple Coast ice streams to retreat to near their present positions by 5,000–10,000 years ago. Equation (9) then would lead us to predict deltas some 40–80 km long at the mouths of ice stream B and other ice streams. As discussed below, recent mapping of the grounding line [Shabtaie and Bentley, 1987] is entirely consistent with the existence of such deltas.

4. Till deltas

4.1. Ice-till coupling

For the calculations in section 3 we assumed perfect coupling between ice and till, but we now consider the possibility of partial decoupling across a water film at the ice-till interface. At Blue Glacier, Engelhardt et al. [1978] observed almost complete ice-till coupling in most boreholes; however, in some instances, drilling of boreholes introduced large volumes of excess water to the bed, causing ice-till separation and rapid sliding between ice and till for a short time before the excess water drained away. At Breidamerkurjökull, Boulton [1979] measured relative motion equal to about 10% of the basal ice velocity between a marker placed in the base of the ice and another placed in the top of the till. Because of the finite size of the markers and the depth variation of velocity in the till, this shows that true sliding of ice over till accounts for ≤10% of the basal ice velocity but does not rule out the possibility of partial decoupling. It thus is reasonable to suppose that sliding over an immobile bed and full coupling of ice to a deforming bed represent end-members of a spectrum of possible behaviors.

As argued in sections 2.3 and 3.1, our understanding of the water balance of ice stream B indicates that Weertman sliding is unlikely to be significant near the head of the ice stream, and geophysical data, combined with glaciological calculations, show that more than half of the ice velocity at UpB arises from till deformation. However, Weertman and Birchfield [1982] show that the water generated beneath the ice stream is sufficient to allow rapid sliding near the grounding line if all the water flows in a Weertman film, and Jenssen et al. [1985] show model results indicating that rapid sliding is possible for at least the lower half of the ice stream.

Classical sliding theory has been developed for a rigid, consolidated bed and may require modification for a deformable bed. For example, Weertman [1972] argues that under many circumstances, water beneath a glacier resting on a rigid, impermeable bed will flow preferentially in a film that lubricates sliding rather than in channels that have little effect on sliding. However, this analysis assumes physical limitations (such as the assumption that erosional deepening of tributary Nye channels is slow) that are valid for rigid beds but that probably require reassessment in relation to an unconsolidated, deforming bed.

Also, it seems likely that the unconsolidated nature of a till bed will limit the thickness of a Weertman film. To see this, notice that a water film of thickness δ carries a quantity of water, q_δ, per unit width given by [Weertman and Birchfield, 1982]

$$q_\delta = \frac{\delta^3(\rho_i g \tan \alpha)}{12\eta_w} \tag{10}$$

where ρ_i is the density of ice, g is the gravitational acceleration, $\tan \alpha$ is the surface slope, the quantity $\rho_i g \tan \alpha$ is the pressure gradient driving water flow, and $\eta_w = 1.8 \times 10^{-3}$ Pa s is the viscosity of water. The mean flow velocity in the film, u_δ, is

$$u_\delta = \frac{q_\delta}{\delta} \tag{11}$$

With $\tan \alpha$ ranging from 0.001 to 0.003 on ice stream B [Shabtaie et al., this issue] and taking $\delta = 5$ mm, the minimum for fast sliding [Weertman, 1969], this gives $u_\delta \approx 10$ to 30 mm s^{-1}.

From Hjulström's curves [Hjulström, 1935; 1939; Sundborg, 1956, pp. 177–180], flows in this range of velocities are rapid enough to transport clay, silt, and fine sand, although such flows are not rapid enough to mobilize lodged sediment. Deformation of till beneath an ice stream might serve to loosen fine material for transport, aided by higher flow velocities caused by local channeling of water around clasts. Silt and clay thus could be winnowed out of the upper surface of the till, creating a high-permeability zone of sand and gravel there. Such a high-permeability zone would conduct some water, thinning the Weertman film, and would appear rough to the glacier, increasing ice-till coupling. (Some sand and gravel lenses observed in tills may have such an origin.)

We know of no direct demonstration of this winnowing effect on a deforming bed, but other evidence tends to support the idea. Hallet [1979] shows, from examination of deposits on recently deglaciated surfaces, that fine-grained material was transported by a water film involved in the regelation process. Boulton and Dent [1974] have shown that significant amounts of clay and silt are removed from till by water percolation during as little as one year of subaerial exposure at Breidamerkurjökull. Clarke *et al.* [1984] argue that high-permeability flow paths exist within till beneath Trapridge Glacier and are maintained by water flow but that they are largely destroyed by periodic surges.

It thus seems likely that a winnowing mechanism does place some limit on the thickness of a Weertman film over a deformable bed and that under some circumstances (including, perhaps, beneath ice streams) this limit may be small enough to prevent rapid sliding. Combined with the possibility that some excess water flows in channels rather than in a film at the ice-water interface, this suggests that most of the ice velocity along most of the ice stream arises from till deformation rather than from sliding between ice and till. Notice, however, that the value of ΔP at UpB [Blankenship *et al.*, 1987, this issue] is significantly smaller than would occur if all water beneath the ice stream were to flow in a single Röthlisberger channel [Bindschadler, 1983], so channel flow is not as well developed as a single Röthlisberger channel. Also, notice that examination of equations (10) and (11) shows that for a given supply of water, a decrease in the pressure gradient driving water flow will increase the thickness of any water film present and will decrease the water flow velocity in this film and thus the winnowing effect, both of which will increase sliding between ice and till as a percentage of total basal velocity [Weertman and Birchfield, 1982]. As discussed below, this becomes important in our understanding of ice flow over till deltas.

From the arguments presented in section 3.1 to the effect that a till layer up to 15 m thick would deform at UpB but that the till layer there averages only about 6 m thick, we can calculate that partial ice-till decoupling must occur over till deltas. Suppose that ice and till are fully coupled along the entire length of the ice stream. Model results presented in the following paper [Alley *et al.*, this issue] then show that the till layer thins slightly downstream but that the ice velocity (and thus the potential maximum thickness of deforming till) increases downstream. If full coupling were to continue from the ice stream onto a thick till delta, then the deforming till layer would thicken onto the delta. The till flux entering the delta from the ice stream then would be less than the till flux across the delta, and the delta would be eroded. Such erosion could not proceed far enough to produce total ice-till decoupling by floating the ice off the delta because some ice-till coupling must be maintained to transport till from the ice stream to the front of the growing delta; nonetheless, such erosion would have to occur until partial decoupling developed.

The weak, thick deforming till on a delta would support only small basal shear stress and thus small slope of the ice-air surface. The pressure gradient driving water flow at the ice-till interface arises largely from the slope of the ice-air surface, so a small slope would allow partial ponding of water and partial ice-till decoupling. We expect that small surface slope, low basal shear stress, partial ponding of water, and partial ice-till decoupling develop in harmony at the front of the delta as it prograded.

4.2. Evidence for till deltas

Morainal banks (or till deltas) deposited subaqueously at a grounding line are commonly observed glacial features [Sugden and John, 1976, pp. 324–325; Powell, 1984; Oldale, 1985] and should be expected in many situations. Such morainal banks generally are smaller than the features proposed here, but often were deposited over relatively short periods of time at unstable glacial margins and may have had deposition rates at least as large as those proposed here [Oldale, 1985].

The till deltas proposed here have not been observed directly, and direct observation is unlikely in the near future. Nonetheless, recent mapping of the grounding line by Shabtaie and Bentley [1987] strongly suggests the presence of such till deltas. Their new grounding line map, based on detailed studies using satellite translocation and airborne radar, shows lobate regions of grounded ice tens of kilometers long at the mouths of ice streams A and B; ice-air surface slopes in these regions are intermediate between ice stream and ice shelf values, as predicted by our till delta model. The new mapping clearly shows grounded ice in regions where Robertson [1975] (also in the work by Greischar and Bentley [1980]) reported water layers of 50–100 m based on seismic measurements. It is likely that the "water layer" detected seismically in these regions actually is unconsolidated, water-saturated till [Shabtaie and Bentley, 1987]; further seismic experiments clearly are warranted.

4.3. Possible implications of till deltas

The possibility of extensive till deltas at the grounding line has implications for the dynamics of the ice sheet and for the sedimentary record produced by the ice sheet. Because the basal shear stress supported by a delta is small, the delta should make only a small contribution to the force balance of the ice sheet. The "coupling line" at the head of the delta (the upstream limit of till too thick to deform to bedrock; Figure 1), which marks the downstream limit of significant basal shear stress and ice-air surface slope, may be more important dynamically to the ice sheet than the grounding line at the front of the delta. (This coupling line essentially corresponds to the grounding line as mapped by Rose [1979].) The delta does serve to stabilize the ice sheet somewhat, however, because it causes the ice thickness at the coupling line to exceed slightly the thickness that can be floated by the existing sea level and so provides a buffer against coupling line retreat.

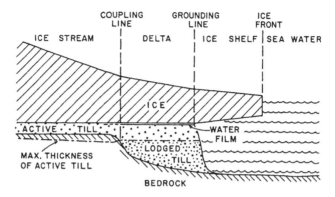

Figure 1 Diagram of our ice stream/delta model.

It is worth noting that the coupling line would advance in "conveyor belt" fashion [Powell, 1984] by erosion of the upglacier end of the delta and deposition on the marine face if the ice-air surface slope were to increase over the head of the delta, thus increasing the water flux and consequently the ice-till coupling there. Such an increase in surface slope over the head of the delta could be caused in several ways; for example, a drop in sea level would increase interaction of the ice shelf with pinning points, which would increase the backstress on the ice sheet and cause the grounded ice to thicken and steepen to maintain force balance.

As discussed in section 4.1, full ice-till coupling at the head of the delta would cause rapid erosion of till there and would increase the till flux to the grounding line. Based on the model results presented in the following paper [Alley *et al.*, this issue] and on the discussion in section 3.1, we estimate that till just upstream of the delta on ice stream B may be about 5 m thick but that full ice-till coupling there would deform more than 20 m of till. If full coupling were established over the head of the delta, the till flux across the delta and the rate of grounding line advance would increase by a factor of 4 or more until the till thickness at the head of the delta was reduced to about 5 m. If sea level were to fall steadily, then continuous erosion of the head of the delta and deposition at the grounding line would cause the grounding line and coupling line to advance rapidly across the continental shelf. The grounded ice behind this advancing delta would be characterized by a low, ice stream profile (see also Thomas [1979]) and would be lubricated by a thin (≈5 m) till layer; the entire till thickness would be deforming and eroding subjacent bedrock.

Subsequent sea level rise and grounding line retreat of such an expanded, till-lubricated ice sheet would leave a distinctive sedimentologic signature consisting of a basal unconformity (either in strips corresponding to ice streams, or regional if the fast moving zones of ice coalesced) covered by a thin layer of till. Morainal banks or till deltas would mark the terminal position (and possibly recessional positions) of the grounding line unless the terminal position corresponded with the outer limit of the continental shelf, in which case the glacially transported debris would be moved to the base of the continental slope by mass-wasting processes. The similarity between this model and the observed sedimentary features in the Ross Embayment [Johnson et al., 1982; Anderson et al., 1984; Grosval'd, 1984a, b; Powell, 1984; Karl et al., 1987] is suggestive.

5. Conclusions

Analysis of seismic data from UpB on ice stream B, West Antarctica [Blankenship et al., 1986, this issue; Rooney et al., this issue] has allowed us to develop a model of ice stream behavior and its relation to glacial geological processes that is consistent with recent mapping of the grounding line by Shabtaie and Bentley [1987] and with other data sets. We propose the following:

1.   The ice stream velocity arises from deformation throughout a subglacial till that averages 6 m thick.
2.   The till is eroding an angular unconformity, which is fluted parallel to ice flow, on subjacent bedrock.
3.   The till flux at UpB is equivalent to a steady state erosion rate of about 0.4 mm yr^{-1} averaged over the catchment area of the ice stream.
4.   The till flux has caused the post-Wisconsinan deposition of a till delta tens of kilometers long at the mouth of ice stream B (and adjacent ice streams).
5.   The till delta is characterized by partial ice-till decoupling across a water film and thus by small ice-air surface slope.
6.   During the Wisconsinan, falling sea level caused rapid "conveyor belt" advance of till deltas into the Ross Embayment, forming a grounded, till-lubricated ice sheet with a low, ice stream profile.

Clearly, much of the material presented here is somewhat speculative, and numerous further studies are needed. Further observation of the bed of ice stream B should receive top priority, using seismic techniques, drilling, and radar if possible. Theoretical, laboratory, and field studies of ice-water-till interactions also are needed.

Notation

g	magnitude of gravitational acceleration, m s^{-2}.
$\dot{g}$	grounding line advance rate, m s^{-1}.
$2G$	gradient in longitudinal deviatoric force per unit width, Pa.

h	ice thickness, m.
h_b	till thickness, m.
h_d	till thickness on delta, m.
l	length of ice stream, m.
n	till porosity.
n_d	till porosity on delta.
(o)	subscript indicating head of ice stream.
q_δ	water flux per unit width in Weertman film, $m^2 \, s^{-1}$.
Q_r	rock flux in till beneath ice stream B2 at UpB, $m^3 \, s^{-1}$.
Q_{rg}	rock flux in till beneath ice stream B at grounding line, $m^3 \, s^{-1}$.
Q_w	water flux in till beneath ice stream B2 at UpB, $m^3 \, s^{-1}$.
S	average basal stress on ice stream from side drag.
$\tan \alpha$	ice-air surface slope.
$\tan \phi_d$	internal friction of till.
$\bar{u}_b$	till velocity averaged over depth at UpB, $m \, s^{-1}$.
u_δ	water velocity averaged over depth in Weertman film, $m \, s^{-1}$.
w	ice stream width, m.
w_d	delta width, m.
x	horizontal coordinate axis parallel to flow.
δ	water layer thickness, m.
ΔP	excess of overburden pressure over water pressure in till, Pa.
$\dot{\varepsilon}$	simple strain rate in till, s^{-1}.
$\dot{\varepsilon}_1$	simple strain rate required to dilate till, s^{-1}.
η_w	water viscosity, Pa s.
λ_r	erosion rate of rock in catchment, $m \, s^{-1}$.
λ_w	generation rate of water by melt in catchment, $m \, s^{-1}$.
ρ_i	density of ice, $kg \, m^{-3}$.
$\bar{\sigma}'_x$	depth-averaged longitudinal deviatoric stress, Pa.
τ	driving stress for ice flow, Pa.
τ_b	basal shear stress, Pa.
τ_d	shear strength of dilated till, Pa.
τ_s	depth-averaged side shear stress, Pa.

Acknowledgments

We wish to thank S. Shabtaie for access to unpublished data, G.K.C. Clarke, H.F. Engelhardt, R.H. Thomas, and I.M. Whillans for helpful comments on the manuscript, and A.N. Mares and P.B. Dombrowski for manuscript and figure preparation. This work was supported in part by the U.S. National Science Foundation under grant DPP84-12404. Geophysical and Polar Research Center, University of Wisconsin-Madison, contribution 455.

References

Alley, R.B., A non-steady ice-sheet model incorporating longitudinal stresses, *Rep. 84*, The Ohio State Univ. Inst. of Polar Stud., Columbus, 1984.

Alley, R.B., D.D. Blankenship, C.R. Bentley, and S.T. Rooney, Deformation of till beneath ice stream B, West Antarctica, *Nature*, *322*, 57–59, 1986.

Alley, R.B., D.D. Blankenship, S.T. Rooney, and C.R. Bentley, Till beneath ice stream B, 4, A coupled ice-till flow model, *J. Geophys. Res.*, this issue.

Anderson, J.B., D.F. Brake, and N.C. Myers, Sedimentation on the Ross Sea continental shelf, Antarctica, *Mar. Geol.*, *57*, 295–333, 1984.

Bentley, C.R., and J.W. Clough, Seismic refraction shooting in Ellsworth and Queen Maud Land, in *Antarctic Geology and Geophysics*, edited by R.J. Adie, pp. 683–691, Universitetsforlaget, Oslo, 1972.

Bindschadler, R.A., The importance of pressurized subglacial water in separation and sliding at the glacier bed, *J. Glaciol.*, *29*, 3–19, 1983.

Blankenship, D.D., C.R. Bentley, S.T. Rooney, and R.B. Alley, Seismic measurements reveal a saturated, porous layer beneath an active Antarctic ice stream, *Nature*, *322*, 54–57, 1986.

Blankenship, D.D., S. Anandakrishnan, J. Kempf, and C.R. Bentley, Microearthquakes under and alongside ice stream B, detected by a new passive seismic array, *Ann. Glaciol.*, *9*, in press, 1987.

Blankenship, D.D., C.R. Bentley, S.T. Rooney, and R.B. Alley, Till beneath ice stream B, 1, Properties derived from seismic travel times, *J. Geophys. Res.*, this issue.

Boulton, G.S., Processes of glacier erosion on different substrata, *J. Glaciol.*, 23, 15–38, 1979.

Boulton, G.S., and D.L. Dent, The nature and rates of post-depositional changes in recently deposited till from south-east Iceland, *Geogr. Ann.*, *56A*, 121–134, 1974.

Boulton, G.S., and A.S. Jones, Stability of temperate ice caps and ice sheets resting on beds of deformable sediment, *J. Glaciol.*, *24*, 29–43, 1979.

Boulton, G.S., and M.A. Paul, The influence of genetic processes on some geotechnical properties of glacial tills, *Q.J. Eng. Geol.*, *9*, 159–194, 1976.

Boulton, G.S., D.L. Dent, and E.M. Morris, Subglacial shearing and crushing, and the role of water pressures in tills from south-east Iceland, *Geogr. Ann.*, *56A*, 135–145, 1974.

Budd, W.F., D. Jenssen, and U. Radok, Derived physical characteristics of the Antarctic ice sheet, *ANARE Rep. 120, Ser. A, IV, Glaciol.*, Aust. Natl. Antarct. Res. Exped., Melbourne, Australia, 1971.

Budd, W.F., D. Jenssen, and I.N. Smith, A three-dimensional time-dependent model of the West Antarctic ice sheet, *Ann. Glaciol.*, *5*, 29–36, 1984.

Clarke, G.K.C., Subglacial till: A physical framework for its properties and processes, *J. Geophys. Res.*, this issue.

Clarke, G.K.C., S.G. Collins, and D.E. Thompson, Flow, thermal structure, and subglacial conditions of a surge-type glacier, *Can. J. Earth Sci.*, *21*, 232–240, 1984.

Clayton, L., J.T. Teller, and J.W. Attig, Surging of the southwestern part of the Laurentide ice sheet, *Boreas*, *14*, 235–241, 1985.

Engelhardt, H.F., W.D. Harrison, and B. Kamb, Basal sliding and conditions at the glacier bed as revealed by bore-hole photography, *J. Glaciol.*, *20*, 469–508, 1978.

Goldthwait, R.P., Giant grooves made by concentrated basal ice streams, *J. Glaciol.*, *23*, 297–307, 1979.

Gow, A.J., H.T. Ueda, and D.E. Garfield, Antarctic ice sheet: Preliminary results of first core hole to bedrock, *Science*, *161*, 1011–1013, 1968.

Gow, A.J., S. Epstein, and W. Sheehy, On the origin of stratified debris in ice cores from the bottom of the Antarctic ice sheet, *J. Glaciol.*, *23*, 185–192, 1979.

Greischar, L.L., and C.R. Bentley, Isostatic equilibrium grounding line between the West Antarctic inland ice sheet and the Ross Ice Shelf, *Nature*, *283*, 651–654, 1980.

Grosval'd, M.G., Glaciation of the continental shelves (Part I), *Polar Geogr. Geol.*, *8*, 194–258, 1984a.

Grosval'd, M.G., Glaciation of the continental shelves (Part II), *Polar Geogr. Geol.*, *8*, 287–351, 1984b.

Hallet, B., Subglacial regelation water film, *J. Glaciol.*, *23*, 321–334, 1979.

Hjulström, F., Studies of the morphological activity of rivers as illustrated by the River Fyris, *Bull. Geol. Inst. Univ. Uppsala*, *25*, 221–527, 1935.

Hjulström, F., Transportation of detritus by moving water, in *Recent Marine Sediments*, edited by P.D. Trask, pp. 5–31, American Association of Petroleum Geologists, Tulsa, Okla., 1939.

Hughes, T.J., West Antarctic ice streams, *Rev. Geophys.*, *15*, 1–46, 1977.

Jenssen, D., W.F. Budd, I.N. Smith, and U. Radok, On the surging potential of polar ice streams, Part II, Ice streams and physical characteristics of the Ross Sea drainage basin, West Antarctica, Append. B, Meteorol. Dep., Univ. of Melbourne and Coop. Inst. for Res. in Environ. Sci., Univ. of Colo., Boulder, 1985.

Jezek, K.C., A modified theory of bottom crevasses used as a means for measuring the buttressing effect of ice shelves on inland ice sheets, *J. Geophys. Res.*, *89*, 1925–1931, 1984.

Johnson, G.L., J.R. Vanney, and D. Hayes, The Antarctic continental shelf, in *Antarctic Geoscience*, edited by C. Craddock, pp. 995–1002, University of Wisconsin Press, Madison, 1982.

Karl, H.A., E. Reimnitz, and B.D. Edwards, Extent and nature of Ross Sea unconformity in the western Ross Sea, Antarctica, in *Geology and Geophysics of the Western Ross Sea*, *Earth Sci. Ser.*, vol. 5B, edited by A.K. Cooper and F.J. Davey, Circumpacific Council for Energy and Mineral Resources, Houston, Tex., in press, 1987.

MacClintock, P., and A. Dreimanis, Reorientation of till fabric by overriding glacier in the St. Lawrence valley, *Am. J. Sci.*, *262*, 133–142 1964.

Oldale: R.N., Upper Wisconsinan submarine end moraines off Cape Ann, Massachusetts, *Quat. Res.*, *24*, 187–196, 1985.

Paterson, W.S.B., *The Physics of Glaciers*, 2nd ed., Pergamon, New York, 1981.

Powell, R.D., Glacimarine processes and inductive lithofacies modelling of ice shelf and tide-water glacier sediments based on Quaternary examples, *Mar. Geol.*, *57*, 1–52, 1984.

Robertson, J.D., Geophysical studies on the Ross Ice Shelf, Antarctica. Ph.D. thesis, 214 pp., Univ. of Wis., Madison, 1975.

Rooney, S.T., D.D. Blankenship, and C.R. Bentley, Seismic refraction measurements of crustal structure in West Antarctica, in *Gondwana Six: Structure, Tectonics, and Geophysics, Geophys. Monogr. Ser.*, vol. 40, edited by G.D. McKenzie, AGU, Washington, D.C., in press, 1987.

Rooney, S.T., D.D. Blankenship, R.B. Alley, and C.R. Bentley, Till beneath ice stream B. 2. Structure and continuity, *J. Geophys. Res.*, this issue.

Rose, K.E., Characteristics of ice flow in Marie Byrd Land, Antarctica, *J. Glaciol.*, *24*, 63–75, 1979.

Shabtaie, S., and C.R. Bentley, West Antarctic ice streams draining into the Ross Ice Shelf: Configuration and mass balance, *J. Geophys. Res.*, *92*, 1311–1336, 1987.

Shabtaie, S., I.M. Whillans, and C.R. Bentley, The morphology of ice streams A, B, and C, West Antarctica, and their environs, *J. Geophys. Res.*, this issue.

Sugden, D.E., and B.S. John, *Glaciers and Landscape: A Geomorphological Approach*, John Wiley, New York, 1976.

Sundborg, A., The River Klarälven, a study of fluvial processes, *Geogr. Ann.*, 38, 125–316, 1956.

Thomas, R.H., West Antarctic ice sheet: Present-day thinning and Holocene retreat of the margins, *Science*, 205, 1257–1258, 1979.

Thomas, R.H., and C.R. Bentley, A model for Holocene retreat of the West Antarctic ice sheet, *Quat. Res.*, 10, 150–170, 1978.

Thomas, R.H., and D.R. MacAyeal, Derived characteristics of the Ross Ice Shelf, Antarctica, *J. Glaciol.*, 28, 397–412, 1982.

Vornberger, P.L., and I.M. Whillans, Surface features of ice stream B, Marie Byrd Land, West Antarctica, *Ann. Glaciol.*, 8, 168–170, 1986.

Walker, B.F., Vane shear stress testing, in *In-Situ Testing for Geotechnical Investigations*, edited by M.C. Ervin, pp. 65–72, A.A. Balkema, Rotterdam, 1983.

Wardlaw, N.C., M.R. Stauffer, and M. Hoque, Striations, giant grooves, and superposed drag folds, Interlake area, Manitoba, *Can. J. Earth Sci.*, 6, 577–593, 1969.

Weertman, J., Water lubrication mechanism of glacier surges, *Can. J. Earth Sci.*, 6, 929–942, 1969.

Weertman, J., General theory of water flow at the base of a glacier or ice sheet, *Rev. Geophys.*, 10, 287–333, 1972.

Weertman, J., Penetration depth of closely spaced water-free crevasses, *J. Glaciol.*, 18, 37–46, 1977.

Weertman, J., and G.E. Birchfield, Subglacial water flow under ice streams and West Antarctic ice-sheet stability, *Ann. Glaciol.*, 3, 316–320, 1982.

Whillans, I.M., Force balance of ice sheets, in *The Dynamics of the West Antarctic Ice Sheet*, edited by J. Oerlemans and C.J. Van der Veen, pp. 17–36, D. Reidel, Hingham, Mass., 1987.

Young, N.W., Responses of ice sheets to environmental changes, *IASH Pub.*, 131, 331–360, 1981.

21

DRUMLIN FIELDS, DISPERSAL TRAINS, AND ICE STREAMS IN ARCTIC CANADA

A.S. Dyke and T.F. Morris

Source: *Canadian Geographer* 32(1) (1988): 86–90.

Drumlin fields, or swarms of hills streamlined in the direction of glacier flow, are common elements of glaciated landscapes. Most are believed to have formed a short distance behind the ice margin just prior to deglaciation and therefore record the final direction of ice movement. *Dispersal trains* are long and relatively narrow belts of glacial debris of some distinctive composition transported down-ice from a source region. *Ice streams* are linear zones of rapidly flowing ice within an ice sheet flanked by ice moving much more slowly. They are important features of the present Greenland and Antarctic ice sheets, where they account for more than half the ice sheet drainage (Hughes, Denton, and Fastook 1985). They are now being recognized in the former Laurentide Ice Sheet.

There is not necessarily any interconnection between these three features. Dispersal trains can be formed by slow as well as by rapidly flowing ice, and ice streams can totally evacuate all debris from their beds and leave areas of bare bedrock rather than drumlin fields.

Prince of Wales Island, in the central Canadian Arctic, has three intersecting drumlin fields as well as a spectacular dispersal plume and a peculiar 'lateral shear moraine,' a feature unknown elsewhere. These features are believed to have been formed by large ice streams comparable in size to Hudson Strait. They are well illustrated on Landsat imagery (Figure 1). Here we briefly describe them and call attention to certain implications for interpretation of dispersal trains and drumlin fields. All three drumlin fields are composed of till.

The oldest of the 'drumlin fields' on southern Prince of Wales Island (1, Figure 1) is oriented northwestward, parallel to the west coast. It is made conspicuous by inter-drumlin lakes and troughs partly filled with vegetated

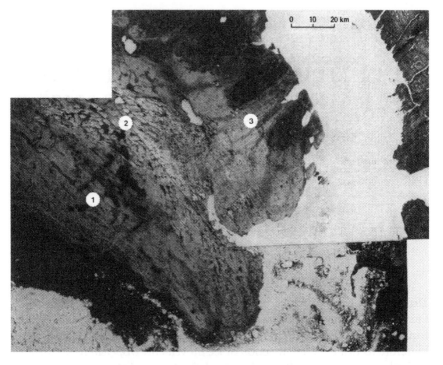

Figure 1 Three intersecting drumlin fields (numbered 1, 2 and 3, from oldest to youngest) and eastward-oriented dispersal plumes on southeastern Prince of Wales Island. The large channel in the upper right is Peel Sound (oriented north-south), with Somerset Island and Boothia Peninsula east of the sound. 1:1,000,000 scale Landsat image. North to top. Straight line SE of 2 is image boundary.

(dark-toned) raised marine sediments. Individual drumlinoid forms are several kilometres long but are very low.

A younger drumlin field (2, Figure 1) trends northwestward in an arc across the first with a counter-clockwise curving flowline. Features in this field are classic 'drumlins,' 10–30 m high, with distinct tails. Individual drumlins or drumlin clusters stand out as light-toned 'islands' on the image. The western side of this drumlin field is delineated by a single narrow ridge of till, which, with minor gaps, is 68 km long, while possible extensions occur 25 km to the south. This ridge is much lower and narrower than the drumlins of field 2, to which it is related, but individual segments have drumlinoid forms and the feature as a whole has an ice-moulded appearance.

What is the origin of this feature? Is it an attenuated drumlin, or is it a peculiar sort of left-lateral moraine, as its position at the side of drumlin field 2 might suggest? It is difficult to envisage what might cause a single body of drift to become excessively attenuated at the *edge* of a drumlin field

656

rather than nearer the axis; therefore, we propose that it is a sort of lateral moraine with two possible modes of formation.

(1) If the area west of drumlin field 2 was ice-free at the time of formation of the drumlin field and ridge, the ridge probably marks the side of a large readvance lobe. The perfect curving parallelism of the drumlins to the bounding ridge indicates that there could have been no splaying flow within this lobe, which might suggest that the features were formed by rapid, but brief, flow; i.e. a surge. The fact that drumlin field 2 and ridge arc across the older drumlins (1) could indicate that the area to the west was ice free, because no remoulding or modification of the older drumlins occurred. However, this is not a satisfactory line of argument, as will become apparent in discussing the youngest drumlin field, 3.

(2) If the area of drumlin field 1 was still ice covered when drumlin field 2 and ridge were formed, the ridge most likely marks the boundary between rapidly flowing ice to the east and ice experiencing no *basal* flow to the west. In other words, the ridge marks a shear zone at the side of an ice stream, in which case it could be referred to as a 'lateral shear moraine.' Ice to the west of the moraine must have become cold based after formation of drumlin field 1 in order to account for lack of remoulding.

The youngest drumlin field, 3, was formed by eastward-flowing ice (Figure 1). On southeastern Prince of Wales Island that flow completely obliterated the older flows, if they ever affected that area. Further west the eastward flow is superimposed on the northward flow, forming distinct tails of till on the eastern sides of the older drumlins, 2. Still further up-ice the eastward-oriented bedforms dwindle in size until they disappear and the older bedforms are left entirely unmodified by the youngest flow in the region where the ice sheet was thickest. This youngest flow affected all of eastern Prince of Wales Island and extended across Peel Sound to Somerset Island and Boothia Peninsula (Dyke 1984).

On southeastern Prince of Wales Island, the youngest flow formed a remarkable 10.5-km-wide plume of light, limestone/dolomite-charged till, spread eastward across much darker, red clastic sediments that underlie the island's east side. That plume has very abrupt and straight sides; other weaker plumes can be seen in the image (Figure 1) north and south of it. The sharp sides of the plume and the much longer transport of material within it than on either side indicate that the plume was formed by ice flowing much more rapidly than ice on either side – i.e. by an ice stream. Drumlins within it are parallel to the sides of the plume and are longest near its axis. The smaller drumlins and till tails at the head of the drumlin field indicate a strong flow convergence that likely was the triggering mechanism for the ice stream: ice streams in Antarctica and Greenland originate in such zones. The weaker plumes have slighter flow convergences at their heads.

These plumes on Prince of Wales Island are parts of a much larger dispersal train. Sedimentary rock debris from Prince of Wales Island is spread eastward across Precambrian metasedimentary rocks of Somerset Island and Boothia Peninsula, east of Peel Sound and into the Gulf of Boothia. This light-toned debris on the dark Precambrian Shield is conspicuous on Figure 1. The dispersal train that crosses the Shield there is 150 km wide. The rock terrain within it is heavily ice-scoured and moulded into extensive swarms of rock drumlins. The ice flow pattern is much more heavily imprinted than on Prince of Wales Island (up-ice) or on areas to the north and south (Dyke 1984).

The spectacular plume on southeastern Prince of Wales Island (Figure 1) is aligned with the central zone of the much larger dispersal train that crosses the Precambrian rocks to the east. Further, it is similar in size to the axial plume of another large train that crosses southern Boothia Peninsula (Dyke 1984). It is likely, therefore, that it formed in the central zone of a much wider ice stream rather than by a discrete, independent, smaller ice stream. It differs from the axial plume of the southern Boothia dispersal train, however, in two important respects: it has more abrupt sides, and it is one of several plumes, although much the best developed one, rather than the sole axial plume. We believe that these differences are due to the more headward location of the Prince of Wales Island plumes within the large dispersal train. Figure 2 attempts to summarize the glaciological conditions that might have been responsible for the association of dispersal features and landforms in the 'northern Boothia dispersal train.' Near the head of the train several plumes were set up by flow convergences separated by islands of weak or no basal flow (flow divergences). These plumes became amalgamated into a single axial plume further down ice.

Not all dispersal trains were formed by ice streams. Figure 3 contrasts those formed by ice streams (the Boothia type) with those formed under normal, or even under slower than average, regional flow ('stream flow' as opposed to 'sheet flow' of Hughes, Denton, and Fastook 1985). In the Boothia type, debris is spread down-ice from a small part of a relatively large source area. Because ice flowed across the entire region in the same direction and for the same duration, the greater transport of material in the dispersal train must indicate a zone of more rapid ice flow (an ice stream). In the other type of dispersal train, debris is spread down-ice from a relatively restricted source area, but ice-flow rate may have been uniform across the entire region. Most small dispersal trains are of this type, as are some large ones such as the giant Dubawnt dispersal train of central Keewatin (Shilts, Cunningham, and Kaszycki 1979). Boothia-type dispersal trains so far have been recognized only in the northern parts of the former Laurentide Ice Sheet. Several occur around Foxe Basin (Dyke, Dredge and Vincent 1982). Discovery of other Boothia-type dispersal trains from other parts of the Laurentide Ice Sheet would enhance our reconstruction of ice-sheet dynamics.

These observations have some implications for how drumlins relate to deglacial history and for where they form. (1) Not all drumlin fields date

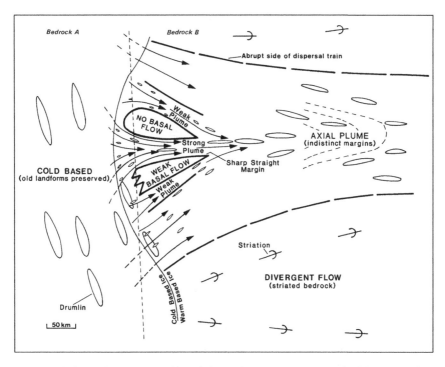

Figure 2 Schematic representation of the major components of the large 'northern Boothia dispersal train' and probable glaciological conditions that formed them.

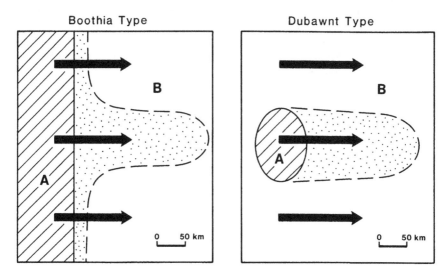

Figure 3 Two types of large dispersal trains: the Boothia type, formed by ice streams, and the Dubawnt type, formed under normal but sustained regional flow. A and B are two distinctly different rock types; arrows indicate ice-flow direction; dots indicate glacial dispersal of debris from rock type A.

from the final stages of deglaciation, or necessarily from deglaciation at all. We believe, for instance, that the eastward flow in the central Arctic dates from the last glacial maximum (Dyke 1984); if so, two older drumlin fields must date from the last glacial build-up or earlier. (2) The morphological zonation of streamlined landforms both across and along large dispersal trains (ice streams), and the strong flow convergence indicated by drumlinoid forms at the heads of former ice streams, near the former ice divide, mean that drumlin formation occurred along the entire ice stream and was not restricted to the marginal zone of the ice sheet. (3) Likely areas for preservation of glacial landforms ante-dating the last glacial maximum are areas that lay beneath ice divides. There, the horizontal component of flow is low and basal sliding may not occur.

Note

This article is Geological Survey of Canada Contribution 53186.

References

DYKE, A.S. 1984 'Quaternary geology of Boothia Peninsula and northern District of Keewatin, central Canadian Arctic' Geological Survey of Canada, Memoir 407

DYKE, A.S., DREDGE, L.A., and VINCENT, J-S. 1982 'Configuration and dynamics of the Laurentide Ice Sheet during the late Wisconsin maximum' *Géographie physique et Quaternaire* 36, 5–14

HUGHES, T.J., DENTON, G.H., and FASTOOK, J.L. 1985 'The Antarctic Ice Sheet: an analog for Northern Hemisphere paleo-ice sheets' in M.J. Woldenberg ed *Models in Geomorphology* (Boston: Allen and Unwin) 25–72

SHILTS, W.W., CUNNINGHAM, C.M., and KASZYCKI, C.A. 1979 'Keewatin Ice Sheet – re-evaluation of the traditional concept of the Laurentide Ice Sheet' *Geology* 7, 531–7

22

STRUCTURE AND STABILITY OF THE FORMER SUBGLACIAL DRAINAGE SYSTEM OF THE GLACIER DE TSANFLEURON, SWITZERLAND

M. Sharp, J.C. Gemmell and J.-L. Tison

Source: *Earth Surface Processes and Landforms* 14 (1989): 119–34.

Abstract

A well-developed subglacial drainage system consisting of large cavities developed in the lee of bedrock steps connected together by a network of Nye channels is exposed on an area of recently deglaciated limestone bedrock in front of Glacier de Tsanfleuron, Switzerland. This system covers some 51 per cent of the bedrock surface area, and is believed to have transported the bulk of supraglacially-derived meltwaters through the glacier. Using the cavity hydraulics model of Kamb (1987), it is shown that the geometry of the system rendered it stable against collapse by meltback of channel roofs into a tunnel-dominated system. For likely combinations of glacier geometry and meltwater discharge, the steady state water pressure in this system would have been only a small fraction of that required for flotation, and for discharges of less than about 0.5–5 m^3 s^{-1} water would have flowed at atmospheric pressure. The system appears to have adjusted to varying discharges by a combination of varying water pressure and changing the total cross-sectional area of flow by altering the number of active channels connecting cavities. Glacier sliding velocity would have been independent of meltwater discharge for discharges at which water flowed at atmospheric pressure, but would have risen with increasing discharge for higher flows. Velocities on the order of 0.1 m d^{-1} are predicted for a realistic range of discharges and effective pressures, and these are believed to be plausible. Episodes of enhanced sliding in glaciers with similar drainage systems could be triggered by a rise in meltwater discharge across the threshold between flows at atmospheric pressure and flow under pressure from the glacier.

Introduction

In recent years, the realization that the sliding velocity of glaciers is sensitively dependent upon the subglacial effective pressure (ice overburden pressure minus subglacial water pressure) (Iken, 1981; Kamb et al., 1985; Iken and Bindschadler, 1986) has focussed attention on our ability to predict the effective pressure as a function of meltwater discharge, glaciological conditions, and drainage network properties (Röthlisberger, 1972; Bindschadler, 1983; Walder, 1986; Kamb, 1987). Early work concentrated upon the relationships between water pressure and discharge in tunnels incised upwards into the base of glaciers (Röthlisberger, 1972), and predicted an inverse relationship under steady state conditions. One consequence of this relationship is that large conduits should capture water from smaller ones with a lower discharge, and should therefore grow at their expense, with the drainage converging on a single channel in which water pressure is relatively low (and effective pressure therefore relatively high). Water transfer through such a system should be efficient and rapid, as suggested by the results of many dye injection experiments carried out on glaciers (Collins, 1982; Burkimsher, 1983). In glaciers with tunnel-dominated drainage systems, there should, under steady-state conditions, be an inverse relationship between meltwater discharge and glacier sliding velocity. In reality, however, this relationship tends to break down under conditions of rapidly rising discharge (Spring and Hutter, 1982), such that over the short term water pressure and sliding velocity may both increase with discharge.

More recently, attention has been directed towards the analysis of the behaviour of drainage systems made up of networks of cavities which are located downstream from bedrock obstacles and are connected together by multiple narrow orifices. The existence of such systems has been postulated to explain field observations made at Findelengletscher, Switzerland (Iken and Bindschadler, 1986) and during the 1982–83 surge of Variegated Glacier, Alaska (Kamb et al., 1985). These observations reveal the existence of subglacial drainage systems through which water transit velocities are very low, in which injected dye undergoes considerable spatial and temporal dispersion, and in which measured water pressures are too high to be accounted for by the single conduit model. In such systems, slow transit speeds are accounted for by the throttling of water flow through orifices which results in the temporary storage of water in cavities, while dye dispersion results from the existence of multiple interconnections between cavities within the drainage network. Analyses of the hydraulics of such systems indicate a positive relationship between the subglacial water pressure and meltwater discharge (and hence a negative relationship between discharge and effective pressure) (Walder, 1986; Kamb, 1987). They have, however, been based upon idealizations of drainage network geometry which, though necessary, are not well constrained by field observations.

Subglacial cavities and orifices form on the leeside of bedrock eminences as a glacier slides over a rough bed. Their geometry is initially controlled by the subglacial water pressure and the glacier sliding velocity, higher values of either parameter producing greater ice–bedrock separation. This geometry can, however, be modified by roof melting induced by viscous heat dissipation in water flowing through the cavities and orifices. In the model proposed by Kamb (1987), it is assumed that head loss is confined to the orifices, in which the local hydraulic gradient is given by:

$$(\alpha/\omega)(L_c/L_o) \tag{1}$$

where α is the overall hydraulic gradient (taken as the surface slope of the glacier), ω is the average tortuosity of flow paths through the drainage system, and (L_c/L_o) is a 'head concentration factor' which expresses the ratio between the lengths of that part of the drainage path which is confined to cavities (L_c) and that part which is confined to orifices, (L_o). Kamb defines a 'melting stability parameter', Ξ, which measures the relative importance of viscous heating as an influence on orifice geometry. Where $\Xi \geq 1$ the system is dominated by the effects of viscous heating and the orifices are unstable against rapid growth in response to increases in water pressure or orifice size above the steady state values. Under these conditions, some orifices may grow into tunnels with steady state water pressures below those prevailing in the remaining cavities. Such tunnels will capture water from the cavities and the cavity system will not survive as a stable drainage configuration capable of conducting the bulk of through-flowing meltwaters. When $\Xi \leq 1$, however, the sliding velocity of the glacier renders the orifices stable against such perturbations, and they will persist. It is then possible to determine the steady state effective pressure as a function of the meltwater discharge and the geometry of the cavity network, and to use it as an input to effective pressure-dependent sliding laws in order to investigate the relationship between water discharge and sliding velocity. Since the sliding velocity is sensitively dependent upon the water pressure, the cross-sectional area of individual orifices increases rapidly as water pressure rises, resulting in an increase in discharge with water pressure.

In principle, Kamb's model can be used to explain a number of glaciological phenomena. Kamb himself argues that glacier surges can be explained by a transition between a tunnel-dominated drainage system which exists during the quiescent phase of the surge cycle and a cavity-dominated system with lower steady-state effective pressures which exists during surges. A permanently stable cavity system which yields low steady-state effective pressures might explain the equilibrium fast flow behaviour of some ice streams, while permanently stable cavity systems, which presumably yield high steady-state effective pressures, have been described from beneath non-surging glaciers by Walder and Hallet (1979) and Hallet and Anderson (1980). It is therefore of interest to determine whether these different situations can be explained in terms of variations in meltwater discharge and drainage system geometry.

Study site and aims

The present study was carried out at the Glacier de Tsanfleuron, a 4 km^2 plateau glacier located at elevations between 2420–2850 m a.s.l. on the northern side of the Rhône valley above Sion, Switzerland (Figure 1). The glacier flows from west to east and, except for a 700 m long narrow tongue, it terminates on a plateau composed of massive Cretaceous limestones. Since about 1860, it has been retreating from a moraine ridge which is now located about 1.6 km from the ice front. As it has done so, it has exposed about 2.5 km^2 of smoothed and polished bedrock (Figure 2) on which are preserved subglacial calcite precipitates (Lemmens *et al.*, 1982; Souchez and Lemmens, 1985) and the vestiges of an interconnected cavity drainage system. We have mapped a representative 30 × 30 m tract of this bedrock in

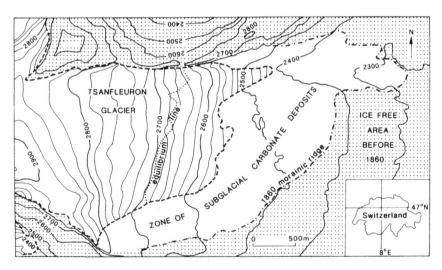

Figure 1 Map showing the location of Glacier de Tsanfleuron, and the geometry of the glacier and its proglacial area.

Figure 2 Recently deglaciated limestone bedrock surface in front of Glacier de Tsanfleuron—the location of the area mapped in detail in this study.

considerable detail (Figure 3), and have used the map to determine appropriate values for such drainage system properties as the tortuosity of flow paths, the number of independent orifices found along a transverse section across the glacier (N_o), and the proportion of flow paths in cavities and orifices, all of which are required as input to the Kamb model. We have then used these values to assess the stability of the observed system against roof meltback, and to determine the form of the effective pressure–discharge relationship for the system. Finally, for a realistic range of meltwater discharges, we have examined the implications of the predicted steady state effective pressures for the sliding velocity of the glacier. As a result of this study we are able to identify a number of differences between the cavity system which we have mapped and the idealized cavity geometry assumed by Kamb (1987).

Mapping technique

The topography and micro-scale geomorphology of a 30×30 m area were mapped as follows. A regular 2×2 m grid of primary control points was laid out across the study area, and located by plane table and alidade survey. Elevations of the control points, and locations and elevations of a number of other points located at major breaks of slope were determined by

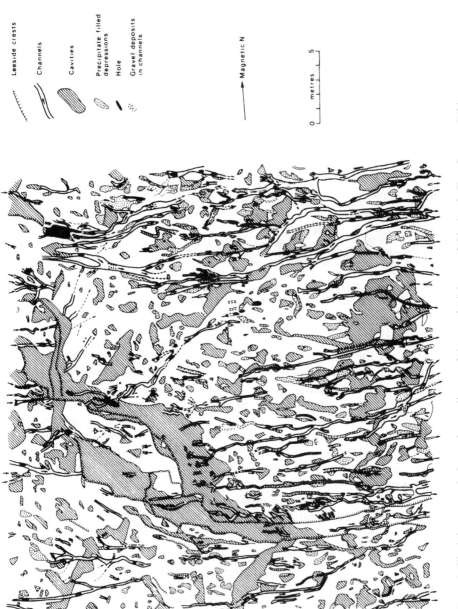

Figure 3 Glacial geomorphology of a small area of the proglacial area of Glacier de Tsanfleuron. White areas represent areas of intimate ice–bedrock contact.

Leeside crests

Channels

Cavities

Precipitate filled depressions

Hole

Gravel deposits in channels

Magnetic N

0   metres   5

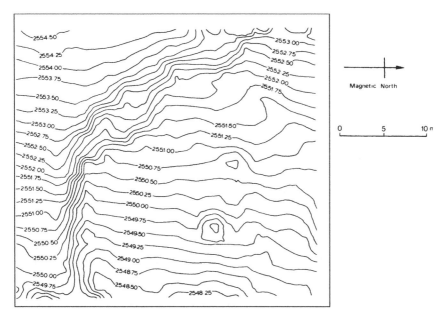

Figure 4 Topography of the area mapped in Figure 3.

surveying with a Wild T2 theodolite and Kern DM102 electrooptical distance meter. The topography of the area is shown in Figure 4. Geomorphological detail within each 2 × 2 m grid square was added by sketching directly onto the plane table map. Accuracy in the location of major geomorphological contacts is believed to be better than 0.2 m, and is comparable to that obtained by Hallet and Anderson (1980).

Geomorphology of the mapped area

To provide a basis for mapping, the surface was divided into a number of units, each of which reflects the action of a distinctive suite of geomorphic processes. The surface properties of these geomorphic units were very different in character from those of strongly weathered limestone surfaces located immediately outside the 1860 moraine, and they are therefore believed to be primarily subglacial in origin. The units defined were essentially the same as those recognized by Walder and Hallet (1979): (a) areas of intimate ice–bedrock contact; (b) channels incised into the bedrock (Nye channels); (c) cavities on the leeside of bedrock obstacles; (d) surface depressions filled with calcite precipitates; and (e) karst sinkholes. The proportions of the total mapped area taken up by each terrain type is shown in Table I, together with comparable figures for Blackfoot Glacier (Walder and Hallet,

667

Table I Proportions of deglaciated bedrock surfaces at Glacier de Tsanfleuron, Blackfoot Glacier and Castleguard Glacier occupied by various geomorphic elements.

Geomorphic element	Tsanfleuron	% cover Blackfoot	Castleguard
Intimate ice–bedrock contact	47.2 ⎫	70	
Nye Channel	16.1 ⎭		
Cavities	28.3 ⎫		⎫
Precipitate-filled depressions	7.1 ⎭	20	30 ⎬
Karst sinkholes	1.3		⎭

1979) and Castleguard Glacier (Hallet and Anderson, 1980). These figures suggest that the cavity system (including precipitate-filled depressions) at Glacier de Tsanfleuron is relatively more extensive than at the other two glaciers.

Areas of intimate ice–bedrock contact

The former existence of close contact between ice and bedrock is indicated by (i) striated bedrock, which indicates that sliding ice was dragging clasts across the glacier bed; (ii) solution furrows on the upstream side of bedrock bumps, which indicate dissolution of subglacial bedrock by subglacial waters produced by pressure melting, and (iii) subglacial calcite precipitates, which often consist of spicular crystals aligned in the former direction of ice movement. These precipitates are believed to form by the rejection of calcite from saturated regelating waters where they refreeze on the downstream side of bedrock bumps (Hallet, 1976) (Figure 5a).

Nye channels

Linear channels of dimensions 0.1–0.2 m wide and *c*. 0.1 m deep are spaced at intervals of approximately 1.5 m across the glacier bed (Figure 5b). They are aligned parallel to the former ice flow direction at 080–100°E. They form an anastomosing rather than a dendritic network, and about 75 per cent of them have blind terminations. Where such terminations occur, however, it is often possible to detect evidence of more limited water flow across the bed to connect with other channel segments or cavities further downglacier. This suggests that the blind channels were active on an irregular basis. The Nye channels function as a link between major cavities located on the downstream side of vertical steps in the bedrock surface. It therefore appears that these cavities and their interconnecting Nye channels provide the major route by which waters drain across the glacier bed. Tracing of potential drainage paths across the mapped area suggests that approximately 64 per cent

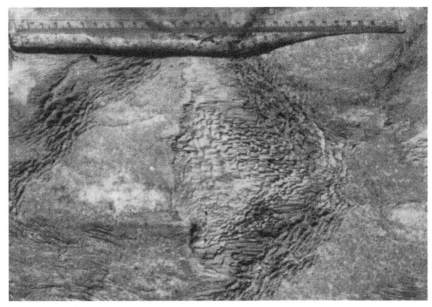

A

B

Figure 5 (a) Detail of an area of intimate ice–bedrock contact, showing striated bedrock, solution furrows and subglacially-precipitated calcites. Ice flow was from left to right; (b) Network of Nye channels incised into bedrock.

669

C

D

Figure 5 (c) Leeside cavity, showing water-polished bedrock surface; (d) Precipitate-filled depression at the margin of a Nye channel.

670

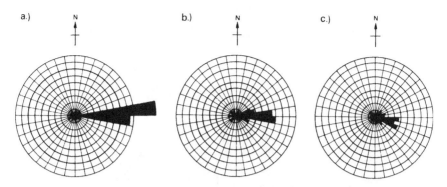

Figure 6 Rose diagrams showing the orientation of (a) bedrock striations, (b) Nye channels, and (c) long axes of cavities. Each infilled section represents one observation.

of the length of these drainage paths is in Nye channels. The geometry of this drainage system is similar in many respects to that assumed in Kamb's 'step-cavity' model, so this model is used in subsequent analysis, with each Nye channel treated as an orifice. The main geometric differences between the system which we have mapped and that envisaged by Kamb are (i) that the orifices at Glacier de Tsanfleuron are aligned parallel to rather than normal to the direction of ice flow (Figure 6), (ii) that the measured mean tortuosity of potential drainage paths (1.07) is significantly less than assumed by Kamb, and (iii) that there is evidence, in the form of channels with blind terminations and infills of precipitate, that the number of active orifices may vary with discharge or water pressure (or both), rather than remain constant as assumed by Kamb. This suggests that to accommodate increasing discharges the system may enlarge the total cross-sectional area of active orifices by opening up additional orifices rather than by enlarging existing orifices, as assumed by Kamb. This process is, however, also likely to be activated by an increase in cavity water pressure as discharge rises, such that increased glacier-bed separation at the downstream end of cavities opens the connection to new orifices. We consider that the similarity between this process and that described by Kamb is sufficient to justify the use of Kamb's model for a first analysis of the behaviour of the system.

Leeside cavities

Cavities, defined by areas of bedrock showing evidence of water washing and solution rather than of glacial abrasion, are extensively developed on the downstream side of bedrock obstacles (Figure 5c). Some are associated with the active fracture of leeside surfaces as evidenced by signs of recent block removal, while others seem to be associated with the preservation of preglacial surfaces which are inclined too steeply downglacier for the glacier

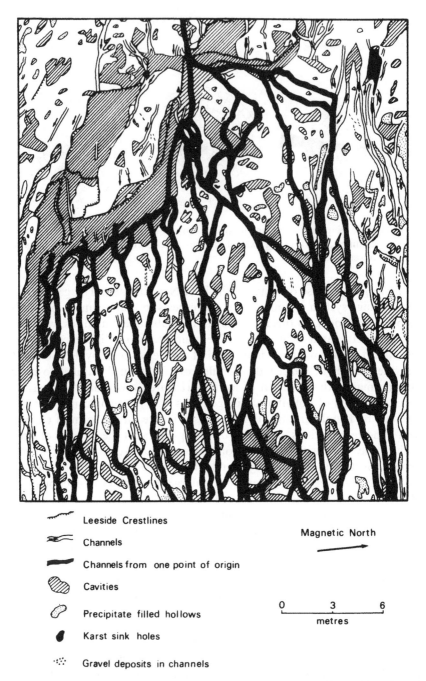

Leeside Crestlines

Channels

Channels from one point of origin

Cavities

Precipitate filled hollows

Karst sink holes

Gravel deposits in channels

Magnetic North

0   3   6

metres

Figure 7 Potential flow routes of water entering the mapped area in one Nye channel located near the centre of the area, to show the dispersing effect of flow through cavities.

to remain in contact with the bed along them. Such surfaces often preserve evidence of karren features which are extensively developed on limestone outcrops outside the 1860 moraine, but which have generally been removed by glacial erosion within it. Three main types of cavity have been identified: (i) large cavities aligned at high angles to the ice flow direction which are developed in the lee of steps in the bedrock surface. These are fully integrated into the main drainage system and are connected together by Nye channels. They provide considerable potential for the lateral dispersion of subglacial meltwaters (Figure 7), indicating that it is in the cavities rather than the channels where most such dispersion occurs; (ii) cavities elongated in the direction of ice flow, which appear to be linearly continuous with Nye channels; and (iii) small cavities developed in the lee of wave-like bumps on the glacier bed, which appear to be isolated from the step-cavity/Nye channel drainage system, and which frequently contain subglacial calcite precipitates. The presence of these precipitates suggests that these cavities form part of a drainage system which transports regelating waters rather than bulk waters derived from the glacier surface. The extent to which this drainage system interacts with the main cavity system is unknown and needs to be investigated further.

Precipitate-filled depressions

Depressions almost totally infilled with calcite precipitates occur quite widely at the head of and along the margins of Nye channel segments (Figure 5d). Their location suggests that they may reflect the temporary incursion of glacier ice and associated regelating waters into the step-cavity/Nye channel system at times of low discharge. This provides evidence that the number of active orifices may vary over time.

Karst sinkholes

There are a number of well-developed sinkholes within the mapped area, almost all of which are fed directly by Nye channels or elongated cavity segments. This indicates that there is an interconnection between the subglacial drainage system and the subterranean karst drainage system. Interestingly, a number of the sinkholes also have Nye channels leading away from them in a downglacier direction, suggesting either that at times of high discharge they must fill up and overflow, or that they periodically become infilled with ice.

The geometry and stability of the subglacial drainage system

The above results suggest that, as at Blackfoot and Castleguard Glaciers, the subglacial drainage system of Glacier de Tsanfleuron has two principal

Table II Values assigned to parameters in the cavity hydraulics model of Kamb (1987) in comparison to those used in this study.

	Symbol	Tsanfleuron	Kamb (1987)
Hydraulic gradient	α	0.13	0.1
Sinuosity	ω	1.07	4
Number of orifices	N_o	1,000	50
Step height	h	0.1 m	0.05 m
Head concentration factor	L_c/L_o	0.57	10

components. These are a system of large step cavities connected together by Nye channels and elongated cavity segments, and a system of isolated cavities connected by the regelation water film (*cf.* Hallet, 1979). These two components respectively drain primary supraglacially- and primarily subglacially-derived waters. In subsequent analysis, we consider only the step-cavity/Nye channel part of the system, and we treat the Nye channels as 'orifices'.

From our knowledge of the geometry of Glacier de Tsanfleuron and analysis of our map, we have derived typical values for those properties of the drainage system which enter into Kamb's model—namely the hydraulic gradient (α) (taken as the surface slope of the glacier), the tortuosity of the drainage paths (ω), the number of independent orifices across the glacier (N_o) (extrapolated from their density in the mapped area), the height of the steps which initiate orifices (h), and the 'head concentration factor' (L_c/L_o). Values for these parameters are listed in Table II, where they are compared to the values assumed by Kamb (1987) in his analysis of the drainage system of Variegated Glacier. The Glacier de Tsanfleuron system has many more orifices, higher step heights, much less tortuous drainage paths, and a much lower head concentration factor than assumed by Kamb.

The stability of this drainage system against collapse by roof-melting was analysed by calculation of Kamb's melting stability parameter, Ξ:

$$\Xi = \frac{2^{1/3}}{\pi^{1/2}} \frac{(\alpha(L_c/L_o)/\omega)^{3/2}}{DM} \frac{(\eta_R)^{1/2}}{(v_R\sigma_R)} h^{7/6} \frac{(\tau_R)^{3/2}}{\tau} \tag{2}$$

In this calculation the following parameter values, consistent with Kamb (1987), are used:

$$D = 31,000 \text{ m} \qquad \rho_i H/\rho_w g$$
$$M = 0.1 \text{ m}^{-1/3}\,\text{s}^{-1} \qquad \text{Manning roughness}$$
$$\eta_R = 10 \text{ kPa yr}^{-1} \qquad \text{Reference ice viscosity}$$
$$\sigma_R = 600 \text{ kPa} \qquad \text{Reference effective pressure}$$
$$v_R = 1 \text{ m d}^{-1} \qquad \text{Reference sliding velocity}$$

Here ρ_i and ρ_w are the densities of ice and water, g is the acceleration due to gravity, and H is the latent heat of melting. The three reference parameters are derived by substitution from equations 7 and 36 of Kamb (1987) which describe the dependence of ice viscosity and sliding velocity on effective stress, and which are important in relating the orifice geometry to the sliding process. As in Kamb's calculations, it is assumed that the shear stress, τ, is equal to the reference shear stress, τ_R (the result is insensitive to this decision). The resulting value of Ξ, 0.057, is significantly lower than Kamb's value of 0.18, and much less than the critical value of 1 at which a melting instability develops. Examination of the parameter values used in the calculation reveals that the Glacier de Tsanfleuron drainage system derives its stability from the low value of the head concentration factor. The relatively great length of orifices reduces the local hydraulic gradient and hence minimizes the tendency of the orifices to enlarge by roof melting.

Effective pressure–discharge relationships

Having demonstrated that the orifices are stable against melting instabilities, it is now possible to determine the relationship between effective pressure (σ) and discharge in the drainage system. According to equation 38a of Kamb (1987):

$$Q_w = \frac{2^{4/3}}{\pi^{1/2}} \frac{N_o}{M} \left(\frac{\alpha(L_c/L_o)}{\omega} \right)^{1/2} \left(\frac{\eta_R V_R}{\sigma_R} \right)^{1/2} h^{13/6} \phi \left(\frac{\tau}{\tau_R} \right)^{3/2} \left(\frac{\sigma_R}{\sigma} \right)^3 \tag{3}$$

This relation is based upon the Manning equation for turbulent flows, and relates the meltwater discharge to the effective pressure via the latter's influence on the cross-sectional area of the orifices and the velocity of water flow. In this calculation, a value of 0.417 is taken for the 'flux factor', ϕ, as read from Figure 7 in Kamb's paper. In the light of the evidence that about 75 per cent of the orifices have blind terminations, two sets of calculations were carried out. In the first it was assumed that all orifices were active simultaneously ($N_o = 1{,}000$), whereas in the second it was assumed that only those without blind terminations were active ($N_o = 267$). Using the stated parameter values, Equation 3 reduces to:

(a) $\qquad\qquad Q_w = 26.27(6/\sigma)^3 \quad (N_o = 1{,}000)$

(b) $\qquad\qquad Q_w = 7.01(6/\sigma)^3 \quad (N_o = 267)$

These relationships are plotted in Figure 8, along with the corresponding relationship derived by Kamb. The result is that, whatever the number of active orifices, effective pressures for a given discharge are considerably

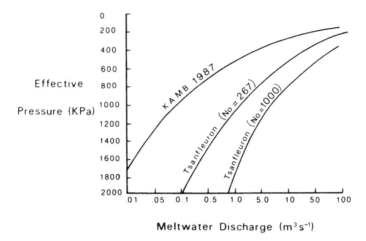

Meltwater Discharge (m³ s⁻¹)

Figure 8 Relationship between meltwater discharge and effective pressure for the two cases $N_o = 1,000$ and $N_o = 267$. The relation determined for Variegated Glacier by Kamb (1987) is shown for comparison.

higher than for the cavity configuration assumed by Kamb for Variegated Glacier. This difference arises because of the larger number of active orifices, the larger step height, the lower tortuosity of the drainage paths, and the steeper overall hydraulic gradient at Glacier de Tsanfleuron, all of which facilitate drainage of water from the glacier.

To determine the likely values of effective pressure beneath Glacier de Tsanfleuron, it is necessary to estimate the average meltwater discharge and the number of active orifices. Discharge can be estimated by considering the runoff which would be generated by different ablation rates acting over a glacier 4 km² in area (Table III). For ablation rates of less than 0.1 m d⁻¹, mean discharges will be less than 5 m³ s⁻¹. A discharge of 1 m³ s⁻¹ would

Table III Relationships between ablation rates, mean discharge, effective pressure, and sliding velocity.

Ablation rate $(m\ d^{-1})$	Mean discharge $(m^3\ s^{-1})$	Effective pressure $(N_o = 1,000)$ (kPa)	Sliding velocity $(\tau_b = 150\ kPa)$ $(m\ d^{-1})$	Effective pressure $(N_o = 267)$ (kPa)	Sliding velocity $(\tau_b = 150\ kPa)$ $(m\ d^{-1})$
0.002	0.1	3,700	0.004	2,454	0.015
0.011	0.5	2,220	0.019	1,436	0.073
0.022	1.0	1,770	0.039	1,140	0.146
0.108	5.0	1,040	0.193	671	0.710
0.216	10.0	830	0.383	533	1.430
2.160	100.0	390	3.750	250	13.820

correspond to an effective pressure of 1,100–1,800 kPa, depending upon the number of active orifices. Since ice thicknesses at Glacier de Tsanfleuron are unlikely to exceed 200 m (equivalent to a normal pressure of 1,766 kPa or less), subglacial water pressures must be only a small fraction of that for flotation (0–38 per cent for a 200 m thick glacier and 1 m^3 s^{-1} discharge).

If we make the assumption that the adjustment to varying discharge takes place solely by varying the number of active orifices, the number of orifices required to transmit a given discharge can be estimated by using the Manning equation to calculate the velocity of flow through an orifice of 0.01 m^2 cross-sectional area (comparable to the dimensions of Nye channels at Glacier de Tsanfleuron). The Manning equation has the form:

$$u = \left[\frac{A_c}{4\pi}\right]^{1/3}\left[\frac{((\alpha/\omega)(L_c/L_o))^{1/2}}{M}\right] \tag{4}$$

Here U is the velocity of water flow, $((\alpha/\omega)(L_c/L_o))$ is the local hydraulic gradient, A_c is the cross-sectional area of the channel, and M is the Manning roughness. This predicts a velocity of 0.25 m s^{-1}, an order of magnitude higher than the transit velocities measured at Variegated Glacier during its surge (Kamb *et al.*, 1985), but still at the lower end of the range of velocities determined from dye tracing experiments on non-surging glaciers (Collins, 1982). With a discharge through each orifice of 0.0025 m^3 s^{-1}, 400 orifices would be needed to transport a discharge of 1 m^3 s^{-1} and 2,000 to transport 5 m^3 s^{-1}. A discharge of around 0.67 m^3 s^{-1} could be transported by those continuous orifices which are believed to have been permanently active, while additional orifices would be needed to transport larger discharges. Examination of Figure 8, however, suggests that increasing discharges are not accommodated solely by an increase in the number of active orifices. The steady-state effective pressures for 267 orifices transmitting 0.67 m^3 s^{-1} and 1,000 orifices transmitting 2.5 m^3 s^{-1} would both be around 1,500 kPa. If effective pressure did not fall with rising discharge, however, there would be no mechanism available to increase the number of active orifices. Adjustment to varying discharges must therefore involve a combination of varying orifice dimensions and varying orifice numbers. The water pressure–discharge relationship will then depend upon the rate at which cavity growth is able to activate new orifices, and it will be strongly influenced by the geometry of the glacier bed. Future analyses of cavity hydraulics should consider this effect explicitly.

Implications for glacier sliding

The above considerations suggest that for meltwater discharges of less than 5 m^3 s^{-1} (corresponding to ablation rates of <0.1 m d^{-1}) effective pressures would range from 600 to 3,800 kPa, depending upon the number of active orifices. Although the thickness of the glacier is unknown, it can be estimated from a knowledge of its surface slope (0.13) and an assumed value for the basal shear stress using the relation:

$$\tau = f\rho_i gh \sin \alpha \tag{5}$$

where f is a shape factor, ρ_i is the density of ice, g is gravity, h is the ice thickness and α the slope angle. Taking $f = 0.8$, this gives thickness values of 109 m for a basal shear stress of 100 kPa and of 163 m for 150 kPa. The maximum possible effective pressures for these ice thicknesses are 962 kPa and 1,439 kPa respectively. Given the effective pressure–discharge relationships predicted earlier, this suggests that, depending upon the number of active orifices, for discharges of less than 0.5–5.0 m^3 s^{-1} maximum possible effective pressures may be less than those predicted by the theory. Under these conditions, subglacial water flow would be at atmospheric pressure (open channel flow). Only at higher discharges would pressurized flow occur, so only then would effective pressure vary and affect the sliding velocity. This result is also consistent with our estimate of the discharges which could be carried by the permanently active orifices alone, and suggests that flow in these orifices may often have been at atmospheric pressure.

Kamb (1987) suggested the following empirical relationship between glacier sliding velocity (v) and effective pressure:

$$V = V_R \left[\frac{\tau}{\tau_R} \cdot \frac{\sigma_R}{\sigma} \right]^3 \tag{6}$$

Using this relationship and the parameter values given earlier, and assuming that $\tau = \tau_R$, then if $Q_w = 1$ m^3 s^{-1}, $N_o = 267$ and $\sigma = 1,140$ kPa, $v = 0.146$ m d^{-1}. If $Q_w = 5$ m^3 s^{-1}, $N_o = 1,000$ and $\sigma = 1,038$ kPa, $v = 0.193$ m d^{-1} (Figure 9; Table III). Although we have no direct measurements of the sliding velocity of Glacier de Tsanfleuron, these values are not greatly dissimilar (given the empirical nature of the sliding law) from the rate of 0.03 m d^{-1} measured over similar bedrock topography beneath Grinnell Glacier, Montana, by Anderson et al. (1982). They are considerably larger, however, than value of 0.0005 m d^{-1} calculated for Glacier de Tsanfleuron from measured surface roughnesses with the sliding theory of Nye (1969) on the assumption of no cavitation and a 100 kPa basal shear stress. They are also sufficiently low to confirm our earlier suggestion that stable cavity systems with low steady-state water pressures can exist beneath non-surging glaciers. Since all warm-based glaciers resting on bedrock substrates slide over irregular beds, we suggest that a network of leeside cavities will exist beneath all such glaciers, although its geometry may vary considerably with bedrock type and with

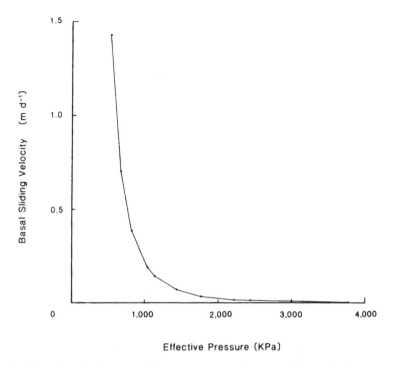

Figure 9 Relationship between effective pressure and glacier sliding velocity predicted by the sliding law given in Equation 5.

changes in the orientation of ice flow relative to the strike of bedding. This network is likely to be the main route by which surface-derived meltwaters pass through the glacier unless the particular combination of network and glacier geometry is such as to render the system unstable against roof meltback. If this is the case a tunnel-dominated drainage system is likely to develop at the expense of the cavity system, and will drain the bulk of supraglacially-derived meltwaters. In such cases, meltwater erosion is unlikely to enhance the development of the cavity orifice system and it may therefore be hard to detect on deglaciated bedrock surfaces. Cavity-dominated systems may develop particularly commonly on carbonate substrates in which permanently active orifices (i.e. small Nye channels) can be created by active solution of the substrate (existing mapped cavity systems are all from carbonate terrains).

Conclusions

Surface-derived meltwaters are evacuated from Glacier de Tsanfleuron via a network of step cavities linked by Nye channels spaced at approximately

1.5 m intervals across the bed. This network covers approximately 51 per cent of the glacier bed. The channels are aligned subparallel to the ice flow direction and drainage paths have low tortuosity. From 51 to 83 per cent of the length of individual drainage paths occurs in channels, but the cavities play the major role in causing lateral dispersion of flow. The step cavity drainage system model of Kamb (1987) predicts that this drainage system is stable against meltback instability of channel roofs, and that for meltwater discharges of less than about $0.5-5$ m^3 s^{-1} water flow will be at atmospheric pressure. Effective pressures at the glacier bed are then a direct function of the ice thickness. For higher discharges, water flows under pressure and effective pressures are reduced, though in this case only slightly for what appears to be a realistic range of discharges. Sliding velocities calculated for appropriate values of effective pressure using an empirical sliding law seem realistic. The Kamb model can, therefore, successfully predict the stability of a cavity-channel system in which subglacial water pressures are low. It can thus account for the low sliding velocities which occur beneath glaciers like Glacier de Tsanfleuron.

Three differences are, however, identified between Kamb's step cavity model and the cavity system mapped here. These are:

1. That the orifices which provide the linkages between cavities are aligned subparallel to the ice flow rather than at right angles to it as assumed by Kamb.
2. As a result of (1), the drainage system may partly adjust to changing discharges by varying the number of active orifices, rather than solely by enlargement of those which are already active. Activation of additional orifices will, however, require rising water pressures to increase the extent of glacier-bed separation and open up connections between cavities and new orifices. In this respect, the process of adjustment may be very similar to that envisaged by Kamb, and may justify continued use of his model.
3. In relatively thin glaciers, like Glacier de Tsanfleuron, the ice thickness may set an upper limit to the effective pressure which is below the values predicted by Kamb's theory. This implies the existence of a threshold meltwater discharge, below which water flows at atmospheric pressure and sliding velocity is independent of discharge. The crossing of such a threshold could account for the initiation of rapid motion events in such glaciers, and would allow them to be explained without recourse to changes in flow routing or drainage system structure.

Acknowledgements

We gratefully acknowledge financial support from Fitzwilliam College, Cambridge, and Christ Church, Oxford. We are indebted to Professor Roland

Souchez for introducing us to the geomorphology of the proglacial area of Glacier de Tsanfleuron. Pam Nickell assisted greatly with the fieldwork. Dr K.S. Richards provided a helpful review of an earlier draft.

References

Anderson, R.S., Hallet, B., Walder, J.S., and Aubry, B.F. 1982. 'Observations in a cavity beneath Grinnell Glacier', *Earth Surface Processes and Landforms*, **7**, 63–70.

Bindschadler, R.A. 1983. 'The importance of pressurized water in sliding and separation at the glacier bed', *Journal of Glaciology*, **29**, 3–19.

Burkimsher, M. 1983. 'Investigations of glacial hydrological systems using dye tracer techniques: observations at Pasterzengletscher, Austria', *Journal of Glaciology*, **29**, 403–416.

Collins, D.N. 1982. 'Flow-routing of meltwater in an Alpine glacier as indicated by dye tracer tests', *Beitrage zur Geologie der Schweiz—Hydrologie*, **Bd 28 II**, 523–534.

Hallet, B. 1976. 'Deposits formed by subglacial precipitation of $CaCO_3$', *Geological Society of America Bulletin*, **87**, 1003–1015.

Hallet, B. 1979. 'Subglacial regelation water film', *Journal of Glaciology*, **23**, 321–334.

Hallet, B. and Anderson, R.S. 1980. 'Detailed glacial geomorphology of a proglacial bedrock area at Castleguard Glacier, Alberta, Canada', *Zeitschrift fur Gletscherkunde und Glazialgeologie*, **16**, 171–184.

Iken, A. 1981. 'The effect of the subglacial water pressure on the sliding velocity of a glacier in an idealised numerical model', *Journal of Glaciology*, **27**, 407–421.

Iken, A. and Bindschadler, R.A. 1986. 'Combined measurements of subglacial water pressure and surface velocity at Findelengletscher, Switzerland: conclusions about drainage system and sliding mechanism', *Journal of Glaciology*, **32**, 101–119.

Kamb, W.B. 1987. 'Glacier surge mechanism based on linked cavity configuration of the basal water conduit system', *Journal of Geophysical Research*, **92**, B9, 9083–9100.

Kamb, W.B., Raymond, C.F., Harrison, W.D., Engelhardt, H., Echelmeyer, K.A., Humphrey, N., Brugman, M.M., and Pfeffer, T. 1985. 'Glacier surge mechanism: 1982–1983 surge of Variegated Glacier, Alaska', *Science*, **227**, 469–479.

Lemmens, M., Lorrain, R., and Haren, J. 1982. 'Isotopic composition of ice and subglacially-precipitated calcite in an alpine area', *Zeitschrift für Gletscherkunde und Glazialogeologie*, **18**, 151–159.

Nye, J.F. 1969. 'A calculation on the sliding of ice over a wavy surface using a Newtonian viscous approximation', *Proceedings of the Royal Society of London*, **A311**, 445–467.

Röthlisberger, H. 1972. 'Water pressure in intra- and subglacial channels', *Journal of Glaciology*, **11**, 177–203.

Souchez, R.A. and Lemmens, M. 1985. 'Subglacial carbonate deposition: an isotopic study of a present day case', *Palaegeography, Palaeoclimatology and Palaeoecology*, **51**, 357–364.

Spring, U. and Hutter, K. 1982. 'Conduit flow of a fluid through its solid phase and its application to intraglacial channel flow', *International Journal of Engineering Science*, **20**, 327–363.

Walder, J.S. 1986. 'The hydraulics of subglacial cavities', *Journal of Glaciology*, **32**, 439–445.

Walder, J.S. and Hallet, B. 1979. 'Geometry of former subglacial water channels and cavities', *Journal of Glaciology*, **23**, 335–346.

23

CONSTRAINTS ON THE PRESERVATION OF DIAMICT FACIES (MELT-OUT TILLS) AT THE MARGINS OF STAGNANT GLACIERS

M.A. Paul and N. Eyles

Source: *Quaternary Science Reviews* 9(1) (1990): 51–69.

This paper reviews the formation and preservation of diamict facies (melt-out till) generated *in situ* by the downwasting of stagnant ice at the margins of continental glaciers. Melt-out tills are commonly envisaged as being the product of the top (supraglacial) or bottom (subglacial) melt of stagnant debris-rich glacier ice but such deposits are relatively uncommon in modern glacier environments where they are thin, laterally discontinuous and have a low preservation potential. Theoretical considerations of the distribution of englacial debris in continental ice sheets show that only the glacier margin is likely to contain the necessary volumes of englacial debris for melt-out to be a significant diamict-forming process. However the application of thaw–consolidation theory to the formation of such deposits shows that substantial porewater pressures and self-sediment deformations can be induced during melt-out, especially in matrix-rich debris. The classic melt-out mechanism widely invoked in the literature since 1875 and involving passive *in situ* aggregation from thawing englacial debris can be seen to operate under restricted conditions. Diamict facies produced by melt-out are likely to form spatially disjunct and laterally discontinuous elements within complex and ice marginal stratigraphics; the melt-out process is not a mechanism whereby regionally extensive diamict(ite) units can be deposited across glaciated basins.

Introduction

The formation of glacial diamict facies by the *in situ* melting of debris-rich stagnant ice was first discussed in detail by Goodchild (1875). Since then this depositional model has been widely employed in the interpretation of massive or crudely-stratified, sometimes thick and laterally extensive Pleistocene and Pre-Pleistocene glacial diamict(ites) sequences (e.g. Upham, 1891; Chamberlin, 1894; Wordie, 1947; Carruthers, 1939, 1943, 1944, 1953; Elson, 1961, 1988; Boulton, 1970, 1971, 1976, 1977; Flint, 1971; Spencer, 1971, 1985; Mickelson, 1973; Dreimanis, 1976, 1988; Cegla *et al.*, 1976; Minell, 1977, 1977a; Eyles and Slatt, 1977; Shaw, 1979, 1982; Lawson, 1979; Haldersen and Shaw, 1982; Madgett and Catt, 1978; Catt and Madgett, 1981; Edwards, 1981; Drozdowski, 1983, 1984; Stephan and Ehlers, 1983; Dardis and McCabe, 1984; Bouchard *et al.*, 1984; Dowdeswell *et al.*, 1985; Drewry, 1986).

It is the purpose of this paper to examine the geotechnical conditions under which melt-out occurs, both below and on top of stagnant glacier ice. The present analysis is based on geotechnical modelling of the decay of perennially-frozen ground in Arctic areas, which is directly applicable to stagnant glaciers, a few published descriptions of melt-out processes at modern glaciers, and observations during the course of fieldwork in Iceland, Spitsbergen, and Canada. We will show that the melt-out process is not a simple process and that there is a process/facies continuum between lodgement and melt-out with severe constraints on the formation and preservation of undisturbed melt-out tills. Diamict facies produced by melt-out are likely to form thin, coarse-grained and spatially disjunct elements within complex stratigraphies deposited along the ice margin; the melt-out process is not a mechanism whereby thick regionally extensive diamict(ite) units can be deposited and melt-out till(ite)s should be only rare components of ancient terrestrial glacial deposits.

A note on terminology

Before proceeding, the meaning of certain terms should be clarified. The term 'debris' is used according to the definition stated by Boulton and Eyles (1979, p. 12) for 'a dispersion of particles in the glacier or on its surface acquired by no matter what means'. The term 'till' is used according to the definition of Boulton (1972a) for an aggregate 'whose components are brought together and deposited by the direct agency of glacier ice and which, although it may suffer postdepositional deformation, does not undergo subsequent disaggregation and redeposition' (see Eyles, *et al.*, 1984 and discussion therein). The term 'melt-out till' is defined as a diamict that has accumulated *in situ* by the aggregation of englacial debris as a result of the melting of the upper and/or lower surfaces of debris-laden stagnant ice. This definition will also carry the rider identified above regarding disaggregation and redeposition.

The common visualization of the meltout process is a 'grain-by-grain' accumulation as each particle or particle aggregate is released from ice (Fig. 1). It is important to note that the key to the definition is *in situ* aggregation; newly released melt-out debris is commonly subject to syn- and postdepositional processes (most commonly mass flow) which act to redeposit and restructure such materials. It is generally agreed that this resedimented material is not 'till'; sediment gravity flow facies may be transported great distances by debris flow and associated processes and deposited within a variety of subaerial and subaqueous fan settings. Postdepositional resedimentation may moreover, continue for many thousands of years after regional deglaciation.

Melting of stagnant debris-laden ice: a thaw consolidation problem

The slow melting of debris-laden glacier ice and the aggregation of newly-thawed debris after its release is in essence a thaw consolidation problem of the type studied by Morgenstern and Nixon (1971). Their work was originally developed for the analysis of the thawing and consolidation of perennially frozen ground by *in situ* melting around pipelines but the results can equally be applied to conditions during the downwasting of stagnant, debris-rich glacier ice (Lawson, 1982). The linear form of their theory applies to ice-debris mixtures that are characterized by numerous intergranular contacts but for an ice-rich mixture some non-linear behaviour may be expected due to variation in englacial debris contents within the glacier and the heterogeneity of freshly exposed debris. Additional analysis by the above authors (Nixon and Morgenstern, 1973) has shown that inclusion of non-linear behaviour by the thawing debris does not change the basic conclusions, but serves to increase estimates of porewater pressure developed during thawing. Since the arguments presented in this paper emphasize the sedimentological importance of high porewater pressures during melt-out, the more simple linear theory is employed to establish a lower limiting condition for sediment behaviour during thawing.

The process of thaw consolidation creates an elevated pore pressure within the thawed sediment, the magnitude of which is determined by the balance between the rate of meltwater production and the rate of consolidation of the sediment. This balance is measured by the 'thaw consolidation ratio'; which is a parameter that reflects both the material properties of the sediment and its environmental situation. The higher the value of this ratio, the higher will be the excess pore pressures during melt-out.

The thaw consolidation ratio (R) that is likely to have obtained during melt-out is estimated by:

$$R = \frac{\alpha}{2\sqrt{C_v}}$$

Figure 1 Basal melt-out at the margin of wet based glaciers: (a) Stagnant ice-cores at margin of Kviarjokull, Iceland being overridden by active ice to form folded basal debris layers (b). (c) Large mass of downwasting dirty basal ice, trapped against bedrock high. Active ice above moving right to left. Width of photography is about 10 m. Berendon Glacier, British Columbia. (d) Dirty basal ice about 8 m thick, trapped against bedrock high (to left). The mass has been detached from active ice moving overhead from right to left. Morasrjokull, Iceland. (e) The margin of Athabaska Glacier, Alaska, Canada. Photograph (f) shows the result of *in situ* downwasting of basal ice; basal debris has been lowered onto the substrate to form a thin rubbly melt-out deposit one or two clasts thick. Trowel for scale. *In situ* basal debris can be seen behind trowel.

where α is the 'thaw constant' that describes the non-linear rate of penetration of the thawing boundary into the debris and has units of length/(time)$^{1/2}$. The parameter C_v is the coefficient of consolidation of the sediment after thawing, and describes the rate of which the pore pressure can fall during consolidation as a result of drainage and associated volume changes.

In order to estimate R, values of C_v and α must be estimated for a range of thawed sediment types. C_v is a property of the material itself, and is related to its grading, density and fabric. The principal control on C_v is the permeability of the thawing sediment which is controlled by grain-size. The value of C_v can vary greatly in practice from about 1 m^2 year^{-1}, for a clayrich diamict, to about 100 m^2 year^{-1} or more for coarse-grained, free-draining facies. The values of C_v increased by such structural features as fissures and sand laminations that improve internal drainage, and is decreased by a clayey matrix.

The thaw constant α is controlled by both the thermal properties of the diamict and by the ambient air temperature during the thaw period. McRoberts (1975) has studied in detail the effect of variations in these parameters and has calculated that the range 0.015–0.095 cm sec$^{-1/2}$ (0.84–5.3 m year$^{-1/2}$) encompasses the range of ice contents (5 to 70%) and temperatures (2 to 20°C) that are likely to occur in practice with regard to permafrozen sediments of a wide range of origins. With regard to glaciers, the ice content in englacial debris is greater than that identified by McRoberts for perennially frozen-ground, and temperature is the major variable. Therefore, if the ambient temperature during melt-out can be estimated, the appropriate value of α can be determined. In the case of subglacial melt-out where the ice base is not open to the atmosphere, an ambient air temperature value close to 0°C is appropriate, whereas for supraglacial melt-out the range should extend to perhaps 25°C (see below). The corresponding values of α thus covers the range 0 to 2.5 m year$^{-1/2}$. Figure 5 shows the thaw consolidation ratios that correspond to various combinations of thaw rate and coefficient of consolidation over the ranges mentioned. This figure emphasizes that in many melt-out situations the thaw consolidation ratio will have a value between 0.05 and 1.25, and will be governed more by the consolidation characteristics of the debris than by the thaw rate.

Morgenstern and Nixon's (1971) analysis allows solutions to be obtained for two limiting cases: that of an entirely weightless material that is thawing beneath a fixed confining load, and that of a material consolidating under its own self weight. These extremes correspond respectively to the geological situations of subglacial melt-out, under a substantial confining load of stagnant glacier ice, and supraglacial melt-out with a much reduced confining load. The more complex case of subglacial release where the ice is thin relative to the sediment thickness can be obtained by superposition of the solutions for the two cases. This analysis is by necessity, grossly simplified since the boundary conditions can vary enormously and the analysis presented here is therefore likely to be conservative (see below).

For the above conditions of *subglacial* and *supraglacial* melt-out, depth profiles of porewater pressures within the newly thawed diamict can be determined (Figs 6a,b). Both depth and pore pressure are normalized by an appropriate scaling factor, which for depth is the total thawed depth and for pressure is a measure of the ambient stress (this measure is different for the supraglacial and subglacial cases). The significant features of the solution are the increase of pore pressure towards the thaw plane, and the strong dependence of the maximum pressure on the thaw consolidation ratio in both the subglacial and supraglacial cases. This is emphasized in Fig. 6c, which illustrates that the pore pressure can carry a very substantial fraction of the applied stress, and hence that the intergranular forces can be greatly reduced in a sediment while it is melting out. Figure 6c shows that in the supraglacial case this reduction can be about 90% of the long term value, and that in theory in the subglacial case the intergranular forces can be reduced to zero.

Pressure gradients are obtained by differentiation of the function shown in Figs 6a,b,c (Fig. 6d). In the case of supraglacial melt-out the normalized gradient is constant with depth, and although it increases with increasing thaw consolidation ratio, it is always less than unity. In the subglacial case, by contrast, the gradient is depth dependent, and this dependence becomes increasingly pronounced as R increases. The maximum values of normalized gradient occur close to the drainage boundary, and may be in excess of unity.

The sedimentological consequences of these data are significant. If the pore pressure gradient is high, the drag forces on individual particles may be sufficient to overcome frictional or self-weight forces. The particles are then able to move relative to one another, thus disturbing any clast or matrix fabric inherited from original englacial debris and leading to visible water escape structures. If the overall porewater pressure is high, the sediment will have a low shear strength and can be easily mobilized, also modifying or destroying any relict englacial structure developed in the parent ice mass. These processes of deformation are significant for environmental reconstructions and facies modelling in that they produce diamict lithofacies whose character may obscure the primary origin of the deposit by melt-out. It would appear therefore, that the preservation of undisturbed sediments of melt-out origin is heavily constrained by the conditions which govern the onset of syndepositional deformation under or on the buried ice.

Syndepositional deformation during *in situ* melt-out

During deposition, the presence of both excess pore pressures and strong pressure gradients can cause deformation by two distinct processes which we shall term *shear instability* and *hydraulic instability*, that lead respectively to deformation by *shear failure* and by *water escape*. These two processes are independent and are considered separately.

Shear instability occurs when the available strength of the sediment is exceeded by the applied shear stress. The strength is lowered by the elevated pore water pressure, as shown in Fig. 6e. The loss of strength increases rapidly with increasing thaw consolidation ratio and can be substantial. In the subglacial environment applied shear stresses may be produced as a result of (for example) any steep slopes on the underlying bedrock surface or by differential settlement around substrate irregularities. In the supraglacial environment the applied shear stress usually develops as a result of deposition on a slope such as on the flank of an icecored ridge.

Hydraulic instability arises when seepage forces are sufficient to overcome the self-weight and frictional contact forces that act on the individual sedimentary particles. These forces result from the pore pressure gradient, and thus are at a maximum where the gradients are greatest. In the case of subglacial sediments the maximum gradient occurs at the base of the sequence and the magnitude of the gradient increases rapidly with thaw consolidation ratio; in the supraglacial case the gradient is constant with depth, and the dependence on thaw consolidation ratio is less strong (Fig. 6d). Hydraulic instability will also be produced by low effective stresses, since interparticle forces will be reduced and therefore particles will be less constrained by their neighbours. Under these conditions there will be a critical hydraulic gradient which will just enable interparticle movement to occur, and once this critical gradient is exceeded, water escape structures may be expected to form in the sediment.

The onset of hydraulic instability is analogous to the well known 'quicksand' condition of upward flow ('fluidization'; Lowe, 1975; Postma, 1983). This implies a normalized hydraulic gradient of unity and Fig. 6d shows that, in principle this condition will not be produced by simple supraglacial thaw consolidation. However, it should be noted that supraglacial deposition also involves conditions of rapid undrained loading by sediment gravity flows (see below) which can increase local pore pressure gradients sufficiently for water escape structures to develop (Hutchinson and Bhandari, 1971; Lawson, 1982; see below). In the subglacial case, the situation is more complex as a result of the confined setting of the sediment, and the fact that the water flow is generally downwards. However, if it is assumed that some lateral movement of the sediment is possible (for example) it can be shown that when the thaw consolidation ratio lies in the approximate range 0.5 to 1.0, the combination of pore pressure and pressure gradient is sufficient to allow interparticle movement to begin. The precise value is determined by the extent of lateral restraint and the internal friction of the sediment; for fine grained, matrix-dominant diamict facies the value will lie at the lower end of the range, and vice versa for clast-dominated facies.

The two processes of shear and hydraulic instability have so far been considered as independent of one another but they also affect sediments in combination. Consequently, it is possible in principle to identify four deformational environments defined by (1) the level of shear stress relative to the available strength of the sediment and (2) the level of the pore pressure gradient relative to the critical hydraulic gradient necessary for fluidization. This is shown in Fig. 7 which generalizes, and adds geological detail to Fig. 6e.

The boundaries between the zones are defined by two sets of conditions. The shear failure boundary is simply the failure envelope for the sediment, and its position on the diagram is determined by the material properties of the sediment and also by the porewater pressure, which in turn is determined by the thaw consolidation ratio as indicated in the figure. In practice it may well be preferable to define this boundary in terms of the maximum allowable imposed shear stress (or some geologically meaningful expression of this, based, for example, on surface of the ice slope or bedrock configuration below the glacier). The hydraulic disruption boundary is that value of the thaw consolidation ratio at which the level of effective stress and hydraulic gradient permit sufficient local grain separation to allow the development of internal structures. In practice, it is useful to express this boundary directly in terms of a limiting thaw consolidation ratio as was shown on Fig. 6e.

Diamict facies originating by melt-out may be expected to reflect the deformational environment in which they form, and so each situation will give rise to a characteristic deposit or deposits. We anticipate that these will be broadly as follows:

ZONE I: In this zone the shear stress on the thawing debris is always below that required to produce deformation. Similarly, the pore pressure conditions are below those required for fluidization. As a result, the accumulating diamict undergoes no deformation other than vertical settlement as particles move into contact following the loss of interstitial ice. Melt-out in this subenvironment corresponds to the 'classic' meltout model that has been frequently used in the literature involving 'grain-by-grain' aggregation of englacial debris. The characteristics of melt-out tills have been summarized by Lawson (1981) and include a preferred clast orientation parallel to iceflow (see below), the presence of discontinuous lenses and of poorly sorted sediment, inherited from pre-existing stratified englacial debris, which may have a textural, composition or colour distinct from the surrounding diamict. Such lenses may be draped over the larger clasts as a result of differential compaction (Shaw, 1982; Bouchard *et al.*, 1984). In general, clasts are not greatly reorientated compared with that of englacial debris only to the extent of a flatter dip and increased scatter around the mean as a result of compaction (Lawson, 1979; Dowdeswell and Sharp, 1986; Eyles *et al.*, 1987). The bulk textures of the parent englacial debris will remain unaltered during melt-out although Haldersen (1983) has demonstrated that melt-out tills can be deficient in medium to fine silt grades when compared to lodgement tills reflecting the increased abrasion that englacial debris undergoes during the lodgement process.

ZONE II: In this zone, values of shear stress are sufficient to produce failure in newly melted-out debris, although the pore pressure and seepage conditions are insufficient to generate fluidization. Since the highest pore pressures occur at the thaw interface, it is here that repeated failure will occur as the englacial debris thaws and thus the sediment may be expected to show evidence of shearing throughout its final thickness. The sediment-ary features identified in zone I (above) would be eliminated and replaced by sedimentary and geotechnical properties in common with lodgement till (e.g. Eyles *et al.*, 1982; Sladen and Wrigley, 1984). Supraglacial sediment would be deposited as sediment gravity flows as a result of the earlier remobilization of newly-thawed and saturated debris.

ZONE III: In this zone pore pressure and seepage conditions are suffi-cient to generate fluidization, but there is no shear failure of the sediment. Porewater will escape from the sediment and cause loss of fines coupled with disruption of clastic and matrix fabrics. This disruption may be pervasive, but is more likely to be localized to well defined water escape structures or zones of piping (Lowe, 1975, 1976; Lafleur *et al.*, 1982; Postma, 1983). If the sediment were initially to contain lenses or pods of contrasting texture, colour or composition, the lowered effective stress might assist the genera-tion of a variety of 'load structures' formed by the differential settlement of gravitationally unstable coarser-grain sizes or aggregates following the creation of reverse-density gradients (e.g. Butrym *et al.*, 1964; Anketell *et al.*, 1970). Bouchard *et al.* (1984) describe coarse-grained, stratified basal tills, showing differential compaction around clasts, water escape structures and diapiric deformations, that probably formed under such conditions.

ZONE IV: In this zone, both shear failure and fluidization can occur. The result will be a diamict with a wide variety of deformation structures; in particular some degree of textural sorting as a result of shear segregation may be apparent. The latter may provide a means of generating a crude layering or shear-banding, from originally poorly-sorted englacial debris.

It is clear from this brief examination that a broad continuum of diamict facies can be formed by the simple removal of interstitial ice as a result of varying syndepositional deformation. Simple grain-by-grain lowering, which is the mechanism widely invoked in the literature (classic melt-out; zone I) comprises one particular case only. The key idea to emerge is that of a deformation field defined in terms of thaw consolidation ratio and level of applied shear stress which greatly constrain the formation and preservation of melt-out till. The classic model of melt-out can be seen to be applicable only under well-defined and very restricted boundary conditions. Outside this zone, melt-out tills will undergo extensive syn- and postdepositional deformation.

Constraints on the distribution of englacial debris in ice sheets

In order to apply the foregoing theory to the analysis of melt-out tills, it is necessary to consider the spatial distribution of englacial debris in ice sheets. Many authors have modelled notional ice sheets, resting on relatively low relief beds, and illustrated the pattern of ice flow that is to be expected (Boulton, 1972; Boulton *et al.*, 1977, 1985; Sugden, 1977; Denton and Hughes, 1981; Hughes, 1987; Andrews, 1983, 1987 and references therein). Over most of the ice sheet, flow vectors are subparallel to the underlying bed whereas at the margin of the ice sheet deceleration causes movement towards the ice surface.

An important contrast in basal conditions occurs between those areas of the ice sheet which are at the pressure melting point, and therefore undergoing *basal melting*, and those which are below the melting point and are areas of net *basal freezing* or are *frozen*. This contrast develops in response to the changing balance between the production and discharge of heat at the glacier sole and is therefore dynamic in nature both spatially and temporally (Weertman, 1961; Boulton, 1972). The exact configuration of these thermal zones beneath both theoretical models and ancient ice sheets, whose beds are now exposed, is a matter of considerable debate (Moran *et al.*, 1980; Denton and Hughes, 1981; Punkari, 1984; Boulton *et al.*, 1985; Drewry, 1986; Hughes, 1987) but they are generally pictured as concentric belts, arranged parallel to the margin, whose boundaries may migrate with time.

In areas of *basal melting* the thickness of the basal debris zone is limited by bedrock topography, since incorporation of debris cannot take place above the crest height of bed obstacles and further dispersion by clast-to-clast collisions limited (Weertman, 1968; Boulton, 1975). At the ice-bed interface debris is held in a zone of traction where it is frictionally retarded relative to the moving ice. Depending upon the ice velocity and bed/clast contact forces the debris may either be lodged against the bed by pressure melting (lodgement) or be swept across the bed with resultant abrasion and the generation of comminution products (Boulton, 1975, 1978).

Within the zone of basal melting, the opportunities for melt-out till units to be deposited are strictly limited. Suitable conditions are only likely to occur where portions of the moving ice base stagnate in the region of bedrock irregularities (e.g. Boulton, 1970; Bouchard *et al.*, 1984; Fig. 1). Due to the low overall concentration of debris in the basal suspension zone the thickness of the melted-out deposit will be very much less (perhaps by an order of magnitude) than the height of the obstruction. Such units will obviously be limited in their lateral extent, and will be confined to areas of moderate to strong bedrock relief or obstructions caused by glacetectonic deformations (e.g. Hillefors, 1973; Haldorsen, 1982, 1983; Shaw, 1982). It can be argued therefore that in areas of basal melting, which probably occupied the larger part of the beds of ancient ice sheets, melt-out till(ite)s are not significant elements within the subglacial sedimentary record.

Zones of *basal freezing* are commonly associated with a thickening of the basal debris layer due to repeated freezing on of debris and meltwater and quarrying of the substrate (Weertman, 1961; Boulton, 1972; Moran *et al.*, 1980; Denton and Hughes, 1981). This thickened layer will, if it undergoes stagnation and *in situ* melt, produce an increased volume of melted-out deposits. It is not possible for melt-out to occur while the basal temperature is below freezing however, and yet, if the basal regime changes to one of melting (due to the dynamic nature of the thermal zones), the thickened debris layer will likely be thinned and dispersed by ice flow.

Melt-out deposits are most likely to accumulate at the ice margin during glacial retreat, when bodies of stagnant cold-based ice become detached from the retreating glacier (Figs 2–4). Ice sheet margins are tectonically complex as a result of flow deceleration and give rise to compression and thickening of the basal debris zone (Figs 1–4). This deceleration may be the result of diverging flow (Shaw, 1979), topographic obstacles (Clayton and Moran, 1974; Paul, 1983) or overriding of stagnant ice (Attig *et al.*, 1985; Fig. 1). In the case of surging glaciers large volumes of meltwater and debris may be incorporated within the glacier by freezing (e.g. Clapperton, 1975; Lawson, 1979a). In all these cases, overfolding and thrusting at the ice margin causes stratigraphic repetition of the basal debris zone, giving the appearance of a system of multiple debris bands, often with a steep upglacier

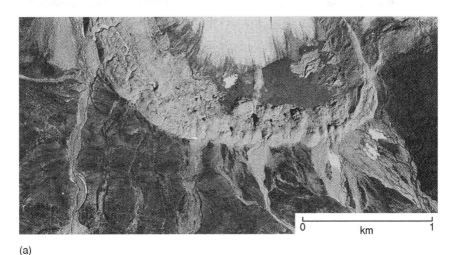

(a)

Figure 2 (a) Broad zone of stagnant ice with supraglacial sediment cover at the margin of Spitsbergen outlet glacier. Subsurface structure of ice margin as in Figs 9, 11. Glacier is thermally-complex and is surge-prone. Part of photograph 569 2675, reproduced by permission of Norsk Polarinstitutt.

(b)

(c)

(d)

(e)

Figure 2 (cont'd) (b) Outcrop of basal debris zone at ice margin composed of alternating debrisrich and debris-poor bands, ice axe for scale. (c) Steeply dipping debris band outcropping as a dyke. Sediment gravity flows move newly melted-out debris down the slope of the surrounding debris-mantled (and therefore insulated) ice-cored slope. Kviarjokull, Iceland. (d) 1.5 m thick cover of supraglacial overlying dead ice with englacial debris bands dipping from right to left. Supraglacial melt-out is occurring at the ice/ sediment interface. Supraglacial sediment cover consists mostly of melt-out till that has been resedimented downslope by gravity flow. Aavatsmarkbreen, Spitsbergen. (e) Contact of supraglacial melt-out till (about 1 m thick) and underlying dirty ice. Note that ice is completely stratified but overlying melt-out till is massive.

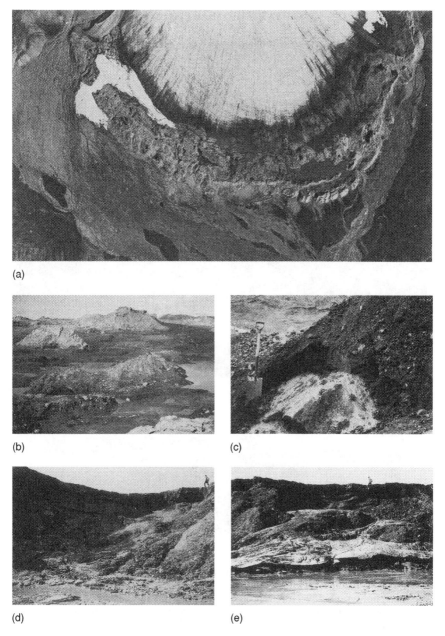

Figure 3 (a) Broad zone of stagnant ice with supraglacial sediment cover at Elisebreen, Spitsbergen. For map of area see Paul (1983). Part of photograph 569 2188, reproduced with permission of Norsk Polarinstitutt. Scale as in Fig. 2a. (b) Ice-cored hummocks with cover of supraglacial diamict; highs are progressively being lowered as supraglacial sediment is stripped off by sediment gravity flow (Fig. 4c). Matanuska Glacier, Alaska. (c) Ice-core below supraglacial sediment cover; Kviarjokull, Iceland. (d,e) 2 m thick supraglacial sediment cover (below figure) overlying stagnant ice with englacial debris bands. The supraglacial cover is being stripped by sediment gravity flows.

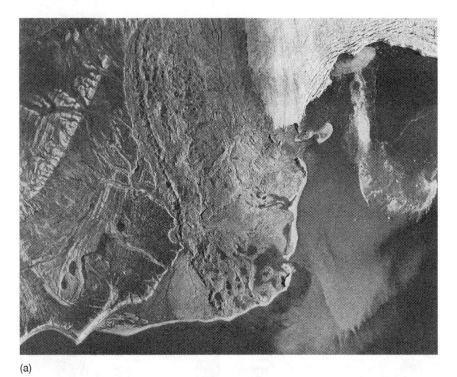

(a)

(b)

Figure 4 (a) Broad zone of ice-cored hummocks and troughs (e.g. Figs 9,11) at Aavatsmarkbreen, Spitsbergen (for map of area, see Paul, 1983). Part of photograph 569 2188, reproduced with permission of Norsk Polarinstitutt. Scale as in Fig. 2a. (b) View of outer ice-cored margin showing ice-cored high and supraglacial lake. Lake overflowing waters have cut into the ice-core and triggered a large lobate sediment gravity flow. Back scarp of flow is about 8 m high.

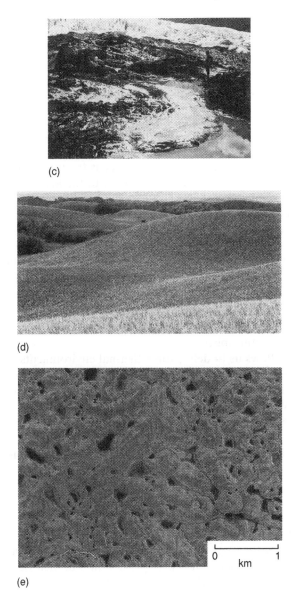

(c)

(d)

(e)

Figure 4 (*cont'd*) (c) Lobate sediment gravity flow moving to right; Matanuska Glacier, Alaska (cf. Lawson, 1982). (d) Ground and air. (e) photographs of typical hummocky 'disintegration' topography deposited as in Figs 9, 11; near Drumheller, Alberta. Cattle for scale. Air photograph part of photograph A7629-74, Alberta Government.

dip (Fig. 2); the downwasting ice surface intersects this structure and leads to a complex outcrop pattern of the debris layers as a series of dykes at the glacier surface (Fig. 2).

Application to subglacial environments

The aim of this section is to apply the above theory to discuss the way in which melted out units can be distributed in the subglacial environment, and to what extent they may suffer syndepositional disturbance. Although our model is of necessity conjectural, its value is to emphasize the potential complexity of melted out deposits, in contrast to the more simplistic model of 'classic' melt-out.

We consider a simple situation. We suppose that a body of stagnant ice is resting on an irregular surface of sediment and bedrock, and that meltwater is in places able to circulate beneath the ice through the sediment or at the ice-rock interface. Such water is a common feature beneath glaciers, and is believed on mechanical grounds to be above 0°C due to kinetic heating and mixing with groundwater (Rothlisberger, 1972; Vivian, 1980; Lliboutry, 1983; Hooke, 1984). It is also supposed for illustration that in places there are open cavities below the ice, and that some of these are in communication with the external atmosphere.

This model allows us to define three thermal environments. These are (1) areas which receive only geothermal heat, (2) areas where there is additional heat from meltwater advection and (3) areas where heat is gained from the atmosphere. We note that in order for syndepositional disturbance to occur, there is also a kinematic condition to be satisfied, since the sediment must be able to be displaced in some direction, in order for shear structures to develop. If this condition is not satisfied, the shear stress cannot rise above its 'at rest' (K) value, and failure is prohibited. Displacement will be allowed if either there is an open cavity into which the sediment can move, or consolidation over an irregular bed causes lateral strains to develop as a result of differential vertical compression.

We consider separately the effects of melt-out in each of these thermal environments for the cases of a fine-grained clay till and a coarser textured silty till. In order to discuss the likely preservation of melt-out till, we must estimate the likely value of the thaw consolidation ratio during deposition, and also establish whether the shear stress has become limiting. This enables the depositional environment to be positioned on a diagram such as Fig. 7.

Table 1 has been constructed for some illustrative values of heat flow and coefficient of consolidation: these are believed to be typical, but it is stressed that they are not prescriptive. At the ice front the air temperature has been taken to be 5°C, and the basal water temperature has been taken to be 1°C. The geothermal heat flow is typical for intraplate areas. The example

Table 1 Examples of thaw consolidation conditions for subglacial melt-out.

		Example calculations					
		Clay till *C = 1 ma*			*Silty till* *C = 10 ma*		
Heat source	Heat flow	R	U/		R	U/	
Geothermal heat only	~ 50 mW/m = 0.06 ma	0.03	<2%	13	0.01	<1%	19
Geothermal heat plus meltwater advection	~ 100 mW/m = 0.12 ma	0.06	<5%	12	0.02	<1%	19
Geothermal heat plus air advection at 5°C	~ 1000 mW/m = 1.2 ma	0.60	45%	8	0.19	9%	18

calculations show (a) the thaw consolidation ratio that will arise in each case, (b) the excess pore water pressure (expressed as a percentage of over-burden) that will be generated by thaw consolidation and (c) an indication of the bed slope on which the shear stress will reach a limiting value for the type of till in question, assuming the free percolation of meltwater in the downslope direction. This is not the only suitable measure of the onset of shear instability, but it provides an easily visualized example.

Consider first the simplest case, that of an isolated area of bed that is receiving only geothermal heat. In this case the thaw consolidation ratio will not exceed a value of around 0.03 even in fine-grained sediments. Correspondingly, the excess pore water pressure will rise to around 2% of overburden. This value is well below that required to generate water escape structures. Similarly, the raised pore pressure hardly affects the stability of the sediment. If kinematic conditions allow, shear failure will occur on a bed slope around one half of the angle of friction. Since, however, this corresponds to the angle of 15 to 20°, we suggest that much sediment will remain undisturbed: this is therefore a situation in which melt-out till will be preserved. If we consider the addition of advected heat from meltwater, we see from Table 1 that the thaw consolidation ratio rises to around 0.06, and the excess pore pressure, to around 5%. These are still low values, and the above arguments apply. Thus we conclude that the presence of percolating meltwater does not have a major influence on the preservation potential of subglacial melt-out till, and that sediments which accumulate under this regime will fall mainly into our zone I, and possibly into zone II.

The advection of heat from the atmosphere, for example through open cavities, causes a very different situation. In the case of a clay till, the thaw consolidation ratio will be around 0.6 and the excess pore pressure will reach around 45% of overburden pressure. These are maximum values which will decline inwards from the margin. They are however, substantial:

the thaw consolidation ratio implies the onset of water escape from the sediment, and the pore pressure would cause shear failure on bed slopes of around 8. Thus sediments that melt-out under these conditions will be expected to fall into either of our zones III or IV.

Distribution of subglacial melt-out tills

A basic division can be made between marginal areas and those in the interior of the stagnant ice. In the interior, restricted heat flow will favour the accumulation of melt-out till: the spatial distribution will thus reflect the distribution of the englacial sediment at the time of stagnation, and in particular the continuity or discontinuity of the debris layers, as discussed in an earlier section.

Near the ice front, where higher heat flows allow elevated pore pressures to be generated within melted-out diamict, water escape and loading structures may occur (e.g. Zone III, Fig. 7). Substrate sediments may also show features such as clastic dykes and other injection structures, resulting from high local pore pressures and fluidized sediment. Several authors have briefly alluded to the existence of such higher porewater pressure and the likelihood of deformation during melt-out (Boulton, 1970; Holdsworth, 1973; Boulton and Paul, 1976; Shaw, 1982; Drozdowski, 1983; Bleuer, 1983; Brodzikowski and Van Loon, 1983). Since shear strength is reduced by high porewater pressures, diamicts that melt-out on areas of a sloping subglacial bed may be expected to be completely remoulded and be redeposited as debris flows (Zone IV, Fig. 7).

Further in from the ice margin where heat flow is reduced and the rate of melting is lower the pore pressure generated during melt-out is reduced. In areas of sloping substrates, where the shear stress exceeds the shear strength of the sediment, shear failure will occur by relatively simple movements on discrete failure surfaces throughout the sediment body (Zone II, Fig. 7); this will destroy or severely modify any relict englacial structure inherited from the basal debris zone. In areas of low stress ratio (i.e. bed hollows) melted-out sediments can accumulate without structural disturbance (Zone I, Fig. 7) corresponding to the classic melt-out situation described in the literature (Boulton, 1970; Shaw, 1979; Haldorsen, 1982; Bouchard et al., 1984).

In an idealized case there may be a simple upglacier succession from zones III/IV to zones I/II in response to the reduction in heat flow. In reality, two qualifications must be made. First, the heat flow at the terminus must be sufficient to generate the necessary porewater pressure to cause hydraulic instability. If this is not so, zones I and II will extend right to the terminus.

The second qualification concerns the replacement of one zone by another when the ice thins and the margin recedes. When the ice withdraws, heat flow presumably increases and as a result the stress ratio and pore pressure will rise. Thus sediments that originally melted out under zone I or II conditions may be overlain by sediments deposited under conditions of zones III or IV. The deformation that will affect the latter increments may also penetrate into the earlier sediment and thus partially or totally overprint any pre-existing structure. It can thus be anticipated that most fine-grained melted-out sediments will be modified and restructured by syndepositional disturbance; the term melt-out till cannot be applied to these sediments (see above).

Application to supraglacial environments

Descriptions of depositional processes occurring within complex supra-glacial environments (e.g. Figs 1–4) have been published by many workers (Clayton, 1964; Boulton, 1972; Boulton and Paul, 1976; Eyles, 1979, 1983; Wright, 1980; Driscoll, 1980; Lawson, 1982; Sharp, 1985; Drozdowski, 1985). Englacial debris is released as the ice surface downwastes and either remains *in situ* as a supraglacial melt-out till, or is resedimented by debris flow processes to be variably reworked by meltstreams and interbedded with other facies.

Ice-cored slopes are subject to frequent landslips (e.g. Fig. 4) which lead to disaggregation and resedimentation of supraglacial sediments, and thus the preservation of melt-out tills requires that the ice surface on which they accumulate should remain stable. As a result, *slope processes* are critical to the preservation of supraglacial melt-out tills.

There are many processes that control slope stability in the supraglacial environment such as ambient air temperature, summer rainfall, density and thickness of debris, presence of supraglacial lakes, vegetation, activity of meltstreams etc. In this paper we stress two processes that potentially lead to slope failure in ice-cored terrain; both involve the elevation of pore pressures above its long term value. The first process is that of melt-out itself; the second is the rapid loading of the lower part of the slope by the arrival of a sediment gravity flow from debris higher upslope. We shall examine these two processes separately.

(1) Instability due to melt-out

The stability of thawing slopes in perennially-frozen ground has been analyzed by McRoberts and Morgenstern (1974). Table 2 shows, in an analogous manner to Table 1, the effect of supraglacial melt-out on sediment stability. The significant feature of this table is the higher heat flows that occur in the supraglacial environment, and the correspondingly higher thaw consolidation ratios and pore water pressures that are generated. These elevated pore water pressures have an adverse effect on the stability of the sediment cover on ice-cored slopes (see also Dyke and Egginton, 1988). Figure 9a is based

Table 2 Examples of thaw consolidation conditions for supraglacial melt-out.

		Example calculations					
		Clay till C = 1 ma			Silty till C = 10 ma		
Heat source	Heat flow	R	U/		R	U/	
Atmosphere at 5°C: that at surface	Step temp 5°C = 1.67 ma	0.83	60%	6	0.26	15%	17
Atmosphere at 5°C: thaw at depth of 2 m	~1000 mW/m	0.60	45%	8	0.19	9%	18

on their analysis and shows the relationship between the thaw consolidation ratio and the critical angle of stability. It can be seen that this angle rapidly reduces to a very low value as the thaw consolidation ratio increases. As discussed earlier, in any particular case the thaw consolidation ratio is determined by the relative rates of melting and consolidation; in the supraglacial environment the rate of thaw is governed principally by the summer air temperature. From Fig. 5 we see that in a matrix rich sediment the thaw consolidation ratio may reach 1.0, whereas in a coarse-grained sediment (under the same conditions) it is unlikely to exceed 0.2. Reference to Fig. 9a shows that the matrix-rich sediment will become unstable on a slope of 5° or less, whereas the coarse diamict will remain stable on a slope of almost 20°. It is thus apparent that when melting-out from an ice-core fine-grained sediment is very unlikely to remain *in situ* and will be disaggregated by downslope movement as a debris flow (e.g. Fig. 4).

Since melted-out sediment accumulates on a slope whose maximum angle can be predicted, we can determine the maximum thickness of the melted-out unit for a given concentration of debris. Figure 9b shows this information in terms of the thaw consolidation ratio under which the slope is stable and the type of sediment of which it is composed. Reference to this figure illustrates that fine-grained sediments are not likely to exceed two metres in maximum thickness per hundred metres of slope width, and from simple geometry the units may be expected to be approximately lens shaped over their width.

(2) Instability due to superimposition of sediment gravity flows

In the case of a supraglacial diamict sequence formed from superimposed debris flows, the episodic arrival of flows from upslope can impose an additional load which may lead to the downslope failure of the sequence.

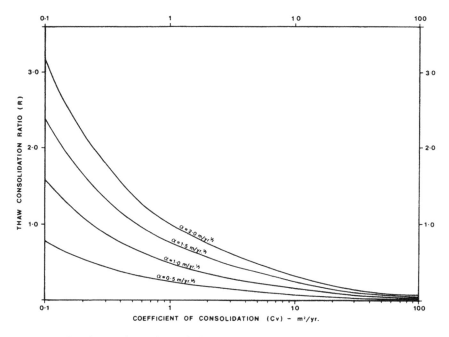

Figure 5 Graphs to show the variations of thaw consolidation ratio (R) with the coefficient of consolidation (C_v) and the thaw rate (α). The graphs indicate that in the case of coarse-grained sediments (high C_v) variations in either C_v or α are relatively unimportant, since the thaw ratio will in practice always have a low value. The thaw consolidation ratio will in practice always have a low value. The thaw consolidation ratios of fine-grained sediments, on the other hand, are relatively large, and variations in either of these parameters may have an important influence on the behaviour of the sediment.

Slope stability under these conditions can be analyzed using the undrained loading model of Hutchinson and Bhandari (1971).

Suppose that the initial thickness of the sediment 'carpet' is z, and that it undergoes a rapid increase in thickness by an amount z' due to the superimposition of a new flow unit. There will be an increase in pore pressure within the sediment body by an amount close to the increment of geostatic pressure $\gamma z'$, where γ is the unit weight of the sediment. The pore pressure ratio (the ratio of pore pressure to overburden pressure) in the sediment will rise by an amount close to z'/z and so the critical angle at which the slope becomes unstable will be reduced. Figure 9c shows the critical slope angles appropriate to various values of z'/z, assuming an initial groundwater condition of flow parallel to and coincident with the ground surface.

Observations made by the authors at the margins of glaciers in Spitsbergen and Iceland suggest that those debris flows that possess sufficient fluidity

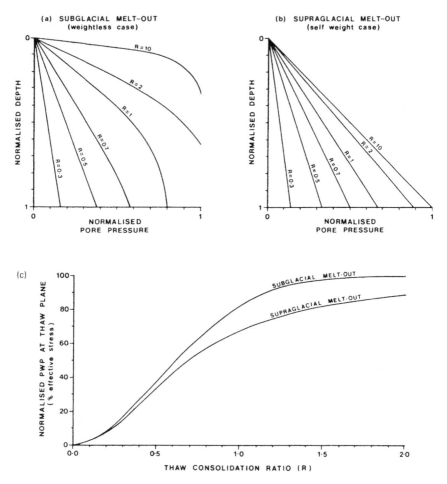

Figure 6 Graphs based on the work of Morgenstern and Nixon (1971), to show the variation of pore pressure and pressure gradient with depth and thaw consolidation ratio during (a) subglacial and (b) supraglacial melt-out. Figure (c) shows the pressure expressed as a proportion of the effective stress that would exist in the absence of thaw; this proportion of the load is carried by the pore pressure during thaw and leads to a reduction in the frictional forces between particles.

and mobility to extend over large areas rarely exceed 03–0.5 m in thickness (Fig. 4) (see also Boulton, 1968; Lawson, 1982). If a sequence of debris flows is built up by superimposition, the ratio z'/z will fall in the approximate sequence 1, 0.5, 0.33 etc. The corresponding critical slope angles from Fig. 9c shows that in its early stages the carpet will only remain stable on very low angled slopes, but that as its thickness builds up the carpet can

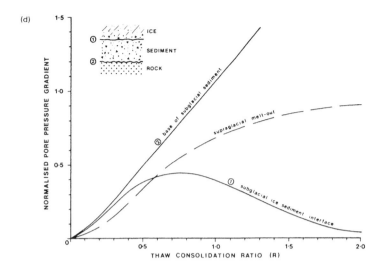

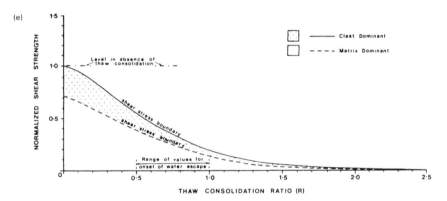

Figure 6 (cont'd) (d) shows pore pressure gradients that correspond to the pressure profiles of (a) and (b). In the supraglacial case the gradient is constant with depth; in the subglacial case the gradients at the sediment base and the thaw plane are shown separately. Figure (e) identifies the shear deformation boundary based on the shear strength of the sediment, taking into account the increase in pore pressure that occurs with increasing thaw consolidation ratio. The hydraulic deformation boundary is based on the approximate thaw consolidation ratio at which interparticle movement starts to become possible as a result of seepage forces. The boundary between matrix dominant and clast dominant sediments is taken to be a sand content of 40% (Sladen and Gilroy, 1983).

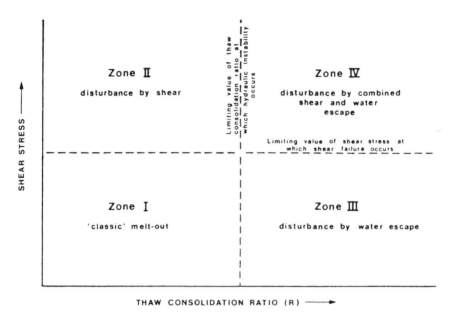

Figure 7 Schematic representation of deformation zones based on the occurrence of shear instability and/or hydraulic instability during melt-out from stagnant ice.

remain stable on increasingly steep slopes, up to the limit of stability for the groundwater conditions described above. For example, on a slope of 5° supraglacial diamict would fail under loading by superimposed flows until the ratio z'/z had fallen to around 0.3 or less; i.e. until it had achieved a thickness of at least three or four flow increments. This would correspond to an actual thickness of around one to two metres, and is in agreement with the common observation that supraglacial materials are extensively remoulded during their early phases of accumulation, and that the sediment cover only stabilizes when it reaches about one metre in thickness.

The above argument implies that supraglacial melt-out tills will only be preserved if they accumulate beneath a stable sediment gravity flow sequence or other sediments such as lacustrine or outwash facies.

Distribution of supraglacial melt-out tills

The distribution of ice-cored ridges within any ice margin is a fundamental control on the spatial distribution of melt-out tills deposited from the ice surface. A distinction can be drawn between those ice-cored ridges that are orientated subparallel to the ice margin ('controlled'; Gravenor and Kupsch, 1959) and that are unorientated with regard to ice flow direction ('uncontrolled'). This reflects the orientation and distribution of englacial debris bands

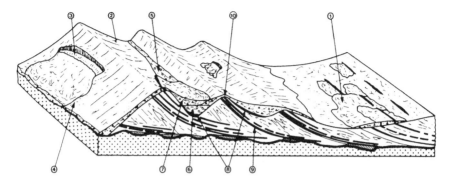

Figure 8 Schematic block diagram to show typical processes and morphological features associated with supraglacial deposition. This diagram is based on the cross section illustrated in Fig. 9. (1) Carpet of superimposed sediment gravity flows derived from debris outcrops. (2) Continuous carpet concealing large volumes of buried ice. (3) Local failure of the carpet exposing the ice-core. (4) Remobilization of the carpet by undrained loading. (5) Glaciofluvial sediment in trough, grading laterally into (6) glaciolacustrine sediment and buried (7) by further sediment gravity flows. (8) Accumulation of melted-out sediment on the proximal flanks of ridges from (9) englacial debris. On the distal flanks, the carpet is derived at depth from debris which crops out as linear bands and has moved downslope (10). Compare with Figs 2, 3 and 4.

and in turn the tectonic structure of the ice margin. It is usually observed that the trend of an ice-cored slope is parallel to the trend of the debris band(s) it contains. Since the latter dip upglacier, often steeply, they will subcrop below the supraglacial carpet as narrow belts on the distal flanks of downwasting ice cores, having a wider subcrop on the proximal flanks (Fig. 8). During meltdown of the ice core, the distal outcrops will contribute material from a series of point sources which are unlikely to coalesce to form a continuous supraglacial cover. On the other hand, material that melts out of the proximal slope will form a more continuous unit which can be added incrementally to the supraglacial carpet. Consequently melt-out tills are likely to accumulate preferentially on the proximal flanks of the ridges. Therefore the final spatial arrangement of the melt-out units will be dictated by the disposition of the former ice bodies, with a tendency for such units to lie on the proximal side of the hollows recording the former positions of buried ice masses. As a result, supraglacial melt-out tills will usually be confined to low points in the topography since hummocks are most likely, because of topographic inversion, to be underlain by the fillings of supraglacial troughs such as sediment gravity flow, and waterlain sediments. In these low points supraglacial lakes may assist the preservation of fine-grained melt-out tills by slowing the thaw rate.

(a)

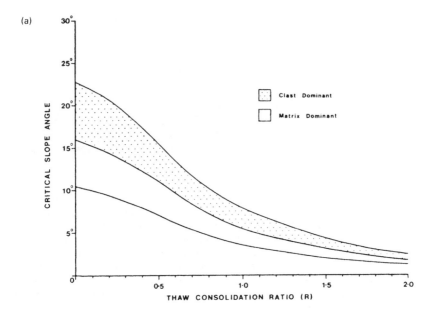

(b)

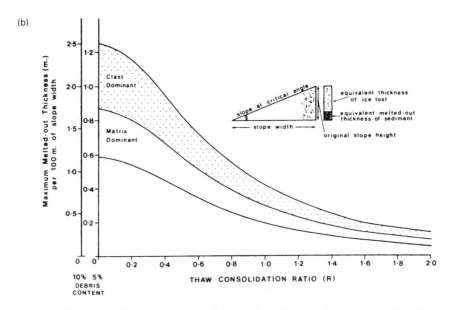

Figure 9 Graphs to illustrate the stability of the sediment carpet on thawing, ice-cored slopes. Figure (a) shows the basic relationship between the critical angle of stability and the thaw consolidation ratio. Higher values of the ratio lead to increased pore pressures during thaw, and hence to a reduction in the slope angle. Figure (b) indicates likely values of thickness of melted-out sediment if slopes of these angles subsequently melted-down undisturbed.

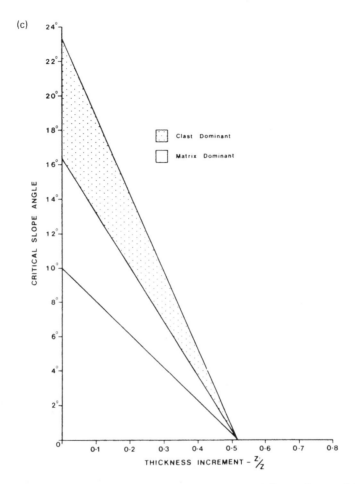

Figure 9 (*cont'd*) (c) shows the critical slope angle for stability under conditions of undrained loading. Envelopes for clast dominant and matrix dominant facies are based on limiting values for angle of internal friction within most facies will fall, i.e. 20–30° for matrix dominant and 30–40° for clast dominant.

In the case of melt-out below a supraglacial lake, heat flow into the sediment is controlled by water temperature. As noted by Driscoll (1980) this can rise above the summer air temperature in the case of a shallow lake, although in a deep lake the water temperature (at depth) will normally be below the air temperature. The presence of the water body will depress the permafrost surface and so will allow the melting-out of an increased thickness of sediment. However, the preservation of this sediment is still controlled by slope stability considerations, and the mere presence of the water body does not guarantee its survival.

Discussion

The analysis presented in this paper provides quantitative support for the oft-quoted idea that a process/facies continuum exists between lodgement and melt-out below the margins of glaciers. This continuum (Fig. 7) will be evident not only spatially across the bed of the glacier at any one time but will also occur during the deposition of a single till sequence as a result of 'overprinting'. Boulton *et al.* (1974) for example has clearly shown the postdepositional deformation that lodgement till undergoes as a result of changing ice thickness during glacial retreat; this is also implicit in discussion of deforming beds below ice sheets involving the widespread remoulding of substrate sediments (Boulton and Jones, 1979; Boulton, 1987; Boulton and Hindmarsh, 1987; Clark, 1987; Elson, 1988). Comprehensive discrimination of the range of subglacially-produced diamict facies will ultimately be reliant on detailed outcrop, clast, matrix and palaeomagnetic studies of depositional sequences combined with more sophisticated instrumentation of modern glaciers. At this point in time, field criteria for the recognition of lodgement and melt-out till(ite)s are available and in many cases identification of a 'lodgement till complex' or a 'melt-out till complex' may be possible. In many cases however where the dominant depositional process is unclear, use of the umbrella term 'basal till complex' may be more appropriate.

Clearly, the data presented in this paper impact on discussions regarding the definition of 'till', in particular how much syn- and postdepositional deformation and remobilization is allowed (see above). Only those diamicts deposited in Zone I on Fig. 7 can be identified as melt-out tills *sensu stricto*. Within the terrestrial glacial environment the opportunity for melt-out tills *sensu stricto* to accumulate *in situ* and be preserved is restricted. Sufficient englacial debris is only likely to be present within glacier margins but it is in this area where the geotechnical and sedimentary conditions that lead to subglacial deformation, remoulding and restructuring are most intense. With regard to supraglacial settings the distribution and volume of englacial debris is likely to be extremely irregular promoting differential ablation of the ice margin and leading to a system of ice-cored highs and intervening troughs. Slope processes will promote stripping and destruction of the ice-cored highs, the topographic 'inversion' of topography, and the mechanical disturbance of melted-out facies by sediment gravity flow processes (Lawson, 1988).

The importance of melt-out has probably been overstated in many interpretations of ancient diamict(ite) deposits that are thick and laterally extensive. Observations in the modern environment and the geotechnical arguments assembled here indicate that melt-out is of only local significance, and do not support the hypothesis that it is a mechanism by which regionally extensive, tabular diamict units of many metres thickness can be deposited. These observations, based principally on modern glacier margins

710

of restricted extent are equally valid when applied to ancient sequences deposited by large continental ice sheets. In North America, within the area affected by the Laurentide Ice Sheet belts of hummocky disintegration topography deposited by stagnant ice (e.g. Fig. 4d) extend over several thousands of square kilometres (Moran *et al.*, 1980); but even here, though their subsurface sedimentology is not well documented as yet, it is clear that individual facies are spatially discontinuous stratigraphic elements within the former marginal zone (e.g. Paul, 1983). Large-scale stagnation fails to produce regionally extensive diamict units by melt-out essentially because of complexity within the depositional system created by repeated incremental stagnation (e.g. Fig. 8).

Acknowledgements

The authors wish to thank J. Shaw, E.C. McRoberts and C.H. Eyles for review of previous drafts of the manuscript and J.D. Peacock for Figs 2d, 3d,e. N. Eyles is grateful to the donors of the Petroleum Research Fund of the American Chemical Society and the Natural Sciences and Engineering Research Council of Canada for their generous support.

References

Andrews, J.T. (1983). On the reconstruction of Pleistocene ice sheets: A review. *Quaternary Science Reviews*, **1**, 1–30.

Andrews, J.T. (1987). The Late Wisconsin glaciation and deglaciation of the Laurentide Ice Sheet. *In*: Ruddiman, W.F. and Wright, H.E., Jr. (eds), *North America and Adjacent Oceans During the Last Deglaciation*, Volume K-3, pp. 13–38. Geological Society of America, Geology of North America.

Anketell, J.M., Cegla, J. and Dzulynski, S. (1970). On the deformational structures in systems with reversed density gradients. *Annals de la Société géologique de Pologne*, **XL**, 3–29.

Attig, J.W., Clayton, L. and Mickelson, D.M. (1985). Correlation of late Wisconsin glacial phases in the western Great Lakes area. *Geological Society of America Bulletin*, **96**, 1585–1593.

Bleuer, N.K. (1983). Load and slide deformation of Wisconsinan substrates in the central Indiana till plain. *Geological Society of America Abstracts with Programs*, **15**, p. 21.

Bouchard, M.A., Cadieux, B. and Goutier, F. (1984). L'Origine et les caracteristiques des lithofacies du till dans le secteur nord of Lac Albanel, Quebec: Une etude de la dispersion glaciare clastiques. Chibougamou — Stratigraphy and Mineralization. Canadian Institute Mining, Spec. Pub. **34**, 244–260.

Boulton, G.S. (1968). Flow tills and related deposits on some Vestspitsbergen glaciers. *Journal of Glaciology*, **7**, 391–412.

Boulton, G.S. (1970). On deposition of subglacial and melt-out tills at the margins of certain Svalbard glaciers. *Journal of Glaciology*, **9**, 231–246.

Boulton, G.S. (1971). Till genesis and fabric in Svalbard, Spitsbergen. *In*: Goldthwait, R.P. (ed.), *Till: a Symposium*, pp. 41–72. Columbus, Ohio State University Press.

Boulton, G.S. (1972). The role of thermal regime in glacial sedimentation. Inst. British Geographers, Spec. Pub. No. 4, 1–20.

Boulton, G.S. (1972a). Modern Arctic glaciers as depositional models for former ice sheets. *Journal of the Geological Society of London*, **128**, 361–393.

Boulton, G.S. (1975). Processes and patterns of subglacial sedimentation: a theoretical approach. *In*: Wright, A.E. and Moseley, F. (eds), *Ice Ages: Ancient and Modern*, pp. 7–42. Seel House Press, Liverpool.

Boulton, G.S. (1976). A genetic classification of tills and criteria for distinguishing tills of different origin. *In*: *Tills — Genesis and Diagenesis. Geografia*, **12**, 65–80.

Boulton, G.S. (1977). A multiple till sequence by a late Devensian Welsh ice-cap: Glanllynnau, Gwynedd. *Cambria*, **4**, 10–31.

Boulton, G.S. (1978). Boulder shapes and grain-size distribution of debris as indicators of transport paths through a glacier and till genesis. *Sedimentology*, **25**, 773–799.

Boulton, G.S. (1987). A theory of drumlin formation by subglacial sediment deformation. *In*: Menzies, J. and Rose, J. (eds), *Drumlin Symposium*, pp. 25–80. Balkema, Rotterdam.

Boulton, G.S., Dent, D.L. and Morris, E.M. (1974). Subglacial shearing and crushing and the role of water pressures in tills from south-east Iceland. *Annales de géographie*, **56**, 135–145.

Boulton, G.S. and Paul, M.A. (1976). The influence of genetic processes on some geotechnical properties of glacial tills. *Quarterly Journal of Engineering Geology*, **9**, 159–194.

Boulton, G.S. and Eyles, N. (1979). Sedimentation by valley glaciers: Model and genetic classification. *In*: Schluchter, Ch. (ed.), *Moraines and Varves: Origin/ Genesis/Classification*, pp. 11–23. Balkema, Rotterdam.

Boulton, G.S., Jones, A.S., Clayton, K.M. and Kenning, M.J. (1977). A British ice sheet model and patterns of glacial erosion and deposition in Britain. *In*: Shotton, F.W. (ed.), *British Quaternary Studies: Recent Advances*, pp. 231–246. Oxford University Press.

Boulton, G.S., Smith, G.D., Jones, A.S. and Newsome, J. (1985). Glacial geology and glaciology of the last mid-latitude ice sheets. *Journal of the Geological Society of London*, **142**, 447–474.

Boulton, G.S. and Hindmarsh, R.C.A. (1987). Sediment deformation beneath glaciers: Rheology and geological consequences. *Journal of Geophysical Research*, **92**, 9059–9082.

Brodzikowski, K. and Van Loon, A.J. (1983). Sedimentology and deformational history of unconsolidated Quaternary sediments of the Jaroszow zone (Sudetic Foreland). *Geologia Sudetica*, **18**, 124–195.

Butyrym, J., Cegla, J., Dzulynski, S. and Nakonieczny, S. (1964). New interpretation of periglacial structures. *Folia Quaternaria*, **17**, 1–34.

Carruthers, R.G. (1939). On northern glacial drifts: some peculiarities and their significance. *Quarterly Journal of the Geological Society of London*, **379**, 299–333.

Carruthers, R.G. (1943). The secret of the glacial drifts. *Proceedings of the Yorkshire Geological Society*, **27**, 43–57.

Carruthers, R.G. (1944). The secret of the glacial drifts. Part II. Applications to Yorkshire. *Proceedings of the Yorkshire Geological Society*, **27**, 129–172.

Carruthers, R.G. (1953). *Glacial Drift and the Undermelt Theory.* Harold Hill, Newcastle-upon-Tyne, England.

Catt, J.A. and Madgett, P.A. (1981). The work of W.S. Bisat F.R.S. on the Yorkshire coast. *In:* Neale, J. and Flenley, J. (eds), *The Quaternary in Britain*, pp. 114–136. Pergamon Press, Oxford.

Cegla, J.C., Dzulynski, S. and Rzechowski, J. (1976). Experiments on gradational contacts between varves and till. *Geografia*, **12**, 161–165.

Chamberlin, T.C. (1894). Proposed genetic classification of Pleistocene glacial formations. *Journal of Geology*, **2**, 517–538.

Clapperton, C.M. (1975). The debris content of surging glaciers in Svalbard and Iceland. *Journal of Glaciology*, **14**, 395–406.

Clarke, G.K.C. (1987). Subglacial till: A physical framework for its properties and processes. *Journal of Geophysical Research*, **92**, 9023–9036.

Clayton, L. (1964). Karst topography on stagnant glaciers. *Journal of Glaciology*, **5**, 107–112.

Claryton, L. and Moran, S.R. (1974). A glacial process-form model. *In:* Coates, D.R. (ed.), *Glacial Geomorphology*, pp. 89–120. New York State University.

Dardis, G.F., McCabe, A.M. and Mitchell, W.I. (1984). Characteristics and origin of lee-side stratification sequences in late Pleistocene drumlins, Northern Ireland. *Earth Surface Processes and Landforms*, **9**, 409–424.

Denton, G.H. and Hughes, T.J. (eds). *The Last Great Ice Sheets.* John Wiley, New York, 310 pp.

Dowdeswell, J.A., Hambrey, M.J. and Wu, R. (1985). A comparison of clast fabric and shape in late Precambrian and modern glacigenic sediments. *Journal of Sedimentary Petrology*, **55**, 691–704.

Dowdeswell, J.A. and Sharp, M. (1986). Characterization of pebble fabrics in modern terrestrial glacigenic sediments. *Sedimentology*, **33**, 699–710.

Dreimanis, A. (1976). Tills, their origin and properties. *In:* Leggett, R.F. (ed.), *Glacial Till*, pp. 11–49. Special Publication No. 12, Royal Society of Canada, Ottawa.

Dreimanis, A. (1988). Tills: Their genetic terminology and classification. *In:* Goldthwait, R.P. and Matsch, C. (eds), *Genetic Classification of Glacigenic Deposits*, pp. 17–84. Balkema, Rotterdam.

Drewry, D.J. (1986). *Glacial Geologic Processes.* Arnold, Canada, 274 pp.

Driscoll, F.G. (1980). Wastage of the Klutlan ice-cored moraines, Yukon Territory, Canada. *Quaternary Research*, **14**, 31–49.

Drozdowski, E. (1983). Load deformation in melt-out till and underlying laminated till: An example from northern Poland. *In:* Evenson, E.B., Schlüchter, Ch and Rabassa, J. (eds), *Tills and Related Deposits*, pp. 119–124. Balkema, Rotterdam.

Drozdowski, E. (1984). Subglacial meltout till as a deposit indicative of areal deglaciation. Abstracts of Papers, 25th Intn. Geographical Congress, Paris, France, **1**, Th. 1.22.

Dorzdowski, E. (1985). On the effects of bedrock protuberances upon the depositional and relief-forming processes in different marginal environments of Spitsbergen Glaciers. *Palaeogeography, Palaeoclimatology, Palaeoecology*, **51**, 397–415.

Dyke, L. and Egginton, P. (1988). Till behaviour and its relationship to active-layer hydrology, District of Keewatin, Northwest Territories. *Canadian Geotechnical Journal*, **25**, 167–172.

Edwards, C.A. (1981). The tills of Filey Bay. *In*: Neale, J. and Flenley, J. (eds), *The Quaternary in Britain*, pp. 108–118. Pergamon Press, Oxford.

Elson, J.A. (1961). *Geology of Tills*. National Research Council of Canada. Technical Memorandum No. 69, pp. 5–17.

Elson, J.A. (1988). Glacetectonite: Brecciated sediments and cataclastic sedimentary rocks formed subglacially. *In*: Goldthwait, R.P. and Matsch, C.L. (eds), *Genetic Classification of Glacigenic Deposits*, pp. 89–91. Balkema, Rotterdam.

Eyles, N. (1979). Facies of supraglacial sedimentation on Icelandic and Alpine temperate glaciers. *Canadian Journal of Earth Sciences*, **16**, 1341–1361.

Eyles, N. (1983). Modern Icelandic glaciers as depositional models for 'hummocky moraine' in the Scottish Highlands. *In*: Evenson, E.B., Schlüchter, Ch. and Rabasa, J. (eds), *Tills and Related Deposits*, pp. 47–59. Balkema, Rotterdam.

Eyles, N. and Slatt, R.M. (1977). Ice-marginal sedimentary, glacitectonic and morphologic features of Pleistocene drift; an example from Newfoundland. *Quaternary Research*, **8**, 267–281.

Eyles, N., Sladen, J.A. and Gilroy, S. (1982). A depositional model for stratigraphic complexes and facies superimposition in lodgement till. *Boreas*, **11**, 317–333.

Eyles, N., Eyles, C.H. and Miall, A.D. (1984). Lithofacies types and vertical profile models: An alternative approach to the description and environmental interpretation of glacial diamict and diamictite sequences. Reply to Comments. *Sedimentology*, **31**, 891–898.

Eyles, N., Day, T.E. and Gavican, A. (1987). Depositional controls on the magnetic characteristics of lodgement tills and other glacial diamict facies. *Canadian Journal of Earth Sciences*, **24**, 2436–2458.

Flint, R.F. (1971). *Glacial and Quaternary Geology*. John Wiley, New York.

Gravenor, C.P. and Kupsch, W.O. (1959). Ice-disintegration features in western Canada. *Journal of Geology*, **67**, 48–64.

Goodchild, J.G. (1875). The glacial phenomena of the Eden Valley and the western part of the Yorkshire Dales district. *Quarterly Journal of the Geological Society of London*, **31**, 55–99.

Haldorsen, S. (1982). The genesis of tills from Astadalen, southeastern Norway. *Norsk Geologisk Tidsskrift*, **62**, 17–38.

Haldorsen, S. (1983). Mineralogy and geochemistry of basal till and their relationship to fill-forming processes. *Norsk Geologisk Tidsskrift*, **63**, 15–25.

Haldrosen, S. and Shaw, J. (1982). The problem of recognizing melt-out till. *Boreas*, **11**, 261–277.

Hillefors, A. (1973). The stratigraphy and genesis of stoss and leeside moraines. *Bulletin of the Geological Institute, Uppsala University*, **5**, 139–154.

Holdsworth, G. (1973). Ice deformation and moraine formation at the margin of an ice cap adjacent to a proglacial lake. *In*: Fahey, B.D. and Thompson, R.D. (eds), *Research in Polar and Alpine Geomorphology*, pp. 187–199. Geo-Books, Norwich.

Hooke, R. LeB. (1984). On the role of mechanical energy in maintaining subglacial water conduits at atmospheric pressure. *Journal of Glaciology*, 30, 180–187.

Hughes, T. (1987). Ice dynamics and deglaciation models when ice sheets collapsed. *In*: Ruddiman, W.F. and Wright, H.E., Jr. (eds), *North America and Adjacent Oceans During the Last Deglaciation*, Volume K-3, pp. 183–220. Geological Society of America, Geology of North America.

Hutchinson, J.N. and Bhandari, R.K. (1971). Undrained loading, a fundamental mechanism of mudflows and other mass movement. *Geotechnique*, **21**, 353–358.

Lafleur, J., Cummins, A. and Chiche, S. (1982). Self-filtration of tills submitted to hydraulic gradients. Proceedings 35th Canadian Geotechnical Society Conference, Montreal, pp. 50–62.

Lawson, D.E. (1979). A comparison of the pebble orientation in ice and deposits of the Matanuska Glacier, Alaska. *Journal of Geology*, **87**, 629–645.

Lawson, D.E. (1979a). Sedimentological analysis of the western terminus region of the Matanuska Glacier, Alaska. U.S. Army Cold Regions Research and Engineering Laboratory, Report 79-9, 103 p.

Lawson, D.E. (1981). Sedimentological characteristics and classification of depositional processes and deposits in the glacial environment: U.S. Army Cold Regions Research and Engineering Laboratory, Report 81-27, 16 p.

Lawson, D.E. (1982). Mobilization, movement and deposition of active subaerial sediment flows. Matanuska Glacier, Alaska. *Journal of Geology*, **90**, 279–300.

Lliboutry, L. (1983). Modifications to the theory of intraglacial waterways for the case of subglacial ones. *Journal of Glaciology*, **29**, 216–226.

Lowe, D.R. (1975). Water escape structures in coarse-grained sediments. *Sedimentology*, **22**, 157–204.

Lowe, D.R. (1976). Subaqueous liquefied and fluidized sediment flows and their deposits. *Sedimentology*, **23**, 285–308.

Madgett, P.A. and Catt, J.A. (1978). Petrography, stratigraphy and weathering of Late Pleistocene tills in East Yorkshire, Lincolnshire and North Norfolk. *Proceedings of Yorkshire Geological Society*, **42**, 55–108.

McRoberts, E.C. (1975). Field observations of thawing in soils. *Canadian Geotechnical Journal*, **12**, 126–130.

McRoberts, E.C. and Morgenstern, N.R. (1974). The stability of thawing slopes. *Canadian Geotechnical Journal*, **11**, 447–469.

Mickelson, D.M. (1973). Nature and rate of basal till deposition in a stagnating ice mass, Burroughs Glacier, Alaska. *Arctic Alpine Research*, **5**, 17–27.

Minell, H. (1977). Hog's-back moraines in Harrejokko, northern Sweden. *Geologiska Foreningens Stockholm Forhandlingar*, **99**, 264–270.

Minell, H. (1977a). Transverse moraine ridges of basal origin in Harjedalen. *Geologiska Foreningens Stockholm Forhandlingar*, **99**, 271–277.

Moran, S.R., Clayton, L., Hooke, R. LeB., Fenton, M.M. and Andriashek, L.D. (1980). Glacier-bed landforms of the Prairie region of North America. *Journal of Glaciology*, **25**, 457–476.

Morgenstern, N.R. and Nixon, J.F. (1971). One dimensional consolidation of thawing soils. *Canadian Geotechnical Journal*, **8**, 558–565.

Nixon, J.F. and Morgenstern, N.R. (1973). Practical extensions to a theory of consolidation for thawing soils: Proc. 2nd Int. Conf. Permafrost. North American Contribution, pp. 369–377.

Paul, M.A. (1983). The supraglacial land system. *In*: Eyles, N. (ed.), *Glacial Geology: An Introduction to Engineers and Earth Scientists*, pp. 71–90. Pergamon Press, Oxford.

Postma, G. (1983). Water escape structures in the context of a depositional model of a mass flow dominated conglomerate fan-delta (Abrioja Formation, Pliocene, Almeria Basin, SE Spain). *Sedimentology*, **30**, 91–103.

Punkari, M. (1984). The relations between glacial dynamics and tills in the eastern part of the Baltic Shield. *Striae*, **20**, 49–54.

Rothlisberger, H. (1972). Water pressure in intra and subglacial channels. *Journal of Glaciology*, **11**, 177–203.

Sharp, M. (1985). Sedimentation and stratigraphy at Eyjabakkajokull, — an Icelandic surging glacier. *Quaternary Research*, **24**, 268–284.

Shaw, J. (1979). Genesis of the Sveg tills and Rogen moraines of central Sweden: a model of basal melt-out. *Boreas*, **8**, 409–426.

Shaw, J. (1982). Melt-out till in the Edmonton area, Alberta, Canada. *Canadian Journal of Earth Sciences*, **19**, 1548–1569.

Sladen, J.A. and Wrigley, W. (1984). Geotechnical properties of lodgement till — a review. *In*: Eyles, N. (ed.), *Glacial Geology*, pp. 184–212. Pergamon Press, Oxford.

Spencer, A.M. (1971). Late Precambrian glaciation in Scotland. *Geological Society of London, Memoir No. 6*.

Spencer, A.M. (1985). Mechanisms and environments of deposition of Late Precambrian geosynclinial tillites: Scotland and East Greenland. *Palaeogeography, Palaeoclimatology, Palaeoecology*, **51**, 143–158.

Stephan, H.-J. and Ehlers, J. (1983). North German till types. *In*: Ehlers, J. (ed.), *Glacial Deposits in North-West Europe*, pp. 239–248. Balkema, Rotterdam.

Sugden, D.E. (1977). Reconstruction of the morphology, dynamics and thermal characteristics of the Laurentide Ice Sheet at its maximum. *Arctic Alpine Research*, **9**, 21–47.

Upham, W. (1891). Criteria of englacial and subglacial drift. *American Geologist*, **8**, 376–385.

Vivian, R. (1980). The nature of the ice-rock interface: the results of investigations on 20,000 m^2 of the rockbed of temperate glaciers. *Journal of Glaciology*, **25**, 267–277.

Weertman, J. (1961). Mechanism for the formation of inner moraines found near the edge of cold ice caps and ice sheets. *Journal of Glaciology*, **3**, 965–978.

Weertman, J. (1968). Diffusion law for the dispersion of hard particles in an ice matrix that undergoes simple shear deformation. *Journal of Glaciology*, **7**, 161–165.

Wordie, J.M. (1947). Discussion on the origin of glacial drifts. *Journal of Glaciology*, **1**, 430–440.

Wright, H.E., Jr. (1980). Surge moraines of the Klutlan Glacier, Yukon Territory, Canada: Origin, wastage, vegetation succession, lake development and application to the late-glacial of Minnesota. *Quaternary Research*, **14**, 2–18.

24

POTENTIAL EFFECTS OF SUBGLACIAL WATER-PRESSURE FLUCTUATIONS ON QUARRYING

N.R. Iverson

Source: *Journal of Glaciology* 37(125) (1991): 27–36.

Abstract

Water-pressure fluctuations beneath glaciers may accelerate rock fracture by redistributing stresses on subglacial bedrock and changing the pressure of water in bedrock cracks. To study the potential influence of water-pressure fluctuations on the fracture of subglacial bedrock, ice flow over a small bedrock step with a water-filled cavity in its lee is numerically modeled, and stresses on the bedrock surface are calculated as a function of transient water pressures in the cavity. Stresses on the bed are then used to calculate principal stress differences within the step. Rapid reductions in cavity water pressure increase principal stress differences in the bed, increasing the likelihood of crack growth in the step and the formation of predominantly vertical fractures. Relatively impermeable bedrock may be most susceptible to fracturing during water-pressure reductions because high water pressure in cracks within the rock can be maintained, as water pressure decreases in cavities. These results, when considered in conjunction with the strong likelihood that increases in water pressure accelerate the removal of rock fragments looened from the bed, suggest that in zones of ice–bed separation where water-pressure fluctuations typically are large, rates of quarrying may be higher than along other parts of glacier beds.

Introduction

It is well established that where ice separates from bedrock beneath glaciers large temporal variations in subglacial water pressure can occur. Basal water pressures have been measured directly in boreholes and moulins, and have sometimes been found to fluctuate diurnally through more than 0.5 MPa (Kamb and others, 1985; Iken and Bindschadler, 1986; Hooke and others, 1987, 1989). In addition to affecting the sliding velocity of glaciers, these pressure variations may play an important role in quarrying subglacial bedrock (Robin, 1976; Röthlisberger and Iken, 1981). The intent of this study was to explore this idea further by analyzing how the stresses exerted by ice and water on a small part of an idealized glacier bed are influenced by fluctuations in the water pressure within a nearby cavity.

As noted by Röthlisberger and Iken (1981), quarrying comprises several processes. Bedrock blocks must be first loosened from the bed by either subglacial rock fracture or by the presence of preglacial joints. They then have to be removed from the bed and entrained by basal ice. Here, the focus will be on the effects of water-pressure fluctuations on rock fracture. Stresses along the surface of a bedrock step are calculated for steady and transient water pressures in a subglacial cavity with the finite-element method. The calculated stress distributions are used to determine stresses at depth within the bedrock, and the likelihood of fracture propagation in the rock is assessed with the Griffith failure criterion.

Rock fracture

Background

Quarrying can only occur if subglacial bedrock is sufficiently fractured. Many authors have noted the role of preglacial joints in the process (Matthes, 1930; Crosby, 1945; Zumberge, 1955; Addison, 1981; Rastas and Seppala, 1981). It is unlikely, however, that preglacial joints will always be dense and continuous enough to completely isolate bedrock blocks so that removal requires that only frictional forces be overcome. One might argue that preglacial weathering processes could create sufficiently fragmented bedrock, but this does not explain the erosion of valleys and fiords that are thousands of meters deep. Subglacial fracture must occur, therefore, to some extent. This is supported by observations of groups of boulders with quarried lee surfaces that are firmly embedded in lodgement tills (Sharp, 1982).

Morland and Morris (1977) examined the possibility that sliding ice might exert a sufficiently large viscous drag on bedrock obstacles to induce fracture. Using Nye's (1969) theoretical pressure distribution over the surface of a wavy bed, they solved for the stress field in an asymmetric, elastic bedrock hump. Using the Coulomb failure criterion and including the effect of pore-water pressure, they concluded that differences between the two effective principal stresses in the hump were generally not large enough to induce failure in intact bedrock.

The analysis presented in this section is similar to that of Morland and Morris (1977) in that stress fields in an elastic bedrock irregularity are

718

calculated for a range of conditions, and the resultant tendency for fracture is assessed. However, important differences exist. Here, stresses are analyzed in an idealized bedrock step with a cavity in its lee. The effect of water-pressure variations in the cavity on the ice flow and on the pressure distribution along the rock surface is determined using a method similar to that used by Iken (1981). Because the macroscopic fracture of rock depends sensitively on poorly constrained variables like rock microstructure and environmental factors associated with sub-critical crack propagation (e.g. Atkinson, 1984), the main objective of this study is not to predict the exact conditions under which failure will occur. Instead, changes in the propensity for rock fracture will be evaluated for specified fluctuations in water pressure, assuming lesser-known variables remain constant.

Ice flow over a bedrock step

A finite-element program developed by Hanson (1985) is used to model ice flow over a step on the bed that is 1 m in height (Fig. 1). The chosen geometry, although simplified, is not unrealistic; planar bedrock steps have

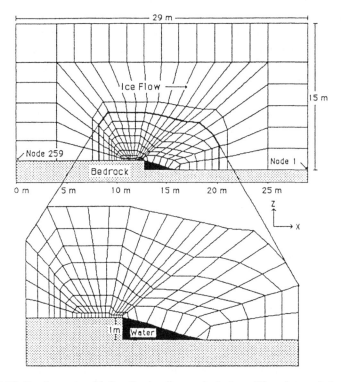

Figure 1 Finite-element grid for the ice-flow calculation. The plane of the grid is vertical and parallel to the flow direction.

frequently been observed (Anderson and others, 1982; Hallet and Anderson, 1982). Given the necessary boundary conditions, the program calculates vertical and horizontal velocities at each node and the mean pressure within each element. Strain rates normal to the flow plane are assumed to be zero (plane strain). Equations solved are for continuity of an incompressible fluid and conservation of momentum when accelerations are negligible. The familiar Glen-type power law for flow is assumed:

$$\dot{\varepsilon} = (\tau/B)^n \qquad (1)$$

where $\dot{\varepsilon}$ and τ are the effective strain rate and shear stress, respectively, n is a constant taken to be equal to 3, and B is a temperature-dependent viscosity coefficient.

At nodes along the lower boundary of the domain, where the ice is in contact with the bed, the vertical ice velocity is fixed at zero. This constrains the direction of ice movement to be parallel to the ice–bed interface. The basal ice is assumed to contain no debris, so the effect of ice convergence with the bed on fragment–bed stresses (Hallet, 1979) is not considered. At nodes along the zone of ice–bed separation in the lee of the step, where the glacier sole is supported by water under pressure, a force normal to the cavity roof is specified. These are identical to the conditions specified by Iken (1981) in her numerical simulation of hydraulically jacked sliding over a wavy bed. The method used to determine the size of the separated zone is discussed below.

A normal stress and shear stress act on the upper boundary. A normal stress of 2.57 MPa was chosen, equal to the pressure beneath 285 m of ice. A shear stress of 0.09 MPa was applied to the upper boundary, consistent with a glacier surface slope of 4° and assuming that the basal shear stress is reduced by one-half to account approximately for the fraction of basal shear stress supported by valley walls (Nye, 1965).

Along the ends of the modeled section, forces are prescribed at each node commensurate with a pressure normal to the boundary equal to $\rho g H$ where ρ is the ice density, g is the gravitational acceleration, and H is the overlying ice thickness. This is strictly true only for a fluid at rest sustaining no shear stresses. However, because the glacier surface slope is small, it is a good approximation. In addition, stresses normal to the boundary will deviate from $\rho g H$ if a velocity gradient across the length of the domain exists, implying a gradient in the longitudinal stress. Thus, for this to be a valid boundary condition, horizontal velocities calculated along the up-glacier side of the domain must approximately equal those calculated along the domain's down-glacier side. To maintain mechanical equilibrium, shear stresses must act on the ends of the modeled section. These shear stresses are not fixed as boundary conditions, but are produced during the balancing of forces during a given computation, as implied by velocity gradients in the resultant flow solutions. Finally, at the nodes located at the two lower outside corners of the section (nodes 1 and 259; Fig. 1), a reasonable sliding velocity parallel to the bed is specified. Note that the specification of a sliding velocity at the lower outside corners of the section and a shear stress along the upper boundary implies a rough bed. The roughness can be thought of as resulting from small-scale roughness elements on the planar surface.

Modeling of a steady cavity

The cavity that forms down-glacier from the bedrock step is steady if it is neither shrinking nor expanding at a given water pressure. This means that a parcel of ice on the cavity roof has a net velocity parallel to the cavity surface. A primary goal of this modeling is to determine the pressure distribution against the rock upstream from a steady cavity and then to examine how it is affected by transients in the cavity water pressure. In contrast, Iken (1981), who modeled water-pressurized cavities in the lee of sinusoidal bed undulations, was principally concerned with the effect of transients in cavity water pressure on the basal sliding velocity. The iterative method used to model steady cavities in this analysis therefore is different in some respects from the one used by Iken.

The iteration is begun by specifying a sliding velocity at each end of the domain (nodes 1 and 259), a water pressure in the cavity, and a reasonable cavity geometry. The program is then run and the velocity field examined. The velocity calculated at nodes along the cavity roof usually shows that the cavity is either expanding or shrinking, or expanding at some places and shrinking at others. Two changes are therefore made before the next iteration: (1) the geometry of the cavity is changed, as suggested by the velocity field, and (2) the velocity at the two lower, outside corners of the domain is increased if the cavity is shrinking or decreased if the cavity is expanding. By this iterative procedure the geometry of the cavity is adjusted until velocity vectors at nodes along the cavity wall are within 3° the cavity surface. Thus, a cavity is obtained that has a size and geometry commensurate with the specified cavity water pressure and sliding velocity. The effect on the cavity geometry of viscous heat dissipation by subglacial water flow is not incorporated into the calculation, due to its dependence on a number of poorly constrained hydrological parameters. This effect, however, may be important, increasing the length of a steady-state cavity by as much as a factor of 3 (Kamb, 1987).

A steady cavity with an internal water pressure, P_w, of 2.1 MPa is shown in Figure 2 (solid line). The sliding velocity at the ends of the domain was 5.2 m year^{-1}. As expected, the ice pressure is concentrated on the area of the bed immediately upstream from the cavity. Pressure on the rock in this area is just over 4.0 MPa or about 50% greater than the ambient mean pressure in the basal ice. Velocities at nodes along the up-glacier end of the domain

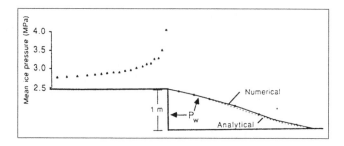

Figure 2 Geometry of a steady cavity (solid line) and the mean ice pressure in elements
along the bed. $P_w = 2.1$ MPa, $B = 0.28$ MPa year, the sliding velocity is
5.2 m year^{-1}, and the ice thickness is 300 m. Dotted line is Kamb's (1987)
analytical solution for the cavity geometry using a linear rheology, and
assuming no melting occurs.

are within 4% of velocities at nodes an equal distance above the bed along
the down-glacier side of the domain. Gradients in the longitudinal strain
rate and stress over the domain length are therefore small. A value of
$B = 0.28$ MPa year$^{1/3}$ was used in the calculation. The solution proved to be
stable against a factor of 2 reduction in the size of the domain, and was also
nearly identical to one obtained from Kamb's (1987) analytical solution for
the geometry of a step cavity using a linear rheology with appropriately
scaled viscosities (Fig. 2, dotted line).

A steady cavity was also calculated with a B value of 0.176 MPa year$^{1/3}$,
which is probably more appropriate for temperate ice (Hooke, 1981, fig. 2).
In this case, a sliding velocity of about 19 m year^{-1} was necessary to produce
a steady cavity of the same dimensions as that calculated with the larger B
value. This is consistent with the power-law relation for flow, which predicts
a four-fold increase in the effective strain rate as a result of reducing B from
0.28 to 0.176 MPa year$^{1/3}$. The stresses in the ice are the same in both cases,
however, and thus the pressure distribution along the bed upstream from
the cavity remains unchanged.

Transient water pressures

Any sudden decrease in the cavity water pressure will be accompanied by
an increase in ice pressure against the bed, because the pressure averaged
over the base of the glacier must remain constant. In the vicinity of the
cavity, the magnitude of the change in ice pressure will depend on the geometry of the cavity and on the magnitude of the water-pressure change.

To examine the effects of transient water pressures, forces at nodes along
the wall of the steady cavity ($P_w = 2.1$ MPa) were specified that are commensurate with water pressures of 2.5, 1.9 and 1.5 MPa. Other than these
changes, boundary conditions were unchanged from the last iteration of the

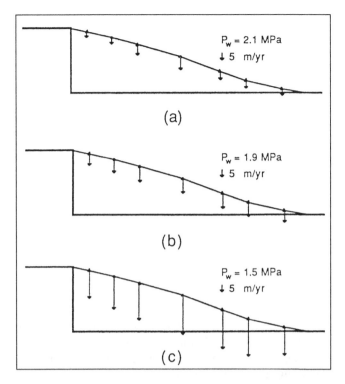

Figure 3 The vertical component of the ice velocity at nodes along the cavity ceiling when the cavity is (a) steady at a water pressure of 2.1 MPa, and when the water pressure is instantaneously reduced to (b) 1.9 MPa and (c) 1.5 MPa. When P_w is raised to 2.5 MPa, downward vertical velocities are all less than 0.7 m year^{-1}.

steady-cavity calculation. Keeping the sliding velocity at the domain ends at the steady value is incorrect, however, because increases or decreases in the cavity water pressure should affect the far-field sliding velocity (Iken, 1981). Horizontal components of the ice velocity are therefore not accurate in the transient calculations. However, the vertical component of the ice velocity at each node and, more importantly, the pressure distribution against the bed, are accurate because they are independent of the horizontal velocity specified at the domain ends.

Figure 3b and c illustrate the effect of reductions in the cavity water pressure on the vertical component of the ice velocity at the cavity wall. Reducing the water pressure 0.6 MPa results in about a five-fold increase in the downward component of the ice velocity relative to the steady case (Fig. 3a). Figure 4 shows the distribution of ice pressure over 1.5 m of the bed immediately upstream from the cavity for three different transient water pressures and for the steady-state water pressure. In the element nearest the

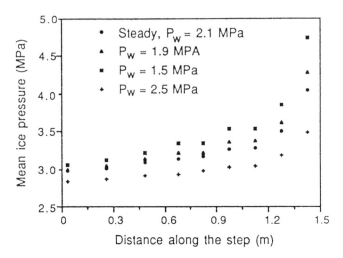

Figure 4 Mean ice pressures in elements along the bed immediately upstream from the cavity (cavity edge at 1.5 m) for steady and transient cavity water pressures.

cavity, a decrease in water pressure from the steady value results in an increase in ice pressure of roughly the same magnitude as the water-pressure decrease. The steps in the curves are products of the geometry of the elements chosen along the bedrock surface. Solutions using different element configurations are only slightly different, with steps occurring at different positions along the bed.

The magnitude of the increase in ice pressure should depend on the extent of ice–bed separation (Robin, 1976, p. 192). If, for example, two cavities had been included at the base of the section rather than a single cavity, a given reduction in water pressure would have shifted a larger fraction of the weight of the glacier to zones of ice-bed contact, resulting in greater increases in ice pressure against the bed.

Bedrock stresses

For the steady cavity modeled above, reductions in the cavity water pressure result in increases in ice pressure against the bed upstream from the cavity, whereas increases in water pressure reduce this pressure (Fig. 4). To determine the effect of the pressure changes on the propensity for bedrock fracture, stress fields within the bedrock step must be calculated. A program developed by Gustafson and others (1977) is used to obtain a finite-element approximation of the stress field. Linear elasticity and plane strain are assumed. A vertical plane through the bedrock step is discretized into triangular elements (Fig. 5) and, with appropriate boundary forces and displacements, the program calculates the magnitude and orientation of the

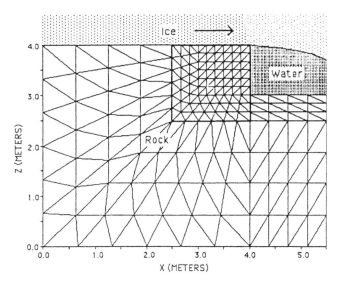

Figure 5 Finite-element grid for calculating bedrock stresses in a vertical plane through the step, parallel to the direction of ice flow.

principal stresses within each element. Elements are concentrated in a 1.5 m square where bedrock fracture associated with quarrying is expected. At nodes along the upper surface of the bed, normal forces are prescribed using the distributions of basal ice pressure calculated above. At nodes along the bed contained within the cavity, a normal force is specified that is commensurate with the water pressure in the cavity. Although a shear stress of 0.09 MPa is supported along the bed in the ice-flow calculation, assigning boundary conditions along the ends of the modeled section is greatly simplified if a shear stress is not specified in the bedrock stress calculation. Neglecting the shear stress only slightly influences bedrock fracture because the shear stress is small relative to the difference between the normal stresses applied to the step's upper and lee surfaces. Along ends of the section, zero horizontal displacement is specified, which is equivalent to assuming that the horizontal stress, σ_x, at the ends of the section equals $v(1 - v)^{-1}\rho gH$, where v is Poisson's ratio (Jaeger and Cook, 1979, p. 372). At nodes along the lower boundary, zero horizontal and vertical displacement are assumed.

Contours of the two principal stresses in a part of the step adjacent to the steady cavity ($P_w = 2.1$ MPa) are plotted as solid lines in Figure 6a and b. Both principal stresses are compressive (positive); therefore, the greatest principal stress (σ_1) is the one that is most compressive. σ_1 increases toward the cavity edge, reflecting the increase in ice pressure toward the cavity. σ_2 becomes less compressive away from the upper corner of the step where the ice pressure is highest. Gradients in both σ_1 and σ_2 are greatest near the

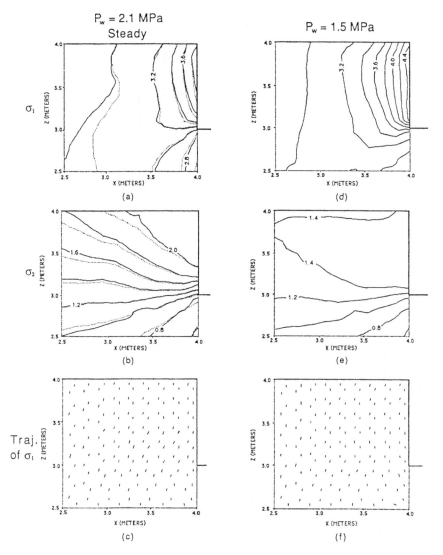

Figure 6 Contours of (a) σ_1 and (b) σ_2 when the cavity is steady at a water pressure of 2.1 MPa (solid lines). Dotted contours were calculated with an expanded modeling domain. (c) Trajectories of σ_1 when the cavity is steady. Contours of (d) σ_1 and (e) σ_2 when the cavity water pressure is reduced to 1.5 MPa. (f) Trajectories of σ_1 when the cavity water pressure is reduced to 1.5 MPa. Step surface is delineated with a bold line. Stresses are in MPa.

base of the step. Trajectories (orientations) of σ_1 are nearly vertical or steeply dipping up-glacier (Fig. 6c). The same calculation using a domain that was 4 m longer and 3 m deeper yielded a similar stress distribution (Fig. 6a and b; dotted lines). Applying a 0.1 MPa shear stress to the step surface upstream from the cavity likewise had only a small effect. Apparent deviations from applied boundary stresses along the step surface result from the finiteelement approximation, which limits the distance over which stress gradients can be resolved to the width of an element.

To calculate stress fields in the step for transient water pressures in the cavity, stresses along the upper surface of the domain were changed in accordance with the change in water pressure and the resultant variation in ice pressure against the bed (Fig. 4). The focus here will be on the effect of an instantaneous decrease in cavity water pressure from the steady value of 2.1 to 1.5 MPa. Contours of σ_1 and σ_2 for $P_w = 1.5$ MPa are shown in Figure 6d and e. The reduction in water pressure increases σ_1 and decreases σ_2 over most of the step, resulting in larger principal stress differences than those produced under steady water pressure. This is due to the increase in ice pressure against the step's upper surface, together with the reduction in water pressure against its lee surface. As the water pressure in the lee of the step falls and ice pressure normal to the step's upper surface increases, vertical compressive stresses in the rock dominate horizontal compressive stresses, rotating the trajectory of the most compressive principal stress more nearly vertical (Fig. 6f). With higher water pressure in the cavity, normal pressures against the upper and lee surfaces of the step are more nearly equal and trajectories of σ_1 tend to dip more uniformly up-glacier.

Fracture criterion

The theory of Griffith (1924) will be used to assess the likelihood of fracture from the bedrock stresses. Unlike the Coulomb criterion, it directly addresses the mechanism of crack growth in brittle materials. In addition, numerically intensive alternatives involving the calculation of stress intensities for mixed-mode crack propagation under compressive conditions (e.g. Ingraffea and Heuze, 1980) are avoided.

The Griffith (1924) criterion defines the state of stress necessary to induce fracture growth from the tips of minute, pre-existing cracks in a material. Such cracks pervade all rocks in the Earth's upper crust. Under compressive conditions, tensile stresses may develop locally near the tips of favorably oriented cracks. These stresses are largest at crack surfaces, their magnitude depending upon the curvature and length of the crack tip, as well as on the nature of the remote stresses. Fracture is initiated at or near the tips of cracks when tensile stresses exceed some critical value. The critical stress for fracture propagation can be equated with the uniaxial tensile strength of the rock, T_0, determined from uniaxial tests. The fracture criterion, therefore, is expressed in terms of T_0 and the principal stresses and is of the form:

$$\frac{(\sigma_1 - \sigma_2)^2}{8(\sigma_1 + \sigma_2)} = T_0 \quad \text{if } \sigma_1 + 3\sigma_2 > 0 \tag{2a}$$

$$\sigma_2 = -T_0 \qquad \qquad \text{if } \sigma_1 + 3\sigma_2 < 0. \tag{2b}$$

The condition $\sigma_1 + 3\sigma_2 > 0$ defines a stress state that is essentially compressive (Paterson, 1978). Unstable crack growth begins when criterion (2a) or (2b) is satisfied, but much slower stable (sub-critical) crack growth will occur when tensile stresses are less than T_0 depending upon the crack length, temperature, pressure and chemical environment (see Atkinson (1984) for a review). These conditions are poorly constrained and thus predicting the state of stress under which fracture will occur is difficult. However, the criterion is adequate for comparing the propensity for fracture under contrasting stress fields.

Unfortunately, the Griffith criterion provides little insight into the orientation of macroscopic failure surfaces. As fractures grow from the tips of micro-cracks, local stress patterns around individual cracks begin to interfere with each other, rendering Griffith's treatment invalid. Final failure surfaces presumably result from the coalescence of these cracks (Murrell, 1971; Hallbauer and others, 1973).

Despite the local re-orientation of stresses that occurs as cracks propagate and coalesce, a substantial amount of empirical evidence suggests that fracture paths can be approximated from the stress state immediately prior to crack propagation. Studies of contact-induced fracture in glass (Frank and Lawn, 1967) and rocks (Lindquist and others, 1984) show that fractures lie approximately parallel to the most compressive principal stress of the pre-fracture stress field. In triaxial compression tests, fractures often form $20–30°$ from the most compressive principal stress, with the size of the angle being proportional to the confining pressure on the specimen (Paterson, 1978, p. 17). It is probably safe to assume, therefore, that fractures will lie at a relatively small angle ($<30°$) to σ_1 of the pre-fracture stress field. More specific predictions cannot be made, however, particularly in the light of the structural anisotropy in many rocks. Trajectories of σ_1 were approximately vertical in both the steady and transient cases considered here (Fig. 6c and f), and thus predominantly vertical fractures are expected. A shear stress applied to the bedrock surface rotates the dip of σ_1 trajectories down-glacier. However, this effect is small unless the shear stress against the bed is comparable in magnitude to the normal stress.

In order to evaluate the likelihood of rock fracture from the principal stresses, a fracture index, T_g, representing the tensile stress near favorably oriented cracks that result from the prescribed principal stresses, can be substituted for T_0 in the fracture criterion (Equations (2)) and treated as a dependent variable (Johnson, 1975). Principal stresses are used to determine

T_g from the criterion equation for each element in the modeled section. Large values of T_g denote a greater propensity for rock fracture and, where $T_g > T_0$, unstable crack propagation is expected. Note that T_g becomes larger as principal stress differences increase and becomes smaller as mean stress increases (Equation (2a)). T_g is contoured in Figure 7a and b for the

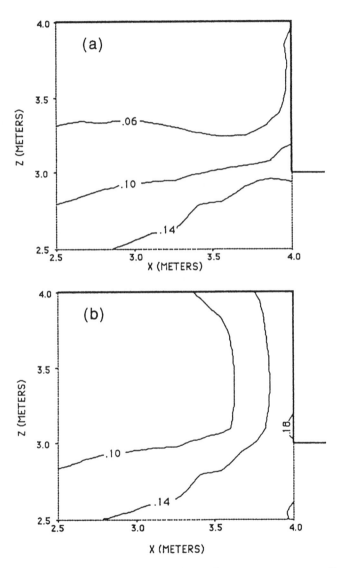

Figure 7 Contours of T_g when (a) the cavity is steady at a water pressure of 2.1 MPa, and (b) the cavity water pressure is reduced to 1.5 MPa. Step surface is delineated with a bold line. Values of T_g are in MPa.

steady cavity $P_w = 21$ bar) and for the case of the rapid decline in water pressure to 1.5 MPa. No internal water pressure is assumed to exist within micro-cracks. Throughout most of the step the tendency for crack propagation is increased by the reduction in water pressure. In the outer 0.5 m closest to the cavity where quarrying might be considered to be most likely, values of T_g are over twice as large in the transient case. There is also a tendency in both the steady and transient cases for T_g to increase with depth in the rock. This results mainly from the general decrease in the mean compressive stress with depth.

Values of T_g in Figure 7 are smaller than values of T_0 obtained from uniaxial tensile tests, which for most rocks exceed 1.0 MPa (Jaeger and Cook, 1979, p. 190), and thus in this example unstable fracture propagation is unlikely. However, if there are pre-existing, centimeter-scale fractures in the rock (e.g. Peng, 1975), values of T_0 may be reduced by more than an order of magnitude, and crack extension should occur in some sandstones, limestones and marbles. In addition, much slower, subcritical fracture propagation is expected due to stress corrosion, which may reduce values of T_g necessary to cause fracture by as much as a factor of 5 (Atkinson, 1984).

Effect of water pressure within cracks

Subglacial bedrock is likely to be saturated with water. Because subglacial water pressure may frequently be a significant fraction of the ice-overburden pressure, internal water pressure within cracks, P_c, may be large and play an important role in fracturing bedrock. The same failure criterion applies, except that σ_1 and σ_2 are replaced by the effective principal stresses σ_1' and σ_2' where

$$\sigma_1' = \sigma_1 - P_c \tag{3a}$$

and

$$\sigma_2' = \sigma_2 - P_c. \tag{3b}$$

It is assumed that cracks are sufficiently isolated from each other so that remote stresses applied to an individual crack are not affected by other cracks in the rock. It is a good assumption for crystalline rocks in the light of their low porosities but probably is a poor assumption for certain sedimentary rocks. The assumption that cracks are widely spaced and thus do not interfere with each other was also made by Griffith (1924) when the local stress field around an individual arbitrarily oriented crack was determined in the derivation of Equations (2).

Values of T_g were recalculated taking into consideration the effect of water pressure in cracks by substituting Equations (3) into Equations (2) and calculating effective principal stresses. For the steady cavity, it is assumed that the crack water pressure has had sufficient time to equilibrate with the water pressure in the cavity. Thus, $P_c = P_w = 2.1$ MPa. Values of T_g (Fig. 8a) indicate that by reducing the effective mean compressive stress on microcracks in the step, the water pressure in cracks results in local tensile stresses at crack tips that are up to ten times greater than in the case of unsaturated bedrock (Fig. 7a). In this case, values of T_g are large enough to cause sub-critical crack extension in most sedimentary rocks and some crystalline rocks, particularly if there are centimeter-scale pre-existing cracks in the rock (e.g. Peng, 1975; Labuz and others, 1985).

A critical problem is estimating how the water pressure within cracks changes in response to variations in subglacial water pressure. An abrupt reduction in the cavity water pressure from the steady value should result in a time-dependent change in the water pressure within cracks in the step. Rigorous analysis of the rate of this decay is difficult due to the step geometry and the dependence of the rock's hydraulic properties on the change in ice pressure that accompanies the decline in P_w. The main concern here, however, is grossly constraining the length of time that water pressure in cracks can remain out of equilibrium with the subglacial water pressure. A reasonable approximation is that the time necessary for water-pressure relaxation in cracks scales with h^2/D, where h is the step height and D is the hydraulic diffusivity of the rock (see Carslaw and Jaeger (1959) for analogous treatments of transient heat flow). Values of D vary widely for different rock types, primarily as a function of differences in rock permeability. Unfractured igneous and metamorphic rocks may have diffusivities as low as 7×10^{-7} m^2 s^{-1}, whereas a typical diffusivity for sedimentary rocks is 10 m^2 s^{-1} (Freeze and Cherry, 1979). Therefore, the time required for P_c to equilibrate with changes in subglacial water pressure throughout the step may be on the order of several weeks or fractions of a second. Thus, for relatively impermeable rocks, typical diurnal water-pressure fluctuations beneath glaciers (Kamb and others, 1985; Iken and Bindschadler, 1986; Hooke and others, 1987, 1989) may produce large gradients in P_c very near cavities, without significantly affecting P_c in bedrock that is slightly removed from cavities. In more permeable rocks, however, P_c throughout subglacial bedrock may vary roughly simultaneously with subglacial water pressure.

In order to analyze the effect of a rapid decrease in P_w, on rock fracture, two end members of the continuum of crack water-pressure responses are considered: (1) the rock is sufficiently impermeable so that P_c remains at essentially the steady value throughout the step after a reduction in P_w, and (2) the rock is sufficiently permeable so that changes in P_c, throughout the step occur simultaneously with changes in P_w. Note that in both of these ideal cases P_c is uniform throughout the step, and therefore the possible effect of transient hydraulic gradients on rock fracture (Rice and Cleary, 1976) is not considered. If the water pressure in cracks throughout the step

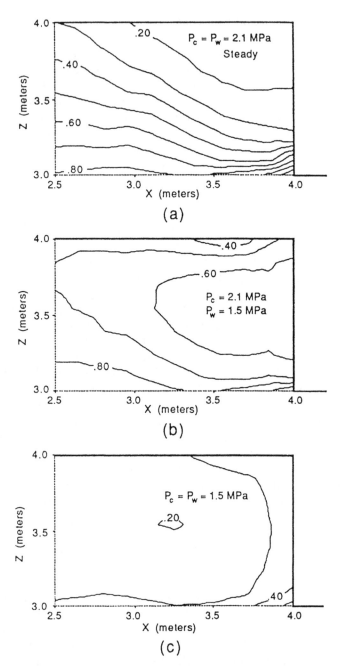

Figure 8 Contours of T_g when (a) the cavity is steady at a water pressure of 2.1 MPa and $P_c = 2.1$ MPa, (b) the cavity water pressure is reduced to 1.5 MPa and $P_c = 2.1$ MPa, and (c) the cavity water pressure is reduced to 1.5 MPa and $P_c = 1.5$ Mpa. Step surface is delineated with a bold line. Values of T_g are in Mpa.

remains at 2.1 MPa while the cavity water pressure falls to 1.5 MPa, the tendency for crack propagation, relative to the steady case when the rock is water-saturated, is increased over most of the step, in some areas by as much as a factor of 3 (Fig. 8b). This is consistent with earlier calculations when it was assumed that $P_c = 0$ (Fig. 7). However, if P_c varies in phase with P_w and both simultaneously fall from 2.1 to 1.5 MPa, the effect on crack propagation is less distinct (Fig. 8c). T_g is slightly increased relative to the steady case in the upper corner of the step, but elsewhere T_g is slightly less than in the steady case. In this situation, the decrease in the crack water pressure increases the effective mean compressive stress on cracks sufficiently to counteract the effect of the increase in $\sigma_1 - \sigma_2$ that occurs when P_w is reduced. Thus, the increase in the propensity for bedrock fracture that occurs with transient decreases in subglacial water pressure should be greater in less permeable rocks.

Conclusions

Subglacial water pressure influences rock fracture in two ways: it affects the distribution and magnitude of stress against the bedrock surface and the internal water pressure within fractures and microscopic cracks in the rock.

When water pressure falls in a subglacial cavity, some of the weight of the glacier formerly supported by pressurized water is shifted to the bedrock surface, which increases the ice pressure against the bed up-glacier from the cavity. This effect has been illustrated for the bedrock step considered here (Fig. 4) and for sinusoidal bed undulations (Röthlisberger and Iken, 1981). The increase in ice pressure against the bed, together with the decrease in water pressure against the lee surface of the bedrock irregularity, increases the difference between principal stresses in the rock. The result is an increase in the tendency for growth of pre-existing cracks in a direction roughly parallel to the most compressive principal stress, producing approximately vertical fractures in the case of a bedrock step. This effect may be partly or wholly counteracted if the rock permeability is sufficiently high so that the water pressure in cracks is reduced roughly simultaneously with the water pressure in cavities. In this case, the effective mean compressive stress on cracks increases as the crack water pressure is reduced, inhibiting crack growth.

Macroscopic fractures produced subglacially, probably acting in conjunction with preglacial joints and bedding planes, will isolate some rock fragments from the surrounding bedrock. Unlike subglacial rock fracture, the removal of loosened rock fragments is expected to be accelerated during periods of increasing subglacial water pressure and low effective stress. During periods of increasing water pressure, pressure-release freezing between ice and parts of the bed may occur as glacier sliding is accelerated, thereby increasing bed-parallel stresses on detached rock fragments (Robin, 1976;

Röthlisberger and Iken, 1981). Accelerated glacier sliding should also increase the drag on rough fragment surfaces exposed to ice. Concurrently, effective stresses should be reduced between fragments and adjacent rock as the water pressure in fractures increases, reducing the frictional resistance to fragment dislodgement. This effect should be reinforced by the reduction in ice pressure against the bed that occurs as water pressure in cavities increases.

Thus, fluctuations in subglacial water pressure may result in periods of accelerated crack propagation in intact bedrock that alternate with periods of accelerated rock-fragment removal from the bed. This suggests that as water pressure fluctuates beneath glaciers, bedrock in zones of ice–bed separation subject to large waterpressure fluctuations will be quarried faster than in other areas where, due to differences in the local drainage system, water-pressure fluctuations are smaller.

Acknowledgements

I am grateful to R. Hooke for valuable guidance throughout the course of this study and for reviewing a preliminary version of this paper. Suggestions made by B. Hallet, A. Iken and an anonymous reviewer substantially improved the paper. J. Paetz helped solve numerous computer problems. Funds for computing were provided, in part, by the Minnesota Supercomputer Institute. I also thank P. Hudleston, H. Wright, J. Labuz and W. Gerberich for reviewing a preliminary version of this paper as a section of my doctoral thesis at the University of Minnesota.

References

Addison, K. 1981. The contribution of discontinuous rock-mass failure to glacier erosion. *Ann. Glaciol.*, **2**, 3–10.

Anderson, R.S., B. Hallet, J. Walder and B.F. Aubry. 1982. Observations in a cavity beneath Grinnell Glacier. *Earth Surface Processes and Landforms*, **7**(1), 63–70.

Atkinson, B.K. 1984. Subcritical crack growth in geological materials. *J. Geophys. Res.*, **89**(B6), 4077–4114.

Carslaw, H.S. and J.C. Jaeger. 1959. *Conduction of heat in solids*. Oxford, Clarendon Press.

Crosby, I.B. 1945. Glacial erosion and the buried Wyoming valley of Pennsylvania. *Bull. Geol. Soc. Am.*, **56**, 389–400.

Frank, F.C. and B.R. Lawn. 1967. On the theory of Hertzian fracture. *Proc. R. Soc. London*, **299**, 291–306.

Freeze, R.A. and J.A. Cherry. 1979. *Groundwater*. Englewood Cliffs, NJ, Prentice-Hall.

Griffith, A.A. 1924. The theory of rupture. *In* Biezeno C.B. and J.M. Burgers, *eds. Proc. First Inter. Congr. Appl. Mech.* Delft, J. Waltman, Jr.

Gustafson, R.J., D.J. Hansen and D.A. Folen. 1977. *Finite element method package*. St. Paul, MN, University of Minnesota.

Hallbauer, D.K., H. Wagner and N.G.W. Cook. 1973. Some observations concerning the microscopic and mechanical behavior of quartzite specimens in stiff, triaxial compression tests. *Int. J. Rock Mech. Min. Sci. Geomech. Abstr.* **10**, 713–726.

Hallet, B. 1979. A theoretical model of glacial abrasion. *J. Glaciol.*, 23(89), 39–50.

Hallet, B. and R.S. Anderson. 1982. Detailed glacial geomorphology of a proglacial bedrock area at Castleguard Glacier, Alberta, Canada. *Z. Gletscherkd. Glazialgeol.*, **16**(2), 1980, 171–184.

Hanson, B. 1985. *Climate sensitivity of a numerical model of ice sheet dynamics and thermodynamics.* Cooperative thesis No. 91, National Center for Atmospheric Research, Boulder, CO.)

Hooke, R. LeB. 1981. Flow law for polycrystalline ice in glaciers: comparison of theoretical predictions, laboratory data, and field measurements. *Rev. Geophys. Space Phys.*, **19**(14), 664–672.

Hooke, R. LeB., P. Holmlund and N.R. Iverson. 1987. Extrusion flow demonstrated by bore-hole deformation measurements over a riegel, Storglaciären, Sweden. *J. Glaciol.*, **33**(113), 72–78.

Hooke, R. LeB., P. Calla, P. Holmlund, M. Nilsson and A. Stroeven. 1989. A 3 year record of seasonal variations in surface velocity, Storglaciären, Sweden. *J. Glaciol.*, **35**(120), 235–247.

Iken, A. 1981. The effect of the subglacial water pressure on the sliding velocity of a glacier in an idealized numerical model. *J. Glaciol.*, **27**(97), 407–421.

Iken, A. and R.A. Bindschadler. 1986. Combined measurements of subglacial water pressure and surface velocity at Findelengletescher, Switzerland: conclusions about drainage system and sliding mechanism. *J. Glaciol.*, **32**(110), 101–119.

Ingraffea, A.R. and F.E. Heuze. 1980. Finite element models for rock fracture mechanics. *Int. J. Numer. Anal. Methods Geomech.*, **4**, 25–43.

Jaeger, J.C. and N.G.W. Cook. 1979. *Fundamentals of rock mechanics.*, New York, Chapman and Hall.

Johnson, C.B. 1975. *Characteristics and mechanics of formation of glacial arcuate abrasion cracks.* (Ph.D. thesis, Pennsylvania State University.)

Kamb, W.B. 1987. Glacier surge mechanism based on linked cavity configuration of the basal water conduit system. *J. Geophys. Res.*, **92**(B9), 9083–9100.

Kamb, W.B. *and 7 others.* 1985. Glacier surge mechanism: 1982–1983 surge of Variegated Glacier, Alaska. *Science*, **227**(4686), 469–479.

Labuz, J.F., S.P. Shah and C.H. Downing. 1985. Experimental analysis of crack propagation in granite. *Int. J. Rock Mech. Min. Sci. Geomech. Abstr.*, **22**(2), 85–98.

Lindquist, P.A., H.H. Lai and O. Alm. 1984. Indentation fracture developments in rock continuously observed with a scanning electron microscope. *Int. J. Rock Mech. Min. Sci. Geomech. Abstr.*, **21**(4), 165–182.

Matthes, F.E. 1930. Geological history of the Yosemite Valley. *U.S. Geol. Surv. Prof. Pap.* 160.

Morland, L.W. and E.M. Morris. 1977. Stress fields in an elastic bedrock hump due to glacier flow. *J. Glaciol.*, **18**(78), 67–75.

Murrell, S.A.F. 1971. Micromechanical basis of the deformation and fracture of rocks. *In* Te'eni, M., *ed. Structure, solid mechanics, and engineering design.* London, Wiley-Interscience.

Nye, J.F. 1965. The flow of a glacier in a channel of rectangular, elliptic or parabolic cross-section. *J. Glaciol.*, **5**(41), 661–690.

Nye, J.F. 1969. A calculation on the sliding of ice over a wavy surface using a Newtonian viscous approximation. *Proc. R. Soc. London, Ser. A*, **311**, 445–467.

Paterson, M.S. 1978. *Experimental rock deformation — the brittle field.* Berlin, etc., Springer-Verlag.

Peng, S.S. 1975. A note on fracture propagation and time-dependent behavior of rocks in uniaxial tension. *Int. J. Rock Mech. Min. Sci. Geomech. Abstr.*, **12**, 125–127.

Rastas, J. and M. Seppälä. 1981. Rock jointing and abrasion forms on *roches moutonnées*, SW Finland. *Ann. Glaciol.*, **2**, 159–163.

Rice, J.R. and M.P. Cleary. 1976. Some basic diffusion solutions for fluid-saturated elastic media with compressible constituents. *Rev. Geophys. Space Phys.*, **14**(2), 227–241.

Robin, G, de Q. 1976. Is the basal ice of a temperate glacier at the pressure melting point? *J. Glaciol.*, **16**(74), 183–196.

Röthlisberger, H. and A. Iken. 1981. Plucking as an effect of water-pressure variations at the glacier bed. *Ann. Glaciol*, **2**, 57–62.

Sharp, M. 1982. Modifications of clasts in lodgement tills by glacial erosion. *J. Glaciol.*, **28**(100), 475–481.

Zumberge, J.H. 1955. Glacial erosion in tilted rock layers. *J. Geol.*, **63**(2), 149–158.

25

A PLOUGHING MODEL FOR THE ORIGIN OF WEAK TILLS BENEATH ICE STREAMS

A qualitative treatment

S. M. Tulaczyk, R. P. Scherer and C. D. Clark

Source: *Quaternary International* 86 (2001): 59–70.

Abstract

Glaciological studies of West Antarctic ice streams have shown that weak sub-ice-stream tills provide the basal lubrication that makes fast ice streaming possible under low driving stresses. Given the significant current interest in time-dependent ice stream behavior, there is a clear need for a conceptual model of weak sub-ice-stream tills that treats in a simple, but physically correct, way the coupling between evolution of till properties and ice stream dynamics. As a possible alternative to the previous, viscous-bed model, we propose a ploughing model that is consistent with the experimentally determined Coulomb-plastic rheology of sub-ice-stream till. In the ploughing model, the till is a several-meters-thick layer of sedimentary material that is disturbed and transported by ploughing that occurs during sliding of a bumpy ice base. The thickness of the till layer is determined in the ploughing model by the amplitude of the largest roughness elements ("ice keels"). There is no direct proof for the existence of ice bumps and ice keels beneath the modern West Antarctic ice streams but bedforms (e.g. megalineations and bundle structures) left behind by Pleistocene ice streams strongly support our assumption that an ice stream base is irregular. Generation of new till material occurs when ice keels protrude through the existing till layer and erode the top of the sub-till preglacial sediments. Based on a single tethered stake measurement of Engelhardt and Kamb (J. Glaciol.

44 (1998) 223) made at the UpB camp, Ice Stream B, West
Antarctica (Fig. 1), we estimate that the till flux due to sliding
with ploughing is there <88 $m^3 yr^{-1}$ per meter width. To balance
the estimated till flux in the UpB area, substrata erosion by ice
keels would have to take place at a high, but not unreasonable,
non-dimensional rate of <1.7×10^{-4} (assuming 1% contact
area). In the case of the West Antarctic ice streams, erosion of
sub-till materials by ice keels may be particularly fast and
unimpeded because these ice streams are overriding unlithified
preglacial (Tertiary) sediments. The most significant implication
of the proposed ploughing model is that it permits treating
basal resistance to ice motion as being velocity independent
(plastic till rheology) while allowing subglacial transport of
till as in the viscous-bed model. Models of ice streams with
a plastic bed exhibit a greater potential for unstable behavior
than models of ice streams with viscous beds. © 2001 Elsevier
Science Ltd and INQUA. All rights reserved.

1. Introduction

Discovery of a high-porosity till layer beneath one of the fast-moving West
Antarctic ice streams (Blankenship et al., 1986, 1987; Rooney et al., 1987;
Engelhardt et al., 1990) provided an impetus for a significant paradigm shift
in glaciology and glacial geology (Boulton, 1986). This finding demonstrated
that fast ice motion may be accommodated not only by sliding of ice over
rigid bedrock, as it has been assumed in the classical glaciological models,
but also by shearing of weak subglacial till (Weertman, 1957; Kamb, 1970;
Lliboutry, 1979; Alley et al., 1986, 1987; Boulton and Hindmarsh, 1987).
Over the last decade, an increasing interest in stability of the West Antarctic
ice sheet and in the dynamic behavior of Pleistocene ice sheets has further emphasized the need for a thorough understanding of the role of
weak tills in ice stream mechanics and dynamics (MacAyeal, 1989, 1992;
Hughes, 1992; Clark et al., 1996; Jenson et al., 1996; Marshall et al., 1996;
Bindschadler, 1997, 1998). Ice motion associated with shearing of subglacial
till has also become important in glacial geology where it is the favored
mechanism used to explain high fluxes of glacial sediments (Alley, 1991;
Hooke and Elverhøi, 1996; Alley et al., 1998).

The initial approach to modelling the role of subglacial tills in fast ice
streaming was based on application of the viscous till model developed by
Boulton and Hindmarsh (1987) for the margin of Breidamerkurjokull glacier
in Iceland (Alley et al., 1986, 1987). In this approach, the sub-ice-stream till
is treated as a viscous fluid that deforms pervasively throughout its thickness and replenishes itself by erosion of the underlying geologic substratum.
The viscous till model of ice streaming assumed that the gravitational
shear stress, which drives ice stream motion, is fully balanced by the basal

shear stress arising during viscous till deformation (Alley *et al.*, 1987). Influence of stresses due to ice deformation in ice stream shear margins and due to longitudinal extension or compression of an ice stream was presumed to be negligible. Thus, ice stream velocity was thought to be fully determined by the magnitude of the gravitational driving stress (assumed equal to the basal shear stress), till thickness, and till viscosity.

Recent studies of the West Antarctic ice streams paint a more complex picture of ice stream mechanics. Firstly, samples of the sub-ice-stream till recovered in several locations on Ice Stream B have shown practically no viscous behavior in laboratory tests (Kamb, 1991; Tulaczyk *et al.*, 2000a). Rather, the tested till samples behave as a Coulomb-plastic material whose strength is linearly dependent on effective stress but practically independent of strain rate. Moreover, analysis of ice stream force balance indicate that rapid ice deformation in the shear margins contributes significantly more resistive stress than deformation of subglacial till (Echelmeyer *et al.*, 1994; Jackson and Kamb, 1997). This conclusion is consistent with direct measurements of till strength (Kamb, 1991) which have demonstrated that the strength of the sub-ice-stream material, and the magnitude of the basal shear stress that it can support, is several times smaller than the magnitude of the gravitational driving stress acting on the ice stream (~2 kPa vs. ~14–20 kPa, Jackson and Kamb, 1997).

The emerging picture of ice streams suggests that they share some important characteristics of ice shelves and grounded ice masses. Whereas ice sheets are firmly coupled to their base, in the sense that the gravitational driving stress is equal to the basal shear stress, in the floating ice shelves basal stresses are zero and the gravitational driving stress is supported by marginal and longitudinal stresses (Paterson, 1994). Ice streams seem to fall in between these two cases. They do not float in water but they do move over till so weak that it cannot support the gravitational driving stress. The weaker the till is, the faster the ice stream motion because a greater fraction of the driving stress must be supported by ice deformation in shear margins (Raymond, 1996, Eq. (39); Tulaczyk *et al.*, 2000b).

These new glaciological findings represent a significant departure from the viscous till model, mainly because they imply a less active role of the till layer in determining ice stream velocity. Similar conclusion can be drawn from the results of recent studies that have failed to find viscous deforming till beneath mountain glaciers (Blake, 1992; Iverson *et al.*, 1995; Hooke *et al.*, 1997; Iverson *et al.*, 1997).

If the sub-ice-stream till layers are not made of viscous fluids that deform actively and erode the substrata to replenish themselves, then how should one account for their existence and properties in the new framework of ice stream mechanics? Here, we propose a new, ploughing model for the origin of weak sub-ice-stream tills and discuss implications of this model for ice-stream stability and for geologic models of glacial sediment fluxes.

2. Characteristics of sub-ice-stream tills

In this section, we review the properties of sub-ice-stream tills to provide background for further discussion of our model of till generation.

Considering the logistical challenges facing glaciological research in West Antarctica, a significant effort has been invested into studying the weak till layers underlying the West Antarctic ice streams. This effort yielded a number of important observations made using remote sensing combined with inverse modeling, geophysical surveys, borehole experiments, or direct analysis of subglacial till cores (Alley *et al.*, 1986; Blankenship *et al.*, 1986, 1987; Rooney *et al.*, 1987; Engelhardt *et al.*, 1990; Kamb, 1991; MacAyeal *et al.*, 1995; Jacobel *et al.*, 1996; Jackson and Kamb, 1997; Engelhardt and Kamb, 1997, 1998; Scherer *et al.*, 1998). Much of the existing knowledge comes from research on Ice Stream B, especially in the area of the so-called UpB camp (Fig. 1).

Both geophysical and sedimentological investigations suggest strongly that the weak sub-ice-stream tills are developed over sedimentary basins that are buried beneath parts of the West Antarctic ice sheet (Rooney *et al.*, 1991; Scherer, 1991; Anandakrishnan *et al.*, 1998; Bell *et al.*, 1998; Tulaczyk *et al.*, 1998). These basins likely contain Mid to Late Cenozoic sediments similar to the ones forming the glacimarine sedimentary sequence that fills the Ross Sea sedimentary basin (Barrett, 1975; Hayes and Frakes, 1975). This inference is strongly corroborated by three lines of evidence: (1) seismic properties of the sedimentary package underlying the UpB area (Rooney *et al.*, 1991);

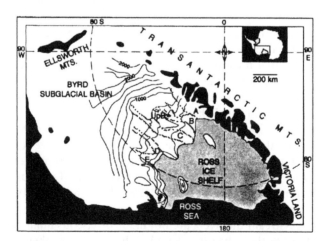

Figure 1 Location map shows outlines of the ice streams flowing through the Ross Sea Section of West Antarctica. Letters A through E denote the individual ice streams. Ice elevation is indicated by contour lines at 250 m intervals. Major mountain ranges are shown in black (after Tulaczyk *et al.*, 1998, Fig. 1).

(2) textural properties of the till from beneath Ice Stream B that suggest glacial recycling from parent sediments analogous to those found in the Ross Sea (Tulaczyk *et al.*, 1998); and (3) finding that the Ice Stream B till samples contain microfossil assemblages whose age ranges correspond to the age ranges of Ross Sea sequences (Scherer, 1991; Scherer *et al.*, 1998). In contrast, subglacial presence of hard bedrock appears to hinder development of efficient basal lubrication necessary for ice streaming conditions (Anandakrishnan *et al.*, 1998; Bell *et al.*, 1998).

The till layer beneath Ice Stream B is fairly continuous and has thickness up to several meters (Blankenship *et al.*, 1986, 1987; Rooney *et al.*, 1987; Engelhardt *et al.*, 1990; Tulaczyk *et al.*, 1998). In places, it may be punctuated by 'sticky spots' (Rooney *et al.*, 1987; Alley, 1993) although more abundant evidence for their presence comes from studies on the stopped Ice Stream C (Anandakrishnan and Bentley, 1993; Anandakrishnan and Alley, 1997). Geophysical investigations in the UpB area yielded evidence for drumlinization or large-scale fluting of the till layer (Rooney *et al.*, 1987; Novick *et al.*, 1994). Vertical and lateral variations in microfossil abundance, till porosity and composition, and in seismic properties, suggest that the several-meter-thick layers of sub-ice-stream till may in fact represent packets of till units (Rooney, 1988 and the next section of this paper).

One of the most consistent physical characteristics of the till samples recovered from beneath the West Antarctic ice streams is their high porosity ranging from ~39% to ~45% (Tulaczyk *et al.*, 2000a, 2001). This water-rich, muddy till has very low strength of only a few kPa (Kamb, 1991). Mechanical behavior of the sub-ice-stream till samples has been examined using several different apparatuses: shear box, triaxial device, ring shear device. The results of these tests have shown repeatedly that the strength of this till is practically independent of strain magnitude or shear strain rate but it depends linearly on effective stress (Kamb, 1991; Tulaczyk, 1998; Tulaczyk *et al.*, 2000a).

Very little is known about the character of deformation that the sub-ice-stream tills experience in situ. The most relevant record available is that obtained with a tethered stake employed at the bottom of a borehole in the UpB area (Engelhardt and Kamb, 1998). With the tethered stake it is possible to determine what fraction of ice stream surface velocity is accommodated below and above the depth of emplacement of this device in the till, in this case <0.3 m. The record obtained in the UpB area indicated that over 26 days only 31% of ice stream velocity was accommodated below the shallow depth of emplacement. This result favors basal sliding and/or shallow till deformation as the predominant mode of ice stream motion. Perhaps the most surprising feature of the tethered stake record was its very high temporal variability. The observed local contribution of sliding and shallow till deformation to ice stream motion ranges from ca. 5% to 100%.

3. The need for a new model

In addition to the general need for a new look at sub-ice-stream tills resulting from the recent shift in understanding of ice-stream mechanics (as discussed in the introduction), we consider three lines of evidence gathered through studies of the West Antarctic ice streams to be inconsistent with the viscous-bed model of ice stream mechanics (Alley *et al.*, 1986, 1987; Alley, 1989a, b): (1) results of geotechnical tests showing that rheology of sub-ice-stream tills is not viscous but Coulomb-plastic (Kamb, 1991; Tulaczyk, 1998; Tulaczyk *et al.*, 2000a), (2) the tethered stake record from Ice Stream B demonstrating predominance of basal sliding and/or shallow deformation (Engelhardt and Kamb, 1998), and (3) presence of intra-till inhomogeneities in porosity, composition, and microfossils. Since the first two lines of evidence listed above have been discussed in previous publications we will explain only briefly their importance to our current line of argumentation. The third point is developed here.

One of the most straightforward ways to test the viscous-till model of ice streaming is to verify whether the samples of sub-ice-stream till that have been obtained by drilling, do indeed have the nearly linearly viscous rheology assumed in this model (e.g. Alley *et al.*, 1987). First such verification has been performed using strain-rate-controlled and stress-controlled shear-box tests (Kamb, 1991). This test yielded a negative result in that the examined till samples demonstrated nearly plastic rheology with almost no dependence of strength on strain rate. This result was criticized on the basis of the assumption that a relevant till rheology can only be obtained if one shears till samples to very high strains (100s or 1,000s) whereas shear-box test is capable of accumulating only relatively small strains (~10). A second testing program was designed to examine the potential influence of such variables like strain magnitude and effective stress on till rheology (Tulaczyk, 1998). This program involved tests in a triaxial device and in a ring-shear device. The results of these tests confirmed fully the conclusion reached on the basis of the earlier shear box tests. At different levels of strain (1 to ca. 1,000) and at different magnitudes of effective stress (~5–500 kPa), the samples of the UpB till behave like a Coulomb-plastic material with no significant strain rate dependence of strength (Tulaczyk, 1998; Tulaczyk *et al.*, 2000a). These results are inconsistent with the viscous-till model because the model assumed that till viscosity is the fundamental parameter that determines the velocity of ice stream flow and that the strain-rate dependence of till strength forces the till to deform continuously throughout its thickness. Since it is not possible to observe this strain-rate dependence of strength and to measure till viscosity, the viscous-till model loses its appeal as a predictive and an explanatory tool.

The tethered stake record from the UpB area (Engelhardt and Kamb, 1998) is also inconsistent with the predictions of the viscous-till model. Deep deformation distributed throughout several meters of till thickness should

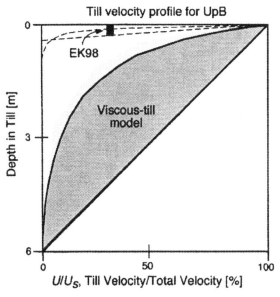

Figure 2 Hypothetical vertical distribution of strain in the till beneath the UpB camp, West Antarctica. The shaded area represents the best estimate for strain distribution under the assumption of viscous till rheology (Alley *et al.*, 1989, Fig. 5 for a till of 6 m thickness). The solid square located near the *x*-axis represents the result of the tethered stake experiment performed by Engelhardt and Kamb (1998) at the UpB camp. Dashed thin lines give two possible scenarios for strain distribution under ploughing conditions with an assumption of a much thinner 'active' portion of till.

accommodate almost all of the ice stream velocity (Fig. 2) (Alley *et al.*, 1989, their Fig. 5). In the observed record, it is sliding and/or shallow (<0.3 m) deformation that account for most of the motion. Equally important is the fact that the tethered stake record is so highly variable. Large variability of strain rates is characteristic for deformations taking place in highly non-linear and plastic materials (Hindmarsh, 1997).

The third line of evidence that we consider to be inconsistent with the predictions of the viscous-till model for ice stream beds is the presence of inhomogeneities in porosity, composition, and microfossil content that we have detected in the till cores recovered from beneath Ice Stream B near the UpB camp (Fig. 3). These inhomogeneities are real in the sense that their magnitudes are greater than the analytical uncertainties associated with the applied data collection methods. The viscous-till model predicted that in the UpB area the sub-ice-stream till layer should be deforming pervasively and continuously to the thickness of ca. 6 m (Alley *et al.*, 1989). Such continuous deformation persisting as long as the ice stream is active (at least 1,000s of years, Bentley, 1997) should result in a homogenization of the till

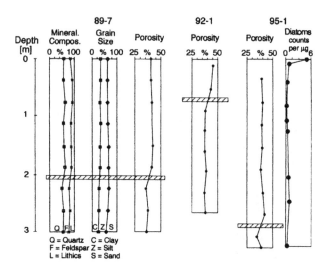

Figure 3 Vertical changes in till properties observed in the three longest cores recovered from the UpB area, 89-7, 92-1, 95-1, West Antarctica (see Tulaczyk *et al.*, 2000a, for core locations). Between depths 1.9–2.3 m, core 89-7 displayed significant changes in three different properties: (1) mineralogical composition of the sand fraction (increase in lithic fragments), (2) grain-size distribution (increase in sand abundance), and (3) till porosity. The upper part of the core 92-1 had higher porosity than the till samples taken from below ca. 0.9 m in this core (this quantitative observation was corroborated by a more 'wet' appearance of the upper part of the core when it was handled in the lab). Core 95-1 showed a significant drop in porosity below 2.9 m. The same core had also much greater abundance of diatom fragments in the uppermost layer than in the main body of the core (data from Scherer *et al.*, 1998, Table 1). All porosities were determined by the weight-loss method (Bowles, 1992). Mineralogical composition of the sand fraction and diatom abundance were derived by point counting (Tulaczyk *et al.*, 1998). Combination of sieving and pipetting was used to analyze the abundance of sand, silt, and clay in the till samples.

layer (Piotrowski and Tulaczyk, 1999). The most striking is the distribution of microfossils, which are found in any appreciable abundance only near the top of the till cores from UpB (Fig. 3; Scherer *et al.*, 1998; Tulaczyk *et al.*, 1998). We interpret this observation as a sign of preferential till transport in a thin layer immediately adjacent to the ice base.

UpB diatoms are too rare to perform accurate estimates of whole fossils, but abundance estimates of small diatom fragments (>2 μm) in the <250 μm fraction of these sediments are highly reproducible, using the method of Scherer (1994), despite the fact that the number of fragments generated from a single diatom can vary from a few to more than several hundred. The data demonstrate a distinct difference between sediments recovered by piston coring several cm to several m beneath the ice, and sediment samples that include the particles in closest proximity to the ice. The uppermost sediments contain a significantly higher concentration of diatoms and diatom fragments than those below. Core samples from beneath the uppermost several cm contain a mean concentration of 1.4×10^5 fragments per gram dry sediment (fr/g) ($n = 14$), whereas samples from closer to the ice base at the central part of the ice stream at UpB contain a mean of 1.8×10^6 fr/g ($n = 11$) (Scherer and Tulaczyk, 1997). Furthermore, the samples known to contain Quaternary diatoms contain as much as two orders of magnitude higher total concentration of diatom fragments, with a mean of 2.3×10^8 fr/g ($n = 4$) (Scherer et al., 1998).

Significant variation in diatom abundance and assemblages is also noted normal to ice flow, across a few km, but diatom abundance and diatom species groupings tend to be comparable parallel to flow, up to tens of km downstream in the Upstream B region (Scherer et al., 1998). We interpret these observations as further evidence of ploughing with ice flow, which may create distinct lateral till packets. The highest concentration of diatom fragments, including whole, well-preserved diatoms, is in a region identified by seismic properties as possessing an unusually thin or absent till layer (Rooney et al., 1987). We speculate that this may be a region recently excavated by ice keel ploughing. Well-preserved Quaternary diatoms may have been carried from upstream sedimentary deposits in suspension, in the cavity formed in the lee of an ice keel (Fig. 6B).

The distribution of porosity in the three longest cores recovered at UpB also suggests that till deformation does not extend everywhere to depths of several meters. In these cores, there is a significant drop in till porosity between ~0.5 and ~2.5 m depth (Fig. 3). Since till strength decreases with increasing till porosity (Tulaczyk et al., 2000a), the upper, more porous parts of the till cores must be weaker than the lower, less porous sections. It is mechanically more favorable for any type of deformation (viscous or plastic) to concentrate in the weaker material. Therefore, the thickness of the more porous sections of the cores probably reflects the depth to which till deformation is distributed in situ. Seismic data collected in the UpB area also suggest that there is a weaker, 0.5–1.5 m thick layer on top of the several-meter-thick packet of till that was presumed to have been deforming pervasively in the viscous-till model (Rooney, 1988). This seismic subunit was detected over most of the length of the seismic profiles. The seismic evidence indicates that the upper core sections with higher porosity are not local aberrations but that they may be representative of a continuous weak layer to which most of till deformation should be confined.

4. Sliding and ploughing by an uneven ice base

To avoid the problematic assumptions of viscous till rheology and continuous till deformation, we propose a 'ploughing' model for a weak, sub-ice-stream till layer. In this model, the till is a few-meter-thick layer of sub-till material

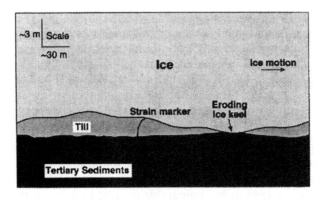

Figure 4 Schematic representation of the ploughing model.

disturbed and transported by ploughing that occurs during sliding of a bumpy ice base (Fig. 4). The model is similar to these presented before by Brown *et al.* (1987) and Beget (1986). However, those authors emphasized the importance of clast ploughing and we stress deformation of till around 'bumps' in the ice base itself. This change in emphasis is driven by the observation that basal ice in the West Antarctic ice streams is devoid of debris (Scherer *et al.*, 1998) and the underlying tills are remarkably fine-grained and clast-poor (Tulaczyk *et al.*, 1998). Our treatment of till formation by sliding and ploughing is similar to conceptual models for generation of fault gouge layers (Tchalenko, 1970; Eyles and Boyce, 1998). In this analogy the ice base represents a rigid upper fault plate with asperities. Ploughing by these asperities is assumed to be the most important process for generation of the weak, deformable till (i.e. fault gouge layer) from the underlying strata (i.e. the lower fault plane in the analogous fault gouge model).

Recent investigations of modern and fossil beds of ice streams provide abundant evidence for presence of basal ice bumps. For instance, large bumps in the base of Ice Stream B have been inferred from radar data (Novick *et al.*, 1994) and studies of sea-bottom topography in the Ross Sea indicated that the base of the Pleistocene West Antarctic ice sheet left a very irregular imprint after deglaciation (Anderson *et al.*, 1992; Shipp and Anderson, 1997a, b; Shipp *et al.*, 1999). Perhaps the best insight into the geometry of an ice stream base is provided by the topography of bedforms left behind by Pleistocene ice streams in North America and elsewhere (Clark, 1993, 1994; Canals *et al.*, 2000; Clark and Stokes, 2001). Analysis of geophysical data, satellite imagery and air photos acquired over the areas of former ice streams show elongated bedforms of varying amplitude, wavelength, and width. Fig. 5 shows an example of such bedforms generated by the M'Clintock Channel ice stream that is discussed in detail in Clark and Stokes (2001). Rather than the ridges having formed by some relief

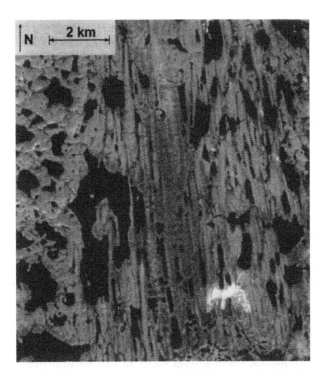

Figure 5 Landsat TM image (band 3) positioned across the boundary of the M'Clintock paleo ice stream, Arctic Canada, that is running here approximately in the south–north direction. Image is centered on 72°:14′:21″N, 106°:11′:22″W. Western third of the image is characterized by approximately equidimensional hummocks, presumably formed beneath slow-moving ice, whereas the central and eastern parts of the image are dominated by strongly elongated bedforms thought to be indicative of paleo-ice streams (Clark and Stokes, 2001). Amplitude and transverse spacing of the elongated bedforms have been measured using 1 : 50,000 topographic maps with contour interval of 10 m and air photos from the same area as the one shown on the satellite image.

amplification process based around viscous deformation of sediment (Boulton, 1987; Hindmarsh, 1997), in our model we view them as residual accumulations of sediment whose form is a consequence of the carving of intervening grooves. Whilst this does not explain the pattern of all bedforms such as drumlins and rogen moraine, it may explain the observed form of megalineations (Clark, 1993) or bundle structures (Canals *et al.*, 2000).

Is it realistic to propose, as we do here, that an ice base can be treated as the (more or less) rigid element that deforms and molds the underlying substratum? In classical hard-bed models of ice motion, it was the substratum that was the rigid element in a glacial system. Irregularities in the bed forced ice to deform and, thus, shaped the ice base geometry (e.g. Weertman, 1957). The rigid-ice-base assumption seems realistic as long as the substratum is made of very weak till. Ice strength at strain rates common in nature is of the order of 100 kPa (Paterson, 1994) and the strength of the sub-ice-stream till is two orders of magnitude smaller (Kamb, 1991). For instance, elimination of an ice bump with some arbitrary amplitude A and wavelength $\sim 10A$ through ice shearing would require accumulation of ~ 0.1 shear strain. Since ice strain rates at the low stress level of 1 kPa are of the order of 10^{-7} yr^{-1} (Glen flow law with flow-law constant of 5.3×10^{-15} (s kPa)$^{-1}$, Paterson, 1994, Table 3.3), the time required to eliminate the ice bump ($\sim 10^6$ yr) would be much longer than the residence time of ice in an ice stream ($\sim 10^3$ yr; ice residence time can be estimated by dividing the typical length of an ice stream, 100s of km, by the typical ice stream velocity, 100s of m yr^{-1}).

If it exists, differential basal melting represents a more potent way of eliminating ice bumps. Melting may concentrate on ploughing ice bumps because stresses acting on them during ploughing should be several times greater than stresses associated with a flat ice bed sliding over till or with shear on intra-till planes (Brown et al., 1987; Tulaczyk, 1999). The additional melting due to this increased shear heating may be of the order of 1 mm yr^{-1} (taking ice stream velocity ~ 100 m yr^{-1}, till strength ~ 1 kPa, and large stress concentration factor of ~ 10). Under these conditions, ice bumps with initial amplitude smaller than ca. 1 m can be destroyed by melting during the estimated ice residence time in an ice stream. However, the fast motion of the ice stream base may help to minimize the influence of the localized increased shear heating, thus diminishing the importance of the differential shear heating. Thorsteinsson and Raymond (2001) consider this problem for a wavy ice base moving over till of viscous rheology. They conclude that differential melting should affect mainly those roughness elements whose wavelength is relatively small (~ 1 m or less). Thus, it is reasonable to expect that an ice stream base will be relatively smooth at small scale and that it is more likely to have bumps whose wavelength is of the order of meters or more. Nevertheless, the fact that bedforms left behind by paleo-ice streams are characterized by presence of very long, continuous megalineations (Clark, 1993, 1994; Shipp et al., 1999; also bundle structure of Canals et al., 2000) indicates that at least the large-scale ice base irregularities survive over distances of 10–100s of kilometers.

The main stage of ice bump generation takes place before or when ice enters an ice stream, at the time when it is still in contact with a rigid bed. If an ice stream is covered with weak, continuous till, the last contact of ice with hard bedrock should largely determine the geometry of the ice base. This speculation is consistent with the geometry of the bundle structure from Antarctic Peninsula (Canals et al., 2000, Fig. 3). The exact geometry of the basal ice bumps, thus also bedform geometry, will then result from combination of the initial ice base roughness (governed by the geometry of the rigid bed) and later melting and large-scale stretching that ice experiences during and after entering an ice stream. Unfortunately, basal geometry for the West Antarctic ice streams has not been examined with sufficient resolution to provide the observational constraints that would be necessary for a quantitative treatment of our ploughing model of till generation. This fact forces us to stress qualitative aspects of our ploughing model and to invoke data from areas affected by paleo-ice streams (e.g. Fig. 5) to justify our assumption that an ice stream base is not flat.

If an ice stream base is uneven and bumpy, this fact will cause distribution of deformation in sub-ice-stream tills. In the past, it has been explicitly or implicitly assumed that only till of linear or mildly non-linear rheology could experience distributed deformation (e.g. Alley, 1993). However, this presumption is correct only if one is considering a perfectly flat ice base moving over a smooth fluid with no effective-stress dependence of strength (Iverson *et al.*, 1998; Tulaczyk, 1999; Tulaczyk *et al.*, 2000a). In the presence of basal roughness elements (clasts or ice bumps), deformation is distributed because till must move from the zone of compression in front of each roughness element to the zone of extension behind it. This process of till flow around clasts and bumps generates a viscous-like distribution of deformation in a plastic till (Tulaczyk, 1999).

In support of our ploughing model of sub-ice-stream tills, we show qualitatively that interaction of basal ice bumps with the UpB till provides a plausible explanation for the significant fluctuations of sliding velocity observed in the tethered stake record obtained by Engelhardt and Kamb (1998). During an observation period lasting ca. 26 days, the sliding velocity measured beneath ISB experienced several significant fluctuations (Fig. 6A). Engelhardt and Kamb (1998) infer that at least at the initial stages of the experiment, the tethered stake was located within centimeters of the ice base. They also propose that the biggest and longest lasting dip in the measured sliding velocity occurred because the tethered stake was dragged by a clast or an ice bump protruding down from the ice base. Here we hypothesize that the other, smaller and shorter lasting velocity slow-downs may have occurred when the tethered stake found itself in a deformation zone surrounding ploughing bumps (Fig. 6B). In this zone, till is dragged in the direction of ice motion and a tethered stake imbedded in the till will record this as a slow-down. The recorded fluctuations in sliding velocity last typically one or two days and have a similar repeat interval. Given an ice base velocity of ca. 1.2 m day^{-1}, the roughness elements generating the sliding fluctuations should have a wavelength of a few meters. According to our conceptual model (Fig. 6B), the magnitude of the apparent slow-downs is controlled by the relative amplitude of basal protuberances with respect to the depth at which the tethered stake is employed. If we accept the inference that this depth was <0.3 m (Engelhardt and Kamb, 1998), the amplitude of most of the basal bumps should have been smaller than

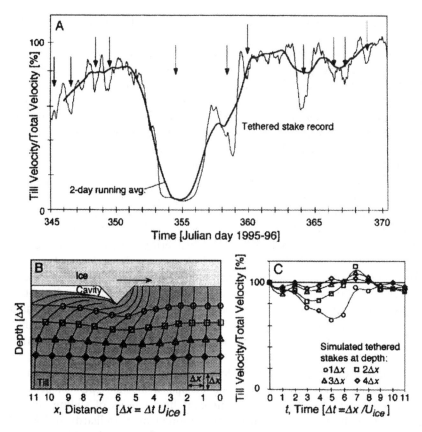

Figure 6 (A) Record of basal sliding obtained beneath Ice Stream B in the UpB area, West Antarctica, using a tethered stake; modified from Engelhardt and Kamb (1998, Fig. 4). The non-dimensional sliding velocity is obtained by dividing the measured sliding rate by the ice surface velocity U_{ice} = 440 m yr^{-1} observed in the UpB area by Whillans and van der Veen (1993). Vertical arrows point out major departures of the measured sliding velocity from the surface velocity. (B) and (C) illustrate how such departures may result from clasts or ice protrusions ploughing the underlying till. Panel (B) shows an example of the pattern of deformation that may be caused by till ploughing (modified from Tulaczyk, 1999, Fig. 9). Taking a reference frame moving with the ice (at speed U_{ice}), we can equivalently assume that the vertical grid markers in (B) represent progressive changes in the position and shape of one originally vertical marker which experiences incremental deformation due to passage of a ploughing clast or ice protuberance. Open symbols indicate consecutive positions of four simulated tethered stakes which were initially emplaced at different depths on the perfectly vertical grid marker (initial position $x = 0$). If these four simulated tethered stakes were to experience a passage of a ploughing protrusion as shown in (B), they would yield the sliding records shown in (C). We determined these synthetic sliding records by measuring off the horizontal distance of each symbol from $x = 0$ in (B) and dividing it by the distance at which this symbol would be located if the initially vertical grid marker had not deformed.

750

this because they have caused only moderate apparent slow-downs of ~20% of total velocity. Each slow-down in sliding velocity is matched by a corresponding increase in distributed till deformation (Fig. 6C).

Just as in the case of the viscous-till model (Alley et al., 1986), ploughing by ice bumps is associated with distributed till deformation that produces net till flux in the direction of ice stream motion. However, the tethered stake experiment and the evidence indicating active till layer thinner than 6 m (Fig. 3; Rooney, 1988) suggest that till flux in the ploughing model is much smaller than that under the assumption of viscous till. For instance (Fig. 2), extrapolating linearly from 100% of deformation at the ice base and through average of 31% at the depth of emplacement of the tethered stake (<0.3 m, Engelhardt and Kamb, 1998), one obtains an estimate of $H < 0.4$ m as the thickness of the till layer that deforms during ploughing in the UpB area. This assumed geometry gives <88 m^3 yr^{-1} per meter width of an ice stream as the total till flux, Q_t [obtained from $Q_t = 0.5HU_s$, where U_s is the ice stream velocity of ca. 440 m yr^{-1} at UpB (Whillans and van der Veen, 1993)]. Whereas this till flux is approximately an order of magnitude lower than the flux predicted by the viscous-till model (Alley et al., 1986, 1989), it is still a significant flux that requires relatively high till generation rates for a steady state. If there is a steady state and if till is generated by erosion of substrata beneath the ice stream itself, as we will argue below, the till flux estimated above would translate into an average till generation rate of <0.73 mm yr^{-1} for the ca. 120 km long flowline upstream of the UpB camp.

Our presumption that till generation takes place mainly beneath the ice stream itself is driven largely by the observation that in the UpB area there was no evidence for significant debris in the basal ice (Scherer et al., 1998; Tulaczyk et al., 1998). If this observation is representative for the ice stream in general, influx of debris with ice entering the ice stream can be assumed negligible. As illustrated in Fig. 4, in our model till generation takes place also through ploughing, but only when and where a ploughing ice bump has an amplitude that exceeds the thickness of the till itself and protrudes into the underlying substrata. In this framework, till generation is performed just by the largest bumps (several meters?) whereas the smaller bumps contribute only to till transport. Since we hypothesize that till generation is localized rather than aerially extensive (as it would be for abrasion of substratum by sliding till, e.g. Cuffey and Alley, 1996), it should be expected that when a localized, large ploughing bump does erode the substratum, it is doing it at a relatively fast rate.

At this point, the specific geologic setting of the West Antarctic ice streams over young sedimentary basins may become important. Weak glacial and glaciomarine diamictons may represent the material that is beneath the currently 'active' till layer. Samples of this sub-till material have not been recovered, but Rooney et al. (1991) and Tulaczyk et al. (1998) suggested that these diamictons are similar to those studied previously in the Ross Sea (e.g. Hayes and Frakes, 1975; Anderson *et al.*, 1984). Microfossil ages suggest that the till is derived mostly from upper Miocene sediments (Scherer, 1991; Scherer *et al.*, 1998). Samples of Miocene sediments taken from the top several hundred meters of the Ross Sea sequence were unlithified and had high porosity of 34.6% to 43.5% (Barrett and Froggatt, 1978, Table 3) comparable to that of the sub-ice-stream tills themselves. Thus, the substratum-ploughing ice bumps that are responsible for generation of till in our model may indeed be efficient at generating new till because they do not have to erode competent bedrock but simply dislodge unlithified sediments.

A snapshot of the sub-ice-stream zone beneath the UpB camp obtained with seismic surveys indicated that over most of the area (~98%) the weak till layer separates the ice stream base from the underlying sedimentary sequence (Rooney *et al.*, 1987). If at any single moment the contact area between the largest ice bumps and the sequence is indeed of the order of 1%, the local rate of till generation must be of the order of <73 mm yr^{-1} for the areally averaged till generation rate to reach the previously estimated steady-state value of <0.73 mm yr^{-1}. Although the local erosion rate of <73 mm yr^{-1} is high, it is not unreasonably high given the fast ice stream velocity and (presumed) weak nature of the eroded substrata. If expressed in non-dimensional terms, i.e. the dimensional erosion rate of <73 mm yr^{-1} divided by the ice velocity of 440 m yr^{-1}, the local erosion rate is $<1.7 \times 10^{-4}$. This is comparable to the non-dimensional erosion rate reported by Humphrey and Raymond (1994) for bedrock erosion by the Variegated glacier during surge, $\sim 10^{-4}$.

It is important to note that, at least qualitatively, there is a potential for a stabilizing feedback between the till thickness and the till generation rate. For instance, if for some arbitrary reason the till layer becomes thicker, then fewer ice bumps will be protruding through the layer and eroding the substratum, which should counteract the initial tendency of the till to thicken. Similarly, if the till becomes thinner due to some arbitrary disturbance, more ice bumps will span its whole thickness and the areally averaged till generation rate should go up, pushing the system back toward a steady-state thickness of the till that is consistent with the (as of yet unknown) roughness of the ice stream base.

5. Consequences for ice stream stability

Qualitatively, incorporation of plastic till rheology into models of ice stream motion introduces a greater potential for an unstable ice stream behavior as compared to the past models that considered sub-ice-stream tills as nearly linearly viscous fluid (Alley, 1989a, b vs. Kamb, 1991). In the plastic case, there is simply no direct feedback between ice stream velocity and the resistance provided by the subglacial till against ice stream motion because till strength does not depend on shear rates. On the other hand, assumption of viscous till implied greater ice stream stability because variations in ice stream velocity would face strong 'quenching' due to the strain-rate-dependence of viscous till strength.

Recently, Hindmarsh (1998) argued that plastic till rheology may not be appropriate for models of ice motion over deformable till beds because ice stream velocity is finite and confined to a relatively narrow range of values ($<1,000$s m yr^{-1}). According to this argument, such behavior is more consistent with viscous till rheology, whose 'quenching' of velocity variations provides a simple explanation for this relatively low variability of ice velocities. However, the example of ice shelves clearly demonstrates that finite ice velocities with a relatively low range of variations can be achieved without any significant viscous basal control (Thomas, 1973a, b). It is the requirement of ice continuity coupled with the mildly non-linear rheology of ice itself that force ice velocities to fall within a finite and reasonably narrow range. In ice streams, as in ice shelves, weak or no basal resistance against ice motion is compensated by shifting the support for the gravitational driving stress to the side margins and by expanding part of the driving stress on ice stretching (Echelmeyer et al., 1994; Jackson and Kamb, 1997; MacAyeal et al., 1995; Whillans and van der Veen, 1997). Raymond (1996, Eq. (39)) and Tulaczyk et al. (2000b) have shown that ice stream models assuming plastic till beds can generally reproduce the observed ice stream velocities.

The plastic rheology of sub-ice-stream tills will have especially far-reaching implications for modeling of transient behavior of ice streams (e.g. initiation or cessation of ice streaming). Because the plastic till strength is highly sensitive to till water content (e.g. Tulaczyk et al., 2000a), transient behavior of ice streams is likely to be caused by an increase or decrease in water storage in till (Tulaczyk et al., 2000b). Dynamics of subglacial water balance is thus of primary importance in considerations of changes in ice stream behavior. Advection of water in till transported during ploughing is likely to be a significant component of this water balance. Using our previous estimate of till flux at the UpB camp, <88 m^{-3} yr^{-1} per meter width, we can estimate the magnitude of water advection at <36 m^{-3} yr^{-1} (assuming 40% till porosity). In a steady state, basal melting would have to be <0.3 mm yr^{-1} along the ca. 120 km long flowline upstream of the UpB area to compensate for the water loss due to advection. Given the fact that basal melting (or freezing) rate is expected to be of the order of 1 mm yr^{-1} or less in West Antarctica (Tulaczyk et al., 2000b), water advection in till is likely to represent one of the primary components of sub-ice-stream water balance. This component will help determine the appropriate timescales for significant changes in ice stream velocity.

6. Conclusions

Recent developments in research on ice-stream mechanics and on till rheology call for a new model for generation of weak, sub-ice-stream tills which previously have been thought to represent an actively deforming and eroding layer of viscous till. The ploughing model proposed here represents a variation on a previous model (e.g. Beget, 1986; Brown *et al.*, 1987) that invoked ploughing of till by clasts. In the framework of our ploughing model, weak sub-ice-stream tills represent simply a wet and slippery boundary layer caught between bumps in the ice stream base. We conclude that such bumps exist based on several lines of evidence provided by geophysical remote sensing of modern and fossil ice stream beds. Till is transported subglacially as it deforms around the bumps in a spatially and temporally variable fashion. The carving of numerous parallel grooves may explain the form of some commonly observed subglacial bedforms, particularly megalineations and bundle structures. Till transport rate should scale with the (as of yet unknown) amplitude and wavelength of these bumps. The largest of the basal ice bumps (referred to as 'ice keels') protrude through the whole till thickness and regenerate the weak till by scrapping material off the top of the underlying sequence of unlithified older glacial or glaciomarine sediments and by adding basal meltwater in the process. In spite of many uncertainties regarding the geometry of the basal ice bumps and the kinematics of till deformation around them, we estimate the till flux at the UpB camp to be <88 m^3 yr^{-1} per meter width of the ice stream (extrapolated from a single tethered stake experiment performed by Engelhardt and Kamb, 1998). In a steady state this flux would require that the ice keels erode the sub-till material at a reasonable non-dimensional rate of $<1.7 \times 10^{-4}$ (assuming ~1% of contact area). A great advantage of the ploughing model for generation of sub-ice-stream tills is that it allows for significant till transport while dropping the problematic expectation that till has viscous rheology. If sub-ice-stream tills behave plastically in situ, as they do in laboratory tests, this suggests that ice streams may behave in an unstable manner (e.g. turn on and off without significant changes in driving stress) over relatively short time periods.

Acknowledgements

Partial financial support for this research was provided by the National Science Foundation through grant EAR-9819346 to S. T., and the Swedish Natural Sciences Research Council (NFR) to R.S. The principal author acknowledges also financial support from the Office of Vice-Chancellor for Research and Graduate Studies at the University of Kentucky that subsidized presentation of this research at the XV Congress of INQUA in Durban, South Africa, August 1999.

References

Alley, R.B., 1989a. Water-pressure coupling of sliding and bed deformation 1. Water system. Journal of Glaciology 35, 108–118.

Alley, R.B., 1989b. Water-pressure coupling of sliding and bed deformation 2. Velocity–depth profiles. Journal of Glaciology 35, 119–129.

Alley, R.B., 1991. Deforming-bed origin for southern Laurentide till sheets? Journal of Glaciology 37, 67–76.

Alley, R.B., 1993. How can low-pressure channels and deforming tills coexist subglacially? Journal of Glaciology 38, 200–207.

Alley, R.B., Blankenship, D.D., Bentley, C.R., Rooney, S.T., 1986. Deformation of till beneath Ice Stream B, West Antarctica. Nature 322, 57–59.

Alley, R.B., Blankenship, D.D., Bentley, C.R., Rooney, S.T., 1987. Till beneath Ice Stream B, 4. A coupled ice-till flow model. Journal of Geophysical Research 92, 8921–8929.

Alley, R.B., Blankenship, D.D., Rooney, S.T., Bentley, C.R., 1989. Water-pressure coupling of sliding and bed deformation: III. Application to Ice Stream B, Antarctica. Journal of Glaciology 35, 130–139.

Alley, R.B., Cuffey, K.M., Evenson, E.B., Strasser, J.C., Lawson, D.E., Larson, G.J., 1998. How glaciers entrain and transport basal sediments: physical constraints. Quaternary Science Reviews 16, 1017–1038.

Anandakrishnan, S., Alley, R.B., 1997. Tidal forcing of basal seismicity of Ice Stream C, West Antarctica, observed far inland. Journal of Geophysical Research 102, 15183–15196.

Anandakrishnan, S., Bentley, C.R., 1993. Micro-earthquakes beneath Ice Stream B and Ice Stream C, West Antarctica. Journal of Glaciology 39, 455–462.

Anandakrishnan, S., Blankenship, D.D., Alley, R.B., Stoffa, P.L., 1998. Influence of subglacial geology on the position of a West Antarctic ice stream from seismic observations. Nature 394, 62–65.

Anderson, J.B., Brake, C.F., Myers, N.C., 1984. Sedimentation on the Ross Sea continental shelf, Antarctica. Marine Geology 57, 295–333.

Anderson, J.B., Shipp, S.S., Bartek, L.R., Reid, D.E., 1992. Evidence for a grounded ice sheet on the Ross Sea continental shelf during the late Pleistocene and preliminary paleodrainage reconstructions. In: Elliot, D.H. (Ed.), Contribution to Antarctic Research III. Antarctic Research Series 57, American Geophysical union, Washington D.C., pp. 39–62.

Barrett, P.J., 1975. Textural characteristics of Cenozoic preglacial and glacial sediments at site 270, Ross Sea, Antarctica. In: Hayes, D.E., Frakes, L.A. (Eds.), Initial Reports of the Deep Sea Drilling Project, 28. US Government Printing Office, Washington, DC, pp. 757–767.

Barrett, P.J., Froggatt, P.C., 1978. Densities, porosities, and seismic velocities of some rocks from Victoria Land, Antarctica. New Zealand Journal of Geology and Geophysics 21, 175–187.

Beget, J., 1986. Influence of till rheology on Pleistocene glacier flow in the southern Great Lakes area. USA Journal of Glaciology 32, 235–241.

Bell, R.E., Blankenship, D.D., Finn, C.A., Morse, D.L., Scambos, T.A., 1998. Influence of subglacial geology on the onset of a West Antarctic ice stream from aerogeophysical observations. Nature 394, 58–62.

Bentley, C.R., 1997. Rapid sea-level rise soon from West Antarctic Ice Sheet collapse. Science 275, 1077–1078.

Bindschadler, R., 1997. West Antarctic Ice-Sheet collapse. Science 276, 662–663.

Bindschadler, R., 1998. Monitoring ice sheet behavior from space. Reviews of Geophysics 36, 79–104.

Blake, E.W., 1992. The deforming bed beneath a surge-type glacier: measurement of mechanical and electrical properties. Unpublished Ph.D. Thesis, University of British Columbia, Vancouver, 179pp.

Blankenship, D.D., Bentley, C.R., Rooney, S.T., Alley, R.B., 1986. Seismic measurements reveal a saturated porous layer beneath an active Antarctic ice stream. Nature 322, 54–57.

Blankenship, D.D., Bentley, C.R., Rooney, S.T., Alley, R.B., 1987. Till beneath Ice Stream B, 1. Properties derived from seismic travel times. Journal of Geophysical Research 92, 8903–8911.

Boulton, G.S., 1986. Geophysics—a paradigm shift in glaciology. Nature 322, 18.

Boulton, G.S., 1987. A theory of drumlin formation by subglacial sediment deformation: In: Menzies, J., Rose, J. (Eds.), Drumlin Symposium. A.A. Balkema, Rotterdam, pp. 25–80.

Boulton, G.S., Hindmarsh, R.C.A., 1987. Sediment deformation beneath glaciers— rheology and geological consequences. Journal of Geophysical Research 92, 9059–9082.

Bowles, J.E., 1992. Engineering Properties of Soils and their Measurement. McGraw-Hill, New York, 291pp.

Brown, N.E., Hallet, B., Booth, D.B., 1987. Rapid soft bed sliding of the Puget glacial lobe. Journal of Geophysical Research 92, 8985–8997.

Canals, M., Urgeles, R., Calafat, A.M., 2000. Deep sea-floor evidence of past ice streams off the Antarctic Peninsula. Geology 28, 31–34.

Clark, C.D., 1993. Mega-scale glacial lineations and cross-cutting ice-flow landforms. Earth Surface Processes and Landforms 18, 1–29.

Clark, C.D., 1994. Large scale ice-molded landforms and their glaciological significance. Sedimentary Geology 91, 253–268.

Clark, C.D., Stokes, C.R., 2001. Extent and basal characteristics of the M'Clintock Channel Ice Stream. Quaternary International, this issue.

Clark, P.U., Licciardi, J.M., MacAyeal, D.R., Jenson, J.W., 1996. Numerical reconstruction of a soft-bedded Laurentide ice-sheet during the last glacial maximum. Geology 24, 679–682.

Cuffey, K., Alley, R.B., 1996. Is erosion by deforming subglacial sediments significant? (toward till continuity). Annals of Glaciology 22, 17–24.

Echelmeyer, K.A., Harrison, W.D., Larsen, C., Mitchell, J.E., 1994. The role of the margins in the dynamics of an active ice stream. Journal of Glaciology 40, 527–538.

Engelhardt, H.F., Kamb, B., 1997. Basal hydraulic system of a West Antarctic ice stream: constraints from borehole observations. Journal of Glaciology 43, 207–230.

Engelhardt, H.F., Kamb, B., 1998. Sliding velocity of Ice Stream B. Journal of Glaciology 44, 223–230.

Engelhardt, H.F., Humphrey, N., Kamb, B., Fahnestock, M., 1990. Physical conditions at the base of a fast moving Antarctic ice stream. Science 248, 57–59.

Eyles, N., Boyce, J.J., 1998. Kinematics indicators in fault gouge: tectonic analog for soft-bedded ice sheets. Sedimentary Geology 116, 1–12.

Hayes, D.E., Frakes, L.A., 1975. General synthesis. In: Hayes, D.E., Frakes, L.A. (Eds.), Initial Reports of the Deep Sea Drilling Project, 28. US Government Printing Office, Washington, DC, pp. 919–942.

Hindmarsh, R.C.A., 1997. Deforming beds: viscous and plastic scales of deformation. Quaternary Science Reviews 16, 1039–1056.

Hindmarsh, R.C.A., 1998. Drumlinisation and drumlin-forming instabilities: viscous till mechanics. Journal of Glaciology 44, 293–314.

Hooke, R.L., Elverhøi, A., 1996. Sediment flux from a fjord during glacial periods, Isfjorden, Spitsbergen. Global and Planetary Change 12, 237–249.

Hooke, R.L., Hanson, B., Iverson, N.R., Jansson, P., Fischer, U.H., 1997. Rheology of till beneath Storglaciären, Sweden. Journal of Glaciology 43, 172–179.

Hughes, T., 1992. Abrupt climatic change related to unstable ice-sheet dynamics: towards a new paradigm. Palaeogeography Palaeoclimatology Palaeoecology 97, 203–234.

Humphrey, N., Raymond, C.F., 1994. Hydrology, erosion and sediment production in a surging glacier, Variegated Glacier, Alaska, 1982–1983. Journal of Glaciology 40, 539–552.

Iverson, N.R., Hanson, B., Hooke, R.L., Jansson, P., 1995. Flow mechanics of glaciers on soft beds. Science 267, 80–81.

Iverson, N.R., Baker, R.W., Hooyer, T.S., 1997. A ring shear device for the study of till deformation: tests on tills with contrasting clay content. Quaternary Science Reviews 16, 1057–1066.

Iverson, N.R., Hooyer, T.S., Baker, R.W., 1998. Ring-shear studies of till deformation and an hypothesis for reconciling Coulomb-plastic behavior with distributed strain in glacier beds. Journal of Glaciology 44, 634–642.

Jackson, M., Kamb, B., 1997. The marginal shear stress of Ice Stream B. Journal of Glaciology 43, 415–426.

Jacobel, R.W., Scambos, T.A., Raymond, C.F., Gades, A.M., 1996. Changes in the configuration of ice stream flow in the West Antarctic Ice Sheet. Journal of Geophysical Research 101, 5499–5504.

Jenson, J.W., MacAyeal, D.R., Clark, P.U., Ho, C.L., Vela, J.C., 1996. Numerical modeling of subglacial sediment deformation—implications for the behavior of the Lake-Michigan lobe, Laurentide Ice-Sheet. Journal of Geophysical Research 101, 8717–8728.

Kamb, B., 1970. Sliding motion of glaciers: theory and observations. Reviews of Geophysics and Space Physics 8, 673–728.

Kamb, B., 1991. Rheological nonlinearity and flow instability in the deforming-bed mechanism of ice stream motion. Journal of Geophysical Research 96, 16585–16595.

Lliboutry, L., 1979. Local friction laws for glaciers: a critical review and new openings. Journal of Glaciology 23, 67–95.

MacAyeal, D.R., 1989. Large-scale ice flow over a viscous basal sediment—theory and application to Ice Stream B, Antarctica. Journal of Geophysical Research 94, 4071–4087.

MacAyeal, D.R., 1992. Irregular oscillations of the West Antarctic ice sheet. Nature 359, 29–32.

MacAyeal, D.R., Bindschadler, R., Scambos, T.A., 1995. Basal friction of Ice Stream E, West Antarctic. Journal of Glaciology 41, 247–262.

Marshall, S.J., Clarke, G.K.C., Dyke, A.S., Fisher, D.A., 1996. Geologic and topographic controls on fast flow in the Laurentide and Cordilleran ice sheets. Journal of Geophysical Research 101, 17827–17839.

Novick, A.N., Bentley, C.R., Lord, N., 1994. Ice thickness, bed topography and basal-reflection stregths from radar sounding, Upstream B, West Antarctica. Annals of Glaciology 20, 128–152.

Paterson, W.S.B., 1994. The Physics of Glaciers. Pergamon, Oxford, UK, 480pp.

Piotrowski, J.A., Tulaczyk, S., 1999. Subglacial conditions under the last ice sheet in northwest Germany: ice-bed separation and enhanced basal sliding? Quaternary Science Reviews 18, 737–751.

Raymond, C., 1996. Shear margins in glaciers and ice sheets. Journal of Glaciology 42, 90–102.

Rooney, S.T., 1988. Subglacial geology of Ice Stream B, West Antarctica. Unpublished Ph.D. Thesis, University of Wisconsin-Madison, Madison, 188pp.

Rooney, S.T., Blankenship, D.D., Alley, R.B., Bentley, C.R., 1987. Till beneath Ice Stream B, 2. Structure and continuity. Journal of Geophysical Research 92, 8913–8920.

Rooney, S.T., Blankenship, D.D., Alley, R.B., Bentley, C.R., 1991. Seismic reflection profiling of a sediment-filled graben beneath Ice Stream B, West Antarctica. In: Thomson, M.R., Crame, J.A., Thomson, J.W. (Eds.), Geological Evolution of Antarctica. British Antarctic Survey, Cambridge, UK, pp. 261–265.

Scherer, R.P., 1991. Quaternary and tertiary microfossils from beneath Ice Stream B: evidence for a dynamic West Antarctic Ice Sheet history. Palaeogeography, Palaeoclimatology, Palaeoceanography 90, 395–412.

Scherer, R.P., 1994. A new method for the determination of absolute abundance of diatoms and other silt-sized sedimentary particles. Journal of Paleolimnology 12, 171–180.

Scherer, R.P., Tulaczyk, S., 1997. Diatoms in subglacial sediments yield clues regarding West Antarctic ice-sheet history and ice-stream processes. Antarctic Journal of the United States 32, 32–34.

Scherer, R.P., Aldahan, A., Tulaczyk, S., Kamb, B., Engelhardt, H., Possnert, G., 1998. Pleistocene collapse of the West Antarctic ice sheet. Science 281, 82–85.

Shipp, S., Anderson, J.B., 1997a. Drumlin field of the Ross Sea continental shelf, Antarctica. In: Davis, T.A., et al. (Ed.), Glaciated Continental Margins: an Atlas of Acoustic Images. Chapman & Hall, London, pp. 54–55.

Shipp, S., Anderson, J.B., 1997b. Lineations on the Ross Sea continental shelf, Antarctica. In: Davis, T.A., et al. (Ed.), Glaciated continental margins: an atlas of acoustic images. Chapman & Hall, London, pp. 378–381.

Shipp, S., Anderson, J.B., Domack, E., 1999. Late Pleistocene-Holocene retreat of the West Antarctic ice-sheet system in the Ross Sea: Part I—geophysical results. Geological Society of America Bulletin 111, 1486–1516.

Tchalenko, J.S., 1970. Similarities between shear zones of different magnitudes. Geological Society of America Bulletin 81, 1625–1640.

Thomas, R.H., 1973a. The creep of ice shelves; theory. Journal of Glaciology 12, 45–53.

Thomas, R.H., 1973b. The creep of ice shelves; interpretation of observed behavior. Journal of Glaciology 12, 55–70.

Thorsteinsson, T., Raymond C.F., 2001. Sliding versus till deformation in the motion of an ice stream over deformable till. Journal of Glaciology 46, 633–640.

Tulaczyk, S., 1998. Basal mechanics and geologic record of ice streaming, West Antarctica. Unpublished Ph.D. thesis, California Institute of Technology, Pasadena, California, 359pp.

Tulaczyk, S., 1999. Ice sliding over weak, fine-grained tills: dependence of ice-till interactions on till granulometry. In: Mickelson, D.M., Attig J. (Eds.), Glacial Processes: Past and Modern, Geological Society of America Special Paper 337, Geological Society of America, Boulder, Colorado, pp. 159–177.

Tulaczyk, S., Kamb, B., Scherer, R., Engelhardt, H.F., 1998. Sedimentary processes at the base of a West Antarctic ice stream: constraints from textural and compositional properties of subglacial debris. Journal of Sedimentary Research 68, 487–496.

Tulaczyk, S., Kamb, B., Engelhardt, H., 2000a. Basal mechanics of Ice Stream B.I. Till mechanics. Journal of Geophysical Research 105, 463–481.

Tulaczyk, S., Kamb, B., Engelhardt, H., 2000b. Basal mechanics of Ice Stream B.II Plastic-undrained-bed model. Journal of Geophysical Research 105, 483–494.

Weertman, J., 1957. On sliding of glaciers. Journal of Glaciology 3, 33–38.

Whillans, I.M., van der Veen, C.J., 1993. New and improved determinations of velocity of Ice Stream B and C, West Antarctica. Journal of Glaciology 39, 483–490.

Whillans, I.M., van der Veen, C.J., 1997. The role of lateral drag in the dynamics of Ice Stream B, Antarctica. Journal of Glaciology 43, 231–237.

Printed and bound by CPI Group (UK) Ltd, Croydon, CR0 4YY

01/11/2024

01782632-0020